MULTIVARIABLE CALCULUS WITH ENGINEERING AND SCIENCE APPLICATIONS

MULTIVARIABLE CALCULUS WITH ENGINEERING AND SCIENCE APPLICATIONS

Preliminary Version

Philip M. Anselone
Oregon State University

John W. Lee
Oregon State University

Prentice Hall
Upper Saddle River, NJ 07458

Library of Congress Cataloging-in-Publication Data

Anselone, Philip M.
Multivariable calculus with engineering and science applications / Philip M.
Anselone, John W. Lee   Prelim. version.
p.   cm.
Includes index.
ISBN 0-13-045279-3
1.Calculus.   I. Lee, John W.   II. Title.
QA303.A5215   1995   95-33009
515'.84—dc20   CIP

Acquisitions Editor: George Lobell
Editorial Production/Supervision: Rachel J. Witty, Letter Perfect, Inc.
Manufacturing Buyer: Alan Fischer
Marketing Manager: Frank Nicolazzo
Cover Designer: Jayne Conte
Cover Photo: Concert hall. *Source*: Timothy Hursley.
Editorial Assistant: Gale A. Epps

Printed in the United States of America

10 9 8 7 6 5 4 3 2 1

ISBN 0-13-045279-3

Prentice-Hall International (UK) Limited, *London*
Prentice-Hall of Australia Pty. Limited, *Sydney*
Prentice-Hall Canada, Inc., *Toronto*
Prentice-Hall Hispanoamericano, S.A., *Mexico*
Prentice-Hall of India Private Limited, *New Delhi*
Prentice-Hall of Japan, Inc., *Tokyo*
Simon & Schuster Asia Pte. Ltd., *Singapore*
Editora Prentice-Hall do Brasil, Ltda., *Rio de Janeiro*

Contents

CHAPTER 1 SEQUENCES AND SERIES 1

1.1 Sequences 2
1.2 Monotone Sequences and Successive Approximations 16
1.3 Infinite Series 29
1.4 Series with Nonnegative Terms and Comparison Tests 44
1.5 Absolute and Conditional Convergence; Alternating Series 59
1.6 The Ratio and Root Tests 69

Chapter Highlights 81
Chapter Project: Dynamical Systems 82
Chapter Review Problems 84

CHAPTER 2 POWER SERIES 88

2.1 Taylor Polynomials 88
2.2 Taylor Series and Power Series 103
2.3 Differentiation and Integration of Power Series 114
2.4 Power Series and Differential Equations; The Binomial Series 126

Chapter Highlights 133
Chapter Project: Random Walks 134
Chapter Review Problems 136

CHAPTER 3 VECTORS 139

3.1 Rectangular Coordinates in 3-Space 139
3.2 Vectors 142
3.3 The Dot Product 153
3.4 The Cross Product 164
3.5 Lines and Planes 177

Chapter Highlights 188
Chapter Project: Friction 189
Chapter Review Problems 191

CHAPTER 4 VECTOR CALCULUS 194

4.1 Parametric Curves 194
4.2 Vector Functions, and Curve Length 207
4.3 Velocity, Speed, and Acceleration 220
4.4 Curvature; Tangential and Normal Components of Acceleration 234
4.5 Motion in Polar Coordinates 245

Chapter Highlights 260
Chapter Project: Pursuit Problems 262
Chapter Review Problems 264

CHAPTER 5 DIFFERENTIAL CALCULUS FOR FUNCTIONS OF TWO AND THREE VARIABLES 267

5.1 Functions and Graphs 267
5.2 Limits and Continuity 280
5.3 Partial Derivatives 289
5.4 Tangent Planes, Linear Approximations, and Differentials 298
5.5 Chain Rules and Directional Derivatives 310
5.6 Gradients 328

Chapter Highlights 343
Chapter Project: Curves of Steepest Descent and Ascent 344
Chapter Review Problems 346

CHAPTER 6 MAX-MIN PROBLEMS FOR FUNCTIONS OF TWO AND THREE VARIABLES 350

6.1 Maximum and Minimum Values 350
6.2 Higher Order Partial Derivatives and the Second Partials Test 361
6.3 Constrained Max-Min Problems and Lagrange Multipliers 373

Chapter Highlights 383
Chapter Project: Optimal Location 384
Chapter Review Problems 387

CHAPTER 7 INTEGRAL CALCULUS FOR FUNCTIONS OF TWO AND THREE VARIABLES 389

7.1 Double Integrals in Rectangular Coordinates 389
7.2 Triple Integrals in Rectangular Coordinates 406
7.3 Double and Triple Integrals in Polar and Cylindrical Coordinates 420
7.4 Triple Integrals in Spherical Coordinates 433
7.5 Further Applications of Double and Triple integrals 445

Chapter Highlights 465
Chapter Project: Numerical Integration 467
Chapter Review Problems 470

CHAPTER 8 ELEMENTS OF VECTOR ANALYSIS 474

8.1 Scalar and Vector Fields 474
8.2 Line Integrals 488
8.3 The Fundamental Theorem for Line Integrals 504
8.4 Green's Theorem and Applications 518
8.5 Surface Area and Surface Integrals 533
8.6 The Divergence Theorem (Gauss' Theorem) and Applications 544
8.7 Stokes' Theorem and Applications 558

Chapter Highlights 569
Chapter Project: Heat Conduction 572
Chapter Review Problems 575

APPENDICES A 1–A 27

APPENDIX A RADIUS OF CONVERGENCE OF A POWER SERIES A 1

APPENDIX B DIFFERENTIATION AND INTEGRATION OF POWER SERIES A 3

APPENDIX C ANSWERS TO SELECTED ODD-NUMBERED PROBLEMS A 6

INDEX A 29

Preface

This calculus book is designed primarily for students aiming at careers in science, engineering, or mathematics. In order to reach a wide audience, we have adopted an informal conversational style that students should find appealing and a mode of exposition we call *intuition before rigor.* When presenting a new topic, we begin with a special case or an elementary example that illustrates salient geometric or numerical features of the general situation. Then precision and appropriate rigor are added. Some of the more difficult passages are deferred to ends of sections and some are starred, which means that they can be skipped without adverse effect on later topics. Thus, a high level of rigor is maintained, but it does not get in the way of the flow of ideas. We believe that putting intuition before rigor conveys a deeper understanding to a wider range of students than the alternative practice of first presenting rigorous definitions and theorems, and then explaining and applying them. The trend in some current calculus books to present intuitive ideas, but not much more, seriously handicaps abler students in their future courses and careers.

To serve students who will go on in science or engineering, we provide a considerable variety of interesting applications throughout the book. Those drawn from the physical sciences usually involve motion, force, work, or energy. Applications that illustrate these important topics are treated seriously, both from the physical and mathematical points of view, with the result that students better understand and appreciate both the physical concepts and the related mathematics. Applications modeled by differential equations appear early and often. Necessary background for the applications is provided in the text.

A do-it-yourself project on a topic in science, engineering, or probability is given at the end of each chapter. Particular projects include the optimal placement of a power plant, dynamical systems and chaos, quality control, pursuit problems, and random walks. A typical project uses the methods of the chapter (or earlier chapters) and appropriate graphing utilities, a CAS, or other software. The projects are laid out in a "problem format" to help guide the student. We recommend that a student be asked to write up each project in the form of a report. The report, which expands the outline, should develop the material in the student's own words and should incorporate the problems and their solutions.

The text is technology friendly but is not technology dependent. This is true both for technology that facilitates graphing and for technology that enables students to perform significant numerical calculations. We believe that the geometric ideas of calculus and modern graphing technology complement and reinforce each other when used to best advantage. For example, use of a graphics utility to obtain the graph of a function often gives valuable initial information. However, often significant features of a graph may be obscured by, distorted by, or even be absent from the default graphing window. Knowledge obtained from calculus such as the location of critical points helps students determine appropriate viewing windows and generally leads to more effective use of graphics technology. The intensity of use of technology in calculus varies substantially from school to school and even within sections taught at a particular institution. There is similar diversity in hardware and software. Hardware ranges from graphing calculators through PCs and MACs to sophisticated work stations, and software includes such familiar products as Derive, Maple, Mathcad, Mathematica, and Matlab. Each hardware and software platform has its own local implementation with site-specific protocols. We have responded to this diversity by restricting our discussion of technology primarily to general issues regarding the appropriate and effective use of technology in calculus. Occasionally, we have included short examples of Matlab code when that seemed helpful. Finally, our discussion of technology normally comes at the ends of appropriate sections. This placement enables us to discuss more effectively the interplay between the use of technology and the calculus concepts just covered.

Our book contains a substantial number of innovations in choice of material, order of topics, derivations, and applications. Some of the more significant innovations are described in the following chapter-by-chapter summary. Although the book is mainly about multivariable calculus, two chapters on sequences and series are included in order to provide for flexibility in curriculum planning.

Chapter 1 begins with sequences and limits. Several special limits are derived and illustrated. Monotone sequences are discussed. Applications to Newton's method and successive substitutions help students appreciate the usefulness of sequences and the need for general convergence theorems. Series come next. Most of the results are standard. We have also included optional treatments of numerical approximations of sums and error bounds. This material is particularly suitable for courses that emphasize the use of technology.

Chapter 2 is on power series. The emphasis is on the approximation of functions by Taylor polynomials, the closely related topic of Taylor series, and the convenient computational properties of convergent power series that facilitate their use in applications. The characterization of the radius of convergence of a power series and the justification of term-by-term differentiation and integration are deferred to appendices. The chapter ends with the interplay between first-order differential equations and power series solutions. The ideas are used to derive the binomial series.

Vectors are introduced in Chapter 3. Two- and three-dimensional vectors are treated in tandem. We believe that students will understand three-dimensional concepts better if they are linked closely with the corresponding two-dimensional concepts. For example, the direction of a vector in 3-space is determined by the direction of its projection onto the xy–plane and its angle of elevation relative to the xy–plane. This idea helps students to visualize the orientation of a line in 3–space.

Parametric curves in 2–space and 3–space are introduced in Chapter 4. Then they are recast in vector form. A method of visualizing space curves is adapted from topographic maps. A curve C in 3–space is determined by its projection, C_0, onto the xy–plane and its "elevation function," $z(t)$. Our treatment of curvature is more strongly based on geometry and easier to understand than the usual presentations. It is preceded by discussion of angular velocity. The same ideas developed for angular velocity, when applied to the unit tangent vector, yield formulas for curvature. We have included an optional section on motion in polar coordinates. Although not essential for topics covered later in the text, this material is used frequently in upper division science and engineering courses dealing with mechanics. Furthermore, it provides the foundation for an accessible derivation of Kepler's laws of planetary motion that is given in the text.

Partial derivatives are introduced in Chapter 5. The idea that partial derivatives are particular ordinary derivatives, with all but one variable held fixed, is constantly reiterated along with the message that virtually everything that was learned about ordinary derivatives carries over to partial derivatives. All the various chain rules are presented as variants on one basic chain rule that is easy to remember. Tangent planes and directional derivatives are presented first in scalar forms and then recast in vector forms with the aid of the gradient.

Max-min problems are studied in Chapter 6. Several examples show how to identify maxima and minima by direct examinations of surfaces before the second partials test is introduced. All too often, the second partials test is applied in a rote manner, without motivation. We explain its natural geometric content before applying it. Lagrange multipliers are preceded by a generous discussion, with examples, of constrained max-min problems.

Chapter 7 is on double and triple integrals. The order is dictated by geometry rather than by dimension. Thus, double and triple integrals in rectangular coordinates are treated consecutively. Integrals in polar and cylindrical coordinates are treated in the same section. This chapter includes a considerable variety of physical applications of double and triple integrals. For example, spherical coordinates are used to extend Newton's law of gravitation to homogeneous spherical bodies. We also show that a solid body moves in accordance with Newton's second law of motion as if it were concentrated at its center of mass.

Chapter 8, on vector analysis, presents the integral theorems of Green, Gauss, and Stokes. The emphasis is on physical motivations and applications, particularly steady fluid flow. The section on Green's theorem can serve as a conclusion to a shortened introduction to vector analysis; both the divergence theorem and Stokes' theorem in the plane are formulated here. Surface areas and surface integrals are treated in this chapter in order to be nearer to the divergence theorem and Stokes' theorem in 3–space.

We acknowledge the following reviewers for their comments and suggestions:

Andre Adler, *Illinois Institute of Technology*
Daniel D. Anderson, *University of Iowa, Iowa City*
Leon Chen, *Manhattan College*
Kathy Davis, *University of Texas at Austin*
Moshe Dror, *University of Arizona at Tucson*
John H. Ellison, *Grove City College*
Said Fariabi, *University of Texas at San Antonio*
George Giordano, *Ryerson Polytechnic University*
Cecilia Knoll, *Florida Institute of Technology*
James R. Retherford, *Louisiana State University*
Nathan P. Ritchey, *Youngstown State University*
J.T. Sheick, *Ohio State University*
Susan Sitton, *Illinois Institute of Technology*
David J. Sprows, *Villanova University*
Charles Swartz, *New Mexico State University*

Philip M. Anselone
John W. Lee

MULTIVARIABLE CALCULUS WITH ENGINEERING AND SCIENCE APPLICATIONS

CHAPTER 1
SEQUENCES AND SERIES

In everyday usage a sequence is an infinite list of numbers in a definite order. Sequences of approximations to desired quantities that cannot be determined exactly play an important role in mathematics and its applications to engineering and other fields. In fact, the need to find practical approximations for complicated functions and solutions to challenging problems was the primary motivating force behind the development of the material we are about to study. And the subject is still rich with promising new applications. For example, the colorful fractal patterns you may have seen in recent years are generated by special sequences, and sequential reasoning helps us understand fractal behavior. After learning more about sequences, we move on to the closely related field of infinite series. Roughly speaking, infinite series are sums with infinitely many terms. Such series have many applications in calculus, other branches of mathematics, computer science, the physical sciences, economics, and, more recently, the biological sciences.

Sections 1.1 and 1.2 deal with sequences. Limits of sequences are introduced in Sec. 1.1. Much of the discussion is reminiscent of basic properties of limits of functions. In addition, several new limits that occur frequently in calculus and beyond are introduced. Section 1.2 focuses on convergence and limits of sequences with increasing or decreasing terms. Such sequences have many applications, as we shall see. The first two sections set the stage for the study of infinite series, which begins in Sec. 1.3. We learn that some series have sums while others do not and we are led naturally to the notions of convergence and divergence for series. A number of tests for convergence and divergence of series are developed in the rest of the chapter. There are comparison tests (based on area comparisons) for series with nonnegative terms, an alternating series test that applies to many series whose terms alternate in sign, and very useful ratio and root tests. Refinements of several of these tests also enable us to estimate the sum of a convergent series to within a prescribed accuracy.

The more powerful computer algebra systems, such as Maple and Mathematica, can find the limits of many sequences and sums of many series exactly. Often a CAS returns the limit or sum in a familiar form. Sometimes the limit or sum is expressed in a less convenient way, such as when the sum of a series is given by an internal CAS function that is defined by an improper integral. In the latter case, a CAS can provide a numerical approximation of the sum. If you have access to a CAS, we recommend that you become familiar with its procedures for finding limits and sums and use them to extend your ability to work effectively with sequences and series. As a first step, you may want to check some of your work using a CAS.

We close these introductory remarks by mentioning some standing conventions that are used throughout the book. Occasionally a section or subsection is marked with an asterisk. The asterisk means that the topic is not essential for material covered later in the book. However, such topics always have important applications elsewhere. In order to make clear and precise mathematical statements, it is very helpful to use the logical symbols:

$\Rightarrow$ which means *implies,*
$\Leftrightarrow$ which means *if and only if* or *is equivalent to.*

For example,

$$a > 0, b > 0 \;\Rightarrow\; ab > 0,$$
$$y = \sqrt{x} \;\Leftrightarrow\; x = y^2 \quad \text{for } x, y \geq 0.$$

In fact, you should recognize that the second statement is the definition of the square root of x. We shall use the symbols $\Rightarrow$ and $\Leftrightarrow$ frequently in coming chapters.

1.1 Sequences

Sequences and limits play central roles in mathematics and its applications. For example, many problems cannot be solved exactly. Instead, successive approximate solutions are calculated, which are designed to approach the true solution with greater and greater accuracy. The approximate solutions form a sequence whose limit is the exact solution to the problem at hand. The approximate solutions just mentioned are usually generated by computer programs. Typical methods of scientific computing are based on sequences and sequential (algorithmic) reasoning. Such algorithms generate the sequences that produce the intricate and beautiful fractal patterns that you have probably seen.

In this section, we first define sequences and limits. Then we present algebraic limit laws and squeeze laws that help us to find limits of sequences efficiently. These basic limit laws for sequences are very much like corresponding limit laws for functions. Therefore, much of this section will be easy going and we shall be brief. Along the way, you will learn that sequences are special kinds of functions. Consequently, it should come as no surprise that techniques used earlier in calculus to find limits of functions often can be applied (virtually without change) to find limits of sequences. Finally, we introduce several special limits that play important roles in mathematics and its applications.

Sequences and Graphs

A sequence is an infinite list of numbers in a definite order. More precisely, if to each positive integer $n = 1, 2, 3, \ldots$ there corresponds a particular real number a_n, then the numbers

$$a_1, a_2, a_3, \ldots, a_n, \ldots$$

form a **sequence**. Each of the numbers a_1, a_2, a_3, ... is called a **term** of the sequence. The first term is a_1, the second term is a_2, and the nth term is a_n. The dots indicate terms of the sequence that are not explicitly written. For example,

$$1, \frac{1}{2}, \frac{1}{3}, \ldots, \frac{1}{n}, \ldots,$$

is a sequence with nth term $a_n = 1/n$.

In general, a sequence

$$a_1, a_2, a_3, \ldots, a_n, \ldots,$$

is expressed more compactly with the brace notation $\{a_n\}_{n=1}^{\infty}$ or $\{a_n\}$. Occasionally the braces are dropped and the sequence is denoted simply by a_n. For example, the sequence 1, 1/2, 1/3, $\cdots$ also can be expressed by $\{1/n\}_{n=1}^{\infty}$, $\{1/n\}$, or simply by $1/n$. A few other examples are

$$\left\{1, \frac{1}{2^2}, \frac{1}{3^2}, \frac{1}{4^2}, \ldots\right\} = \left\{\frac{1}{n^2}\right\}, \qquad \{1, -1, 1, -1, \ldots\} = \{(-1)^{n+1}\},$$

$$\{1, 4, 9, 16, \ldots\} = \{n^2\}, \qquad \left\{1, -\frac{1}{4}, \frac{1}{9}, -\frac{1}{16}, \ldots\right\} = \left\{\frac{(-1)^{n+1}}{n^2}\right\}.$$

There are two useful graphical representations of a sequence $\{a_n\}$. The first is a plot of the terms $a_1, a_2, a_3, \ldots$ on a number line. Figure 1 is such a plot for the sequence $a_n = 1/n$. The second is a plot of the points (n, a_n) in a coordinate plane. Such a plot is shown in Fig. 2 for the sequence 2/1, 3/2, 4/3, 5/4, ... whose nth term is $a_n = (n+1)/n$ for $n =$ 1, 2, 3, ...

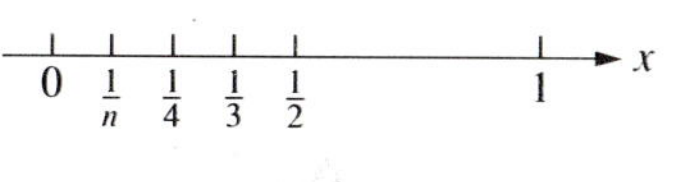

FIGURE 1

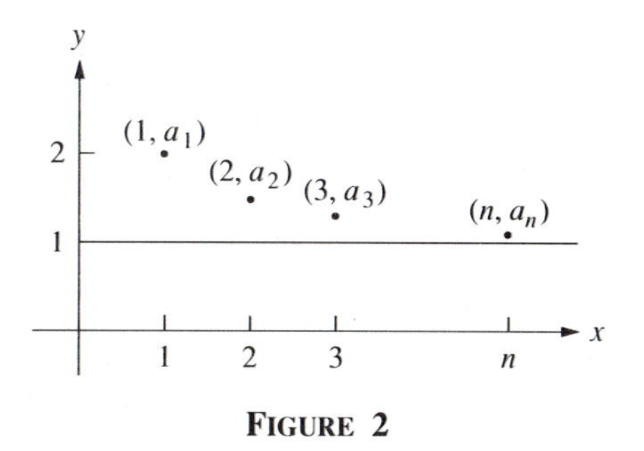

FIGURE 2

A graph such as Fig. 2 reveals that, from another point of view, a sequence $\{a_1, a_2, a_3, \ldots, a_n, \ldots\}$ is a function with domain the set of positive integers n and range the set of values a_n. The notation

$$a(n) = a_n$$

expresses this idea. For example, the sequence graphed in Fig. 2 can be expressed either by $a_n = (n+1)/n$ or $a(n) = (n+1)/n$.

Although the letter n is commonly used for the index of a sequence, other letters such as m or k are often used as well. On occasion, it is convenient to start indexing a sequence with an initial value other than 1. A frequent starting index is 0. For example, $\{(-1)^k/(k+1)\}_{n=o}^{\infty} = \{1, -1/2, 1/3, -1/4, \ldots\}$.

Limits of Sequences

The notion of limit of a sequence is analogous to limit of a function. Not surprisingly, the notation and language used for limits of sequences are very much like the corresponding notation and language used for limits of functions. We begin with the intuitive idea of limit, as applied to sequences.

Often, as n increases, the nth term of a sequence seems to approach a particular number called its limit. As we mentioned earlier, sequences are often constructed to have limits that solve applied problems.

The sequence $a_n = 1/n$ will help us to better understand the idea of a limit. Note that

$$0 < \frac{1}{n} < \frac{1}{10} \text{ if } n > 10, \qquad 0 < \frac{1}{n} < \frac{1}{100} \text{ if } n > 100, \qquad 0 < \frac{1}{n} < \frac{1}{1000} \text{ if } n > 1000,$$

and so on. Evidently, $1/n$ is as near to 0 as we like for all n sufficiently large. We say that the sequence $\{1/n\}$ has limit 0, which is expressed in symbols by

$$\frac{1}{n} \to 0 \text{ as } n \to \infty \text{ or } \lim_{n \to \infty} \frac{1}{n} = 0.$$

The arrows are read as "tends to". The notation $n \to \infty$ is a convenient shorthand which means that n gets larger and larger, growing without bound.

Based on the foregoing discussion, we are ready to define what it means for a real number L to be the limit of a sequence $\{a_n\}$.

Definition *Finite Limit of a Sequence*
A real number L is the **limit** of a sequence $\{a_n\}$ and we write

$$a_n \to L \text{ as } n \to \infty \text{ or } \lim_{n \to \infty} a_n = L$$

if for any $\varepsilon > 0$ no matter how small there is a corresponding number N such that

$$|a_n - L| < \varepsilon \text{ for } n > N.$$

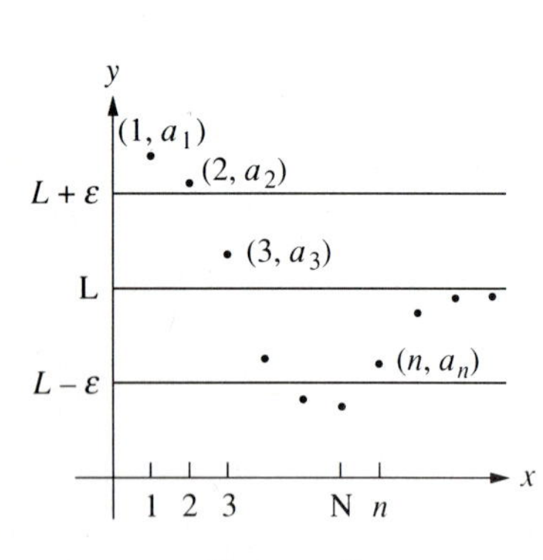

FIGURE 3

If a sequence has a finite limit, we say the sequence **converges** or is **convergent**. The definition of limit is illustrated in Fig. 3, where the points (n, a_n) for $n > N$ lie between the horizontal lines $y = L + \varepsilon$ and $y = L - \varepsilon$. Since the number N depends on ε, we often write $N(\varepsilon)$. Typically, the smaller that ε is taken, the larger that $N(\varepsilon)$ must be.

It should be apparent that a sequence cannot have two different limits. In other words, if a sequence has a limit, it is unique, and we are justified in speaking of "the" limit. It follows immediately from the definition of limit that the constant sequence $a_n = c$ for all n has limit c:

$$\lim_{n \to \infty} c = c \text{ for any constant } c.$$

It is important to realize that not all sequences have limits. For example, $\{(-1)^{n+1}\}_{n=1}^{\infty} = \{1, -1, 1, -1, \ldots, 1, -1, \ldots\}$ does not have a limit; its terms oscillate back and forth between 1 and -1.

We can express the definition of limit, $a_n \to L$ as $n \to \infty$, informally by saying a_n is as near to L as we like for all n large enough. When working with sequences, we are often guided by this intuitive idea of a limit. Nevertheless, it is worthwhile to begin getting used to the precise mathematical definition that is the foundation upon which all limit and convergence properties of sequences rest.

EXAMPLE 1. Use the definition of a limit to verify that

$$a_n = \frac{2n^2 + 1}{n^2} \to 2 \text{ as } n \to \infty.$$

Solution. First, write a_n in the form $a_n = 2 + 1/n^2$. Then

$$|a_n - 2| = 1/n^2.$$

For any number $\varepsilon > 0$,

$$|a_n - 2| = \frac{1}{n^2} < \varepsilon \quad \Leftrightarrow \quad n^2 > \frac{1}{\varepsilon} \quad \Leftrightarrow \quad n > \frac{1}{\sqrt{\varepsilon}}.$$

Let $N(\varepsilon) = 1/\sqrt{\varepsilon}$ to obtain

$$|a_n - 2| < \varepsilon \text{ for } n > N(\varepsilon),$$

which verifies that $a_n \to 2$ as $n \to \infty$. Calculations show that a_n approaches 2 rather rapidly. For example,

$$|a_n - 2| = \frac{1}{n^2} < \frac{1}{100} \quad \text{for } n > 10,$$

$$|a_n - 2| = \frac{1}{n^2} < \frac{1}{10{,}000} \quad \text{for } n > 100. \ \square$$

In applications, rates of convergence are very important. Rapidly convergent sequences of approximate solutions to problems are highly desirable. They save time and money in the calculation of accurate approximations.

Two useful consequences of the definition of a limit are

$$a_n \to L \quad \Leftrightarrow \quad |a_n - L| \to 0,$$

$$a_n \to 0 \quad \Leftrightarrow \quad |a_n| \to 0,$$

where it is understood that the limits are taken as $n \to \infty$. For example, $a_n = (n+1)/n = 1 + 1/n \to 1$ because $|a_n - 1| = 1/n \to 0$ and $b_n = (-1)^n/n \to 0$ because $|b_n| = 1/n \to 0$.

Infinite limits come up naturally for sequences, just as they do for functions. An example of an infinite limit is

$$a_n = \frac{4n^2 + 3}{2n - 1} = \frac{4n^2(1 + 3/4n^2)}{2n(1 - 1/2n)} = 2n \cdot \frac{1 + 3/4n^2}{1 - 1/2n} \to \infty \text{ as } n \to \infty,$$

which means that a_n is as large as we like for all n large enough. The precise mathematical definition follows.

Definition *Infinite Limit of a Sequence*
A sequence $\{a_n\}$ has **limit** ∞ and we write

$$a_n \to \infty \text{ as } n \to \infty \text{ or } \lim_{n \to \infty} a_n = \infty$$

if for any $M > 0$ no matter how large there is a corresponding number N such that

$$a_n > M \text{ for } n > N.$$

The definition of $a_n \to -\infty$ as $n \to \infty$ is similar. If $a_n \to \infty$ or $a_n \to -\infty$, we say that the sequence **diverges** to ∞ or $-\infty$.

EXAMPLE 2. Use the definition of an infinite limit to confirm that

$$a_n = \frac{4n^2 + 3}{2n - 1} \to \infty \text{ as n} \to \infty.$$

Solution. It follows from the remarks preceding the definition that

$$a_n = 2n \cdot \frac{1 + 3/4n^2}{1 - 1/2n} > 2n \text{ for } n = 1, 2, 3, \ldots .$$

Therefore, given any $M > 0$,

$$a_n > 2n > M \text{ if } n > N = M/2,$$

which proves that $a_n \to \infty$ as $n \to \infty$. □

As we have said, a sequence converges if it has a finite limit. All other sequences **diverge** or are **divergent**. A divergent sequence may diverge to $\pm\infty$ or it may have no limit at all. It should be clear from the definition of a limit that the convergence or divergence of a sequence $\{a_n\}$ depends only on the terms a_n for arbitrarily large n. Consequently, any finite number of terms can be changed, removed, or added without having any effect on the convergence or divergence of a sequence or on the value of its limit if it has one. This simple observation will save a good deal of work as we go along.

Limit Laws for Sequences

Limit laws for sequences are strictly analogous to their counterparts for functions. We summarize those laws here and give a few illustrations. The **algebraic limit laws** come first.

$$\text{Assume that } a_n \to a,\ b_n \to b, \text{ and } c \text{ is any number. Then}$$
$$ca_n \to ca, \qquad a_n \pm b_n \to a \pm b,$$
$$a_n b_n \to ab, \qquad \frac{a_n}{b_n} \to \frac{a}{b} \text{ if } b \neq 0.$$

For example, since $1/n \to 0$ and the constant sequence $\{1\}$ has limit 1, $a_n = (n+1)/n = 1 + 1/n \to 1$ as $n \to \infty$.

A useful **squeeze law** for sequences is

$$a_n \leq b_n \leq c_n, \qquad a_n \to L, \quad c_n \to L \qquad \Rightarrow \qquad b_n \to L.$$

The squeeze law is often applied in the special cases with $a_n = L$ or $c_n = L$ for all n. For example, for any fixed $p \geq 1$,

$$0 \leq \frac{1}{n^p} \leq \frac{1}{n} \qquad \text{and} \qquad \frac{1}{n} \to 0$$

imply that

$$\lim_{n\to\infty} \frac{1}{n^p} = 0 \text{ for any } p \geq 1.$$

Another useful limit law, which should seem obvious, is

$$a_n \leq b_n, \quad a_n \to a, \quad b_n \to b \qquad \Rightarrow \qquad a \leq b.$$

In applications of this limit law, it may happen that $a_n < b_n$ for all n, but $\lim a_n = \lim b_n$. For example,

$$a_n = 1 - \frac{1}{n}, \quad b_n = 1 + \frac{1}{n}, \qquad \Rightarrow \qquad a_n < b_n \qquad \text{and} \qquad \lim a_n = \lim b_n = 1.$$

Very often a sequence $\{a_n\}$ is the restriction of a function $a(x)$ to the positive integers n. This means that $a_n = a(n)$. In such cases, familiar methods used to find limits of functions often help to determine limits of sequences. The next two examples illustrate how l'Hôpital's rule for limits of functions can be used to find limits of sequences. Recall that if $f(x) \to 0$ and $g(x) \to 0$ as $x \to \infty$, then

$$\lim_{x\to\infty} \frac{f(x)}{g(x)} = \lim_{x\to\infty} \frac{f'(x)}{g'(x)}$$

whenever the limit on the right exists.

EXAMPLE 3. Show that

$$a_n = \frac{\ln n}{n} \to 0 \text{ as } n \to \infty.$$

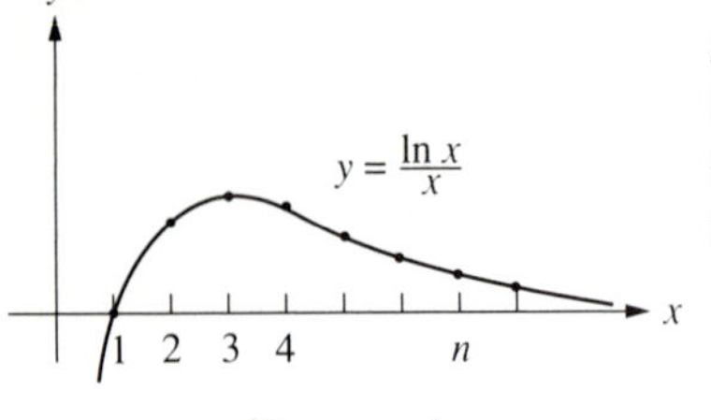

FIGURE 4

Solution. The given sequence $\{a_n\}$ is the restriction of the function $a(x) = \ln x/x$ to the positive integers n. Figure 4 illustrates the situation. The graph of the sequence $\{a_n\}$ consists of the points (n, a_n) on the graph of $a(x)$. From l'Hôpital's rule,

$$\lim_{x \to \infty} \frac{\ln x}{x} = \lim_{x \to \infty} \frac{1/x}{1} = 0.$$

It is apparent from Fig. 4, and follows easily from the definition of a limit, that

$$\frac{\ln x}{x} \to 0 \text{ as } x \to \infty \quad \Rightarrow \quad \frac{\ln n}{n} \to 0 \text{ as } n \to \infty. \ \square$$

Reasoning as in Ex. 3 yields the useful limit law:

$$\boxed{\begin{array}{c} a(x) \to L \text{ as } x \to \infty \\ \Rightarrow \\ a_n = a(n) \to L \text{ as } n \to \infty. \end{array}}$$

Here $L = \pm\infty$ are allowed.

EXAMPLE 4. Find the limit of $a_n = e^n/n$ as $n \to \infty$.

Solution. By l'Hôpital's rule, $e^x/x \to \infty$ as $x \to \infty$. Therefore, $e^n/n \to \infty$ as $n \to \infty$. $\square$

Often we can take advantage of continuity to evaluate the limit of a sequence.

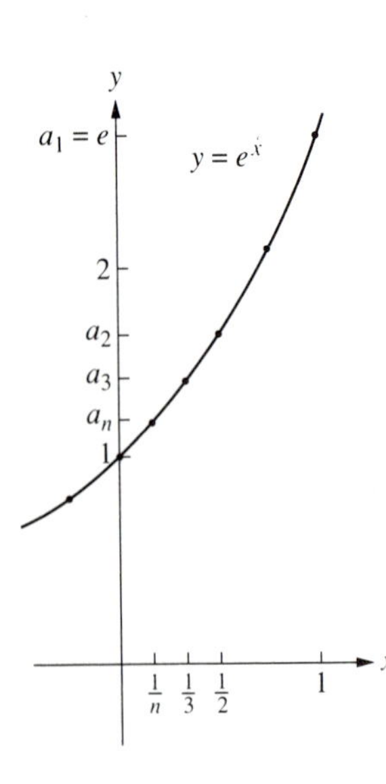

FIGURE 5

EXAMPLE 5. Find the limit of $e^{1/n}$ as $n \to \infty$.

Solution. Since the function e^x is continuous,

$$e^x \to e^0 = 1 \text{ as } x \to 0.$$

It follows that

$$e^{1/n} \to e^0 = 1 \text{ as } n \to \infty$$

because $1/n \to 0$ as $n \to \infty$. See Fig. 5. $\square$

Example 5 illustrates the use of and the reason behind one of two powerful substitution rules.

> **Theorem 1** *Substitution Rule: Continuity Form*
> Assume that $f(x)$ is continuous at $x = L$.Then
> $a_n \to L$ as $n \to \infty \quad \Rightarrow \quad f(a_n) \to f(L)$ as $n \to \infty$.

Let $f(x) = e^x$ and $a_n = 1/n$ to recover Ex. 5.

A more general substitution rule is illustrated next.

EXAMPLE 6. Find the limit of $n \ln(1 + 1/n)$ as $n \to \infty$.

Solution. By l'Hôpital's rule,

$$\frac{\ln(1+x)}{x} \to 1 \text{ as } x \to 0.$$

Since $1/n \to 0$ as $n \to \infty$, it follows that

$$n \ln\left(1 + \frac{1}{n}\right) = \frac{\ln(1 + 1/n)}{1/n} \to 1 \text{ as } n \to \infty. \ \square$$

> **Theorem 2** *Substitution Rule: General Form*
> Assume that $f(x) \to c$ as $x \to L$. Then
> $a_n \neq L, \quad a_n \to L$ as $n \to \infty \quad \Rightarrow \quad f(a_n) \to c$ as $n \to \infty$.

In this rule, $L = \pm\infty$ are allowed.

Let $a_n = 1/n$ and $f(x) = (1/x) \ln(1 + x)$ to recover Ex. 6. Notice that the continuity form of the substitution rule cannot be used in Ex. 6 because $f(x)$ is not continuous at $x = 0$; it is not even defined there.

Special Limits

Next we derive several special limits that come up frequently in calculus and beyond. The derivations are based in large part on properties of exponential functions, particularly

$$e^x \to \infty \text{ as } x \to \infty, \qquad e^x \to 0 \text{ as } x \to \infty,$$

$$a^b = e^{b \ln a} \text{ for } a > 0 \text{ and } -\infty < b < \infty.$$

An example of the first special limit is $2^n \to \infty$ as $n \to \infty$. More generally,

> $x^n \to \infty$ as $n \to \infty$ for any $x > 1$.

Here and in the other limits that follow, x is fixed. Since $x > 1$, we have $\ln x > 0$. Then $x^n = e^{n \ln x} \to \infty$ because $n \ln x \to \infty$ as $n \to \infty$.

The next special limit is similar. An example is $\left(\frac{1}{2}\right)^n \to 0$ as $n \to \infty$. In general,

$$x^n \to 0 \text{ as } n \to \infty \text{ if } |x| < 1.$$

If $x = 0$, the limit is obvious. Fix $x \neq 0$ with $|x| < 1$. Then $\ln|x| < 0$ and $|x^n| = |x|^n = e^{n \ln|x|} \to 0$ because $n \ln|x| \to -\infty$ as $n \to \infty$.

The third special limit is illustrated by $2^{1/n} \to 1$ as $n \to \infty$. It has the form

$$x^{1/n} \to 1 \text{ as } n \to \infty \text{ for any } x > 0.$$

In this case, $x^{1/n} = e^{\frac{1}{n}\ln x} \to e^0 = 1$ as $n \to \infty$.

The next special limit is illustrated by $n^{1/2} = \sqrt{n} \to \infty$ as $n \to \infty$. In general,

$$n^x \to \infty \text{ as } n \to \infty \text{ for any } x > 0.$$

By the now familiar reasoning, $n^x = e^{x \ln n} \to \infty$ as $n \to \infty$.

The following special limit may surprise you:

$$n^{1/n} \to 1 \text{ as } n \to \infty.$$

Some numerical evidence for this limit is given by

$$100^{1/100} \approx 1.05, \qquad 1000^{1/1000} \approx 1.007.$$

To derive the limit, rewrite $n^{1/n}$ as

$$n^{1/n} = e^{\frac{1}{n}\ln n} = e^{\frac{\ln n}{n}}.$$

By Ex. 3, $(\ln n)/n \to 0$ as $n \to \infty$. Therefore, $n^{1/n} \to e^0 = 1$ as $n \to \infty$.

Many calculus texts define e, the base of the natural logarithms, by the limit relation

$$\left(1 + \frac{1}{n}\right)^n \to e \text{ as } n \to \infty,$$

which is a special case of the following more general result:

$$\left(1 + \frac{x}{n}\right)^n \to e^x \text{ as } n \to \infty \text{ for any } x.$$

There are several ways to establish this special limit. Here is one of them. First, the limit is obvious if $x = 0$. Fix $x \neq 0$ and let

$$a_n = \left(1 + \frac{x}{n}\right)^n.$$

We must show that $a_n \to e^x$ as $n \to \infty$. Note that a_n is indeterminate in the form 1^∞ as $n \to \infty$. Such indeterminate forms for functions are often evaluated by first taking logarithms and then applying l'Hôpital's rule. So, let's take logarithms:

$$\ln a_n = n \ln\left(1 + \frac{x}{n}\right) = x\left(\frac{\ln(1 + x/n)}{x/n}\right),$$

$$\ln a_n = x\left(\frac{\ln(1+t)}{t}\right), \qquad t = \frac{x}{n}.$$

Let $n \to \infty$. Since x is fixed, $t \to 0$ and an application of l'Hôpital's rule gives

$$\frac{\ln(1+t)}{t} \to 1.$$

Hence, $\ln a_n \to x$ as $n \to \infty$ and

$$a_n = e^{\ln a_n} \to e^x \text{ as } n \to \infty.$$

The Ratio Test for Sequences

Sequences that converge to zero come up frequently. Often it is obvious that the limit is zero, but sometimes it is not. In more subtle cases, the following two very useful ratio tests for convergence to zero are often applicable.

A special case will reveal what is involved. Suppose that a given sequence $\{a_n\}$ satisfies

$$a_n > 0 \text{ and } \frac{a_{n+1}}{a_n} \leq \frac{1}{2} \text{ for } n = 0, 1, 2, \ldots .$$

Then $a_{n+1} \leq \frac{1}{2} a_n$ for every n. For example, a_n could be the amount of a chemical substance remaining after n hours of a chemical reaction. Since $a_{n+1} \leq \frac{1}{2} a_n$, the amount present is cut at least in half every hour. It should seem obvious that $a_n \to 0$ as $n \to \infty$. To confirm our expectation, note that

$$0 < a_1 \leq \frac{1}{2} a_0, \qquad 0 < a_2 \leq \frac{1}{2} a_1 \leq \frac{1}{2^2} a_0, \qquad 0 < a_3 \leq \frac{1}{2} a_2 \leq \frac{1}{2^3} a_0.$$

By induction, $0 < a_n \leq a_0/2^n$. The squeeze law and $a_0/2^n \to 0$ give $a_n \to 0$ as $n \to \infty$.

A similar argument gives

Theorem 3 *The Ratio Test for Sequences*
If $a_n \neq 0$ and

$$\left|\frac{a_{n+1}}{a_n}\right| \leq r < 1 \text{ for all } n \text{ large enough,}$$

then $a_n \to 0$ as $n \to \infty$.

Only small modifications in the previous argument are needed to derive this test. First, recall that the convergence of a sequence $\{a_n\}$ depends only upon a_n for n arbitrarily large. Therefore, we can adjust the initial terms in the sequence as needed so that $n = 0, 1, 2, 3, \ldots$ and

$$a_n \neq 0, \left|\frac{a_{n+1}}{a_n}\right| \leq r < 1 \text{ for all } n.$$

Then $|a_{n+1}| \leq r|a_n|$ for all n. Let $n = 0, 1, 2$ to obtain

$$0 < |a_1| \leq r|a_0|, \quad 0 < |a_2| \leq r|a_1| \leq r^2|a_0|, \quad 0 < |a_3| \leq r|a_2| \leq r^3|a_0|.$$

By induction,

$$0 < |a_n| \leq r^n|a_0| \text{ for } n = 0, 1, 2, \ldots.$$

Since $0 < r < 1$, the second special limit gives $r^n \to 0$ as $n \to \infty$. By the squeeze law, $|a_n| \to 0$ as $n \to \infty$. Hence, $a_n \to 0$ as $n \to \infty$. The smaller the number r, the faster a_n converges to zero.

A special case of the ratio test for sequences is

Theorem 4 *The Ratio Test for Sequences: Limit Form*
If $a_n \neq 0$ for all large n and

$$\left|\frac{a_{n+1}}{a_n}\right| \to \rho < 1 \text{ as } n \to \infty,$$

then $a_n \to 0$ as $n \to \infty$.

To establish this test, select and fix any number r with $\rho < r < 1$. Since $|a_{n+1}/a_n|$ can be made as near to ρ as we like by taking n large enough, we must have $|a_{n+1}/a_n| \leq r$ for n large enough. Hence, by Th. 3, $a_n \to 0$ as $n \to \infty$.

The smaller the number ρ, the faster a_n converges to zero. If $\rho = 0$ so that

$$\frac{a_{n+1}}{a_n} \to 0 \text{ as } n \to \infty,$$

then $a_n \to 0$ very rapidly. Our next example will exhibit this behavior. It involves factorials.

Factorials usually are encountered first in connection with the binomial theorem. They also come up in many other areas, including combinatorics and probability. The first few factorials are

$$0! = 1, \quad 1! = 1, \quad 2! = 2 \cdot 1 = 2, \quad 3! = 3 \cdot 2 \cdot 1 = 6.$$

For $n = 1, 2, 3, \ldots$, **n − factorial** is defined by

$$\boxed{n! = n(n-1)(n-2) \cdots (2)(1).}$$

Notice that

$$n! = n(n-1)! \text{ and } (n+1)! = (n+1)n!.$$

Arithmetic with factorials is easy but requires a little care. For example,

$$\frac{9!}{7!} = \frac{9 \cdot 8 \cdot 7!}{7!} = 9 \cdot 8 = 72,$$

$$\frac{10!}{7!\,3!} = \frac{10 \cdot 9 \cdot 8 \cdot 7!}{7!\,3!} = \frac{10 \cdot 9 \cdot 8}{3 \cdot 2 \cdot 1} = 10 \cdot 3 \cdot 4 = 120,$$

$$\frac{(n+2)!}{(n-1)!} = \frac{(n+2)(n+1)n(n-1)!}{(n-1)!} = (n+2)(n+1)n.$$

EXAMPLE 7. Find the limit of $a_n = 4^n/n!$ as $n \to \infty$.

Solution. In the following table, the first five values a_n are exact.

n	0	1	2	3	4	5	10	12	15
a_n	1	4	8	$10\frac{2}{3}$	$10\frac{2}{3}$	8.5	.29	.035	.0008

After increasing initially, the terms of the sequence appear to decrease and converge to zero. To verify the limit, observe that

$$\frac{a_{n+1}}{a_n} = a_{n+1} \cdot \frac{1}{a_n} = \frac{4^{n+1}}{(n+1)!} \cdot \frac{n!}{4^n} = \frac{4}{n+1} \to 0 \text{ as } n \to \infty.$$

By the limit form of the ratio test, $a_n = 4^n/n! \to 0$ as $n \to \infty$. So we should expect and the table suggests rapid convergence to zero. A further calculation yields $a_{20} \approx 0.0000005$. □

Example 7 is a particular case of an important special limit:

$$\boxed{\frac{x^n}{n!} \to 0 \text{ as } n \to \infty \text{ for any } x.}$$

The limit is obvious if $x = 0$. Otherwise, we proceed just as in Ex. 7. Let $a_n = x^n/n!$ for $x \neq 0$. Then

$$\left|\frac{a_{n+1}}{a_n}\right| = \left|\frac{x^{n+1}}{(n+1)!} \cdot \frac{n!}{x^n}\right| = \frac{|x|}{n+1} \to 0 \text{ as } n \to \infty.$$

By the limit form of the ratio test, $x^n/n! \to 0$ as $n \to \infty$. This limit will play a prominent role in our study of infinite series.

PROBLEMS

In Probs. 1–32, the nth term of a sequence is given. (a) Determine decimal approximations for the terms with $n = 5, 10, 20, 50, 100, 500, 1000$. (b) Based on the results in (a), and perhaps further calculator experiments, state whether you think the sequence converges or diverges. If you believe the sequence has a limit state what it is. (c) Finally, use the methods of this section to determine whether the sequence converges or diverges and to find the limit, if there is one.

1. $\dfrac{n-1}{n+1}$
2. $\dfrac{3n-4}{n+1}$
3. $1 + (-1)^n$
4. $\dfrac{(-1)^n}{2}$
5. $\sin\dfrac{n\pi}{2}$
6. $\sin(4n+1)\dfrac{\pi}{2}$
7. $e^{-1/n}$
8. e^{-n}
9. $(-1)^n e^{-1/n}$
10. $(-1)^n e^{-n}$
11. $\tan\left(\dfrac{n\pi}{4n-1}\right)$
12. $\arctan\left(\dfrac{2-n}{n+3}\right)$
13. $(-1)^{n+1}\tan\left(\dfrac{n\pi}{4n-1}\right)$
14. $(-1)^{n+1}\arctan\left(\dfrac{2-n}{n+3}\right)$
15. $1/3^n$
16. $(-1/3)^n$
17. $\dfrac{2^n+3^n}{4^n}$
18. $(-1)^n\dfrac{4^n}{2^n+3^n}$
19. $\dfrac{2^{1/n}+3^{1/n}}{4^{1/n}}$
20. $\dfrac{2^{1/n}+(-1)^n 3^{1/n}}{4^{1/n}}$
21. $5^{1/n}$
22. $5^{-1/n}$
23. $n2^{-n}$
24. $n^2 2^{-n}$
25. $3^n/n$
26. $3^n/n^2$
27. $n!/10^n$
28. $10^n/n!$
29. $\sqrt[2n]{n}$
30. $\sqrt[3n]{n^4}$
31. $(5/2n)^{1/n}$
32. $\sqrt[n]{n^2+n}$

In Probs. 33–70, the nth term of a sequence is given. Use the methods of this section to determine whether the sequence converges or diverges and to find the limit, if there is one.

33. $(-4)^n/n!$

34. $(-4)^n/3^n n!$

35. $(n!)^2/(2n)!$

36. $(n!)^3/(2n)!$

37. $\dfrac{\ln n}{\sqrt{n}}$

38. $\dfrac{\ln n^2}{\sqrt{n}}$

39. $\cos\left(\dfrac{4n^2+1}{n-2n^2}\pi\right)$

40. $\sin\left(\dfrac{n^3-1}{2n^3+5n}\pi\right)$

41. $\sin\left(n\sin\dfrac{1}{n}\right)$

42. $\cos\left(\dfrac{\ln(2n)}{n}\pi\right)$

43. $\sin\left(2n\sin\dfrac{1}{n}\right)$

44. $\tan\left(n\sin\dfrac{1}{2n}\right)$

45. $\left(\sin\dfrac{1}{n}\right)^{1/n}$

46. $\left(\sin\dfrac{1}{n^2}\right)^{1/n}$

47. $\left(1+\dfrac{x}{2n}\right)^n$, x fixed

48. $\left(1+\dfrac{1}{2n}\right)^{nx}$, x fixed

49. $\dfrac{\ln 2^n}{\sqrt{n}}$

50. $\dfrac{\ln 2^{-n}}{\sqrt{n}}$.

51. $\arctan n$

52. $\arcsin(e^{1/n})$

53. $\dfrac{n^n}{(n+1)^n}$

54. $\dfrac{n^{n+1}}{(n+1)^n}$

55. $\left(1+\dfrac{2}{n}\right)^n$

56. $\left(1-\dfrac{1}{2n}\right)^n$

57. $n^2(1-\cos(1/n))$

58. $n^3(1-\cos(1/n))$

59. $n^2\sin(1/n^2)$

60. $n^2\sin^2(1/n)$

61. $\dfrac{2^n}{4^n+1}$

62. $\dfrac{2^{-n}}{4^{-n}+1}$

63. $\ln n - \ln(2n)$

64. $\ln n - \ln(n+2)$

65. $\dfrac{(\ln n)^2}{n}$

66. $\left(1-\dfrac{2}{n}\right)^{n^2}$

67. $\dfrac{n-2\sin n}{n+2\cos n}$

68. $\dfrac{n-2\sinh n}{n+2\cosh n}$

69. $(n+1)e^{-n}$

70. $\sqrt{n+1}-\sqrt{n}$

In Probs. 71–78 illustrate different rates of convergence. (a) First, verify that the sequence has a finite limit. (b) Then find the smallest integer N such that $|a_n - L| < .001$ for $n > N$. You may use a calculator to help with (b).

71. $1/n^2$

72. $1/\sqrt{n}$

73. $1/2^n$

74. $1/3^n$

75. $n^4/2^n$

76. $n^4/3^n$

77. $\arctan n$

78. $\tanh n$

In Probs. 79–84, use the definition of a limit to establish the following limits as $n \to \infty$.

79. $\dfrac{n+1}{n} \to 1$

80. $\dfrac{n-3}{2n+1} \to \dfrac{1}{2}$

81. $\dfrac{1}{2^n} \to 0$

82. $e^{1/n} \to 1$

83. $\dfrac{n^2}{2n-1} \to \infty$

84. $\ln\left(\dfrac{1}{n}\right) \to -\infty.$

85. Fix x with $0 < x < 1$. Use the ratio test for sequences to show that as $n \to \infty$ (a) $nx^n \to 0$, (b) $n^2x^n \to 0$, (c) $n^px^n \to 0$ for any fixed p.

86. Prove that a sequence $\{a_n\}$ can have at most one limit. *Hint.* If $\{a_n\}$ has two limits L and M, take $\varepsilon < \frac{1}{2}|L - M|$ in the definition of a limit to arrive at a contradiction. A graph like Fig. 2 or Fig. 3 may help.

1.2 Monotone Sequences and Successive Approximations

Knowledge of a few basic limits and use of various limit laws and other techniques applied in Sec. 1.1 enable us to evaluate many limits. Despite these successes, there are many important sequences that converge but their limits are unknown and cannot be found exactly by any available means. Indeed, this is frequently the situation for practical applications of sequences. When we can't find the exact solution to a problem, often we are able to calculate a sequence of approximate solutions that converges to the exact solution. Three steps are involved. First, calculate the successive approximate solutions by some procedure, such as Newton's method. Second, show that the approximate solutions converge to a limit. Third, show that the limit is the solution we seek.

For the time being, we shall focus on the second step, expressed more broadly:

> Given a sequence $\{a_n\}$, how can you tell, just from the behavior of its terms, whether the sequence converges or diverges?

In many cases of interest, the sequences are increasing or decreasing. We shall find that every increasing or decreasing sequence has a limit. The limit is finite if the sequence is bounded. The limit is infinite if the sequence is unbounded.

The foregoing results will be applied to the approximate solution of equations. They will be used to establish the convergence of Newton's method for typical problems. Then the method of successive substitutions will be introduced and similar convergence results will be presented.

Bounded Sequences

The boundedness properties of sets and sequences are essentially the same. We shall concentrate on sequences. A sequence $\{a_n\}$ is **bounded above** if there is a real number M such that

$$a_n \leq M \text{ for all } n.$$

We call M an **upper bound** for $\{a_n\}$. Similarly, $\{a_n\}$ is **bounded below** if there is a real number m such that

$$m \leq a_n \text{ for all } n.$$

We call m a **lower bound** for $\{a_n\}$. A sequence is **bounded** if it is bounded above and below.

For example, the sequence

$$\{1, 2, 3, 1, 2, 3, 1, 2, 3, \ldots\}$$

is bounded. An upper bound is 3 and a lower bound is 1. Moreover, every number $M \geq 3$ is an upper bound and every number $m \leq 1$ is a lower bound. All the terms of this sequence lie in the interval $[1, 3]$ on the real line. In general, a sequence $\{a_n\}$ is bounded if all its terms lie in an interval $[m, M]$.

Here is a very useful observation.

Every convergent sequence is bounded.

To see why, suppose that $a_n \to L$ as $n \to \infty$. Since a_n is arbitrarily near L for n large enough, we must have $L - 1 < a_n < L + 1$ if $n > N$ for some integer N. Only a finite number of terms a_n lie outside the interval $[L - 1, L + 1]$ in Fig. 1. So all the terms lie in some interval $[m, M]$ and $\{a_n\}$ is bounded.

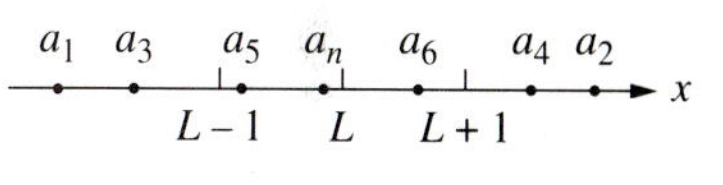

FIGURE 1

EXAMPLE 1. The familiar sequence

$$\{a_n\} = \left\{1, \frac{1}{2}, \frac{1}{3}, \ldots, \frac{1}{n}, \ldots\right\}$$

is graphed in Fig. 2. Clearly, $a_n \to 0$ and $\{a_n\}$ is bounded. Any number $M \geq 1$ is an upper bound for $\{a_n\}$. The least (smallest) upper bound for $\{a_n\}$ is $M = 1$, which is also the maximum term of the sequence. Any number $m \leq 0$ is a lower bound for $\{a_n\}$. The greatest (largest) lower bound for $\{a_n\}$ is $m = 0$, which is not a term of the sequence. This sequence has no minimum term. □

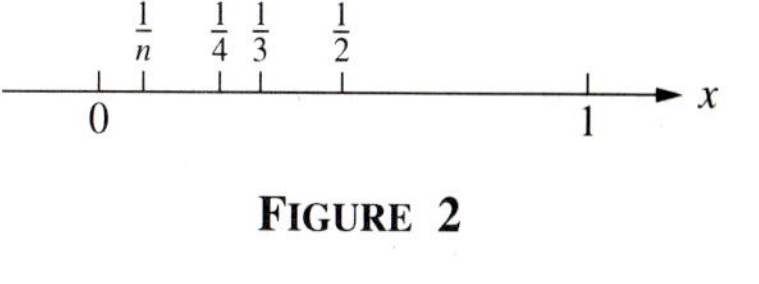

FIGURE 2

EXAMPLE 2. The sequence

$$\{a_n\} = \left\{0, \frac{1}{2}, \frac{2}{3}, \ldots, \frac{n-1}{n}, \ldots\right\}$$

is graphed in Fig. 3. Note that $a_n \to 1$ and $\{a_n\}$ is bounded. Just as in Ex. 1, any $M \geq 1$ is an upper bound and any $m \leq 0$ is a lower bound. The least upper bound

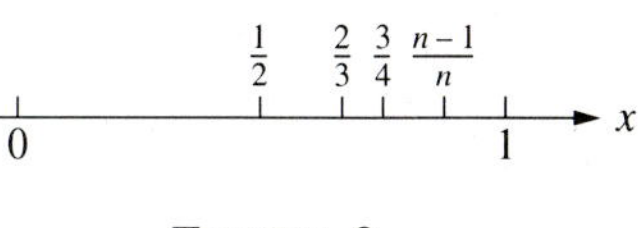

FIGURE 3

for $\{a_n\}$ is $M = 1$, which is not a term of the sequence. This sequence has no maximum term. The greatest lower bound for $\{a_n\}$ is $m = 0$, which is also the minimum term of the sequence. □

If all the rational numbers are plotted on a real number line many points remain unmarked – no rational number corresponds to such a point. For example, the point at distance $\sqrt{2}$ from the origin does not correspond to any rational number. On the other hand, when all the real numbers are plotted on a number line all points are marked with a unique real number. This is a fundamental property of the real number system, called **completeness**. We express it informally by saying "there are no holes in the number line". An equivalent formulation of the completeness property is as follows:

> Every sequence that is bounded above has a least upper bound.
> Every sequence that is bounded below has a greatest lower bound.

Increasing and Decreasing Sequences

A sequence $\{a_n\}$ is **increasing** if

$$a_{n+1} > a_n \text{ for all } n.$$

The sequence is **decreasing** if

$$a_{n+1} < a_n \text{ for all } n.$$

For example,

$$\{a_n\} = \left\{0, \frac{1}{2}, \frac{2}{3}, \frac{3}{4}, \ldots, \frac{n-1}{n}, \ldots\right\} \text{ is increasing,}$$

$$\{a_n\} = \left\{1, \frac{1}{2}, \frac{1}{3}, \frac{1}{4}, \ldots, \frac{1}{n}, \ldots\right\} \text{ is decreasing.}$$

Most of the sequences you will meet are either increasing or decreasing after some finite number of terms. We call such sequences **eventually increasing** or **eventually decreasing**.

EXAMPLE 3. Let $a_n = 4^n/n!$ for $n = 0, 1, 2, \ldots$. Calculations made in Ex. 7 of Sec. 1.1 suggest that $\{a_n\}$ is decreasing for $n \geq 4$. Confirm this observation.

Solution. As we saw in Sec. 1.1,

$$\frac{a_{n+1}}{a_n} = \frac{4^{n+1}}{(n+1)!} \cdot \frac{n!}{4^n} = \frac{4}{n+1}.$$

If $n \geq 4$ then

$$\frac{a_{n+1}}{a_n} < 1 \text{ and } a_{n+1} < a_n.$$

Thus, $\{a_n\}$ is decreasing for $n \geq 4$. □

The next example shows how calculus can help to determine whether a sequence is increasing or decreasing.

EXAMPLE 4. In Sec. 1.1 we proved that

$$a_n = \frac{\ln n}{n} \to 0 \text{ as } n \to \infty.$$

Show that $\{a_n\}$ is decreasing for $n \geq 3$.

Solution. Let

$$a(x) = \frac{\ln x}{x} \text{ for } x > 0.$$

By the quotient rule,

$$a'(x) = \frac{1 - \ln x}{x^2}.$$

If $x > e \approx 2.7$, then $\ln x > 1$ and $a'(x) < 0$, so that $a(x)$ is decreasing. Since $a_n = a(n)$, it follows that $\{a_n\}$ is decreasing for $n \geq 3$. □

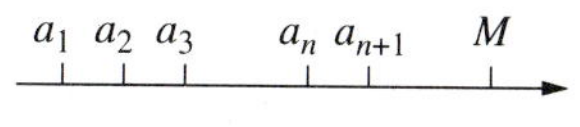

FIGURE 4

A bounded increasing sequence $\{a_n\}$ and its least upper bound M are illustrated in Fig. 4. It should seem evident from the figure that $a_n \to M$ as $n \to \infty$. A bounded decreasing sequence $\{b_n\}$ and its greatest lower bound m are shown in Fig. 5. Apparently, $b_n \to m$ as $n \to \infty$. These observations are special cases of the following fundamental theorem.

m b_{n+1} b_n b_3 b_2 b_1

FIGURE 5

Theorem 1 *Convergence of Bounded Increasing or Decreasing Sequences*
Every bounded increasing or decreasing sequence converges.
If $\{a_n\}$ is increasing and M is the least upper bound for $\{a_n\}$, then $a_n \to M$ as $n \to \infty$.
If $\{b_n\}$ is decreasing and m is the greatest lower bound for $\{b_n\}$, then $b_n \to m$ as $n \to \infty$.

We shall indicate how the proof of Th. 1 goes for the case of a bounded increasing sequence $\{a_n\}$ with least upper bound M. The argument is short and instructive. Since M is an upper bound for the sequence,

$$a_n \leq M \text{ for all } n.$$

Let $\varepsilon > 0$. Since M is the *least* upper bound of $\{a_n\}$, $M - \varepsilon$ is *not* an upper bound for the sequence. Consequently, at least one member of the sequence, say a_N, must be greater than $M - \varepsilon$. That is,

$$a_N > M - \varepsilon \text{ for some } N.$$

Since the sequence is increasing,

$$a_n > a_N > M - \varepsilon \text{ for all } n > N.$$

Combine these observations to obtain

$$M - \varepsilon < a_n \leq M \text{ and } |a_n - M| < \varepsilon \text{ for } n > N.$$

By the definition of a limit, $a_n \to M$ as $n \to \infty$. The proof for a bounded decreasing sequence is very similar.

Unbounded increasing or decreasing sequences have obvious limit properties. For example, the sequence $\{n^2\}$ is increasing and unbounded. It has limit ∞. The sequence $\{-n^2\}$ is decreasing and unbounded. It has limit $-\infty$. These two cases are typical. Every unbounded increasing sequence diverges to ∞. Every unbounded decreasing sequence diverges to $-\infty$.

Monotone Sequences

It is useful to generalize the notions of increasing and decreasing sequences a bit. A sequence $\{a_n\}$ is **nondecreasing** (not decreasing) if

$$a_{n+1} \geq a_n \text{ for all } n.$$

The sequence is **nonincreasing** (not increasing) if

$$a_{n+1} \leq a_n \text{ for all } n.$$

For example,

$$\left\{\frac{1}{2}, \frac{1}{2}, \frac{2}{3}, \frac{2}{3}, \frac{3}{4}, \frac{3}{4}, \ldots\right\} \text{ is nondecreasing,}$$

$$\left\{\frac{1}{2}, \frac{1}{2}, \frac{1}{3}, \frac{1}{3}, \frac{1}{4}, \frac{1}{4}, \ldots\right\} \text{ is nonincreasing.}$$

Clearly, any increasing sequence is nondecreasing, and any decreasing sequence is nonincreasing.

A sequence is **monotone** if it is either nondecreasing or nonincreasing. A sequence is **eventually monotone** if it is monotone after a finite number of terms.

The convergence and divergence properties of increasing and decreasing sequences carry over to monotone sequences in a straightforward manner. There is nothing really new. The following theorem extends our previous conclusions.

> **Theorem 2** *Limits of Monotone Sequences*
> Every monotone sequence $\{a_n\}$ has a limit.
> The limit is finite if $\{a_n\}$ is bounded.
> The limit is infinite if $\{a_n\}$ is unbounded.

It suffices for the sequence to be eventually monotone.

Successive Approximations

Theorems 1 and 2 may seem to be only of theoretical interest. This is far from the truth. In fact, these theorems are of great practical value. It often happens that successive approximate solutions of a problem form a bounded monotone sequence $\{a_n\}$ and, moreover, that the limit L of the sequence guaranteed by Th. 1 or 2 is the exact solution of the problem. Even though we do not know the exact solution L, the approximate solutions a_n for n not too large may be accurate enough for practical purposes.

We shall illustrate the foregoing approximation procedure in the context of rootfinding, first with Newton's method and then with the method of successive substitutions. In either case, $\{a_n\}$ is a sequence of approximations to a solution r of a given equation. Ordinarily, we do not know r exactly. So we generate the sequence $\{a_n\}$ to obtain successively more accurate approximations for r.

Newton's Method

Newton's method first comes up early in calculus where it is treated rather informally. Now, after a little review, Th. 1 will be used to show that Newton's method converges in typical cases. We seek a root r of a twice differentiable function $f(x)$ graphed in Fig. 6. An initial guess for r is x_0. The point x_1, where the tangent line T_0 at $(x_0, f(x_0))$ crosses the x–axis, is a better approximation for r than x_0 is. A formula for x_1 is

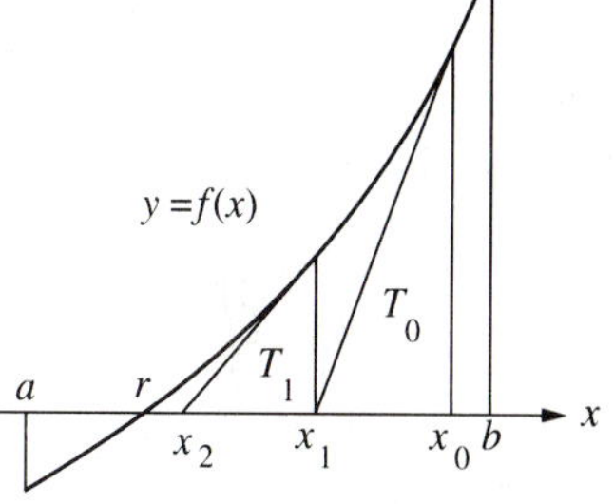

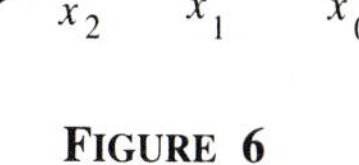

FIGURE 6

$$x_1 = x_0 - \frac{f(x_0)}{f'(x_0)}.$$

Repeated application of the same procedure gives successive Newton approximations $x_0, x_1, x_2, \dots, x_n, \dots$ defined by

$$\boxed{x_{n+1} = x_n - \frac{f(x_n)}{f'(x_n)}.}$$

Usually $f(x)$ is either increasing or decreasing and either concave up or concave down near a root. Since the four cases are very similar, we shall treat just one of them. Figure 6 illustrates the case with

$$f(a) < 0, \qquad f(b) > 0,$$
$$f' > 0, \quad f \text{ increasing} \quad \text{on } [a, b],$$
$$f'' > 0, \quad f \text{ concave up} \quad \text{on } [a, b].$$

By the intermediate value theorem, f has a root r in (a, b). Since f is increasing, r is the only root of f in (a,b).

Let $r < x_0 \leq b$. Since f is concave up, the tangent line T_0 lies below the curve except at $(x_0, f(x_0))$. It follows that $x_0 > x_1 > r$. For the same reason,

$$x_0 > x_1 > x_2 > \cdots > x_n > x_{n+1} > \cdots > r.$$

Thus, $\{x_n\}$ is decreasing and bounded below. By Th. 1,

$$x_n \to s \quad \text{as} \quad n \to \infty$$

for some s with $r \leq s \leq b$. It is not hard to show that $s = r$. Since f and f' are continuous, $f(x_n) \to f(s)$ and $f'(x_n) \to f'(s)$ as $n \to \infty$. Let $n \to \infty$ in the Newton iteration formula to obtain

$$s = s - \frac{f(s)}{f'(s)}, \qquad f(s) = 0.$$

Since r is the only root of f in $[a, b]$, $s = r$ and $x_n \to r$ as $n \to \infty$.

EXAMPLE 5. Show that $f(x) = x - 1 + \dfrac{1}{2} e^{-x}$ has a unique positive root r and approximate r by Newton's method.

Solution. Note that

$$f'(x) = 1 - \frac{1}{2} e^{-x} > 0, \qquad f(x) \text{ is increasing} \quad \text{for} \quad x \geq 0,$$

$$f''(x) = \frac{1}{2} e^{-x} > 0, \qquad f(x) \text{ is concave up} \quad \text{for} \quad x \geq 0.$$

Since $f(x)$ is increasing for $x \geq 0$, f can have at most one positive root. Use of a calculator gives

$$f(0.5) \approx -0.20, \qquad f(1) \approx 0.18.$$

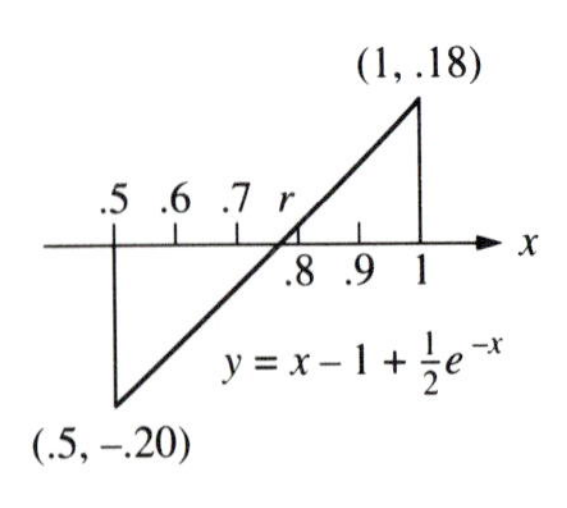

FIGURE 7

Therefore, f has a unique positive root r and $0.5 < r < 1$. A graph of $f(x)$ for $.5 \leq x \leq 1$ is shown in Fig. 7. Since the curve is nearly straight, Newton's method should converge very rapidly. Let $x_0 = 1$. From the discussion preceding Ex. 5, we know that the Newton iterates decrease and converge to r. A few calculations yield

$$\begin{aligned} x_0 &= 1.000, & f(x_0) &\approx 0.184, \\ x_1 &\approx 0.775, & f(x_1) &\approx 0.005, \\ x_2 &\approx 0.768, & f(x_2) &\approx 0.00003. \end{aligned}$$

One more calculation yields $f(0.767) \approx -0.0008$. Since $f(0.767) < 0 < f(0.768)$, the intermediate value theorem tells us that $0.767 < r < 0.768$. □

Successive Substitutions

The method of successive substitutions is another iterative procedure for calculating approximate solutions of equations. It applies to equations of the form $x = g(x)$ with $g(x)$ continuous. Before describing the method, let's dwell briefly on the kind of equation it is designed for. Often, an equation comes naturally expressed in the form $x = g(x)$, particularly in population dynamics. More about that a little later. Any equation can be put in the form $x = g(x)$ by simple algebra. The equation $x - 1 + \frac{1}{2}e^{-x} = 0$ for $x \geq 0$ in Ex. 5 has the equivalent form $x = g(x)$ with $g(x) = 1 - \frac{1}{2}e^{-x}$. The functions $y = x$ and $y = g(x)$ are graphed in Fig. 8. The unique positive solution r of $x = g(x)$ is the x–coordinate of the point of intersection of the two graphs. Thus, $r = g(r)$. Since r is unchanged when the function g is applied to it, r is called a *fixed point* of the function g.

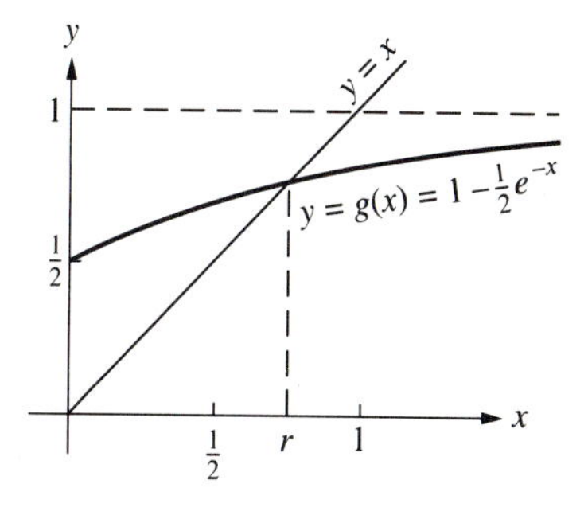

FIGURE 8

As we have said, the method of successive substitutions yields approximate solutions of an equation $x = g(x)$ with $g(x)$ continuous. Let x_0 be an initial approximation for a solution. Substitute $x = x_0$ into $g(x)$ to obtain $x_1 = g(x_0)$. Then substitute $x = x_1$ into $g(x)$ to obtain $x_2 = g(x_1)$. Continue in the same way to obtain the **successive substitutions** $x_0, x_1, x_2, \ldots, x_n, \ldots$ determined by

$$\boxed{x_{n+1} = g(x_n) \text{ for } n = 1, 2, 3, \ldots.}$$

Suppose that we can manage to show that $x_n \to r$ as $n \to \infty$ for some r. Let $n \to \infty$ in $x_{n+1} = g(x_n)$ to obtain $r = g(r)$. We have learned an important fact about the successive substitutions x_n.

$$\boxed{\begin{array}{l}\text{If } x_n \to r \quad \text{as} \quad n \to \infty, \text{ then} \\ r \text{ is a solution of } x = g(x).\end{array}}$$

EXAMPLE 6. Apply the method of successive approximations to the equation

$$x = g(x) \text{ with } g(x) = 1 - \frac{1}{2}e^{-x} \text{ for } x \geq 0.$$

Solution. As we already observed, this equation has the unique solution r indicated in Fig. 8. Newton's method was used in Ex. 5 to obtain the approximation $r \approx 0.768$. Let's see what we get with the method of successive substitutions. Let $x_0 = 0$. Then

$$x_{n+1} = g(x_n) = 1 - \frac{1}{2}e^{-x_n} \text{ for } n = 0, 1, 2, \ldots.$$

Carry out the first few calculations to obtain

$$x_1 = 0.5, \qquad x_2 \approx 0.697, \qquad x_3 \approx 0.751, \qquad x_4 \approx 0.764.$$

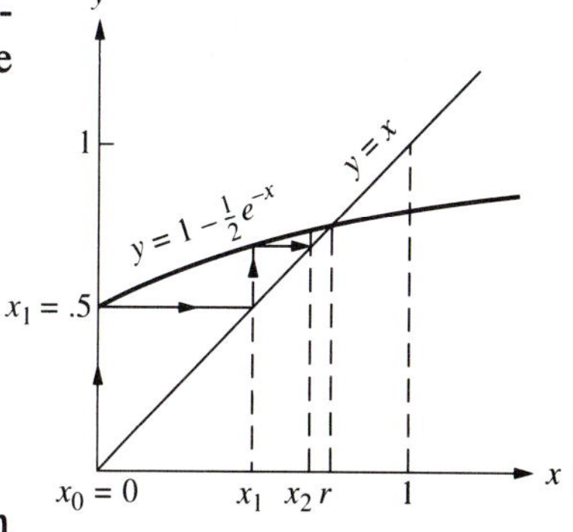

FIGURE 9

The successive substitutions x_n seem to be converging to the solution $r \approx 0.768$. Geometric evidence for this conclusion is presented in Fig. 9. In the figure, $x_1 = g(x_0) = 0.5$ is the height up to the graph of g at $x_0 = 0$. To transfer x_1 to the x–axis, move horizontally over to the diagonal line $y = x$ and then vertically down to the x–axis. Now $x_2 = g(x_1)$ is the height up to the graph of g at x_1. Transfer x_2 to the x–axis by moving horizontally to the line $y = x$ and then vertically to the x–axis. Notice the staircase pattern in Fig. 9. From the geometry in the figure, the sequence $\{x_n\}$ is increasing and $x_n < r$ for all n. By Th. 1, $x_n \to s$ as $n \to \infty$ for some s. Then $s = g(s)$ by an argument made a moment ago. Since r is the unique positive solution of $x = g(x)$, we have $s = r$ and $x_n \to r$ as $n \to \infty$. □

The equation $x = 1 - \frac{1}{2}e^{-x}$ is typical of a class of equations $x = g(x)$ for which the method of successive substitutions is successful. Assume that the function g satisfies

$$g(a) > a, \qquad g(b) < b,$$

$$0 < g'(x) < 1 \quad \text{for} \quad a \le x \le b.$$

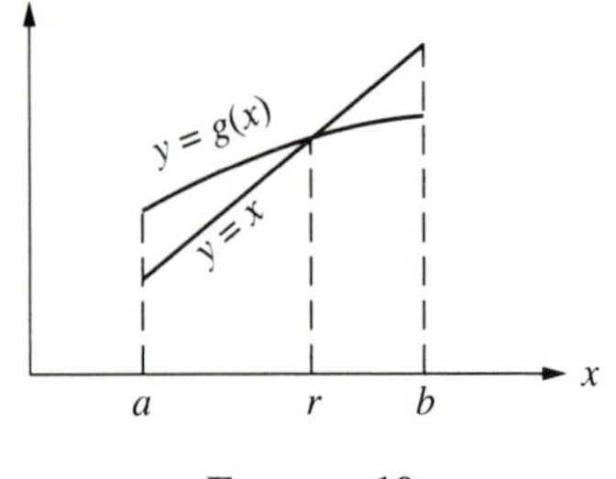

FIGURE 10

Such a function $g(x)$ is graphed in Fig. 10. It should be obvious from the figure that the equation $x = g(x)$ has a unique solution r in (a, b). Choose any initial approximation x_0 with $a \le x_0 < r$. Then the method of successive substitutions, $x_{n+1} = g(x_n)$, yields an increasing sequence of approximations x_n for r such that $x_n \to r$ as $n \to \infty$.

In the next example, the method of successive substitutions is applied to an equation $x = g(x)$ with $-1 < g'(x) < 0$. Once again the successive substitutions converge to the unique solution of $x = g(x)$. In this case the geometry is more interesting.

EXAMPLE 7. Apply the method of successive substitutions to the equation

$$x = g(x) \text{ with } g(x) = 1 + 1/x \text{ for } x > 0.$$

Solution. Since $x = g(x) \Leftrightarrow x^2 - x - 1 = 0$, the quadratic formula gives the unique positive solution

$$r = \frac{1}{2}(1 + \sqrt{5}) \approx 1.618.$$

This time, let $x_1 = 1$ be our initial approximation for r. Then

$$x_{n+1} = g(x_n) = 1 + \frac{1}{x_n} \text{ for } n = 0, 1, 2, \ldots.$$

Carry out the first several calculations to obtain

$$x_2 = 2 = 2.000, \qquad x_5 = \frac{8}{5} = 1.600,$$

$$x_3 = \frac{3}{2} = 1.500, \qquad x_6 = \frac{13}{8} = 1.625,$$

$$x_4 = \frac{5}{3} \approx 1.667, \qquad x_7 = \frac{21}{13} \approx 1.615$$

The first four approximations are graphed in Fig. 11. In this case we have a spiral pattern rather than a staircase pattern. From the graph,

$$x_1 < x_3 < x_5 < \cdots < r < \cdots < x_6 < x_4 < x_2.$$

The figure suggests that $x_n \to r$ as $n \to \infty$. Since $\{x_n\}$ is not monotone, we cannot apply Th. 1 directly to $\{x_n\}$ to reach this conclusion. To show that $x_n \to r$, we could apply Th. 1 to the "even" and "odd" subsequences of $\{x_n\}$ and then show that they have the same limit. We shall not take the time to complete the argument. □

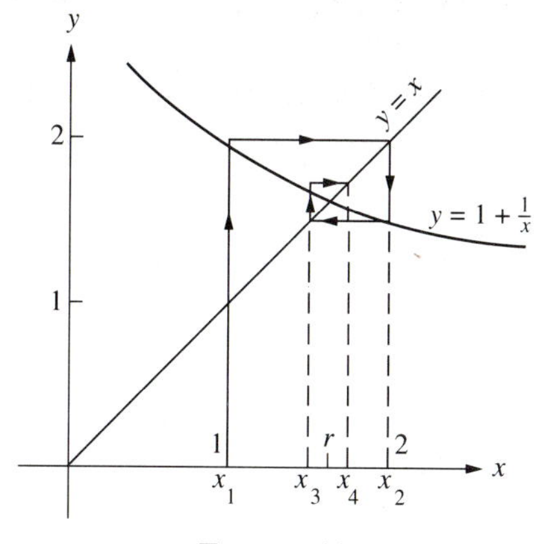

FIGURE 11

Here is an interesting application of Ex. 7. An algebra problem of the thirteenth century led to an extraordinarily useful sequence of numbers, called **Fibonacci numbers**. The first few Fibonacci numbers are

$$1, 1, 2, 3, 5, 8, 13, 21, 34, 45, 79, \cdots.$$

You may have noticed that, after the first two terms, each term of the sequence is the sum of the preceding two terms. Thus, the sequence is determined by

$$F_1 = 1, \qquad F_2 = 1, \qquad F_{n+1} = F_n + F_{n-1}.$$

In the old algebra problem, F_n is the number of pairs of rabbits that are present after n months of breeding. Evidently, the rabbit population grows rapidly. But the monthly rate of increase, which is given by $x_n = F_{n+1}/F_n$, stabilizes as n increases. In fact,

$$x_n = \frac{F_{n+1}}{F_n} \to \frac{1}{2}(1 + \sqrt{5}) \text{ as } n \to \infty.$$

To verify this limit, note that $x_1 = 1$ and

$$x_{n+1} = \frac{F_{n+2}}{F_{n+1}} = \frac{F_n + F_{n+1}}{F_{n+1}} = \frac{F_n}{F_{n+1}} + 1,$$

$$x_{n+1} = \frac{1}{x_n} + 1.$$

This is the same iteration formula we encountered in Ex. 7, where we learned that $x_n \to \frac{1}{2}(1 + \sqrt{5})$ as $n \to \infty$.

The iteration scheme $x_{n+1} = g(x_n)$ that was used in the preceding examples to find approximate solutions of equations has other significant applications. The project at the end of the chapter gives an introduction to how such schemes are used to study the evolution of physical and biological systems. From this point of view, the iteration scheme describes the evolution of a dynamical system.

PROBLEMS

In Probs. 1–20 do the following: (a) Show that the sequence $\{a_n\}$ is increasing or decreasing, or is eventually increasing or decreasing. (b) Show that $\{a_n\}$ is bounded, and find the least upper bound and greatest lower bound. (c) Find the limit of the sequence.

1. $\left\{\dfrac{4+n}{2n}\right\}_{n=1}^{\infty}$

2. $\left\{\dfrac{3n}{5+n}\right\}_{n=0}^{\infty}$

3. $\left\{\dfrac{n^2}{4+3n^2}\right\}_{n=0}^{\infty}$

4. $\left\{\dfrac{n-n^2}{3+4n^2}\right\}_{n=0}^{\infty}$

5. $\left\{\tan\left(\dfrac{\pi n^2-1}{n^2+4}\right)\right\}_{n=0}^{\infty}$

6. $\{\arctan n\}_{n=0}^{\infty}$

7. $\left\{\dfrac{10^n}{n!}\right\}_{n=0}^{\infty}$

8. $\left\{\dfrac{(n+2)!}{(2n)!}\right\}_{n=0}^{\infty}$

9. $\{n2^{-n}\}_{n=1}^{\infty}$

10. $\{n^{10}2^{-n}\}_{n=1}^{\infty}$

11. $\left\{\sin\dfrac{6}{n}\right\}_{n=1}^{\infty}$

12. $\left\{\cos\dfrac{6}{n}\right\}_{n=1}^{\infty}$

13. $\left\{n\sin\dfrac{1}{n}\right\}_{n=1}^{\infty}$

14. $\{ne^{-n}\}_{n=1}^{\infty}$

15. $\left\{\dfrac{\ln(n+2)}{n}\right\}_{n=1}^{\infty}$

16. $\left\{\dfrac{\ln(1+2/n)}{n}\right\}_{n=1}^{\infty}$

17. $\left\{\dfrac{(\ln n)^2}{n}\right\}_{n=1}^{\infty}$

18. $\{n^{1/n}\}_{n=1}^{\infty}$

19. $\left\{\sqrt{n}\sin\dfrac{1}{\sqrt{n}}\right\}_{n=1}^{\infty}$

20. $\left\{n\left(1-\cos\dfrac{1}{n}\right)\right\}_{n=1}^{\infty}$

In Probs. 21–26, use Newton's method, as follows, to approximate the root r of the given function in the given interval. (a) Show that f changes sign, is increasing or decreasing, and is concave up or down on the given interval $a \le x \le b$. (b) Choose $x_0 = a$ or b in order to obtain a sequence of Newton iterates that is monotone and has limit r. (c) Calculate the Newton iterates until $|x_{n+1} - x_n| < 10^{-4}$ for the first time. Report n and x_{n+1}.

21. $f(x) = x^2 - 5$ for $2 \le x \le 3$

22. $f(x) = xe^x - 3$ for $1 \le x \le 2$

23. $f(x) = e^{-x} - \ln x$ for $1 \le x \le e$

24. $f(x) = x^3 + 7x - 6$ for $0 \le x \le 1$

25. $f(x) = 2x \arctan x - 1$ for $0 \le x \le 1$

26. $f(x) = \cos x - x \sin x$ for $0 \le x\ \pi/3$

In Probs. 27–32, use successive substitutions to approximate a solution of the equation $x = g(x)$ for the given function $g(x)$ and starting value x_0. Proceed as follows: (a) Use a graphical analysis as in Fig. 9 to show that the equation has a unique root r. (b) Plot the successive iterates x_0, x_1, x_2, and x_3 on a graph as in Figs. 9 and 11. (c) Calculate the successive iterates until $|x_{n+1} - x_n| < 10^{-4}$ for the first time. Report n and x_{n+1}.

27. $g(x) = \frac{1}{2}\cos x$ for $x \ge 0$, $\quad x_0 = 0$

28. $g(x) = 2 + \ln x$ for $x \ge 1$, $\quad x_0 = 2$

29. $g(x) = e^{-x}$ for $x \ge 0$, $\quad x_0 = 1$

30. $g(x) = 1 + \arctan x$ for $x \ge 0$, $\quad x_0 = 0$

31. $g(x) = 2 + \tanh x$ for $x \ge 0$, $\quad x_0 = 0$

32. $g(x) = 1/\sqrt{x+1}$ for $x \ge 0$, $\quad x_0 = 3$

In the preceding problems the method of successive substitutions works very well. However, as indicated in the text, success depends on suitable properties of the function $g(x)$.

We cannot go into details here, but Probs. 33–35 hint at topics explored in numerical analysis courses.

33. Show that the equation $x^2 - 6 = 0$ can be expressed as $x = g(x)$ for (a)

    $g(x) = x^2 + x - 6$ (b) $g(x) = 6/x$, (c) $g(x) = (1/2)(x + 6/x)$, and (d) $g(x) = 1 + x - (1/6)x^2$.

34. In what follows $g(x)$ is from the corresponding part of the previous problem. Use successive substitutions with the given initial guesses and calculate several iterates in an attempt to approximate the solution $r = \sqrt{6} \approx 2.4495$ of the equation $x^2 - 6 = 0$. What do you conclude about convergence of the given successive substitution scheme?

    (a) $x_0 = 2$, $x_0 = 3$, $x_0 = 2.4495$
    (b) $x_0 = 2, 4, 7$ or any positive value except $\sqrt{6}$.
    (c) x_0 is any positive value.
    (d) $x_0 = 0, 1, 2, \ldots, 10$.

35. Use the following steps, motivated by successive substitutions, to show that the following interesting sequence converges and to find its limit:

$$\sqrt{2}\quad \sqrt{2+\sqrt{2}},\quad \sqrt{2+\sqrt{2+\sqrt{2}}},\quad \sqrt{2+\sqrt{2+\sqrt{2+\sqrt{2}}}},\ \cdots$$

(a) Let $x_0 = 0$ and $g(x) = \sqrt{2+x}$. Show that the preceding sequence is generated by the successive substitution scheme $x_{n+1} = g(x_n)$.
(b) Show that $\{x_n\}$ is bounded above. *Hint.* Show that $x < 2 \Rightarrow g(x) < 2$.
(c) Show that $\{x_n\}$ is increasing. *Hint.* Show that $g(x) > x$ on $[0,2]$.
(d) Now explain why the sequence $\{x_n\}$ converges to a limit L with $0 < L \leq 2$.
(e) Explain why $L = \sqrt{2+L}$ and solve this equation to find L, which is the limit of the sequence we started with.

36. Use reasoning similar to that in the previous problem to show that the sequence

$$\sqrt{6},\quad \sqrt{6+\sqrt{6}},\quad \sqrt{6+\sqrt{6+\sqrt{6}}},\quad \sqrt{6+\sqrt{6+\sqrt{6+\sqrt{6}}}},\ \cdots$$

converges and to find its limit.

37. Fix a positive number c. Define $x_1 = \sqrt{c}$ and $x_{n+1} = \sqrt{c + x_n}$. *Assuming* that $x_n \to L$ as $n \to \infty$, find L.

38. Prove that the sequence in the preceding problem is increasing and bounded above and hence does have a limit.

39. Let $x_0 = 1$ and $x_{n+1} = 1 - x_n^2$. (a) *Assuming* that $x_n \to L$ as $n \to \infty$, find L. (b) Calculate $x_1, x_2, x_3 \ldots$. What do you conclude about the assumption in (a)?

40. Let

$$x_0 = 1 \text{ and } x_{n+1} = \frac{1}{2}\left(x_n + \frac{4}{x_n}\right) \text{ for } n = 0, 1, 2, \ldots.$$

(a) *Assuming* that $x_n \to L$ as $n \to \infty$, find L. (b) Show that $f(x) = \frac{1}{2}(x + 4/x)$ for $x > 0$ has minimum $f(2) = 2$ and that $f(x) \leq x$ for all $x \geq 2$. (c) Conclude that $\{x_n\}_{n=1}^{\infty}$ is bounded below, nonincreasing, and, hence, convergent. (So the assumption in (a) is justified.)

41. Let

$$x_0 = 1 \text{ and } x_{n+1} = \frac{1}{2}\left(x_n - \frac{4}{x_n}\right) \text{ for } n = 0, 1, 2, \ldots.$$

(a) *Assuming* that $x_n \to L$ as $n \to \infty$, find L. (b) Can you say whether or not the sequence converges?

42. Fix a number $a > 0$. Let $x_0 > 0$ with $x_0 \neq \sqrt{a}$ an initial guess at $\sqrt{a}$. Show that the Newton sequence for finding a root of $x^2 - a = 0$ is determined by

$$x_{n+1} = \frac{1}{2}\left(x_n + \frac{a}{x_n}\right), \text{ for } n = 1, 2, \cdots.$$

(a) Prove that

$$x_1 > x_2 > \cdots > x_n > \cdots > \sqrt{a}.$$

Hint. See Prob. 40. (b) Conclude that the Newton sequence converges and find its limit.

1.3 Infinite Series

In everyday language, an infinite series is a sum with an infinite number of terms. Infinite series arose early in the development of calculus. Solutions to a great many mathematical and physical problems are represented as sums of infinite series. For example, the shape of a vibrating string can be expressed as an infinite series involving sines and cosines. Such series are called Fourier series. In this section we cover basic properties of infinite series. We shall learn that some series have sums and others do not. Along the way, we shall meet a number of special series, such as the harmonic series, geometric series, and series with telescoping sums. Further properties of infinite series will be developed later in this chapter and in the following chapter.

Convergence, Divergence, and Sums of Series

Before making general definitions, let's look at an example that leads naturally to an infinite series and the meaning of its sum.

Take a mile walk along a straight road, pictured as the x–axis in Fig. 1. Walk from 0 toward 1 in stages as follows. Walk halfway and stop; then walk half the remaining distance and stop; then walk half the remaining distance and stop; and so on and on. After many stages, you are very close to 1.

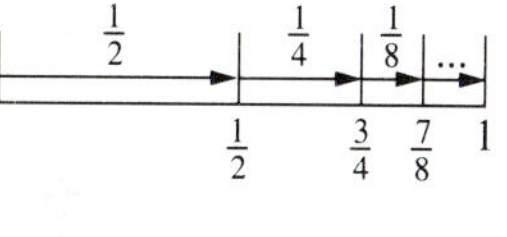

FIGURE 1

Figure 1 suggests that

$$\frac{1}{2}+\frac{1}{4}+\frac{1}{8}+\frac{1}{16}+\cdots=1.$$

The expression on the left is called an infinite series. Our intuition suggests that its sum should be 1. However, there is a problem. The rules of arithmetic assign a sum only when a finite number of terms are added. What sense can we make of a sum with an infinite number of terms?

The mode of travel suggests what to do. The distances traveled after completion of the successive stages of the trip in Fig. 1 are

$$S_1=\frac{1}{2}=1-\frac{1}{2},\qquad S_2=\frac{1}{2}+\frac{1}{4}=\frac{3}{4}=1-\frac{1}{4},\qquad S_3=\frac{1}{2}+\frac{1}{4}+\frac{1}{8}=\frac{7}{8}=1-\frac{1}{8},$$

and so forth. In general,

$$S_n=\frac{1}{2}+\frac{1}{4}+\frac{1}{8}+\cdots+\frac{1}{2^n}=1-\frac{1}{2^n}.$$

One way to derive the formula for S_n is by induction. A quicker argument uses

$$S_n=\frac{1}{2}+\frac{1}{4}+\frac{1}{8}+\cdots+\frac{1}{2^{n-1}}+\frac{1}{2^n},$$

$$2S_n=1+\frac{1}{2}+\frac{1}{4}+\cdots+\frac{1}{2^{n-2}}+\frac{1}{2^{n-1}}.$$

Subtract the formulas for $2S_n$ and S_n to obtain $S_n=1-1/2^n$. (Most of the terms subtract out.) The distance traveled during the entire trip ought to be the limit of S_n as $n \to \infty$, which is given by

$$\lim_{n \to \infty} S_n = \lim_{n \to \infty} \left(1 - \frac{1}{2^n}\right) = 1.$$

In view of the definition of S_n,

$$\lim_{n \to \infty} \left(\frac{1}{2} + \frac{1}{4} + \frac{1}{8} + \cdots + \frac{1}{2^n}\right) = 1.$$

This limit statement is what is meant by

$$\frac{1}{2} + \frac{1}{4} + \frac{1}{8} + \cdots + \frac{1}{2^n} + \cdots = 1$$

or, more briefly, in summation notation by

$$\sum_{n=1}^{\infty} \frac{1}{2^n} = 1.$$

In general, an expression of the form

$$\sum_{n=1}^{\infty} a_n = a_1 + a_2 + a_3 + \cdots + a_n + \cdots$$

is called an **infinite series**, or just a **series**. The ***n*th term** of the series is a_n. The ***n*th partial sum** is

$$S_n = a_1 + a_2 + \cdots + a_n = \sum_{k=1}^{n} a_k.$$

If $\lim_{n \to \infty} S_n = S$ exists (finite or infinite), then S is the **sum** of the series and we write

$$\sum_{n=1}^{\infty} a_n = S \text{ or } a_1 + a_2 + a_3 + \cdots = S.$$

We say that a series **converges** if it has a finite sum. From the introductory discussion,

$$\sum_{n=1}^{\infty} \frac{1}{2^n} = 1.$$

So this series converges.

A series **diverges** if it has an infinite sum or no sum at all. A series with the sum ∞ (or $-\infty$) is said to *diverge to* ∞ (or $-\infty$). Examples of divergent series are

$$\sum_{n=1}^{\infty} n^2 = 1 + 4 + 9 + 16 + \cdots = \infty,$$

$$\sum_{n=0}^{\infty}(-1)^n = 1 - 1 + 1 - 1 + \cdots.$$

The partial sums of the last series are 1, 0, 1, 0, ..., which obviously have no limit. So the series $\sum_{n=0}^{\infty}(-1)^n$ has no sum.

In the last example, the summation index n started with $n = 0$ instead of $n = 1$. From time to time, we shall use other starting values and other summation indices besides n. The symbol used for the index of summation really doesn't matter because it is a dummy variable, much like a variable of integration.

Next we give several examples of infinite series with finite or infinite sums. Our first example illustrates the close connection between unending decimals and infinite series.

EXAMPLE 1. The unending decimal for 1/3 is $0.3333\cdots$, which means that

$$\frac{1}{3} = \lim_{n\to\infty}(0.333\cdots33) \text{ with } n - \text{place decimals.}$$

Observe that

$$0.333 = 0.3 + 0.03 + 0.003 = \frac{3}{10} + \frac{3}{10^2} + \frac{3}{10^3}.$$

Similarly, for n–place decimals,

$$0.333\cdots33 = 0.3 + 0.03 + 0.003 + \cdots + 0.000\cdots03 = \frac{3}{10} + \frac{3}{10^2} + \frac{3}{10^3}\cdots + \frac{3}{10^n}.$$

Thus, the partial sums of the infinite series $\sum_{n=1}^{\infty} 3/10^n$ are the n–place decimals $0.333\cdots33$, which converge to 1/3 as $n\to\infty$. It follows that

$$\sum_{n=1}^{\infty}\frac{3}{10^n} = \frac{1}{3}. \ \square$$

EXAMPLE 2. Verify that

$$\sum_{n=1}^{\infty}\frac{1}{\sqrt{n}} = \infty.$$

Solution. The nth partial sum of the series is

$$S_n = 1 + \frac{1}{\sqrt{2}} + \frac{1}{\sqrt{3}} + \cdots + \frac{1}{\sqrt{n}}.$$

Since the smallest term on the right is $1/\sqrt{n}$,

$$S_n > \frac{1}{\sqrt{n}} + \frac{1}{\sqrt{n}} + \frac{1}{\sqrt{n}} + \cdots + \frac{1}{\sqrt{n}} = n\cdot\frac{1}{\sqrt{n}} = \sqrt{n}.$$

It follows that $S_n \to \infty$ as $n \to \infty$. So the series diverges to ∞. □

The conclusion that the series

$$\sum_{n=1}^{\infty} \frac{1}{\sqrt{n}} = 1 + \frac{1}{\sqrt{2}} + \frac{1}{\sqrt{3}} + \cdots + \frac{1}{\sqrt{n}} + \cdots$$

diverges to ∞ is often troubling at first to students who notice that the general term of the series tends to zero:

$$\frac{1}{\sqrt{n}} \to 0 \text{ as } n \to \infty.$$

It is tempting to believe that a series must converge if its nth term tends to 0. This general statement is FALSE, as Ex. 2 shows. Loosely speaking, an infinite series converges if its nth term tends to zero fast enough. However, there is no criterion that distinguishes all convergent series from divergent series. The following important series provides a further illustration of the danger of guessing whether a series converges or diverges just by looking at the size of the nth term.

The **harmonic series** is

$$\sum_{n=1}^{\infty} \frac{1}{n} = 1 + \frac{1}{2} + \frac{1}{3} + \cdots + \frac{1}{n} + \cdots.$$

The nth term is $1/n$. It tends to zero very much faster than the nth term of the series $\sum_{n=1}^{\infty} 1/\sqrt{n}$ because

$$\frac{1/n}{1/\sqrt{n}} = \frac{1}{\sqrt{n}} \to 0 \quad \text{as} \quad n \to \infty.$$

Nevertheless, the nth term of the harmonic series does not tend to zero *fast enough* to make the series converge. The harmonic series diverges to ∞:

$$\boxed{\sum_{n=1}^{\infty} \frac{1}{n} = \infty.}$$

To prove that the harmonic series diverges, we use an argument similar to the one used in Ex. 2, but a little more complicated. Consider the partial sums

$$S_n = 1 + \frac{1}{2} + \frac{1}{3} + \cdots + \frac{1}{n}.$$

Clearly, $\{S_n\}$ is increasing. Observe that

$$\frac{1}{3} + \frac{1}{4} > \frac{1}{4} + \frac{1}{4} = \frac{1}{2}, \qquad \frac{1}{5} + \frac{1}{6} + \frac{1}{7} + \frac{1}{8} > \frac{1}{8} + \frac{1}{8} + \frac{1}{8} + \frac{1}{8} = \frac{1}{2}.$$

It follows that

$$S_1 = 1,$$

$$S_2 = 1 + \left(\frac{1}{2}\right),$$

$$S_4 = 1 + \left(\frac{1}{2}\right) + \left(\frac{1}{3} + \frac{1}{4}\right) > 1 + 2\left(\frac{1}{2}\right),$$

$$S_8 = 1 + \left(\frac{1}{2}\right) + \left(\frac{1}{3} + \frac{1}{4}\right) + \left(\frac{1}{5} + \frac{1}{6} + \frac{1}{7} + \frac{1}{8}\right) > 1 + 3\left(\frac{1}{2}\right).$$

Since S_n increases as n increases

$$S_n > S_4 > 1 + 2\left(\frac{1}{2}\right) \text{ for } n > 4 = 2^2,$$

$$S_n > S_8 > 1 + 3\left(\frac{1}{2}\right) \text{ for } n > 8 = 2^3.$$

The pattern continues. For any $m = 1, 2, 3, \cdots,$

$$S_n > 1 + m\left(\frac{1}{2}\right) \text{ for } n > 2^m.$$

Therefore, $S_n \to \infty$ as $n \to \infty$ and the harmonic series diverges to infinity.

Convergence or Divergence?

A central question in the study of infinite series is whether a given series converges or diverges. We present a few simple tests for convergence or divergence here. Additional tests will be given in the coming sections.

We begin with a simple but useful observation. Consider the two series

$$1 + \frac{1}{2} + \frac{1}{4} + \cdots + \frac{1}{2^n} + \cdots \quad \text{and} \quad 5 + \frac{1}{2} + \frac{1}{4} + \cdots + \frac{1}{2^n} + \cdots.$$

We know that the first series converges. What about the second series? Each partial sum of the second series is four more than the corresponding partial sum of the first series. It follows that the second series also converges. By the same reasoning, the series

$$1 + \frac{1}{2} + \frac{1}{3} + \cdots + \frac{1}{n} + \cdots \quad \text{and} \quad 5 + \frac{1}{2} + \frac{1}{3} + \cdots + \frac{1}{n} + \cdots$$

both diverge. Similar arguments show that, for any m,

$$\sum_{n=1}^{\infty} a_n \text{ and } \sum_{n=m}^{\infty} a_n \text{ both converge or both diverge.}$$

A very important conclusion follows.

> Any finite number of terms of a series can be changed, removed, or added without any effect on the convergence or divergence of the series.

We often express an infinite series by Σa_n without indicating the starting value of n. Usually the starting value is clear from the context. Often it does not matter because the starting index has no effect on the convergence or divergence of a series.

The following properties are direct consequences of corresponding properties for finite sums and algebraic limit laws.

> Assume that Σa_n and Σb_n converge and c is any number. Then Σca_n and $\Sigma(a_n \pm b_n)$ converge and
> $$\Sigma ca_n = c\,\Sigma a_n,$$
> $$\Sigma(a_n \pm b_n) = \Sigma a_n \pm \Sigma b_n.$$

For example, since

$$\sum_{n=1}^{\infty} \frac{1}{2^n} = 1 \quad \text{and} \quad \sum_{n=1}^{\infty} \frac{3}{10^n} = \frac{1}{3},$$

it follows that

$$\sum_{n=1}^{\infty} \frac{1}{10^n} = \sum_{n=1}^{\infty} \frac{1}{3} \cdot \frac{3}{10^n} = \frac{1}{3} \sum_{n=1}^{\infty} \frac{3}{10^n} = \frac{1}{3}\left(\frac{1}{3}\right) = \frac{1}{9},$$

$$\sum_{n=1}^{\infty} \left(\frac{5}{2^n} - \frac{18}{10^n}\right) = 5(1) - 18\left(\frac{1}{9}\right) = 3.$$

It should seem plausible that the nth term of a series must tend to zero for there to be any hope that the series converges. To confirm this statement, suppose that the series Σa_n converges to S and that the indexing starts with $n = 1$. Then

$$S_n = a_1 + a_2 + \cdots + a_{n-1} + a_n = S_{n-1} + a_n,$$

$$a_n = S_n - S_{n-1} \to S - S = 0 \quad \text{as} \quad n \to \infty.$$

We have established the following very important result.

> If Σa_n converges, then $a_n \to 0$ as $n \to \infty$.

For example, the series $\Sigma 1/2^n$ converges and $1/2^n \to 0$ as $n \to \infty$. It is worth pointing out again that $a_n \to 0$ does *not* imply that Σa_n converges. Remember that the harmonic series $\Sigma 1/n$ diverges, even though $1/n \to 0$ as $n \to \infty$.

Suppose that $a_n \not\to$, which means that a_n does not converge to zero. Thus, a_n has no limit or has a limit different from 0. Then Σa_n cannot converge because, if it did, then $a_n \to 0$ according to the preceding displayed statement. This simple observation gives us a very useful test for divergence.

Basic Divergence Test
If $a_n \not\to 0$, then Σa_n diverges.

EXAMPLE 3. The series $\Sigma e^{1/n}$ diverges because $e^{1/n} \to 1 \neq 0$ as $n \to \infty$. □

Geometric Series

The most important series in calculus and its applications is the **geometric series**

$$\sum_{n=0}^{\infty} x^n = 1 + x + x^2 + \cdots + x^n + \cdots.$$

The following theorem gives its essential convergence properties.

Theorem 1 *The Geometric Series*

$$\sum_{n=0}^{\infty} x^n = \frac{1}{1-x} \quad \text{if} \quad |x| < 1.$$

The series diverges if $|x| \geq 1$.

For example, if $x = 1/2$ then $\Sigma_{n=0}^{\infty} 1/2^n = 2$, and if $x = 2$ then $\Sigma_{n=0}^{\infty} 2^n = 1 + 2 + 4 + 8 + \cdots = \infty$.

To prove Th. 1 and obtain some important related results, we begin with the partial sums

$$S_n = 1 + x + x^2 + \cdots + x^n.$$

Then

$$xS_n = x + x^2 + x^3 + \cdots + x^n + x^{n+1},$$
$$S_n - xS_n = 1 - x^{n+1},$$
$$S_n = \frac{1 - x^{n+1}}{1-x} \quad \text{for } x \neq 1.$$

It follows that

$$1 + x + x^2 + \cdots + x^n = \frac{1 - x^{n+1}}{1-x} \quad \text{for } x \neq 1,$$

$$\frac{1}{1-x} = 1 + x + x^2 + \cdots + x^n + \frac{x^{n+1}}{1-x} \quad \text{for } x \neq 1.$$

If $|x| < 1$, then $x^{n+1} \to 0$ as $n \to \infty$ and, hence,

$$1 + x + x^2 + \cdots + x^n \to \frac{1}{1-x} \quad \text{for any} \quad |x| < 1.$$

Equivalently,

$$\sum_{n=0}^{\infty} x^n = \frac{1}{1-x} \quad \text{if} \quad |x| < 1,$$

which proves the first assertion in Th. 1. If $|x| \geq 1$, then $x^n \not\to 0$ and, hence, Σx^n diverges, which proves the second assertion in Th. 1.

A more general geometric series is

$$\sum_{n=0}^{\infty} a r^n$$

where $a \neq 0$ and r are given numbers. The number r is the **common ratio** of two successive terms of the series. Evidently,

$$\sum_{n=0}^{\infty} a r^n = \frac{a}{1-r} \quad \text{if} \quad |r| < 1$$

and the series diverges if $|r| \geq 1$.

EXAMPLE 4.

$$\sum_{n=0}^{\infty} \frac{4}{5^n} = \sum_{n=0}^{\infty} 4\left(\frac{1}{5}\right)^n = \frac{4}{1-1/5} = \frac{5 \cdot 4}{5-1} = 5.$$

Consequently,

$$\sum_{n=1}^{\infty} \frac{4}{5^n} = \left(\sum_{n=0}^{\infty} \frac{4}{5^n}\right) - 4 = 5 - 4 = 1. \ \square$$

EXAMPLE 5.

$$\sum_{n=0}^{\infty} \frac{3^n - 2}{5^n} = \sum_{n=0}^{\infty} \left(\frac{3}{5}\right)^n - 2\sum_{n=0}^{\infty} \left(\frac{1}{5}\right)^n = \frac{1}{1-3/5} - \frac{2}{1-1/5}$$

$$= \frac{5}{5-3} - \frac{10}{5-1} = \frac{5}{2} - \frac{10}{4} = 0\,! \ \square$$

EXAMPLE 6. Express the unending periodic decimals $0.2222\cdots$ and $.5373737\cdots$ in fractional form.

Solution. First,

$$0.2222\cdots = \frac{2}{10} + \frac{2}{10^2} + \frac{2}{10^3} + \cdots = \frac{2}{10}\left(1 + \frac{1}{10} + \frac{1}{10^2} + \cdots\right) = \frac{2}{10} \cdot \frac{1}{1-1/10} = \frac{2}{9}.$$

Next,

$$0.5373737\cdots = 0.5 + \frac{37}{10^3} + \frac{37}{10^5} + \frac{37}{10^7} + \cdots = \frac{1}{2} + \frac{37}{10^3}\left(1 + \frac{1}{10^2} + \frac{1}{10^4} + \cdots\right)$$

$$= \frac{1}{2} + \frac{37}{10^3}\left(\frac{1}{1 - 1/10^2}\right) = \frac{1}{2} + \frac{37}{990} = \frac{1064}{1980}.$$

Evaluate 1064/1980 on a calculator to recover several of the periodic decimal places. □

Variants of Geometric Series

The geometric series

$$\sum_{n=0}^{\infty} x^n = 1 + x + x^2 + x^3 + \cdots = \frac{1}{1-x} \quad \text{for } |x| < 1$$

is often used with other indices and with various changes of variable. Such variants of the geometric series will become increasingly important as we go along. For the moment, we just give a few examples to illustrate the possibilities.

First, observe that

$$\sum_{n=1}^{\infty} x^n = x + x^2 + x^3 + \cdots = x(1 + x + x^2 + \cdots) = \frac{x}{1-x} \quad \text{for } |x| < 1.$$

What is the sum if the starting index is $n = 2$?

Replace x by $-x$ in the geometric series and use the fact that $|-x| < 1$ if and only if $|x| < 1$ to obtain

$$\sum_{n=0}^{\infty} (-1)^n x^n = 1 - x + x^2 - x^3 + \cdots = \frac{1}{1+x} \quad \text{for } |x| < 1.$$

For example,

$$\sum_{n=0}^{\infty} \frac{(-1)^n}{3^n} = 1 - \frac{1}{3} + \frac{1}{3^2} - \frac{1}{3^3} + \cdots = \frac{1}{1 + (1/3)} = \frac{3}{4}.$$

In the series for $1/(1-x)$ and $1/(1+x)$ let $x = t^2$ and use the fact that $|t^2| < 1$ if and only if $|t| < 1$ to obtain

$$\sum_{n=0}^{\infty} t^{2n} = 1 + t^2 + t^4 + t^6 + \cdots = \frac{1}{1-t^2} \quad \text{for } |t| < 1,$$

$$\sum_{n=0}^{\infty} (-1)^n t^{2n} = 1 - t^2 + t^4 - t^6 + \cdots = \frac{1}{1+t^2} \quad \text{for } |t| < 1.$$

There is no need to memorize formulas such as these. They can be recovered from the geometric series by simple changes of variable whenever you need them.

Applications of Geometric Series

It's time for a little diversion. We shall give two applications of geometric series. The first could be called "follow the bouncing ball", and the second could be called "who wins?".

EXAMPLE 7. A ball is dropped from a height of h feet. It bounces repeatedly but loses energy so that on each bounce the height it reaches is only a fixed fraction r times the previous height, where $0 < r < 1$. What is the total distance traveled by the ball? What if $h = 10$ feet and $r = 3/4$?

Solution. The motion of the bouncing ball is described in the following table.

distance up		rh	r^2h	r^3h	$\cdots$
distance down	h	rh	r^2h	r^3h	$\cdots$

Indeed, the ball first falls a distance h, then bounces up a distance rh, then falls the same distance rh, then bounces up a distance $r(rh) = r^2h$, and so on. The total distance traveled downward is

$$S_d = \sum_{n=0}^{\infty} hr^n = \frac{h}{1-r}.$$

The total distance traveled upward is

$$S_u = \sum_{n=1}^{\infty} hr^n = h\left(\sum_{n=1}^{\infty} r^n\right) = \frac{hr}{1-r}.$$

Thus, the total distance traveled by the bouncing ball is

$$S = S_d + S_u = \frac{h}{1-r} + \frac{hr}{1-r} = h\left(\frac{1+r}{1-r}\right).$$

In particular, if $h = 10$ feet and $r = 3/4$, then $S = 70$ feet. □

EXAMPLE 8. George and Martha take turns tossing a possibly biased coin. Martha tosses first. The game is won by the player who first tosses a head. What is the probability that Martha will win?

Solution. We use a few laws of probability that should seem reasonable. Let

$$h = \text{ the probability of heads and } 0 < h < 1.$$

Then

$$t = 1 - h \text{ is the probability of tails and } 0 < t < 1.$$

Martha wins precisely when one of the following events occurs:

Martha wins on her first toss,
Martha wins on her second toss,
Martha wins on her third toss,

and so on. These events are mutually exclusive, which means that no two of them can occur. Let

p_n = the probability that Martha wins on her nth toss,
p = the probability that Martha wins the game.

Then

$$p = p_1 + p_2 + p_3 + \cdots = \sum_{n=1}^{\infty} p_n.$$

The probabilities add because the events of winning on a particular toss are mutually exclusive. Martha wins on her nth toss if both Martha and George throw tails on their first $n - 1$ tosses and Martha throws a head on her nth toss. The probability of $2n - 2$ tails and one head is

$$p_n = t^{2n-2}h = t^{2(n-1)}h.$$

The probabilities multiply because the events of a head or tail on separate tosses of the coin are independent. The outcome on any one toss has no effect on the other tosses. Then, because $0 < t < 1$ and $h = 1 - t$,

$$p = \sum_{n=1}^{\infty} p_n = \sum_{n=1}^{\infty} ht^{2(n-1)} = h(1 + t^2 + t^4 + t^6 + \cdots) = \frac{h}{1 - t^2} = \frac{1}{1 + t}.$$

The probability that Martha wins the game is

$$p = \frac{1}{1 + t} = \frac{1}{2 - h}$$

and the probability that George wins is

$$q = 1 - p.$$

Since $0 < t < 1$,

$$\frac{1}{2} < p < 1 \quad \text{and} \quad 0 < q < \frac{1}{2}.$$

We conclude that Martha, who plays first, always has a better chance of winning the game. (Can you explain why this must be so without any calculations?) Of course, the solution gives more information. It gives the probability that Martha will win. For example, if the coin is fair, then $h = t = 1/2$ and $p = 2/3$. Notice that if h is very near 0 and, correspondingly, t is very near 1, then both p and q are very near 1/2. Why should you expect such an outcome? □

Telescoping Sums

A glance back through this section will confirm that the only convergent series whose sums we found were particular geometric series. For such series, we were able to derive a convenient form for the nth partial sum, which enabled us to find the sum of the series. This pleasant state of affairs happens only for certain special series, which include series with telescoping sums.

EXAMPLE 9. Show that the series

$$\sum_{n=1}^{\infty} \frac{1}{n(n+1)} = \frac{1}{1 \cdot 2} + \frac{1}{2 \cdot 3} + \frac{1}{3 \cdot 4} + \cdots$$

converges and find its sum.

Solution. The nth partial sum is

$$S_n = \frac{1}{1 \cdot 2} + \frac{1}{2 \cdot 3} + \frac{1}{3 \cdot 4} + \cdots + \frac{1}{n(n+1)} = \sum_{k=1}^{n} \frac{1}{k(k+1)}.$$

A new index of summation k is used because n serves another purpose. Notice the similarity between

$$\frac{1}{k(k+1)} \quad \text{and} \quad \frac{1}{x(x+1)}.$$

This suggests the partial fractions expansion

$$\frac{1}{k(k+1)} = \frac{1}{k} - \frac{1}{k+1}.$$

Then

$$S_n = \sum_{k=1}^{n} \left(\frac{1}{k} - \frac{1}{k+1}\right) = \left(\frac{1}{1} - \frac{1}{2}\right) + \left(\frac{1}{2} - \frac{1}{3}\right) + \left(\frac{1}{3} - \frac{1}{4}\right) + \cdots + \left(\frac{1}{n} - \frac{1}{n+1}\right).$$

Notice the cancellation of all the pairs of "interior" terms in the sum on the right. It is called a *telescoping sum.* We conclude that

$$S_n = 1 - \frac{1}{n+1} \to 1 \quad \text{as} \quad n \to \infty,$$

$$\sum_{n=1}^{\infty} \frac{1}{n(n+1)} = 1. \ \square$$

Although telescoping sums are very special, they do crop up in unexpected places and it is good to be on the lookout for them.

A Look Ahead

Two basic questions come up in typical problems involving infinite series.

1. Does a given series converge or diverge?
2. What is its sum, if there is one?

In some cases, such as for the geometric series and for series with telescoping sums, the nth partial sum S_n of the series can be expressed in a simple form that enables us to find its limit S. Then both questions are answered at once. However, often there is no convenient expression for S_n and other means are needed to answer the two questions. As we mentioned earlier, several tests for convergence or divergence are developed in later sections.

In general, it is much harder to find the sum of a series than to show whether the series converges or diverges. In certain cases, the sum of a series can be discovered by ingenious mathematical arguments. For example, in about 1736 Euler showed that

$$\sum_{n=1}^{\infty} \frac{1}{n^2} = \frac{\pi^2}{6}.$$

However, no simple formula is known for the partial sums of this series. On the other hand, as we shall see, the series

$$\sum_{n=1}^{\infty} \frac{1}{n^3}$$

converges, but nobody knows its exact sum. The best we can do is to estimate the sum for such a series. Often the straightforward approach works: Just approximate the sum S by a partial sum S_n for some large n. This works reasonably well for $\sum_{n=1}^{\infty} 1/n^3$ because the series converges rather rapidly, meaning that $S_n \to S$ rather rapidly. The same approach is a disaster for the convergent series $\sum_{n=2}^{\infty} 1/n(\ln n)^2$ because the series converges very slowly. Some other ways to estimate sums of series will be given as we go along. Stay tuned!

PROBLEMS

In Probs. 1–27 determine whether the series converges, diverges to ∞ or $-\infty$, or has no sum. In case of convergence, find the sum.

1. $\sum_{n=1}^{\infty} 4/10^n$
2. $\sum_{n=1}^{\infty} (-1)^{n-1} 4/10^n$
3. $\sum_{n=0}^{\infty} 2^n/3^n$
4. $\sum_{n=0}^{\infty} (\sqrt{3})^{-n}$
5. $\sum_{n=0}^{\infty} 2^{n-1}/3^n$
6. $\sum_{n=0}^{\infty} 2^n/3^{n-1}$
7. $\sum_{n=2}^{\infty} (-1)^n 2^n/3^n$
8. $\sum_{n=2}^{\infty} (-1)^{n-1} 2^{n-1}/3^{n+1}$
9. $\sum_{n=0}^{\infty} \dfrac{3 \cdot 2^n + 4 \cdot 3^n}{5^n}$
10. $\sum_{n=0}^{\infty} \dfrac{5/2^n - 3/4^n}{1/3^n}$
11. $1 + \dfrac{1}{\sqrt{2}} + \dfrac{1}{\sqrt[3]{3}} + \dfrac{1}{\sqrt[4]{4}} + \cdots + \dfrac{1}{\sqrt[n]{n}} + \cdots$
12. $1 + 2^{-1/2} + 2^{-1/3} + 2^{-1/4} + \cdots + 2^{-/n} + \cdots$
13. $1 - e^{-\pi} + e^{-2\pi} - e^{-3\pi} + e^{-4\pi} - + \cdots$
14. $1 - 2 + 3 - 4 + 5 - 6 + - \cdots$
15. $\ln\left(\frac{2}{1}\right) + \ln\left(\frac{3}{2}\right) + \ln\left(\frac{4}{3}\right) + \ln\left(\frac{5}{4}\right) + \ln\left(\frac{6}{6}\right) + \cdots$
16. $\sum_{n=1}^{\infty} n \sin(1/n)$
17. $\sum_{n=0}^{\infty} \arctan n$

18. $\sum_{n=0}^{\infty} \sin n\pi$

19. $\sum_{n=0}^{\infty} \cos n\pi$

20. $\sum_{n=1}^{\infty} 1/n^{(1-1/n)}$

21. $\sum_{n=1}^{\infty} \ln n/n$

22. $\sum_{n=2}^{\infty} \frac{1}{n^2-1}$

23. $\sum_{n=0}^{\infty} \frac{1}{4n^2-1}$

24. $\sum_{n=2}^{\infty} 1/(n^2-n)$

25. $\sum_{n=1}^{\infty} 1/(n^2+5n+6)$

26. $\sum_{n=1}^{\infty} \left(\frac{n-1}{n}\right)^n$

27. $\sum_{n=1}^{\infty} \left(\frac{n^2-1}{n^2}\right)^n$

In Probs. 28–31, express the following unending decimals as fractions.

28. $0.4444\cdots$

29. $0.43434343\cdots$

30. $0.12340404040\cdots$

31. $0.101101101101\cdots$

32. Explain why

$$\sum_{n=1}^{\infty} \left(\frac{3}{n} - \frac{4}{5^n}\right)$$

diverges.

33. (a) Write out the first five terms of the series

$$\sum_{n=1}^{\infty} \frac{1}{2n-1} \quad \text{and} \quad \sum_{k=0}^{\infty} \frac{1}{2k+1}.$$

(b) Make the change of variable $k = n - 1$ in the first series to obtain the second series.

34. (a) Write out the first five terms of the series

$$\sum_{n=3}^{\infty} \frac{(-1)^{n-1}}{(n-2)(n+3)} \quad \text{and} \quad \sum_{p=0}^{\infty} \frac{(-1)^p}{(p+1)(p+6)}.$$

(b) Make the change of variable $p = n - 3$ in the first series to obtain the second series.

35. (a) Show that

$$\sum_{n=1}^{\infty} \frac{3}{n(n+3)} = 1 + \frac{1}{2} + \frac{1}{3}.$$

(b) Let p be a fixed positive integer. What do you suspect the sum of the series $\sum_{n=1}^{\infty} p/n(n+p)$ to be?
(c) Prove your guess. *Hint.* Let $S_N = \sum_{n=1}^{N} p/n(n+p)$. Use partial fractions. Note that $\sum_{n=1}^{N} 1/(n+p) = \sum_{k=p+1}^{p+N} 1/k = \sum_{n=p+1}^{p+N} 1/k$.

In Probs. 36–41, find series expansions, similar to the geometric series expansion of $1/(1-x)$, for the following functions. State the set of x values for which the expansions are valid.

36. $\frac{x}{1-x}$

37. $\frac{x}{1-2x}$

38. $\dfrac{1}{1 + 9x^2}$

39. $\dfrac{x^2}{1 - 4x^2}$

40. $\dfrac{x - x^2}{(1 - x)(1 + 2x)}$

41. $\dfrac{x^3 - 3x^2}{x^2 - 4x + 3}$

In Probs. 42–51, find the sum of the indicated series and state the set of x values for which the equality holds. (You may wish to write out several terms of the series.)

42. $\sum_{n=2}^{\infty} x^n$

43. $\sum_{n=3}^{\infty} x^n$

44. $\sum_{n=0}^{\infty} x^{2n}$

45. $\sum_{n=0}^{\infty} x^{3n-1}$

46. $\sum_{n=0}^{\infty} \dfrac{(-1)^n}{4^n} x^n$

47. $\sum_{n=0}^{\infty} \dfrac{(-1)^n x^{2n}}{2^n}$

48. $\sum_{n=1}^{\infty} (-1)^{n+1}(x + 2)^n$

49. $\sum_{n=0}^{\infty} (2x - 1)^n$

50. $\sum_{n=0}^{\infty} \dfrac{x^n}{(1 - x)^n}$

51. $\sum_{n=0}^{\infty} 2^n \sin^n x$

In Probs. 52–59, (a) make the indicated change of variable $x = f(t)$ in the geometric series $\sum_{n=0}^{\infty} x^n = 1/(1 - x)$ for $|x| < 1$ to obtain a related series in t. (b) State the range of t values for which the new series is valid. (c) Write out the first five terms in the new series.

52. $x = -2t$

53. $x = t/2$

54. $x = \sqrt{t}$

55. $x = 2 \cos t$

56. $x = t/(1 + t)$

57. $x = 2t/(1 - t)$

58. $x = \ln t$

59. $x = e^{-t}$

60. In Ex. 8, George and Martha toss a fair six–sided die instead of a biased coin. This time, the game is won by the first person to throw a 3. Martha tosses first. Find the probability that she wins.

61. George and Martha toss a pair of fair dice. Martha tosses first, as usual. The game is won by the first person to throw a 7; that is the total of the pips showing is 7. Find the probability that Martha wins.

62. An urn contains 100 red balls and 200 blue balls. Jack and Jackie take turns drawing balls from the urn. Jack goes first. Each player draws a ball, records its color, and then replaces the ball. The urn is stirred between plays to keep the balls thoroughly mixed. Under these conditions, the probability of drawing a red ball is 1/3 and the probability of drawing a blue ball is 2/3. The first person to draw a red ball wins. Find the probability that Jackie wins.

63. Start with a square with side 2. Join the midpoints of its sides to form a new square. Join the midpoints of the sides of the new square to form yet another square, and so on forever. (a) Make a sketch of this procedure. (b) Find the sum of the areas of all the squares.

64. Start with any triangle. Let A be its area and P be its perimeter. Join the midpoints of its sides to form a new triangle. Join the midpoints of the sides of the new triangle to form yet another new triangle, and so on forever. (a) Make a *careful* sketch of this procedure. (b) Find the sum of the areas of all the triangles. (c) Find the sum of all the perimeters of the triangles.

65. *(The Multiplier Effect in Economics)* A new, primary industry moves into your community. Its annual payroll is \$10 million. Assume each employee of the new company spends a fixed fraction r of his or her income in the community. The merchants in the community who take in the new income, in turn, spend the same fraction r of their added revenues. This spending goes to others in the community who, in turn, spend the fraction r of their added income, and so on. Find the total secondary spending in the community generated by the \$10 million in new primary income.

1.4 Series with Nonnegative Terms and Comparison Tests

In this section we consider only series with nonnegative terms. The question of convergence or divergence is settled more easily for series of this kind. We shall find that every series with nonnegative terms has a finite or infinite sum. The series converges if the sum is finite and the series diverges if the sum is infinite.

We shall give three comparison tests that tell us whether a series Σa_n with $a_n \geq 0$ for all n converges or diverges. The first test, called the basic comparison test, compares Σa_n term by term with another series Σb_n that is known to converge or diverge. The second test, called the limit comparison test, compares a_n and b_n as $n \to \infty$. The third comparison test, called the integral test, compares Σa_n with an improper integral $\int_1^\infty a(x)dx$, where a_n and $a(x)$ are related by $a(n) = a_n$.

Series with Nonnegative Terms

Two series with nonnegative terms are

$$\sum_{n=1}^{\infty} \frac{1}{2^n} = 1 \quad \text{and} \quad \sum_{n=1}^{\infty} \frac{1}{\sqrt{n}} = \infty.$$

The first series has a finite sum and it converges. The second has an infinite sum and it diverges. These series are typical of the general situation.

Consider a series Σa_n with every $a_n \geq 0$. For convenience, let $n = 1, 2, 3, \ldots$. Then the nth partial sum is $S_n = a_1 + a_2 + \cdots + a_{n-1} + a_n$. Since $S_n = S_{n-1} + a_n \geq S_{n-1}$, the sequence $\{S_n\}$ is nondecreasing. By Th. 2 in Sec 1.2,

$$S_n \text{ has a finite limit if } \{S_n\} \text{ is bounded,}$$
$$S_n \to \infty \text{ if } \{S_n\} \text{ is unbounded.}$$

Since the limit of S_n is the sum of the series, we have established

> **Theorem 1** *Series with Nonnegative Terms*
> Every series Σa_n with $a_n \geq 0$ for all n has a finite or infinite sum. Furthermore,
>
> $$\Sigma a_n < \infty \text{ if } \{S_n\} \text{ is bounded,}$$
> $$\Sigma a_n = \infty \text{ if } \{S_n\} \text{ is unbounded.}$$

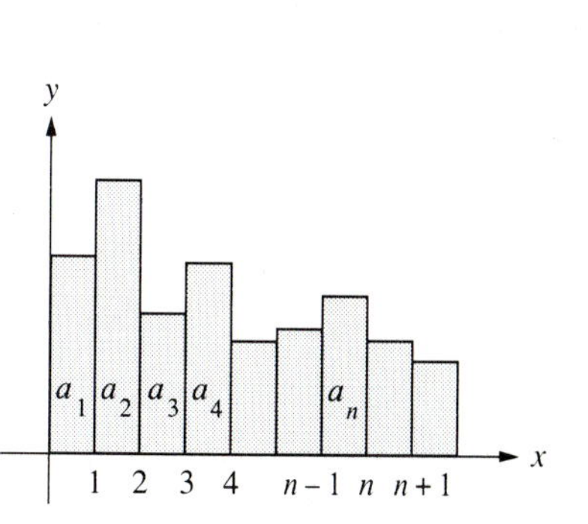

FIGURE 1

Theorem 1 and most of the other results in this section have simple area interpretations. Figure 1 shows how to interpret the sum of a series $\Sigma_{n=1}^{\infty} a_n$ with $a_n \geq 0$ as an area. The nth rectangle has height a_n, width 1, and area a_n. The sum S of the series is the shaded area (finite or infinite) of the system of rectangles.

The area in Fig. 1 can be regarded as the value of the improper integral of the function $a(x)$ whose graph is made up of the tops of the rectangles in the figure. Thus,

$$\sum_{n=1}^{\infty} a_n = \int_0^{\infty} a(x)dx,$$

where

$$a(x) = a_n \quad \text{for} \quad n-1 < x \leq n \quad \text{and} \quad n = 1, 2, 3, \cdots.$$

Most of the results in this section could be deduced from known properties of improper integrals. However, it is more instructive to work directly with series.

Theorem 1 enables us to extend two handy algebraic properties of convergent series to series that may be divergent, provided all terms are nonnegative.

> If $a_n \geq 0$, $b_n \geq 0$, and $c \geq 0$, then
>
> $$\Sigma(a_n + b_n) = \Sigma a_n + \Sigma b_n,$$
> $$\Sigma(ca_n) = c\,\Sigma a_n.$$

These results should seem reasonable. The easy proofs are outlined in the problems.

The Basic Comparison Test

As we said earlier, the basic comparison test compares two series term by term. Let's look at an example. Since

$$\frac{1}{3^n} \leq \frac{1}{2^n} \quad \text{for} \quad n = 0, 1, 2, \ldots,$$

each partial sum of $\Sigma_{n=1}^{\infty} 1/3^n$ is less than or equal to the corresponding partial sum of $\sum_{n=0}^{\infty} 1/2^n$. Consequently, the limits of the partial sums, which are the sums of the series, satisfy

$$\sum_{n=0}^{\infty} \frac{1}{3^n} \le \sum_{n=0}^{\infty} \frac{1}{2^n}.$$

In the same way,

$$\boxed{0 \le a_n \le b_n \quad \text{for all} \quad n \quad \Rightarrow \quad \Sigma a_n \le \Sigma b_n,}$$

where the sums may be finite or infinite. An immediate consequence is

Theorem 2 *The Basic Comparison Test*
Let $0 \le a_n \le b_n$ for all n. Then

$$\Sigma b_n < \infty \quad \Rightarrow \quad \Sigma a_n < \infty,$$
$$\Sigma a_n = \infty \quad \Rightarrow \quad \Sigma b_n = \infty.$$

In other words,

if $0 \le a_n \le b_n$ for all n, then
Σb_n converges $\Rightarrow \Sigma a_n$ converges,
Σa_n diverges $\Rightarrow \Sigma b_n$ diverges.

The basic comparison test remains valid if $0 \le a_n \le b_n$ for all n large enough, because a finite number of terms of a series has no effect on its convergence or divergence.

Here are two simple examples of the basic comparison test. First,

$$\frac{1}{2^n + n} \le \frac{1}{2^n} \quad \text{and} \quad \sum_{n=0}^{\infty} \frac{1}{2^n} \quad \text{converges} \quad \Rightarrow \quad \sum_{n=0}^{\infty} \frac{1}{2^n + n} \quad \text{converges.}$$

Second, since the harmonic series diverges and

$$1 + \frac{1}{3} + \frac{1}{5} + \cdots > \frac{1}{2} + \frac{1}{4} + \frac{1}{6} + \cdots = \frac{1}{2}\left(1 + \frac{1}{2} + \frac{1}{3} + \cdots\right),$$

the series on the left diverges.

The next three examples begin to develop the convergence properties of the so-called ***p*-series**,

$$\sum_{n=1}^{\infty} \frac{1}{n^p} = 1 + \frac{1}{2^p} + \frac{1}{3^p} + \frac{1}{4^p} + \cdots,$$

where p is a fixed number. For example, the p–series with $p = 2$ and $p = 1/2$ are

$$\sum_{n=1}^{\infty} \frac{1}{n^2} = 1 + \frac{1}{2^2} + \frac{1}{3^2} + \frac{1}{4^2} + \cdots,$$

$$\sum_{n=1}^{\infty} \frac{1}{n^{1/2}} = 1 + \frac{1}{\sqrt{2}} + \frac{1}{\sqrt{3}} + \frac{1}{\sqrt{4}} + \cdots.$$

EXAMPLE 1. Show that the series $\sum_{n=1}^{\infty} 1/n^2$ converges.

Solution. Notice that

$$\frac{1}{2^2} = \frac{1}{2 \cdot 2} \leq \frac{1}{2 \cdot 1}, \qquad \frac{1}{3^2} = \frac{1}{3 \cdot 3} \leq \frac{1}{2 \cdot 3}, \qquad \frac{1}{4^2} = \frac{1}{4 \cdot 4} \leq \frac{1}{3 \cdot 4}.$$

In general

$$\frac{1}{n^2} \leq \frac{1}{(n-1)n} \quad \text{for} \quad n \geq 2.$$

From Ex. 9 in Sec. 1.3,

$$\sum_{n=2}^{\infty} \frac{1}{(n-1)n} = 1.$$

Therefore,

$$\sum_{n=2}^{\infty} \frac{1}{n^2} \leq \sum_{n=2}^{\infty} \frac{1}{(n-1)n} = 1,$$

$$\sum_{n=1}^{\infty} \frac{1}{n^2} = 1 + \sum_{n=2}^{\infty} \frac{1}{n^2} \leq 2.$$

We have shown that the series $\sum_{n=1}^{\infty} 1/n^2$ converges and has sum at most 2. It is much harder to find the exact sum of this series. As we mentioned earlier, Euler proved that $\sum_{n=1}^{\infty} 1/n^2 = \pi^2/6 \approx 1.645$. □

EXAMPLE 2. Show that the p–series, $\sum 1/n^p$, converges for any $p \geq 2$.

Solution. Since the p–series converges for $p = 2$, and

$$p > 2 \quad \Rightarrow \quad \frac{1}{n^p} \leq \frac{1}{n^2} \quad \text{for} \quad n = 1, 2, 3, \ldots,$$

the p–series converges for any $p > 2$ by the basic comparison test. □

EXAMPLE 3. Show that the p–series, $\sum 1/n^p$, diverges for any $p \leq 1$.

Solution. The p–series for $p = 1$ is the harmonic series, which diverges. Since

$$p < 1 \quad \Rightarrow \quad \frac{1}{n^p} \geq \frac{1}{n} \quad \text{for} \quad n = 1, 2, 3, \ldots,$$

the basic comparison test implies that the p–series diverges for any $p < 1$. □

Examples 1–3 settle the behavior of the p–series for $p \le 1$ and $p \ge 2$. What do you think happens for $1 < p < 2$? Stay tuned. In the meantime, we give some other applications of the basic comparison test.

EXAMPLE 4. Does the series $\sum_{n=0}^{\infty} 1/n!$ converge or diverge?

Solution. Since $n! = 1 \cdot 2 \cdot 3 \cdot \cdots \cdot n \ge 1 \cdot 2 \cdot 2 \cdot \cdots \cdot 2 = 2^{n-1}$ for $n \ge 1$,

$$\sum_{n=1}^{\infty} \frac{1}{n!} \le \sum_{n=1}^{\infty} \frac{1}{2^{n-1}} = 1 + \frac{1}{2} + \frac{1}{2^2} + \cdots = 2,$$

$$\sum_{n=0}^{\infty} \frac{1}{n!} = 1 + \sum_{n=1}^{\infty} \frac{1}{n!} \le 3.$$

So $\sum_{n=0}^{\infty} 1/n!$ converges. We shall show later that the sum is $e \approx 2.718$. □

EXAMPLE 5. Does $\sum n!/n^n$ converge or diverge?

Solution. For $n \ge 2$,

$$\frac{n!}{n^n} = \frac{1 \cdot 2 \cdot 3 \cdot \cdots \cdot n}{n \cdot n \cdot n \cdot \cdots \cdot n} = \frac{2}{n^2} \cdot \left(\frac{3}{n} \cdot \frac{4}{n} \cdot \cdots \cdot \frac{n}{n}\right) \le \frac{2}{n^2},$$

$$\sum \frac{n!}{n^n} \le 2 \sum \frac{1}{n^2} < \infty,$$

and the given series converges. □

EXAMPLE 6. Does $\sum (n-1)/(n^3 + 2n + 3)$ converge or diverge?

Solution. Since

$$\frac{n-1}{n^3 + 2n + 1} \le \frac{n}{n^3} = \frac{1}{n^2} \quad \text{and} \quad \sum \frac{1}{n^2} < \infty,$$

the given series converges by the basic comparison test. □

The strategy in Ex. 6 is based on two observations. First, the dominant terms in the numerator and denominator are n and n^3, so we expect that

$$\frac{n-1}{n^3 + 2n + 1} \approx \frac{n}{n^3} = \frac{1}{n^2} \quad \text{for all large n.}$$

This suggests a comparison with the series $\sum 1/n^2$. To carry out the comparison and with an eye toward establishing convergence, the numerator of the original fraction is increased and its denominator is decreased. This step gets rid of the lower-order terms. A similar strategy is used in the next example, but this time we expect divergence, so the numerator is decreased and the denominator is increased.

EXAMPLE 7. Does $\Sigma(4n^2 + 3n + 5)/(4n^3 - 2n - 1)$ converge or diverge?

Solution. Since

$$\frac{4n^2 + 3n + 5}{4n^3 - 2n - 1} \approx \frac{4n^2}{4n^3} = \frac{1}{n} \quad \text{for all large } n$$

and $\Sigma 1/n = \infty$, we expect divergence. Now,

$$\frac{4n^2 + 3n + 5}{4n^3 - 2n - 1} \geq \frac{4n^2}{4n^3} = \frac{1}{n}.$$

So the given series diverges, as anticipated. □

The Limit Comparison Test

Often the basic comparison test is hard to execute because inequalities of the type used in Exs. 6 and 7 are not readily available. For example, if the plus and minus signs are interchanged in those examples, then inequalities of the desired type are harder to get. (Try to find them.) The next example, which illustrates the limit comparison test in a particular case, shows how routine limit evaluations can replace tricky work with inequalities.

EXAMPLE 8. Does $\Sigma(4n + 1)/(2n^3 - 1)$ converge or diverge?

Solution. Since

$$\frac{4n + 1}{2n^3 - 1} = \frac{n(4 + 1/n)}{n^3(2 - 1/n^3)} = \frac{1}{n^2} \cdot \frac{4 + 1/n}{2 - 1/n^3} \approx \frac{2}{n^2} \quad \text{for large } n$$

and $\Sigma 2/n^2$ converges, we suspect that the given series converges. But there does not seem to be a simple way to make direct use of the basic comparison test. A slightly more sophisticated argument does the trick. Since

$$\frac{4 + 1/n}{2 - 1/n^3} \to 2 \quad \text{as} \quad n \to \infty,$$

the definition of a limit tells us that

$$1 < \frac{4 + 1/n}{2 - 1/n^3} < 3$$

for n large enough, say, for $n \geq N$. (In fact, $N = 2$ will do, but we do not need to know that.) Consequently,

$$\frac{4n + 1}{2n^3 - 1} = \frac{1}{n^2} \cdot \frac{4 + 1/n}{2 - 1/n^3} < \frac{1}{n^2} \cdot 3 = \frac{3}{n^2} \quad \text{for} \quad n \geq N.$$

Since $\Sigma 3/n^2$ converges, the basic comparison test implies that the given series converges. □

The reasoning used in Ex. 8 can be applied to many other series Σa_n. The basic idea is to express a_n as a product, $a_n = b_n c_n$, where Σb_n is known to converge or diverge and c_n approaches a finite limit.

Theorem 3 *The Limit Comparison Test*
Let $a_n \geq 0$ and $b_n \geq 0$.
If $a_n = b_n c_n$ and $c_n \to L$ with $0 < L < \infty$, then Σa_n and Σb_n both converge or both diverge.

Proof. Since $c_n \to L$, there is an integer N such that

$$\frac{1}{2}L \leq c_n \leq 2L \quad \text{for} \quad n \geq N.$$

Multiply by b_n and use $a_n = b_n c_n$ to obtain

$$\frac{1}{2}Lb_n \leq a_n \leq 2Lb_n \quad \text{for} \quad n \geq N.$$

Then

$$\frac{1}{2}L \sum_{n=N}^{\infty} b_n \leq \sum_{n=N}^{\infty} a_n \leq 2L \sum_{n=N}^{\infty} b_n.$$

Since $0 < L < \infty$, the sums $\sum_{n=N}^{\infty} a_n$ and $\sum_{n=N}^{\infty} b_n$ are both finite or both infinite. Since a finite number of terms has no effect on the convergence or divergence of a series, Σa_n and Σb_n both converge or both diverge. □

Incidentally, for the convergence of Σa_n it is enough for $0 \leq L < \infty$, and for the divergence of Σa_n it is enough for $0 < L \leq \infty$. These extensions of Th. 3 are occasionally useful.

EXAMPLE 9. Does $\Sigma (n^2 + 3n)/(n^5 + 4)$ converge or diverge?

Solution. Let a_n be the nth term of the series. Then

$$a_n = \frac{n^2 + 3n}{n^5 + 4} = \frac{n^2(1 + 3/n)}{n^5(1 + 4/n^5)} = \frac{1}{n^3} \cdot \frac{1 + 3/n}{1 + 4/n^5},$$

$$a_n = b_n c_n, \qquad b_n = \frac{1}{n^3}, \qquad c_n = \frac{1 + 3/n}{1 + 4/n^5}.$$

Since $c_n \to 1$ and $\Sigma b_n = \Sigma 1/n^3$ converges, the limit comparison test implies that the given series converges. □

It should be clear from Exs. 8 and 9 that the convergence or divergence of a series $\Sigma P(n)/Q(n)$, where $P(n)$ and $Q(n)$ are polynomials, is determined by the convergence or divergence of the corresponding series in which only the highest powers in $P(n)$ and $Q(n)$ are retained.

EXAMPLE 10. Does $\sum_{n=1}^{\infty} \sin(1/n)$ converge or diverge?

Solution. The series has all positive terms. (Why?) Since $\sin x \approx x$ for $x \approx 0$, we have $\sin(1/n) \approx 1/n$ for large n. So perhaps the series diverges. To prove that it does, observe that

$$\sin\left(\frac{1}{n}\right) = \frac{1}{n} \cdot \frac{\sin(1/n)}{1/n}.$$

Since

$$\frac{\sin(1/n)}{1/n} \to 1 \quad \text{as} \quad n \to \infty \quad \text{and} \quad \sum \frac{1}{n} \quad \text{diverges,}$$

the series $\sum \sin(1/n)$ diverges by the limit comparison test. □

The Integral Test

We turn now to a different type of comparison, in which an infinite series and an improper integral are compared. Suppose that $f(x) \geq 0$ and $f(x)$ is continuous for $a \leq x < \infty$. Then

$$\int_a^{\infty} f(x)\,dx = \lim_{b \to \infty} \int_a^b f(x)dx.$$

The value of the integral, finite or infinite, is represented by the shaded area in Fig. 2.

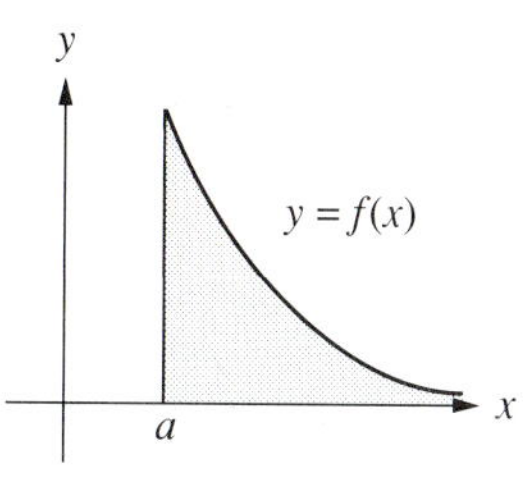

FIGURE 2

We observed in Sec. 1.1 that the terms of a sequence $\{a_n\}$ are often given by a familiar function $a(x)$ defined for $1 \leq x < \infty$; that is, $a_n = a(n)$. For example, if $a_n = 1/n^2$ then $a_n = a(n)$ with $a(x) = 1/x^2$. The integral test, which compares a series of the form $\sum a_n = \sum a(n)$ and the improper integral of $a(x)$, is basically an area comparison.

Theorem 4 *The Integral Test*
Let $a(x)$ be a continuous, positive, decreasing function for $1 \leq x < \infty$.
Let $a_n = a(n)$. Then

$$\sum_{n=1}^{\infty} a_n \quad \text{and} \quad \int_1^{\infty} a(x)dx$$

are both finite (both converge) or both infinite (both diverge).

Figures 3 and 4 show the graph of such a function $a(x)$ and two systems of rectangles that display the partial sums $S_n = a_1 + a_2 + \cdots + a_n$ of the series

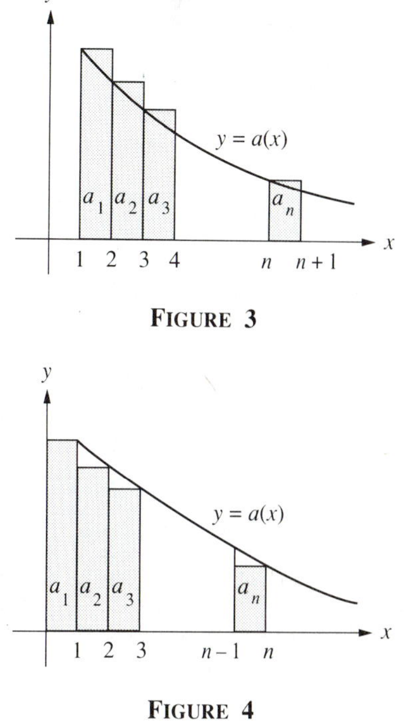

FIGURE 3

FIGURE 4

Σa_n as areas. Area comparisons, first in Fig. 3 and then in Fig. 4, yield the two inequalities

$$\int_1^{n+1} a(x)dx \le S_n \le a_1 + \int_1^n a(x)dx.$$

Let $n \to \infty$ to obtain

$$\int_1^{\infty} a(x)dx \le \sum_{n=1}^{\infty} a_n \le a_1 + \int_1^{\infty} a(x)dx.$$

It follows from these inequalities that the series and the improper integral are both finite or both infinite, which is the conclusion of the integral test.

EXAMPLE 11. The integral test gives another way to prove that the harmonic series $\Sigma_{n=1}^{\infty} 1/n$ diverges. To see this, let $a(x) = 1/x$ for $1 \le x < \infty$. Then $a(x)$ is continuous, positive, decreasing, and $a(n) = 1/n$. Since

$$\int_1^{\infty} \frac{1}{x} dx = [\ln x]_1^{\infty} = \infty,$$

the integral test tells us that $\Sigma_{n=1}^{\infty} 1/n = \infty$. □

Let $a(x) = 1/x$ in the general inequality

$$\int_1^{n+1} a(x)dx \le S_n \le a_1 + \int_1^n a(x)dx$$

to obtain the following estimates for the partial sums S_n of the harmonic series:

$$\ln(n+1) \le S_n \le 1 + \ln n.$$

These inequalities measure how fast $S_n \to \infty$. For example, let $n = 1{,}000$ and $n = 1{,}000{,}000$ to obtain

$$6.91 \le S_{1{,}000} \le 7.91 \quad \text{and} \quad 13.82 \le S_{1{,}000{,}000} \le 14.82.$$

The partial sums of the harmonic series grow very slowly indeed! About how many terms must be added to make $S_n \ge 20$? Further information about the harmonic series may be found in the problem set.

EXAMPLE 12. The integral test provides an easy means to see that the p-series, $\Sigma 1/n^p$, converges for any $p > 1$. We verified the convergence for $p \ge 2$ in Ex. 2. What is new is that the series also converges for $1 < p < 2$. We apply the integral test with $a(x) = 1/x^p$, which is continuous, positive, decreasing, and satisfies $a(n) = 1/n^p$. Since

$$\int_1^{\infty} \frac{1}{x^p} dx = \left[\frac{1}{-p+1} x^{-p+1}\right]_1^{\infty} = \frac{1}{p-1} \quad \text{if} \quad p > 1,$$

the integral test implies that the p-series converges for any $p > 1$. □

Now we know the full story about convergence of the p–series:

$$\sum_{n=1}^{\infty} \frac{1}{n^p} = 1 + \frac{1}{2^p} + \frac{1}{3^p} + \cdots \text{ converges if } p > 1$$

and the series diverges if $p \leq 1$.

The integral test was stated for series with $n = 1, 2, 3, \ldots$ and improper integrals over $1 \leq x < \infty$. The same reasoning establishes corresponding integral tests with

$$\sum_{n=m}^{\infty} a_n \quad \text{and} \quad \int_m^{\infty} a(x)dx,$$

where $a(x)$ is continuous, positive, and decreasing for $m \leq x < \infty$. In most applications, $m = 0$ or 1 or 2.

EXAMPLE 13. Test $\sum_{n=2}^{\infty} 1/n(\ln n)^2$ for convergence or divergence.

Solution. We apply the integral test with

$$a(x) = \frac{1}{x(\ln x)^2} \quad \text{for} \quad 2 \leq x < \infty.$$

Use the change of variables $u = \ln x$ to find that

$$\int_2^{\infty} \frac{1}{x(\ln x)^2} dx = \int_{\ln 2}^{\infty} \frac{1}{u^2} du = \left[-\frac{1}{u} \right]_{\ln 2}^{\infty} = \frac{1}{\ln 2} < \infty.$$

The given series converges by the integral test. □

Often more than one test can be used to show that a particular series converges. For practice, show that $\sum 1/(n^2 + 1)$ converges by the integral test, the basic comparison test, and the limit comparison test.

*Approximate Sums Based on the Integral Test**

Starred material is not essential for topics that follow.

Slight modifications in the derivation of the integral test lead to effective means for approximating the sums of certain series. Assume that $a(x)$ is continuous, positive, and decreasing for $1 \leq x < \infty$, $a_n = a(n)$, and $S = \sum_{n=1}^{\infty} a_n < \infty$. Express the sum in the form

$$S = \sum_{n=1}^{N} a_n + \sum_{n=N+1}^{\infty} a_n = S_N + \sum_{n=N+1}^{\infty} a_n.$$

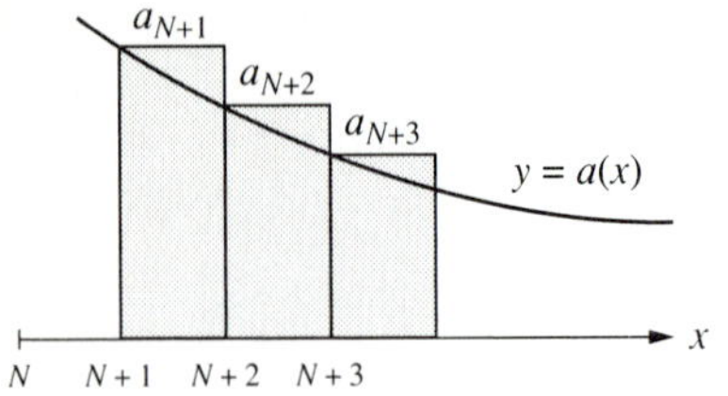

FIGURE 5

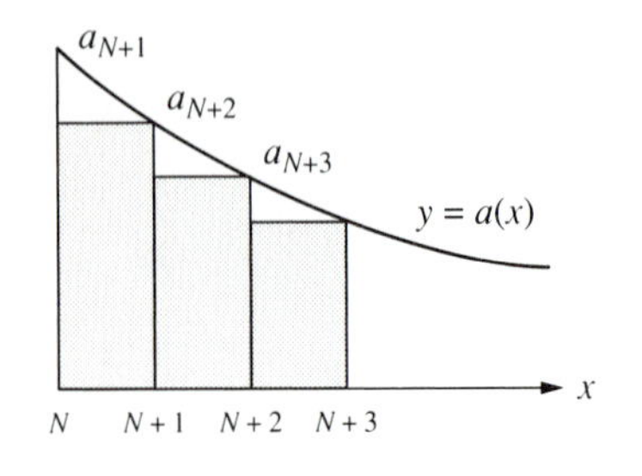

FIGURE 6

Then

$$S - S_N = \sum_{n=N+1}^{\infty} a_n.$$

Area comparisons, first in Fig. 5 and then in Fig. 6, yield the two inequalities

$$\int_{N+1}^{\infty} a(x)dx < S - S_N < a_{N+1} + \int_{N+1}^{\infty} a(x)dx.$$

These inequalities indicate both how good and how bad S_N can be as an approximation for S.

If the series converges very slowly, then the partial sums S_N have little practical value as approximations for S. Fortunately, we can do better with really no further effort by exploiting the estimates of $S - S_N$ given previously. Let

$$T_N = S_N + \int_{N+1}^{\infty} a(x)dx.$$

Add S_N to each member of the inequalities for $S - S_N$ to obtain

$$T_N < S < T_N + a_{N+1} \quad \text{and} \quad 0 < S - T_N < a_{N+1}.$$

Since $a_{N+1} \to 0$, the squeeze law gives $T_N \to S$ as $N \to \infty$. Figure 7 compares S_N and T_N as approximations for S. In the figure, S is the total area of all the rectangles, S_N is the area of the first N rectangles, and T_N is the shaded area. Clearly, T_N is better than S_N as an approximation for S.

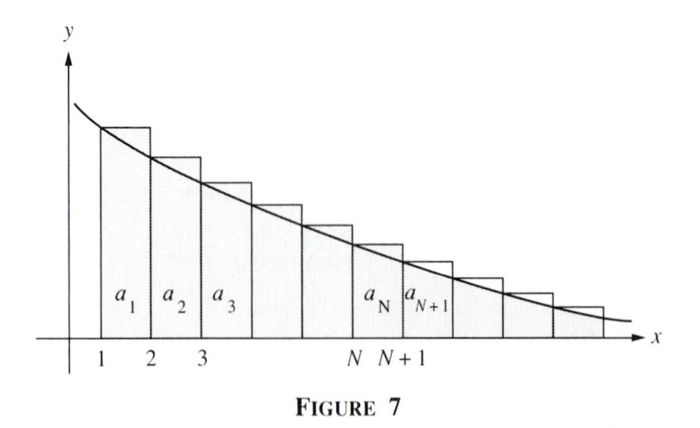

FIGURE 7

EXAMPLE 14. Find $S = \sum_{n=1}^{\infty} 1/n^3$ accurate to within 0.001.

Solution. The series converges; it is the p–series with $p = 3$. We mentioned in Sec. 1.3 that the exact value of S is not known. We shall use T_N to approximate

S. Now $a_n = 1/n^3$, $a(x) = 1/x^3$, and, from the preceding displayed results,

$$T_N = S_N + \int_{N+1}^{\infty} \frac{1}{x^3} dx = S_N + \frac{1}{2(N+1)^2},$$

$$0 \leq S - T_N \leq \frac{1}{(N+1)^3}.$$

So T_N approximates S to within 0.001 if

$$\frac{1}{(N+1)^3} < 0.001 \quad \Leftrightarrow \quad (N+1)^3 > 10^3 \quad \Leftrightarrow \quad N > 9.$$

Let $N = 10$. Then $0 < S - T_{10} < 0.001$. Use a calculator to find

$$S_{10} = \sum_{n=1}^{10} \frac{1}{n^3} \approx 1.1975, \qquad T_{10} = S_{10} + \frac{1}{2(11)^2} \approx 1.2016.$$

The approximation $S \approx 1.2016$ is accurate to within 0.001. □

In Ex. 14, we also could use a partial sum S_N to approximate S. To see the advantage of using T_N as we did, apply the general inequality

$$S - S_N > \int_{N+1}^{\infty} a(x) dx$$

with $a(x) = 1/x^3$ to obtain

$$S - S_N > \frac{1}{2(N+1)^2}.$$

So $S - S_N \geq = 0.001$ at least as long as

$$\frac{1}{2(N+1)^2} \geq 0.001 \quad \Leftrightarrow \quad (N+1)^2 \leq 500 \quad \Leftrightarrow \quad N \leq 22.$$

To attain $S - S_N < 0.001$ we must take N at least equal to 23 and possibly greater. Thus, more than twice as many terms are needed to approximate S by S_N to within 0.001 as were needed for T_N in Ex. 14.

The next example is even more striking. The partial sums S_N are of no practical value as approximations to S for any N, no matter how large.

EXAMPLE 15. Approximate $S = \sum_{n=2}^{\infty} 1/n(\ln n)^2$ to within 0.01.

Solution. The series converges by Ex. 13. In this case, $a_n = 1/n(\ln n)^2$ and $a(x) = 1/x(\ln x)^2$. Then

$$T_N = S_N + \int_{N+1}^{\infty} \frac{dx}{x(\ln x)^2} = S_N + \frac{1}{\ln(N+1)},$$

$$0 < S - T_N < a_{N+1} = \frac{1}{(N+1)[\ln(N+1)]^2}.$$

A little experimentation with a calculator gives

$$0 < S - T_N < 0.01 \quad \text{for} \quad N \geq 14.$$

Let $N = 14$. Further calculations yield

$$S_{14} = \sum_{n=1}^{14} \frac{1}{n(\ln n)^2} \approx 1.736 \quad \text{and} \quad T_{14} = S_{14} + \frac{1}{\ln 15} \approx 2.105.$$

The approximation $S \approx 2.105$ is accurate to within 0.01. □

Now let's try to approximate S by S_N in Ex. 15. The inequality

$$S - S_N > \int_{N+1}^{\infty} a(x)dx$$

with $a(x) = 1/x(\ln x)^2$ yields

$$S - S_N > \frac{1}{\ln(N+1)}.$$

Hence, $S - S_N \geq .01$ as long as

$$\frac{1}{\ln(N+1)} \geq .01 \quad \Leftrightarrow \quad \ln(N+1) \leq 100 \quad \Leftrightarrow \quad N \leq e^{100} - 1 \approx 2.7 \times 10^{43}.$$

To achieve $S - S_N < 0.01$ requires a choice of $N > 2 \times 10^{43}$. Calculating a partial sum with this many terms is beyond the capability of any modern computer. Among other reasons, there just is not enough time. The age of the universe is about 15 billion years or 5×10^{17} sec. To add 2×10^{43} terms, starting with the beginning of time, the computer would have to perform 4×10^{25} additions per second. No machine even comes close!

PROBLEMS

In Probs. 1–6, use a comparison test to determine whether the given series converges or diverges.

1. $\sum \frac{1}{2n+1}$
2. $\sum \frac{\sqrt{n}}{3n+2}$
3. $\sum \frac{1 + \sin n}{n^2}$
4. $\sum \frac{\arctan n}{n^2 + 4}$
5. $\sum \frac{1}{\sqrt{n(n+1)}}$
6. $\sum \frac{1}{n^2(n+1)}$

In Probs. 7–12, use the integral test to determine whether the given series converges or diverges. Be sure to check that the function you use satisfies the conditions in the integral test.

7. $\sum \dfrac{1}{n^2 + 4}$
8. $\sum \dfrac{n}{n^2 + 4}$
9. $\sum n e^{-n^2}$
10. $\sum n e^{-n}$
11. $\sum \dfrac{1}{n(n + 2)}$
12. $\sum \dfrac{n^2 + 3}{n^2(n^2 + 1)}$

In Probs. 13–44, test the given series for convergence or divergence. You may use any test you have learned. (Often more than one test will apply. It is a good idea to try several tests. This will help you to learn which tests are likely to apply or be easier to apply to particular classes of series.)

13. $\sum \dfrac{1}{n^2 + 1}$
14. $\sum \dfrac{1}{(n + 1)^2}$
15. $\sum \dfrac{1}{n + 1}$
16. $\sum \dfrac{1}{2n - 1}$
17. $\sum \dfrac{n}{n + 1}$
18. $\sum \dfrac{\sqrt{n}}{n + 4}$
19. $\sum n 2^{-n}$
20. $\sum n^2 2^{-n}$
21. $\sum \dfrac{1}{\sqrt{4n^3 + 1}}$
22. $\sum \dfrac{n}{\sqrt{4n^3 + 1}}$
23. $\sum \dfrac{1}{n \ln n}$
24. $\sum \dfrac{\ln n}{n^{3/2}}$
25. $\sum \dfrac{n}{\sqrt{4n^2 + 1}}$
26. $\sum \left(\dfrac{n}{2n + 1}\right)^n$
27. $\sum \dfrac{n^{1/n}}{\sqrt{n^2 + 4}}$
28. $\sum \dfrac{n^{-1/n}}{\sqrt{n^2 + 4}}$
29. $\sum \dfrac{1}{n^3 + n}$
30. $\sum \dfrac{n \ln n}{n^3 + n}$
31. $\sum \dfrac{\tanh n}{n^2 + 1}$
32. $\sum \operatorname{sech}^2 n$
33. $\sum \sin \dfrac{1}{n^2}$
34. $\sum n \sin \dfrac{1}{n^2}$
35. $\sum \sinh \dfrac{1}{n}$
36. $\sum \sinh \dfrac{1}{n^2}$
37. $\sum \left(1 - \cos \dfrac{1}{n}\right)$
38. $\sum \left(1 - \cos \dfrac{1}{\sqrt{n}}\right)$
39. $\sum \dfrac{1}{n^n}$
40. $\sum \dfrac{1}{n^{\ln n}}$

41. $\Sigma \dfrac{n!}{(2n)!}$

42. $\Sigma \dfrac{(n!)^2}{(2n)!}$

43. $\Sigma \dfrac{n^2 - 3n + 2}{5n^4 + 2n - 7}$

44. $\Sigma \dfrac{n + 5}{4n^3 - 3}$

45. Determine whether the series $\Sigma (\ln n)/n^{5/2}$ converges or diverges. *Hint.* $\ln x/x^{1/2} \to 0$ as $x \to \infty$.

46. Determine whether the series $\Sigma (\ln n)/n^{3/2}$ converges or diverges.

47. Determine the values of p for which the series $\Sigma_{n=1}^{\infty} 1/(2n - 1)^p$ converges and diverges.

48. Determine the values of p for which the series $\Sigma_{n=2}^{\infty} 1/n(\ln n)^p$ converges and diverges.

49. Determine the values of p for which the series $\Sigma_{n=1}^{\infty} (\ln n)/n^p$ converges and diverges.

50. The p–series converges for each fixed $p > 1$ and diverges for each fixed $p \leq 1$. Test the series (a) $\Sigma 1/n^{(1 + 1/n)}$ and (b) $\Sigma 1/n^{(1 - 1/n)}$ for convergence or divergence. Notice that the series in (a) and (b) are *not* p–series because the exponents in the series are not constant; they vary with n.

51. Assume $a_n \geq 0$ and that Σa_n converges. (a) Show that Σa_n^2 also converges. (b) Give a specific example that shows that Σa_n^2 can converge while Σa_n diverges.

In Probs. 52–55, the series converges by the integral test. Use the approximation $T_N = S_N + \int_{N+1}^{\infty} a(x)dx$, developed in the text, to approximate the sum of the series to within 0.001.

52. $\displaystyle\sum_{n=1}^{\infty} \frac{1}{n^4}$

53. $\displaystyle\sum_{n=1}^{\infty} \frac{1}{n^5}$

54. $\displaystyle\sum_{n=1}^{\infty} \frac{1}{n^2}$

55. $\displaystyle\sum_{n=1}^{\infty} ne^{-n^2}$

The next problem, which builds on Ex. 11 and the discussion following it, takes an even closer look at the relation between the partial sums of the harmonic series and $\ln n$.

56. Let $S_n = 1 + 1/2 + 1/3 + \cdots + 1/n$. We found following Ex. 11 that

$$\ln(n + 1) \leq S_n \leq 1 + \ln n.$$

(a) Show that $c_n = S_n - \ln n$ satisfies $0 \leq c_n \leq 1$.
(b) Show that

$$\ln(n + 1) - \ln n = \int_n^{n+1} \frac{1}{x} dx > \frac{1}{n + 1}$$

and illustrate the inequality with an area comparison.
(c) Use (b) to show that $\{c_n\}$ is decreasing.
(d) Explain why $\{c_n\}$ converges. It is customary to denote its limit, called **Euler's constant**, by γ.

$$\gamma = \lim_{n \to \infty} \left[\left(1 + \frac{1}{2} + \frac{1}{3} + \cdots + \frac{1}{n}\right) - \ln n\right] \approx 0.5772156649.$$

Euler's constant crops up in numerous places in both theoretical and applied mathematics. No one knows whether γ is rational or is irrational.
(e) Show that the result in (d) can be put in the often more convenient form

$$1+\frac{1}{2}+\frac{1}{3}+\cdots+\frac{1}{n}=\ln n+\gamma+\varepsilon_n,$$

where $\varepsilon_n \to 0$ as $n \to \infty$.

57. Prove the algebraic laws displayed after Th. 1. *Hint.* Let $\{S_N\}$ and $\{T_N\}$ be the sequences of partial sums for $\sum a_n$ and $\sum b_n$. Then $S_N \to S = \sum a_n$ and $T_N \to T = \sum b_n$ with $0 \leq S_N \leq S \leq \infty$ and $0 \leq T_N \leq T \leq \infty$. Now use properties of limits of sequences.

1.5 Absolute and Conditional Convergence; Alternating Series

In the previous section we studied series with nonnegative terms. Now we consider series with terms of mixed signs. The cancellation that occurs in the partial sums of such series tends to promote convergence. For example, no matter how you change the signs of the terms in a convergent series with nonnegative terms, the resulting series will still converge. On the other hand, if you start with a convergent series having terms of mixed signs and you replace each term by its absolute value, then the resulting series may converge or diverge. If the series of absolute values converges, we call the original series absolutely convergent. If the series of absolute values diverges, we call the original series conditionally convergent. It turns out that absolutely convergent series behave like finite sums under most familiar algebraic operations while conditionally convergent series often behave quite differently. We shall discuss this matter more fully at the end of the section. Meanwhile, we shall learn that every absolutely convergent series is convergent. This fact enables us to use convergence tests originally designed for series with nonnegative terms to show that many series with terms of mixed signs are convergent.

The most important series with terms of mixed signs that occur in applications have consecutive terms which alternate in sign. Such series are called alternating series. We shall derive a very handy test which tells us that a large class of alternating series converges. Moreover, the test also yields error bounds for the approximation of the sum of such a series by its partial sums. Alternating series typically arise from approximation procedures that successively overshoot and undershoot the exact answer to a problem.

Absolute Convergence

A series $\sum a_n$ **converges absolutely** if $\sum |a_n|$ converges. Our first goal is to prove that every absolutely convergent series is convergent. If this were not true, the terminology just introduced would be very misleading.

EXAMPLE 1. (a) Show that the series

$$\sum_{n=1}^{\infty} a_n=\sum_{n=1}^{\infty}\frac{(-1)^{n+1}}{n^2}=1-\frac{1}{2^2}+\frac{1}{3^2}-\frac{1}{4^2}+\frac{1}{5^2}-\frac{1}{6^2}+\cdots.$$

converges absolutely. (b) Then show that the series converges.

Solution. (a) The given series converges absolutely because we know that the series

$$\sum_{n=1}^{\infty} |a_n| = \sum_{n=1}^{\infty} \frac{1}{n^2} = 1 + \frac{1}{2^2} + \frac{1}{3^2} + \frac{1}{4^2} + \frac{1}{5^2} + \frac{1}{6^2} + \cdots$$

converges. (b) To show that $\Sigma(-1)^{n+1}/n^2$ is convergent, we use the basic comparison test twice: Both of the series

$$1 + 0 + \frac{1}{3^2} + 0 + \frac{1}{5^2} + 0 + \cdots,$$

$$0 + \frac{1}{2^2} + 0 + \frac{1}{4^2} + 0 + \frac{1}{6^2} + \cdots$$

converge by comparison with the series $\Sigma 1/n^2$. Therefore, the difference of these two series,

$$1 - \frac{1}{2^2} + \frac{1}{3^2} - \frac{1}{4^2} + \frac{1}{5^2} - \frac{1}{6^2} + \cdots,$$

which is the given series $\Sigma(-1)^{n+1}/n^2$, converges. □

The foregoing reasoning applies to any absolutely convergent series.

Theorem 1 *Absolute Convergence Implies Convergence*
$\Sigma |a_n|$ converges $\Rightarrow \Sigma a_n$ converges.

Proof. Assume $\Sigma |a_n|$ converges. Define

$$b_n = \left\{ \begin{array}{r} a_n = |a_n| \text{ if } a_n \geq 0 \\ 0 \text{ if } a_n < 0 \end{array} \right\}, \qquad c_n = \left\{ \begin{array}{r} 0 \text{ if } a_n \geq 0 \\ -a_n = |a_n| \text{ if } a_n < 0 \end{array} \right\}.$$

Then $a_n = b_n - c_n$ and

$$0 \leq b_n \leq |a_n|, \qquad 0 \leq c_n \leq |a_n| \qquad \text{for all } n.$$

Since $\Sigma |a_n|$ converges, the basic comparison test implies that Σb_n and Σc_n converge. Consequently, $\Sigma a_n = \Sigma (b_n - c_n)$ converges. □

The converse of Th. 1 is false. This means that there are series that converge but do not converge absolutely. Such series are called conditionally convergent. In other words, a series **converges conditionally** if Σa_n converges but $\Sigma |a_n|$ diverges. We shall meet examples of such series shortly, but for the moment we concentrate on absolute convergence.

An important consequence of Th. 1 is that it enables us to use the comparison tests of Sec. 1.4 to show that various series are absolutely convergent and, hence, convergent. Here are a few examples.

EXAMPLE 2. Show that the series

$$\sum_{n=1}^{\infty} a_n = \sum_{n=1}^{\infty} \sin\left(\frac{n\pi}{2}\right) \cdot \frac{1}{2^n} = 1 - \frac{1}{2^3} + \frac{1}{2^5} - \frac{1}{2^7} + \cdots$$

converges absolutely.

Solution. Since $|\sin x| \leq 1$ for all x, $|a_n| \leq 1/2^n$. By the basic comparison test, $\Sigma |a_n|$ converges. So Σa_n converges absolutely and, hence, converges. □

Whenever the conclusion is reached that a particular series converges absolutely, it should be realized immediately that the series converges. We usually do not bother to say so.

Just as for ordinary convergence, for any m,

$$\Sigma_{n=1}^{\infty} a_n \text{ converges absolutely} \iff \Sigma_{n=m}^{\infty} a_n \text{ converges absolutely.}$$

This observation is useful in the next example.

EXAMPLE 3. Show that the series

$$\sum_{n=1}^{\infty} a_n = \sum_{n=1}^{\infty} (-1)^n \frac{n^2 - 4n + 1}{n^5 - 2n^3 + 4}$$

converges absolutely.

Solution. By easy arguments, the numerators and denominators in the series are positive for $n \geq 4$. Hence,

$$\sum_{n=4}^{\infty} |a_n| = \sum_{n=4}^{\infty} \frac{n^2 - 4n + 1}{n^5 - 2n^3 + 4}.$$

This series converges by limit comparison with $\Sigma 1/n^3$ or, more quickly, by the observation made in Sec. 1.4 that the convergence or divergence of a series $\Sigma P(n)/Q(n)$, where $P(n)$ and $Q(n)$ are polynomials, is determined by the convergence or divergence of the corresponding series in which only the leading terms of $P(n)$ and $Q(n)$ are kept. It follows that $\Sigma_{n=4}^{\infty} a_n$ converges absolutely and, hence, so does the given series. □

EXAMPLE 4. Show that the series

$$\sum_{n=1}^{\infty} \frac{(-1)^{n+1}}{n^2 + 1}$$

converges absolutely.

Solution. We must show that $\Sigma_{n=1}^{\infty} 1/(n^2 + 1)$ converges. There are several ways to reach this conclusion, including the basic comparison test. For variety, we shall use the integral test. Let $a(x) = 1/(x^2 + 1)$. Then $a(x)$ is continuous, positive, decreasing, and

$$\int_1^{\infty} \frac{dx}{x^2 + 1} = [\arctan x\ ]_1^{\infty} = \frac{\pi}{2} - \frac{\pi}{4} = \frac{\pi}{4}.$$

By the integral test, $\sum_{n=1}^{\infty} 1/(n^2+1)$ converges. Therefore, the given series converges absolutely. □

Alternating Series

Any series with consecutive terms of opposite signs is called an **alternating series**. We shall give a very convenient test for convergence of many alternating series. The reasoning leading to the test also provides practical means for estimating the sums of such alternating series. In order to better understand why many alternating series converge, we begin by discussing two informative examples.

At the beginning of Sec. 1.3, the partial sums of the series

$$\frac{1}{2}+\frac{1}{4}+\frac{1}{8}+\frac{1}{16}+\cdots$$

were illustrated by a mile walk along the x–axis. We walked halfway from 0 toward 1, then half the remaining distance, then half the remaining distance, and so on. The limiting position as the number of stages of the walk tends to infinity is the sum of the series $S = 1$.

Let's change things a bit. Consider the series

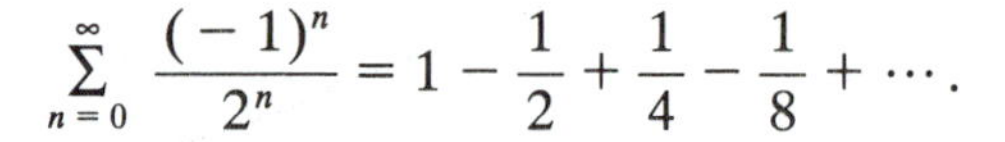

$$\sum_{n=0}^{\infty} \frac{(-1)^n}{2^n} = 1-\frac{1}{2}+\frac{1}{4}-\frac{1}{8}+\cdots.$$

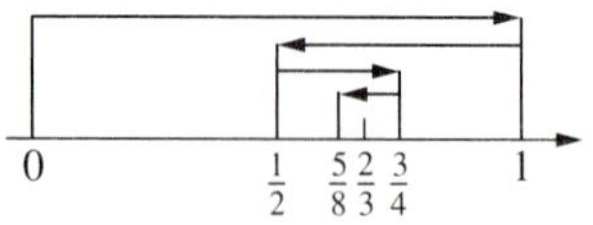

This is the geometric series $\sum_{n=0}^{\infty} x^n$ with $x = -1/2$. Its sum is $S = 2/3$. On the x–axis in Fig. 1, first walk forward 1 mile from 0 to 1, then backward 1/2 mile, then forward 1/4 mile, then backward 1/8 mile, and so on. The positions after the successive stages of the walk are

FIGURE 1

$$1,\quad 1-\frac{1}{2}=\frac{1}{2}=0.5,\quad 1-\frac{1}{2}+\frac{1}{4}=\frac{3}{4}=0.75,\quad 1-\frac{1}{2}+\frac{1}{4}-\frac{1}{8}=\frac{5}{8}=0.625,$$

and so on. These are the first four partial sums of the series $\sum_{n=0}^{\infty}(-1)^n/2^n$. The partial sums squeeze in on the point $S = 2/3$ in Fig. 1. The limiting position, as the number of stages of the walk tends to infinity, is $S = 2/3$.

Let's try another walk where we do not know the limiting position. Consider the **alternating harmonic series**

$$1-\frac{1}{2}+\frac{1}{3}-\frac{1}{4}+\cdots.$$

On the x–axis, starting at $x = 0$, first walk forward 1 mile, then backward 1/2 mile, then forward 1/3 mile, then backward 1/4 mile, and so on. The positions after the successive stages of this walk are the partial sums of the alternating harmonic series. The first six partial sums are

$$S_1 = 1, \qquad S_2 = 1-\frac{1}{2},$$

$$S_3 = 1-\frac{1}{2}+\frac{1}{3}, \qquad S_4 = 1-\frac{1}{2}+\frac{1}{3}-\frac{1}{4},$$

$$S_5 = 1-\frac{1}{2}+\frac{1}{3}-\frac{1}{4}+\frac{1}{5}, \quad S_6 = 1-\frac{1}{2}+\frac{1}{3}-\frac{1}{4}+\frac{1}{5}-\frac{1}{6}.$$

The values of $S_1, S_2, \cdots, S_6$ are displayed in Fig. 2, which shows that

$$S_2 < S_4 < S_6 < S_5 < S_3 < S_1.$$

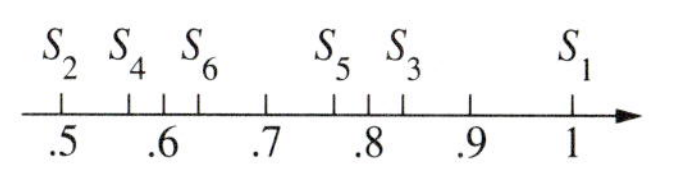

FIGURE 2

These inequalities are easy to confirm, either with a calculator or by algebra. The algebra gives more insight. For example,

$$S_5 = S_3 - \frac{1}{4} + \frac{1}{5} < S_3, \quad S_6 = S_4 + \frac{1}{5} - \frac{1}{6} > S_4, \quad S_6 = S_5 - \frac{1}{6} < S_5.$$

The pattern continues with partial sums

$$S_n = 1 - \frac{1}{2} + \frac{1}{3} - \frac{1}{4} + \cdots + \frac{1}{n}, \qquad n \text{ odd},$$

$$S_n = 1 - \frac{1}{2} + \frac{1}{3} - \frac{1}{4} + \cdots - \frac{1}{n}, \qquad n \text{ even}.$$

Notice that each of the "odd partial sums" $S_1, S_3, S_5, \ldots$ ends with an addition and each of the "even partial sums" $S_2, S_4, S_6, \ldots$ ends with a subtraction. Reason as before to obtain

$$S_2 < S_4 < S_6 < \cdots S_n < S_{n-1} < \cdots < S_5 < S_3 < S_1, \qquad n \text{ even}.$$

Thus, the even partial sums increase and the odd partial sums decrease. Since any bounded increasing or decreasing sequence converges, the even and odd partial sums converge. Furthermore,

$$S_{n-1} - S_n = \frac{1}{n} \to 0 \text{ as } n \to \infty \text{ with } n \text{ even}.$$

It follows that the even and odd partial sums tend to a common limit S and

$$S_2 < S_4 < S_6 < \cdots S_n < S < S_{n-1} < \cdots < S_5 < S_3 < S_1, \qquad n \text{ even}.$$

Figure 3 illustrates these inequalities.

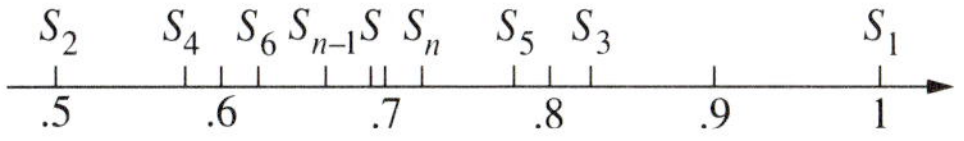

S_2 S_4 S_6 S_{n-1} S S_n S_5 S_3 S_1
.5 .6 .7 .8 .9 1

FIGURE 3

Since the even and odd partial sums of the alternating harmonic series have the common limit S, it follows that the full sequence of partial sums also has limit S. We have proved that the alternating harmonic series converges and has sum S:

$$S = \sum_{n=1}^{\infty} (-1)^{n+1} \frac{1}{n} = 1 - \frac{1}{2} + \frac{1}{3} - \frac{1}{4} + \cdots.$$

We shall discover later that $S = \ln 2 \approx 0.69$. So the walk according to the alternating harmonic series has the limiting position $\ln 2$. In other words, the net distance walked is $\ln 2$. The total distance walked is the sum of the harmonic series $\sum_1^\infty 1/n$, which is infinite. Apparently, the walking back and forth made for a lot of cancellation! Since the harmonic series diverges, the alternating harmonic series does not converge absolutely. It converges conditionally. This is our first example of a conditionally convergent series.

The reasoning that led to the convergence of the alternating harmonic series depended on just three properties of the terms $(-1)^{n+1}/n$:

The terms alternate in sign
The terms decrease in magnitude;
The nth term has limit zero.

he following theorem tells us that any series with these properties converges.

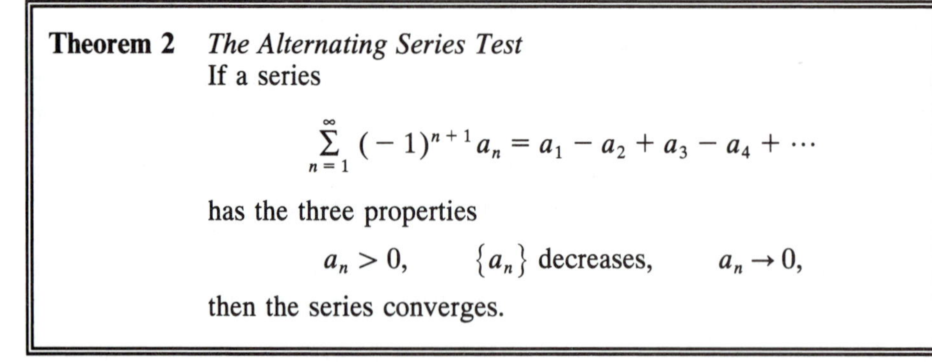

Theorem 2 *The Alternating Series Test*
If a series

$$\sum_{n=1}^{\infty} (-1)^{n+1} a_n = a_1 - a_2 + a_3 - a_4 + \cdots$$

has the three properties

$$a_n > 0, \qquad \{a_n\} \text{ decreases}, \qquad a_n \to 0,$$

then the series converges.

Proof. Since the reasoning is virtually the same as for the alternating harmonic series, we only hit the highlights. For n even (hence, n − 1 odd),

$$S_{n-1} = a_1 - a_2 + a_3 - a_4 + \cdots + a_{n-1},$$

$$S_n = a_1 - a_2 + a_3 - a_4 + \cdots + a_{n-1} - a_n.$$

An examination of the successive partial sums shows that, for n even,

$$S_2 < S_4 < S_6 < \cdots S_n < S_{n-1} < \cdots < S_5 < S_3 < S_1,$$

$$S_{n-1} - S_n = a_n \to 0 \quad \text{as} \quad n \to \infty.$$

As before, these relations and the completeness property imply that the full sequence of partial sums has a limit S, which is the sum of the series. Thus, the series converges. □

A by-product of the proof is

$$S_2 < S_4 < S_6 < \cdots S_n < S < S_{n-1} < S_5 < S_3 < S_1, \qquad n \text{ even}.$$

We shall use this chain of inequalities shortly to approximate S.

EXAMPLE 5. Both of the series

$$\sum_{n=1}^{\infty} \frac{(-1)^{n+1}}{2n-1} = 1 - \frac{1}{3} + \frac{1}{5} - \frac{1}{7} + \cdots,$$

$$\sum_{n=0}^{\infty} \frac{(-1)^n}{\sqrt{n+1}} = 1 - \frac{1}{\sqrt{2}} + \frac{1}{\sqrt{3}} - \frac{1}{\sqrt{4}} + \cdots$$

converge by the alternating series test. We learned earlier that the corresponding series with all plus signs diverge. So neither series converges absolutely. Both series converge conditionally. □

EXAMPLE 6. Does the alternating series

$$\sum_{n=1}^{\infty} (-1)^{n+1} \frac{2n+1}{3n^2+2}$$

converge or diverge?

Solution. Clearly, the terms alternate in sign. Let

$$a_n = \frac{2n+1}{3n^2+2}.$$

Then $a_n \to 0$ as $n \to \infty$. To show that $\{a_n\}$ is decreasing, let $a(x) = (2x+1)/(3x^2+2)$. Differentiate to find that $a'(x) < 0$ and $a(x)$ is decreasing for $x > 1$. Therefore, $\{a_n\}$ is decreasing and the series converges by the alternating series test. □

EXAMPLE 7. Does the series $\sum_{n=1}^{\infty}(-1)^{n+1}(n+1)/2n$ converge or diverge?

Solution. Here

$$a_n = \frac{n+1}{2n} = \frac{1}{2} + \frac{1}{2n} > 0$$

and $\{a_n\}$ is decreasing. But $a_n \to 1/2 \neq 0$. So the alternating series test gives no information about the series. However, since $a_n \not\to 0$, the series diverges. □

When trying to decide whether a series $\sum a_n$ converges or diverges, you should check first whether $a_n \to 0$ as $n \to \infty$. If not, the series diverges and there is no need to try any of the convergence tests we have learned. It is just a waste of time.

Approximating Sums of Alternating Series

Assume that $S = \sum_{n=1}^{\infty}(-1)^{n+1}a_n$, where the series converges by the alternating series test. As we have seen, all the even partial sums are smaller than S and all the odd partial sums are larger than S. Also, $S_{n+1} = S_n \pm a_n$, depending on whether n is even or odd. Thus,

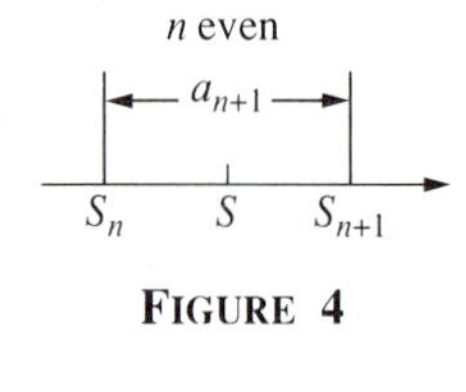

FIGURE 4

$$S_n < S < S_{n+1} \quad \text{and} \quad S_{n+1} - S_n = a_{n+1} \quad \text{for} \quad n \text{ even,}$$

$$S_{n+1} < S < S_n \quad \text{and} \quad S_n - S_{n+1} = a_{n+1} \quad \text{for} \quad n \text{ odd.}$$

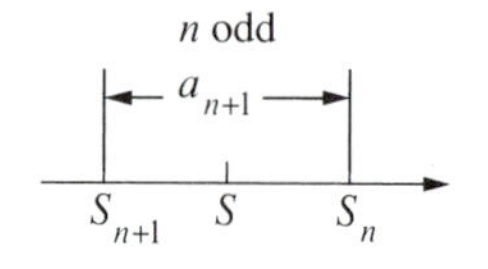

FIGURE 5

See Figs. 4 and 5.

It follows that for any n, even or odd,

$$\boxed{|S_n - S| < a_{n+1}.}$$

Thus, the error of the approximation of S by S_n is less than the magnitude of the next term in the series. For example, the sum of the alternating harmonic series is $S = \ln 2 \approx 0.693$ and

$$S_9 = 1 - \frac{1}{2} + \frac{1}{3} - \frac{1}{4} + \frac{1}{5} - \frac{1}{6} + \frac{1}{7} - \frac{1}{8} + \frac{1}{9} \approx 0.746.$$

Since the next term in the series is $-1/10$, the error in S_9 as an approximation for S satisfies $|S_9 - S| < 1/10 = 0.1$. The actual error using $S = \ln 2$ satisfies $|S_9 - S| \approx 0.053$.

EXAMPLE 8. We know that the alternating series $\sum_{n=1}^{\infty}(-1)^{n+1}/n^2$ converges. Determine its sum S to within 0.001.

Solution. Let $a_n = 1/n^2$ in the foregoing general error estimate to obtain

$$|S_n - S| < \frac{1}{(n+1)^2} \quad \text{for all} \quad n.$$

Then S_n approximates S to within 0.001 if

$$\frac{1}{(n+1)^2} < 0.001 \qquad \Leftrightarrow \qquad (n+1)^2 > 10^3 \qquad \Leftrightarrow \qquad n+1 > 10\sqrt{10} > 31.$$

Let $n = 31$. Then $|S_{31} - S| < 0.001$. Calculations yield

$$S_{31} = \sum_{n=1}^{31} \frac{(-1)^{n+1}}{n^2} \approx 0.82297.$$

As a point of information, the sum of the full series is known to be $S = \pi^2/12 \approx 0.82247$. □

Concluding Remarks

The alternating harmonic series

$$1-\frac{1}{2}+\frac{1}{3}-\frac{1}{4}+\cdots$$

is only conditionally convergent while the series

$$1-\frac{1}{2^2}+\frac{1}{3^2}-\frac{1}{4^2}+\cdots$$

is absolutely convergent. Why all the fuss about this distinction? After all, we are primarily interested in the given series and not the corresponding series of absolute values.

The distinction is important because conditionally convergent series and absolutely convergent series behave quite differently under familiar algebraic operations. First, the good news: The terms of an absolutely convergent series can be added in any order without changing the sum of the series or the fact that it converges. You say of course, but read on.

The bad (but intriguing) news: The sum of a conditionally convergent series may change and the series may even become divergent if its terms are rearranged. For example,

$$1-\frac{1}{2}+\frac{1}{3}-\frac{1}{4}+\frac{1}{5}-\frac{1}{6}+\cdots=\ln 2,$$

$$1+\frac{1}{3}-\frac{1}{2}+\frac{1}{5}+\frac{1}{7}-\frac{1}{4}+\cdots=\frac{3}{2}\ln 2.$$

See the problems.

The famous German mathematician Bernhard Riemann (1826–1866) proved that given *any* real number S, the terms of a conditionally convergent series can be rearranged so that the rearranged series converges and has sum S. Moreover, other rearrangements diverge to ∞ and $-\infty$.

PROBLEMS

In Probs. 1–22, determine whether the given series converges absolutely (hence, converges), converges conditionally, or diverges. Be sure to check all the hypotheses of any test you use.

1. $\sum_{n=1}^{\infty}(-1)^{n+1}/(4n-3)$
2. $\sum_{n=1}^{\infty}(-1)^{n-1}/\sqrt{2n+1}$
3. $\sum_{n=1}^{\infty}(-1)^{n-1}/n^2$
4. $\sum_{n=1}^{\infty}(-1)^{n+1}/(n^2-\ln n)$
5. $\sum_{n=0}^{\infty}(-1)^n\frac{3n}{n^2+4}$
6. $\sum_{n=0}^{\infty}\frac{(-1)^n}{n^2+4}$
7. $\sum_{n=0}^{\infty}\frac{\cos n\pi}{\sqrt{n}}$
8. $\sum_{n=2}^{\infty}\frac{\sin(2n+1)\pi/2}{\ln n}$

9. $\sum_{n=2}^{\infty} (-1)^n/n \ln n$

10. $\sum_{n=2}^{\infty} (-1)^n/n(\ln n)^2$

11. $\sum_{n=1}^{\infty} (-1)^{n-1}\frac{\ln n}{n}$

12. $\sum_{n=2}^{\infty} (-1)^n\frac{n}{\ln n}$

13. $\sum_{n=1}^{\infty} (-1)^{n-1}\frac{\sin(1/n)}{n}$

14. $\sum_{n=1}^{\infty} (-1)^{n+1}\sin\frac{1}{n}$

15. $\sum_{n=1}^{\infty} (-1)^{n-1}\left(1-\cos\frac{1}{n}\right)$

16. $\sum_{n=1}^{\infty} (-1)^{n+1}\left(1-\cos\frac{1}{\sqrt{n}}\right)$

17. $\sum_{n=1}^{\infty} \left(\frac{\sin n}{n}\right)^n$

18. $\sum_{n=1}^{\infty} \left(\frac{\ln n}{n}\right)^n$

19. $\sum_{n=1}^{\infty} \frac{(-1)^{n+1}}{\sqrt{n(n+1)}}$

20. $\sum_{n=1}^{\infty} (-1)^{n-1}\frac{n!-n}{n(n!)}$

21. $\sum_{n=1}^{\infty} (-1)^{n-1}(\sqrt{n+1}-\sqrt{n})$

22. $\sum_{n=1}^{\infty} (-1)^{n-1}(\ln(n+1)-\ln n)$

In Probs. 23–26, determine the values of x for which the series converges.

23. $\sum_{n=1}^{\infty} \frac{x^n}{n^2}$

24. $\sum_{n=1}^{\infty} \frac{x^n}{n}$

25. $\sum_{n=1}^{\infty} \frac{x^n}{2^n n^2}$

26. $\sum_{n=1}^{\infty} \frac{\sin nx}{n^2}$

In Probs. 27–30, use the error estimate $|S_n - S| < a_{n+1}$ associated with the alternating series test: (a) Find the smallest n such that the error estimate guarantees that S_n will approximate S to within 0.001. (b) Then calculate the corresponding S_n to four decimal places.

27. $\sum_{n=1}^{\infty} (-1)^{n+1}/n^3$

28. $\sum_{n=1}^{\infty} (-1)^{n+1}/n^n$

29. $\sum_{n=1}^{\infty} \frac{(-1)^{n+1}\sin(1/n)}{n}$

30. $\sum_{n=2}^{\infty} \frac{(-1)^n}{n(\ln n)^2}$

31. Show that the series $\sum_{n=1}^{\infty}(-1)^{n+1}a_n$ with $a_n = 1/\sqrt{n} + (-1)^{n+1}/n$ satisfies $a_n \geq 0$ and $a_n \to 0$, but the series diverges. Why doesn't this contradict the alternating series test?

32. Find the values of p such that $\sum_{n=1}^{\infty}(-1)^{n+1}/n^p$ converges absolutely, converges conditionally, and diverges.

33. If $\sum a_n$, and $\sum b_n$ are absolutely convergent and c is any constant, show that $\sum(a_n + b_n)$ and $\sum c\,a_n$ are absolutely convergent.

34. If $\sum a_n$ is absolutely convergent, show that $|\sum a_n| \leq \sum |a_n|$.

35. If $\sum a_n^2$ and $\sum b_n^2$ are convergent, show that $\sum a_n b_n$ is absolutely convergent. *Hint.* $(a-b)^2 \geq 0$.

36. The improper integral $\int_0^\infty [(\sin x)/x]\,dx$ plays an important role in Fourier series and Fourier integrals. It turns out that the integral converges, as you are about to show, and has value $\pi/2$. Use the following steps to show that the improper integral converges. First sketch a graph of $y = (\sin x)/x$; it will help you see what is going on.
(a) For any $b > 0$, explain why there are consecutive integers n and $n + 1$ such that $n\pi \le b < (n + 1)\pi$. Then show that
$$\left|\int_0^b \frac{\sin x}{x}\,dx - \int_0^{n\pi} \frac{\sin x}{x}\,dx\right| \le \ln\left(\frac{n+1}{n}\right).$$
(b) Conclude from (a) that the improper integral converges if and only if
$$\int_0^{n\pi} [(\sin x)/x]\,dx \to L \text{ as } n \to \infty.$$
(c) Show that
$$\int_0^{n\pi} \frac{\sin x}{x}\,dx = \sum_{k=1}^{n} (-1)^{k+1} \int_{(k-1)\pi}^{k\pi} \left|\frac{\sin x}{x}\right| dx.$$
Conclude that the integral on the left will have a limit as $n \to \infty$ if the series $\sum_{k=1}^\infty (-1)^{k+1} a_k$ with $a_k = \int_{(k-1)\pi}^{k\pi} [(\sin x)/x]\,dx$ converges.
(d) Show that the series converges.

37. Given that the alternating harmonic series has sum $\ln 2$, show that the rearrangement of the series $1 + 1/3 - 1/2 + 1/5 + 1/7 - 1/4 + \cdots$ has sum $(3/2)\ln 2$. *Hint.* (a) Explain why $1/2 - 1/4 + 1/6 - 1/8 + \cdots = (1/2)\ln 2$. (b) Now insert zero terms into this series (including before the first term) in such a way that the sum of the augmented series and the harmonic series is the rearranged series. (c) Explain briefly why inserting zeros as in (b) will not destroy the convergence of the series or change its sum.

If you have solved Prob. 54 in Sec. 1.4, you can verify that the alternating harmonic series has sum ln 2 by solving the next problem. Otherwise, stay tuned.

38. Let S_n be the nth partial sum of the alternating harmonic series and T_n the nth partial sum of the harmonic series. (a) Show that $S_{2n} = T_{2n} - T_n$. *Hint.* Simplify $T_{2n} - T_n$ in the case $n = 3$ to see what is happening. Then note that $T_n = 2(\frac{1}{2} + \frac{1}{4} + \frac{1}{6} + \cdots + \frac{1}{2n})$ to get the general result. (b) Use Prob. 54 in Sec. 1.4 to show that $S_{2n} \to \ln 2$. (c) Show that $S_{2n+1} \to \ln 2$ by a simple argument based on (b). (d) Conclude $S_n \to \ln 2$ as $n \to \infty$.

1.6 The Ratio and Root Tests

The chapter ends with the ratio and root tests, which are two of the most widely applicable tests for the convergence or divergence of a series. Both tests are based on comparisons with geometric series. Further comparisons yield numerical approximations for sums of series.

The Ratio Test

The ratio test is a little simpler for series with positive terms. So we restrict the discussion to that case for a while.

We begin with a few comments about the geometric series $\sum r^n$ with $r > 0$. The geometric series converges if $r < 1$ and diverges if $r \ge 1$. Notice that the common ratio of two consecutive terms of the series is $r = r^{n+1}/r^n$. Therefore, the geometric series $\sum r^n$ converges if the common ratio satisfies $r < 1$ and diverges if the common ratio satisfies $r \geq 1$. A very similar result holds for any series of positive terms.

> Let $a_n > 0$. Then
> $$\sum a_n \text{ converges if } \frac{a_{n+1}}{a_n} \leq r < 1 \text{ for large } n,$$
> $$\sum a_n \text{ diverges if } \frac{a_{n+1}}{a_n} \geq r \geq 1 \text{ for large } n.$$

For convenience, let $n = 0, 1, 2, \ldots$. First, we establish the test for convergence. Since the convergence of $\sum a_n$ depends only on a_n for n arbitrarily large, we can change the initial terms of the series so that

$$\frac{a_{n+1}}{a_n} \leq r < 1 \text{ for } n = 0, 1, 2, \ldots.$$

Then, just as in the proof of the ratio test for sequences given in Sec. 1.1,

$$a_1 \leq ra_0, \qquad a_2 \leq ra_1 \leq r^2 a_0, \qquad a_3 \leq ra_2 = r^3 a_0,$$

and, in general,

$$a_n \leq a_0 r^n.$$

Since $0 < r < 1$, $\sum a_0 r^n$ converges and the basic comparison test implies that $\sum a_n$ converges. The proof of the test for divergence amounts to reversing the inequalities and noting that $\sum a_0 r^n$ diverges for $r \geq 1$.

EXAMPLE 1. Does $\sum_{n=1}^{\infty} n^2/2^n$ converge or diverge?

Solution. Let $a_n = n^2/2^n$. Then

$$\frac{a_{n+1}}{a_n} = a_{n+1} \cdot \frac{1}{a_n} = \frac{(n+1)^2}{2^{n+1}} \cdot \frac{2^n}{n^2} = \frac{1}{2}\left(\frac{n+1}{n}\right)^2 = \frac{1}{2}\left(1 + \frac{1}{n}\right)^2,$$

so that a_{n+1}/a_n decreases as n increases. The first three ratios are

$$\frac{a_2}{a_1} = 2, \qquad \frac{a_3}{a_2} = \frac{9}{8}, \qquad \frac{a_4}{a_3} = \frac{8}{9}.$$

Therefore, $a_{n+1}/a_n \leq 8/9 < 1$ for $n \geq 3$ and the series $\sum n^2/2^n$ converges. □

Another way to proceed in Ex. 1 is to notice that

$$\frac{a_{n+1}}{a_n} = \frac{1}{2}\left(1 + \frac{1}{n}\right)^2 \to \frac{1}{2} \text{ as } n \to \infty.$$

Therefore, a_{n+1}/a_n can be made as near to 1/2 as we like by taking n large enough. Since $1/2 < 3/4$, it follows that

$$\frac{a_{n+1}}{a_n} \leq \frac{3}{4} \text{ for all } n \text{ large enough.}$$

Consequently, $\sum n^2/2^n$ converges. This line of reasoning leads to

The Ratio Test for Series with Positive Terms

Let $a_n > 0$ and assume that

$$\frac{a_{n+1}}{a_n} \to \rho \text{ (finite or infinite) as } n \to \infty.$$

Then

$\sum a_n$ converges if $\rho < 1$,
$\sum a_n$ diverges if $\rho > 1$,
the test fails if $\rho = 1$.

Proof. First, assume that $\rho < 1$. Choose any number r with $\rho < r < 1$. Since a_{n+1}/a_n can be made as near to ρ as we like by taking n large enough, we must have

$$\frac{a_{n+1}}{a_n} \leq r < 1 \text{ for all } n \text{ large enough.}$$

Therefore, $\sum a_n$ converges. The reasoning when $\rho > 1$ is almost the same. When $\rho = 1$, the series may converge or diverge. Two examples that show this are

$$\sum \frac{1}{n} \text{ diverges and } \frac{a_{n+1}}{a_n} = \frac{1/(n+1)}{1/n} = \frac{n}{n+1} \to 1,$$

$$\sum \frac{1}{n^2} \text{ converges and } \frac{a_{n+1}}{a_n} = \frac{1/(n+1)^2}{1/n^2} = \left(\frac{n}{n+1}\right)^2 \to 1.$$

So the test fails to distinguish between convergence and divergence if $\rho = 1$. □

EXAMPLE 2. Does $\sum_{n=0}^{\infty} 5^n/n!$ converge or diverge?

Solution. Let $a_n = 5^n/n!$. Then

$$\frac{a_{n+1}}{a_n} = a_{n+1} \cdot \frac{1}{a_n} = \frac{5^{n+1}}{(n+1)!} \cdot \frac{n!}{5^n} = \frac{5}{n+1} \to 0 \text{ as } n \to \infty.$$

The given series converges by the ratio test. In the same way, $\sum_{n=0}^{\infty} x^n/n!$ converges for any fixed x. We shall find in Sec. 2.2 that the sum of the series is e^x. □

EXAMPLE 3. Does $\sum n^n/n!$ converge or diverge?

Solution. Let $a_n = n^n/n!$. Then

$$\frac{a_{n+1}}{a_n} = \frac{(n+1)^{n+1}}{(n+1)!} \cdot \frac{n!}{n^n} = \frac{(n+1)^{n+1}}{(n+1)n^n} = \frac{(n+1)^n}{n^n} = \left(1+\frac{1}{n}\right)^n \to e>1 \text{ as } n \to \infty,$$

and the series diverges. (Can you give a simpler way to show that the series diverges?) □

Much the same reasoning applies to the series $\sum n!/n^n$, whose terms are the reciprocals of those in Ex. 3. In this case, $a_{n+1}/a_n \to 1/e < 1$, so that $\sum n!/n^n$ converges. The convergence was proved by a comparison test in Ex. 5 of Sec. 1.4. Once again, more than one test does the job. Which is easier?

EXAMPLE 4. Test the series $\sum n!/(2n)!$ for convergence or divergence.

Solution. Let $a_n = n!/(2n)!$. Then (watch the arithmetic carefully)

$$\begin{aligned}\frac{a_{n+1}}{a_n} &= \frac{(n+1)!}{[2(n+1)]!} \cdot \frac{(2n)!}{n!} = \frac{(n+1)n!}{(2n+2)(2n+1)(2n)!} \cdot \frac{(2n)!}{n!} \\ &= \frac{n+1}{2(n+1)(2n+1)} = \frac{1}{2(2n+1)} \to 0 \text{ as } n \to \infty,\end{aligned}$$

so the series converges. □

Examples 2 – 4 illustrate that the ratio test is often useful when the general term of a series involves factorials and/or powers in combination. For practice, show that each of the following series converges for any fixed x with $0 < x < 1$:

$$\sum x^n, \quad \sum nx^n, \quad \sum n^2x^n, \quad \sum n^3x^n, \quad \ldots$$

The general ratio test, for series with terms of possibly mixed signs, comes next. It includes the ratio test for series of positive terms as a special case.

Theorem 1 The Ratio Test
Let $a_n \neq 0$ and

$$\left|\frac{a_{n+1}}{a_n}\right| \to \rho \text{ (finite or infinite) as } n \to \infty.$$

Then
$\sum a_n$ converges absolutely if $\rho < 1$,
$\sum a_n$ diverges if $\rho > 1$,
the test fails if $\rho = 1$.

Proof. First assume

$$\frac{|a_{n+1}|}{|a_n|} = \left|\frac{a_{n+1}}{a_n}\right| \to \rho < 1.$$

Then $\sum |a_n|$ converges by the ratio test for series with positive terms. Thus, $\sum a_n$ converges absolutely. This proves the first assertion of the theorem. Now let

$$\frac{|a_{n+1}|}{|a_n|} = \left|\frac{a_{n+1}}{a_n}\right| \to \rho > 1.$$

Since $|a_{n+1}|/|a_n|$ can be made as near to ρ as we like by taking n large enough,

$$\frac{|a_{n+1}|}{|a_n|} \geq 1 \text{ and } |a_{n+1}| \geq |a_n| \text{ for large } n.$$

Since $|a_n| > 0$ and is nondecreasing for large n, a_n cannot tend to 0. So $\sum a_n$ must diverge and the second assertion of the theorem is proved. Finally, the test fails if $\rho = 1$, as illustrated earlier by the series $\sum 1/n^2$ and $\sum 1/n$. □

EXAMPLE 5. For what values of x does the series

$$\sum_{n=1}^{\infty} (-1)^n \frac{x^{2n}}{n^2}$$

converge?

Solution. Let $a_n = (-1)^n x^{2n}/n^2$. If $x = 0$, it is obvious that the series converges absolutely. For $x \neq 0$,

$$\left|\frac{a_{n+1}}{a_n}\right| = \frac{|x|^{2(n+1)}}{(n+1)^2} \cdot \frac{n^2}{|x|^{2n}} = |x|^2 \left(\frac{n}{n+1}\right)^2 \to |x|^2.$$

By the ratio test, the given series

$$\text{converges absolutely if } |x|^2 < 1 \iff |x| < 1,$$
$$\text{and diverges if } |x|^2 > 1 \iff |x| > 1.$$

If $|x| = 1$, that is if $x = \pm 1$, the ratio test gives no information. However, for $x = \pm 1$, the series is

$$\sum_{n=1}^{\infty} \frac{(-1)^n}{n^2},$$

which we know converges absolutely. In summary, $\sum (-1)^n x^{2n}/n^2$ converges absolutely for $|x| \leq 1$ and diverges for $|x| > 1$. □

The Root Test

As we mentioned earlier, the root test comes from the basic comparison test using comparisons with geometric series. The reasoning is very similar to that used for the ratio test.

To begin with, consider a series $\sum a_n$ with every $a_n \geq 0$. Compare $\sum a_n$ with the geometric series $\sum r^n$ to obtain

$$\sum a_n \text{ converges if } a_n \leq r^n \text{ for large } n \text{ and } r < 1,$$

$$\sum a_n \text{ diverges if } a_n \geq r^n \text{ for large } n \text{ and } r \geq 1.$$

These results are expressed more conveniently in the following forms.

> Let $a_n \geq 0$. Then
> Σa_n converges if $a_n{}^{1/n} \leq r < 1$ for large n,
> Σa_n diverges if $a_n{}^{1/n} \geq r \geq 1$ for large n.

EXAMPLE 6. Does $\Sigma a_n = \Sigma \left(\dfrac{n+2}{2n-1}\right)^n$ converge or diverge?

Solution. Since

$$a_n^{1/n} = \frac{n+2}{2n-1} = \frac{1}{2}\left(\frac{2n+4}{2n-1}\right) = \frac{1}{2}\left(\frac{2n-1+3}{2n-1}\right) = \frac{1}{2}\left(1 + \frac{3}{2n-1}\right),$$

$a_n^{1/n}$ decreases as n increases. (Alternatively, differentiate $f(x) = (x+2)/(2x-1)$ and check that $f'(x) < 0$ for $x \geq 1$). The first four values of $a_n^{1/n}$ are

$$a_1 = 3, \qquad a_2^{1/2} = \frac{4}{3}, \qquad a_3^{1/3} = 1, \qquad a_4^{1/4} = \frac{6}{7}.$$

It follows that $a_n^{1/n} \leq 6/7$ for $n \geq 4$ and the given series converges. □

Another way to proceed in Ex. 6 makes use of

$$a_n^{1/n} = \frac{n+2}{2n-1} \to \frac{1}{2} \text{ as } n \to \infty.$$

Consequently, $a_n^{1/n} \leq \dfrac{3}{4}$ for large n and the series in Ex. 6 converges. The reasoning just used stands behind

> **The Root Test for Series with Nonnegative Terms**
> Let $a_n \geq 0$ and assume
> $a_n^{1/n} \to \rho$ (finite or infinite) as $n \to \infty$.
> *Then*
> Σa_n converges if $\rho < 1$,
> Σa_n diverges if $\rho > 1$,
> the test fails if $\rho = 1$.

Proof. Assume that $\rho < 1$. Choose any number r with $\rho < r < 1$. Since $a_n^{1/n}$ can be made as near to ρ as we like by taking n large enough,

$$a_n^{1/n} \leq r < 1 \text{ for large } n$$

and the series Σa_n converges. The reasoning for the case $\rho > 1$ is almost the same. When $\rho = 1$, the series may converge or diverge. For example,

$$\Sigma \frac{1}{n} \text{ diverges and } a_n^{1/n} = \frac{1}{n^{1/n}} \to 1,$$

$$\Sigma \frac{1}{n^2} \text{ converges and } a_n^{1/n} = \frac{1}{(n^2)^{1/n}} = \frac{1}{(n^{1/n})^2} \to 1.$$

So the test fails if $\rho = 1$. □

We observed earlier that the series

$$\Sigma x^n, \qquad \Sigma n x^n, \qquad \Sigma n^2 x^n, \qquad \Sigma n^3 x^n, \qquad \ldots$$

with fixed $0 < x < 1$ converge by the ratio test. The root test gives the convergence even more easily.

EXAMPLE 7. Does $\Sigma_{n=2}^{\infty} n/(\ln n)^n$ converge or diverge?

Solution. Let $a_n = n/(\ln n)^n$. Then $a_n^{1/n} = n^{1/n}/\ln n$. Since $n^{1/n} \to 1$ from Sec. 1.1, $a_n^{1/n} \to 0$ *as* $n \to \infty$ and the series converges. □

EXAMPLE 8. In Ex. 1 we found that $\Sigma n^2/2^n$ converges by the ratio test. It also converges by the root test because

$$\left(\frac{n^2}{2^n}\right)^{1/n} = \frac{(n^{1/n})^2}{2} \to \frac{1}{2} < 1 \text{ as } n \to \infty. \ \square$$

Next comes the general root test for series with terms of possibly mixed signs. It includes the root test for series with nonnegative terms as a special case.

Theorem 2 The Root Test
Assume that

$$|a_n|^{1/n} \to \rho \text{ (finite or infinite) as } n \to \infty.$$

Then
Σa_n converges absolutely if $\rho < 1$,
Σa_n diverges if $\rho > 1$,
the test fails if $\rho = 1$.

The proof is very much like that for the general ratio test. We leave it for the problems.

EXAMPLE 9. Show that the series $\Sigma(-1)^n (\ln n/\sqrt{n})^n$ converges absolutely.

Solution. Let $a_n = (-1)^n (\ln n/\sqrt{n})^n$. Then, from known properties of $\ln x$,

$$|a_n|^{1/n} = \frac{\ln n}{\sqrt{n}} \to 0 \text{ as } n \to \infty.$$

So the series converges absolutely by the root test. □

Approximating Sums of Series *

Refinements of the comparisons with geometric series used to establish the ratio and root tests lead to practical numerical approximations for sums of series that converge by these tests. For simplicity, the discussion is restricted to series with nonnegative terms. There are similar results for series with mixed signs.

A few preliminary observations about the geometric series $\sum r^k$ with $0 < r < 1$ will prepare the way for the approximations that follow. Note that

$$\sum_{k=0}^{\infty} r^k = \sum_{k=0}^{n} r^k + \sum_{k=n+1}^{\infty} r^k,$$

$$\sum_{k=n+1}^{\infty} r^k = r^{n+1}(1 + r + r^2 + r^3 + \cdots) = \frac{r^{n+1}}{1-r},$$

$$\sum_{k=0}^{\infty} r^k = \sum_{k=0}^{n} r^k + \frac{r^{n+1}}{1-r}.$$

Thus, $r^{n+1}/(1-r)$ is the error in the approximation of the sum of the geometric series by the Nth partial sum.

Error Bounds Based on the Root Test

We begin with the root test because the error in approximation of the sum of a series by a partial sum is a little easier to estimate in this case. Suppose that the root test is used to show that a series $\sum_{n=1}^{\infty} a_n$ with $a_n \geq 0$ converges. Thus, $a_n^{1/n} \to \rho$ as $n \to \infty$ and $\rho < 1$. Let $\rho < r < 1$. Then

$$a_n^{1/n} \leq r \text{ and } a_n \leq r^n \text{ for } n > N$$

with N large enough. Let

$$S = \sum_{k=1}^{\infty} a_k, \qquad S_n = \sum_{k=1}^{\infty} a_k.$$

Then, for $n > N$,

$$0 \leq S - S_n = \sum_{k=n+1}^{\infty} a_k \leq \sum_{k=n+1}^{\infty} r^k = \frac{r^{n+1}}{1-r}.$$

In summary,

> if $a_n \geq 0$ and $a_n^{1/n} \leq r < 1$ for $n > N$, then
> $$0 \leq S - S_n \leq \frac{r^{n+1}}{1-r} \text{ for } n > N.$$

In applications, we try to select an index N of moderate size for which r is "reasonably" less than 1 in order to make the bound $r^{N+1}/(1-r)$ for $S - S_N$ small enough for practical purposes. The choices of N and r involve a balancing act, as the next example shows.

EXAMPLE 10. Find $S = \sum_{n=2}^{\infty} n/(\ln n)^n$ accurate to within 0.001.

Solution. Let $a_n = n/(\ln n)^n$. The series converges by Ex. 7, where the root test was used with

$$a_n^{1/n} = \frac{n^{1/n}}{\ln n} \to 0 \text{ as } n \to \infty.$$

It is easy to show, using logarithmic differentiation, that $f(x) = x^{1/x}$ is decreasing for $e \le x < \infty$. Hence, $n^{1/n}$ decreases for $n \ge 3$. Since $\ln n$ increases, it follows that $a_n^{1/n}$ decreases for $n \ge 3$. Calculations yield

$$a_{10}^{1/10} \approx 0.547, \qquad a_{100}^{1/100} \approx 0.227.$$

Therefore, $a_n^{1/n} \le 0.55$ for $n \ge 10$ and $a_n^{1/n} \le 0.23$ for $n \ge 100$. We could take $N \ge 100$ and $r = 0.23$ in the bound for $S - S_N$. But that would require us to use partial sums with at least 100 terms. A more practical option is to take $N \ge 10$ and $r = 0.55$. Then

$$0 \le S - S_N \le \frac{r^{N+1}}{1-r} \text{ for } N \ge 10 \text{ and } r = 0.55.$$

The desired accuracy of 0.001 is achieved if

$$\frac{r^{N+1}}{1-r} < 10^{-3} \Leftrightarrow (N+1)\ln r < \ln[10^{-3}(1-r)] \Leftrightarrow N > \frac{\ln[10^{-3}(1-r)]}{\ln r} - 1.$$

(The inequality reversed because $\ln r < 0$.) Set $r = 0.55$ to find $N > 11.8$. For $r = 0.55$ and $N = 12$, we obtain

$$0 \le S - S_{12} \le \frac{r^{13}}{1-r} < 0.0009.$$

Use of a calculator gives

$$8.25262 \le S_{12} \le 8.25263.$$

Add the last two inequalities to obtain

$$8.25262 \le S < 8.25353 \text{ or } 0 \le S - 8.25262 < 0.00091.$$

So the approximation $S \approx 8.25262$ is accurate to within 0.001. □

Error Estimates Based on the Ratio Test

The corresponding story for the ratio test is a little more complicated. Suppose that a series $\sum a_n$ with positive terms converges by the ratio test. Thus, $a_{n+1}/a_n \to \rho < 1$ as $n \to \infty$. Let $\rho < r < 1$. Then

$$\frac{a_{n+1}}{a_n} \le r \text{ and } a_{n+1} \le r a_n \text{ for } n \ge m$$

with some m. It follows that

$$a_{m+1} \le r a_m, \quad a_{m+2} \le r a_{m+1} \le r^2 a_m, \quad a_{m+3} \le r a_{m+2} \le r^3 a_m.$$

In general,

$$a_n \le a_m r^{n-m} \text{ for } n \ge m.$$

For any $N \ge m$,

$$0 \le S - S_N = \sum_{n=N+1}^{\infty} a_n \le \sum_{n=N+1}^{\infty} a_m r^{n-m},$$

$$0 \le S - S_N \le a_m \frac{r^{N+1-m}}{1-r}.$$

In summary,

> if $a_n > 0$ and $a_{n+1}/a_n \le r < 1$ for $n \ge m$, then
> $$0 \le S - S_N \le a_m \frac{r^{N+1-m}}{1-r} \text{ for } N \ge m.$$

EXAMPLE 11. Find $S = \sum_{n=1}^{\infty} n^2/2^n$ accurate to within 0.001.

Solution. Let $a_n = n^2/2^n$. From Ex. 1, the series converges,

$$\frac{a_{n+1}}{a_n} = \frac{1}{2}\left(1 + \frac{1}{n}\right)^2,$$

and a_{n+1}/a_n decreases as n increases. Numerical experiments yield

$$\frac{a_{n+1}}{a_n} \le \frac{a_{11}}{a_{10}} = .605 \text{ for } n \ge 10.$$

Let $N \ge m = 10$ and $r = 0.605$. Then

$$0 \le S - S_N \le a_{10} \frac{r^{N-9}}{1-r}.$$

Proceed as in Ex. 10 to obtain

$$a_{10}\frac{r^{N-9}}{1-r} < 0.001 \qquad \Leftrightarrow \qquad N > 9 + \frac{\ln[10^{-3}(1-r)/a_{10}]}{\ln r}.$$

Use $r = 0.605$ and $a_{10} = 10^2/2^{10}$ to find $N > 19.97$. So $N = 20$ will do the job and

$$S_{20} = \sum_{n=1}^{20} \frac{n^2}{2^n} \approx 5.9995$$

approximates the sum of the series to within 0.001 . You may suspect that the sum of the series is 6. It is. Stay tuned. □

PROBLEMS

In Probs. 1–40 test the following series for convergence or divergence. In the case of convergence, state whether the series is absolutely or conditionally convergent. (As usual, more than one test may do the job. It is a good idea to try to apply several methods to each series and to compare the effort required when more than one method is successful.)

1. $\sum n^3/2^n$
2. $\sum n^5/3^n$
3. $\sum(-1)^n 3^n/n2^n$
4. $\sum(-1)^n 3^n/n^5 2^n$
5. $\sum 10^n/n!$
6. $\sum \sqrt{10^n/n!}$
7. $\sum n^2 e^{-n}$
8. $\sum (n+1)^2 n!/e^n$
9. $\sum \dfrac{[(2n)!]^2}{(4n)!}$
10. $\sum \dfrac{n!(2n)!}{(3n)!}$
11. $\sum \left(1+\dfrac{2}{n}\right)^{-n}$
12. $\sum \left(1-\dfrac{2}{n}\right)^{n}$
13. $\sum \left(1-\dfrac{1}{n^2}\right)^{n}$
14. $\sum \left(1-\dfrac{1}{n}\right)^{n^2}$
15. $\sum(-1)^{n+1}\dfrac{(\ln n)3^n}{n!}$
16. $\sum(-1)^{n+1}\dfrac{2^n e^{-n}}{\ln n}$
17. $\sum \dfrac{2^n n!}{n^n}$
18. $\sum \dfrac{3^n n!}{n^n}$
19. $\sum n^n/(2n)!$
20. $\sum n^n/(n!)^2$
21. $\sum n\left(\dfrac{n-3}{2n+1}\right)^n$
22. $\sum n^2\left(\dfrac{n-3}{2n+1}\right)^{n/2}$
23. $\sum n \sin^n\left(\dfrac{1}{n}\right)$
24. $\sum \dfrac{1}{n}\cos^n\left(\dfrac{1}{n}\right)$
25. $\sum \left(\dfrac{n}{n+5}\right)^n$
26. $\sum \left(\dfrac{n}{2n+5}\right)^n$

27. $\sum \frac{2^n (n!)^2}{(2n)!}$

28. $\sum \frac{5^n (n!)^2}{(2n)!}$

29. $\sum \frac{n^{2n}}{(2n)!}$

30. $\sum \frac{(\ln n)^n}{n!}$

31. $\sum \frac{(\ln n)^2}{(\ln 2)^n}$

32. $\sum \frac{(\ln n)^3}{(\ln 3)^n}$

33. $\sum (\sqrt{n} - 1)^{n/2}$

34. $\sum 2^n \left(\frac{n}{n+1}\right)^n$

35. $\sum 2^n \left(\frac{n}{n+1}\right)^{n^2}$

36. $\sum 3^n \left(\frac{n}{n+1}\right)^{n^2}$

37. $1 + \frac{1 \cdot 4}{1 \cdot 3} \cdot \frac{1}{1^2} + \frac{1 \cdot 4 \cdot 7}{1 \cdot 3 \cdot 5} \cdot \frac{1}{2^2} + \frac{1 \cdot 4 \cdot 7 \cdot 10}{1 \cdot 3 \cdot 5 \cdot 7} \cdot \frac{1}{3^2} + \frac{1 \cdot 4 \cdot 7 \cdot 10 \cdot 13}{1 \cdot 3 \cdot 5 \cdot 7 \cdot 9} \cdot \frac{1}{4^2} + \cdots$

38. $\frac{2}{1} + \frac{2^2 \cdot 3}{2!} + \frac{2^3 \cdot 3^2}{3!} + \frac{2^4 \cdot 3^3}{4!} + \cdots$

39. $\frac{2^2 \cdot 3}{4^3} + \frac{2^3 \cdot 3^2}{4^4 \cdot 5} + \frac{2^4 \cdot 3^3}{4^5 \cdot 5^2} + \frac{2^5 \cdot 3^4}{4^6 \cdot 5^3} + \cdots$

40. $1 + \frac{7^2}{2^2} + \frac{7^4}{(2 \cdot 4)^2} + \frac{7^6}{(2 \cdot 4 \cdot 6)^2} + \frac{7^8}{(2 \cdot 4 \cdot 6 \cdot 8)^2} + \ldots$

41. (a) Explain why the root test cannot be used to test the series

$$\frac{1}{2} + \frac{1}{3^2} + \frac{1}{2^3} + \frac{1}{3^4} + \frac{1}{2^5} + \frac{1}{3^6} + \cdots$$

for convergence or divergence. (b) Show that the displayed test that leads to the root test does settle the issue.

42. (a) Explain why the ratio test cannot be used to test for convergence or divergence of the series

$$1 + \frac{1}{2} + \frac{1}{2 \cdot 3} + \frac{1}{2^2 \cdot 3} + \frac{1}{2^2 \cdot 3^2} + \frac{1}{2^3 \cdot 3^2} + \frac{1}{2^3 \cdot 3^3} + \cdots,$$

where the terms after the leading one are obtained by alternately multiplying the previous term by 1/2 or 1/3. (b) Show that the displayed test that leads to the ratio test does settle the issue.

43. Show that the ratio test and the root test both are inconclusive for the p-series, for any p.

44. Let $p > 0$. Determine the values of p for which

$$\sum_{n=1}^{\infty} \frac{(\ln n)^p}{(\ln p)^n}$$

converges or diverges.

For each series in Probs. 45–50, show that

$$\text{either } a_n^{1/n} \leq r \text{ or } \frac{a_{n+1}}{a_n} \leq r \text{ for } n \geq m,$$

for some m and for some r with $0 < r < 1$. Use the corresponding error estimate for $S - S_N$ to calculate S accurate to within 0.001.

45. $\sum_{n=1}^{\infty} n/2^n$

46. $\sum_{n=1}^{\infty} n^2/3^n$

47. $\sum_{n=1}^{\infty} \left(\frac{n}{2n+1}\right)^n$

48. $\sum_{n=0}^{\infty} \left(\frac{n+5}{2n+1}\right)^n$

49. $\sum_{n=0}^{\infty} 2^n/n!$

50. $\sum_{n=1}^{\infty} n!/n^n$

Chapter Highlights

Sequences have important practical applications, particularly to methods of successive approximations for the solutions of problems, as illustrated by Newton's method and the method of successive substitutions. Sequences and their convergence properties are essential for the study of infinite series. Several special limits, including $x^{1/n} \to 1$ as $n \to \infty$ if $x > 0$, will be used frequently in Ch. 2. A very important property of sequences is that every monotone sequence has a limit. The limit is finite if the sequence is bounded and the limit is infinite if the sequence is unbounded.

Most of the chapter is concerned with infinite series. There are two main questions. Does the series converge or diverge? If the series converges, what is its sum? Tests for convergence or divergence help to answer the first question. The second question can be answered in relatively few cases. Often we have to resort to numerical methods to approximate the sum of a series.

There are a few standard series you should recognize immediately. One is the geometric series,

$$\sum_{n=0}^{\infty} x^n = \frac{1}{1-x} \quad \text{for} \quad |x| < 1.$$

Another is the harmonic series,

$$\sum_{n=1}^{\infty} \frac{1}{n} = \infty.$$

The harmonic series is the special case with $p = 1$ of the p-series

$$\sum_{n=1}^{\infty} \frac{1}{n^p},$$

which converges if $p > 1$ and diverges if $p \leq 1$.

Several tests for convergence, absolute convergence, or divergence are presented in the text. You should know what these tests are and how to apply them. First came three tests for series with nonnegative terms,

the basic comparison test,
the limit comparison test,
the integral test.

Then we met series with terms of mixed sign and

the alternating series test.

The chapter ends with two tests for absolute convergence or divergence,

the ratio test and the root test.

The first three tests also can be used to test for absolute convergence. Just replace each term in the given series by its absolute value.

When faced with a particular series $\sum a_n$ to examine for convergence or divergence, first check whether $a_n \to 0$ as $n \to \infty$. If not, the series diverges by the basic divergence test. Sometimes the basic comparison test reveals at a glance that the series converges or diverges by comparison with a standard series. But don't struggle with the basic comparison test; often the limit comparison test is easier to apply. For a series with alternating terms, it may be obvious that the conditions of the alternating series test are satisfied, so that the series converges. We leave to the last the tests that may require more work. You should try the integral test if $a_n = a(n)$, where $a(x)$ is decreasing for $1 \leq x < \infty$, $a(x) \to 0$ as $x \to \infty$, and the integral of $a(x)$ is not too difficult. The ratio test is often successful if a_n involves nth powers or factorials. Finally, the root test is worth trying if a_n involves nth powers but does not involve factorials.

The project for this chapter concerns the evolution of a dynamical system. The emphasis is on the logistic equation $x_{n+1} = \mu x_n(1 - x_n)$, which is often used to model the development of a population, say of deer in a forest, at discrete times. For smaller values of μ the population size x_n approaches a limiting equilibrium value as n increases. For larger values of μ, x_n oscillates chaotically. In recent years, chaotic behavior has been a focus of investigations by many mathematicians, scientists, and engineers. You may have seen reports about it in the news. Areas under active investigation include the transition from well-organized flow patterns to chaotic behavior in water waves and in the atmosphere; chaotic behavior in certain chemical reactions; the chaotic dynamics of nonlinear electrical circuits such as the Van der Pol oscillator; and the erratic swings in population observed in many biological populations. It has even been suggested that chaotic neural activity may play a role in the way the brain processes information.

Chapter Project: Dynamical Systems

A *dynamical system* is a physical or biological system that changes over time according to a well–determined rule of evolution. Mathematical models of such systems are either continuous or discrete, meaning that the system is modeled for all time in some relevant interval or the system is observed only at particular (usually equally spaced) instants in time. This project is a brief introduction to discrete dynamical systems. To be definite, a biological model is assumed. At time zero, a number x_0 called the *state* of the system is given. At time 1, the next instant of time when the system is observed the state is x_1. Likewise, $x_2, x_3, \ldots$ give the subsequent states. The rule of evolution of the system is described by

$$x_{n+1} = g(x_n)$$

for $n = 0, 1, 2, \ldots$, where g is a function determined by the particular biological situation.

One of the most useful population models, first used in 1845 by P. F. Verhulst, is the so–called *logistic equation*

$$x_{n+1} = \mu x_n(1 - x_n).$$

Here $\mu > 0$ and x_n is a normalized population characteristic. Thus, $0 \leq x_n \leq 1$. For example, x_n could be the percentage of the population infected with a particular disease or it could be the ratio of population size to the carrying capacity (maximum possible population), given the resources available.

Use a programmable calculator or computer as needed in the following problems. The problems refer to the logistic equation unless the contrary is explicitly stated. The emphasis is on the long–term behavior of the system. That is, pay special attention to the behavior of x_n for large n. To be definite, assume x_n is a (normalized) population and think in those terms.

Problem 1. Let $\mu = 2$. Then $x_{n+1} = 2x_n(1 - x_n)$. Take several different values for the initial population x_0 in $[0,1]$ and calculate step by step the values $x_1, x_2, \ldots, x_n$ until nothing new of interest happens. What do the experiments suggest about the long–term behavior of the population? Be specific.

Problem 2. (a) What function $g(x)$ describes the logistic equation as $x_{n+1} = g(x_n)$? (b) Now let $\mu = 2$ and use a graph, as in the part of Sec. 1.2 dealing with successive substitutions, to explain the results in Prob. 1. This time be sure to try $x_0 = 0$.

In general, if $x_{n+1} = g(x_n)$ describes a discrete dynamical system, a point a is an *equilibrium point (fixed point)* of the system (or of g) if $a = g(a)$.

Problem 3. Let $x_{n+1} = g(x_n)$ be any discrete dynamical system with g continuous. Assume that $x_n \to L$ as $n \to \infty$ for some L. Show that L is an equilibrium point.

Problem 4. Find the equilibrium point(s) of the logistic equation when $\mu = 2$.

Problem 5. Find the equilibrium point(s) of the logistic equation for any $\mu > 0$.

An equilibrium point a of a dynamical system is *stable* if $x_n \to a$ for all choices of x_0 close enough to a; otherwise, a is *unstable.* An equilibrium point a is *attracting* for a point c if the sequence $\{x_n\}$ with $x_0 = c$ has limit a.

Problem 6. Determine with graphical analysis whether the equilibrium point(s) of the logistic equation with $\mu = 2$ are stable or unstable. Also determine the attracting set(s).

There is an important criterion for stability that you may use without proof.

If a is an equilibrium point for $x_{n+1} = g(x_n)$
and g has a continuous derivative near a, then

$$|g'(a)| < 1 \quad \Rightarrow \quad a \text{ stable,}$$
$$|g'(a)| > 1 \quad \Rightarrow \quad a \text{ unstable.}$$

Problem 7. Apply the stability criterion to the equilibrium point(s) of the logistic equation when (a) $\mu = 2$, (b) $\mu = 3$, and (c) $\mu = 4$. What can you conclude in each case?

Problem 8. Determine the stability of the equilibrium points of the logistic equation for any $\mu > 0$, $\mu \neq 3$.

Problem 9. Use numerical experiments and graphical analysis to try and determine the stability of the equilibrium points of the logistic equation when $\mu = 3$. What do you believe happens?

Problem 10. Let $\mu = 3.3$ in the logistic equation. Use numerical experiments to examine the long–run behavior of the population. Try several different x_0 in $[0,1]$ (perhaps chosen by a random number generator). What do you observe about the long–run behavior? Be specific.

Problem 11. Repeat Prob. 10 for $\mu = 3.5$.

It turns out that the population dynamics of the logistic equation become increasing complex (as hinted at by the results in Probs. 8–11) as μ increases from 0 to 4 (and beyond).

Problem 12. Repeat Prob. 10 for $\mu = 4$.

Surprisingly, the erratic behavior exhibited in Prob. 12 and in several other dynamical systems can be described in very precise mathematical terms. This leads to the notion of *chaotic dynamics.* The logistic equation exhibits chaotic dynamics for $\mu = 4$. It would take us too far afield to describe fully the transition to chaos in the logistic equation. However, the basic facts are as follows. For $1 < \mu < 3$ the origin is an unstable equilibrium point and $a = (\mu - 1)/\mu$ is stable; in fact, a attracts any x_0 in $(0,1)$ and the population settles down toward the equilibrium state a. As μ passes through 3, the fixed point $a = (\mu - 1)/\mu$ becomes unstable and the population begins to cycle between two stable points of period 2. (A point b has period 2 if the sequence starting with $x_0 = b$ is b, c, b, c, b, ..., where $c = g(b)$ and $b = g(c)$.) As μ increases further, the points of period 2 become unstable and four points of period 4 appear, then period 8 points appear. As μ continues to increase, points of all other periods eventually appear and the erratic behavior associated with chaos sets in. Problems 10 and 11 give numerical confirmation that points of period 2 and 4 appear as μ increases. Further insight into this transition to chaos is provided by the next problem.

The last problem for this project requires good plotting resources and the ability to use them. If you have the knowledge and resources, the results are gratifying.

Problem 13. Take $x_0 = 1/2$. For each of 1,500 equally spaced values of μ between 0 and 4, calculate the first 500 points in the logistic sequence. Then plot μ on the horizontal axis and vertically above μ plot the last 400 points in the logistic sequence. Only the last 400 points are plotted to avoid any transient behavior near the beginning. (Very similar plots are obtained for any choice of x_0 in $(0,1)$.)

CHAPTER REVIEW PROBLEMS

In Probs. 1–16, use the algebraic limit laws and special limits to determine whether the sequence whose nth term is given converges or diverges. In the case of convergence, find the limit. (If you don't see what to expect, do some experimenting with a calculator or computer.)

1. $\dfrac{n^2 + 3}{\sqrt{4n + 9n^2}}$

2. $\dfrac{2/n + 3}{9 - 1/n^2}$

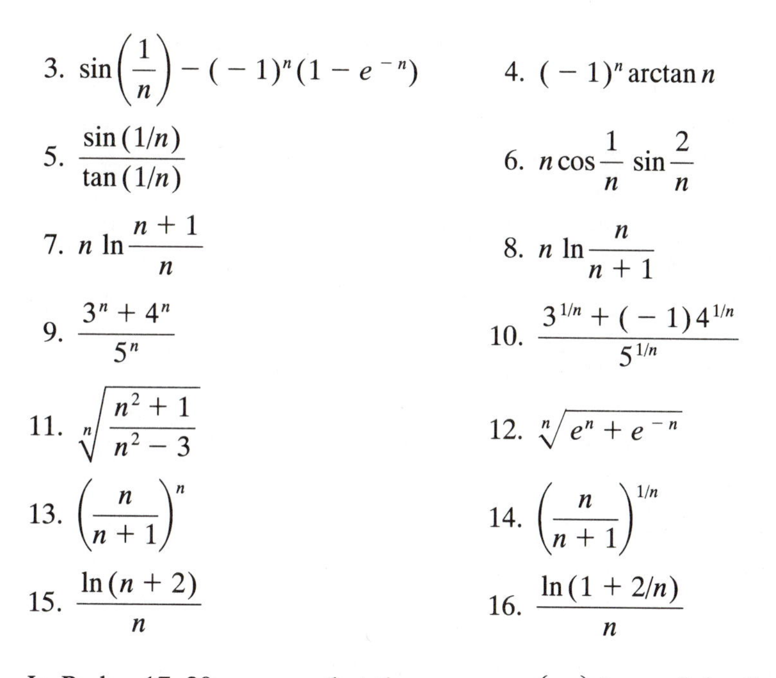

3. $\sin\left(\frac{1}{n}\right) - (-1)^n(1 - e^{-n})$

4. $(-1)^n \arctan n$

5. $\frac{\sin(1/n)}{\tan(1/n)}$

6. $n \cos\frac{1}{n} \sin\frac{2}{n}$

7. $n \ln\frac{n+1}{n}$

8. $n \ln\frac{n}{n+1}$

9. $\frac{3^n + 4^n}{5^n}$

10. $\frac{3^{1/n} + (-1)4^{1/n}}{5^{1/n}}$

11. $\sqrt[n]{\frac{n^2+1}{n^2-3}}$

12. $\sqrt[n]{e^n + e^{-n}}$

13. $\left(\frac{n}{n+1}\right)^n$

14. $\left(\frac{n}{n+1}\right)^{1/n}$

15. $\frac{\ln(n+2)}{n}$

16. $\frac{\ln(1 + 2/n)}{n}$

In Probs. 17–20, assume that the sequence $\{a_n\}$ has a finite limit L as $n \to \infty$. Find the limit if the terms of the sequence satisfy $a_1 = 1$ and the recurrence formula that follows.

17. $a_{n+1} = \frac{1}{n} a_n$

18. $a_{n+1} = 3 + \frac{(-1)^n}{\sqrt{n}} a_n$

19. $a_{n+1} = \frac{a_n(a_n^2 + 12)}{3a_n^2 + 4}$

20. $a_{n+1} = a_n + 1 - \ln a_n$

21. Show that the sequence in Prob. 17 is bounded and monotone. Thus, the limit assumed in Prob. 17 does exist.

22. Show that the sequence defined by $x_0 = 1$ and $x_{n+1} = 1 + \sqrt{x_n}$ for $n \geq 1$ has a limit and find it. *Hint.* (a) Show that $x_n \geq 1$ and that $x_n \leq 4$ for all n. (b) Show step by step that $x_{n+1} > x_n$ for all n by using the mean value theorem and

$$x_{n+1} - x_n = g(x_n) - g(x_{n-1})$$

where $g(x) = 1 + \sqrt{x}$ for $x \geq 1$.

In Probs. 23–49, determine the convergence properties of the given series. That is, determine if the series converges conditionally, converges absolutely, or does not converge. (Assume that $n \geq 1$ unless otherwise specified.)

23. $\Sigma \frac{n}{\sqrt{n^2+4}}$

24. $\Sigma \frac{1}{n\sqrt{n^2+4}}$

25. $\Sigma \frac{2^{-n}}{3^n}$

26. $\Sigma \frac{3^n}{2^n + 3^n}$

27. $\Sigma \ln\frac{n+1}{3n+4}$

28. $\Sigma(-1)^n \ln\left(\frac{1}{n}\right)$

29. $\frac{1}{2} - \frac{1}{\sqrt{2}} + \frac{1}{\sqrt[3]{2}} - \frac{1}{\sqrt[4]{2}} + \cdots$

30. $1 - \frac{3}{1 \cdot 2} + \frac{5}{3 \cdot 2} - \frac{7}{4 \cdot 3} + \frac{9}{5 \cdot 4} - \cdots$

31. $\sqrt{3} + \frac{3}{\sqrt{2}} + \frac{3^{3/2}}{\sqrt{3}} + \frac{3^2}{\sqrt{4}} + \frac{3^{5/2}}{\sqrt{5}} + \cdots$

32. $1 + \frac{1}{\sqrt{2}} + \frac{1}{2\sqrt{3}} + \frac{1}{3\sqrt{4}} + \frac{1}{4\sqrt{5}} + \cdots$

33. $2 - \frac{3}{2\sqrt{2}} + \frac{4}{3\sqrt{3}} - \frac{5}{4\sqrt{4}} + \cdots$

34. $\frac{2^2}{1 + 2^3} + \frac{2^3}{1 + 3^3} + \frac{2^4}{1 + 4^3} + \frac{2^5}{1 + 5^3} + \cdots$

35. $\sum \sin \frac{\pi}{n^2}$

36. $\sum \tan \frac{\pi}{n^2}$

37. $\sum \frac{\sqrt{n}}{n^2 + 1}$

38. $\sum \frac{\arctan n}{n^2 + 1}$

39. $\sum \frac{\cos n\pi}{\sqrt{n}}$

40. $\sum \frac{\cos n\pi}{n + 1}$

41. $\sum \frac{1}{n^2}\left(1 + \frac{1}{n}\right)^{-n}$

42. $\sum \frac{1}{n} \ln\left(1 + \frac{1}{n}\right)$

43. $\sum_{n=2}^{\infty} (\ln n)^{-2}$

44. $\sum_{n=2}^{\infty} \frac{(-1)^n}{n \ln(\ln n)}$

45. $\sum \frac{n^e}{e^n}$

46. $\sum \frac{e^n}{n^e}$

47. $\sum \frac{(-1)^n \ln n}{n^{1/3}}$

48. $\sum \frac{(-1)^{n+1} \ln 2n}{n^{4/3}}$

49. $\sum (-1)^n \frac{\sin(1/n)}{\sqrt{n}}$

50. $\sum \frac{\sin(1/n)}{\sqrt{n}}$

51. $\sum (-1)^n \ln \frac{n}{n + 1}$

52. $\sum \ln \frac{n}{n + 1}$

In Probs. 53–60, determine the values of x (if any) for which the series converges (conditionally or absolutely).

53. $1 - x + x^2 - x^3 + x^4 - x^5 + \cdots$

54. $1 + x^3 + x^6 + x^9 + x^{12} + \cdots$

55. $\sum_{n=0}^{\infty} \frac{n!x^n}{3^n}$

56. $\sum_{n=0}^{\infty} \frac{3^{n/2}x^n}{(n + 1)2^n}$

57. $\sum_{n=0}^{\infty} \frac{(2n)!x^{2n}}{2^{2n}(n!)^2}$

58. $\sum_{n=0}^{\infty} \frac{\ln n}{n} x^{3n}$

59. $\sum_{n=0}^{\infty} \frac{(-1)^n(x+1)^n}{\sqrt[n]{2}}$

60. $\sum_{n=0}^{\infty} \frac{x^{3n}}{2^n \sqrt[n]{n}}$

In Probs. 61–64, find the sum of the series.

61. $\sum_{n=0}^{\infty} e^{-n}$

62. $\sum_{n=0}^{\infty} \frac{(-1)^{n+1} e^n}{\pi^n}$

63. $\sum_{n=1}^{\infty} \frac{1}{n(n+2)}$

64. $\sum_{n=3}^{\infty} \frac{1}{n^2+3n+2}$

In Probs. 65–69, find the sum of the series accurate to within 0.001.

65. $\sum_{n=0}^{\infty} \frac{(-1)^n}{n!}$

66. $\sum_{n=1}^{\infty} \frac{(-1)^{n+1}n}{e^n}$

67. $\sum_{n=0}^{\infty} \frac{1}{n^2+1}$

68. $\sum_{n=1}^{\infty} \frac{1}{(2n-1)^2}$

CHAPTER 2
POWER SERIES AND TAYLOR SERIES

An infinite series of the form

$$a_0 + a_1x + a_2x^2 + \cdots + a_nx^n + \cdots$$

with any constants a_n is called a power series in x. Such series occur in many branches of mathematics and its applications. For example, power series are used to solve some of the most important differential equations of science and engineering and they are used frequently in numerical approximation methods. In addition, most of the functions you know have power series representations, which means that the function is a sum of a particular power series. In this context, the power series also is called the Taylor series for the function, after the English mathematician Brook Taylor (1685-1731). Sometimes these series are called Maclaurin series after Colin Maclaurin (1698-1746). For simplicity, we shall refer to them only as Taylor series. The partial sums of the Taylor series for a function f, which are polynomials, are called Taylor polynomials for f.

There are two interrelated themes in this chapter. One concerns general properties of power series. The other concerns the representation of functions by Taylor series. Taylor polynomials are studied in Sec. 2.1. We return to power series and Taylor series in Sec. 2.2. Questions of convergence will be explored and Taylor series for a variety of functions will be derived. Power series look like "infinite–degree polynomials" and, very often, they behave like polynomials. We show in Sec. 2.3 that power series can be differentiated and integrated term by term just as if they were polynomials.

As mentioned above, power series methods enable us to solve differential equations when more elementary solution methods fail. The basic ideas used to find power series solutions are illustrated in Sec. 2.4. The binomial series, which generalizes the binomial theorem of algebra, is obtained as a by–product.

Taylor polynomials and Taylor series in powers of $x–a$ will be considered rather briefly. Their properties are inherited from Taylor polynomials and Taylor series in powers of x by a simple change of variable.

Computer algebra systems are very useful for finding the first few terms in a power series or Taylor series, especially when there is no simple pattern followed by the coefficients in the series. Likewise, a CAS can be used to calculate the first few terms in a power series solution to a differential equation. If you have access to a CAS, we recommend that you become acquainted with the procedures your system uses for carrying out the computations just mentioned. You can use the system as a check on your work as you study this chapter.

2.1 Taylor Polynomials

The Taylor polynomials for a given function $f(x)$ are successive approximations for $f(x)$. The successive Taylor polynomials match $f(x)$ and more and more of its derivatives at a given point $x = a$. Typically the Taylor polynomials approximate the function rather well near $x = a$ and the accuracy increases as the number of derivatives matched goes up. We shall learn how to find the Taylor polynomials for a given function and to estimate how accurately they approximate the function. We shall deal mainly with the case with $a = 0$ and return to the more general case at the end of the section. Thus, for the time being, the Taylor polynomials for a function match the function and successive derivatives of the function at $x = 0$.

Preview

Before turning to formal definitions and the details of the error estimation just mentioned, we pause for a brief preview of Taylor polynomials and related Taylor series (coming in Sec. 2.2). The nth Taylor polynomial, denoted by $P_n(x)$, has at most degree n. A typical example will show how Taylor polynomials are found. The function $f(x) = 1/(1 - x)$ with $x < 1$ is graphed in Fig. 1.

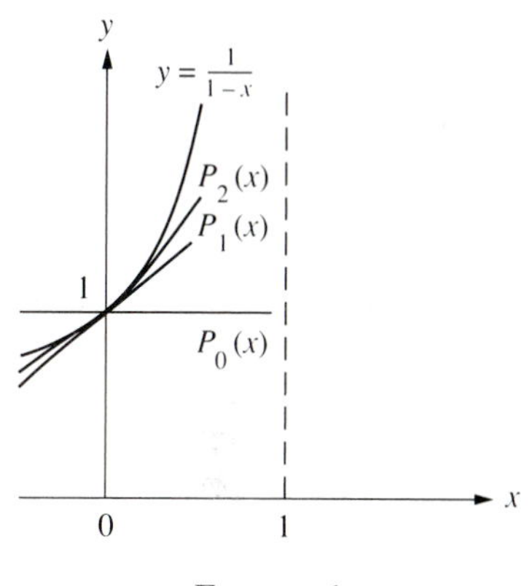

FIGURE 1

To find the first three Taylor polynomials, $P_0(x)$, $P_1(x)$, and $P_2(x)$, for $f(x)$, note that

$$f(x) = \frac{1}{1-x}, \qquad f(0) = 1,$$

$$f'(x) = \frac{1}{(1-x)^2}, \qquad f'(0) = 1,$$

$$f''(x) = \frac{2}{(1-x)^3}, \qquad f''(0) = 2.$$

The Taylor polynomial $P_0(x)$ is constant and satisfies $P_0(0) = f(0) = 1$. Therefore, $P_0(x) = 1$ for all x, as indicated in Fig. 1. The Taylor polynomial $P_1(x)$ is linear and satisfies $P_1(0) = f(0) = 1$ and $P_1'(0) = f'(0) = 1$. Thus, the graph of $P_1(x)$ is the tangent line to the graph of $f(x)$ at $x = 0$. By the point-slope equation for a line, $P_1(x) = 1 + x$. The Taylor polynomial $P_2(x)$ has the form $P_2(x) = a + bx + cx^2$ and satisfies

$$P_2(0) = f(0) = 1, \qquad P_2'(0) = f'(0) = 1, \qquad P_2''(0) = f''(0) = 2.$$

The graphs of $P_2(x)$ and $f(x) = 1/(1 - x)$ have the same values, slopes, and bending at $x = 0$. A little algebra yields $P_2(x) = 1 + x + x^2$. The pattern continues. The nth Taylor polynomial $P_n(x)$ and $f(x) = 1/(1 - x)$ have the same values and the same first n derivatives at $x = 0$. It turns out that

$$P_n(x) = 1 + x + x^2 + \cdots + x^n.$$

Observe that $P_n(x)$ is the nth partial sum for the geometric series for $1/(1 - x)$. In Sec. 1.3 we showed that $P_n(x) \to 1/(1 - x)$ as $n \to \infty$ for $|x| < 1$ and, consequently,

$$\sum_{n=0}^{\infty} x^n = \frac{1}{1-x} \quad \text{for} \quad |x| < 1.$$

The geometric series on the left is the Taylor series for the function $f(x) = 1/(1-x)$.

Taylor polynomials $P_n(x)$ for other functions $f(x)$ are determined in the same manner. Under reasonable conditions on $f(x)$, the successive Taylor polynomials are better and better approximations for $f(x)$ near $x = 0$. We shall derive useful estimates for the error in the approximation of $f(x)$ by $P_n(x)$. These error estimates will be used in Sec. 2.2 to show that, for many familiar functions $f(x)$, $P_n(x) \to f(x)$ as $n \to \infty$ for x in some interval about 0. It will follow that $f(x)$ is represented by a Taylor series in that interval.

Before getting down to business, we need to augment the prime notation for derivatives:

$$f'(x), \quad f''(x), \quad f'''(x), \quad f''''(x), \; \ldots .$$

The prime notation becomes cumbersome after two or three derivatives. A convenient alternative is to denote the successive derivatives of $f(x)$ by

$$f^{(1)}(x), \quad f^{(2)}(x), \quad f^{(3)}(x), \quad f^{(4)}(x), \quad \ldots, \quad f^{(n)}(x), \quad \ldots .$$

If we also define $f^{(0)}(x) = f(x)$, a number of summation formulas take simpler forms. With this notation, $f(x) = \sin x$ has derivatives

$$f^{(1)}(x) = \cos x, \quad f^{(2)}(x) = -\sin x, \quad f^{(3)}(x) = -\cos x, \quad f^{(4)}(x) = \sin x, \quad \ldots .$$

New Formulas for Polynomials

The key to deriving formulas for Taylor polynomials (about 0) lies in the relationship between the coefficients of a polynomial and the derivatives of the polynomial at $x = 0$. Since this relationship is of independent interest, it is developed next, after which we shall return to Taylor polynomials.

The case of a polynomial of degree 4 will make the general pattern clear:

$$\begin{aligned}
P(x) &= a_0 + a_1x + a_2x^2 + a_3x^3 + a_4x^4, & P(0) &= a_0, \\
P'(x) &= a_1 + 2a_2x + 3a_3x^2 + 4a_4x^3, & P'(0) &= a_1, \\
P''(x) &= 2a_2 + 3 \cdot 2a_3x + 4 \cdot 3a_4x^2, & P''(0) &= 2a_2, \\
P^{(3)}(x) &= 3 \cdot 2a_3 + 4 \cdot 3 \cdot 2a_4x, & P^{(3)}(0) &= 3 \cdot 2a_3, \\
P^{(4)}(x) &= 4 \cdot 3 \cdot 2a_4, & P^{(4)}(0) &= 4 \cdot 3 \cdot 2a_4.
\end{aligned}$$

Solve for a_0, a_1, a_2, a_3, a_4 to obtain

$$a_0 = P(0), \qquad a_1 = P'(0), \qquad a_2 = \frac{P''(0)}{2}, \qquad a_3 = \frac{P^{(3)}(0)}{3!}, \qquad a_4 = \frac{P^{(4)}(0)}{4!}.$$

Since $0! = 1$, $1! = 1$, and $2! = 2$,

$$a_k = \frac{P^{(k)}(0)}{k!} \quad \text{for} \quad k = 0, 1, 2, 3, 4\,.$$

It follows that

$$P(x) = P(0) + P'(0)x + P''(0)\frac{x^2}{2} + P^{(3)}(0)\frac{x^3}{3!} + P^{(4)}(0)\frac{x^4}{4!}.$$

Thus, any polynomial of degree 4 is determined by $P(0)$ and the first four derivatives of $P(x)$ at $x = 0$.

EXAMPLE 1. The unique polynomial $P(x)$ of degree 4 with

$$P(0) = 3, \qquad P'(0) = 4, \qquad P''(0) = 6, \qquad P^{(3)}(0) = -30, \qquad P^{(4)}(0) = 48$$

is

$$P(x) = 3 + 4x + 6 \cdot \frac{x^2}{2} - 30 \cdot \frac{x^3}{3!} + 48 \cdot \frac{x^4}{4!} = 3 + 4x + 3x^2 - 5x^3 + 2x^4. \ \square$$

The reasoning for the case with $n = 4$ extends to polynomials of any degree and yields:

Theorem 1 *Formulas for Polynomials*
The coefficients $a_0, a_1, \ldots, a_n$ of any polynomial

$$P(x) = a_0 + a_1 x + a_2 x^2 + \cdots + a_n x^n$$

of degree n satisfy

$$a_k = \frac{P^{(k)}(0)}{k!} \quad \text{for} \quad k = 0, 1, \ldots, n.$$

Consequently, $P(x)$ can be expressed as

$$P(x) = \sum_{k=0}^{n} P^{(k)}(0)\frac{x^k}{k!}$$

$$= P(0) + P'(0)x + P''(0)\frac{x^2}{2!} + \cdots + P^{(n)}(0)\frac{x^n}{n!}.$$

EXAMPLE 2. A car travels along a straight road. Its position at time t is $s(t)$. At time $t = 0$, its position, velocity, and acceleration are given by $s_0 = s(0) = 5$, $v_0 = s'(0) = 30$, and $a_0 = s''(0) = 4$. Suppose that $s(t)$ is a quadratic polynomial in t. Then

$$s(t) = s(0) + s'(0)t + s''(0)\frac{t^2}{2!} = 5 + 30t + 2t^2. \ \square$$

Taylor Polynomials

Remember the basic idea behind Taylor polynomials: Successive polynomials match more and more derivatives of a given function at $x = 0$. An immediate consequence of Th. 1 is

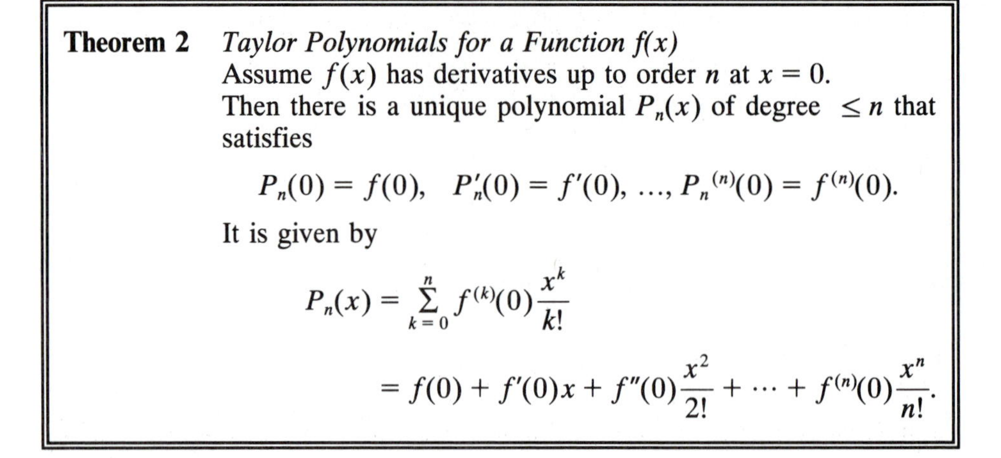

Theorem 2 *Taylor Polynomials for a Function f(x)*
Assume $f(x)$ has derivatives up to order n at $x = 0$.
Then there is a unique polynomial $P_n(x)$ of degree $\leq n$ that satisfies

$$P_n(0) = f(0), \quad P_n'(0) = f'(0), \ldots, P_n{}^{(n)}(0) = f^{(n)}(0).$$

It is given by

$$P_n(x) = \sum_{k=0}^{n} f^{(k)}(0)\frac{x^k}{k!}$$

$$= f(0) + f'(0)x + f''(0)\frac{x^2}{2!} + \cdots + f^{(n)}(0)\frac{x^n}{n!}.$$

The unique polynomial $P_n(x)$ in Th. 2 is the **nth Taylor polynomial for f about 0**. Taylor polynomials about any point $x = a$ are discussed at the end of the section. Since we are dealing mainly with Taylor polynomials about 0, we refer to them simply as Taylor polynomials whenever no confusion will result.

For example, return to $f(x) = 1/(1 - x)$. Now,

$$f'(x) = 1/(1 - x)^2, \qquad f''(x) = 2/(1 - x)^3, \qquad f^{(3)}(x) = 3!/(1 - x)^4,$$

and, in general, $f^{(n)}(x) = n!/(1 - x)^{n+1}$. Therefore, $f^{(n)}(0) = n!$ and the nth Taylor polynomial for $f(x) = 1/(1 - x)$ is

$$P_n(x) = 1 + x + x^2 + \cdots + x^n,$$

just as we stated earlier.

EXAMPLE 3. Find the Taylor polynomials for $f(x) = e^x$.

Solution. Now $f(0) = 1$ and

$$f'(x) = e^x, \qquad f''(x) = e^x, \ldots, \qquad f^{(n)}(x) = e^x,$$
$$f'(0) = 1, \qquad f''(0) = 1, \ldots, \qquad f^{(n)}(0) = 1.$$

By Th. 2,

$$P_0(x) = 1, \qquad P_1(x) = 1 + x, \qquad P_2(x) = 1 + x + \frac{x^2}{2},$$

and, in general,

$$P_n(x) = \sum_{k=0}^{n} \frac{x^k}{k!} = 1 + x + \frac{x^2}{2} + \cdots + \frac{x^n}{n!}. \quad \square$$

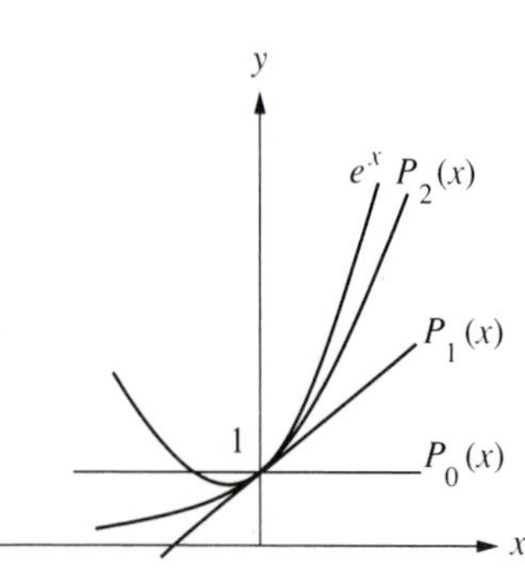

FIGURE 2

Figure 2 displays graphs of e^x and the first three Taylor polynomials for e^x. The graph of $P_1(x)$ is the tangent line to the graph of e^x at $x = 0$. The graph of $P_2(x)$ has the same value, slope, and bending at $x = 0$ as the graph of e^x.

EXAMPLE 4. Find the Taylor polynomials for $f(x) = \sin x$.

Solution. In this case,

$$f(x) = \sin x, \qquad f'(x) = \cos x, \qquad f''(x) = -\sin x, \qquad f^{(3)}(x) = -\cos x,$$
$$f^{(4)}(x) = \sin x, \qquad f^{(5)}(x) = \cos x, \qquad f^{(6)}(x) = -\sin x, \qquad f^{(7)}(x) = -\cos x,$$

and the pattern continues. It follows that

$$f(0) = 0, \qquad f'(0) = 1, \qquad f''(0) = 0, \qquad f^{(3)}(0) = -1,$$
$$f^{(4)}(0) = 0, \qquad f^{(5)}(0) = 1, \qquad f^{(6)}(0) = 0, \qquad f^{(7)}(0) = -1,$$

and so on. Hence, the first few Taylor polynomials for $\sin x$ are

$$P_0(x) = 0$$
$$P_1(x) = P_2(x) = x,$$
$$P_3(x) = P_4(x) = x - \frac{x^3}{3!},$$
$$P_5(x) = P_6(x) = x - \frac{x^3}{3!} + \frac{x^5}{5!}.$$

The Taylor polynomials for $\sin x$ involve only odd powers of x and the signs alternate. A general formula for the Taylor polynomials for $\sin x$ is

$$P_n(x) = P_{n+1}(x) = \sum_{k=0}^{(n-1)/2} (-1)^k \frac{x^{2k+1}}{(2k+1)!}, \quad \text{for } n \text{ odd.}$$

What are $P_7(x)$ and $P_8(x)$? □

EXAMPLE 5. Let $f(x) = \cos x$. Calculations similar to those for $\sin x$ (see the problems) lead to

$$P_0(x) = P_1(x) = 1,$$
$$P_2(x) = P_3(x) = 1 - \frac{x^2}{2!},$$
$$P_4(x) = P_5(x) = 1 - \frac{x^2}{2!} + \frac{x^4}{4!},$$

and, in general,

$$P_n(x) = P_{n+1}(x) = \sum_{k=0}^{n} (-1)^k \frac{x^{2k}}{(2k)!}, \quad \text{for } n \text{ even.} \ \square$$

EXAMPLE 6. Find the nth Taylor polynomial for $f(x) = \ln(1 + x)$.

Solution. Routine calculations give

$$\begin{aligned} f(x) &= \ln(1+x), & f(0) &= 0, \\ f'(x) &= (1+x)^{-1} & f'(0) &= 1, \\ f''(x) &= (-1)(1+x)^{-2} & f''(0) &= -1, \\ f^{(3)}(x) &= (-1)(-2)(1+x)^{-3}, & f^{(3)}(0) &= 2!, \\ f^{(4)}(x) &= (-1)(-2)(-3)(1+x)^{-4}, & f^{(4)}(0) &= -3!, \\ &\cdots & &\cdots \\ f^{(k)}(x) &= (-1)^{k-1}(k-1)!(1+x)^{-k}, & f^{(k)}(0) &= (-1)^{k-1}(k-1)!. \end{aligned}$$

Since $f(0) = 0$ and

$$\frac{f^{(k)}(0)}{k!} = \frac{(-1)^{k-1}(k-1)!}{k!} = \frac{(-1)^{k-1}}{k} \quad \text{for} \quad k \geq 1,$$

the nth Taylor polynomial for $f(x) = \ln(1 + x)$ is

$$P_n(x) = \sum_{k=1}^{n} \frac{(-1)^{k-1}}{k} x^k = x - \frac{x^2}{2} + \frac{x^3}{3} - \frac{x^4}{4} + \cdots + (-1)^{n-1}\frac{x^n}{n}. \ \square$$

Error Estimation for Taylor Polynomials

Now we address the question: How accurately do the Taylor polynomials for a given function approximate the function? Our goal is to find practical estimates for the **error** or **remainder**

$$R_n(x) = f(x) - P_n(x)$$

in the approximation of $f(x)$ by the nth Taylor polynomial $P_n(x)$. We assume that $f(x)$ is defined for $-c \leq x \leq c$, with $c > 0$, and that $f(x)$ has as many derivatives as we need.

If $n = 1$, $P_1(x)$ is the linear approximation of $f(x)$ whose graph is the straight line that is tangent to the graph of $f(x)$ at $(0, f(0))$. Since the second derivative of a function is a measure of the tendency of the graph of the function to bend away from any of its tangent lines, it is natural to expect that the error $R_1(x) = f(x) - P_1(x)$ is related to the magnitude of the second derivative of f. At the end of the section, we shall prove that

$$|f''(x)| \leq B \quad \text{for} \quad -c \leq x \leq c \Rightarrow |R_1(x)| \leq B\frac{x^2}{2} \quad \text{for} \quad -c \leq x \leq c.$$

Essentially the same reasoning used to obtain this estimate of $R_1(x)$ yields the similar estimates for $R_n(x) = f(x) - P_n(x)$.

Theorem 3 *Error in Taylor Approximation*
If $|f^{(n+1)}(x)| \leq B$ for $-c \leq x \leq c$, then

$$|R_n(x)| \leq B\frac{|x|^{n+1}}{(n+1)!} \leq B\frac{c^{n+1}}{(n+1)!} \quad \text{for} \quad -c \leq x \leq c.$$

There are one-sided versions of Th. 3 that deserve mention. If $|f^{(n+1)}(x)| \leq B$ for $0 \leq x \leq c$ or if $|f^{(n+1)}(x)| \leq B$ for $-c \leq x \leq 0$, then the conclusion of Th. 3 holds for $0 \leq x \leq c$ or for $-c \leq x \leq 0$. In fact, Theorem 3 is an immediate consequence of the two one-sided versions.

As indicated earlier, we defer the proof of Th. 3 to the end of the section in order to focus first on applications of the estimate for $R_n(x) = f(x) - P_n(x)$. We shall find that, for some familiar functions $f(x)$, $P_n(x)$ is a very good approximation for $f(x)$ even with n quite small.

Usually $f^{(n+1)}(x)$ is continuous. Then a logical candidate for B in Th. 3 is

$$B = \max |f^{(n+1)}(x)| \quad \text{for} \quad -c \leq x \leq c,$$

or some constant B not too much greater than this maximum.

There is a sharper connection between $f^{(n+1)}(x)$ and $R_n(x)$ than what is given in Th. 3. The French mathematician Joseph Louis Lagrange (1736-1813) discovered that

$$R_n(x) = f^{(n+1)}(z)\frac{x^{n+1}}{(n+1)!} \quad \text{for some } z \text{ between 0 and } x.$$

This is called the **Lagrange form of the remainder**. The number z is somewhere between 0 and x but Lagrange's result does not tell us any more about z. Consequently, the main practical use of the Lagrange form of the remainder is to observe that if $|f^{(n+1)}(z)| \leq B$ for *all* z between 0 and x, then $|R_n(x)| \leq B|x|^{n+1}/(n+1)!$, which is the essential conclusion of Th. 3.

The next example gives a direct application of Th. 3. A more refined use of Th. 3 comes afterward.

EXAMPLE 7. Let $f(x) = e^x$. From Ex. 3,

$$P_5(x) = 1 + x + \frac{x^2}{2!} + \frac{x^3}{3!} + + \frac{x^4}{4!} + \frac{x^5}{5!},$$

$$R_5(x) = e^x - P_5(x), \qquad f^{(6)}(x) = e^x.$$

Let $n = 5$, $c = 1$, and $B = e = \max e^x$ for $-1 \leq x \leq 1$ in Th. 3 to obtain

$$|R_5(x)| \leq e\frac{|x|^6}{6!} \leq \frac{e}{720} < 0.004 \quad \text{for} \quad -1 \leq x \leq 1.$$

Thus, the fifth degree Taylor polynomial $P_5(x)$ gives a very accurate approximation to e^x for $-1 \leq x \leq 1$. We do not provide a sketch of e^x and $P_5(x)$ because their graphs would appear to coincide. Theorem 3 suggests that the error is largest at $x = 1$. In fact, for $x = 1$,

$$f(1) = e \approx 2.7183, \qquad P_5(1) \approx 2.7167,$$
$$R_5(1) = e - P_5(1) \approx 0.0016,$$

which is consistent with the global estimate $|R_5(x)| \leq 0.004$ for $-1 \leq x \leq 1$. □

EXAMPLE 8. What degree Taylor polynomial for e^x about $x = 0$ is needed to approximate e^x for $-1 \leq x \leq 1$ to within $10^{-4} = 0.0001$?

Solution. Let $c = 1$ in Th. 3. Since $f^{(n+1)}(x) = e^x$ for all n, we could use $B = e = \max e^x$ for $-1 \leq x \leq 1$. For simplicity, let's take $B = 3$ instead. Then Th. 3 gives

$$|R_n(x)| < 3\frac{|x|^{n+1}}{(n+1)!} < \frac{3}{(n+1)!} \quad \text{for} \quad -1 \leq x \leq 1.$$

The required accuracy is attained if n satisfies

$$\frac{3}{(n+1)!} < 10^{-4} \iff (n+1)! > 30{,}000.$$

Since $7! = 5{,}040$ and $8! = 40{,}320$, we can take $n = 7$. Then the Taylor polynomial $P_7(x) = \sum_{k=0}^{7} x^k/k!$ provides the required accuracy. □

EXAMPLE 9. You are camping in the redwoods and spot a giant tree. Around midday you pace off the shadow of the tree and estimate it to be 70 feet. You also estimate the angle of elevation of the treetop measured from the tip of the shadow as 80°. Estimate the height of the redwood.

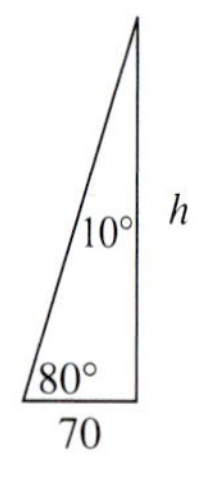

FIGURE 3

Solution. Figure 3 shows the geometric situation with h the height of the redwood. Suppose that you only have a primitive calculator that has no trigonometric functions. So you attempt to make use of the Taylor approximations

$$\sin x \approx x - \frac{x^3}{3!} \quad \text{and} \quad \cos x \approx 1 - \frac{x^2}{2!} \quad \text{for} \quad x \approx 0.$$

To take advantage of these approximations, express h in terms of the smaller angle, $10° = \pi/18$ radians in Fig. 3. Thus,

$$h = 70\cot(\pi/18) = 70\frac{\cos(\pi/18)}{\sin(\pi/18)},$$

where

$$\cos\left(\frac{\pi}{18}\right) \approx 1 - \frac{1}{2}\left(\frac{\pi}{18}\right)^2 \approx 0.985, \qquad \sin\left(\frac{\pi}{18}\right) \approx \frac{\pi}{18} - \frac{1}{6}\left(\frac{\pi}{18}\right)^3 \approx 0.174.$$

Therefore,

$$h \approx 70 \cdot \frac{985}{174} \approx 396\ ft.$$

Check this result against $h = 70 \cot(\pi/18)$ with a calculator. By the way, you may have found a record tree. The tallest confirmed redwood is about 368 feet high. □

General Taylor Polynomials

Up until now, we have approximated functions by their Taylor polynomials about 0. The story is essentially the same for approximations about any other point a, called a **base point**. So we shall be brief.

Let $f(x)$ be defined on an open interval containing the point a and assume that $f(x)$ has at least n derivatives at a. The ***n*th Taylor polynomial of *f* about *a*** is the unique polynomial of degree at most n that satisfies

$$P_n(a) = f(a), \qquad P_n'(a) = f'(a), \ldots, \qquad P_n^{(n)}(a) = f^{(n)}(a).$$

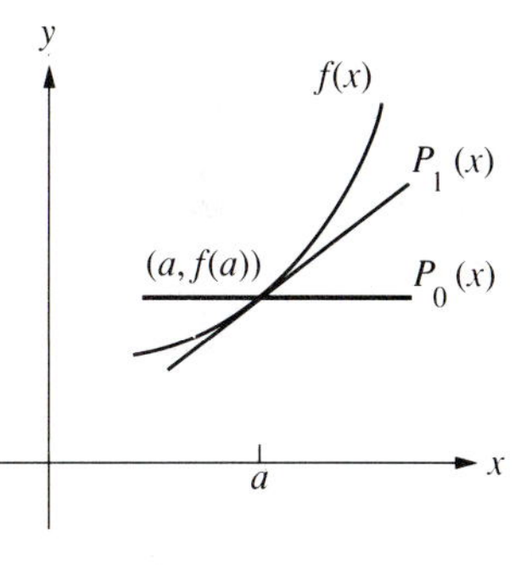

FIGURE 4

In Fig. 4, $P_0(x) = f(a)$ is constant and graphs as a horizontal line through the point $(a, f(a))$, and $P_1(x) = f(a) + f'(a)(x - a)$ is the linear approximation to f at a. It graphs as the tangent line to $y = f(x)$ at the point $(a, f(a))$.

An analogue of Th. 1 gives a convenient representation for Taylor polynomials about a. We call

$$P(x) = b_0 + b_1(x - a) + \cdots + b_n(x - a)^n$$

a *polynomial in* $x - a$. (By expanding out and collecting terms, $P(x)$ also could be written as an ordinary polynomial in x.) The reasoning leading to Th. 1 shows that

$$b_k = k!P^{(k)}(a) \quad \text{for} \quad k = 0, 1, \cdots, n,$$

$$P(x) = \sum_{k=0}^{n} P^{(k)}(a)\frac{(x - a)^k}{k!}.$$

It follows that the nth Taylor polynomial of f about a is

$$\boxed{\begin{aligned} P_n(x) &= \sum_{k=0}^{n} f^{(k)}(a)\frac{(x-a)^k}{k!} \\ &= f(a) + f'(a)(x-a) + f''(a)\frac{(x-a)^2}{2!} + \cdots + f^{(n)}(a)\frac{(x-a)^n}{n!}. \end{aligned}}$$

Set $a = 0$ to recover the Taylor polynomials about 0.

As before, the error $R_n(x) = f(x) - P_n(x)$ for $-c \le x - a \le c$ can be estimated in terms of $|f^{(n+1)}(x)|$. See the problems.

EXAMPLE 10. Find the Taylor polynomials for $f(x) = \ln x$ about $a = 1$.

Solution. Before going ahead, observe that we could not find Taylor polynomials about 0 because $\ln x$ is not defined for $x = 0$. Now,

$$\begin{aligned}
f(x) &= \ln x, & f(1) &= 0,\\
f'(x) &= x^{-1} & f'(1) &= 1,\\
f''(x) &= (-1)x^{-2} & f''(1) &= -1,\\
f'''(x) &= (-1)(-2)x^{-3}, & f'''(1) &= 2!,\\
f^{(4)}(x) &= (-1)(-2)(-3)x^{-4}, & f^{(4)}(1) &= -3!,\\
&\cdots & &\cdots\\
f^{(k)}(x) &= (-1)^{k-1}(k-1)!x^{-k}, & f^{(k)}(1) &= (-1)^{k-1}(k-1)!.
\end{aligned}$$

Therefore,

$$\begin{aligned}
P_0(x) &= 0,\\
P_1(x) &= (x-1)\\
P_2(x) &= (x-1) - \frac{(x-1)^2}{2}\\
P_3(x) &= (x-1) - \frac{(x-1)^2}{2} + \frac{(x-1)^3}{3}\\
&\cdots\\
P_n(x) &= (x-1) - \frac{(x-1)^2}{2} + \frac{(x-1)^3}{3} - \cdots + (-1)^{n-1}\frac{(x-1)^n}{n}.
\end{aligned}$$

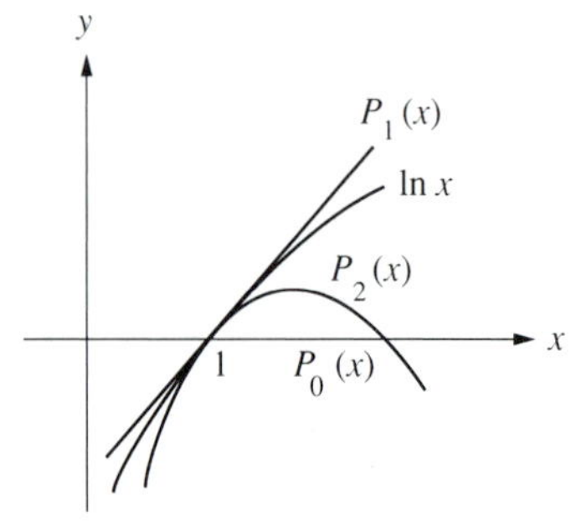

FIGURE 5

Figure 5 displays graphs of the first three Taylor polynomials for $\ln x$ about $a = 1$. □

You may have noticed the similarity between Ex. 6 and Ex. 10. In Ex. 6, $a = 0$ and

$$\ln(1+x) \approx x - \frac{x^2}{2} + \frac{x^3}{3} - \cdots + (-1)^{n-1}\frac{x^n}{n} \quad \text{for} \quad x \text{ near } 0.$$

In Ex. 10, $a = 1$ and

$$\ln x \approx (x-1) - \frac{(x-1)^2}{2} + \frac{(x-1)^3}{3} - \cdots + (-1)^{n-1}\frac{(x-1)^n}{n} \quad \text{for } x \text{ near } 1.$$

The two polynomial approximations are related by a simple change of variables. Let $x = t - 1$ and $t = 1 + x$ in the approximation for $\ln(1 + x)$ and notice that t is near 1 when x is near 0 to obtain

$$\ln t \approx (t-1) - \frac{(t-1)^2}{2} + \frac{(t-1)^3}{3} - \cdots + (-1)^{n-1}\frac{(t-1)^n}{n} \quad \text{for} \quad t \text{ near } 1.$$

This is just the approximation about 1. (Replace t by x if you prefer.)

Other Taylor polynomials in powers of $x - a$ can be found by similar changes of variable. First, let $t = x - a$. Then use Th. 2 to derive the Taylor polynomials in powers of t. Finally, replace t by $x - a$. With a little practice (see the problems), the procedure goes very smoothly.

Proof of Theorem 3*

In Th. 3, $-c \leq x \leq c$. As we mentioned earlier, Th. 3 is an immediate consequence of two one-sided versions of the theorem in which $-c \leq x \leq 0$ and $0 \leq x \leq c$. So it suffices to prove the one-sided versions of Th. 3. Since the proofs of the one-sided versions are very similar, we shall restrict to the case with $0 \leq x \leq c$. Furthermore, essentially the same argument works for any n so we give the proof only for $n = 1$. The benefit of dealing with a particular value of n, such as $n = 1$, is that there is less clutter.

The proof will be based on the geometrically obvious proposition,

$$g(0) = 0, \quad g'(x) \geq 0 \quad \text{for} \quad 0 \leq x \leq c \quad \Rightarrow \quad g(x) \geq 0 \quad \text{for} \quad 0 \leq x \leq c,$$

which follows easily from the mean value theorem. See the problems. Actually, we shall need a "higher order" version of this proposition. Replace $g(x)$ by $g'(x)$ to obtain (again, see the problems)

$$g(0) = g'(0) = 0, \quad g''(x) \geq 0 \quad \text{for} \quad 0 \leq x \leq c \quad \Rightarrow \quad g(x) \geq 0 \quad \text{for} \quad 0 \leq x \leq c.$$

Now we are ready to prove Th. 3 for $n = 1$, which is stated as follows.

> If $|f''(x)| \leq B$ for $-c \leq x \leq c$, then
>
> $$|R_1(x)| \leq B\frac{x^2}{2!} \leq B\frac{c^2}{2!} \quad \text{for} \quad -c \leq x \leq c.$$

Assume that $|f''(x)| \leq B$ for $0 \leq x \leq c$. Equivalently,

$$-B \leq f''(x) \leq B \quad \text{for} \quad 0 \leq x \leq c.$$

Now $P_1(x) = P_1(0) + P_1'(0)x$,

$$P_1(0) = f(0), \qquad P_1'(0) = f'(0), \qquad P_1''(x) = 0 \text{ for all } x.$$

Since $R_1(x) = f(x) - P_1(x)$,

$$R_1(0) = R_1'(0) = 0, \qquad R_1''(x) = f''(x).$$

Let

$$g(x) = B\frac{x^2}{2!} - R_1(x) \quad \text{for} \quad 0 \leq x \leq c.$$

It is easy to check (do it) that

$$g(0) = g'(0) = 0, \qquad g''(x) = B - f''(x) \geq 0 \quad \text{for} \quad 0 \leq x \leq c.$$

As we have seen earlier, it follows that $g(x) \geq 0$ for $0 \leq x \leq c$. From the definition of $g(x)$,

$$R_1(x) \leq B\frac{x^2}{2!} \quad \text{for} \quad 0 \leq x \leq c.$$

Essentially the same argument, with

$$g(x) = B\frac{x^2}{2!} + R_1(x) \quad \text{for} \quad 0 \leq x \leq c,$$

yields

$$-B\frac{x^2}{2!} \leq R_1(x) \quad \text{for} \quad 0 \leq x \leq c.$$

The two inequalities just obtained for $R_1(x)$ are equivalent to

$$|R_1(x)| \leq B\frac{x^2}{2!} \quad \text{for} \quad 0 \leq x \leq c.$$

This completes the proof of the one-sided version of Th. 3 for the case with $n = 1$. As we said, the proof for the general case is essentially the same except for more clutter.

PROBLEMS

In Probs. 1–4, find the unique polynomial of degree n that has the given derivatives.

1. $P(0) = 1, \quad P'(0) = 2, \quad P''(0) = 3, \quad P'''(0) = 4, \quad n = 3$
2. $P(0) = 1, \quad P'(0) = -1, \quad P''(0) = 1, \quad P'''(0) = -1, \quad P^{(4)}(0) = 1, \quad n = 4$
3. $P(0) = 3, \quad P'(0) = 0, \quad P''(0) = -2, \quad P'''(0) = 18, \quad P^{(4)}(0) = -24, \quad n = 4$
4. $P(0) = 0, \quad P'(0) = 1, \quad P''(0) = 0, \quad P'''(0) = -6, \quad n = 3$
5. A car travels down a straight road. At time $t = 0$ the car has position $s(0) = 50$ ft, velocity $v(0) = -30$ ft/sec, and acceleration $a(0) = 16$ ft/sec^2. Find an appropriate Taylor polynomial that approximates $s(t)$ for t near 0.
6. A weather balloon rises vertically from the ground. One minute after launch at time $t = 1$, the balloon has position $s(1) = 125$ ft, velocity $v(1) = 12$ ft/sec, and acceleration $a(1) = 4$ ft/sec^2 . Find an appropriate Taylor polynomial that approximates the position of the balloon $s(t)$ for t near 1.
7. The motion of a pendulum bob with mass m is governed by the differential equation

$$mL\theta'' + mg\sin\theta = 0,$$

where L is the length of the pendulum arm, g is the acceleration due to gravity, and θ is the angle between the pendulum arm and the vertical. (Forces other than gravity have been neglected.) At time $t = 0$, the pendulum bob is set in motion with $\theta(0) = 1$ and $\theta'(0) = -3$. Assume that $L = 16$ ft and $g = 32$ ft/sec^2. Find the second degree Taylor polynomial that approximates the angular position $\theta(t)$ of the bob near $t = 0$. The Taylor polynomial provides an approximation to the solution of the nonlinear differential equation for the pendulum.

8. In the previous problem, find the third order Taylor polynomial that approximates the motion $\theta(t)$ for t near 0.

9. If a frictional force (expressed per unit of mass) proportional to velocity and a driving force (also per unit of mass) of $4\cos 2t$ are assumed in the pendulum motion of Prob. 7, then the differential equation of motion changes to

$$mL\theta'' + mk\theta' + mg\sin\theta = m\cos 2t,$$

Assume $k = 4$, the pendulum starts at rest so that $\theta(0) = 0$, $\theta'(0) = 0$, and retain the values for L and g from Prob. 7. (a) Find the third order Taylor polynomial that approximates the motion $\theta(t)$ for t near 0. (b) Find the fourth order Taylor polynomial that approximates the motion $\theta(t)$ for t near 0.

In Probs 10–25, find the nth order Taylor polynomial about 0 for the given function and the given n.

10. e^{-x}, $n = 5$
11. e^{2x}, $n = 6$
12. $1/(1 - x^2)$, $n = 4$
13. $1/(1 + x^3)$, $n = 3$
14. $\sqrt{1 + x}$, $n = 5$
15. $(1 + x)^{-1/2}$, $n = 5$
16. $\cos 2x$, $n = 6$
17. $\sin 2x$, $n = 5$
18. $x^3 - 2x^2 + 5x - 1$, $n = 3$
19. $x^3 - 2x^2 + 5x - 1$, $n = 5$
20. $x^2 + 3$, $n = 4$
21. $x^3 + 27$, $n = 7$
22. $\tan x$, $n = 5$
23. $\sec x$, $n = 4$
24. $\arcsin x$, $n = 5$
25. $\arctan x$, $n = 5$

In Probs. 26–39, find the nth Taylor polynomial about 0 for the given function.

26. $1/(1 - x)$
27. $1/(1 + x)$
28. e^{2x}
29. e^{-2x}
30. $\sqrt{1 + x}$
31. $(1 + x)^{-1/2}$
32. $\sqrt{x}$
33. $x^{3/2}$
34. xe^x
35. xe^{-x}
36. $\ln(1 - x)$
37. $\ln[(1 + x)/(1 - x)]$
38. $\sinh x$
39. $\cosh x$

40. Calculate the Taylor polynomials about 0 for $\cos x$ and thereby verify the results in Ex. 5.

41. The Taylor polynomial approximation

$$e^{-x} \approx 1 - x + \frac{x^2}{2} - \frac{x^3}{6}$$

is used for x in the interval $-1 \le x \le 1$. Estimate the error in this approximation as accurately as you can.

42. The Taylor polynomial approximation

$$\sqrt{1+x} \approx 1 + \frac{1}{2}x - \frac{1}{8}x^2$$

is used for x in the interval $-1/4 \le x \le 1/4$. Estimate the error in this approximation as accurately as you can.

43. Suppose, in the previous problem, that you only need the approximation for x in the interval $0 \le x \le 1/4$. Estimate the error in this approximation as accurately as you can.

44. You want to approximate e^{-x} for all x in $-1 \le x \le 1$ to within 0.0001 by a Taylor polynomial about 0. Based on the error estimates in the text, what Taylor polynomial should you use?

45. You want to approximate e^{-x} for all x in $-10 \le x \le 10$ to within 0.0001 by a Taylor polynomial about 0. Based on the error estimates in the text, what Taylor polynomial about 0 should you use? Do you think this is a reasonable way to approximate e^{-x} over the entire interval? Explain briefly.

46. You want to approximate $\sin x$ for all x in $-\pi/2 \le x \le \pi/2$ to within 0.0001 by a Taylor polynomial about 0. Based on the error estimates in the text, what Taylor polynomial should you use? *Hint.* Remember that for n odd, $P_n(x) = P_{n+1}(x) \Rightarrow R_n(x) = R_{n+1}(x)$.

47. You want to approximate $\cos x$ for all x in $-\pi/2 \le x \le \pi/2$ to within 0.0001 by a Taylor polynomial about 0. Based on the error estimates in the text, what Taylor polynomial should you use?

48. Find (a) the fifth and (b) the nth Taylor polynomial about 4 of $\sqrt{x}$.

49. Find (a) the fourth and (b) the nth Taylor polynomial about a of e^x.

50. Find (a) the fifth and (b) the nth Taylor polynomial about $\pi/6$ of $\sin x$.

51. Find (a) the fourth and (b) the nth Taylor polynomial about $\pi/4$ of $\cos x$.

52. (a) Let $P(x) = b_0 + b_1(x-a) + b_2(x-a)^2 + b_3(x-a)^3 + b_4(x-a)^4$. Express $P(a)$, $P'(a)$, ..., $P^{(4)}(a)$ in terms of b_0, b_1, ..., b_4. (b) Let $P(x) = b_0 + b_1(x-a) + \cdots + b_n(x-a)^n$.
Show that $b_k = k!P^{(k)}(a)$ for $k = 0, 1, 2, \ldots, n$.

The next problem extends the idea of a change of variables used to compare the Taylor polynomial approximations about different points for the natural logarithm function to other functions.

53. Let $f(x)$ have n derivatives at $x = a$. Let $t = x - a$, equivalently, $x = t + a$. So t is near 0 when x is near a. Define $g(t) = f(t + a)$. (a) Show that the

nth Taylor polynomial of g about 0 is $\sum_{k=0}^{n} f^{(k)}(a)t^k/k!$. (b) Deduce that the nth Taylor polynomial of f about a can be obtained from the nth Taylor polynomial of g about 0 by making the substitution $t = x - a$ in the Taylor polynomial of g.

54. You know from Ex. 3 that the nth Taylor polynomial about 0 of e^t is $\sum_{k=0}^{n} t^k/k!$. Use the previous problem to write down the nth Taylor polynomial about 3 of e^x.

55. Let $|f^{(n+1)}(x)| \le B$ for $a - c \le x < a + c$. Use the change of variables described earlier to obtain the error estimate corresponding to Th. 3 for $R_n(x) = f(x) - P_n(x)$ on $a - c \le x \le a + c$. Here $P_n(x)$ is the nth Taylor polynomial about a. You will find

$$|R_n(x)| \le B\frac{|x-a|^{n+1}}{(n+1)!} \le B\frac{c^{n+1}}{(n+1)!} \quad \text{for} \quad a - c \le x \le a + c.$$

56. If you solved Prob. 50(a), use the estimate in the previous problem to bound the error in approximating $\sin x$ by its fifth Taylor polynomial about $\pi/6$ for angles between 20° and 40°.

57. Derive the following propositions used in the proof of Th. 3:

(a)
$$g(0) = 0, \qquad g'(x) \ge 0 \quad \text{for} \quad 0 \le x \le c$$
$$\Rightarrow \qquad g(x) \ge 0 \quad \text{for} \quad 0 \le x \le c.$$

Hint. Use the mean value theorem.

(b)
$$g(0) = g'(0) = 0, \qquad g''(x) \ge 0 \quad \text{for} \quad 0 \le x \le c$$
$$\Rightarrow \qquad g(x) \ge 0 \quad \text{for} \quad 0 \le x \le c.$$

Hint. Start by applying (a) with $g(x)$ replaced by $g'(x)$.

2.2 Taylor Series and Power Series

As we stated in the chapter introduction, this chapter has two principal themes. One concerns general properties of power series. The other concerns the representation of particular functions by power series, known as Taylor series for those functions. Actually, as we shall see, the two themes are not all that separate. Each casts light on the other.

For the time being, all series are in powers of x. We begin with Taylor series. Building on the properties of Taylor polynomials developed in Sec. 2.1, we find Taylor series for several common functions and we determine the values of x for which these Taylor series converge. Then we return to power series. The principal result is that, for any power series $\sum a_n x^n$, there is a unique r, $0 \le r \le \infty$, called the radius of convergence, such that the series converges if $|x| < r$ and diverges if $|x| > r$. For $|x| = r$, the series may converge or diverge. Several examples show how to find the radius of convergence in particular cases.

At the end of the section, we briefly consider more general power series in powers of $x - c$.

Taylor Series

Assume that $f(x)$ is defined on an open interval containing 0 and has derivatives of all orders at $x = 0$. The **Taylor series for** f about 0 (also called the **Maclaurin series**) is the power series

$$\sum_{n=0}^{\infty} f^{(n)}(0)\frac{x^n}{n!} = f(0) + f'(0)x + f''(0)\frac{x^2}{2!} + \cdots + f^{(n)}(0)\frac{x^n}{n!} + \cdots.$$

For brevity, this is called simply the Taylor series for f when no misunderstanding is likely. The series may converge or diverge, depending on the function f and the value of x.

For example, let $f(x) = 1/(1-x)$. Then, from Sec. 2.1, $f^{(n)}(0) = n!$. So the Taylor series for $f(x) = 1/(1-x)$ is the familiar geometric series

$$\sum_{n=0}^{\infty} x^n = 1 + x + x^2 + \cdots + x^n + \cdots,$$

which converges for $|x| < 1$ and diverges for $|x| \geq 1$.

The nth partial sum of the Taylor series for $f(x)$ is the nth Taylor polynomial

$$P_n(x) = f(0) + f'(0)x + f''(0)\frac{x^2}{2!} + \cdots + f^{(n)}(0)\frac{x^n}{n!}.$$

Consequently, if we know the Taylor polynomials $P_n(x)$ for a function $f(x)$, then we can write down the Taylor series for $f(x)$ immediately.

Recall that the error or remainder in the approximation of $f(x)$ by $P_n(x)$ is

$$R_n(x) = f(x) - P_n(x).$$

Since the sum of a series is the limit of the partial sums,

$$f(x) = \sum_{n=0}^{\infty} f^{(n)}(0)\frac{x^n}{n!} \quad \Leftrightarrow \quad P_n(x) \to f(x) \text{ as } n \to \infty$$
$$\Leftrightarrow \quad R_n(x) \to 0 \text{ as } n \to \infty.$$

The estimates for $R_n(x)$ derived in Sec. 2.1 will help us show in many cases that $R_n(x) \to 0$ as $n \to \infty$ and, hence, that $f(x)$ is the sum of its Taylor series. From Th. 3 of Sec. 2.1, we have the following estimate for $R_n(x)$.

If $|f^{(n+1)}(x)| \leq B$ for $-c \leq x \leq c$, then

$$|R_n(x)| \leq B\frac{c^{n+1}}{(n+1)!} \quad \text{for} \quad -c \leq x \leq c.$$

We shall use this estimate for $R_n(x)$ to show that e^x, $\sin x$, and $\cos x$, are the sums of their respective Taylor series for $-\infty < x < \infty$. First,

$$e^x = \sum_{n=0}^{\infty} \frac{x^n}{n!} = 1 + x + \frac{x^2}{2!} + \frac{x^3}{3!} + \cdots + \frac{x^n}{n!} + \cdots \text{ for all } x.$$

The proof is as follows. Let $f(x) = e^x$. From Ex. 3 in Sec. 2.1, the nth Taylor polynomial for e^x is

$$P_n(x) = 1 + x + \frac{x^2}{2!} + \frac{x^3}{3!} + \cdots + \frac{x^n}{n!}.$$

Therefore, the displayed series is the Taylor series for e^x. To prove that $R_n(x) = e^x - P_n(x) \to 0$ as $n \to \infty$, fix any $c > 0$ and consider e^x for $-c \le x \le c$. Since $f^{(n)}(x) = e^x$ for all n, we can apply the estimate for $R_n(x)$ with $B = e^c = \max e^x$ for $-c \le x \le c$ and any n to obtain

$$|R_n(x)| \le e^c \frac{c^{n+1}}{(n+1)!} \text{ for } -c \le x \le c.$$

By the last of the special limits in Sec. 1.1, $c^{n+1}/(n+1)! \to 0$ as $n \to \infty$. Therefore, $R_n(x) \to 0$ as $n \to \infty$ and

$$e^x = \sum_{n=0}^{\infty} \frac{x^n}{n!} \text{ for } -c \le x \le c.$$

Since c can be any positive number, there is no restriction on x. So e^x is the sum of its Taylor series for any x. We also say that e^x is **represented** by its Taylor series for any x.

The Taylor series for $\sin x$ is given by

$$\sin x = \sum_{n=0}^{\infty} (-1)^n \frac{x^{2n+1}}{(2n+1)!} = x - \frac{x^3}{3!} + \frac{x^5}{5!} - \frac{x^7}{7!} + \cdots \text{ for all } x.$$

The derivation is similar to that for e^x . Let $f(x) = \sin x$. From Ex. 4 in Sec. 2.1, the Taylor polynomials for $\sin x$ are

$$P_n(x) = x - \frac{x^3}{3!} + \frac{x^5}{5!} - \cdots \pm \frac{x^n}{n!} \text{ for odd } n.$$

So the displayed series is the Taylor series for $\sin x$. It remains to show that the sum of the Taylor series is $\sin x$. Now,

$$R_n(x) = f(x) - P_n(x) = \sin x - P_n(x) \text{ for odd } n.$$

Fix any $c > 0$. For every n, $f^{(n)}(x)$ is either $\pm \sin x$ or $\pm \cos x$ and $|f^{(n)}(x)| \le 1$ for $-c \le x \le c$. By the estimate for $R_n(x)$ with $B = 1$,

$$|R_n(x)| \le 1 \cdot \frac{c^{n+1}}{(n+1)!} \to 0 \text{ as } n \to \infty \text{ with } n \text{ odd.}$$

We conclude that $\sin x$ is represented by its Taylor series for $-c \le x \le c$. Since $c > 0$ is arbitrary, so is x. Therefore, $\sin x$ is represented by its Taylor series for all x.

Similar reasoning, left for the problems, leads to the Taylor series for $\cos x$:

$$\cos x = \sum_{n=0}^{\infty} (-1)^n \frac{x^{2n}}{(2n)!} = 1 - \frac{x^2}{2!} + \frac{x^4}{4!} - \frac{x^6}{6!} + \cdots \text{ for all } x.$$

Power Series

We shall derive Taylor series for several other familiar functions in the next section. For the moment, however, we change our focus and consider an arbitrary power series

$$\sum_{n=0}^{\infty} a_n x^n = a_0 + a_1 x + a_2 x^2 + \cdots + a_n x^n + \cdots.$$

We address the important question: For what values of x does the power series converge? As we have seen, the power series $\sum x^n$ converges for $|x| < 1$, while the power series for e^x, $\sin x$, and $\cos x$ converge for all x. Obviously, every power series $\sum a_n x^n$ converges for $x = 0$ and its sum is a_0. Convergence is an issue only for $x \neq 0$. The following theorem summarizes the possibilities. It shows that the foregoing examples exhibit typical behavior.

Theorem 1 *Convergence and Divergence of Power Series*
For any power series $\sum a_n x^n$ there are three possibilities:
(a) $\sum a_n x^n$ converges only for $x = 0$;
(b) $\sum a_n x^n$ converges absolutely for all x;
(c) there is a number $r > 0$ such that
$\sum a_n x^n$ converges absolutely for $|x| < r$,
$\sum a_n x^n$ diverges for $|x| > r$.

Theorem 1 is proved in Appendix A. Let C (for convergence) be the set of all x for which $\sum a_n x^n$ converges. By Th. 1, C is one of the six intervals

$$[0,0], \quad (-r,r), \quad (-r,r], \quad [-r,r), \quad [-r,r], \quad (-\infty,\infty)$$

for some $r > 0$. We call C the **interval of convergence** of the power series. Define $r = 0$ if $C = [0,0]$ and $r = \infty$ if $C = (-\infty, \infty)$. Then r is called the **radius of convergence** of the power series. (This name comes from properties of series $\sum a_n z^n$ with z a complex number.) The following example shows that all six possibilities for the interval of convergence do occur.

EXAMPLE 1. Show that

(a) $\sum n! x^n$ has $r = 0$ and $C = [0,0]$;
(b) $\sum x^n/n!$ has $r = \infty$ and $C = (-\infty, \infty)$;
(c) $\sum x^n$ has $r = 1$ and $C = (-1,1)$;
(d) $\sum x^n/n$ has $r = 1$ and $C = [-1,1)$;

(e) $\Sigma(-1)^n x^n/n$ has $r = 1$ and $C = (-1, 1]$;

(f) $\Sigma x^n/n^2$ has $r = 1$ and $C = [-1, 1]$.

Solution. Since convergence at $x = 0$ is automatic, we need to examine the series for convergence only for $x \neq 0$.

(a) Apply the ratio test. Since

$$\left|\frac{(n+1)!x^{n+1}}{n!x^n}\right| = (n+1)|x| \to \infty \quad \text{as} \quad n \to \infty \quad \text{for} \quad x \neq 0,$$

$\Sigma n!x^n$ diverges for $x \neq 0$, which confirms (a).

(b) The second statement follows from the ratio test with

$$\left|\frac{x^{n+1}}{(n+1)!} \cdot \frac{n!}{x^n}\right| = \frac{|x|}{n+1} \to 0 \quad \text{as} \quad n \to \infty \quad \text{for} \quad x \neq 0.$$

(c) The third series is the geometric series and we already know that it converges for $|x| < 1$ and diverges for $|x| \geq 1$.

(d) Apply the root test. Since

$$\left|\frac{x^n}{n}\right|^{1/n} = \frac{|x|}{n^{1/n}} \to |x| \quad \text{as} \quad n \to \infty,$$

$\Sigma x^n/n$ converges absolutely for $|x| < 1$ and diverges for $|x| > 1$. So the radius of convergence of the series is 1, and the interval of convergence is one of the four intervals with endpoints -1 and 1. It remains to check for convergence or divergence at $x = \pm 1$:

$$x = 1 \quad \Rightarrow \quad \Sigma\frac{x^n}{n} = \Sigma\frac{1}{n}, \text{ which diverges,}$$

$$x = -1 \quad \Rightarrow \quad \Sigma\frac{x^n}{n} = \Sigma\frac{(-1)^n}{n}, \text{ which converges,}$$

$C = (-1, 1]$.

(e) As in (d), the root test gives the radius of convergence 1 and we must check for convergence or divergence at $x = \pm 1$:

$$x = 1 \quad \Rightarrow \quad \Sigma\frac{(-1)^n x^n}{n} = \Sigma\frac{(-1)^n}{n} \text{ which converges,}$$

$$x = -1 \quad \Rightarrow \quad \Sigma\frac{(-1)^n x^n}{n} = \Sigma\frac{1}{n} \text{ which diverges,}$$

$C = [-1, 1)$.

(f) By the root test with

$$\left|\frac{x^n}{n^2}\right|^{1/n} = \frac{|x|}{(n^{1/n})^2} \to |x| \quad \text{as} \quad n \to \infty,$$

the series converges absolutely for $|x| < 1$ and diverges for $|x| > 1$. Once again, the radius of convergence is 1. Check $x = \pm 1$:

$$x = 1 \quad \Rightarrow \quad \Sigma \frac{x^n}{n^2} = \Sigma \frac{1}{n^2} \text{ which converges,}$$

$$x = -1 \Rightarrow \quad \Sigma \frac{x^n}{n} = \Sigma \frac{(-1)^n}{n^2} \text{ which converges,}$$

$C = [-1,1]$. □

In the problems, we ask you to check that the series

$$\Sigma \frac{x^n}{n^2}, \quad \Sigma \frac{x^n}{n^3}, \quad \Sigma \frac{x^n}{n^4}, \cdots$$

all have radius of convergence 1 and interval of convergence $[-1,1]$. The successive series converge more and more rapidly. Similarly, the series

$$\Sigma nx^n, \quad \Sigma n^2 x^n, \quad \Sigma n^3 x^n, \cdots$$

all have radius of convergence 1 and interval of convergence $(-1,1)$. The successive series converge more and more slowly.

EXAMPLE 2. Find the radius and interval of convergence of the series $\Sigma n! x^n/(2n)!$.

Solution. Let $b_n = n! x^n/(2n)!$. Then, for $x \neq 0$,

$$\left|\frac{b_{n+1}}{b_n}\right| = \left|\frac{(n+1)! x^{n+1}}{[2(n+1)]!} \cdot \frac{(2n)!}{n! x^n}\right| = \frac{(n+1)|x|}{(2n+2)(2n+1)} = \frac{|x|}{2(2n+1)} \to 0 \text{ as } n \to \infty.$$

So the series converges for all x, the radius of convergence is $r = \infty$, and the interval of convergence is $(-\infty, \infty)$. □

Let's try something a little harder.

EXAMPLE 3. Find the radius and interval of convergence of the series $\Sigma(-1)^n x^n/(2^n \ln n)$.

Solution. Let $b_n = (-1)^n x^n/(2^n \ln n)$. Then for $x \neq 0$,

$$\left|\frac{b_{n+1}}{b_n}\right| = \left|\frac{x^{n+1}}{2^{n+1} \ln(n+1)} \cdot \frac{2^n \ln n}{x^n}\right| = \frac{1}{2}|x| \frac{\ln n}{\ln(n+1)}.$$

By l'Hôpital's rule, $\ln x/\ln(x+1) \to 1$ as $x \to \infty$. So $\ln n/\ln(n+1) \to 1$ as $n \to \infty$ and

$$\left|\frac{b_{n+1}}{b_n}\right| \to \frac{1}{2}|x| \text{ as } n \to \infty.$$

By the ratio test, the given series

$$\text{converges absolutely if} \quad \tfrac{1}{2}|x| < 1 \quad \text{or} \quad |x| < 2,$$

$$\text{diverges if} \quad \tfrac{1}{2}|x| > 1 \quad \text{or} \quad |x| > 2.$$

So the given power series has radius of convergence 2. To find the interval of convergence, we must check $x = \pm 2$. First,

$$x = 2 \quad \Rightarrow \quad \Sigma \frac{(-1)^n x^n}{2^n \ln n} = \Sigma \frac{(-1)^n}{\ln n},$$

which converges by the alternating series test. Next,

$$x = -2 \quad \Rightarrow \quad \Sigma \frac{(-1)^n x^n}{2^n \ln n} = \Sigma \frac{1}{\ln n},$$

which diverges by comparison with $\Sigma\, 1/n$ since $\ln x < x$ for $x > 0$. Consequently, the interval of convergence is $(-2, 2]$. □

Algebraic Properties of Power Series

Power series can be added, subtracted, multiplied, and divided in much the same way that polynomials are. We simply summarize these useful facts here.

Let $\Sigma_{n=0}^{\infty} a_n x^n$ be a power series with radius of convergence $r_a > 0$ and let $\Sigma_{n=0}^{\infty} b_n x^n$ be a power series with radius of convergence $r_b > 0$. Let $r = \min(r_a, r_b)$. Then

$$\sum_{n=0}^{\infty} a_n x^n \pm \sum_{n=0}^{\infty} b_n x^n = \sum_{n=0}^{\infty} (a_n \pm b_n) x^n \quad \text{for} \quad |x| < r.$$

If the two power series are multiplied as if they were ordinary polynomials and like powers are collected in the usual way, then the formula obtained for the product is

$$\left(\sum_{n=0}^{\infty} a_n x^n\right)\left(\sum_{n=0}^{\infty} b_n x^n\right) = \sum_{n=0}^{\infty} c_n x^n, \quad \text{where } c_n = \sum_{k=0}^{n} a_k b_{n-k}.$$

This multiplication formula is valid for $|x| < r$.

Division of $\Sigma_{n=0}^{\infty} a_n x^n$ by $\Sigma_{n=0}^{\infty} b_n x^n$ can be carried out formally just as if the power series were polynomials (provided that $b_0 \neq 0$.) The resulting quotient power series converges in some interval about the origin.

In practice, except in special cases, only the first few terms of the product or quotient series can be explicitly calculated. The algebra required to compute even a few terms in the series is often unpleasant and best done with the aid of a CAS.

General Power Series

We call $\Sigma a_n x^n$ a power series in x or a power series about 0. A general power series of the form $\Sigma a_n (x - c)^n$ is called a power series in $x - c$ or a power series about c. The change of variables $t = x - c$ replaces the power series $\sum a_n(x-c)^n$ by the power series $\sum a_n t^n$. In this way, any question about a power series about c can be recast as a question about a power series about 0. In particular, the change of variables enables us to transfer Th. 1 to power series about any point c.

Theorem 2 *Convergence and Divergence of General Power Series*
For any power series $\sum a_n(x-c)^n$ there are three possibilities:
(a) $\sum a^n(x-c)^n$ converges only for $x = c$;
(b) $\sum a_n(x-c)^n$ converges for all x;
(c) there is a number $r > 0$ such that
$\sum a_n(x-c)^n$ converges absolutely for $|x-c| < r$,
$\sum a_n(x-c)^n$ diverges for $|x-c| > r$.

Notice that

$$|x-c| < r \quad \Leftrightarrow \quad c - r < x < c + r.$$

Thus, the interval of convergence C of $\sum a_n(x-c)^n$ is one of the six intervals

$$[c,c],\ (c-r,c+r),\ (c-r,c+r],\ [c-r,c+r),\ [c-r,c+r],\ (-\infty,\infty),$$

for some $r > 0$. Let $r = 0$ if $C = [c,c]$ and $r = \infty$ if $C = (-\infty,\infty)$. Then r is the radius of convergence of the power series.

EXAMPLE 4. Find the radius and interval of convergence of the power series $\sum(-1)^n(x-3)^n/2^n\sqrt{n}$.

Solution. Since

$$\left|\frac{(-1)^n(x-3)^n}{2^n\sqrt{n}}\right|^{1/n} = \frac{|x-3|}{2(n^{1/n})^{1/2}} \to \frac{|x-3|}{2} \quad \text{as } n \to \infty,$$

it follows from the root test that the given series

$$\text{converges absolutely for } \tfrac{1}{2}|x-3| < 1 \quad \text{or} \quad |x-3| < 2,$$

$$\text{diverges if } \tfrac{1}{2}|x-3| > 1 \quad \text{or} \quad |x-3| > 2.$$

The series has radius of convergence 2. Note that $|x-3| < 2 \Leftrightarrow 1 < x < 5$. We check the endpoints of this interval:

$$x = 1 \quad \Rightarrow \quad \sum\frac{(-1)^n(x-3)^n}{2^n\sqrt{n}} = \sum\frac{1}{\sqrt{n}}, \text{ which diverges;}$$

$$x = 5 \quad \Rightarrow \quad \sum\frac{(-1)^n(x-3)^n}{2^n\sqrt{n}} = \sum\frac{(-1)^n}{\sqrt{n}}, \text{ which converges}$$

by the alternating series test. The interval of convergence is $1 < x \leq 5$. □

General Taylor Series

Assume that $f(x)$ has derivatives of all orders at c. The **Taylor series for f about c** is the power series

$$\sum_{n=0}^{\infty} f^{(n)}(c)\frac{(x-c)^n}{n!}.$$

The partial sums are the Taylor polynomials about c. For a large class of functions,

$$f(x) = \sum_{n=0}^{\infty} f^{(n)}(c)\frac{(x-c)^n}{n!}$$

on the interval $(c - r, c + r)$, where r is the radius of convergence of the Taylor series. Sometimes the series converges to $f(x)$ also at one or both of the endpoints $c - r$ and $c + r$. All results related to Taylor series about c can be reduced to questions about related series about the point 0 by the change of variables $t = x - c$. Since there is nothing essentially new, we give just one example.

EXAMPLE 5. Find the Taylor series for $f(x) = \ln x$ about 1.

Solution. From Ex. 10 in Sec. 2.1,

$$f(1) = 0, \quad f^{(n)}(1) = (-1)^{n-1}(n-1)!, \quad \frac{1}{n!} f^{(n)}(1) = \frac{(-1)^{n-1}}{n} \quad \text{for } n \geq 1.$$

So the Taylor series for $\ln x$ about 1 is

$$\sum_{n=1}^{\infty} (-1)^{n-1}\frac{(x-1)^n}{n}.$$

It is easy to show by the ratio or root test that the series has radius of convergence 1. So the Taylor series about 1 converges absolutely for $|x - 1| < 1$. Check the endpoints of this interval to see that the interval of convergence is $0 < x \leq 2$.

Another way to get the Taylor series for $\ln x$ about 1 is to first make the change of variables $t = x - 1$ so that t is near 0 when x is near 1 and $\ln x = \ln(t + 1)$. Next find the Taylor series for $\ln(1 + t)$ about 0. It is

$$\sum_{n=1}^{\infty} (-1)^{n-1}\frac{t^n}{n} = t - \frac{t^2}{2} + \frac{t^3}{3} - \frac{t^4}{4} + \cdots.$$

As before, we find that the series in t has radius of convergence 1 and interval of convergence $-1 < t \leq 1$. Let $t = x - 1$ to obtain the Taylor series in $x - 1$ found previously. □

The change of variables approach may make the calculations a little easier, but it is most useful if there is some additional information to transfer. For example, in Sec. 2.3 we shall prove that

$$\ln(1+t) = \sum_{n=1}^{\infty} (-1)^{n-1}\frac{t^n}{n} = t - \frac{t^2}{2} + \frac{t^3}{3} - \frac{t^4}{4} + \cdots \quad \text{for} \quad -1 < t \le 1.$$

Now the change of variables $x = t + 1$ gives the corresponding series for $\ln x$ about 1 free of charge:

$$\ln x = \sum_{n=1}^{\infty} (-1)^{n-1}\frac{(x-1)^n}{n} =$$

$$= (x-1) - \frac{(x-1)^2}{2} + \frac{(x-1)^3}{3} - \frac{(x-1)^4}{4} + \cdots \quad \text{for} \quad 0 < x \le 2.$$

PROBLEMS

Find the Taylor series about 0 for each function in Probs. 1–14. You may need to write out as many as five to eight derivatives to spot the general pattern that gives $f^{(n)}(x)$.

1. e^{-x}
2. e^{2x}
3. $1/(1+x)$
4. $\ln(1+x)$
5. xe^x
6. x^2e^x
7. $\cos 2x$
8. $\sin^2 x$
9. $x/(1+x)$
10. $x\sin x$
11. $\sinh x$
12. $\cosh x$
13. $(1+x)^{1/2}$
14. $(1+x)^{-1/2}$

Reason as we did in the text for e^x and $\sin x$ to prove that each function in Probs. 15–26 is the sum of its Taylor series about 0 for x in the given interval.

15. e^{-x}, $-\infty < x < \infty$
16. e^{2x}, $-\infty < x < \infty$
17. $1/(1+x)$, $0 \le x < 1$
18. $\ln(1+x)$, $0 \le x < 1$
19. xe^x, $-\infty < x < \infty$
20. x^2e^x, $-\infty < x < \infty$
21. $\cos 2x$, $-\infty < x < \infty$
22. $\sin^2 x$, $-\infty < x < \infty$
23. $x/(1+x)$, $0 \le x < 1$
24. $x\sin x$, $-\infty < x < \infty$
25. $\sinh x$, $-\infty < x < \infty$
26. $\cosh x$, $-\infty < x < \infty$
27. Derive the Taylor series given in the text for $\cos x$.

Find (a) the radius of convergence and (b) the interval of convergence for each series in Probs. 28–53.

28. $\sum x^n/n(n+1)$
29. $\sum(\ln n)x^n/n$

30. $\Sigma x^n/n^3$

31. $\Sigma(-1)^n x^n/n^4$

32. $\Sigma(-1)^{n+1}x^n/\sqrt{n}$

33. $\Sigma x^n/n\sqrt{n}$

34. $\Sigma n x^n$

35. $\Sigma n x^{n-1}$

36. $\Sigma(-1)^n n^2 x^n$

37. $\Sigma(-1)^n\sqrt{n}\,x^n$

38. $2^n x^n/\sqrt{n}$

39. $\Sigma x^n/2^n\sqrt{n}$

40. $\Sigma x^n/(\ln n)$

41. $\Sigma x^n/(\ln n)^2$

42. $\Sigma n^5 x^{2n+1}/e^{2n}$

43. $\Sigma 2^n x^{2n}/(n^2+1)$

44. $\Sigma(-1)^n x^{2n}/3^n(\ln n)$

45. $\Sigma\dfrac{(\ln n)^n x^{2n}}{e^n}$

46. $\Sigma(-1)^n\dfrac{n^2 x^n}{n^2+1}$

47. $\Sigma\dfrac{n^2 x^n}{(\ln 2)^n}$

48. $\Sigma\left(\dfrac{n+1}{n}\right)^{n^2}x^n$

49. $\Sigma\dfrac{1}{n}\left(\dfrac{n+1}{2n}\right)^n x^n$

50. $\Sigma(x-2)^n/3^n$

51. $\Sigma(-1)^n(x-2)^n/(n\ln n)$

52. $\Sigma\dfrac{(2x-3)^n}{n^2+n}$

53. $\Sigma\dfrac{n^3}{n!}\left(\dfrac{x^2+1}{3}\right)^n$

54. Find the radius of convergence of $\Sigma n!\,x^n/n^n$.

55. Find the radius of convergence of $\Sigma(n!)^2 x^n/(2n)!$.

56. Find the radius of convergence of

$$\sum_{n=0}^{\infty}\frac{1\cdot 3\cdot 5\cdots(2n+1)}{n!}x^n.$$

57. Find the radius of convergence of

$$\sum_{n=0}^{\infty}(-1)^n\frac{1\cdot 3\cdot 5\cdots(2n+1)}{4^n n!}x^{2n}.$$

58. Fix any real number p (positive, negative, or zero). (a) Show that the power series $\Sigma x^n/n^p$ has radius of convergence $r=1$. The cases $p=2,3,4,\ldots$ and $p=-1,-2,-3,\ldots$ were mentioned in the text. (b) Show that the interval of convergence C is either $(-1,1)$ or $[-1,1)$ or $[-1,1]$ and determine the values of p that correspond to each of these intervals.

59. Find the radius of convergence of $\Sigma(1+1/2+1/3+\cdots+1/n)x^n$. *Hint.* Use Prob. 54 in Sec. 1.4 and the limit comparison test.

60. Given that $\Sigma a_n x^n$ has radius of convergence r, show that $\Sigma a_n x^{2n}$ has radius of convergence $\sqrt{r}$.

61. (a) Find the Taylor series about 2 of e^x. (b) Show that the series represents e^x for all x.

62. (a) Find the Taylor series about $\pi/6$ of $\sin x$. (b) Show that the series represents $\sin x$ for all x.

63. (a) Find the Taylor series about a of e^x. (b) Show that the series represents e^x for all x.

64. Find the Taylor series about 1 of $\sqrt{x}$.

2.3 Differentiation and Integration of Power Series

In many respects, a power series behaves like an "infinite–degree polynomial" inside its interval of convergence. This is true not only for addition and subtraction but also for differentiation and integration. Any polynomial can be differentiated or integrated term by term. Similarly, any power series can be differentiated or integrated term by term inside its interval of convergence. This is an important practical and theoretical property of power series. It is not true in general that infinite series of functions can be differentiated or integrated term by term. (See the problems.)

The facts concerning term by term differentiation and integration of power series will be stated precisely a little later. We begin with two examples that indicate how term by term differentiation and integration can be justified. These examples illustrate different ways of establishing convergence by showing that remainders tend to zero. Such techniques are frequently effective in other contexts.

EXAMPLE 1. Show that the geometric series

$$\frac{1}{1-x} = 1 + x + x^2 + \cdots + x^n + \cdots \quad \text{for} \quad |x| < 1$$

can be differentiated term by term to obtain

$$\frac{1}{(1-x)^2} = 1 + 2x + 3x^2 + \cdots + nx^{n-1} + \cdots \quad \text{for} \quad |x| < 1.$$

Solution. The starting point is an identity we have met before,

$$\frac{1}{1-x} = 1 + x + x^2 + x^3 + \cdots + x^n + \frac{x^{n+1}}{1-x} \quad \text{for} \quad x \neq 1.$$

It can be verified by multiplying both sides by $1 - x$. Differentiate the identity to get

$$\frac{1}{(1-x)^2} = 1 + 2x + 3x^2 + \cdots + nx^{n-1} + \frac{x^n + (1-x)nx^n}{(1-x)^2} \quad \text{for} \quad x \neq 1.$$

Fix x with $|x| < 1$ and let $n \to \infty$. Then $x^n \to 0$ and, by the ratio test for sequences, $nx^n \to 0$. Therefore,

$$\frac{x^n + (1-x)nx^n}{(1-x)^2} \to 0 \quad \text{as} \quad n \to \infty,$$

which implies that

$$\frac{1}{(1-x)^2} = 1 + 2x + 3x^2 + \cdots + nx^{n-1} + \cdots \quad \text{for} \quad |x| < 1.$$

It is easy to show, using the ratio test, that the series for $1/(1-x)$ and $1/(1-x)^2$ have the same radius of convergence $r = 1$ and the same interval of convergence $(-1,1)$. □

EXAMPLE 2. Replace x by $-x^2$ in the geometric series to get

$$\frac{1}{1+x^2} = 1 - x^2 + x^4 - x^6 + \cdots \quad \text{for} \quad -1 < x < 1.$$

Show that this series can be integrated term by term to obtain

$$\arctan x = \int_0^x \frac{1}{1+t^2}\,dt = x - \frac{x^3}{3} + \frac{x^5}{5} - \frac{x^7}{7} + \cdots \quad \text{for} \quad -1 \le x \le 1.$$

Solution. We begin once again with

$$\frac{1}{1-x} = 1 + x + x^2 + x^3 + \cdots + x^n + \frac{x^{n+1}}{1-x} \quad \text{for} \quad x \ne 1.$$

Let $x = -t^2$ to obtain

$$\frac{1}{1+t^2} = 1 - t^2 + t^4 - t^6 + \cdots + (-1)^n t^{2n} + \frac{(-1)^{n+1} t^{2n+2}}{1+t^2} \quad \text{for all} \quad t.$$

Integrate from 0 to x to find

$$\begin{aligned}\arctan x &= \int_0^x \frac{1}{1+t^2}\,dt \\ &= x - \frac{x^3}{3} + \frac{x^5}{5} - \frac{x^7}{7} + \cdots + (-1)^n \frac{x^{2n+1}}{2n+1} + (-1)^{n+1}\int_0^x \frac{t^{2n+2}}{1+t^2}\,dt.\end{aligned}$$

For $-1 \le x \le 1$,

$$\left|\int_0^x \frac{t^{2n+2}}{1+t^2}\,dt\right| < \int_0^1 t^{2n+2}\,dt = \frac{1}{2n+3} \to 0 \quad \text{as} \quad n \to \infty.$$

It follows that

$$\boxed{\arctan x = x - \frac{x^3}{3} + \frac{x^5}{5} - \frac{x^7}{7} + \cdots \quad \text{for} \quad -1 \le x \le 1.}$$

The series for $1/(1+x^2)$ has radius of convergence $r = 1$ and interval of convergence $(-1,1)$. The integrated series, for $\arctan x$, has the same radius of convergence, $r = 1$, but the larger interval of convergence $[-1,1]$. □

Let $x = 1$ in the series for $\arctan x$ to obtain a remarkable formula,

$$\frac{\pi}{4} = 1 - \frac{1}{3} + \frac{1}{5} - \frac{1}{7} + \cdots.$$

The series on the right converges by the alternating series test. However, it converges very slowly and its partial sums do not provide an effective means for approximating π. You will find such means in the problems.

It is time to formulate some general principles illustrated in the examples and foregoing discussion. Notice one more thing, if you haven't already. The sum of each preceding power series is a continuous function on the interval $(-r,r)$, where r is the radius of convergence. Furthermore, the sum is differentiable infinitely often on $(-r,r)$. All this and more will follow from the next two theorems. In both theorems, the radius of convergence is assumed to be positive, either finite or infinite.

Theorem 1 *Continuity of Sums of Power Series*
Let $f(x) = \sum_{n=0}^{\infty} a_n x^n$ with radius of convergence $0 < r \leq \infty$.
Then $f(x) = \sum_{n=0}^{\infty} a_n x^n$ is continuous on $-r < x < r$.

The proof, which is not difficult, is outlined in the problems.

Theorem 2, which follows, justifies term by term differentiation and integration of any power series inside its interval of convergence.

Theorem 2 *Differentiation and Integration of Power Series*
Let $f(x) = \sum_{n=0}^{\infty} a_n x^n$ with radius of convergence $r > 0$. Then:

(1) $$\int_0^x f(t)\,dt = \sum_{n=0}^{\infty} a_n \frac{x^{n+1}}{n+1} \quad \text{for} \quad -r < x < r.$$

(2) $f(x)$ is differentiable and

$$f'(x) = \sum_{n=1}^{\infty} a_n n x^{n-1} \quad \text{for} \quad -r < x < r.$$

(3) The differentiated and integrated series have the same radius of convergence as the original series.

(4) The interval of convergence for the differentiated series is the same or smaller than for the original series. The interval of convergence for the integrated series is the same or larger than for the original series.

The first three parts of Th. 2 are proved in Appendix B. We shall not use the fourth part.

EXAMPLE 3. By Th. 2, the series

$$\frac{1}{1+x} = 1 - x + x^2 - x^3 + \cdots, \qquad -1 < x < 1,$$

can be integrated term by term to obtain

$$\ln(1+x) = x - \frac{x^2}{2} + \frac{x^3}{3} - \frac{x^4}{4} + \cdots, \qquad -1 < x < 1.$$

Both series have radius of convergence $r = 1$, again by Th. 2. The series for $1/(1+x)$ has interval of convergence $(-1,1)$. Show that the series for $\ln(1+x)$ is valid also at $x = 1$. That is, show that

$$\boxed{\ln 2 = 1 - \frac{1}{2} + \frac{1}{3} - \frac{1}{4} + \cdots.}$$

Solution. We know that the alternating harmonic series converges. It remains to prove that its sum is $\ln 2$. To this end, replace x by $-x$ in the identity for $1/(1-x)$ used in Ex. 1 to obtain

$$\frac{1}{1+x} = 1 - x + x^2 - x^3 + \cdots + (-1)^n x^n + \frac{(-1)^{n+1}x^{n+1}}{1+x}.$$

Integrate from 0 to 1 to obtain

$$\ln 2 = \int_0^1 \frac{1}{1+x}\,dx = 1 - \frac{1}{2} + \frac{1}{3} - \frac{1}{4} + \cdots + \frac{(-1)^n}{n+1} + (-1)^{n+1}\int_0^1 \frac{x^{n+1}}{1+x}\,dx.$$

To obtain the series for $\ln 2$, we must show that the integral on the right tends to 0 as $n \to \infty$. It does:

$$\int_0^1 \frac{x^{n+1}}{1+x}\,dx < \int_0^1 x^{n+1}\,dx = \frac{1}{n+2} \to 0 \quad \text{as} \quad n \to \infty.$$

Therefore, the sum of the alternating harmonic series is $\ln 2$. This substantiates a claim we made some time ago. □

It follows from Ex. 3 that

$$\boxed{\ln(1+x) = x - \frac{x^2}{2} + \frac{x^3}{3} - \frac{x^4}{4} + \cdots \quad \text{for} \quad -1 < x \le 1.}$$

This series for $\ln(1+x)$ gives values of $\ln y$ for $y = 1 + x$ in the interval $0 < y \le 2$. The series converges fairly rapidly for x reasonably close to 0, say $-1/2 \le x \le 1/2$. So the series provides an effective means for approximating $\ln y$ for $1/2 \le y \le 3/2$. On the other hand, the series for $\ln(1+x)$ converges rather slowly for x near -1 or 1. In particular, the series for $\ln 2$ converges very slowly. For example, the approximation of $\ln 2$ by the partial sum with 1000 terms gives $\ln 2 \approx 0.69265$, as compared with the correctly rounded value $\ln 2 \approx 0.69315$.

Replace x by $-x$ in the series for $\ln(1+x)$ and then multiply by -1 to get

$$\boxed{-\ln(1-x) = x + \frac{x^2}{2} + \frac{x^3}{3} + \frac{x^4}{4} + \cdots \quad \text{for} \quad -1 \leq x < 1.}$$

Now add the two logarithmic series to obtain

$$\ln\left(\frac{1+x}{1-x}\right) = 2\left(x + \frac{x^3}{3} + \frac{x^5}{5} + \frac{x^7}{7} + \cdots\right) \quad \text{for} \quad -1 < x < 1.$$

This series is much better for calculating natural logarithms than the preceding series. To help explain the advantage, let

$$y = \frac{1+x}{1-x} = \frac{2}{1-x} - 1 \quad \text{for} \quad -1 < x < 1.$$

Then

$$\frac{dy}{dx} = \frac{2}{(1-x)^2} > 0,$$

y increases as x increases,

$$y \to 0 \text{ as } x \to -1^+, \qquad y \to \infty \text{ as } x \to 1^-.$$

By the intermediate value theorem, y takes on all values with $0 < y < \infty$ as x varies in $-1 < x < 1$. So the series for $\ln y$ with $y = (1+x)/(1-x)$ gives the natural logarithm of *every* positive number. In particular let $x = 1/3$. Then $y = 2$ and

$$\ln 2 = 2\left(\frac{1}{3} + \frac{1}{3} \cdot \frac{1}{3^3} + \frac{1}{5} \cdot \frac{1}{3^5} + \frac{1}{7} \cdot \frac{1}{3^7} + \cdots\right),$$

which converges very rapidly. In fact, the partial sum with only five terms gives $\ln 2 \approx 0.69315$, the correct value to five decimal places. For further results related to practical calculation of logarithms, see the problems.

By Th. 2, the power series obtained by term–by–term differentiation or integration of a given power series has the same radius of convergence as the original series. We can apply Th. 2 repeatedly to show that the power series obtained by differentiating or integrating any number of times has the same radius of convergence.

EXAMPLE 4. By Th. 2, differentiation of the geometric series

$$\frac{1}{1-x} = \sum_{n=0}^{\infty} x^n \quad \text{for} \quad -1 < x < 1$$

gives

$$\frac{1}{(1-x)^2} = \sum_{n=1}^{\infty} nx^{n-1} \quad \text{for} \quad -1 < x < 1.$$

Another differentiation and another application of Th. 2 give

$$\frac{2}{(1-x)^3} = \sum_{n=2}^{\infty} n(n-1)x^{n-2} \quad \text{for} \quad -1 < x < 1.$$

We can continue to differentiate as many times as we like. Alternatively, we can combine these series to get interesting new ones. For example,

$$\frac{x}{(1-x)^2} + \frac{2x^2}{(1-x)^3} = \sum_{n=1}^{\infty} nx^n + \sum_{n=2}^{\infty} n(n-1)x^n = x + \sum_{n=2}^{\infty} n^2x^n,$$

$$\sum_{n=2}^{\infty} n^2x^n = x\left[\frac{1}{(1-x)^2} + \frac{2x}{(1-x)^3} - 1\right].$$

Add x to each side to obtain

$$\sum_{n=1}^{\infty} n^2x^n = x\left[\frac{1}{(1-x)^2} + \frac{2x}{(1-x)^3}\right] = \frac{x(1+x)}{(1-x)^3} \quad \text{for} \quad -1 < x < 1.$$

In particular, the choice $x = 1/2$ gives

$$\sum_{n=1}^{\infty} \frac{n^2}{2^n} = 6.$$

We suspected this result, after approximating the sum of this series, in Ex. 9 of Sec. 1.5. Now, we have proved it. □

Just as in Ex. 4, if $f(x) = \sum a_n x^n$ for $-r < x < r$, then successive applications of Th. 2 show that $f'(x)$, $f''(x)$, $f'''(x)$, ... all exist and are given by the power series obtained by successive term–by–term differentiations of the series for $f(x)$ with $-r < x < r$:

$$f(x) = a_0 + a_1x + a_2x^2 + a_3x^3 + \cdots,$$

$$f'(x) = a_1 + 2a_2x + 3a_3x^2 + 4a_4x^3 + \cdots,$$

$$f''(x) = 2 \cdot 1a_2 + 3 \cdot 2a_3x + 4 \cdot 3a_4x^2 + \cdots,$$

$$f'''(x) = 3 \cdot 2 \cdot 1a_3 + 4 \cdot 3 \cdot 2 \cdot a_4x + \cdots,$$

and, in general,

$$f^{(n)}(x) = n(n-1)\cdots 3 \cdot 2 \cdot 1a_n + (n+1)n\cdots 2a_{n+1}x + \cdots.$$

Set $x = 0$ to find that

$$f(0) = a_0, f'(0) = a_1, f''(0) = 2!a_2, f'''(0) = 3!a_3, \cdots, f^{(n)}(0) = n!a_n, \cdots.$$

Therefore,

$$a_n = f^{(n)}(0)/n! \quad \text{for} \quad n = 0, 1, 2, \ldots,$$

and we have proved the following very important theorem.

Theorem 3 *Uniqueness of Power Series Representations*
Let $f(x) = \sum_{n=0}^{\infty} a_n x^n$ with radius of convergence $r > 0$. Then

$$a_n = \frac{1}{n!} f^{(n)}(0) \quad \text{for all} \quad n,$$

and, hence,

$$f(x) = \sum_{n=0}^{\infty} f^{(n)}(0) \frac{x^n}{n!} \quad \text{for} \quad |x| < r,$$

which is the Taylor series for f about 0.

There are two important conclusions here. The first is in the title of the theorem. It is conceivable that a given function $f(x)$ might have more than one power series representation about 0. Theorem 3 says that this does not happen. If $f(x)$ has a power series representation, then there is only one and it is the Taylor series for $f(x)$. This also means that if we find a power series for a function by any means whatever, the power series we found is the Taylor series for the function. This gives us an important new way to get Taylor series besides calculating the coefficients $f^{(n)}(0)/n!$ directly. For example, all the power series developed earlier in this section are the Taylor series of the functions on the left side. None of these series was found by calculating $f^{(n)}(0)/n!$ directly. To see the advantage, try your luck on a direct calculation with arctan$x$.

Here are a few simple and useful consequences of Th. 3. If $f(x)$ and $g(x)$ have the Taylor series

$$f(x) = \sum_{n=0}^{\infty} a_n x^n, \qquad g(x) = \sum_{n=0}^{\infty} b_n x^n,$$

then the Taylor series for $cf(x)$ and $f(x) \pm g(x)$ are

$$cf(x) = \sum_{n=0}^{\infty} c a_n x^n, \qquad f(x) \pm g(x) = \sum_{n=0}^{\infty} (a_n \pm b_n) x^n.$$

For example, since the Taylor series for e^x and e^{-x} are

$$e^x = 1 + x + \frac{x^2}{2!} + \frac{x^3}{3!} + \frac{x^4}{4!} + \frac{x^5}{5!} + \cdots,$$

$$e^{-x} = 1 - x + \frac{x^2}{2!} - \frac{x^3}{3!} + \frac{x^4}{4!} - \frac{x^5}{5!} + \cdots,$$

the Taylor series for $\sinh x = \frac{1}{2}(e^x - e^{-x})$ and $\cosh x = \frac{1}{2}(e^x + e^{-x})$ are

$$\sinh x = x + \frac{x^3}{3!} + \frac{x^5}{5!} + \cdots,$$

$$\cosh x = 1 + \frac{x^2}{2!} + \frac{x^4}{4!} + \cdots.$$

Both series converge for all x. Direct calculations of the coefficients are easy enough, but take more time.

EXAMPLE 5. In the Taylor series for e^x, replace x by $-x^2$ to obtain the series

$$e^{-x^2} = \sum_{n=0}^{\infty} \frac{(-1)^n x^{2n}}{n!} = 1 - \frac{x^2}{1!} + \frac{x^4}{2!} - \frac{x^6}{3!} + \cdots \quad \text{for all } x.$$

This is the Taylor series for e^{-x^2}. The direct calculation of the coefficients is quite unpleasant. Try it. □

EXAMPLE 6. In the Taylor series for $\sin x$,

$$\sin x = x - \frac{x^3}{3!} + \frac{x^5}{5!} - \frac{x^7}{7!} + \cdots \quad \text{for all } x,$$

divide by x to obtain

$$\frac{\sin x}{x} = 1 - \frac{x^2}{3!} + \frac{x^4}{5!} - \frac{x^6}{7!} + \cdots \quad \text{for all } x \neq 0.$$

This power series actually converges for all x. The continuity of the sum of the series at $x = 0$, guaranteed by Th. 1, gives the familiar limit

$$\lim_{x \to 0} \frac{\sin x}{x} = 1$$

by a new route. □

Similar power series manipulations can be used to evaluate a variety of indeterminate forms, provided you know the power series for the functions that appear. For example, this method works on $(1 - \cos x)/x^2$, which is indeterminate at $x = 0$. See the problems.

The next example shows how Taylor series can be used to approximate integrals that cannot be evaluated exactly by other means.

EXAMPLE 7. Find $\int_0^1 e^{-x^2}\,dx$ to within 10^{-6}.

Solution. This particular integral cannot be evaluated by the fundamental theorem of calculus. So we must settle for accurate approximate values. From Ex. 5,

$$e^{-x^2} = \sum_{n=0}^{\infty} \frac{(-1)^n x^{2n}}{n!} \quad \text{for all } x.$$

Therefore, by Th. 2,

$$\int_0^1 e^{-x^2}\,dx = \sum_{n=0}^{\infty} \int_0^1 \frac{(-1)^n x^{2n}}{n!}\,dx$$

$$= \sum_{n=0}^{\infty} \frac{(-1)^n}{(2n+1)n!} = 1 - \frac{1}{3 \cdot 1!} + \frac{1}{5 \cdot 2!} - \frac{1}{7 \cdot 3!} + \cdots.$$

Since the series on the right satisfies the conditions of the alternating series test, the error made by cutting off the series at $(-1)^n/(2n+1)n!$ is at most the magnitude of the next term in the series. Consequently, we want to choose the smallest n such that

$$\frac{1}{[2(n+1)+1](n+1)!} = \frac{1}{(2n+3)(n+1)!} < 10^{-6}.$$

A little work with a calculator shows that $n = 9$ does the job and that

$$\int_0^1 e^{-x^2}\,dx \approx \sum_{n=0}^{9} \frac{(-1)^n}{(2n+1)n!} = 0.746824,$$

with error less than 10^{-6}. □

An important special case of Th. 3 is that if $\sum_{n=0}^{\infty} a_n x^n = 0$ for $-r < x < r$ with $r > 0$, then $a_n = 0$ for all n. Power series methods for solving differential equations rely on this fact, as we shall see in the next section.

PROBLEMS

In Probs. 1–18, use simple series manipulations, particular power series developed in this and earlier sections, changes of variable, and term–by–term differentiation or integration as appropriate to develop series representations for the given functions. Identify the intervals over which you claim the series representations are valid.

1. $x^2 e^{-x}$
2. xe^{-2x}
3. $1/(1-x)^3$
4. $x/(1-x^{2)2}$
5. $\sin\sqrt{x}$
6. $1-\cos(x^2)$
7. $\dfrac{\ln(1+x)}{x}$
8. $\dfrac{e^{x2}-1}{x}$
9. $\cos^2 x$
10. $\sin^2 x$
11. $\cosh x$
12. $\sinh x$

13. $\displaystyle\int_0^x \frac{\sin t}{t}\,dt$

14. $\displaystyle\int_0^x \frac{e^t - 1}{t}\,dt$

15. $\displaystyle\int_0^{1/2} \frac{dx}{1 + x^3}$

16. $\displaystyle\int_0^1 \frac{\sqrt{x}}{e^x}\,dx$

17. $\operatorname{arctanh} x$

18. $\ln[(1 + x)^2/(1 - x)]$

Review Ex. 4. Then find the sum of each series in Probs. 19–26.

19. $\displaystyle\sum_{n=1}^{\infty} nx^n$

20. $\displaystyle\sum_{n=1}^{\infty} (-1)^n n^2 x^n$

21. $\displaystyle\sum_{n=1}^{\infty} \frac{x^{n+1}}{n(n+1)}$

22. $\displaystyle\sum_{n=1}^{\infty} (-1)^n (n + 1) x^{2n}$

23. $\displaystyle\sum_{n=1}^{\infty} \frac{n^2}{3^n}$

24. $\displaystyle\sum_{n=1}^{\infty} \frac{n}{2^n}$

25. $\displaystyle\sum_{n=1}^{\infty} \frac{(-1)^{n+1}}{n(n+1)}$

26. $\displaystyle\sum_{n=1}^{\infty} \frac{(-1)^n}{n^2 2^n}$

In Probs. 27–32, use series methods, as in Ex. 7, to evaluate the given integrals to within 0.0001. (Remember that $n! \geq 2^{n-1}$ for $n \geq 2$.)

27. $\displaystyle\int_0^1 \frac{\sin x}{x}\,dx$

28. $\displaystyle\int_0^1 \frac{1 - \cos x}{x^2}\,dx$

29. $\displaystyle\int_0^1 \sin(x^2)\,dx$

30. $\displaystyle\int_0^1 e^{x^3}\,dx$

31. $\displaystyle\int_0^1 \frac{e^x - 1}{x}\,dx$

32. $\displaystyle\int_0^{1/2} \frac{\arctan x}{\sqrt{x}}\,dx$

33. Estimate the value of $\int_0^\infty \sqrt{x}\, e^{-x}\,dx$ to within 0.0001 as follows. (a) Write $\int_0^\infty \sqrt{x}\, e^{-x}\,dx$ as the sum of $\int_0^b \sqrt{x}\, e^{-x}\,dx$ and $\int_b^\infty \sqrt{x}\, e^{-x}\,dx$. Use $\sqrt{x}\, e^{-x} \leq x e^{-x}$ for $x \geq 1$ in the last integral to determine a convenient number b so that $\int_b^\infty \sqrt{x}\, e^{-x}\,dx < 0.00005$. (b) Now use series methods to estimate the integral from 0 to b to within 0.00005.

34. Evaluate the value of $\int_0^\infty (e^{-x}/\sqrt{x})\,dx$ to within 0.0001 by following the general approach in the previous problem.

Use known series representations and the fact that every power series is continuous inside its interval of convergence to evaluate the indeterminate forms as $x \to 0$ in Probs. 35–40.

35. $\displaystyle\frac{1 - \cos x}{x^2}$

36. $\displaystyle\frac{1 + 2x - e^{2x}}{x^2}$

37. $\displaystyle\frac{1}{x} - \frac{1}{\sin x}$

38. $\displaystyle\frac{\ln(1 + x^2)}{1 - \cos x}$

39. $\dfrac{x - \arctan x}{x - \sin x}$

40. $\dfrac{\arctan x - \arcsin x}{x^3}$

41. Develop the Taylor series for $-\ln(1-x)$, not as we did in the text, but from scratch, starting with

$$\frac{1}{1-t} = 1 + t + t^2 + t^3 + \cdots + t^n + \frac{t^{n+1}}{1-t} \quad \text{for} \quad t \neq 1.$$

Then make a change of variables to obtain the series for $\ln(1+x)$.

Problems 42–46 show how π can be approximated using the series for the arctangent. The series for $\pi/4$ obtained from the series for $\arctan x$ when $x = 1$ converges too slowly to be of practical value. On the other hand, the series works well if x is reasonably smaller than 1. We shall use a trigonometric identity of Lewis Carroll, author of *Alice in Wonderland,* to take advantage of this. (He was a math teacher.) Efforts along these lines go back at least to Euler.

42. Use the identity (see Prob. 46)

$$\arctan\frac{1}{p} = \arctan\frac{1}{p+q} + \arctan\frac{1}{p+r} \quad \text{where} \quad qr = 1 + p^2$$

to show that

$$\frac{\pi}{4} = \arctan\frac{1}{2} + \arctan\frac{1}{3},$$

$$\arctan\frac{1}{2} = \arctan\frac{1}{3} + \arctan\frac{1}{7},$$

$$\arctan\frac{1}{3} = 2\arctan\frac{1}{7} + \arctan\frac{2}{11}.$$

Combine these results to obtain

$$\frac{\pi}{4} = 3\arctan\frac{1}{7} + 2\arctan\frac{2}{11}.$$

43. Estimate how many terms of the arctangent series with $x = 1$ are needed to approximate $\pi/4$ to within 10^{-6}.

44. Estimate how many terms of the series for arctan 1/2 and arctan 1/3 are needed to approximate $\pi/4$ to within 10^{-6}.

45. Estimate how many terms of the series for 3 arctan 1/7 and 2 arctan 2/11 are needed to approximate $\pi/4$ to within 10^{-6}.

46. Verify the identity of Lewis Carroll, starting with the addition formula for the tangent,

$$\tan(A + B) = \frac{\tan A + \tan B}{1 - \tan A \, \tan B}.$$

Hint. Solve for $A + B$ and then choose $A = \arctan 1/p$ and $B = \arctan 1/(p+q)$.

Problems 47–49 refer to the series in the text for $\ln[(1+x)/(1-x)]$.

47. Give another derivation for the series starting with the power series for $1/(1-x^2)$. *Hint.* Partial fractions.

48. Write

$$\ln\left(\frac{1+x}{1-x}\right) = 2\sum_{k=1}^{n}\frac{x^{2k-1}}{2k-1} + 2\sum_{k=n+1}^{\infty}\frac{x^{2k-1}}{2k-1} = S_n(x) + R_n(x).$$

Show that the remainder satisfies

$$|R_n(x)| < \frac{2|x|^{2n+1}}{(2n+1)(1-x^2)} \quad \text{for} \quad -1 < x < 1,$$

$$0 \le R_n(x) \quad \text{for} \quad 0 \le x < 1.$$

49. This problem shows how to calculate $\ln p$ efficiently using the series for $\ln[(1+x)/(1-x)]$.
(a) Let $x = 1/(2p+1)$ for $p = 1, 2, 3, \ldots$ to obtain

$$\ln\frac{p+1}{p} = 2\sum_{k=1}^{\infty}\frac{1}{(2k-1)(2p+1)^{2k-1}},$$

$$0 \le R_n\left(\frac{1}{2p+1}\right) \le \frac{1}{2(2n+1)(p^2+p)(2p+1)^{2n-1}}.$$

(b) Explain how $\ln 2$, $\ln 3$, $\ln 4$, $\ldots$ can be calculated successively from (a) by setting $p = 1, 2, 3, \ldots$.
(c) Use (a) and (b) to find $\ln 2$, $\ln 3$, $\ln 4$, and $\ln 5$ to within 10^{-6}.
(d) Compare the approximations in (c) against the values provided by a good scientific calculator.

50. This problem shows that the interchange of summation and integration is not always valid for sums with infinitely many terms. Let $S_n(x) = n^2x^n(1-x)$ for $0 \le x \le 1$. Show that the nth partial sum of the series

$$\sum_{k=1}^{\infty} a_k(x) \quad \text{where} \quad a_k(x) = S_k(x) - S_{k-1}(x)$$

is $S_n(x)$. Use this to verify that

$$\int_0^1\left(\sum_{k=1}^{\infty} a_k(x)\right)dx \neq \sum_{k=1}^{\infty}\int_0^1 a_k(x)\,dx.$$

Hint. Evaluate both sides of the inequality.

51. Prove Th. 1 by the following steps. First, write

$$f(x) = S_n(x) + R_n(x), \qquad S_n(x) = \sum_{k=1}^{n} a_k x^k, \qquad R_n(x) = \sum_{k=n+1}^{\infty} a_k x^k.$$

(a) Explain why the series $\sum |a_k x^k|$ converges for $-r < x < r$. Then define

$$g(x) = \sum_{k=1}^{\infty}|a_k x^k| = \tilde{S}_n(x) + \tilde{R}_n(x), \quad \tilde{S}_n(x) = \sum_{k=1}^{n}|a_k x^k|, \quad \tilde{R}_n(x) = \sum_{k=n+1}^{\infty}|a_k x^k|.$$

(b) Fix c with $|c| < r$. Show that

$$|f(x) - f(c)| \le |S_n(x) - S_n(c)| + \tilde{R}_n(x) + \tilde{R}_n(c).$$

Fix $d > 0$ with $|c| < d < r$ and restrict x to the interval $-d \le x \le d$. Now show that

$$|f(x) - f(c)| \le |S_n(x) - S_n(c)| + 2\tilde{R}_n(d) \quad \text{for all} \quad x \quad \text{with} \quad |x| \le d.$$

(c) Use (b) and the verbal definition of a limit to explain why $2\tilde{R}_n(d)$ can be made as near 0 as you like by taking n large enough. Fix such an n. Then explain why $|S_n(x) - S_n(c)|$ can be make as near 0 as you like by taking x close enough to c. Thus, $f(x) \to f(c)$ as $x \to c$, which means that f is continuous at c.

2.4 Power Series and Differential Equations: The Binomial Series*

In this section, we explore some basic connections between power series and differential equations. Knowledge about solutions of particular differential equations can enable us to sum certain power series. Conversely, power series often provide solutions to important differential equations. The interplay between power series and differential equations will be illustrated by the binomial series, which generalizes the binomial theorem you learned in algebra. The binomial series comes up in solutions to a variety of problems in science and engineering.

Power Series and Differential Equations

We begin by taking advantage of our knowledge of solutions to the exponential growth equation $y' = y$ to sum a particular series. The more differential equations you can solve, the more useful is the approach in the next example.

EXAMPLE 1. Find an explicit formula for the sum of the series

$$f(x) = \sum_{n=0}^{\infty} \frac{x^n}{n!} = 1 + x + \frac{x^2}{2!} + \frac{x^3}{3!} + \cdots.$$

Solution. Of course, you know the sum is $f(x) = e^x$. But the point here is the method. So forget you know the answer. The series converges for all x by the ratio test. First, notice that $f(0) = 1$. Differentiate the series for $f(x)$ term by term to obtain

$$f'(x) = 0 + 1 + x + \frac{x^2}{2!} + \frac{x^3}{3!} + \cdots = f(x).$$

The same calculation, expressed in summation notation, is

$$f'(x) = \sum_{n=1}^{\infty} \frac{1}{n!} nx^{n-1} = \sum_{n=1}^{\infty} \frac{x^{n-1}}{(n-1)!} = \sum_{n=0}^{\infty} \frac{x^n}{n!} = f(x),$$

where we replaced n by $n + 1$ in passing from the middle to the last summation. Either way,

$$f'(x) = f(x) \text{ and } f(0) = 1.$$

This is an initial value problem for $f(x)$. We know that the unique solution is $f(x) = e^x$. Thus, we have a new derivation of the power series for e^x. □

A slight alteration of Ex. 1 makes the solution a little more difficult but also more interesting. It points the way to general power series methods for solving differential equations.

EXAMPLE 2. Find a series solution for the initial value problem

$$y'(x) = ky(x), \qquad y(0) = 1,$$

with any constant k.

Solution. Suppose that $y(x)$ is represented by a power series

$$y(x) = \sum_{n=0}^{\infty} a_n x^n$$

that has a positive radius of convergence r. Then

$$y'(x) = \sum_{n=1}^{\infty} n a_n x^{n-1}.$$

Consequently, $y(x)$ is a solution of the initial value problem in Ex. 2 if and only if

$$\sum_{n=1}^{\infty} n a_n x^{n-1} = k \sum_{n=0}^{\infty} a_n x^n, \qquad a_0 = y(0) = 1.$$

To make things easier, write out the first few terms of the two series:

$$a_1 + 2a_2x + 3a_3x^2 + a_4x^3 + \cdots = k(a_0 + a_1x + a_2x^2 + a_3x^3 + \cdots).$$

By the uniqueness theorem for power series, the two series must be equal term by term. Hence,

$$a_1 = ka_0, \quad 2a_2 = ka_1, \quad 3a_3 = ka_3, \quad 4a_4 = ka_3,$$

and, in general, $(n + 1)a_{n+1} = ka_n$. The same result can be obtained from

$$\sum_{n=1}^{\infty} na_n x^{n-1} = k \sum_{n=0}^{\infty} a_n x^n$$

by replacing n by $n + 1$ in the series for y':

$$\sum_{n=0}^{\infty} (n + 1)a_{n+1} x^n = \sum_{n=0}^{\infty} ka_n x^n.$$

Since the coefficients of x^n must match, $(n + 1)\, a_{n+1} = ka_n$ and

$$a_{n+1} = \frac{k}{n+1} a_n.$$

This recurrence formula determines all the coefficients a_n in terms of $a_0 = 1$:

$$a_1 = k, \quad a_2 = \frac{k}{2} a_1 = \frac{k^2}{2}, \quad a_3 = \frac{k}{3} a_2 = \frac{k^3}{3!}, \quad a_4 = \frac{k}{4} a_3 = \frac{k^4}{4!},$$

and, in general, $a_n = k^n/n!$. (A formal proof by induction is easy.) It follows that

$$y(x) = \sum_{n=0}^{\infty} a_n x^n = \sum_{n=0}^{\infty} \frac{(kx)^n}{n!} = e^{kx}.$$

At the beginning, we left open the question as to what the radius of convergence r of $\sum a_n x^n$ is. Now it is clear that $r = \infty$, either by the ratio test or from our knowledge about the Taylor series for the exponential function. In summary, we have solved the initial value problem $y'(x) = ky(x)$, $y(0) = 1$, by deriving a power series solution, which has sum e^{kx}. □

The Taylor series for $\sin x$ and $\cos x$ can be developed in the same way using the fact that all solutions of the equation $y'' + y = 0$ have the form $y = A \cos x + B \sin x$. See the problems.

The Binomial Series

The binomial series is the Taylor series for $(1 + x)^\alpha$, where α is any number. To avoid trivialities, assume $\alpha \neq 0$ throughout our analysis. Particular cases are $(1 + x)^{1/2}$, $(1 + x)^{-1/2}$, and $(1 + x)^{1/3}$. Newton discovered the binomial series. He assumed that its sum was $(1 + x)^\alpha$ and used it for approximate calculations. The series is important not only in science and engineering, but it also is used in convenient derivations of Taylor series for other functions.

We shall derive the Taylor series for $(1 + x)^\alpha$ by a rather roundabout route that saves a lot of work over a direct approach. The first step is to find an initial value problem that $y = (1 + x)^\alpha$ solves. Since $y(0) = 1$ and

$$y' = \alpha(1 + x)^{\alpha - 1} = \frac{\alpha}{1 + x}(1 + x)^\alpha = \frac{\alpha y}{1 + x},$$

$y = (1 + x)^\alpha$ is a solution of the initial value problem

$$\boxed{(1 + x)y' - \alpha y = 0, \qquad y(0) = 1.}$$

We show next that $y = (1 + x)^\alpha$ is the only solution of the initial value problem. To this end, suppose that $y = y(x)$ is any solution. By the quotient rule,

$$\frac{d}{dx} \frac{y}{(1 + x)^\alpha} = \frac{(1 + x)^\alpha y' - \alpha(1 + x)^{\alpha - 1} y}{(1 + x)^{2\alpha}} = \frac{(1 + x)y' - \alpha y}{(1 + x)^{\alpha + 1}} = 0.$$

Hence,

$$\frac{y}{(1+x)^\alpha} = C, \qquad y = C(1+x)^\alpha$$

for some constant C. Since $y(0) = 1$, $C = 1$ and $y = (1+x)^\alpha$. Thus, as claimed, $y = (1+x)^\alpha$ is the only solution of the initial value problem.

Now we come to the second step. We shall show that the displayed initial value problem has a power series solution. Since the problem has only one solution, the power series solution must equal $(1+x)^\alpha$. Thus, we obtain the power series for $y = (1+x)^\alpha$, which is its Taylor series. To this end, let

$$y = \sum_{n=0}^{\infty} a_n x^n.$$

Then $a_0 = y(0) = 1$ and

$$y' = \sum_{n=1}^{\infty} a_n n x^{n-1}.$$

Substitute the series for y and y' into the differential equation $(1+x)y' - \alpha y = 0$ to obtain

$$(1+x)\left(\sum_{n=1}^{\infty} a_n n x^{n-1}\right) - \alpha\left(\sum_{n=0}^{\infty} a_n x^n\right) = 0,$$

$$\sum_{n=1}^{\infty} n a_n x^{n-1} + \sum_{n=1}^{\infty} n a_n x^n - \sum_{n=0}^{\infty} \alpha a_n x^n = 0.$$

Replace n by $n+1$ in the first summation to get

$$\sum_{n=0}^{\infty} (n+1) a_{n+1} x^n + \sum_{n=1}^{\infty} n a_n x^n - \sum_{n=0}^{\infty} \alpha a_n x^n = 0.$$

Notice that the middle sum can be started at $n = 0$ without changing the sum because 0 is added. Then

$$\sum_{n=0}^{\infty} [(n+1)a_{n+1} + (n-\alpha)a_n]x^n = 0,$$

$$(n+1)a_{n+1} + (n-\alpha)a_n = 0, \text{ for } n = 0, 1, 2, \cdots,$$

which yields

$$a_{n+1} = \frac{(\alpha - n)}{n+1} a_n \text{ for } n = 0, 1, 2, \cdots.$$

This recurrence formula allows us to express all the coefficients a_n in terms of $a_0 = 1$:

$$a_0 = 1,$$

$$a_1 = \frac{\alpha}{1} a_0 = \frac{\alpha}{1},$$

$$a_2 = \frac{\alpha - 1}{2} a_1 = \frac{\alpha(\alpha - 1)}{1 \cdot 2},$$

$$a_3 = \frac{\alpha - 2}{3} a_2 = \frac{\alpha(\alpha - 1)(\alpha - 2)}{1 \cdot 2 \cdot 3},$$

and, in general,

$$a_n = \frac{\alpha(\alpha - 1)\cdots(\alpha - n + 1)}{1 \cdot 2 \cdot \cdots \cdot n} \text{ for } n = 1, 2, \cdots,$$

with n factors in the numerator and denominator. Therefore,

$$y = \sum_{n=0}^{\infty} a_n x^n = 1 + \alpha x + \frac{\alpha(\alpha - 1)}{1 \cdot 2} x^2 + \frac{\alpha(\alpha - 1)(\alpha - 2)}{1 \cdot 2 \cdot 3} x^3 + \cdots.$$

If α is a positive integer, then $a_{\alpha+1} = a_{\alpha+2} = a_{\alpha+3} = \cdots = 0$ and the series reduces to a polynomial of degree α. This is a degenerate power series, which has an infinite radius of convergence. If α is *not* a positive integer, then $a_n \neq 0$ for all n and the recurrence formula, expressed as $a_{n+1}/a_n = (\alpha - n)/(n + 1)$, makes it easy to find the radius of convergence of the series: For any $x \neq /0$,

$$\left| \frac{a_{n+1} x^{n+1}}{a_n x^n} \right| = \left| \frac{\alpha - n}{n + 1} \right| |x| \to |x| \text{ as } n \to \infty.$$

By the ratio test, the power series converges absolutely for $|x| < 1$, diverges for $|x| > 1$, and has radius of convergence 1. Therefore,

$$y = \sum_{n=0}^{\infty} a_n x^n = 1 + \alpha x + \frac{\alpha(\alpha - 1)}{1 \cdot 2} x^2 + \frac{\alpha(\alpha - 1)(\alpha - 2)}{1 \cdot 2 \cdot 3} x^3 + \cdots$$

solves the initial value problem $(1 + x)y' - \alpha y = 0, y(0) = 1$ on $-1 < x < 1$.

As we indicated earlier, the sum of this power series must be $(1 + x)^\alpha$ because the initial value problem has a unique solution. Our conclusions are summarized in Th. 1.

Theorem 1 *The Binomial Series*

Let α be any real number and $-1 < x < 1$. Then

$$(1 + x)^\alpha = \sum_{n=0}^{\infty} a_n x^n =$$

$$= 1 + \alpha x + \frac{\alpha(\alpha - 1)}{1 \cdot 2} x^2 + \frac{\alpha(\alpha - 1)(\alpha - 2)}{1 \cdot 2 \cdot 3} x^3 + \cdots$$

where

$$a_n = \frac{\alpha(\alpha - 1)(\alpha - 2)\cdots(\alpha - n + 1)}{1 \cdot 2 \cdot 3 \cdot \cdots \cdot n}.$$

EXAMPLE 3. Find the Taylor series for the function $1/\sqrt{1-x^2}$.

Solution. Since $1/\sqrt{1-x^2} = (1+t)^{-1/2}$ for $t = -x^2$, we can use the binomial series for $(1+t)^{-1/2}$:

$$\frac{1}{\sqrt{1+t}} = 1 + \left(-\frac{1}{2}\right)t + \frac{(-1/2)(-3/2)}{1 \cdot 2}t^2 + \frac{(-1/2)(-3/2)(-5/2)}{1 \cdot 2 \cdot 3}t^3 + \cdots$$

$$= 1 - \frac{1}{2}t + \frac{1 \cdot 3}{2 \cdot 4}t^2 - \frac{1 \cdot 3 \cdot 5}{2 \cdot 4 \cdot 6}t^3 + \cdots \text{ for } |t| < 1.$$

Since $|t| = |-x^2| < 1 \iff |x| < 1$, the substitution $t = -x^2$ gives

$$\frac{1}{\sqrt{1-x^2}} = 1 + \frac{1}{2}x^2 + \frac{1 \cdot 3}{2 \cdot 4}x^4 + \frac{1 \cdot 3 \cdot 5}{2 \cdot 4 \cdot 6}x^6 + \cdots \text{ for } |x| < 1.$$

By Th. 3 in Sec. 2.3, the power series found in this way is the Taylor series for $1/\sqrt{1-x^2}$. □

EXAMPLE 4. Find the Taylor series for the function $\arcsin x$.

Solution. Replace x by u in the series for $1/\sqrt{1-x^2}$ found in Ex. 3 and then integrate from 0 to x to get

$$\arcsin x = \int_0^x \frac{du}{\sqrt{1-u^2}} =$$

$$= x + \frac{1}{2}\frac{x^3}{3} + \frac{1 \cdot 3}{2 \cdot 4}\frac{x^5}{5} + \frac{1 \cdot 3 \cdot 5}{2 \cdot 4 \cdot 6}\frac{x^7}{7} + \cdots \text{ for } |x| < 1. \ \square$$

As we indicated prior to Th. 1, the binomial series reduces to the binomial theorem of algebra if $\alpha = 0, 1, 2, \ldots$ and otherwise the radius of convergence of the series is $r = 1$. Determining the interval of convergence is more difficult. We simply state the results. If α is not a positive integer or zero, then the binomial series converges to $(1+x)^\alpha$ on the interval

$$C = [-1, 1] \text{ if } \alpha > 0, \quad C = (-1, 1] \text{ if } -1 < \alpha \le 0, \quad C = (1, 1) \text{ if } \alpha \le -1.$$

PROBLEMS

1. Show that the series $\sum_{n=0}^{\infty} (-1)^n x^{2n+1}/(2n+1)!$ converges for all x and that its sum, say $y = y(x)$, satisfies the initial value problem

$$y'' + y = 0, \quad y(0) = 0, \quad y'(0) = 1.$$

Use the fact that all solutions to the differential equation $y'' + y = 0$ have the form $y = A\cos x + B\sin x$ for some constants A and B to conclude that the sum of the series is $\sin x$.

2. Show that the series $\sum_{n=0}^{\infty} (-1)^n x^{2n}/(2n)!$ converges for all x and that its sum, say $y = y(x)$, satisfies the initial value problem

$$y'' + y = 0, \quad y(0) = 1, \quad y'(0) = 0.$$

Conclude that the sum of the series is $\cos x$.

3. Show that the series $\sum_{n=0}^{\infty} x^{2n+1}/(2n+1)!$ converges for all x and that its sum, say $y = y(x)$, satisfies the initial value problem

$$y'' - y = 0, \quad y(0) = 0, \quad y'(0) = 1.$$

Given that all solutions of the differential equation $y'' - y = 0$ have the form $y = Ae^x + Be^{-x}$ for some constants A and B, conclude that the sum of the series is $\sinh x = (e^x - e^{-x})/2$.

4. Show that the series $\sum_{n=0}^{\infty} x^{2n}/(2n)!$ converges for all x and that its sum, say $y = y(x)$, satisfies the initial value problem

$$y'' - y = 0, \quad y(0) = 1, \quad y'(0) = 0.$$

Conclude that the sum of the series is $\cosh x = (e^x + e^{-x})/2$.

5. (a) Find a first-order initial value problem satisfied by the sum of the series $\sum_{n=1}^{\infty} x^n/n^2$. (b) Use (a) to show that

$$\sum_{n=1}^{\infty} \frac{x^n}{n^2} = \int_0^x \frac{1}{t} \ln\left(\frac{1}{1-t}\right) dt$$

for $-1 < x < 1$. (This equation also holds for $x = \pm 1$. Using the fact that $\sum_{n=1}^{\infty} 1/n^2 = \pi^2/6$, we obtain an evaluation of the improper integral $\int_0^1 t^{-1} \ln(1-t)^{-1} dt$.)

6. Express the sum of the series $\sum_{n=1}^{\infty} x^n/n(n!)$ as an integral, much as in the previous problem.

In Probs. 7–12, find power series solutions to the following initial value problems. Give the sum of the series if you know it.

7. $y' + 2y = 0, \quad y(0) = 1$

8. $y' - 3y = 0, \quad y(0) = 1/2$

9. $y' + xy = 0, \quad y(0) = 3$

10. $(x+1)y' + (x+2)y = 0,$ $y(0) = 1$

11. $y'' + 4y = 0,\ y(0) = 5,\ y'(0) = -8$

12. $y'' - 9y = 0, \quad y(0) = 1, \quad y'(0) = 2$

In Probs. 13–16, find a power series representation for the given function. Find the radius of convergence of the series.

13. $1/\sqrt[3]{(1+4x)^2}$

14. $\sqrt[3]{2+x}$

15. $1/\sqrt{1+x^2}$

16. $x/\sqrt{1+x^2}$

In Probs. 17–20, use power series methods to approximate the following integrals to within 0.001.

17. $\displaystyle\int_0^{1/2} \sqrt{1+x^4}\, dx$

18. $\displaystyle\int_0^1 \frac{dx}{(8+x^2)^{1/3}}$

19. $\displaystyle\int_0^1 \sqrt{x} \arcsin\left(\frac{x}{3}\right) dx$

20. $\displaystyle\int_0^{1/2} \frac{\arcsin x}{x^{3/2}}\, dx$

21. The length L of the ellipse $x^2/a^2 + y^2/b^2 = 1$ with $0 < b \leq a$ is given by the elliptic integral

$$L = 4a \int_0^{\pi/2} \sqrt{1 - k^2 \sin^2 \theta}\, d\theta, \quad k = \sqrt{1 - b^2/a^2}.$$

The elliptic integral has no elementary antiderivative. The following steps lead to a series for L in terms of the eccentricity k of the ellipse.
(a) Use the binomial series to show that

$$\sqrt{1 - k^2 \sin^2\theta} = 1 - \frac{k^2 \sin^2\theta}{2} - \sum_{n=2}^{\infty} \frac{1 \cdot 3 \cdot \cdots \cdot (2n-3)}{2^n n!} k^{2n} \sin^{2n}\theta \quad \text{for all } \theta.$$

(b) Show that

$$\int_0^{\pi/2} \sin^{2n}\theta\, d\theta = \frac{1 \cdot 3 \cdot \cdots \cdot (2n-1)}{2 \cdot 4 \cdot \cdots \cdot (2n)} \cdot \frac{\pi}{2} \quad \text{for } n \geq 1.$$

(c) Term–by–term integration of the series in (a) is valid. Assume this and derive a power series in k for the length L of an ellipse with eccentricity k.

Chapter Highlights

Every power series $\sum a_n x^n$ has a radius of convergence r, $0 \leq r \leq \infty$. The series converges absolutely if $|x| < r$ and diverges if $|x| > r$. The methods of Ch. 1 can be used to determine whether the series converges or diverges at $x = \pm r$.

Assume that $f(x)$ is defined on an open interval containing 0 and f has derivatives of all orders at $x = 0$. The Taylor series for f (about 0) is the power series

$$\sum_{n=0}^{\infty} f^{(n)}(0) \frac{x^n}{n!}.$$

Most of the functions f that arise in calculus have Taylor series with positive or infinite radius of convergence r and these functions are represented by their Taylor series:

$$f(x) = \sum_{n=0}^{\infty} f^{(n)}(0) \frac{x^n}{n!} \quad \text{for} \quad -r < x < r.$$

Moreover, if f has a power series representation in some open interval about 0, then the power series is the Taylor series for f in that interval.

Three familiar examples of Taylor series, valid for all x, are

$$e^x = 1 + x + \frac{x^2}{2!} + \frac{x^3}{3!} + \cdots,$$

$$\sin x = x - \frac{x^3}{3!} + \frac{x^5}{5!} - \frac{x^7}{7!} + \cdots,$$

$$\cos x = 1 - \frac{x^2}{2!} + \frac{x^4}{4!} - \frac{x^6}{6!} + \cdots.$$

The partial sums of the Taylor series for a function f are the Taylor polynomials $P_n(x)$ for f. The estimates for the remainder $R_n(x) = f(x) - P_n(x)$ developed in the text are of both practical and theoretical value. They tell us how well $f(x)$ is approximated by its Taylor polynomials and they help us find Taylor series for particular functions.

The Taylor series for $f(x)$ can be differentiated or integrated term by term to obtain the Taylor series for $f'(x)$ and $\int_0^x f(t)\,dt$. The radius of convergence does not change. These facts have many applications. In particular, they enable us to find power series solutions of various differential equations. The binomial series in Sec. 2.4 is just one important example.

The chapter project is something like the example in Ch. 1 that we called "follow the bouncing ball." In the project, a marker (it could be a ball) moves back and forth on the real line. Which way it moves is governed by chance. The motion is called a *random walk.* Random walks, particularly in 2–space and 3–space, are important in many branches of science and engineering, including astrophysics and nuclear engineering. An example of a random walk is Brownian motion, in which small particles exposed to molecular shocks dart first one way and then another. A pioneering investigation of Brownian motion was made by Albert Einstein in 1905.

Chapter Project: Random Walks

The random walk takes place on the x–axis. A marker is initially at the origin. It is moved every second. It moves one unit to the right with probability p and one unit to the left with probability $q = 1 - p$. For example, we might have $p = q = 1/2$ or $p = 1/3$ and $q = 2/3$. Our first goal is to determine

$P_1 =$ the probability that the marker will ever land at the position $x = 1$.
Then we seek

$P_k =$ the probability that the marker will ever land at the position $x = k$.
Here $k = 1, 2, 3, \ldots$.

The same mathematical situation occurs in a gambling problem. A gambler plays roulette, always betting \$1 on red at each spin of the wheel. The probability that the ball lands on red is p and the probability that it does not (so it lands on black or green) is $q = 1 - p$. In this case, $p < 1/2$ and $q > 1/2$. The gambler decides to quit playing the first time he or she has a net profit of \$1000. The problem is to determine the probability that the gambler will ever win \$1000. (Of course, in this example, \$1000 can be replaced by any other fixed number of dollars. As a student, your limit might be lower!)

Assume the following facts from probability: The probability of an event is a number in the interval [0,1]. If two or more events are mutually exclusive (no two of them can occur at the same time), then the probability that at least one of the events occurs is the sum of the individual probabilities. If two or more events are independent (the occurrence of any one of the events has no bearing on the occurrence of the others), then the probability that all the events occur is the product of the individual probabilities.

Now return to the random walk. For $n = 1, 2, 3, \ldots$, let

$$x_n = \text{ the position of the marker after the } n\text{th move.}$$

We call x_n a *random variable* because its value depends on chance. Let

$$a_n = \text{the probability that } x_n = 1 \text{ for the first time as a result of the } n\text{th move.}$$

Problem 1 Explain why $a_{2n} = 0$ for all n. Then show by elementary reasoning that $a_1 = p$, $a_3 = qp^2$, and $a_5 = 2q^2 p^3$.

Problem 2 Why should you expect that $P_1 = \sum_{n=1}^{\infty} a_n \leq 1$?

An indirect method, which involves power series, will be used to find the probabilities a_n and P_1. The *generating function* of the sequence $\{a_n\}_{n=1}^{\infty}$ is the power series $A(x) = \sum_{n=1}^{\infty} a_n x^n$. Generating functions are powerful tools for studying sequences. The basic idea is to transform a problem about sequences to a problem about power series, so as to take advantage of useful properties of power series. Generating functions are used extensively in many branches of mathematics, science, and engineering, especially on problems involving probability and statistics.

In the random walk, let

b_n = the probability that $x_n = 2$ for the first time as a result of the nth move.

The probabilities b_n will help you to find the probabilities a_n. The generating function for the sequence $\{b_n\}_{n=1}^{\infty}$ is $B(x) = \sum_{n=1}^{\infty} b_n x^n$.

Problem 3 Show that the series for $A(x)$ and $B(x)$ converge at least for $|x| < 1$.

Problem 4 Explain why $a_n = q b_{n-1}$ for $n = 2, 3, 4, \ldots$.

Problem 5 Deduce that $A(x) = px + qxB(x)$.

The next three problems will yield a surprising relationship between $A(x)$ and $B(x)$.

Problem 6 Let $n > k$. Consider the following compound event: $x_k = 1$ for the first time as a result of the kth move and $x_n = 2$ for the first time as a result of the nth move. Show that the probability of this event is $a_k a_{n-k}$.

Problem 7 Use the laws of probability mentioned earlier to confirm that

$$b_n = a_1 a_{n-1} + a_2 a_{n-2} + \cdots + a_{n-2} a_2 + a_{n-1} a_1.$$

Problem 8 Deduce that $B(x) = A(x)^2$. *Hint.* Recall the formula for multiplying power series in Sec. 2.2.

Problem 9 Infer from Probs. 5 and 8 that

$$A(x) = \frac{1 - (1 - 4pqx^2)^{1/2}}{2qx}.$$

Use the binomial series for $\sqrt{1-t}$ to find formulas for the probabilities a_n.

Problem 10 Explain why

$$P_1 = A(1) = \frac{1 - |p - q|}{2q}.$$

Hint. $(p + q)^2 = 1$.

Problem 11 Deduce from Prob. 10 that

$$P_1 = \left\{ \begin{array}{c} 1 \text{ if } p \geq q, \\ p/q \text{ if } q < p. \end{array} \right\}$$

Notice that $P_1 = 1/2$ if $p = 1/3$ and $q = 2/3$. Does this seem reasonable?

Problem 12 Explain why

$$P_2 = \left\{ \begin{array}{c} 1 \text{ if } p \geq q, \\ (p/q)^2 \text{ if } q < p. \end{array} \right\}$$

Now it is easy to determine generating functions that give the probabilities c_n^k that $x_n = k$ for the first time as a result of the nth move, where $k = 1, 2, 3, \ldots$. Here k is a superscript, not an exponent. Let

$$C^k(x) = \sum_{n=1}^{\infty} c_n^k x^n$$

be the generating function for $\{c_n^k\}_{n=1}^{\infty}$. So $C^1(x) = A(x)$ and $C^2(x) = B(x)$ in the previous notation.

Problem 13 Reason in step–by–step fashion (mathematical induction) much as in Probs. 7 and 8 to show that $C^k(x) = A(x)^k$. Conclude that $P_k = (P_1)^k$ and write out the two–part formula for P_k.

Problem 14 Show that $P_{10} < 0.001$ if $p = 1/3$ and $q = 2/3$. Does this seem reasonable?

Problem 15 If you have access to a computer algebra system, find the first four nonzero probabilities in the sequence $\{c_n^3\}_{n=1}^{\infty}$.

Chapter Review Problems

1. (a) Find the Taylor polynomial $P_2(x)$ for $f(x) = 1/x^2$ about $x = 1$. (b) Express $P_2(x)$ as a polynomial in x.
2. (a) Find the Taylor polynomial $P_4(x)$ for $f(x) = 1/x^3$ about $x = -1$. (b) Express $P_4(x)$ as a polynomial in x.
3. Find the Taylor polynomial $P_5(x)$ for $f(x) = \tan x$.
4. Find the Taylor polynomial $P_6(x)$ for $f(x) = \ln \cos x$.
5. Find the Taylor series for $f(x) = 1/(1 + x)^2$.
6. Find the Taylor series for $f(x) = 1/x$ about $x = 1$.
7. A car travels along a straight road. It has initial position $s(0) = 20$ ft, initial velocity $s'(0) = 80$ ft/sec, and initial acceleration $s''(0) = 12$ ft/sec^2. Find a Taylor polynomial that approximates $s(t)$ for t near 0.
8. A weather balloon rises straight up from the ground. When the balloon is 800 feet above the ground recording devices show that the temperature is 65° and that it is decreasing at 0.05° per foot. Find a Taylor polynomial that

approximates the temperature as a function of height above the ground near the 800 foot level.

In Probs. 9–12, find the sum of the power series.

9. $1 - \dfrac{x^2}{3!} + \dfrac{x^4}{5!} - \dfrac{x^6}{7!} + \cdots$

10. $\dfrac{x^2}{2!} - \dfrac{x^4}{4!} + \dfrac{x^6}{6!} - \cdots$

11. $1 - \dfrac{x}{2} + \dfrac{x^2}{2^2 \cdot 2} - \dfrac{x^3}{2^3 \cdot 3!} + \dfrac{x^4}{2^4 \cdot 4!} - \cdots$

12. $3x - \dfrac{3^2x^2}{2} + \dfrac{3^3x^3}{3} - \dfrac{3^4x^4}{4} + \cdots$

In Probs. 13–22, find the radius and interval of convergence of the power series.

13. $\Sigma \dfrac{x^n}{\sqrt{n(n+1)}}$

14. $\Sigma(-1)^n \dfrac{\ln n}{n}(x-1)^n$

15. $\Sigma \dfrac{(-1)^{n+1}x^n}{2^n \ln n}$

16. $\Sigma \dfrac{x^{2n}}{(3n)!}$

17. $\Sigma \left(1 + \dfrac{2}{n}\right)^{n^2} x^n$

18. $\Sigma \sin\left(\dfrac{1}{n}\right) x^n$

19. $\Sigma(-1)^n \dfrac{x^{2n}}{4^n}$

20. $\Sigma \dfrac{x^n}{2+n^2} x^n$

21. $\Sigma \dfrac{n!}{2^n}(x-3)^n$

22. $\Sigma \dfrac{2^n}{n!}(x-3)^n$

23. Use the Taylor series for e^{-x} to find an approximation for e^{-1} with an error less than 0.01.

24. (a) Find the Taylor polynomial $P_2(x)$ for $f(x) = e^x \sin x$. (b) Estimate the error $|e^x \sin x - P_2(x)|$ for $|x| \leq 0.1$.

25. Find the Taylor series for $\sin^2 x$. *Hint.* Try a half–angle formula.

26. Find the Taylor series for $f(x) = (e^x - 1)/x$. Use it to show that $f(x) \to 1$ as $x \to 0$.

27. Use power series methods to find

$$\lim_{x \to 0} \frac{x - \sin x}{x^3}.$$

28. Show that the improper integral $\int_0^\infty [(\sin x)/x]\,dx$ converges. *Hint.* Look at the graph of $\sin x/x$. Then express the integral as an alternating series.

29. The **error function** is defined by

$$\operatorname{erf}(x) = \frac{2}{\sqrt{\pi}} \int_0^x e^{-t^2}\,dt.$$

It is widely used in probability, statistics, and applied mathematics. Estimate erf(1) with an error less than 0.01.

30. Find a Taylor series for $(1-x)^{-3}$, expressed in the form $\sum_{n=0}^{\infty} a_n x^n$.

31. Use the differentiation formula $(d/dx)\operatorname{arctanh} x = 1/(1-x^2)$ to obtain a power series for $\operatorname{arctanh} x$.

32. Let $f(x) = e^{-1/x^2}$ for $x \neq 0$. (a) Show that $\lim_{x\to 0} f(x) = 0$. (b) Define $f(0) = 0$ and then find the Taylor series for $f(x)$ about $x = 0$. (c) For what values of x does the Taylor series represent the function?

33. Use a power series to estimate $\int_0^1 x^2 \cos x^2\, dx$ to within 0.01.

34. Estimate $\int_0^{1/2} \arctan x dx$ to within 0.001.

35. Verify that the power series $y = \sum_{n=0}^{\infty} x^{2n}/(2^n n!)$ satisfies the differential equation $y' = xy$.

36. Verify that the power series $y = \sum_{n=1}^{\infty} (-1)^{n+1} x^{2n-1}/(2n-1)!$ satisfies the differential equation $y'' = -y$.

37. The **Bessel function** $J_0(x)$ of order zero is defined by

$$J_0(x) = \sum_{n=0}^{\infty} (-1)^n \frac{x^{2n}}{4^n (n!)^2}.$$

(a) Show that the series converges for all x. (b) Show that $y = J_0(x)$ satisfies the differential equation

$$x^2 y'' + xy' + x^2 y = 0.$$

38. (a) Find the Taylor series for $\int_0^x e^{-t^2}\, dt$. (b) Find the radius and interval of convergence.

39. Find the Taylor series about $x = 1$ for $f(x) = x^2 \ln x$. *Hint.* $x^2 = (x-1)^2 + 2(x-1) + 1$.

40. Let $f(x) = 1/\sqrt{1-x}$. (a) Find $P_2(x)$. (b) Find, to three places, $P_2(1/2)$ and $R_2(1/2) = f(1/2) - P_2(1/2)$. (c) Estimate $|R_2(1/2)|$ by Th. 3 in Sec. 2.1.

41. Find $P_3(x)$ for $f(x) = e^{-2x}$. Estimate $|R_3(x)| = |f(x) - P_3(x)|$ for $0 \leq x \leq 1$ in two ways. (a) Use Th. 3 in Sec. 2.1. (b) Justify the use of and use the alternating series error formula.

42. (a) Find a formula for the sum of the series $\sum_{n=1}^{\infty} nx^n$. (b) Evaluate the sum for $x = 2/5$.

43. Find the Taylor series about $x = 0$ for $f(x) = x/(x^2 - 3x + 2)$.

44. Use power series methods to discover the solution to the initial value $y'' + y = 0,\ y(0) = 0,\ y'(0) = 1$.

45. Use power series methods to discover the solution to the initial value $y'' + y = 0,\ y(0) = 1,\ y'(0) = 0$.

46. (a) Use the differentiation formula $(d/dx)\operatorname{arcsinh} x = 1/\sqrt{1+x^2}$ to obtain the first few terms of a power series for $\operatorname{arcsinh} x$. (b) Estimate $\operatorname{arcsinh}(1/4)$ to within 0.001.

In general, the distance of any point $P = (x, y, z)$ from the origin is

$$|OP| = \sqrt{x^2 + y^2 + z^2}.$$

Therefore, the equation for the sphere with center O and radius r i

$$x^2 + y^2 + z^2 = r^2.$$

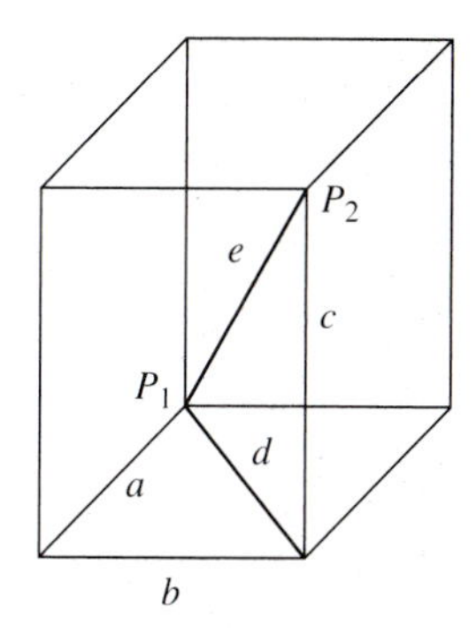

FIGURE 6

Now let $P_1 = (x_1, y_1, z_1)$ and $P_2 = (x_2, y_2, z_2)$ be any two points in 3–space. The case with $x_1 < y_1$, $x_2 < y_2$, $z_1 < z_2$ is shown in Fig. 6, where $a = x_2 - x_1$, $b = y_2 - y_1$, $c = z_2 - z_1$. Reasoning as before, we find that the distance from P_1 to P_2 is $|P_1P_2| = e = \sqrt{a^2 + b^2 + c^2}$. Consequently, the distance between any two points $P_1 = (x_1, y_1, z_1)$ and $P_2 = (x_2, y_2, z_2)$ is

$$\boxed{|P_1P_2| = \sqrt{(x_2 - x_1)^2 + (y_2 - y_1)^2 + (z_2 - z_1)^2}.}$$

For example, if $P_1 = (2, 3, -7)$ and $P_2 = (5, 7, 5)$, then $x_2 - x_1 = 3$, $y_2 - y_1 = 4$, $z_2 - z_1 = 12$, and

$$|P_1P_2| = \sqrt{3^2 + 4^2 + 12^2} = 13.$$

It follows from the distance formula that the equation for the sphere with center $P_0 = (x_0, y_0, z_0)$ and radius r is

$$(x - x_0)^2 + (y - y_0)^2 + (z - z_0)^2 = r^2.$$

For example, the sphere with center $P_0 = (2, 4, -5)$ and radius 3 is given by

$$(x - 2)^2 + (y - 4)^2 + (z + 5)^2 = 9.$$

Verify that the point $P = (1, 2, -3)$ lies on the sphere.

PROBLEMS

In Probs. 1–4 find the distance between the points P and Q.

1. $P = (6, 0, -8)$ and $Q = (0, 0, 0)$.
2. $P = (7, -4, 3)$ and $Q = (-5, 1, 3)$.
3. $P = (2, -2, 5)$ and $Q = (3, 0, 7)$.
4. $P = (3, 7, 12)$ and $Q = (3, 0, -12)$.

In Probs. 5–8 find an equation for the plane through $P = (3, -2, 5)$ which satisfies the given condition.

5. parallel to the *xy*–plane
6. parallel to the *xz*–plane
7. perpendicular to the *x*–axis
8. perpendicular to the *y*–axis

In Probs. 9–10 find an equation for the sphere with center P_0 and radius r.

9. $P_0 = (0, 0, 0)$ and $r = 4$
10. $P_0 = (2, 4, -5)$ and $r = 3$

In Probs. 11–12 find an equation for the sphere with center P_0 and with the point P on the sphere.

11. $P_0=(2,-5,3)$ and $P=(-1,-5,7)$ 12. $P_0=(-1,0,5)$ and $P=(-3,2,6)$

In Probs. 13–14 show that the graph of the given equation is a sphere. Find the center P_0 and the radius r.

13. $x^2+y^2+z^2-4x+6y=3$ 14. $x^2+y^2+z^2+2x-8z=8$

3.2 Vectors

Vectors in 2–space and 3–space are introduced in this section. Vectors have both algebraic and geometric descriptions. The algebraic description enables us to calculate easily with vectors. The geometric description helps with conceptual understanding and with applications, such as problems involving motion or force. From the algebraic point of view, the bottom line is that vectors are added, subtracted, and multiplied by real numbers component by component. In this sense, vectors can be thought of as conceptual prototypes for modern parallel computers. From the geometric point of view, a vector has magnitude (or size) and direction. By contrast, a real number, also called a **scalar**, has only magnitude. It has no direction. Examples of scalars are mass, speed, and temperature.

The algebraic properties of vectors are virtually the same in 2–space and in 3–space. We shall motivate key ideas in the two–dimensional case and add further details when we come to the three–dimensional setting. Our presentation is rather informal, based largely on geometric reasoning.

Vectors in 2–Space

Vectors are represented geometrically as directed line segments. In Fig. 1, the directed line segment from $(0,0)$ to $(3,4)$ represents the vector $\mathbf{v}=<3,4>$. More specifically, the vector $\mathbf{v}$ is *defined* as the ordered pair $<3,4>$. The angle brackets distinguish a vector, in this case $<3,4>$, from the point $(3,4)$. It is common practice to ignore the distinction between a vector defined as an ordered pair and a representation of a vector as a directed line segment. Thus, we think of a vector as having two interpretations, one algebraic and the other geometric.

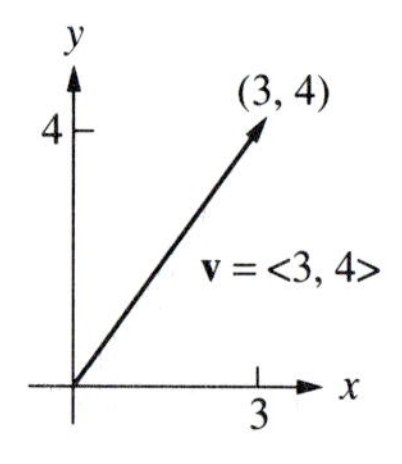

FIGURE 1

Let (a,b) be any point in the xy–plane. Then $\mathbf{v}=<a,b>$ defines a **vector**. The numbers a and b are called **components** of $\mathbf{v}$. Just as for the special vector in Fig. 1, we identify the vector $\mathbf{v}=<a,b>$ with the directed line segment from the origin $(0,0)$ to (a,b) in Fig. 2. The **magnitude** or **length** of $\mathbf{v}$ is

$$||\mathbf{v}||=\sqrt{a^2+b^2},$$

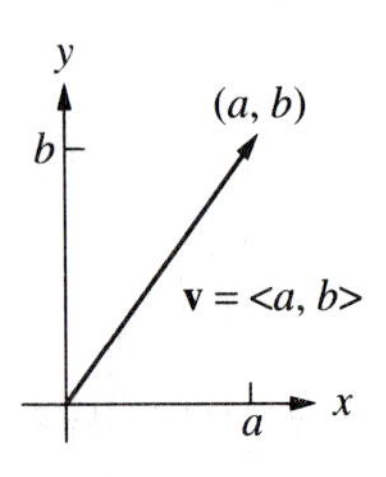

FIGURE 2

which is the length of the directed line segment in Fig. 2. The **zero vector** $\mathbf{0}=<0,0>$ is exceptional; it is represented by the point (0,0) and has zero length and no direction.

Velocities are vectors. To make this more concrete, imagine that an object is moving along a path through the origin in Fig. 1. When it is at the origin, suppose that the object is moving in the direction from $(0,0)$ to $(3,4)$ with speed 5 feet per second (or other units). Then the velocity of the object at that time is the vector $\mathbf{v} = <3,4>$. The direction of $\mathbf{v}$ is the direction of motion and its length $||\mathbf{v}|| = \sqrt{3^2+4^2} = 5$ is the speed. In this context, we refer to $\mathbf{v}$ as a *velocity vector*. The x–coordinate of the object moving with velocity $\mathbf{v} = <3,4>$ is increasing at the rate of 3 ft/sec, which we call the velocity in the x–direction. The velocity in the y–direction is 4 ft/sec.

Forces are vectors. Suppose that an object at the origin in Fig. 1 is subject to a force of magnitude 5 in suitable units acting in the direction from $(0,0)$ to $(3,4)$. Then the force is the vector $\mathbf{v} = <3,4>$, which points in the direction the force acts, and its length $||\mathbf{v}|| = 5$ is the magnitude of the force. In this context, $\mathbf{v}$ is called a *force vector*. Force vectors are denoted more commonly by $\mathbf{F}$ rather than $\mathbf{v}$.

The vector in Fig. 2 is said to be *attached* to the origin. Vectors attached to different points but with the same length and the same direction, as in Fig. 3, are regarded as equal. Strictly speaking, they are different representations of the same vector $\mathbf{v} = <a,b>$. The representation of $\mathbf{v} = <a,b>$, which is attached to the origin in Fig. 2, is said to be in *standard position*. It is also called the **position vector** of the point (a,b). Position vectors are useful for tracking moving objects, as we shall see later.

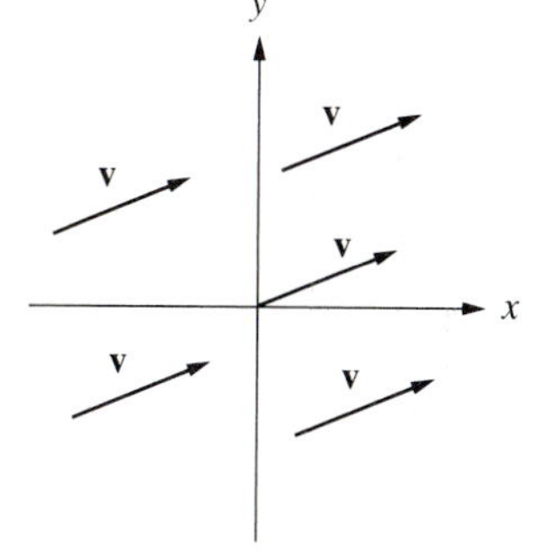

FIGURE 3

In Fig. 4, the two triangles have parallel sides with the same lengths. So the two vectors labeled $\mathbf{v}$ are equal. Both are expressed by $\mathbf{v} = <3,4>$. If $\mathbf{v}$ represents the velocity of a moving object when it is at the point P in Fig. 4, or if $\mathbf{v}$ represents a force acting at P, then it is natural to attach $\mathbf{v}$ to the point P.

The vector $\mathbf{v}$ from $P = (5,2)$ to $Q = (8,6)$ in Fig. 4 is denoted also by $\overrightarrow{PQ}$. Notice that the components of $\overrightarrow{PQ}$ can be found by subtracting the coordinates of P from those of Q. Thus,

$$\overrightarrow{PQ} = <8-5, 6-2> = <3,4>.$$

FIGURE 4

In general, the vector from $P_1 = (x_1, y_1)$ to $P_2 = (x_2, y_2)$ is given by

$$\overrightarrow{P_1P_2} = <x_2 - x_1, y_2 - y_1>.$$

If $\mathbf{v} = \overrightarrow{P_1P_2}$, it is customary to call P_1 the **tail** of $\mathbf{v}$ and P_2 the **head** (or **tip**) of $\mathbf{v}$. This language is used because directed line segments look like arrows.

Scalar Multiples, Sums, and Differences of Vectors

The algebraic rules for vectors are similar to algebraic rules for real numbers. These rules are easy to remember and just as easy to work with. Here we define scalar multiplication, addition, and subtraction of vectors. These algebraic operations have important geometric and physical interpretations that will be discussed as we go along. Other algebraic properties of vectors will be given in the next two sections.

We begin with examples of scalar multiplication of vectors. For any vector $\mathbf{v} = <a,b>$, the vector $-\mathbf{v} = (-1)\mathbf{v}$ is defined by

$$-\mathbf{v} = <-a, -b>,$$

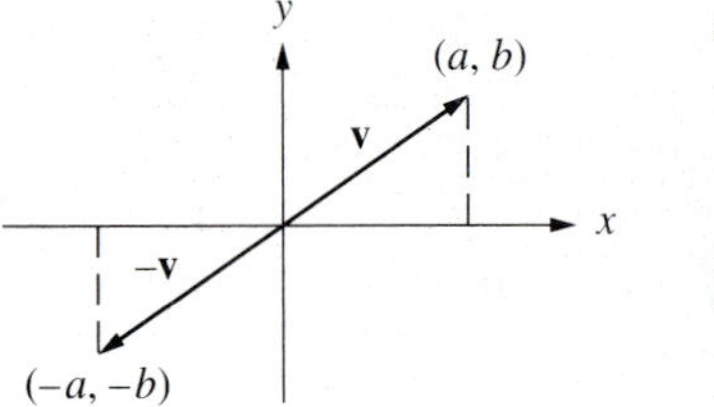

FIGURE 5

as illustrated in Fig. 5. The vectors **v** and $-\mathbf{v}$ have the same length:

$$||-\mathbf{v}|| = ||\mathbf{v}|| = \sqrt{a^2 + b^2}.$$

If $\mathbf{v} \neq \mathbf{0}$, then **v** and $-\mathbf{v}$ point in opposite directions. This is evident from Fig. 5.

Again let $\mathbf{v} = <3,4>$. The vector $2\,\mathbf{v}$ is defined by $2\mathbf{v} = 2<3,4> = <6,8>$. In Fig. 6, **v** and $2\mathbf{v}$ have the same direction and

$$||\mathbf{v}|| = 5, \qquad ||2\mathbf{v}|| = \sqrt{36 + 64} = 10 = 2\,||\mathbf{v}||.$$

The vector $-2\mathbf{v}$ is defined by

$$-2\mathbf{v} = -2 < 3,4 > = < -6, -8 >.$$

In Fig. 7, **v** and $-2\mathbf{v}$ have opposite directions and

$$||-2\mathbf{v}|| = 10 = 2\,||\mathbf{v}||.$$

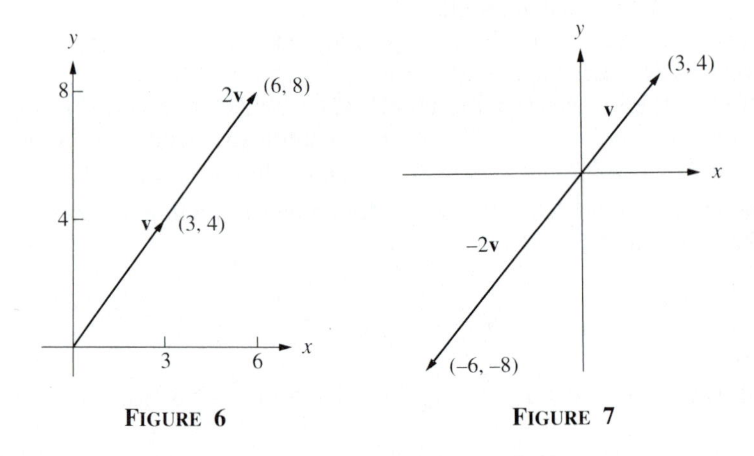

FIGURE 6 FIGURE 7

In general, for any vector $\mathbf{v} = < a, b >$ and any scalar λ, the vector $\lambda\mathbf{v}$, called a **scalar multiple** of **v**, is defined by

$$\lambda\mathbf{v} = \lambda < a, b > = < \lambda a, \lambda b >.$$

It follows easily (see the problems) that

$$||\lambda\mathbf{v}|| = |\lambda|\ ||\mathbf{v}||.$$

Let $\mathbf{v} \neq \mathbf{0}$. Then

$$\mathbf{v} \text{ and } \lambda\mathbf{v} \text{ have the same direction if } \lambda > 0,$$
$$\mathbf{v} \text{ and } \lambda\mathbf{v} \text{ have opposite directions if } \lambda < 0.$$

Two nonzero vectors **v** and **w** are *parallel* if $\mathbf{w} = \lambda\,\mathbf{v}$ for some λ. So **v** and **w** have the same direction or opposite directions. For convenience, we also say that the zero vector is parallel to every vector.

Addition of vectors comes next. We begin with a typical example. If $\mathbf{v} = <3,1>$ and $\mathbf{w} = <2,4>$, then their sum is the vector obtained by adding the corresponding components of $\mathbf{v}$ and $\mathbf{w}$:

$$\mathbf{v} + \mathbf{w} = <3,1> + <2,4> = <3+2, 1+4> = <5,5>.$$

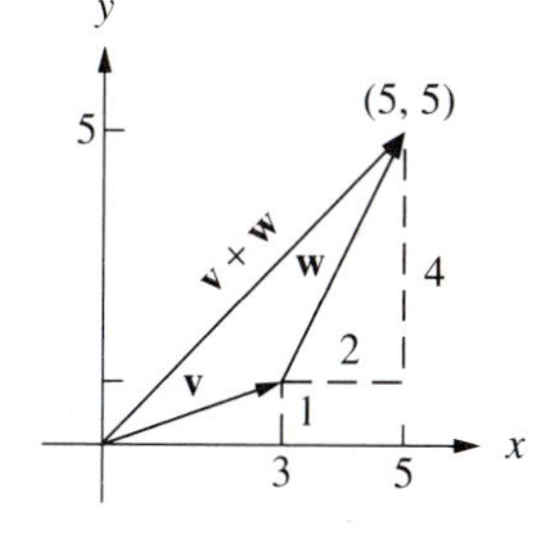

FIGURE 8

Figure 8 gives a picture of $\mathbf{v} + \mathbf{w}$. The vectors $\mathbf{v}$ and $\mathbf{w}$ are plotted end to end. Note that $\mathbf{v}$, $\mathbf{w}$, and $\mathbf{v} + \mathbf{w}$ form a triangle. This geometric construction of $\mathbf{v} + \mathbf{w}$ is called the **triangle law of vector addition.**

In the same way, given any vectors $\mathbf{v} = <v_1, v_2>$ and $\mathbf{w} = <w_1, w_2>$, their **sum** is defined by

$$\mathbf{v} + \mathbf{w} = <v_1 + w_1, v_2 + w_2>.$$

Since $v_1 + w_1 = w_1 + v_1$ and $v_2 + w_2 = w_2 + v_2$,

$$\mathbf{v} + \mathbf{w} = \mathbf{w} + \mathbf{v}.$$

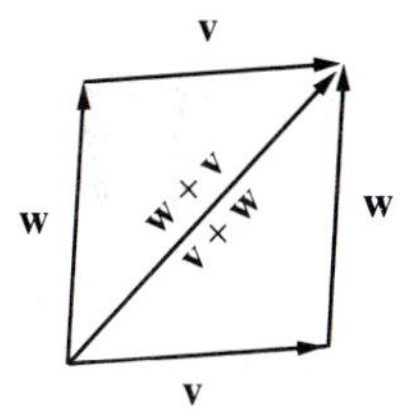

FIGURE 9

Figure 9 gives a geometric interpretation of this equality. Two applications of the triangle law of vector addition give $\mathbf{v} + \mathbf{w} = \mathbf{w} + \mathbf{v}$. Since $\mathbf{v} + \mathbf{w}$ is the diagonal of the parallelogram formed by $\mathbf{v}$ and $\mathbf{w}$, the construction of $\mathbf{v} + \mathbf{w}$ in Fig. 9 is also referred to as the **parallelogram law of vector addition.**

The **difference** $\mathbf{v} - \mathbf{w}$ of the vectors $\mathbf{v} = <v_1, v_2>$ and $\mathbf{w} = <w_1, w_2>$ is defined by

$$\mathbf{v} - \mathbf{w} = <v_1 - w_1, v_2 - w_2>.$$

For example, let $\mathbf{v} = <5,5>$ and $\mathbf{w} = <\mathbf{3},\mathbf{1}>$. Then

$$\mathbf{v} - \mathbf{w} = <5,5> - <3,1> = <5-3, 5-1> = <2,4>.$$

Notice that $\mathbf{v} - \mathbf{w} = \mathbf{v} + (-\mathbf{w})$. Subtraction of vectors reverses the effect of addition in the sense that

$$(\mathbf{v} - \mathbf{w}) + \mathbf{w} = \mathbf{v}.$$

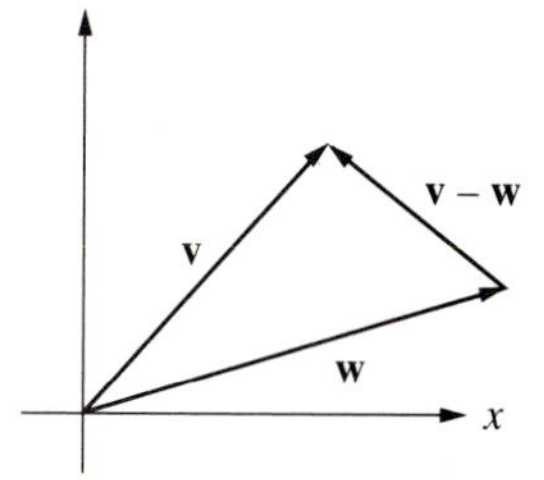

FIGURE 10

Figure 10 shows a geometric construction for $\mathbf{v} - \mathbf{w}$: First, draw $\mathbf{v}$ and $\mathbf{w}$ with the same initial point. Since $(\mathbf{v} - \mathbf{w}) + \mathbf{w} = \mathbf{v}$, it follows from the triangle law of vector addition that $\mathbf{v} - \mathbf{w}$ completes the triangle in Fig. 10.

The definitions of vector addition and subtraction have their origins in the physical world. Suppose that an object at the origin is acted on by two forces $\mathbf{v}$ and $\mathbf{w}$. Experiments show that the joint effect of the two forces on the object is the same as if the single force $\mathbf{v} + \mathbf{w}$ were to act on the object. In this context, $\mathbf{v} + \mathbf{w}$ is called the **resultant** of the forces $\mathbf{v}$ and $\mathbf{w}$. Vector addition is defined precisely so that the resultant of two (or more) forces is the vector sum of the forces. Velocity vectors behave in the same way. For example, if you throw a ball from a moving car, then the velocity of the ball is the vector sum of the velocity of the car plus the velocity you impart to the ball when you throw it.

EXAMPLE 1. Suppose that you are riding in a car with the top down traveling 68 mph along a straight road heading northeast. You throw a ball

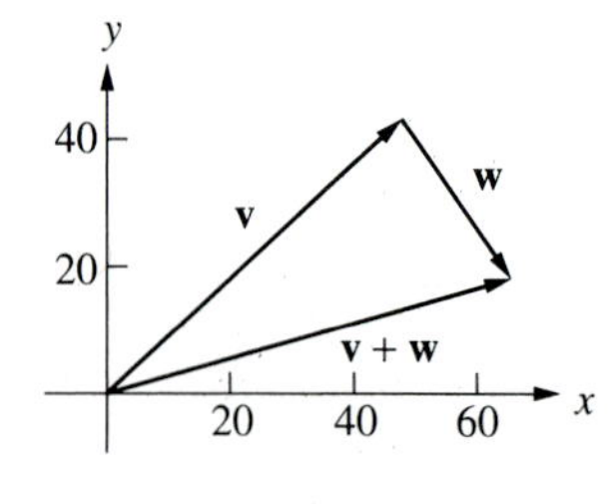

FIGURE 11

toward the southeast at 28 mph. What are the initial velocity and the initial speed of the ball relative to the ground?

Solution. Suppose that the action takes place in the xy–plane with the x–axis pointing east and the y–axis pointing north, as in Fig. 11. Then the velocity of the car is the vector $\mathbf{v}$ with length 68 in the figure. Since $\mathbf{v}$ makes a 45° angle with the x–axis, $\mathbf{v} = < 34\sqrt{2}, 34\sqrt{2} >$. For similar reasons, the velocity you impart to the ball when you throw it is the vector $\mathbf{w} = < 14\sqrt{2}, -14\sqrt{2} >$. The initial velocity of the ball relative to the ground is the vector sum

$$\mathbf{v} + \mathbf{w} = < 48\sqrt{2}, 20\sqrt{2} > = 4\sqrt{2} < 12, 5 >,$$

as shown in Fig. 11. The initial speed of the ball relative to the ground is

$$||\mathbf{v} + \mathbf{w}|| = 4\sqrt{2}\, ||< 12, 5 >|| = 4\sqrt{2} \cdot 13 \approx 73.5 \text{ mph.} \ \square$$

If the vectors $\mathbf{v} = < 34\sqrt{2}, 34\sqrt{2} >$ and $\mathbf{w} = < 14\sqrt{2}, -14\sqrt{2} >$ are forces, then the joint effect or resultant of the two forces is $\mathbf{v} + \mathbf{w} = < 48\sqrt{2}, 20\sqrt{2} >$. Again see Fig. 11.

Direction Angles, Direction Cosines, and Unit Vectors

It is clear from Fig. 12 that the direction of a nonzero vector $\mathbf{v} = < a, b >$ is determined by its components a and b. However, it is often more illuminating to describe the direction of a nonzero vector by means of its direction angles, its direction cosines, or a unit vector with the same direction as the given vector.

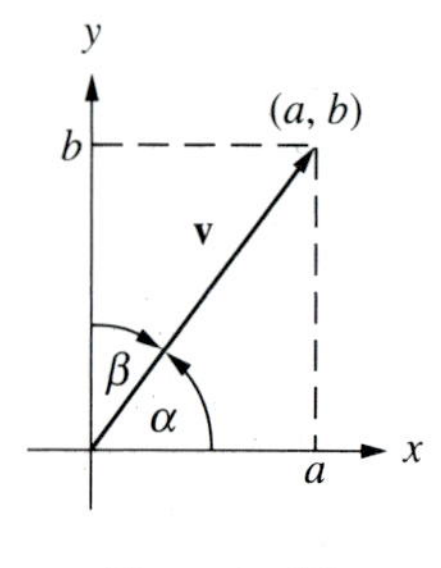

FIGURE 12

The vector $\mathbf{v} = < a, b >$ in Fig. 12 has **direction angles** α and β. As indicated in the figure, α and β are the angles (between 0 and π) that $\mathbf{v}$ makes with the positive x–axis and the positive y–axis. The **direction cosines** of $\mathbf{v}$ are $\cos\alpha$ and $\cos\beta$. They are easy to find from the components of $\mathbf{v}$. If $\mathbf{v} = < a, b >$, then a glance at Fig. 12 reveals that

$$\cos\alpha = \frac{a}{||\mathbf{v}||}, \qquad \cos\beta = \frac{b}{||\mathbf{v}||}.$$

Since $||\mathbf{v}|| = \sqrt{a^2 + b^2}$,

$$\cos^2\alpha + \cos^2\beta = 1.$$

Clearly, the direction of $\mathbf{v}$ is determined by its direction angles or, just as well, by its direction cosines.

EXAMPLE 2. Find the direction cosines and direction angles of the vector $\mathbf{v} = < 3, 4 >$ in Fig. 1.

Solution. Since $\mathbf{v} = < 3, 4 >$ and $||\mathbf{v}|| = 5$, the direction cosines of $\mathbf{v}$ are

$$\cos\alpha = \frac{3}{5}, \qquad \cos\beta = \frac{4}{5}.$$

From a table or a calculator, the direction angles are

$$\alpha \approx 53^\circ, \qquad \beta \approx 37^\circ.$$

So the vector $\mathbf{v} = < 3,4 >$ points into the first quadrant and makes an angle of approximately 53° with the positive x–axis and an angle of approximately 37° with the positive y–axis. □

A **unit vector u** is simply a vector with length $||\mathbf{u}|| = 1$. Another convenient way to describe the direction of a nonzero vector **v** is by means of the unit vector **u** that points in the same direction as **v**. It is easy to find this vector.

EXAMPLE 3. Find the unit vector **u** with the same direction as the vector $\mathbf{v} = < 3,4 >$.

Solution. Recall that **v** has length $||\mathbf{v}|| = 5$. Let

$$\mathbf{u} = \frac{\mathbf{v}}{||\mathbf{v}||} = \frac{1}{5} < 3,4 > \; = \; < \frac{3}{5}, \frac{4}{5} >.$$

Then $||\mathbf{u}|| = 1$, so **u** is a unit vector. Clearly, **u** and **v** have the same direction. So $\mathbf{u} = < 3/5, 4/5 >$ is the unit vector with the same direction as $\mathbf{v} = < 3,4 >$. □

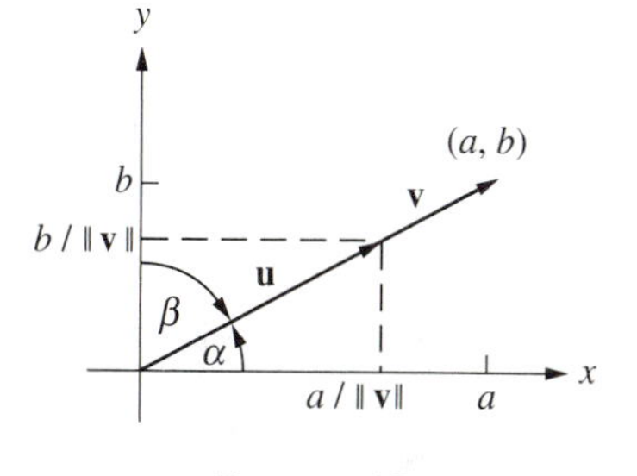

FIGURE 13

The reasoning in Ex. 3 applies to any nonzero vector $\mathbf{v} = < a,b >$, as in Fig. 13. The unit vector with the same direction as **v** is

$$\mathbf{u} = \frac{\mathbf{v}}{||\mathbf{v}||} = \left\langle \frac{a}{||\mathbf{v}||}, \frac{b}{||\mathbf{v}||} \right\rangle.$$

Since the direction cosines of **v** are $\cos\alpha = a/||\mathbf{v}||$ and $\cos\beta = b/||\mathbf{v}||$, another formula for **u** is

$$\mathbf{u} = < \cos\alpha, \cos\beta >.$$

Vectors in 3–Space

Vectors in 3–space are very much like vectors in 2–space. All that is really new is the addition of a third component. Since much of the discussion repeats what we have just done with minor changes, we present the basic facts in a rather concise fashion. At the same time, we give some additional properties of vectors.

Vectors in 3–space are defined in terms of rectangular coordinates. Figure 14 provides a setting. It shows a rectangular box with opposite corners $O = (0,0,0)$ and $P = (a,b,c)$. The distance from O to P is $|OP| = \sqrt{a^2 + b^2 + c^2}$.

A **vector v** in 3–space is an ordered triple of numbers:

$$\mathbf{v} = < a,b,c >.$$

The numbers a, b and c are the **components** of **v**. In Fig. 14, $\mathbf{v} = < a,b,c >$ is represented by the directed line segment from $O = (0,0,0)$ to $P = (a,b,c)$. The **magnitude** or **length** of **v** is

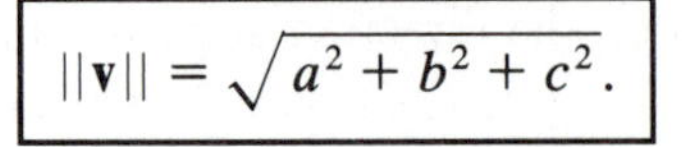

$$||\mathbf{v}|| = \sqrt{a^2 + b^2 + c^2}.$$

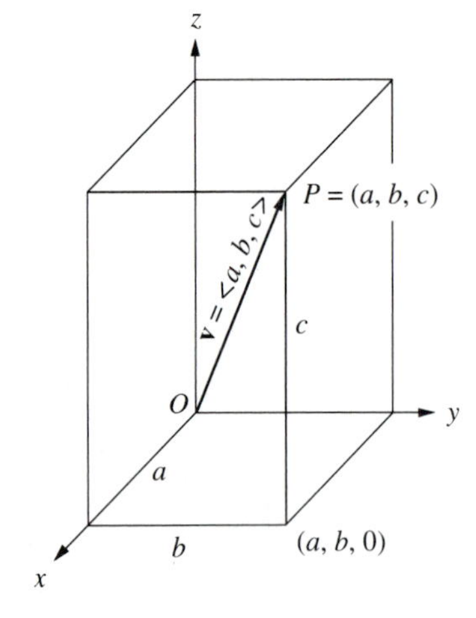

FIGURE 14

We think of a vector in 3–space either as an ordered triple of numbers or as a directed line segment, according to the convenience of the moment.

If $c = 0$, then the vector $\mathbf{v} = < a, b, 0 >$ from $(0, 0, 0)$ to $(a, b, 0)$ lies in the xy–plane. So vectors in the plane are special cases of vectors in space. It really doesn't matter whether we write $\mathbf{v} = < a, b >$ or $\mathbf{v} = < a, b, 0 >$ when we are dealing exclusively with plane vectors.

Any two directed line segments in 3–space with the same direction and the same length represent the same vector. In Fig. 15,

$$P_1 = (x_1, y_1, z_1), \qquad P_2 = (x_2, y_2, z_2),$$
$$a = x_2 - x_1, \qquad b = y_2 - y_1, \qquad c = z_2 - z_1.$$

It should be clear from the geometry that the directed line segments labeled $\mathbf{v} = < a, b, c >$ in Fig. 14 and $\overrightarrow{P_1 P_2}$ in Fig. 15 have the same length and the same direction; so they represent the same vector. Therefore, the vector from $P_1 = (x_1, y_1, z_1)$ to $P_2 = (x_2, y_2, z_2)$ is given by

$$\overrightarrow{P_1 P_2} = < x_2 - x_1, y_2 - y_1, z_2 - z_1 >.$$

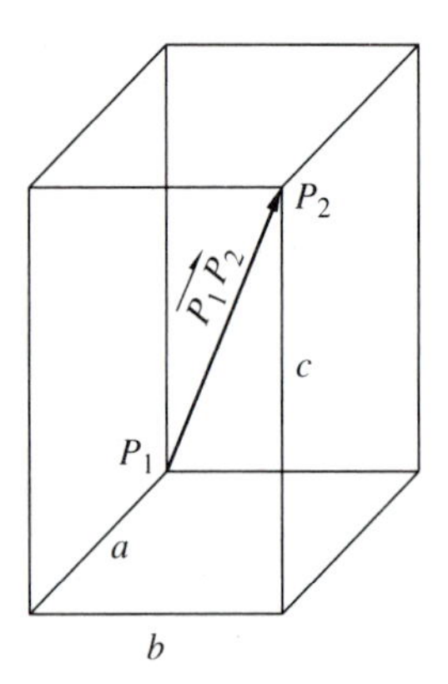

FIGURE 15

The magnitude of $\overrightarrow{P_1 P_2}$ is equal to the distance $|P_1 P_2|$ from P_1 to P_2. From the distance formula in 3–space,

$$||\overrightarrow{P_1 P_2}|| = |P_1 P_2| = \sqrt{(x_2 - x_1)^2 + (y_2 - y_1)^2 + (z_2 - z_1)^2}.$$

EXAMPLE 4. The vector from $P = (3, 2, 1)$ to $Q = (4, 4, 3)$ is

$$\overrightarrow{PQ} = < 4, 4, 3 > - < 3, 2, 1 > = < 1, 2, 2 >,$$

which has length $||\overrightarrow{PQ}|| = \sqrt{1 + 4 + 4} = 3$. □

Basic Vector Algebra

Algebraic operations with vectors are virtually the same in 2–space and in 3–space. For any scalar λ and any vector $\mathbf{v} = < a, b, c >$, the **scalar multiple** $\lambda\,\mathbf{v}$ is defined by

$$\lambda\mathbf{v} = < \lambda a, \lambda b, \lambda c >.$$

A simple calculation yields

$$||\lambda\mathbf{v}|| = |\lambda|\ ||\mathbf{v}||.$$

Notice that $0\mathbf{v} = \mathbf{0}$, $1\mathbf{v} = \mathbf{v}$,

$$-\mathbf{v} = < -a, -b, -c > \quad \text{and} \quad ||-\mathbf{v}|| = ||\mathbf{v}||.$$

Two nonzero vectors $\mathbf{v}$ and $\mathbf{w}$ have the same direction if $\mathbf{w} = \lambda\mathbf{v}$ for some $\lambda > 0$. They have opposite directions if $\mathbf{w} = \lambda\mathbf{v}$ for some $\lambda < 0$. Nonzero vectors with the same or opposite directions are called **parallel**. By convention, the zero vector is parallel to every vector.

Vector addition and subtraction are defined componentwise, just as for vectors in 2–space. Let $\mathbf{v} = < v_1, v_2, v_3 >$ and $\mathbf{w} = < w_1, w_2, w_3 >$. Then the **sum** and **difference** of $\mathbf{v}$ and $\mathbf{w}$ are defined by

$$\mathbf{v} + \mathbf{w} = < v_1 + w_1,\ v_2 + w_2,\ v_3 + w_3 >,$$
$$\mathbf{v} - \mathbf{w} = < v_1 - w_1,\ v_2 - w_2,\ v_3 - w_3 >.$$

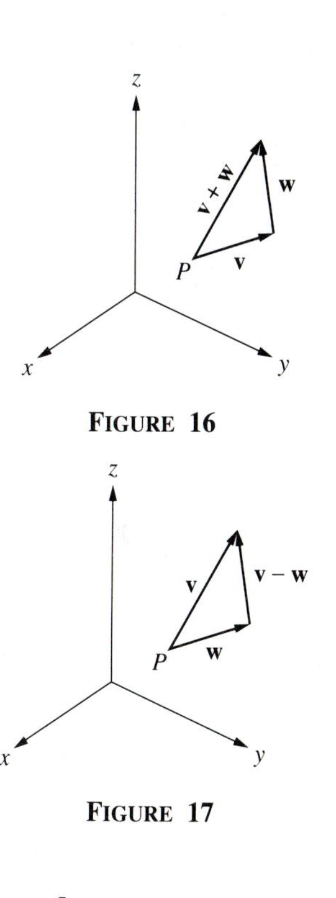

FIGURE 16

FIGURE 17

Note that $\mathbf{v} - \mathbf{w} = \mathbf{v} + (-\mathbf{w})$ and $(\mathbf{v} - \mathbf{w}) + \mathbf{w} = \mathbf{v}$. The triangle laws for vector addition and subtraction are illustrated in Figs. 16 and 17. A similar figure would illustrate the parallelogram law for vector addition.

Since vector addition, subtraction, and scalar multiplication are defined component by component, the following rules of vector algebra should seem obvious (see the problems):

$$\mathbf{v} + \mathbf{w} = \mathbf{w} + \mathbf{v}, \qquad (\mathbf{u} + \mathbf{v}) + \mathbf{w} = \mathbf{u} + (\mathbf{v} + \mathbf{w}),$$
$$\mathbf{v} + \mathbf{0} = \mathbf{v}, \qquad \mathbf{v} + (-\mathbf{v}) = \mathbf{0},$$

$$\lambda(\mathbf{v} + \mathbf{w}) = \lambda\mathbf{v} + \lambda\mathbf{w}, \qquad (\lambda + \mu)\mathbf{v} = \lambda\mathbf{v} + \mu\mathbf{v}, \qquad (\lambda\mu)\mathbf{v} = \lambda(\mu\mathbf{v}) = \mu(\lambda\mathbf{v}).$$

Direction Angles, Direction Cosines, and Unit Vectors

The **direction angles** of a nonzero vector $\mathbf{v}$ are the angles α, β, and γ between 0 and π that $\mathbf{v}$ makes with the three coordinate axes, as illustrated in Fig. 18. The **direction cosines** of $\mathbf{v}$ are

$$\cos\alpha = \frac{a}{||\mathbf{v}||}, \qquad \cos\beta = \frac{b}{||\mathbf{v}||}, \qquad \cos\gamma = \frac{c}{||\mathbf{v}||}.$$

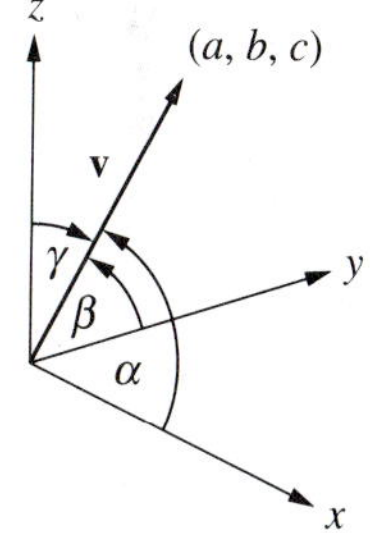

FIGURE 18

In Fig. 14, the direction cosines of $\mathbf{v}$ are the sides a, b, c of the box divided by the length $||\mathbf{v}||$ of the diagonal. The direction of a vector $\mathbf{v}$ is determined by its direction angles or its direction cosines.

As in the plane, a **unit vector u** in space is a vector with length $||\mathbf{u}|| = 1$. A unit vector $\mathbf{u}$ determines a particular direction in space. For this reason, we often refer to a unit vector $\mathbf{u}$ as a **direction.** If $\mathbf{v} = < a, b, c >$ is any nonzero vector, then the unit vector with the same direction as $\mathbf{v}$ is given by

$$\mathbf{u} = \frac{\mathbf{v}}{||\mathbf{v}||} = \left\langle \frac{a}{||\mathbf{v}||}, \frac{b}{||\mathbf{v}||}, \frac{c}{||\mathbf{v}||} \right\rangle.$$

In view of the formulas for the direction cosines of $\mathbf{v}$, another formula for $\mathbf{u}$ is

$$\mathbf{u} = < \cos\alpha,\ \cos\beta,\ \cos\gamma >.$$

Since $||\mathbf{u}|| = 1$,

$$\cos^2\alpha + \cos^2\beta + \cos^2\gamma = ||\mathbf{u}||^2 = 1.$$

The equation

$$\mathbf{v} = ||\mathbf{v}||\ \mathbf{u}$$

expresses any vector $\mathbf{v}$ conveniently in terms of its magnitude $||\mathbf{v}||$ and its direction $\mathbf{u}$.

EXAMPLE 5. The vector $\mathbf{v} = <3, 4, 12>$ has length $||\mathbf{v}|| = 13$ and direction

$$\mathbf{u} = \frac{\mathbf{v}}{||\mathbf{v}||} = \frac{1}{13} < 3, 4, 12 > = \left\langle \frac{3}{13}, \frac{4}{13}, \frac{12}{13} \right\rangle.$$

The direction cosines and direction angles of $\mathbf{v}$ are

$$\cos\alpha = \frac{3}{13}, \qquad \cos\beta = \frac{4}{13}, \qquad \cos\gamma = \frac{12}{13},$$

$$\alpha \approx 77°, \qquad \beta \approx 72°, \qquad \gamma \approx 23°.$$

The direction angles help us visualize the direction of the vector $\mathbf{v} = <3, 4, 12>$ in 3–space. □

Other Notation for Vectors

Three special unit vectors are shown in Fig. 19. They are

$$\mathbf{i} = <1, 0, 0>, \qquad \mathbf{j} = <0, 1, 0>, \qquad \mathbf{k} = <0, 0, 1>.$$

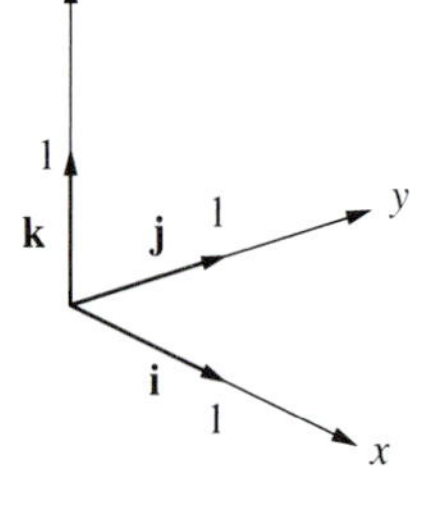

FIGURE 19

We call **i, j, k** the **basic unit vectors**. Let $\mathbf{v} = <a, b, c>$ be any vector. By the algebraic properties of vectors,

$$\begin{aligned} \mathbf{v} &= <a, 0, 0> + <0, b, 0> + <0, 0, c> \\ &= a<1, 0, 0> + b<0, 1, 0> + c<0, 0, 1>. \end{aligned}$$

Hence,

$$\boxed{\mathbf{v} = <a, b, c> = a\mathbf{i} + b\mathbf{j} + c\mathbf{k}.}$$

This shows that any vector can be expressed in either form, $\mathbf{v} = <a, b, c>$ or $\mathbf{v} = a\mathbf{i} + b\mathbf{j} + c\mathbf{k}$. For example, $\mathbf{v} = <1, 2, 2>$ and $\mathbf{v} = \mathbf{i} + 2\mathbf{j} + 2\mathbf{k}$ represent the same vector. Sometimes it is more convenient to express vectors in the angle bracket form $\mathbf{v} = <a, b, c>$ and sometimes the **i j k** form $\mathbf{v} = a\mathbf{i} + b\mathbf{j} + c\mathbf{k}$ is better. We shall use both notations.

When we work with vectors in the plane, we usually drop the third components and the vector **k**. Then the notation simplifies to

$$\mathbf{i} = \langle 1,0 \rangle, \qquad \mathbf{j} = \langle 0,1 \rangle,$$
$$\mathbf{v} = \langle a,b \rangle = a\,\mathbf{i} + b\,\mathbf{j}.$$

For example, $\mathbf{v} = \langle 3,4 \rangle = 3\,\mathbf{i} + 4\,\mathbf{j}$.

Newton's Law of Gravitation

We close this section by expressing Newton's law of gravitation in vector form. Figure 20 provides the setting and the notation. The figure shows two homogeneous spherical masses M and m with centers at P and Q. The distance from P to Q is $r = \|\mathbf{r}\|$. According to Newton's law of gravitation, the magnitude of the gravitational force exerted by M on m is

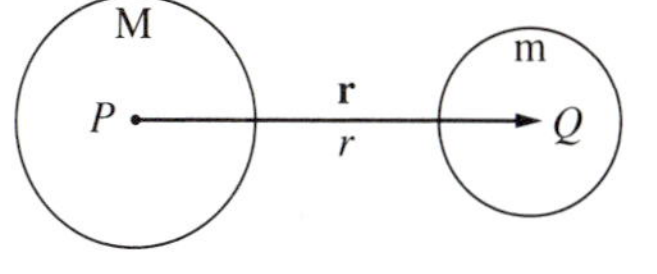

FIGURE 20

$$F = \frac{GMm}{r^2},$$

where G is the universal gravitational constant. Furthermore, the force of M on m is directed from Q to P, which is the direction of the unit vector $-\mathbf{r}/\|\mathbf{r}\|$. Thus, the gravitational force is expressed in vector form by

$$\mathbf{F} = \frac{GMm}{\|\mathbf{r}\|^2}\left(-\frac{\mathbf{r}}{\|\mathbf{r}\|}\right) = -\frac{GMm}{\|\mathbf{r}\|^3}\mathbf{r},$$

which reveals both the magnitude and direction of the gravitational force.

PROBLEMS

In Probs. 1–8 find: (a) $4\mathbf{v} - 3\mathbf{w}$, (b) $\|\mathbf{v}\|$, and (c) a unit vector in the direction of $\mathbf{w}$.

1. $\mathbf{v} = \langle 1,2 \rangle$ $\quad \mathbf{w} = \langle 3,4 \rangle$
2. $\mathbf{v} = \langle -1,2 \rangle$, $\quad \mathbf{w} = \langle 3,4 \rangle$
3. $\mathbf{v} = \langle 7,12 \rangle$, $\quad \mathbf{w} = \langle -5,12 \rangle$
4. $\mathbf{v} = \langle 0,-3 \rangle$, $\quad \mathbf{w} = \langle 5,-12 \rangle$
5. $\mathbf{v} = \langle 2,-1,3 \rangle$, $\quad \mathbf{w} = \langle 5,0,12 \rangle$
6. $\mathbf{v} = \langle 2,-1,1 \rangle$, $\quad \mathbf{w} = \langle -3,1,5 \rangle$
7. $\mathbf{v} = \langle 1,1,-2 \rangle$, $\quad \mathbf{w} = \langle -3,3,2 \rangle$
8. $\mathbf{v} = \langle 1,0,2 \rangle$, $\quad \mathbf{w} = \langle 5,0,12 \rangle$

In Probs. 9–16 find the vector $\overrightarrow{PQ}$: Express your answer both in diamond bracket form and in terms of the basic unit vectors.

9. $P = (1,2)$, $\quad Q = (3,4)$
10. $P = (-1,2)$, $\quad Q = (3,4)$
11. $P = (7,12)$, $\quad Q = (-5,12)$
12. $P = (0,-3)$, $\quad Q = (5,-12)$
13. $P = (2,-1,3)$, $\quad Q = (5,0,12)$
14. $P = (2,-1,1)$, $\quad Q = (-3,1,5)$
15. $P = (1,1,-2)$, $\quad Q = (-3,3,2)$
16. $P = (1,0,2)$, $\quad Q = (5,0,12)$

In Probs. 17–18 answer the questions: (a) Which vectors are parallel? (b) Which vectors have the same direction? (c) Which vectors have the opposite direction?

17. $\mathbf{r} = 3\mathbf{i} - 4\mathbf{j}$, $\mathbf{u} = -3\mathbf{i} + 4\mathbf{j}$, $\mathbf{v} = -21\mathbf{i} + 28\mathbf{j}$, $\mathbf{w} = -6\mathbf{i} + 8\mathbf{j}$

18. $\mathbf{r} = \langle 3, -1, -2 \rangle$, $\mathbf{u} = \langle 2, -1, 1 \rangle$, $\mathbf{v} = \langle 14, -7, 7 \rangle$, $\mathbf{w} = \langle -6, 2, 4 \rangle$

In Probs. 19–20 sketch $\mathbf{v}$ and $\mathbf{w}$. Then use the triangle law of addition to sketch $\mathbf{v} + \mathbf{w}$ and $\mathbf{v} - \mathbf{w}$.

19. $\mathbf{v} = \langle 3, 4 \rangle$, $\mathbf{w} = \langle -1, 2 \rangle$
20. $\mathbf{v} = \langle -2, 3 \rangle$, $\mathbf{w} = \langle 1, 1 \rangle$

21. $\mathbf{v} = \langle 1, 1, 0 \rangle$, $\mathbf{w} = \langle 0, 1, 1 \rangle$
22. $\mathbf{v} = \langle 1, 0, 0 \rangle$, $\mathbf{w} = \langle 1, 1, 2 \rangle$

In Probs. 23–26 find (a) the direction cosines and (b) the direction angles of the given vectors.

23. $\mathbf{v} = \langle 3, -4 \rangle$
24. $\mathbf{v} = -12\mathbf{i} + 13\mathbf{j}$

25. $\mathbf{v} = -3\mathbf{i} + 4\mathbf{j} + 12\mathbf{k}$
26. $\mathbf{v} = \langle 1, 2, -1 \rangle$

27. The vector $\mathbf{v}$ has the same direction as the unit vector $\mathbf{u} = \langle 2/3, 1/3, 2/3 \rangle$. Find the direction cosines of $\mathbf{v}$.

28. The vector $\mathbf{v}$ has the same direction as the vector $\mathbf{w} = \langle 1, -2, 3 \rangle$. Find the direction cosines of $\mathbf{v}$.

29. The vector $\mathbf{v}$ has the opposite direction of the vector $\mathbf{w} = \langle 12, 5 \rangle$. Find the direction cosines of $\mathbf{v}$.

30. Let $\mathbf{v}$, $\mathbf{w}$ be nonparallel vectors in 2–space. (a) What basic fact from Euclidean geometry is expressed by the relation $\|\mathbf{v} + \mathbf{w}\| < \|\mathbf{v}\| + \|\mathbf{w}\|$? Explain briefly. (b) Is the same true if the vectors are in 3–space? Explain briefly.

31. Let $\mathbf{v} = \langle 1, 2 \rangle$. Find all vectors $\mathbf{w}$ with length 2 that are (a) parallel to, (b) have the same direction as, (c) have the opposite direction of $\mathbf{v}$.

32. Let $\mathbf{v} = 3\mathbf{i} - 4\mathbf{j} + 12\mathbf{k}$. Find all unit vectors $\mathbf{u}$ that are (a) parallel to, (b) have the same direction as, (c) have the opposite direction of $\mathbf{v}$.

33. Find all scalars λ such that $\mathbf{v} = \langle 2, -4 \rangle$ and $\mathbf{w} = \langle \lambda, 1 - \lambda \rangle$ are (a) parallel, (b) have the same direction, (c) have the opposite directions.

34. Let P_1 and P_2 be distinct points in space, $\mathbf{v} = \overrightarrow{OP}_1$, and $\mathbf{w} = \overrightarrow{OP}_2$. Explain why the head of the vector $\overrightarrow{OP} = \mathbf{v} + \lambda(\mathbf{w} - \mathbf{v}) = (1 - \lambda)\mathbf{v} + \lambda\mathbf{w}$ for $0 \leq \lambda \leq 1$ traces out the line segment joining P_1 to P_2. Draw a picture that illustrates this situation.

35. Draw a parallelogram with adjacent sides representing vectors $\mathbf{v}$ and $\mathbf{w}$. Express the diagonals of the parallelogram in terms of $\mathbf{v}$ and $\mathbf{w}$.

36. Use vector methods to show that the diagonals of a parallelogram bisect each other. *Hint.* In the previous problem show that the vector from the common initial point of $\mathbf{v}$ and $\mathbf{w}$ to the point where the diagonals intersect can be expressed as $\mathbf{v} + \lambda(\mathbf{w} - \mathbf{v})$ and also as $\mu(\mathbf{v} + \mathbf{w})$ for certain positive scalars λ and μ.

37. Use vector methods to show that the line segment joining the midpoints of two sides of a triangle is parallel to, and half as long as, the third side.

38. A small boat is pulled to shore with a rope inclined at 30° to the horizontal. The tension (force) in the rope is 50 pounds and the water produces a horizontal drag force, opposite to the direction of motion, of 10 pounds. Find the resultant force on the boat.

39. An advertising sign hangs from a support as in Fig. 21. The sign weighs 500 pounds. Find the magnitude of the tension **T** in the cable and the magnitude of the reaction force **R** in the horizontal beam of the support. *Hint.* For the sign to remain at rest the resultant force on it must be zero.

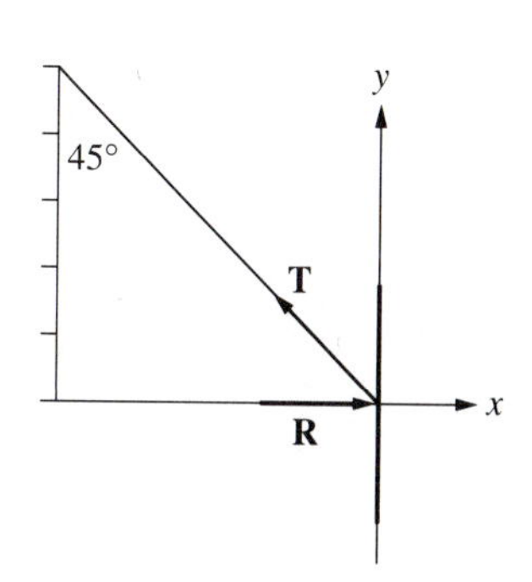

FIGURE 21

40. Let $\mathbf{v}$ and $\mathbf{w}$ be nonparallel, nonzero vectors in 2–space. Show that $a\mathbf{v} + b\mathbf{w} = \alpha\mathbf{v} + \beta\mathbf{w}$ holds if and only if $a = \alpha$ and $b = \beta$.

In Probs. 41–48 prove the given algebraic law by expressing each side in terms of components, simplifying, and comparing the results.

41. $\mathbf{v} + \mathbf{w} = \mathbf{w} + \mathbf{v}$,
42. $(\mathbf{u} + \mathbf{v}) + \mathbf{w} = \mathbf{u} + (\mathbf{v} + \mathbf{w})$
43. $\mathbf{v} + \mathbf{0} = \mathbf{v}$,
44. $\mathbf{v} + (-\mathbf{v}) = \mathbf{0}$
45. $\lambda(\mathbf{v} + \mathbf{w}) = \lambda\mathbf{v} + \lambda\mathbf{w}$,
46. $(\lambda + \mu)\mathbf{v} = \lambda\mathbf{v} + \mu\mathbf{w}$
47. $(\lambda\mu)\mathbf{v} = \lambda(\mu\mathbf{v}) = \mu(\lambda\mathbf{v})$
48. $||\lambda\mathbf{v}|| = |\lambda|\ ||\mathbf{v}||$

3.3 The Dot Product

The dot product (or scalar product) of two vectors plays an essential role in physical applications, particularly in problems involving force or motion. For example, the work done by a force in a particular direction is easily found by using the dot product. Also, the dot product provides the best means for finding the angle between two vectors (or two lines) in space. In particular, there is a convenient dot product test to determine whether two vectors are perpendicular. Finally, as we shall see, components and projections of one vector along another are easily found using dot products.

We begin with the algebraic definition of the dot product and then move on to the geometric description of the dot product that is the key to most applications. Formulas will be developed first for vectors in 3–space. They specialize to 2–space merely by suppressing the third components.

> **Definition** *The Dot Product*
> The **dot product** of $\mathbf{v} = <v_1, v_2, v_3>$ and $\mathbf{w} = <w_1, w_2, w_3>$ is $\mathbf{v} \cdot \mathbf{w} = v_1w_1 + v_2w_2 + v_3w_3$.

The corresponding formula for vectors $\mathbf{v} = <v_1, v_2>$ and $\mathbf{w} = <w_1, w_2>$ in 2–space is

$$\mathbf{v} \cdot \mathbf{w} = v_1w_1 + v_2w_2.$$

Notice that the dot product of two vectors is a *scalar*, not another vector.

EXAMPLE 1. Let $\mathbf{v} = <2, -1, 3>$ and $\mathbf{w} = <1, 3, 2>$. Then

$$\mathbf{v} \cdot \mathbf{w} = (2)(1) + (-1)(3) + (3)(2) = 2 - 3 + 6 = 5. \ \square$$

Algebraic properties of the dot product that follow easily from the definition are

$$\boxed{\begin{aligned} &\mathbf{v} \cdot \mathbf{w} = \mathbf{w} \cdot \mathbf{v} \\ &\mathbf{v} \cdot \mathbf{0} = \mathbf{0} \cdot \mathbf{v} = 0 \\ &(\lambda \mathbf{v}) \cdot \mathbf{w} = \lambda(\mathbf{v} \cdot \mathbf{w}) = \mathbf{v} \cdot (\lambda \mathbf{w}) \\ &\mathbf{u} \cdot (\mathbf{v} + \mathbf{w}) = \mathbf{u} \cdot \mathbf{v} + \mathbf{u} \cdot \mathbf{w} \end{aligned}}$$

These properties can be checked by expressing each side in terms of components, simplifying, and confirming that the final results are the same.

The following dot products involving the basic unit vectors $\mathbf{i} = <1,0,0>$, $\mathbf{j} = <0,1,0>$, $\mathbf{k} = <0,0,1>$ are worth recording:

$$\mathbf{i} \cdot \mathbf{i} = 1, \qquad \mathbf{j} \cdot \mathbf{j} = 1, \qquad \mathbf{k} \cdot \mathbf{k} = 1,$$
$$\mathbf{i} \cdot \mathbf{j} = 0, \qquad \mathbf{i} \cdot \mathbf{k} = 0, \qquad \mathbf{j} \cdot \mathbf{k} = 0.$$

EXAMPLE 2. Let $\mathbf{v} = 3\mathbf{i} - 4\mathbf{j}$ and $\mathbf{w} = 2\mathbf{i} + \mathbf{k}$. Find $\mathbf{v} \cdot \mathbf{w}$.

Solution 1. Since $\mathbf{v} = 3\mathbf{i} - 4\mathbf{j} + 0\mathbf{k} = <3, -4, 0>$ and $\mathbf{w} = 2\mathbf{i} + 0\mathbf{j} + 1\mathbf{k} = <2,0,1>$, the definition of the dot product gives

$$\mathbf{v} \cdot \mathbf{w} = (3)(2) + (-4)(0) + (0)(1) = 6.$$

Solution 2. Use the algebraic properties of the dot product to find

$$\mathbf{v} \cdot \mathbf{w} = (3\mathbf{i} - 4\mathbf{j}) \cdot (2\mathbf{i} + \mathbf{k}) = 6\mathbf{i} \cdot \mathbf{i} + 3\mathbf{i} \cdot \mathbf{k} - 8\mathbf{j} \cdot \mathbf{i} - 4\mathbf{j} \cdot \mathbf{k} = 6 + 0 - 0 - 0 = 6.$$

In most cases the first method of solution is quicker. $\square$

Notice that

$$\mathbf{v} = <a, b, c> \qquad \Rightarrow \qquad \mathbf{v} \cdot \mathbf{v} = a^2 + b^2 + c^2 = ||\mathbf{v}||^2.$$

So the dot product and the length of a vector are related by

$$\boxed{\mathbf{v} \cdot \mathbf{v} = ||\mathbf{v}||^2.}$$

EXAMPLE 3. Let $\mathbf{v} = <2, -5, 1>$. Then $||\mathbf{v}||^2 = \mathbf{v} \cdot \mathbf{v} = 4 + 25 + 1 = 30$. $\square$

Next we prepare for geometric and physical descriptions of the dot product. From the properties of dot products given previously,

$$||\mathbf{v} - \mathbf{w}||^2 = (\mathbf{v} - \mathbf{w}) \cdot (\mathbf{v} - \mathbf{w}) = \mathbf{v} \cdot \mathbf{v} - \mathbf{v} \cdot \mathbf{w} - \mathbf{w} \cdot \mathbf{v} + \mathbf{w} \cdot \mathbf{w},$$

which simplifies to

$$\|\mathbf{v} - \mathbf{w}\|^2 = \|\mathbf{v}\|^2 + \|\mathbf{w}\|^2 - 2\mathbf{v} \cdot \mathbf{w}.$$

The vectors **v**, **w**, and $\mathbf{v} - \mathbf{w}$ form the triangle in Fig. 1. The lengths of the three sides are $\|\mathbf{v}\|$, $\|\mathbf{w}\|$, and $\|\mathbf{v} - \mathbf{w}\|$. In Fig. 1, θ is the angle between the vectors **v** and **w**. In general, the *angle* between two nonzero vectors is the angle θ with $0 \leq \theta \leq \pi$ formed by the vectors when they are drawn from a common point. According to the law of cosines,

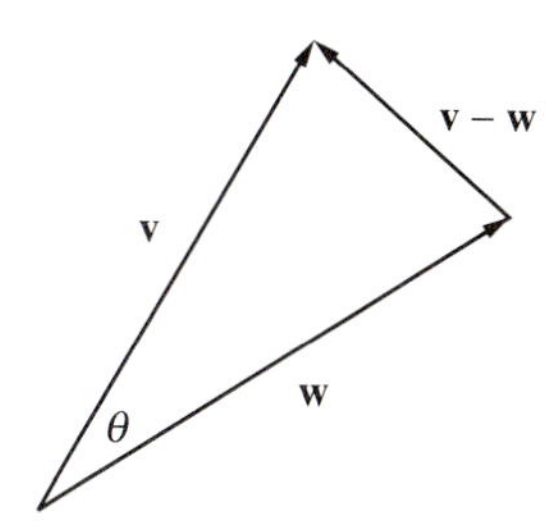

FIGURE 1

$$\|\mathbf{v} - \mathbf{w}\|^2 = \|\mathbf{v}\|^2 + \|\mathbf{w}\|^2 - 2\|\mathbf{v}\|\;\|\mathbf{w}\| \cos\theta.$$

A comparison of the two formulas for $\|\mathbf{v} - \mathbf{w}\|^2$ reveals that

$$\boxed{\mathbf{v} \cdot \mathbf{w} = \|\mathbf{v}\|\;\|\mathbf{w}\| \cos\theta.}$$

Therefore, if **v** and **w** are any nonzero vectors, then

$$\boxed{\cos\theta = \frac{\mathbf{v} \cdot \mathbf{w}}{\|\mathbf{v}\|}\|\mathbf{w}\|.}$$

This formula is very useful for finding angles between vectors. Here is a typical example.

EXAMPLE 4. Let $\mathbf{v} = \langle 1, 2, 3 \rangle$ and $\mathbf{w} = \langle -4, 6, 2 \rangle$. Find the angle θ between **v** and **w**.

Solution. In this case,

$$\cos\theta = \frac{\mathbf{v} \cdot \mathbf{w}}{\|\mathbf{v}\|\;\|\mathbf{w}\|} = \frac{14}{\sqrt{14} \cdot 2\sqrt{14}} = \frac{1}{2}, \qquad \theta = 60^\circ = \frac{\pi}{3} \text{ rad. } \square$$

The dot product provides a quick test to find out whether the angle θ between two nonzero vectors **v** and **w** is acute or obtuse:

$$\mathbf{v} \cdot \mathbf{w} > 0 \quad \Rightarrow \quad \cos\theta > 0 \quad \Rightarrow \quad 0 \leq \theta < \frac{\pi}{2},$$

$$\mathbf{v} \cdot \mathbf{w} < 0 \quad \Rightarrow \quad \cos\theta < 0 \quad \Rightarrow \quad \frac{\pi}{2} < \theta \leq \pi.$$

In Ex. 4, the dot product is positive and the angle is acute.

Orthogonal Vectors

Nonzero vectors **v** and **w** are **orthogonal** if they are perpendicular in the usual sense. This means that the angle θ between **v** and **w** is 90° or $\pi/2$ radians;

hence, $\cos\theta = 0$. Also, by definition, the zero vector is orthogonal to every vector. We write $\mathbf{v} \perp \mathbf{w}$ to indicate that $\mathbf{v}$ is orthogonal to $\mathbf{w}$. For example, $\mathbf{i} \perp \mathbf{j}$, $\mathbf{i} \perp \mathbf{k}$, and $\mathbf{j} \perp \mathbf{k}$.

The dot product provides a convenient test for orthogonality. Since $\mathbf{v} \cdot \mathbf{w} = \|\mathbf{v}\|\,\|\mathbf{w}\| \cos\theta$,

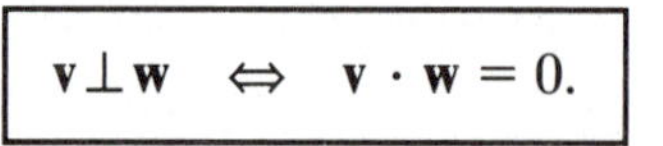

$$\mathbf{v} \perp \mathbf{w} \quad \Leftrightarrow \quad \mathbf{v} \cdot \mathbf{w} = 0.$$

EXAMPLE 5. Let $\mathbf{v} = 3\mathbf{i} - 5\mathbf{j} + 2\mathbf{k}$ and $\mathbf{w} = 4\mathbf{i} + 2\mathbf{j} - \mathbf{k}$. Then

$$\mathbf{v} \cdot \mathbf{w} = 12 - 10 - 2 = 0.$$

So $\mathbf{v} \perp \mathbf{w}$. The angle between $\mathbf{v}$ and $\mathbf{w}$ is 90°. □

EXAMPLE 6. In the xy–plane, let $\mathbf{v} = <2, 3>$ and $\mathbf{w} = <-3, 2>$. Then

$$\mathbf{v} \cdot \mathbf{w} = (2)(-3) + (3)(2) = 0.$$

Hence, $\mathbf{v} \perp \mathbf{w}$. □

Compare the components of the two vectors in Ex. 6. Note the switch and sign change. In the same way, for any a and b,

$$\mathbf{v} = <a, b> \quad \text{and} \quad \mathbf{w} = <-b, a> \quad \Rightarrow \quad \mathbf{v} \perp \mathbf{w}.$$

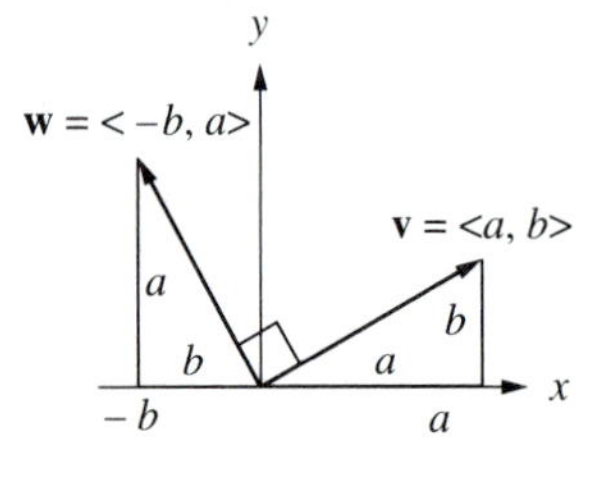

FIGURE 2

Figure 2 illustrates the vectors $\mathbf{v}$ and $\mathbf{w}$ when $\mathbf{v}$ is in the first quadrant. In the figure, and in general, $\mathbf{w}$ is obtained from $\mathbf{v}$ by a counterclockwise rotation of 90°. (A similar idea works in 3–space. See the problems.)

Components of Vectors

Earlier, we called a, b, and c the components of the vector $\mathbf{v} = <a, b, c>$. Now the idea of a component is broadened. We shall define the component of $\mathbf{v}$ along any nonzero vector $\mathbf{w}$.

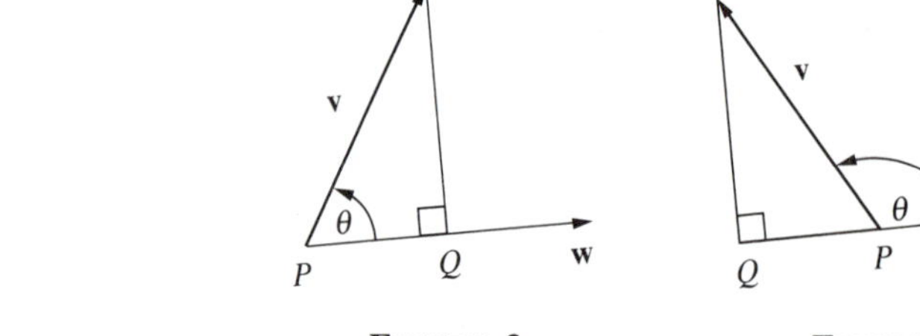

FIGURE 3

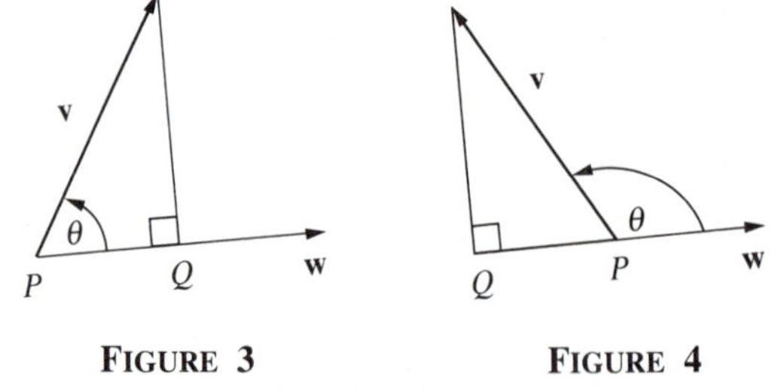

FIGURE 4

Figures 3 and 4 serve as guides and motivate the following definition.

> **Definition** *Component of One Vector Along Another*
> The **component** of $\mathbf{v}$ along $\mathbf{w} \neq \mathbf{0}$ is
> $$\text{comp}_{\mathbf{w}}\, \mathbf{v} = \frac{\mathbf{v} \cdot \mathbf{w}}{\|\mathbf{w}\|} = \|\mathbf{v}\| \cos\theta.$$

The two formulas for the component are equal because $\mathbf{v} \cdot \mathbf{w} = \|\mathbf{v}\|\,\|\mathbf{w}\| \cos\theta$. Observe that

$$\text{comp}_{\mathbf{w}}\, \mathbf{v} = \|\mathbf{v}\| \cos\theta = |PQ| \qquad \text{in Fig. 3,}$$

$$\text{comp}_{\mathbf{w}}\, \mathbf{v} = \|\mathbf{v}\| \cos\theta = -\,\|\mathbf{v}\| \cos(\pi - \theta) = -\,|PQ| \qquad \text{in Fig. 4.}$$

Thus, $\text{comp}_{\mathbf{w}}\, \mathbf{v} > 0$ if θ is acute, and $\text{comp}_{\mathbf{w}}\, \mathbf{v} < 0$ if θ is obtuse.

Let $\mathbf{v} = < a, b, c > = a\mathbf{i} + b\mathbf{j} + c\mathbf{k}$. Then, from Fig. 5 or by a short calculation using the definition of a component,

$$\text{comp}_{\mathbf{i}}\, \mathbf{v} = a, \qquad \text{comp}_{\mathbf{j}}\, \mathbf{v} = b, \qquad \text{comp}_{\mathbf{k}}\, \mathbf{v} = c.$$

So a, b, and c are the components of $\mathbf{v}$ along the basic unit vectors $\mathbf{i}$, $\mathbf{j}$, and $\mathbf{k}$.

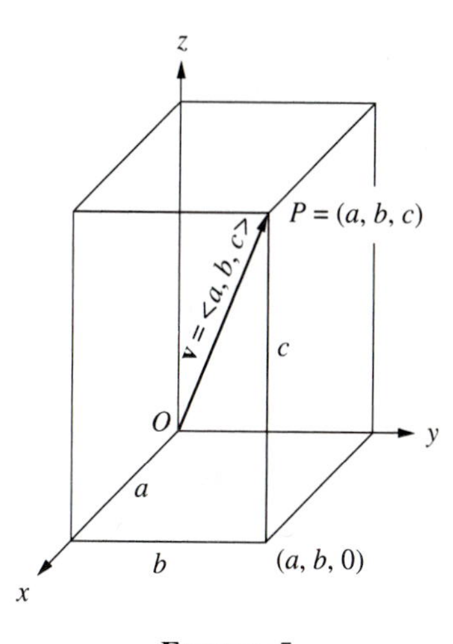

FIGURE 5

EXAMPLE 7. Find the component of $\mathbf{v} = < 4, 5, -2 >$ along $\mathbf{w} = < 4, 0, 3 >$.

Solution. Since $\mathbf{v} \cdot \mathbf{w} = 10$ and $\|\mathbf{w}\| = 5$,

$$\text{comp}_{\mathbf{w}}\, \mathbf{v} = \frac{\mathbf{v} \cdot \mathbf{w}}{\|\mathbf{w}\|} = \frac{10}{5} = 2. \ \square$$

Figures 3 and 4 show that the component of $\mathbf{v}$ along $\mathbf{w}$ depends on the direction of $\mathbf{w}$ but not on its length. The same conclusion follows from

$$\text{comp}_{\mathbf{w}}\, \mathbf{v} = \frac{\mathbf{v} \cdot \mathbf{w}}{\|\mathbf{w}\|} = \mathbf{v} \cdot \frac{\mathbf{w}}{\|\mathbf{w}\|} = \mathbf{v} \cdot \mathbf{u},$$

where $\mathbf{u} = \mathbf{w}/\|\mathbf{w}\|$ is the unit vector in the direction of $\mathbf{w}$. Thus,

$$\boxed{\text{comp}_{\mathbf{w}}\, \mathbf{v} = \text{comp}_{\mathbf{u}}\, \mathbf{v} = \mathbf{v} \cdot \mathbf{u}, \qquad \mathbf{u} = \frac{\mathbf{w}}{\|\mathbf{w}\|}.}$$

For this reason, $\text{comp}_{\mathbf{w}}\, \mathbf{v}$ is also called the component of $\mathbf{v}$ *in the direction of* $\mathbf{w}$.

Components of vectors often have important physical interpretations. Here is an example. An object moves in the xy–plane. Suppose that its velocity is $\mathbf{v} = < 3, 4 >$ when it passes through the point $P = (2, 1)$. In Fig. 6,

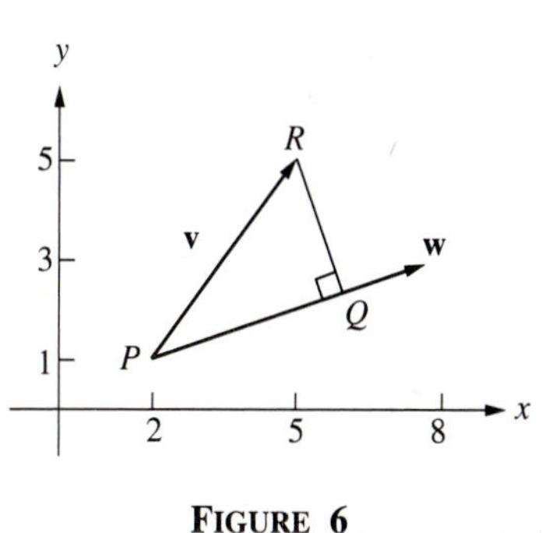

FIGURE 6

$\mathbf{v} = <3,4>$ is represented by the vector $\overrightarrow{PR}$ from $P = (2,1)$ to $R = (5,5)$. The component of $\mathbf{v} = <3,4>$ in the direction of $\mathbf{w} = <6,2>$ is

$$\text{comp}_{\mathbf{w}}\, \mathbf{v} = \frac{\mathbf{v} \cdot \mathbf{w}}{||\mathbf{w}||} = \frac{26}{\sqrt{40}} = \frac{13}{\sqrt{10}} \approx 4.1.$$

If you ride on the object and shine a flashlight perpendicularly onto $\mathbf{w}$, then the spot of light will move along $\mathbf{w}$ at about 4.1 ft/sec as the object passes through the point P in Fig. 6.

Projections of Vectors

Projections and components of vectors are closely related. Projections are components enhanced by directions. So, projections are vectors, whereas components are scalars. There are several ways to define and calculate projections. We begin with the one that has the clearest geometric meaning. The formulas used for most computational purposes come after.

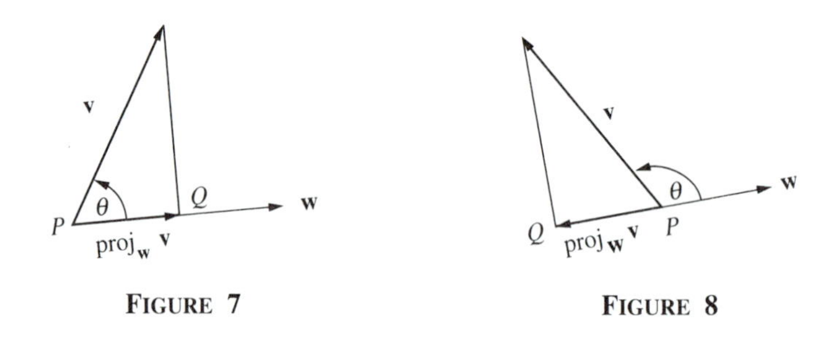

FIGURE 7 FIGURE 8

With the notation in Figs. 7 and 8, the **projection of v along w** (or in the direction of $\mathbf{w}$) is

$$\boxed{\text{proj}_{\mathbf{w}}\, \mathbf{v} = \overrightarrow{PQ}.}$$

If $\mathbf{v} = <a,b,c> = a\mathbf{i} + b\mathbf{j} + c\mathbf{k}$, then from Fig. 5 and the foregoing definition,

$$\text{proj}_{\mathbf{i}}\, \mathbf{v} = a\mathbf{i}, \qquad \text{proj}_{\mathbf{j}}\, \mathbf{v} = b\mathbf{j}, \qquad \text{proj}_{\mathbf{k}}\, \mathbf{v} = c\mathbf{k}.$$

Thus, $\mathbf{v}$ is the sum of its projections in the three coordinate directions.

Figures 7 and 8 reveal the close connection between the component and the projection of $\mathbf{v}$ along $\mathbf{w}$:

$$\boxed{\text{proj}_{\mathbf{w}}\, \mathbf{v} = (\text{comp}_{\mathbf{w}}\, \mathbf{v})\frac{\mathbf{w}}{||\mathbf{w}||}, \qquad ||\text{proj}_{\mathbf{w}}\, \mathbf{v}|| = |\text{comp}_{\mathbf{w}}\, \mathbf{v}|.}$$

Furthermore, since $\text{comp}_{\mathbf{w}}\, \mathbf{v} = \mathbf{v} \cdot \mathbf{w}/||\mathbf{w}||$, we also have

$$\boxed{\text{proj}_{\mathbf{w}} \mathbf{v} = \frac{\mathbf{v} \cdot \mathbf{w}}{||\mathbf{w}||^2} \mathbf{w}.}$$

The last result is expressed more compactly in terms of the unit vector in the same direction as **w**:

$$\boxed{\text{proj}_{\mathbf{w}} \mathbf{v} = (\mathbf{v} \cdot \mathbf{u})\mathbf{u}, \quad \text{where} \quad \mathbf{u} = \frac{\mathbf{w}}{||\mathbf{w}||}.}$$

Each of the formulas for $\text{proj}_{\mathbf{w}} \mathbf{v}$ has advantages in particular situations.

EXAMPLE 8. Find the projection of $\mathbf{v} = <2,5,3>$ along $\mathbf{w} = <1,2,2>$.

Solution. Since $\mathbf{v} \cdot \mathbf{w} = 18$ and $||\mathbf{w}|| = 3$,

$$\text{proj}_{\mathbf{w}} \mathbf{v} = \frac{\mathbf{v} \cdot \mathbf{w}}{||\mathbf{w}||^2} \mathbf{w} = \frac{18}{9} \mathbf{w} = 2\mathbf{w} = <2,4,4>. \ \square$$

Return to Fig. 6 and the object that passes through $P = (2,1)$ with velocity $\mathbf{v} = <3,4>$. The projection of **v** in the direction of $\mathbf{w} = <6,2>$ is

$$\text{proj}_{\mathbf{w}} \mathbf{v} = \frac{\mathbf{v} \cdot \mathbf{w}}{||\mathbf{w}||^2} \mathbf{w} = \frac{26}{40} <6,2> = <3.9,1.3>.$$

If you ride on the object and shine a flashlight perpendicularly onto **w**, then the spot of light will move along **w** with velocity $\text{proj}_{\mathbf{w}} \mathbf{v} = <3.9,1.3>$ as the object passes through the point P in Fig. 6.

Orthogonal Projections

Frequently, we need to express a vector in terms of its components or projections in two orthogonal (perpendicular) directions. We show how to do that next.

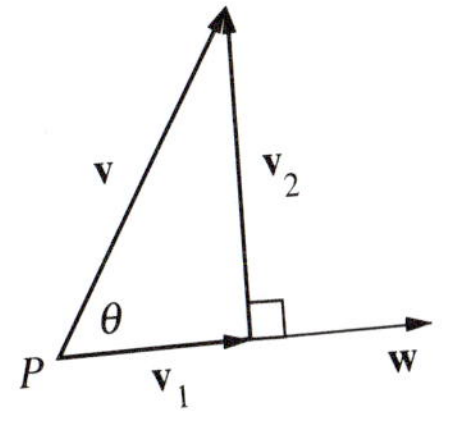

FIGURE 9

Figure 9 shows a vector **v**, its projection $\mathbf{v}_1 = \text{proj}_{\mathbf{w}} \mathbf{v} = (\mathbf{v} \cdot \mathbf{w})\mathbf{w}/||\mathbf{w}||^2$ in the direction of a nonzero vector **w**, and the vector $\mathbf{v}_2 = \mathbf{v} - \mathbf{v}_1$. The figure suggests that $\mathbf{v}_2 \perp \mathbf{w}$. This is true. Since $||\mathbf{w}||^2 = \mathbf{w} \cdot \mathbf{w}$,

$$\mathbf{v}_2 \cdot \mathbf{w} = (\mathbf{v} - \mathbf{v}_1) \cdot \mathbf{w} = \left(\mathbf{v} - \frac{\mathbf{v} \cdot \mathbf{w}}{||\mathbf{w}||^2}\mathbf{w}\right) \cdot \mathbf{w} = \mathbf{v} \cdot \mathbf{w} - \mathbf{v} \cdot \mathbf{w} = 0.$$

We call $\mathbf{v}_2$ the **projection of v orthogonal to w**. The **component of v orthogonal to w** is $||\mathbf{v}_2||$, the length of $\mathbf{v}_2$. Figure 9 displays the basic relations between $\mathbf{v}_1$, $\mathbf{v}_2$, and **v**:

$$\mathbf{v} = \mathbf{v}_1 + \mathbf{v}_2, \quad \mathbf{v}_1 \perp \mathbf{v}_2, \quad ||\mathbf{v}||^2 = ||\mathbf{v}_1||^2 + ||\mathbf{v}_2||^2,$$

$$||\mathbf{v}_1|| = ||\mathbf{v}|| \cos\theta, \quad ||\mathbf{v}_2|| = ||\mathbf{v}|| \sin\theta.$$

EXAMPLE 9. As in Ex. 8, let $\mathbf{v} = < 2,5,3 >$ and $\mathbf{w} = < 1,2,2 >$. Find the projection of $\mathbf{v}$ orthogonal to $\mathbf{w}$ and the component of $\mathbf{v}$ orthogonal to $\mathbf{w}$.

Solution. In Ex. 8 we found that the projection of $\mathbf{v}$ along $\mathbf{w}$ is $\mathbf{v}_1 = < 2,4,4 >$. So the projection of $\mathbf{v}$ orthogonal to $\mathbf{w}$ is

$$\mathbf{v}_2 = \mathbf{v} - \mathbf{v}_1 = < 2,5,3 > - < 2,4,4 > = < 0,1,-1 >.$$

Check that $\mathbf{v}_1 \cdot \mathbf{v}_2 = \mathbf{0}$, as it must be. The component of $\mathbf{v}$ orthogonal to $\mathbf{w}$ is $||\mathbf{v}_2|| = \sqrt{2}$. □

Work

We start with the simplest physical situation. When a constant force of magnitude F moves an object a distance D along a straight path in the direction of the force, then the work done by the force is $W = FD$. If the force is not in the direction of motion, only the component of the force in the direction of motion contributes to the work, as we explain now.

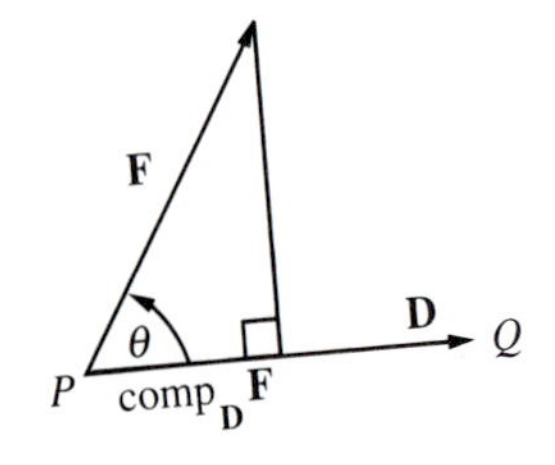

FIGURE 10

In Fig. 10, a force $\mathbf{F}$ acts on an object as it moves along the straight path from P to Q. The vector $\mathbf{D} = \overrightarrow{PQ}$ is called a **displacement vector**. Let θ be the angle between $\mathbf{F}$ and $\mathbf{D}$. In Fig. 10, θ is acute. The component of force in the direction of motion is

$$\text{comp}_{\mathbf{D}}\, \mathbf{F} = \frac{\mathbf{F} \cdot \mathbf{D}}{||\mathbf{D}||}.$$

By definition, the **work** done by $\mathbf{F}$ through the displacement $\mathbf{D}$ is

$$\boxed{W = (\text{comp}_{\mathbf{D}}\, \mathbf{F})\, ||\mathbf{D}|| = \mathbf{F} \cdot \mathbf{D} = ||\mathbf{F}||\, ||\mathbf{D}|| \cos\theta.}$$

This reduces to the familiar formula $W = FD$ with $F = ||\mathbf{F}||$ and $D = ||\mathbf{D}||$ when the force acts in the direction of motion because then $\theta = 0$ and $\cos\theta = 1$. In Fig. 10, where θ is acute, the force acts to assist the motion and the work is positive. If θ is obtuse, then the force acts to oppose the motion and the work is negative.

EXAMPLE 10. A force $\mathbf{F} = < 2,-3,4 >$ moves an object along the straight path from $P = (2,1,3)$ to $Q = (5,-1,4)$. Find the work done. Use English units.

Solution. The force $\mathbf{F} = < 2,-3,4 >$ has units of pounds and the displacement vector $\mathbf{D} = \overrightarrow{PQ} = < 3,-2,1 >$ has units of feet. So the work done by the force is

$$W = \mathbf{F} \cdot \mathbf{D} = 6 + 6 + 4 = 16 \text{ ft–lbs.} \; □$$

Again suppose that a force $\mathbf{F}$ acts on an object as it moves along the straight path from P to Q in Fig. 10. The displacement vector is $\mathbf{D} = \overrightarrow{PQ}$. The projection of $\mathbf{F}$ in the direction of motion is

$$\text{proj}_{\mathbf{D}}\, \mathbf{F} = \frac{\mathbf{F} \cdot \mathbf{D}}{||\mathbf{D}||^2}\mathbf{D}.$$

Dot each side of this equation with $\mathbf{D}$ to find

$$(\text{proj}_{\mathbf{D}}\,\mathbf{F}) \cdot \mathbf{D} = \mathbf{F} \cdot \mathbf{D}.$$

Consequently, the work done by $\mathbf{F}$ is

$$W = \mathbf{F} \cdot \mathbf{D} = (\text{proj}_{\mathbf{D}}\,\mathbf{F}) \cdot \mathbf{D}.$$

For this reason, $\text{proj}_{\mathbf{D}}\,\mathbf{F}$ is called the **effective force** in the direction of motion. The work done by $\mathbf{F}$ depends on $\mathbf{F}$ only through its projection along the direction of motion.

Horizontal and Vertical Projections in 3–Space

Horizontal and vertical projections of vectors are very useful for describing directions of vectors and for visualizing motion in 3–space. Figure 11 shows what we have in mind. As usual, the *xy*–plane is horizontal and the *z*–axis is vertical.

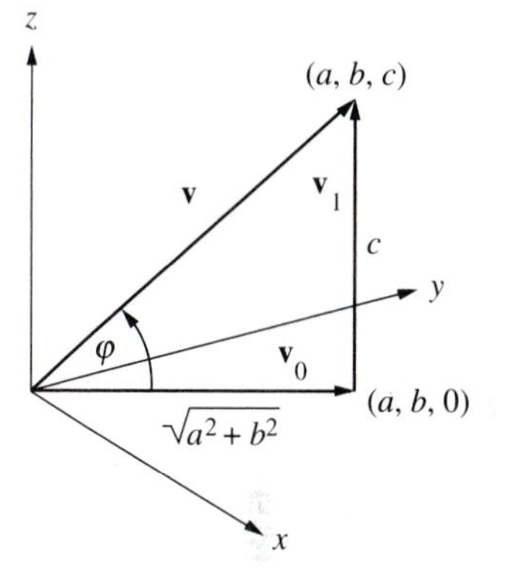

FIGURE 11

The **horizontal projection** of $\mathbf{v} = < a, b, c >$ is $\mathbf{v}_0 = < \mathrm{a}, \mathrm{b}, 0 >$. The **vertical projection** is $\mathbf{v}_1 = < 0, 0, c > = c\mathbf{k}$. Observe that $\mathbf{v}_1$ is the projection of $\mathbf{v}$ along $\mathbf{k}$ and that $\mathbf{v}_0$ is the projection of $\mathbf{v}$ orthogonal to $\mathbf{k}$. In Fig. 11,

$$\mathbf{v} = \mathbf{v}_0 + \mathbf{v}_1, \quad \mathbf{v}_0 \perp \mathbf{v}_1, \qquad ||\mathbf{v}||^2 = ||\mathbf{v}_0||^2 + ||\mathbf{v}_1||^2,$$
$$||\mathbf{v}_0|| = ||\mathbf{v}|| \cos \varphi, \qquad ||\mathbf{v}_1|| = ||\mathbf{v}|| \sin \varphi.$$

If $\mathbf{v}$ is a velocity vector, then $\mathbf{v}_0$ is the horizontal velocity (the velocity relative to the *xy*–plane) and $||\mathbf{v}_0|| = \sqrt{a^2 + b^2}$ is the horizontal speed. Similarly, $\mathbf{v}_1$ is the vertical velocity and $||\mathbf{v}_1|| = |c|$ is the vertical speed.

In Fig. 11, φ is the **angle of inclination** of $\mathbf{v}$. It is the angle between $\mathbf{v}$ and $\mathbf{v}_0$. The **slope (relative to the *xy*–plane)** of $\mathbf{v}$ is

$$m = \tan \varphi = \frac{c}{||\mathbf{v}_0||} = \frac{c}{\sqrt{a^2 + b^2}}.$$

Since this kind of slope is "rise over run," it is similar to the slope of a line in the *xy*–plane. But it is not the same. Rather, we are using slope in the popular sense you might apply to a ladder leaning against a house or a trail or road winding through the mountains. The slope measures steepness relative to the horizontal plane.

EXAMPLE 11. Establish an *xyz*–coordinate system with the *x*–axis pointing east, the *y*–axis pointing north, and the *z*–axis straight up. Describe the direction of flight of a ball as it passes through the origin with velocity $\mathbf{v} = < 1, 1, \sqrt{6}/3 >$.

Solution. The horizontal projection of $\mathbf{v}$ is $\mathbf{v}_0 = < 1, 1, 0 >$ and its vertical projection is $\mathbf{v}_1 = < 0, 0, \sqrt{6}/3 >$. The horizontal projection $\mathbf{v}_0$ points northeast. The slope m and the angle of inclination φ of the ball (relative to the *xy*–plane) satisfy

$$m = \tan \varphi = \frac{(\sqrt{6}/3)}{||\mathbf{v}_0||} = \frac{1}{\sqrt{3}}.$$

Consequently, $\varphi = 30^\circ$ and the ball is moving northeast with angle of inclination of 30°. □

PROBLEMS

In Probs. 1–8 find the dot product of the given vectors.

1. $<1,2>, \quad <3,4>$
2. $<-1,3>, \quad <4,-2>$
3. $3\mathbf{i} - 2\mathbf{j}, \quad -\mathbf{i} + 4\mathbf{j}$
4. $\mathbf{i} + \mathbf{j}, \quad \mathbf{i} - \mathbf{j}$
5. $<2,-1,3>, <-3,6,4>$
6. $<-2,0,5>, \quad <1,1,1>$
7. $3\mathbf{i} - 2\mathbf{j} + 5\mathbf{k}, \quad \mathbf{i} - \mathbf{j} + \mathbf{k}$
8. $3\mathbf{i} - 2\mathbf{j} + 5\mathbf{k}, \quad \mathbf{i} - \mathbf{j} + \mathbf{k}$

In Probs. 9–16 find the angle between the given vectors.

9. $<1,2>, \quad <-3,4>$
10. $<7,-1>, \quad <3,4>$
11. $\sqrt{3}\,\mathbf{i} + \mathbf{j}, \quad \mathbf{i} + \sqrt{3}\,\mathbf{j}$
12. $\mathbf{i} + \mathbf{j}, \quad \mathbf{i} - 2\mathbf{j}$
13. $<5,18,1>, \quad <-20,26,-18>$
14. $<1,0,2>, \quad <3,0,1>$
15. $\mathbf{i} + \mathbf{j} - \mathbf{k}, \quad 5\mathbf{i} - \mathbf{j} + 7\mathbf{k}$
16. $\mathbf{i} + \mathbf{j} + \mathbf{k}, \quad \mathbf{i} + \mathbf{j} + 2\mathbf{k}$
17. A rectangle has dimensions 1 by 2. Find the angles between its sides and a diagonal.
18. A rectangular box has dimensions 1 by 2 by 3. Find the angles between its edges and a diagonal.
19. Find the angle between the diagonal of a cube and any one of its edges.
20. Use a dot product argument to show that the diagonals of a parallelogram are perpendicular if and only if all the sides of the parallelogram are equal. (A parallelogram with equal sides is called a rhombus.)
21. Use a dot product argument to show that an angle inscribed in a semicircle is a right angle. (Such an angle has its vertex on the circle and its sides pass through the ends of a diameter of the circle.)
22. Use a dot product argument to show that an equilateral (all sides equal) triangle also has all angles equal.
23. Let $\overline{AC}$ be the diameter of a sphere and let B be a point on the sphere different from A and C. Show that $\overline{BA}$ is perpendicular to $\overline{BC}$.
24. Let $\mathbf{v}$ and $\mathbf{w}$ be unit vectors along the sides of a given angle. Show that the vector $\mathbf{v} + \mathbf{w}$ bisects the angle.

In Probs. 25–32 find (a) $\text{comp}_{\mathbf{w}}\,\mathbf{v}$ and (b) $\text{proj}_{\mathbf{w}}\,\mathbf{v}$.

25. $<1,2>, \quad <-3,4>$
26. $<7,-1>, \quad <3,4>$
27. $\sqrt{3}\,\mathbf{i} + \mathbf{j}, \quad \mathbf{i} + \sqrt{3}\,\mathbf{j}$
28. $\mathbf{i} + \mathbf{j}, \quad \mathbf{i} - 2\mathbf{j}$
29. $<5,18,1>, \quad <-20,26,-18>$
30. $<1,0,2>, \quad <3,0,1>$
31. $\mathbf{i} + \mathbf{j} - \mathbf{k}, \quad 5\mathbf{i} - \mathbf{j} + 7\mathbf{k}$
32. $\mathbf{i} + \mathbf{j} + \mathbf{k}, \quad \mathbf{i} + \mathbf{j} + 2\mathbf{k}$

33. Let $\mathbf{v} = < 3, -2,4 >$. Find two (nonzero) vectors in space that are orthogonal to $\mathbf{v}$.

34. Let $\mathbf{v} = 2\mathbf{i} - \mathbf{j} + 3\mathbf{k}$. Find two (nonzero) vectors in space that are orthogonal to $\mathbf{v}$.

35. Let $\mathbf{v} = < 1, -2,3 >$ and $\mathbf{w} = < -1,2, -2 >$. Find the projection of $\mathbf{v}$ orthogonal to $\mathbf{w}$.

36. Let $\mathbf{v} = -3\mathbf{i} + 4\mathbf{j}$ and $\mathbf{w} = 7\mathbf{i} - \mathbf{k}$. Find the projection of $\mathbf{v}$ orthogonal to $\mathbf{w}$.

37. Let $\mathbf{v} = < a, b >$. Find the components of the vector $\mathbf{w}$ that is obtained from $\mathbf{v}$ by a clockwise rotation of 90° in the xy–plane.

38. Let $\mathbf{v} = < a, b, c >$ and $\mathbf{w} = < r, s, t >$. Find the projection of $\mathbf{v}$ orthogonal to $\mathbf{w}$.

39. Show: (a) If $\mathbf{v}$ and $\mathbf{w}$ point in the same direction, then $\text{comp}_{\mathbf{w}}\, \mathbf{v} = ||\mathbf{v}||$. (b) If $\mathbf{v}$ and $\mathbf{w}$ are orthogonal, then $\text{comp}_{\mathbf{w}}\, \mathbf{v} = 0$.

40. Express $||\mathbf{v} + \mathbf{w}||^2 + ||\mathbf{v} - \mathbf{w}||^2$ in terms of the lengths of $\mathbf{v}$ and $\mathbf{w}$. The resulting equation is called the **parallelogram law**. Give a geometric interpretation of the result.

41. Establish the **Schwarz inequality** $|\mathbf{v} \cdot \mathbf{w}| \le ||\mathbf{v}||\ ||\mathbf{w}||$.

42. The dot product determines the lengths of vectors through $||\mathbf{v}||^2 = \mathbf{v} \cdot \mathbf{v}$. Show that the dot product is determined by lengths through

$$\mathbf{v} \cdot \mathbf{w} = \frac{1}{4}(||\mathbf{v} + \mathbf{w}||^2 - ||\mathbf{v} - \mathbf{w}||^2).$$

43. For any vector $\mathbf{v} = < a, b, c > = a\mathbf{i} + b\mathbf{j} + c\mathbf{k}$, show that

$$\mathbf{v} \cdot \mathbf{i} = a, \qquad \mathbf{v} \cdot \mathbf{j} = b, \qquad \mathbf{v} \cdot \mathbf{k} = \text{c}.$$

Conclude that

$$\mathbf{v} = (\mathbf{v} \cdot \mathbf{i})\mathbf{i} + (\mathbf{v} \cdot \mathbf{j})\mathbf{j} + (\mathbf{v} \cdot \mathbf{k})\mathbf{k}.$$

44. Suppose a ball has speed of 80 mph, is moving northwest over the xy–plane, and has angle of inclination of 30° relative to that plane. Find the velocity vector $\mathbf{v}$ of the ball and its horizontal and vertical projections. (Use coordinates as in Ex. 11.)

45. Repeat the preceding problem when the angle of inclination is 45° and the ball is thrown in the direction 30° north of west.

46. A wagon is pulled 80 feet along a level sidewalk by a constant force of 5 pounds that acts at (a) 30° and (b) 45° to the horizontal. Find the work done by the force and the effective force in the direction of motion.

47. A boat is pulled 50 feet with a rope inclined at 20° to the water. If the tension (force) in the rope is 8 pounds, find the work done and the effective force in the direction of motion.

48. A constant force $\mathbf{F}$ acts on an object as it makes one circuit around a triangle. Find the work done by $\mathbf{F}$.

49. A person wants to row across a river to a point on the bank directly opposite. The river has straight banks that are 50 yards apart and a constant current (parallel to the banks) that flows at 1/2 ft/sec. The boat can be rowed at 3 ft/sec. At what upstream angle relative to the bank should the boat be

rowed? *Hint.* If **v** is the velocity of the boat through the water and **w** the velocity vector of the current, then it is an experimental fact that the velocity of the boat relative to the river bottom is **v** + **w**.

50. An airliner cruises at 30,000 feet with a speed of 500 mph and is headed southeast when it enters a jet stream blowing at 80 mph toward the northeast. Set up coordinates with the positive x–axis pointing east. At what angle relative to the x–axis should the plane head in order to maintain its southwest course over the ground? *Hint.* In the previous hint replace boat with plane, current with wind speed, and river bottom with ground.

51. If the airliner in the previous problem does not change course upon entering the jet stream, find the velocity of the plane relative to the ground. Find the speed and direction of the plane. Find the speed and direction of the plane.

3.4 The Cross Product

The cross product (or vector product) of two vectors has many applications in mechanics, electromagnetic theory, and other branches of physics and engineering. For example, cross products are used to express various physical properties related to rotary motion. When a rigid body rotates around an axis, the velocity of each particle can be expressed as a cross product. Other applications of cross products pertain to the moment and torque of a force and to various properties of magnetic fields. The Coriolis force, which arises when the motion of a body is described relative to a rotating coordinate system (such as the Earth), is expressed naturally as a cross product. These are but a few of the uses of the cross product.

The formula for the cross product of two vectors looks strange at first. To help us better understand and remember the formula, we begin with some basic properties of determinants.

2 × 2 Determinants

A 2 × 2 *determinant* is a real number that is denoted and defined by

$$\begin{vmatrix} a_1 & a_2 \\ b_1 & b_2 \end{vmatrix} = a_1 b_2 - a_2 b_1,$$

where a_1, a_2, b_1, and b_2 are any real numbers. For example,

$$\begin{vmatrix} 3 & 2 \\ 5 & 4 \end{vmatrix} = 3 \cdot 4 - 2 \cdot 5 = 12 - 10 = 2,$$

$$\begin{vmatrix} 5 & 4 \\ 3 & 2 \end{vmatrix} = 5 \cdot 2 - 4 \cdot 3 = 10 - 12 = -2,$$

$$\begin{vmatrix} 9 & 6 \\ 5 & 4 \end{vmatrix} = 9 \cdot 4 - 6 \cdot 5 = 36 - 30 = 6,$$

$$\begin{vmatrix} 3 & 2 \\ 3 & 2 \end{vmatrix} = 3 \cdot 2 - 2 \cdot 3 = 6 - 6 = 0,$$

$$\begin{vmatrix} 3 & 2 \\ 0 & 0 \end{vmatrix} = 3 \cdot 0 - 2 \cdot 0 = 0 - 0 = 0.$$

These evaluations show that

$$\begin{vmatrix} 3 & 2 \\ 5 & 4 \end{vmatrix} = -\begin{vmatrix} 5 & 4 \\ 3 & 2 \end{vmatrix},$$

$$\begin{vmatrix} 9 & 6 \\ 5 & 4 \end{vmatrix} = \begin{vmatrix} 3 \cdot 3 & 3 \cdot 2 \\ 5 & 4 \end{vmatrix} = 3\begin{vmatrix} 3 & 2 \\ 5 & 4 \end{vmatrix},$$

and illustrate the following rules:

1. If two rows (or columns) are interchanged, then the determinant is multiplied by -1.
2. If a row (or column) is multiplied by a constant, then the determinant is multiplied by that constant.
3. If two rows (or columns) are the same, then the determinant is zero.
4. If there is a zero row (or column), then the determinant is zero.

Determinants arose historically as a systematic means for solving systems of linear equations. Consider the linear system

$$\begin{aligned} ax + by &= e \\ cx + dy &= f \end{aligned}$$

To eliminate y and solve for x, multiply the first equation by d, multiply the second equation by b, and then subtract the resulting equations to obtain

$$(ad - bc)x = ed - bf,$$

$$x = \frac{ed - bf}{ad - bc} = \frac{\begin{vmatrix} e & b \\ f & d \end{vmatrix}}{\begin{vmatrix} a & b \\ c & d \end{vmatrix}}.$$

Similarly,

$$y = \frac{af - ce}{ad - bc} = \frac{\begin{vmatrix} a & e \\ c & f \end{vmatrix}}{\begin{vmatrix} a & b \\ c & d \end{vmatrix}}.$$

These formulas for x and y are valid provided

$$D = \begin{vmatrix} a & b \\ c & d \end{vmatrix} \neq 0.$$

We call D the *determinant of the linear system.* Geometrically, $D \neq 0$ means that the two straight lines determined by the two equations in the system are not parallel; see the problems. If we set

$$D_1 = \begin{vmatrix} e & b \\ f & d \end{vmatrix} \quad \text{and} \quad D_2 = \begin{vmatrix} a & e \\ c & f \end{vmatrix},$$

then the solution of the system can be expressed as

$$x = \frac{D_1}{D}, \qquad y = \frac{D_2}{D}, \quad \text{if} \quad D \neq 0.$$

This is **Cramer's rule.** Notice that D_1 (or D_2) can be obtained from D by replacing its first (or second) column by the column on the right-hand side of the linear system.

EXAMPLE 1. Solve the linear system

$$\begin{aligned} 3x + \ y &= -3, \\ 2x + 4y &= \ \ 8. \end{aligned}$$

Solution. For this system,

$$D = \begin{vmatrix} 3 & 1 \\ 2 & 4 \end{vmatrix} = 10, \quad D_1 = \begin{vmatrix} -3 & 1 \\ 8 & 4 \end{vmatrix} = -20, \quad D_2 = \begin{vmatrix} 3 & -3 \\ 2 & 8 \end{vmatrix} = 30.$$

By Cramer's rule,

$$x = \frac{D_1}{D} = -2, \quad y = \frac{D_2}{D} = 3. \ \square$$

3 × 3 Determinants

The systematic solution of systems of three linear equations in three unknowns leads to 3 × 3 determinants. A 3 × 3 determinant is defined by

$$\begin{vmatrix} a_1 & a_2 & a_3 \\ b_1 & b_2 & b_3 \\ c_1 & c_2 & c_3 \end{vmatrix} = \begin{array}{c} a_1b_2c_3 + a_2b_3c_1 + a_3b_1c_2 \\ - a_1b_3c_2 - a_2b_1c_3 - a_3b_2c_1. \end{array}$$

An equivalent formula for the determinant, which is usually more convenient, is

$$\begin{vmatrix} a_1 & a_2 & a_3 \\ b_1 & b_2 & b_3 \\ c_1 & c_2 & c_3 \end{vmatrix} = a_1 \begin{vmatrix} b_2 & b_3 \\ c_2 & c_3 \end{vmatrix} - a_2 \begin{vmatrix} b_1 & b_3 \\ c_1 & c_3 \end{vmatrix} + a_3 \begin{vmatrix} b_1 & b_2 \\ c_1 & c_2 \end{vmatrix}.$$

A little attention to patterns makes it easy to remember this formula. First, notice the sign pattern: $+ - +$. The numbers a_1, a_2, a_3 in the first row of the 3 × 3 determinant appear in the three terms on the right. The 2 × 2 determinant that multiplies a_1 can be found in the 3 × 3 determinant by covering up the row and column containing a_1. The same observation applies to the 2 × 2 determinants that multiply a_2 and a_3. For example,

$$\begin{vmatrix} 3 & 2 & 1 \\ 1 & -3 & 2 \\ -2 & 1 & -4 \end{vmatrix} = 3 \begin{vmatrix} -3 & 2 \\ 1 & -4 \end{vmatrix} - 2 \begin{vmatrix} 1 & 2 \\ -2 & -4 \end{vmatrix} + 1 \begin{vmatrix} 1 & -3 \\ -2 & 1 \end{vmatrix}.$$

$$= 3(10) - 2(0) + 1(-5) = 25.$$

It is easy, but a little tedious, to show that 3 × 3 determinants obey the four rules given previously for 2 × 2 determinants. Here is another property. Apply the first rule twice, once interchanging rows 2 and 3 and then rows 1 and 2 to obtain

$$\begin{vmatrix} a_1 & a_2 & a_3 \\ b_1 & b_2 & b_3 \\ c_1 & c_2 & c_3 \end{vmatrix} = -\begin{vmatrix} a_1 & a_2 & a_3 \\ c_1 & c_2 & c_3 \\ b_1 & b_2 & b_3 \end{vmatrix} = \begin{vmatrix} c_1 & c_2 & c_3 \\ a_1 & a_2 & a_3 \\ b_1 & b_2 & b_3 \end{vmatrix}.$$

In words, if the first two rows of a 3 × 3 determinant are moved down and the third row put on top, then the value of the determinant is unchanged. Repeat this operation again to find

$$\begin{vmatrix} a_1 & a_2 & a_3 \\ b_1 & b_2 & b_3 \\ c_1 & c_2 & c_3 \end{vmatrix} = \begin{vmatrix} c_1 & c_2 & c_3 \\ a_1 & a_2 & a_3 \\ b_1 & b_2 & b_3 \end{vmatrix} = \begin{vmatrix} b_1 & b_2 & b_3 \\ c_1 & c_2 & c_3 \\ a_1 & a_2 & a_3 \end{vmatrix}.$$

This is sometimes called the *permutation identity.* It can help with some determinant evaluations. For example,

$$\begin{vmatrix} -2 & 4 & 5 \\ 3 & 1 & 2 \\ 1 & 0 & 0 \end{vmatrix} = \begin{vmatrix} 1 & 0 & 0 \\ -2 & 4 & 5 \\ 3 & 1 & 2 \end{vmatrix} = 1\begin{vmatrix} 4 & 5 \\ 1 & 2 \end{vmatrix} = 8 - 5 = 3.$$

There is a version of Cramer's rule for systems of equations with three (or more) unknowns, but we shall not state it. Cramer's rule is an important theoretical tool, but it is seldom a computationally efficient means for finding solutions to systems with more than two unknowns.

Definition and Basic Properties of the Cross Product

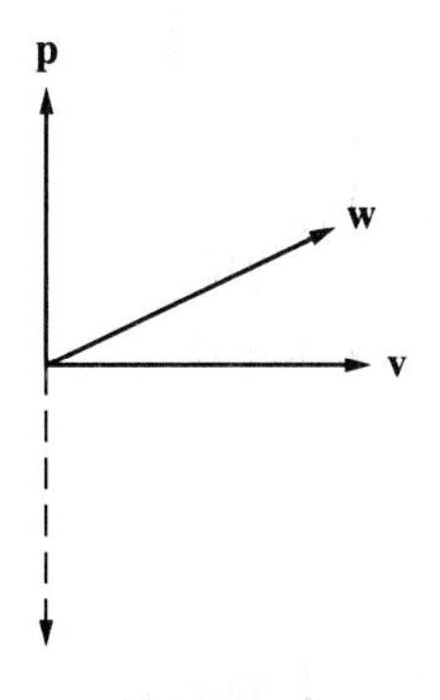

FIGURE 1

The solution of an important geometric problem will lead us to the cross product of two vectors: given two nonparallel vectors **v** and **w**, find a vector **p** that is orthogonal to both **v** and **w.** Figure 1 shows the geometric situation. There are two directions that **p** can point. To be definite, we seek to determine **p** so that **v, w, p** (in that order) obey the right-hand rule (see Fig. 2 in Sec. 3.1). This fixes the direction of **p.** We shall decide on its length later. Let $\mathbf{v} = \langle v_1, v_2, v_3 \rangle$, $\mathbf{w} = \langle w_1, w_2, w_3 \rangle$, and $\mathbf{p} = \langle x, y, z \rangle$, where x, y, and z are unknown. The orthogonality conditions $\mathbf{v} \cdot \mathbf{p} = 0$ and $\mathbf{w} \cdot \mathbf{p} = 0$ give

$$v_1x + v_2y + v_3z = 0,$$
$$w_1x + w_2y + w_3z = 0.$$

These are two equations for the three unknowns. A consequence of the assumption that the vectors $\mathbf{v} = \langle v_1, v_2, v_3 \rangle$ and $\mathbf{w} = \langle w_1, w_2, w_3 \rangle$ are not parallel is that at least one of the determinants

$$\begin{vmatrix} v_1 & v_2 \\ w_1 & w_2 \end{vmatrix}, \quad \begin{vmatrix} v_1 & v_3 \\ w_1 & w_3 \end{vmatrix}, \quad \begin{vmatrix} v_2 & v_3 \\ w_2 & w_3 \end{vmatrix},$$

is nonzero. (See the problems.) Suppose that the first of the three determinants is nonzero. Rewrite the equations for x, y, and z in the form

$$v_1x + v_2y = -v_3z,$$
$$w_1x + w_2y = -w_3z.$$

Use Cramer's rule to solve this 2×2 system for x and y in terms of z:

$$x = \frac{D_1}{D}, \qquad y = \frac{D_2}{D},$$

where

$$D = \begin{vmatrix} v_1 & v_2 \\ w_1 & w_2 \end{vmatrix} \neq 0, \quad D_1 = -z\begin{vmatrix} v_3 & v_2 \\ w_3 & w_2 \end{vmatrix}, \quad D_2 = -z\begin{vmatrix} v_1 & v_3 \\ w_1 & w_3 \end{vmatrix}.$$

Then

$$\mathbf{p} = \langle x, y, z \rangle = \left\langle \frac{D_1}{D}, \frac{D_2}{D}, z \right\rangle = z\left\langle \frac{v_2w_3 - v_3w_2}{D}, \frac{v_3w_1 - v_1w_3}{D}, 1 \right\rangle.$$

The scalar z is still undetermined because we have not fixed the length of **p** or distinguished between the two possible directions for **p** in Fig. 1. It turns out (although it is not obvious and we shall not prove it) that **v**, **w** , and **p** satisfy the right-hand rule if $z = D$ or if z is any positive multiple of D. Let's take $z = D$. Then

$$\mathbf{p} = \langle x, y, z \rangle = \langle v_2w_3 - v_3w_2,\ v_3w_1 - v_1w_3,\ v_1w_2 - v_2w_1 \rangle$$

and **p** is orthogonal to **v** and **w**.

This discussion motivates the following definition.

Definition *The Cross Product*
The **cross product** of $\mathbf{v} = \langle v_1, v_2, v_3 \rangle$ and $\mathbf{w} = \langle w_1, w_2, w_3 \rangle$ is $\mathbf{v} \times \mathbf{w} = \langle v_2w_3 - v_3w_2,\ v_3w_1 - v_1w_3,\ v_1w_2 - v_2w_1 \rangle$.

An equivalent formula for the cross product, which is much easier to remember, is expressed symbolically in the determinant form

$$\mathbf{v} \times \mathbf{w} = \begin{vmatrix} \mathbf{i} & \mathbf{j} & \mathbf{k} \\ v_1 & v_2 & v_3 \\ w_1 & w_2 & w_3 \end{vmatrix} = \mathbf{i}\begin{vmatrix} v_2 & v_3 \\ w_2 & w_3 \end{vmatrix} - \mathbf{j}\begin{vmatrix} v_1 & v_3 \\ w_1 & w_3 \end{vmatrix} + \mathbf{k}\begin{vmatrix} v_1 & v_2 \\ w_1 & w_2 \end{vmatrix}.$$

In view of the discussion preceding the definition of the cross product, if **v** and **w** are not parallel, then $\mathbf{v} \times \mathbf{w} \neq 0$, the vectors **v**, **w**, and $\mathbf{v} \times \mathbf{w}$ obey the right-hand rule, and

v × w is orthogonal to both v and w.

In the problems you are asked to confirm that these orthogonality relations hold for any **v** and **w**, whether parallel or not, by showing that

$$\mathbf{v} \cdot (\mathbf{v} \times \mathbf{w}) = 0, \qquad \mathbf{w} \cdot (\mathbf{v} \times \mathbf{w}) = 0.$$

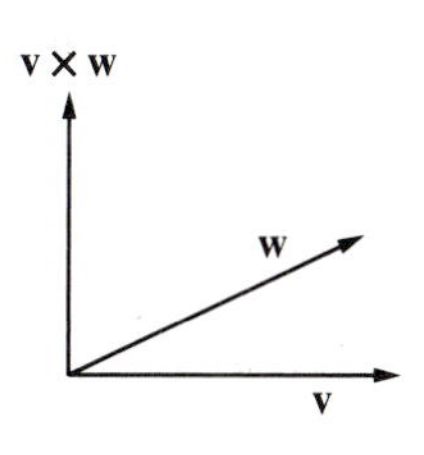

FIGURE 2

Figure 2 illustrates the foregoing properties of **v**, **w**, and **v** × **w**.

The foregoing properties of the cross product are easy to verify for the basic unit vectors **i**, **j**, and **k**. For example, since **i** = $<1,0,0>$ and **j** = $<0,1,0>$,

$$\mathbf{i} \times \mathbf{j} = \begin{vmatrix} \mathbf{i} & \mathbf{j} & \mathbf{k} \\ 1 & 0 & 0 \\ 0 & 1 & 0 \end{vmatrix} = \mathbf{i}\begin{vmatrix} 0 & 0 \\ 1 & 0 \end{vmatrix} - \mathbf{j}\begin{vmatrix} 1 & 0 \\ 0 & 0 \end{vmatrix} + \mathbf{k}\begin{vmatrix} 1 & 0 \\ 0 & 1 \end{vmatrix} = \mathbf{k}.$$

Evidently, **i** × **j** = **k** is orthogonal to both **i** and **j**, and the vectors **i**, **j**, and **i** × **j** = **k** satisfy the right-hand rule. Similar evaluations yield

$$\begin{array}{lll} \mathbf{i} \times \mathbf{j} = \mathbf{k}, & \mathbf{j} \times \mathbf{k} = \mathbf{i}, & \mathbf{k} \times \mathbf{i} = \mathbf{j}, \\ \mathbf{j} \times \mathbf{i} = -\mathbf{k}, & \mathbf{k} \times \mathbf{j} = -\mathbf{i}, & \mathbf{i} \times \mathbf{k} = -\mathbf{j}, \\ \mathbf{i} \times \mathbf{i} = \mathbf{0}, & \mathbf{j} \times \mathbf{j} = \mathbf{0}, & \mathbf{k} \times \mathbf{k} = \mathbf{0}. \end{array}$$

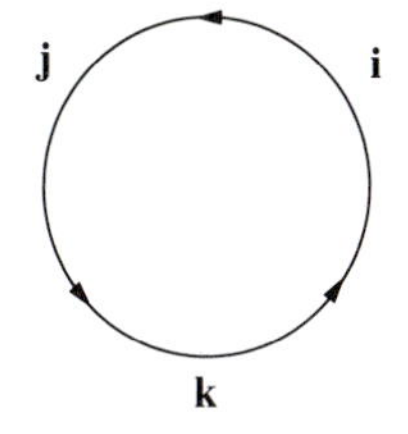

FIGURE 3

Figure 3 provides a handy device for remembering the cross-products of these unit vectors. Start with any one of the vectors **i**, **j**, **k**. If you move counterclockwise along the circle, then the cross product of the first two vectors you meet is the third vector. If you move clockwise, you get the negative of the third vector.

EXAMPLE 2. Find the cross product of **v** = $<2,0,3>$ and **w** = $<1,-2,4>$.

Solution. From the determinant form of the cross product,

$$\mathbf{v} \times \mathbf{w} = \begin{vmatrix} \mathbf{i} & \mathbf{j} & \mathbf{k} \\ 2 & 0 & 3 \\ 1 & -2 & 4 \end{vmatrix} = \mathbf{i}\begin{vmatrix} 0 & 3 \\ -2 & 4 \end{vmatrix} - \mathbf{j}\begin{vmatrix} 2 & 3 \\ 1 & 4 \end{vmatrix} + \mathbf{k}\begin{vmatrix} 2 & 0 \\ 1 & -2 \end{vmatrix}$$

$$= 6\mathbf{i} - 5\mathbf{j} - 4\mathbf{k} = <6, -5, -4>. \ \square$$

The algebraic properties of the cross product stated next follow directly from the determinant form of the definition. The algebraic steps are not informative, so we omit them.

$$\mathbf{v} \times \mathbf{w} = -(\mathbf{w} \times \mathbf{v})$$

$$\mathbf{v} \times \mathbf{v} = \mathbf{0}$$

$$\mathbf{0} \times \mathbf{v} = \mathbf{0} = \mathbf{v} \times \mathbf{0}$$

$$(\lambda\mathbf{v}) \times \mathbf{w} = \lambda(\mathbf{v} \times \mathbf{w}) = \mathbf{v} \times (\lambda\mathbf{w})$$

$$\mathbf{u} \times (\mathbf{v} + \mathbf{w}) = \mathbf{u} \times \mathbf{v} + \mathbf{u} \times \mathbf{w}$$

$$(\mathbf{u} + \mathbf{v}) \times \mathbf{w} = \mathbf{u} \times \mathbf{w} + \mathbf{v} \times \mathbf{w}$$

The cross products of the basic unit vectors displayed earlier illustrate the first two preceding properties.

EXAMPLE 3. Let $\mathbf{v} = 2\mathbf{i} - \mathbf{k}$ and $\mathbf{w} = 3\mathbf{j} + \mathbf{k}$. Find $\mathbf{v} \times \mathbf{w}$.

Solution. We could write $\mathbf{v} = < 2, 0, -1 >$ and $\mathbf{w} = < 0, 3, 1 >$ and then proceed in the manner of Ex. 2, but let's try another way. From the algebraic properties of cross products,

$$\begin{aligned}\mathbf{v} \times \mathbf{w} &= (2\mathbf{i} - \mathbf{k}) \times (3\mathbf{j} + \mathbf{k}) \\ &= 6\mathbf{i} \times \mathbf{j} + 2\mathbf{i} \times \mathbf{k} - 3\mathbf{k} \times \mathbf{j} - \mathbf{k} \times \mathbf{k} \\ &= 6\mathbf{k} - 2\mathbf{j} + 3\mathbf{i} - \mathbf{0} = 3\mathbf{i} - 2\mathbf{j} + 6\mathbf{k}. \ \square\end{aligned}$$

A useful relation between the dot product and the cross product is **Lagrange's identity**:

$$||\mathbf{v} \times \mathbf{w}||^2 = ||\mathbf{v}||^2 ||\mathbf{w}||^2 - (\mathbf{v} \cdot \mathbf{w})^2,$$

Lagrange's identity is easy but time-consuming to prove. Just write out both sides of the equation in terms of the components of $\mathbf{v} = < v_1, v_2, v_3 >$ and $\mathbf{w} = < w_1, w_2, w_3 >$, simplify, and observe that the two sides are identical.

Lagrange's identity leads to a useful and interesting geometric interpretation of $||\mathbf{v} \times \mathbf{w}||$. A physical interpretation comes shortly. For the time being, assume that $\mathbf{v} \neq \mathbf{0}$ and $\mathbf{w} \neq \mathbf{0}$ are not parallel, as in Fig. 4. Then $\mathbf{v}$ and $\mathbf{w}$ determine a parallelogram with base $||\mathbf{v}||$, height $h = ||\mathbf{w}|| \sin\theta$, and area

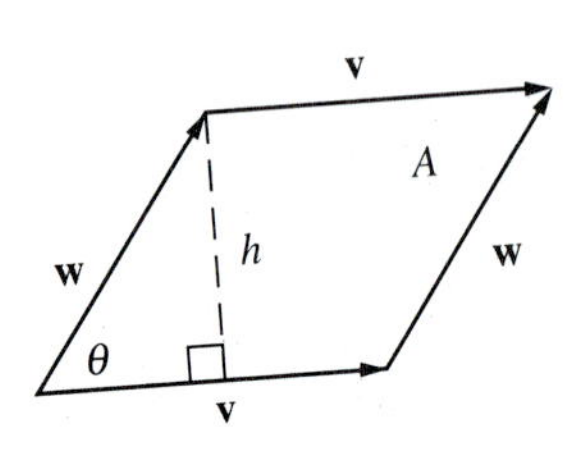

FIGURE 4

$$A = ||\mathbf{v}||\, h = ||\mathbf{v}||\ ||\mathbf{w}|| \sin\theta.$$

From Lagrange's identity and $\mathbf{v} \cdot \mathbf{w} = ||\mathbf{v}||\ ||\mathbf{w}|| \cos\theta$,

$$||\mathbf{v} \times \mathbf{w}||^2 = ||\mathbf{v}||^2 ||\mathbf{w}||^2 (1 - \cos^2\theta) = ||\mathbf{v}||^2 ||\mathbf{w}||^2 \sin^2\theta = A^2.$$

The preceding equation contains two important pieces of information.

The magnitude of the cross product satisfies

$$||\mathbf{v} \times \mathbf{w}|| = ||\mathbf{v}||\ ||\mathbf{w}|| \sin\theta.$$

The area A of the parallelogram determined by $\mathbf{v}$ and $\mathbf{w}$ is

$$A = ||\mathbf{v} \times \mathbf{w}||.$$

If $\mathbf{v}$ and $\mathbf{w}$ are perpendicular, then the parallelogram they determine is a rectangle with area $A = ||\mathbf{v} \times \mathbf{w}|| = ||\mathbf{v}||\ ||\mathbf{w}||$.

The formula $A = ||\mathbf{v} \times \mathbf{w}||$ also makes sense if $\mathbf{v}$ and $\mathbf{w}$ are parallel; then $A = ||\mathbf{v} \times \mathbf{w}|| = 0$. The area interpretation for $||\mathbf{v} \times \mathbf{w}||$ shows that $\mathbf{v} \times \mathbf{w} \neq \mathbf{0}$ if $\mathbf{v}$ and $\mathbf{w}$ are not parallel. Consequently,

$$\boxed{\mathbf{v} \text{ and } \mathbf{w} \text{ are parallel} \quad \Leftrightarrow \quad \mathbf{v} \times \mathbf{w} = \mathbf{0}.}$$

This useful relation can be proved in several other ways.

The foregoing discussion leads to another formula for the cross product, which some authors use for a definition:

$$\mathbf{v} \times \mathbf{w} = ||\mathbf{v}||\ ||\mathbf{w}|| \sin\theta\, \mathbf{n},$$

where $\mathbf{n}$ is a unit vector such that $\mathbf{v}$, $\mathbf{w}$, and $\mathbf{n}$ obey the right-hand rule.

The cross product is also useful for finding areas of triangles. In Fig. 5, $\mathbf{v} = \overrightarrow{PQ}$ and $\mathbf{w} = \overrightarrow{PR}$. Since the area A of the triangle formed by P, Q, and R is half the area of the parallelogram in the figure, $A = \frac{1}{2}||\mathbf{v} \times \mathbf{w}||$.

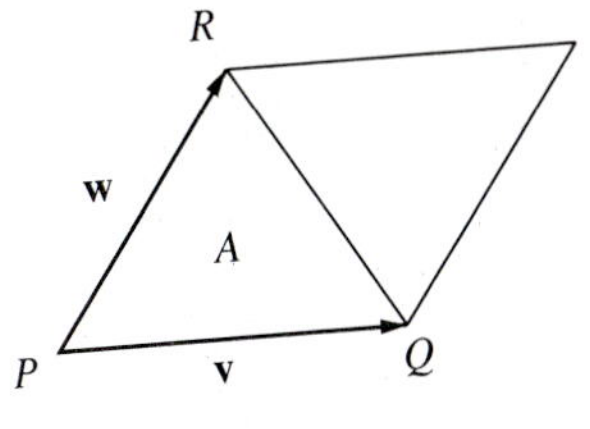

FIGURE 5

EXAMPLE 4. Find the area of the triangle in Fig. 5 if $P = (2, 1, 3)$, $Q = (5, 3, 1)$, and $R = (-4, 1, 5)$.

Solution. In this case, $\mathbf{v} = \overrightarrow{PQ} = <3, 2, -2>$ and $\mathbf{w} = \overrightarrow{PR} = <-6, 0, 2>$. So

$$\mathbf{v} \times \mathbf{w} = \begin{vmatrix} \mathbf{i} & \mathbf{j} & \mathbf{k} \\ 3 & 2 & -2 \\ -6 & 0 & 2 \end{vmatrix} = \mathbf{i}\begin{vmatrix} 2 & -2 \\ 0 & 2 \end{vmatrix} - \mathbf{j}\begin{vmatrix} 3 & -2 \\ -6 & 2 \end{vmatrix} + \mathbf{k}\begin{vmatrix} 3 & 2 \\ -6 & 0 \end{vmatrix}$$

$$= 4\mathbf{i} + 6\mathbf{j} + 12\mathbf{k} = 2 <2, 3, 6>.$$

The area of the triangle is

$$A = \frac{1}{2}||\mathbf{v} \times \mathbf{w}|| = ||<2, 3, 6>|| = \sqrt{4 + 9 + 36} = 7. \ \square$$

The Scalar Triple Product

The **scalar triple product** of $\mathbf{u}$, $\mathbf{v}$, and $\mathbf{w}$ is $\mathbf{u} \cdot (\mathbf{v} \times \mathbf{w})$. This expression comes up rather frequently in vector calculations in science and engineering problems. We shall derive a convenient formula for $\mathbf{u} \cdot (\mathbf{v} \times \mathbf{w})$ and give a geometric interpretation.

Let $\mathbf{u} = <u_1, u_2, u_3>$, $\mathbf{v} = <v_1, v_2, v_3>$, and $\mathbf{w} = <w_1, w_2, w_3>$. First, express $\mathbf{v} \times \mathbf{w}$ as

$$\mathbf{v} \times \mathbf{w} = \mathbf{i}\begin{vmatrix} v_2 & v_3 \\ w_2 & w_3 \end{vmatrix} - \mathbf{j}\begin{vmatrix} v_1 & v_3 \\ w_1 & w_3 \end{vmatrix} + \mathbf{k}\begin{vmatrix} v_1 & v_2 \\ w_1 & w_2 \end{vmatrix}.$$

Then

$$\mathbf{u} \cdot (\mathbf{v} \times \mathbf{w}) = u_1\begin{vmatrix} v_2 & v_3 \\ w_2 & w_3 \end{vmatrix} - u_2\begin{vmatrix} v_1 & v_3 \\ w_1 & w_3 \end{vmatrix} + u_3\begin{vmatrix} v_1 & v_2 \\ w_1 & w_2 \end{vmatrix}.$$

It follows that

$$\mathbf{u} \cdot (\mathbf{v} \times \mathbf{w}) = \begin{vmatrix} u_1 & u_2 & u_3 \\ v_1 & v_2 & v_3 \\ w_1 & w_2 & w_3 \end{vmatrix}.$$

EXAMPLE 5. Let $\mathbf{u} = <2,-3,-4>$, $\mathbf{v} = <3,0,-1>$, $\mathbf{w} = <1,2,2>$. Calculate $\mathbf{u} \cdot (\mathbf{v} \times \mathbf{w})$.

Solution 1. First determine $\mathbf{v} \times \mathbf{w}$:

$$\mathbf{v} \times \mathbf{w} = \begin{vmatrix} \mathbf{i} & \mathbf{j} & \mathbf{k} \\ 3 & 0 & -1 \\ 1 & 2 & 2 \end{vmatrix} = 2\mathbf{i} - 7\mathbf{j} + 6\mathbf{k} = <2,-7,6>.$$

Then

$$\mathbf{u} \cdot (\mathbf{v} \times \mathbf{w}) = (2)(2) + (-3)(-7) + (-4)(6) = 1.$$

Solution 2. By the determinant form of the scalar triple product,

$$\mathbf{u} \cdot (\mathbf{v} \times \mathbf{w}) = \begin{vmatrix} 2 & -3 & -4 \\ 3 & 0 & -1 \\ 1 & 2 & 2 \end{vmatrix} = 2\begin{vmatrix} 0 & -1 \\ 2 & 2 \end{vmatrix} + 3\begin{vmatrix} 3 & -1 \\ 1 & 2 \end{vmatrix} - 4\begin{vmatrix} 3 & 0 \\ 1 & 2 \end{vmatrix}$$

$$= 2(2) + 3(7) - 4(6) = 1.$$

As you can see, the two calculations are nearly identical. □

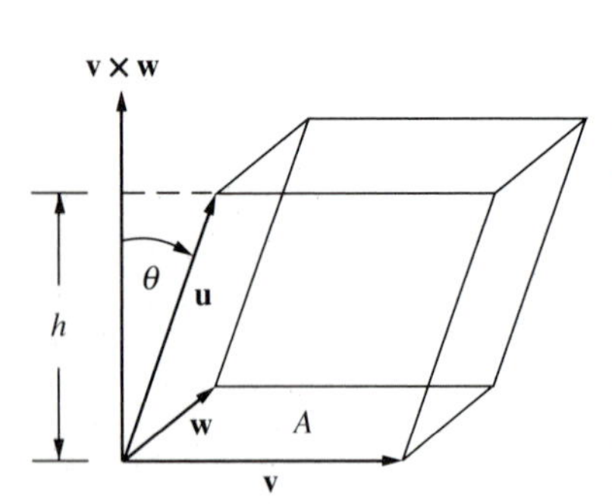

FIGURE 6

Figure 6 gives a geometric interpretation of $\mathbf{u} \cdot (\mathbf{v} \times \mathbf{w})$. Observe that **u**, **v**, **w** obey the right-hand rule if and only if the angle θ between **u** and $\mathbf{v} \times \mathbf{w}$ is acute, as in Fig. 6. Since $\mathbf{u} \cdot (\mathbf{v} \times \mathbf{w}) = \|\mathbf{u}\| \, \|\mathbf{v} \times \mathbf{w}\| \cos\theta$,

u, **v**, and **w** obey the right-hand rule $\Leftrightarrow$ $\mathbf{u} \cdot (\mathbf{v} \times \mathbf{w}) > 0$.

The vectors **u**, **v**, and **w** in Fig. 6 determine a solid figure called a *parallelepiped*. The base has area $A = \|\mathbf{v} \times \mathbf{w}\|$, the height is $h = \|\mathbf{u}\| \cos\theta$, and the volume is

$$V = Ah = \|\mathbf{u}\| \, \|\mathbf{v} \times \mathbf{w}\| \cos\theta = \mathbf{u} \cdot (\mathbf{v} \times \mathbf{w}).$$

In summary,

if **u**, **v**, and **w** obey the right-hand rule, then they determine a parallelepiped with volume
$V = \mathbf{u} \cdot (\mathbf{v} \times \mathbf{w})$.

Similarly, if $\mathbf{u}$, $\mathbf{v}$, and $\mathbf{w}$ obey the left–hand rule, then the angle θ between $\mathbf{u}$ and $\mathbf{v} \times \mathbf{w}$ is obtuse, $\mathbf{u} \cdot (\mathbf{v} \times \mathbf{w}) < 0$, and the volume of the parallelepiped determined by $\mathbf{u}$, $\mathbf{v}$, and $\mathbf{w}$ is $V = |\mathbf{u} \cdot (\mathbf{v} \times \mathbf{w})|$.

The preceding discussion reveals that $\mathbf{u} \cdot (\mathbf{v} \times \mathbf{w}) \neq 0$ whenever $\mathbf{u}$, $\mathbf{v}$, and $\mathbf{w}$ do not lie in a single plane. If they do lie in a single plane, we say that $\mathbf{u}$, $\mathbf{v}$, and $\mathbf{w}$ are *coplanar.* There are three possibilities for coplanar vectors. If at least one of the vectors $\mathbf{u}$, $\mathbf{v}$, or $\mathbf{w}$ is zero, then $\mathbf{u} \cdot (\mathbf{v} \times \mathbf{w}) = 0$. If $\mathbf{v}$ and $\mathbf{w}$ are parallel, then $\mathbf{v} \times \mathbf{w} = \mathbf{0}$ and, hence, $\mathbf{u} \cdot (\mathbf{v} \times \mathbf{w}) = 0$. Finally, if $\mathbf{u}$ lies in the plane of $\mathbf{v}$ and $\mathbf{w}$, as illustrated in Fig. 7, then $\mathbf{u}$ is perpendicular to $\mathbf{v} \times \mathbf{w}$ and, once again, $\mathbf{u} \cdot (\mathbf{v} \times \mathbf{w}) = 0$. All of these cases are summarized by

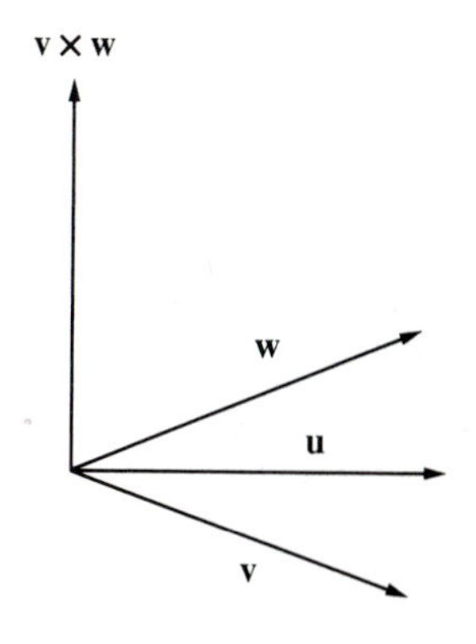

FIGURE 7

$$\boxed{\mathbf{u} \cdot (\mathbf{v} \times \mathbf{w}) = 0 \quad \Leftrightarrow \quad \mathbf{u}, \mathbf{v}, \text{ and } \mathbf{w} \text{ are coplanar.}}$$

EXAMPLE 6. Determine whether or not the vectors $\mathbf{u} = <1, 3, 0>$, $\mathbf{v} = <3, -1, 5>$, and $\mathbf{w} = <2, 0, 3>$ are coplanar.

Solution. We find that

$$\mathbf{u} \cdot (\mathbf{v} \times \mathbf{w}) = \begin{vmatrix} 1 & 3 & 0 \\ 3 & -1 & 5 \\ 2 & 0 & 3 \end{vmatrix} = 1 \begin{vmatrix} -1 & 5 \\ 0 & 3 \end{vmatrix} - 3 \begin{vmatrix} 3 & 5 \\ 2 & 3 \end{vmatrix} - 0 \begin{vmatrix} 3 & -1 \\ 2 & 0 \end{vmatrix}$$

$$= (1)(-3) - (3)(-1) - (0)(2) = 0.$$

Hence, $\mathbf{u}$, $\mathbf{v}$, and $\mathbf{w}$ are coplanar. □

So far, we have expressed the scalar triple product by $\mathbf{u} \cdot (\mathbf{v} \times \mathbf{w})$. Actually, the parentheses are not needed. To understand why, try to do the dot product first and see what happens. In practice, the scalar triple product is expressed by $\mathbf{u} \cdot \mathbf{v} \times \mathbf{w}$, without parentheses.

It follows directly from the determinant formula for the scalar triple product that

$$\boxed{\mathbf{u} \cdot \mathbf{v} \times \mathbf{w} = \mathbf{w} \cdot \mathbf{u} \times \mathbf{v} = \mathbf{v} \cdot \mathbf{w} \times \mathbf{u}.}$$

In fact, this is just a compact way to write the permutation identity for 3×3 determinants given earlier. Other useful vector identities involving the dot and cross products are given in the problems.

Moment of a Force and Torque

We mentioned earlier that cross products have many physical applications. Here is one, related to the twisting effect of a force. We shall meet other physical and geometric applications of the cross product as we go along.

In Fig. 8 a force $\mathbf{F}$ is exerted at the end of a lever arm $\mathbf{r} = \overrightarrow{OP}$. The lever arm could be a broomstick that you hold at O and the force could be the tug of your roommate who wants to do the cleaning. Assume you are strong enough to keep

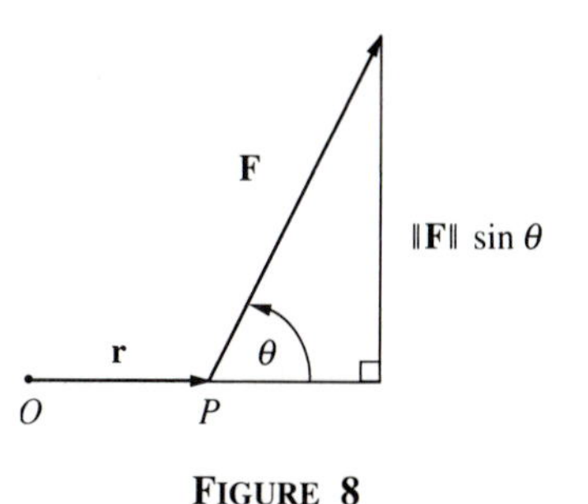

FIGURE 8

your end of the broom at O. That shouldn't be too hard, under the circumstances! Unless the force (exerted by your roommate) is parallel to the lever arm, the arm will tend to rotate about O. Experiments have shown that the tendency of $\mathbf{F}$ to produce rotational motion about O is determined by the component of $\mathbf{F}$ perpendicular to the lever arm $\mathbf{r}$, which is $||\mathbf{F}|| \sin\theta$. The **moment** M of $\mathbf{F}$ about O is defined by

$$M = ||\mathbf{r}||\ ||\mathbf{F}|| \sin\theta.$$

This is a useful quantitative measure of the rotational effect of $\mathbf{F}$. The **torque** of the force $\mathbf{F}$ about O is the vector

$$\boldsymbol{\tau} = \mathbf{r} \times \mathbf{F}.$$

The magnitude of the torque is

$$||\boldsymbol{\tau}|| = ||\mathbf{r} \times \mathbf{F}|| = ||\mathbf{r}||\ ||\mathbf{F}|| \sin\theta,$$

which is just the moment of the force about O. Thus, the magnitude of a cross product of two vectors can be interpreted either as the moment of a force or as the area of a parallelogram.

PROBLEMS

Find $\mathbf{v} \times \mathbf{w}$ in Probs. 1–4.

1. $\mathbf{v} = <1,-2,3>, \quad \mathbf{w} = <0,3,-2>$
2. $\mathbf{v} = <-2,1,-3>, \quad \mathbf{w} = <2,1,1>$
3. $\mathbf{v} = \mathbf{i} - 2\mathbf{k}, \quad \mathbf{w} = 2\mathbf{i} - \mathbf{j} + 4\mathbf{k}$
4. $\mathbf{v} = \mathbf{i} + \mathbf{j} + \mathbf{k}, \quad \mathbf{w} = \mathbf{i} - \mathbf{j} + \mathbf{k}$

In Probs. 5–10 use geometric reasoning to evaluate each expression. Do not compute with components.

5. $\mathbf{i} \cdot (\mathbf{j} \times \mathbf{k})$
6. $\mathbf{i} \times (\mathbf{j} \times \mathbf{k})$
7. $(\mathbf{i} - \mathbf{j}) \times (\mathbf{i} + \mathbf{j})$
8. $(\mathbf{i} - \mathbf{j}) \cdot (\mathbf{i} + \mathbf{j})$
9. $(\mathbf{i} \times \mathbf{j}) \cdot (\mathbf{j} \times \mathbf{k})$
10. $(\mathbf{i} - \mathbf{j}) \times \mathbf{k}$
11. Find a unit vector perpendicular to both $\mathbf{v} = <1,2,-2>$ and $\mathbf{w} = <4,0,-3>$.
12. Find a unit vector perpendicular to both $\mathbf{v} = 3\mathbf{i} - \mathbf{k}$ and $\mathbf{w} = \mathbf{i} - \mathbf{j} + 4\mathbf{k}$.
13. Find a vector perpendicular to the plane determined by $(1,-2,1)$, $(2,-1,-2)$, and $(0,1,0)$.
14. Find a vector perpendicular to the plane determined by the vectors $\mathbf{i} - 3\mathbf{j} + \mathbf{k}$ and $2\mathbf{i} - 2\mathbf{j} - 2\mathbf{k}$.

In Probs. 15–18 find the area of the triangle with the given vertices.

15. $(1,0,0), \quad (0,1,0), \quad (0,0,1)$
16. $(1,1,1), \quad (2,0,-3), \quad (0,1,-4)$

17. $(3,2,1)$, $(2,3,-1)$, $(2,-1,3)$ 18. $(-2,1,0)$, $(0,1,-1)$, $(-3,1,2)$

In Probs. 19–22 find the area of the parallelogram determined by the given vectors.

19. $<-1,1,0>$, $<-1,0,1>$ 20. $<1,-1,-4>$, $<-1,0,-5>$

21. $-\mathbf{i}+\mathbf{j}-2\mathbf{k}$, $\mathbf{i}+3\mathbf{j}+2\mathbf{k}$ 22. $-2\mathbf{i}+\mathbf{k}$, $-3\mathbf{i}+\mathbf{j}+2\mathbf{k}$

Determine whether the vectors in Probs. 23–27 are coplanar.

23. $<1,2,3>$, $<2,-3,1>$, $<3,-1,-2>$

24. $<2,0,3>$, $<1,-1,1>$, $<4,2,7>$

25. $\mathbf{i}+\mathbf{j}+\mathbf{k}$, $2\mathbf{i}-3\mathbf{j}+\mathbf{k}$, $-7\mathbf{i}+21\mathbf{j}+4\mathbf{k}$

26. $2\mathbf{i}+3\mathbf{k}$, $\mathbf{i}-\mathbf{j}+\mathbf{k}$, $4\mathbf{i}+2\mathbf{j}-7\mathbf{k}$

Verify Lagrange's identity for the vectors in Probs. 27–28.

27. $\mathbf{v}=<1,1,-1>$, $\mathbf{w}=<2,0,1>$

28. $\mathbf{v}=2\mathbf{i}-\mathbf{j}+3\mathbf{k}$, $\mathbf{w}=\mathbf{j}+7\mathbf{k}$

29. Use the vectors $\mathbf{u}=\mathbf{i}$, $\mathbf{v}=\mathbf{i}+\mathbf{j}$, and $\mathbf{w}=\mathbf{i}+\mathbf{j}+\mathbf{k}$ to show that the cross product does **not** satisfy the associative law $\mathbf{u}\times(\mathbf{v}\times\mathbf{w})=(\mathbf{u}\times\mathbf{v})\times\mathbf{w}$.

30. For scalars a, b, c with $a\neq 0$, the equation $ab=ac$ implies $b=c$. Exhibit vectors $\mathbf{a}$, $\mathbf{b}$, $\mathbf{c}$ with $\mathbf{a}\neq\mathbf{0}$ such that $\mathbf{a}\times\mathbf{b}=\mathbf{a}\times\mathbf{c}$ but $\mathbf{b}\neq\mathbf{c}$.

31. With reference to the preceding problem, show that if $\mathbf{a}$, $\mathbf{b}$, $\mathbf{c}$ are distinct nonzero vectors, then $\mathbf{a}\times\mathbf{b}=\mathbf{a}\times\mathbf{c}$ if and only if $\mathbf{a}$ is parallel to $\mathbf{b}-\mathbf{c}$.

32. Verify that $(\mathbf{v}+\mathbf{w})\times(\mathbf{v}-\mathbf{w})=2\mathbf{w}\times\mathbf{v}$.

33. Verify that $\mathbf{v}\cdot(\mathbf{v}\times\mathbf{w})=\mathbf{0}$ and $\mathbf{w}\cdot(\mathbf{v}\times\mathbf{w})=\mathbf{0}$.

34. Establish the following properties of the cross product:

    (a) $\mathbf{v}\times\mathbf{w}=\mathbf{w}\times\mathbf{v}$ (b) $\mathbf{v}\times\mathbf{v}=\mathbf{0}$ (c) $\mathbf{0}\times\mathbf{v}=\mathbf{0}$

    (d) $(\lambda\mathbf{v})\times\mathbf{w}=\lambda(\mathbf{v}\times\mathbf{w})=\mathbf{v}\times(\lambda\mathbf{w})$ (e) $\mathbf{u}\times(\mathbf{v}+\mathbf{w})=\mathbf{u}\times\mathbf{v}+\mathbf{u}\times\mathbf{w}$

    Hint. Use the determinant form of the cross product to evaluate each side, simplify, and compare the results.

35. Assume that the vectors $\mathbf{v}=<v_1,v_2,v_3>$ and $\mathbf{w}=<w_1,w_2,w_3>$ are not parallel. Show that at least one of the determinants

$$\begin{vmatrix} v_1 & v_2 \\ w_1 & w_2 \end{vmatrix},\qquad \begin{vmatrix} v_1 & v_3 \\ w_1 & w_3 \end{vmatrix},\qquad \begin{vmatrix} v_2 & v_3 \\ w_2 & w_3 \end{vmatrix},$$

    is nonzero. *Hint.* A 2×2 determinant is zero if and only if one of its rows is a multiple of the other.

36. Prove: If three vectors form a triangle, then the lengths of their pairwise cross products all are equal.

37. Prove Lagrange's identity by following the steps suggested in the text.

38. Check that the identity $\mathbf{u} \cdot \mathbf{v} \times \mathbf{w} = \mathbf{w} \cdot \mathbf{u} \times \mathbf{v} = \mathbf{v} \cdot \mathbf{w} \times \mathbf{u}$ is just a compact way to write the permutation identity for 3×3 determinants. In particular, conclude that

$$\boxed{\mathbf{u} \cdot \mathbf{v} \times \mathbf{w} = \mathbf{u} \times \mathbf{v} \cdot \mathbf{w}.}$$

That is, the dot and cross can be exchanged in the scalar triple product.

39. Show that the determinant D of the 2×2 system S: $ax + by = e$, $cx + dy = f$ is nonzero if and only if the two lines determined by S are not parallel.

40. A force $\mathbf{F}$ with magnitude 50 pounds acts at the point $(1, 2, 3)$ in the direction of the vector $< 1, 2, -1 >$. Find the moment and torque of $\mathbf{F}$ about the origin.

41. Forces of 1, 2, 3, and 4 pounds act at O, A, B, and C and are directed along the sides OA, AB, BC, and CO of a rigid unit square with vertices $O = (0, 0)$, $A = (1, 0)$, $B = (1, 1)$, and $C = (0, 1)$. Find the total moment and torque of these forces about O.

The vector $\mathbf{u} \times (\mathbf{v} \times \mathbf{w})$ is called the **vector triple product** of $\mathbf{u}$, $\mathbf{v}$, and $\mathbf{w}$. It comes up at critical places in various applications.

42. Explain why $\mathbf{u} \times (\mathbf{v} \times \mathbf{w})$ must lie in the plane determined by $\mathbf{v}$ and $\mathbf{w}$ and hence can be expressed as $\mathbf{u} \times (\mathbf{v} \times \mathbf{w}) = \lambda \mathbf{v} + \mu \mathbf{w}$ for certain scalars λ and μ. The next problem identifies the scalars.

43. Prove that

$$\boxed{\mathbf{u} \times (\mathbf{v} \times \mathbf{w}) = (\mathbf{u} \cdot \mathbf{w})\mathbf{v} - (\mathbf{u} \cdot \mathbf{v})\mathbf{w}}$$

by means of the following steps: (a) Show that the **i**–components of both sides are equal by explicit calculation. (b) Conclude, by symmetry, that the **j**–and **k**–components of both sides are the same; hence, the vector identity holds.

44. Determine when $\mathbf{u} \times (\mathbf{v} \times \mathbf{w}) = (\mathbf{u} \times \mathbf{v}) \times \mathbf{w}$ for nonzero vectors $\mathbf{u}$, $\mathbf{v}$, and $\mathbf{w}$. Compare with Prob. 29.

45. Show that $(\mathbf{u} \times \mathbf{v}) \times \mathbf{w} = (\mathbf{u} \cdot \mathbf{w})\mathbf{v} - (\mathbf{v} \cdot \mathbf{w})\mathbf{u}$.

46. Show that $(\mathbf{a} \times \mathbf{b}) \cdot (\mathbf{c} \times \mathbf{d}) = (\mathbf{a} \cdot \mathbf{c})(\mathbf{b} \cdot \mathbf{d}) - (\mathbf{a} \cdot \mathbf{d})(\mathbf{b} \cdot \mathbf{c})$.
Hint. Let $\mathbf{v} = \mathbf{c} \times \mathbf{d}$ and use $\mathbf{a} \times \mathbf{b} \cdot \mathbf{v} = \mathbf{a} \cdot \mathbf{b} \times \mathbf{v}$.

47. Let $\mathbf{v}$ and $\mathbf{w} \neq \mathbf{0}$ be a vectors. Figure 9 in Sec. 3.3 shows $\mathbf{v}_1$, the projection of $\mathbf{v}$ along $\mathbf{w}$, and $\mathbf{v}_2$, the projection of $\mathbf{v}$ orthogonal to $\mathbf{w}$. Establish the following convenient formula for $\mathbf{v}_2$:

$$\mathbf{v}_2 = \frac{\mathbf{w} \times (\mathbf{v} \times \mathbf{w})}{||\mathbf{w}||^2}.$$

Hint. Set $\mathbf{u} = \mathbf{w}$ in the identity in Prob. 43.

3.5 Lines and Planes

Lines and planes in 3–space are described conveniently either by scalar equations or by vector equations. Specialization of these equations to 2–space gives scalar or vector equations for lines in the xy–plane. The new equations for lines in the xy–plane complement the familiar equations for lines you have been using for years. The section also serves as a preview of coming attractions because ideas similar to those presented here will be exploited later to represent more general curves and surfaces.

The vector equations for lines and planes will be expressed in terms of position vectors. Position vectors locate particular points or track a variable point as it moves. In Fig. 1, $\mathbf{r} = \langle x, y \rangle$ is the position vector of the point $P = (x, y)$ in 2–space. Similarly, in Fig. 2, $\mathbf{r} = \langle x, y, z \rangle$ is the position vector of the point $P = (x, y, z)$ in 3–space. Sometimes position vectors of points are called simply positions of the points.

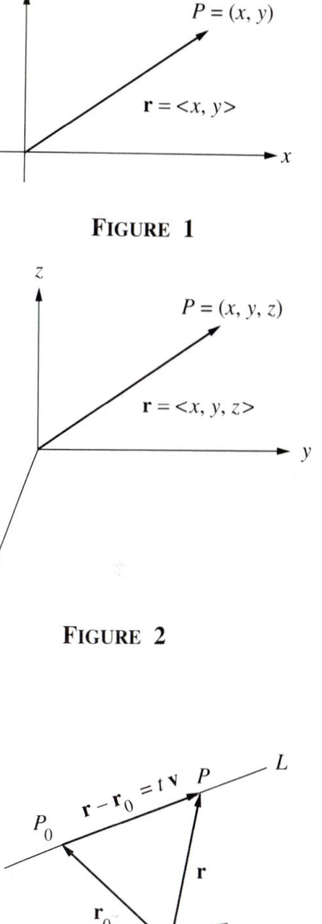

FIGURE 1

FIGURE 2

Lines

Figure 3 will guide us to the vector equation for a line in 3–space. The figure shows a line L that passes through a given point $P_0 = (x_0, y_0, z_0)$ and is parallel to a given *nonzero* vector $\mathbf{v} = \langle a, b, c \rangle$. We call $\mathbf{v}$ a **direction vector** for L. The position vectors for P_0 and a variable point $P = (x, y, z)$ on L are $\mathbf{r}_0 = \langle x_0, y_0, z_0 \rangle$ and $\mathbf{r} = \langle x, y, z \rangle$. Observe that $\mathbf{r} - \mathbf{r}_0 = \overrightarrow{P_0P}$ Clearly P lies on L only when $\overrightarrow{P_0P}$ is parallel to $\mathbf{v}$; that is, $\overrightarrow{P_0P} = t\mathbf{v}$ or $\mathbf{r} - \mathbf{r}_0 = t\mathbf{v}$ for some scalar t. We have established the following **vector equation** for L.

FIGURE 3

> The line L through $P_0 = (x_0, y_0, z_0)$ with direction vector $\mathbf{v} = \langle a, b, c \rangle$ has equation
>
> $$\mathbf{r} = \mathbf{r}_0 + t\mathbf{v}, \qquad -\infty < t < \infty.$$

Figure 4 gives another picture of the line L, with the direction vector $\mathbf{v}$ in L, attached to the point P_0. A physical interpretation of this vector equation will be given shortly.

It is clear from Fig. 4 that as t increases the position vector $\mathbf{r}$ traces out the line in the direction of $\mathbf{v}$, which is called the *positive direction* on L. The opposite direction is called the *negative direction.* Thus, the vector equation $\mathbf{r} = \mathbf{r}_0 + t\mathbf{v}$ for L determines a direction on L. We refer to L as a **directed line** when the direction is significant.

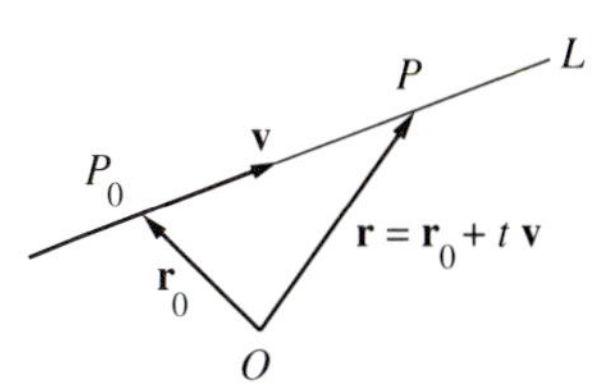

FIGURE 4

Figure 4 illustrates a line either in 2–space or in 3–space. First, let's consider the two–dimensional case. The appropriate notation is:

$$P = (x, y), \qquad P_0 = (x_0, y_0),$$
$$\mathbf{r} = \langle x, y \rangle, \qquad \mathbf{r}_0 = \langle x_0, y_0 \rangle, \qquad \mathbf{v} = \langle a, b \rangle.$$

Since $\mathbf{v} \neq 0$, either $a \neq 0$ or $b \neq 0$ or both. The vector equation $\mathbf{r} = \mathbf{r}_0 + t\mathbf{v}$ for L is satisfied if and only if the components of the two sides of the vector equation are equal. Since $\mathbf{r} = \langle x, y \rangle$ and $\mathbf{r}_0 + t\mathbf{v} = \langle x_0 + at,\ y_0 + bt \rangle$, we must have

$$x = x_0 + at, \qquad y = y_0 + bt \quad \text{for} \quad -\infty < t < \infty.$$

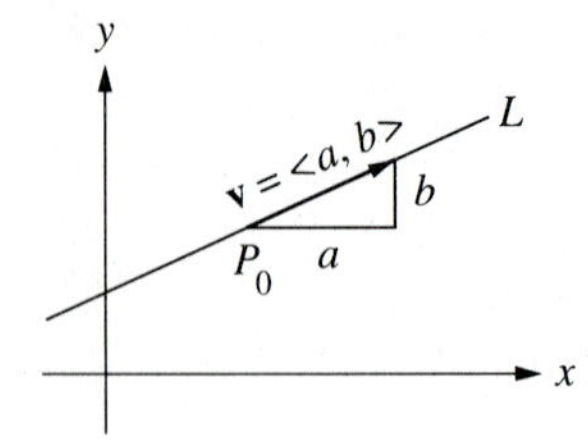

FIGURE 5

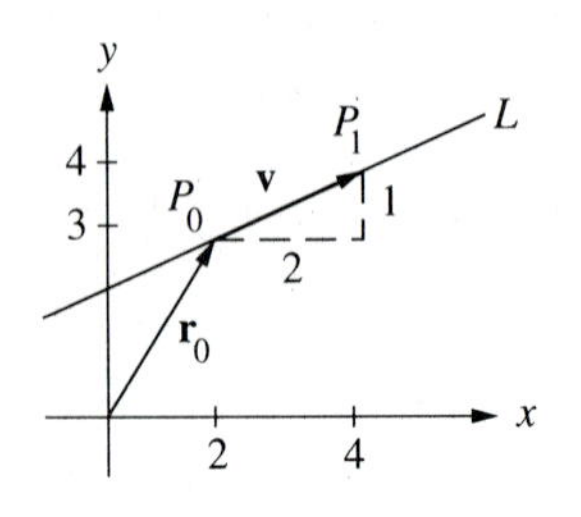

FIGURE 6

These equations are called **parametric equations** for L. The variable t is called a **parameter**. If $b = 0$, then L is the horizontal line $y = y_0$ in the xy–plane. If $a = 0$, then L is the vertical line $x = x_0$. Suppose that $a \neq 0$. Then the slope of L is $m = b/a$, as shown in Fig. 5. The familiar point–slope equation for L can be obtained from the parametric equations by eliminating the parameter t: solve $x = x_0 + at$ for $t = (x - x_0)/a$ and substitute it into $y = y_0 + bt$ to obtain

$$y - y_0 = m(x - x_0).$$

EXAMPLE 1. Let L be the line that passes through the points $P_0 = (2,3)$ and $P_1 = (4,4)$ in Fig. 6. Find vector and parametric equations for L.

Solution. In Fig. 6, the position vector for the point $P_0 = (2,3)$ on L is $\mathbf{r}_0 = <2,3>$ and a direction vector for L is $\mathbf{v} = \overrightarrow{P_0P_1} = <2,1>$. So a vector equation for L is

$$\mathbf{r} = \mathbf{r}_0 + t\mathbf{v} = <2,3> + t<2,1>$$

or, equivalently,

$$\mathbf{r} = <2 + 2t, 3 + t>.$$

Since $\mathbf{r} = <x,y>$, corresponding parametric equations are

$$x = 2 + 2t, \qquad y = 3 + t.$$

For comparison purposes, let's eliminate t. Solve $y = 3 + t$ for $t = y - 3$ and substitute it into $x = 2 + 2t$ to obtain $x = 2y - 4$. Then $y = (1/2)x + 2$. This is the slope–intercept equation for L. It shows that L has slope $m = 1/2$ and y–intercept 2. □

Now suppose that Fig. 4 illustrates a line L in 3–space. Then the notation is:

$$P = (x,y,z), \qquad P_0 = (x_0, y_0, z_0),$$
$$\mathbf{r} = <x,y,z>, \qquad \mathbf{r}_0 = <x_0, y_0, z_0>, \qquad \mathbf{v} = <a,b,c>.$$

Since $\mathbf{v} \neq 0$, at least one of a, b, c is nonzero. A vector equation for L is $\mathbf{r} = \mathbf{r}_0 + t\mathbf{v}$, with $-\infty < t < \infty$. Equate components, just as we did in 2–space, to obtain **parametric equations** for L:

$$x = x_0 + at, \qquad y = y_0 + bt, \qquad z = z_0 + ct, \qquad \text{for} \qquad -\infty < t < \infty.$$

If $c = 0$, then z is constant and L is parallel to the xy–plane. If both $b = 0$ and $c = 0$, then both y and z are constant, so that L is parallel to the x–axis.

If a, b, c are all nonzero, then the three parametric equations can be solved for t to obtain

$$\frac{x - x_0}{a} = \frac{y - y_0}{b} = \frac{z - z_0}{c}.$$

These equations are called the **symmetric equations** for L.

EXAMPLE 2. Find vector and parametric equations for (a) the line L through the points $P_0 = (1,0,2)$ and $P_1 = (3,1,-1)$ and (b) the line segment from $P_0 = (1,0,2)$ to $P_1 = (3,1,-1)$.

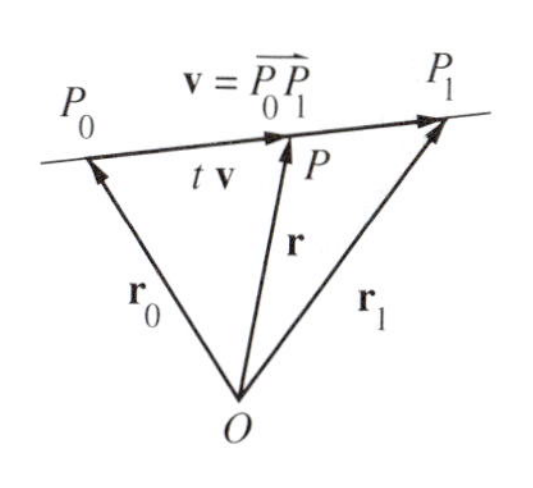

FIGURE 7

Solution. Figure 7 illustrates the situation. (a) Let $\mathbf{r}_0 = <1,0,2>$ and $\mathbf{v} = \overrightarrow{P_0P_1} = <2,1,-3>$ in the vector equation for a line to obtain

$$\mathbf{r} = \mathbf{r}_0 + t\mathbf{v} = <1 + 2t,\ t,\ 2 - 3t>, \qquad -\infty < t < \infty.$$

Since $\mathbf{r} = <x,y,z>$, corresponding parametric equations for L are

$$x = 1 + 2t, \qquad y = t, \qquad z = 2 - 3t, \qquad -\infty < t < \infty.$$

(b) The triangle law for vector addition and a glance at Fig. 7 show that the position vector $\mathbf{r} = \mathbf{r}_0 + t\mathbf{v}$ with $\mathbf{v} = \overrightarrow{P_0P_1}$ traverses the line segment from P_0 to P_1 as t increases from 0 to 1. Thus, the parametric equations in (a) describe the line segment $\overline{P_0P_1}$ when t is restricted to the interval $0 \le t \le 1$. □

An instructive way to consider lines is to imagine that t is time and that the vector equation $\mathbf{r} = \mathbf{r}_0 + t\mathbf{v}$ describes the position of an object at time t as it moves along the line. Let $t_1 < t_2$ and let

$$\mathbf{r}_1 = \mathbf{r}_0 + t_1\mathbf{v}, \qquad \mathbf{r}_2 = \mathbf{r}_0 + t_2\mathbf{v}$$

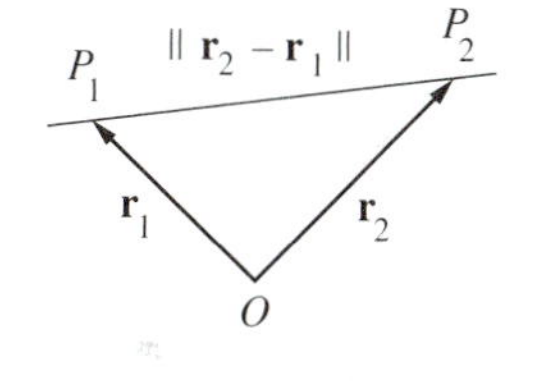

FIGURE 8

be the corresponding positions of the object, as illustrated in Fig. 8. During the time interval from t_1 to t_2, the object travels the distance

$$\|\mathbf{r}_2 - \mathbf{r}_1\| = \|t_2\mathbf{v} - t_1\mathbf{v}\| = (t_2 - t_1)\|\mathbf{v}\|.$$

It follows that the object moves with constant speed $\|\mathbf{v}\|$. Since $\mathbf{v}$ points in the direction of motion, $\mathbf{v}$ is the velocity. From this point of view, the vector equation $\mathbf{r} = \mathbf{r}_0 + t\mathbf{v}$ describes motion along a straight line L with constant velocity $\mathbf{v}$ and constant speed $\|\mathbf{v}\|$.

EXAMPLE 3. Let L be the line with parametric equations

$$x = 5 + 3t, \qquad y = 7 + 4t, \qquad z = 10 + 10t.$$

(a) Give a direction vector $\mathbf{v}$ for L and find a vector equation for L. (b) Find the horizontal and vertical projections of $\mathbf{v}$. Then determine the slope m of $\mathbf{v}$ relative to the xy–plane and its angle of inclination φ. (c) Where does L intersect the xy–plane?

Solution. (a) Compare the given parametric equations with the general parametric equations for a line to see that the coefficients of t determine a direction vector $\mathbf{v} = <3,4,10>$ for L. Let $t = 0$ to obtain the point $P_0 = (5,7,10)$ on L, with position vector $\mathbf{r}_0 = <5,7,10>$. Consequently, a vector equation for L is

$$\mathbf{r} = \mathbf{r}_0 + t\mathbf{v} = <5 + 3t, 7 + 4t, 10 + 10t>.$$

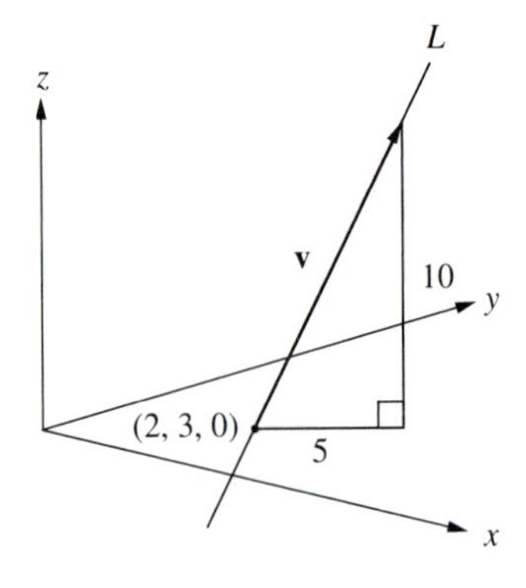

FIGURE 9

(b) The horizontal projection of $\mathbf{v} = <3,4,10>$ is $\mathbf{v}_0 = <3,4,0> = 3\mathbf{i} + 4\mathbf{j}$ and its vertical projection is $\mathbf{v}_1 = <0,0,10> = 10\mathbf{k}$. The slope of $\mathbf{v}$ relative to the *xy*–plane is

$$m = \frac{10}{||\mathbf{v}_0||} = \frac{10}{5} = 2.$$

Since $\tan \varphi = m$, the angle of inclination is

$$\varphi = \arctan 2 \approx 63°.$$

(c) The line L intersects the *xy*–plane when $z = 10 + 10t = 0$. Then $t = -1$, so $x = 2$ and $y = 3$. The point of intersection is $(2,3,0)$. The information in (b) and (c) enables us to graph L in Fig. 9. □

Two lines L_1 and L_2 are parallel if and only if they have parallel direction vectors $\mathbf{v}_1$ and $\mathbf{v}_2$. Since direction vectors are nonzero, this means that $\mathbf{v}_2 = \lambda \mathbf{v}_1$ for some nonzero scalar λ.

EXAMPLE 4. Let L_1 and L_2 be two lines with parametric equations

$$L_1: \quad x = 1 + 2t, \quad y = 2 - 3t, \quad z = 3 + 4t,$$
$$L_2: \quad x = 5 + 4t, \quad y = 7 - 6t, \quad z = 1 + 8t.$$

Are L_1 and L_2 parallel?

Solution. Direction vectors for L_1 and L_2 are

$$\mathbf{v}_1 = <2,-3,4>, \qquad \mathbf{v}_2 = <4,-6,8>.$$

Since $\mathbf{v}_2 = 2\mathbf{v}_1$, L_1 and L_2 are parallel. □

In the *xy*–plane, if two lines are not parallel, they must intersect. But in 3–space, two nonparallel lines may miss each other altogether. For example, every line on the ceiling of a room fails to intersect every line on the floor. The following example shows how to tell whether two lines in space intersect and, if they do, how to find the point of intersection.

EXAMPLE 5. Let L_1 and L_2 be lines with parametric equations

$$L_1: \quad x = 2 + t, \quad y = 1 - t, \quad z = 3 + 2t,$$
$$L_2: \quad x = -1 + s, \quad y = 5 - 2s, \quad z = -s,$$

with $-\infty < t < \infty$ and $-\infty < s < \infty$. Determine whether L_1 and L_2 intersect or not. If they intersect, find the point of intersection.

Solution. Notice that we used different letters for the parameters of the two lines. When dealing with two (or more) lines, it is usually best to use different parameters for each line. We'll have more to say about the reason for this shortly. A point (x,y,z) lies on both lines if and only if

$$x = 2 + t = -1 + s, \qquad y = 1 - t = 5 - 2s, \qquad z = 3 + 2t = -s$$

for some s and t. It follows that L_1 and L_2 intersect if and only if the equations

$$s - t = 3, \qquad 2s - t = 4, \qquad s + 2t = -3$$

for s and t have a solution. Solve the first two equations for $s = 1$ and $t = -2$, which just *happen* to satisfy the third equation. For these values of s and t, the equations for L_1 and L_2 both give $x = 0$, $y = 3$, $z = -1$. So the two lines intersect at the point $(0,3,-1)$. In another problem, if the system of three equations for s and t had no solution, then the lines would not intersect. □

A few comments on Ex. 5 are in order. Suppose we had used the same parameter t for both lines:

$$L_1: \quad x = 2 + t, \qquad y = 1 - t, \qquad z = 3 + 2t,$$
$$L_2: \quad x = -1 + t, \qquad y = 5 - 2t, \qquad z = -t.$$

Equate the x–coordinates to get $2 + t = -1 + t$, an equation with no solution for t. A common mistake is to conclude that the lines L_1 and L_2 do not intersect. Of course, we know that they do intersect at $(0,3,-1)$. But the point $(0,3,-1)$ is obtained when $t = -2$ for L_1 and when $t = 1$ for L_2. There is a simple interpretation in terms of motion along the lines L_1 and L_2. Let t be time, say, measured in seconds. An object that moves along L_1 passes through the point $(0,3,-1)$ at time $t = -2$, while an object moving along L_2 passes through the point $(0,3,-1)$ at time $t = 1$, three seconds later. The point $(0,3,-1)$ is on both trajectories, but the objects pass through the point at different times. Although the trajectories intersect, there is no collision.

EXAMPLE 6. Find the acute angle between the lines L_1 and L_2 in Ex. 5.

Solution. From the parametric equations for L_1 and L_2, direction vectors for L_1 and L_2 are

$$\mathbf{v}_1 = \langle 1,-1,2 \rangle, \qquad \mathbf{v}_2 = \langle 1,-2,-1 \rangle,$$

as indicated in Fig. 10. The angle θ between $\mathbf{v}_1$ and $\mathbf{v}_2$ is given by

$$\cos\theta = \frac{\mathbf{v}_1 \cdot \mathbf{v}_2}{\|\mathbf{v}_1\| \, \|\mathbf{v}_2\|} = \frac{1}{6}, \qquad \theta \cong 80°.$$

The two lines are nearly perpendicular. □

It may happen that the angle between direction vectors $\mathbf{v}_1$ and $\mathbf{v}_2$ for lines L_1 and L_2 is obtuse, as in Fig. 11. In this case $\mathbf{v}_1 \cdot \mathbf{v}_2 < 0$. The acute angle θ between the lines can be obtained by reversing the direction of $\mathbf{v}_2$ or, more quickly, from

$$\cos\theta = \frac{|\mathbf{v}_1 \cdot \mathbf{v}_2|}{\|\mathbf{v}_1\| \, \|\mathbf{v}_2\|}.$$

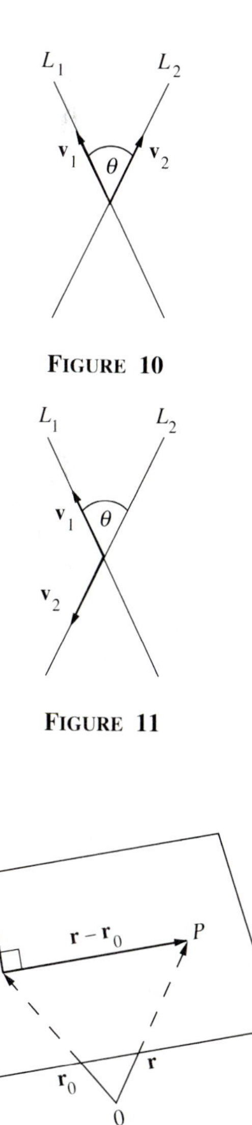

FIGURE 10

FIGURE 11

FIGURE 12

Planes

A plane Π in 3–space is determined by any point $P_0 = (x_0, y_0, z_0)$ on it and any nonzero vector $\mathbf{n} = \langle a,b,c \rangle$ perpendicular to it, as illustrated in Fig. 12.

We say that $\mathbf{n}$ is a **normal vector** for the plane. As usual, the position vector for the fixed point $P_0 = (x_0, y_0, z_0)$ is $\mathbf{r}_0 = < x_0, y_0, z_0 >$ and the position vector for a variable point $P = (x, y, z)$ is $\mathbf{r} = < x, y, z >$.

Figure 12 shows that P lies in the plane Π if and only if $\mathbf{n} \perp \mathbf{r} - \mathbf{r}_0$. Equivalently, $\mathbf{n} \cdot (\mathbf{r} - \mathbf{r}_0) = 0$, which is the **vector equation** for the plane Π.

> The plane Π through $P_0 = (x_0, y_0, z_0)$
> with normal vector $\mathbf{n} = < a, b, c >$ has equation
> $$\mathbf{n} \cdot (\mathbf{r} - \mathbf{r}_0) = 0.$$

Carry out the dot product to obtain a scalar equation for Π:

$$\boxed{a(x - x_0) + b(y - y_0) + c(z - z_0) = 0.}$$

Expand further to obtain another scalar equation for the plane Π:

$$\boxed{ax + by + cz = d,}$$

where $d = ax_0 + by_0 + cz_0$. If a plane is represented in either scalar form, then a normal vector for the plane is $\mathbf{n} = < a, b, c >$.

Since $\mathbf{n} \neq \mathbf{0}$, at least one of a, b, c is nonzero. If $c = 0$, then the equation for the plane reduces to

$$ax + by = d \qquad (\text{any } z).$$

This is a vertical plane, whose intersection with the xy–plane is the line $ax + by = d$, $z = 0$. Similar remarks apply if $a = 0$ or $b = 0$.

EXAMPLE 7. Find the plane Π through the point $P_0 = (5, -1, 2)$ with the normal vector $\mathbf{n} = < 3, 2, -4 >$.

Solution. As usual, $\mathbf{r} = < x, y, z >$ is the position vector of a general point in space. The position vector of P_0 is $\mathbf{r}_0 = < 5, -1, 2 >$. Then $\mathbf{r} - \mathbf{r}_0 = < x - 5, y + 1, z - 2 >$ and a vector equation for the plane is

$$\mathbf{n} \cdot (\mathbf{r} - \mathbf{r}_0) = 3(x - 5) + 2(y + 1) - 4(z - 2) = 0.$$

An equivalent scalar equation for Π is

$$3x + 2y - 4z = 5. \ \square$$

Two planes are parallel if and only if they have parallel normal vectors. For example, the planes

$$3x + 2y - 4z = 1, \qquad 6x + 4y - 8z = 5$$

have the normal vectors

$$\mathbf{n}_1 = < 3, 2, -4 >, \qquad \mathbf{n}_2 = < 6, 4, -8 >.$$

Since $\mathbf{n}_2 = 2\mathbf{n}_1$, the planes are parallel.

Similarly, two planes are perpendicular if and only if they have perpendicular normal vectors, as shown in Fig. 13. For example, the planes

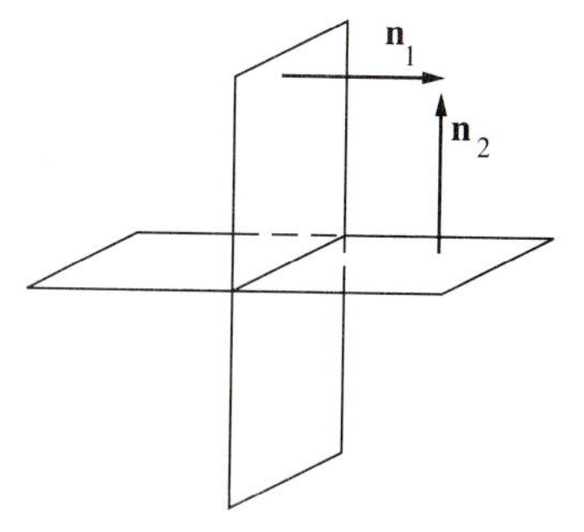

FIGURE 13

$$2x - 3y + 5z = 1, \qquad x + 4y + 2z = 5$$

are perpendicular because the normal vectors $\mathbf{n}_1 = < 2, -3, 5 >$ and $\mathbf{n}_2 = < 1, 4, 2 >$ satisfy $\mathbf{n}_1 \cdot \mathbf{n}_2 = 0$.

Suppose that two planes Π_1 and Π_2 are not parallel. Then they intersect in a line L, as shown in Fig 14. There are normal vectors $\mathbf{n}_1$ to Π_1 and $\mathbf{n}_2$ to Π_2 that meet at an angle θ with $0 \le \theta \le \pi/2$. The figure should help explain why it is reasonable to call θ the **angle between the two planes**. Given any two normal vectors $\mathbf{n}_1$ and $\mathbf{n}_2$ for the planes, the angle between them might be obtuse, as would be the case if $\mathbf{n}_2$ points in the opposite direction in Fig. 14. Then $\mathbf{n}_1 \cdot \mathbf{n}_2 < 0$. If this happens, we could reverse the direction of $\mathbf{n}_2$, which changes the sign of $\mathbf{n}_1 \cdot \mathbf{n}_2$, to find θ. Therefore, in all cases, the angle θ between the two planes is given by

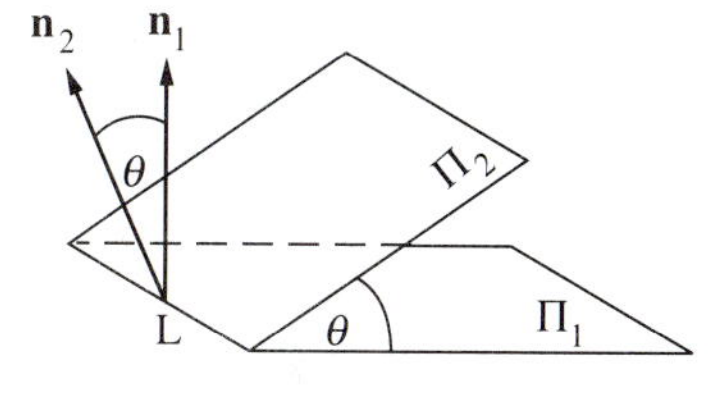

FIGURE 14

$$\cos\theta = \frac{|\mathbf{n}_1 \cdot \mathbf{n}_2|}{||\mathbf{n}_1||\ ||\mathbf{n}_2||}.$$

EXAMPLE 8. Find the angle θ between the planes

$$x + 2y + 3z = 4, \qquad 2x - 3y - z = 5.$$

Solution. Normal vectors for the two planes are

$$\mathbf{n}_1 = < 1, 2, 3 >, \qquad \mathbf{n}_2 = < 2, -3, -1 >.$$

Since $\mathbf{n}_1 \cdot \mathbf{n}_2 = -7$ and $||\mathbf{n}_1|| = ||\mathbf{n}_2|| = \sqrt{14}$,

$$\cos\theta = \frac{|\mathbf{n}_1 \cdot \mathbf{n}_2|}{||\mathbf{n}_1||\ ||\mathbf{n}_2||} = \frac{7}{14} = \frac{1}{2}, \qquad \theta = 60^\circ. \ \square$$

There is a unique plane that contains three distinct points, provided the points do not lie on a straight line. The next two examples show how to find an equation for such a plane.

EXAMPLE 9. Find the plane that contains the three points

$$(2, 0, 0), \qquad (0, 3, 0), \qquad (0, 0, 4).$$

A piece of the plane is the slanting triangle in Fig. 15.

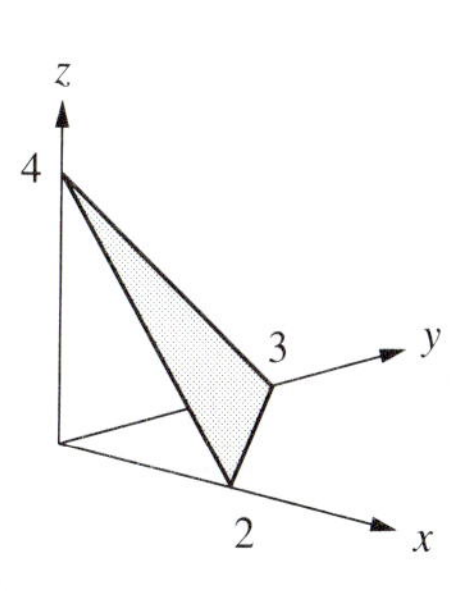

FIGURE 15

Solution. The numbers 2, 3, and 4 are the x–, y–, and z–intercepts of the plane. A scalar equation for the plane with these intercepts can be written down by inspection,

$$\frac{x}{2} + \frac{y}{3} + \frac{z}{4} = 1.$$

Check that this plane contains the three given points. A normal vector for the plane is $\mathbf{n} = \, < 1/2, 1/3, 1/4 >$. □

In the same way, the plane which contains the three points

$$(x_0, 0, 0), \qquad (0, y_0, 0), \qquad (0, 0, z_0),$$

with x_0, y_0, z_0 all nonzero has the equation

$$\frac{x}{x_0} + \frac{y}{y_0} + \frac{z}{z_0} = 1.$$

EXAMPLE 10. Find the plane that contains the three points

$$P_0 = (1, -2, 3), \qquad P_1 = (3, 0, 4), \qquad P_2 = (2, 1, -1).$$

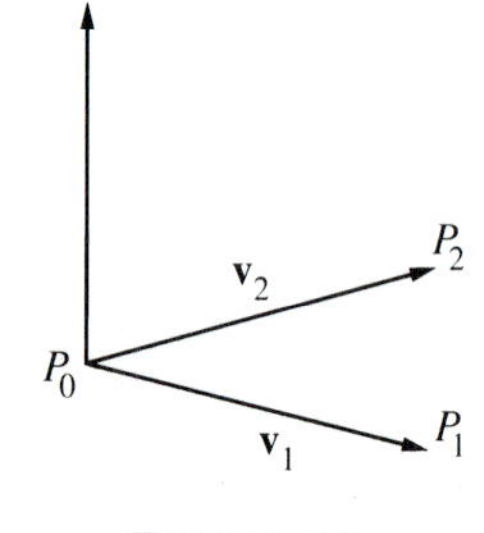

FIGURE 16

Solution. Figure 16 will serve as a guide. The plane we seek contains the point $P_0 = (1, -2, 3)$ and vectors

$$\mathbf{v}_1 = \overrightarrow{P_0P_1} = \, < 2, 2, 1 >, \qquad \mathbf{v}_2 = \overrightarrow{P_0P_2} = \, < 1, 3, -4 >.$$

The vector $\mathbf{n} = \mathbf{v}_1 \times \mathbf{v}_2$ is perpendicular to $\mathbf{v}_1$ and $\mathbf{v}_2$, and, hence, is perpendicular to the plane whose equation we seek. Now,

$$\mathbf{n} = \mathbf{v}_1 \times \mathbf{v}_2 = \begin{vmatrix} \mathbf{i} & \mathbf{j} & \mathbf{k} \\ 2 & 2 & 1 \\ 1 & 3 & -4 \end{vmatrix} = \mathbf{i} \begin{vmatrix} 2 & 1 \\ 3 & -4 \end{vmatrix} - \mathbf{j} \begin{vmatrix} 2 & 1 \\ 1 & -4 \end{vmatrix} + \mathbf{k} \begin{vmatrix} 2 & 2 \\ 1 & 3 \end{vmatrix},$$

$$\mathbf{n} = -11\mathbf{i} + 9\mathbf{j} + 4\mathbf{k} = \, < -11, 9, 4 >.$$

The position vector of P_0 is $\mathbf{r}_0 = \, < 1, -2, 3 >$. The vector equation $\mathbf{n} \cdot (\mathbf{r} - \mathbf{r}_0) = \mathbf{0}$ for the plane yields

$$-11(x - 1) + 9(y + 2) + 4(z - 3) = 0,$$

$$11x - 9y - 4z = 17. \; \square$$

Two nonparallel planes intersect in a line, as in Fig. 14. The next example shows how to find an equation for such a line.

EXAMPLE 11. Find the intersection line L of the planes

$$2x + 2y + z = 6, \qquad x + 3y - 4z = 5.$$

Solution. Refer to Fig. 14 as you go through the solution. Normal vectors for the two planes are

$$\mathbf{n}_1 = <2,2,1>, \qquad \mathbf{n}_2 = <1,3,-4>.$$

Since L lies in both planes, L is perpendicular to both $\mathbf{n}_1$ and $\mathbf{n}_2$. Therefore, L is parallel to $\mathbf{v} = \mathbf{n}_1 \times \mathbf{n}_2$, which is a direction vector for L. From Ex. 10, $\mathbf{v} = <-11,9,4>$. To use the vector formula $\mathbf{r} = \mathbf{r}_0 + t\mathbf{v}$ for L, we also need to find a point P_0 on L. Let's try to find the point $P_0 = (x_0, y_0, 0)$ where L cuts the xy–plane. Since L is on both planes, we set $x = x_0$, $y = y_0$, and $z = 0$ in the equations for the planes to find that

$$2x_0 + 2y_0 = 6,$$
$$x_0 + 3y_0 = 5.$$

These equations are satisfied for $x_0 = 2$ and $y_0 = 1$. Hence, $P_0 = (2,1,0)$ lies on L. Let $\mathbf{r}_0 = <2,1,0>$. Then vector and parametric equations for L are

$$\mathbf{r} = \mathbf{r}_0 + t\mathbf{v} = <2-11t, 1+9t, 4t>,$$
$$x = 2 - 11t, \qquad y = 1 + 9t, \qquad z = 4t.$$

In some cases L will not intersect the xy–plane. Then how can you find a point P_0 on L? □

Finally, we show how to find the (shortest) distance from a point to a plane.

EXAMPLE 12. Find the distance D from $P = (7,24,9)$ to the plane

$$-9x + 12y + 8z = 8.$$

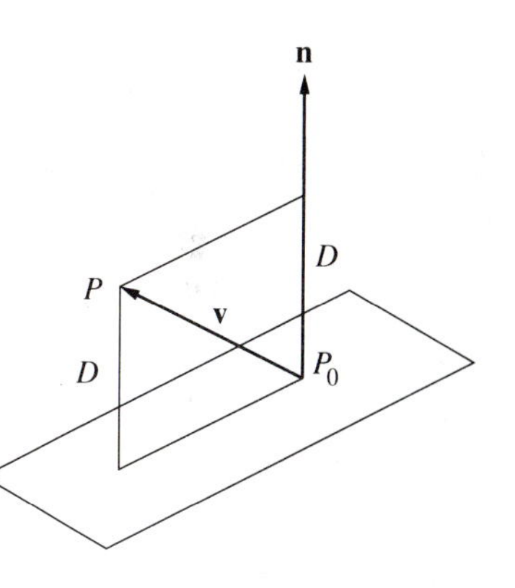

FIGURE 17

Solution. The derivation is guided by Fig. 17. A normal vector for the plane is $\mathbf{n} = <-9,12,8>$. Next we need to find a point P_0 in the plane. This is easy. Let $x = y = 0$ in the equation for the plane and solve for $z = 1$. So $P_0 = <0,0,1>$ lies in the plane. As in Fig. 17, let $\mathbf{v} = \overrightarrow{P_0P} = <7,24,8>$. Then distance from P to the plane is

$$D = |\text{comp}_{\mathbf{n}}\,\mathbf{v}| = \frac{|\mathbf{v}\cdot\mathbf{n}|}{||\mathbf{n}||} = \frac{289}{17} = 17. \ \square$$

The same reasoning leads to a general formula for the distance from a point to a plane. See the problems. However, in particular cases, it is better to reason as in Ex. 12 rather than to memorize a general formula.

PROBLEMS

In Probs. 1–4, find (a) vector and (b) parametric equations for the line in the xy–plane through the given points. (c) Find the slope and y–intercept of the line.

1. $(3,4)$, $(6,6)$
2. $(2,3)$, $(-4,12)$
3. $(1,1)$, $(-2,3)$
4. $(5,0)$, $(0,3)$

In Probs. 5–10, find (a) vector and (b) parametric equations for the line in space with the given properties.

5. Passes through the points $(0,1,-2)$ and $(3,2,-4)$.
6. Passes through the points $(-6,8,3)$ and $(2,5,7)$.
7. Contains the point $(3,1,-1)$ and has direction vector $< 2,1,-3 >$.
8. Contains the point $(3,4,5)$ and has direction vector $< 1,-2,1 >$.
9. Passes through $(1,2,3)$ and is perpendicular to the *yz*–plane.
10. Passes through $(2,4,-5)$ and is parallel to the line with vector equation $\mathbf{r} = < 2 - 3t,\ 4 + t,\ -t >$.

In Probs. 11–14 a line L is described. (a) Find a direction vector $\mathbf{v}$ for L. (b) Write a vector equation for L. (c) Find the point (if any) where L intersects the *xy*–plane. (d) Find the slope (relative to the *xy*–plane) and the angle of inclination for $\mathbf{v}$.

11. $x = 4 - 3t, \quad y = 2t, \quad z = 3 - t$
12. $\mathbf{r} = < 2, 3 - 4t, 5t >$
13. Passes through $(1,0,2)$ and is perpendicular to the plane $x + y - z = 3$.
14. Passes through $(1,1,-1)$ and is perpendicular to the lines $\mathbf{r} = <2,3-4t,5t>$ and $\mathbf{r} = < 15 + 8t, 13 + 6t, 5 + 5t >$.

In Probs. 15–18 determine whether the following lines intersect. If they do, find a) the point of intersection and (b) the smaller angle between them.

15. L_1: $x = 2 + t, y = 2, z = 2 - t$
    L_2: $x = 1 + t, y = 2 - 3t, z = 3 + 2t$
16. L_1: $\mathbf{r} = (3 + 4t)\mathbf{i} + (2 + t)\mathbf{j} + (1 - t)\mathbf{k}$
    L_2: $\mathbf{r} = (1 + t)\mathbf{i} - (4 - 3t)\mathbf{j} - (1 - t)\mathbf{k}$
17. L_1: $\mathbf{r} = < 3 + 4t, 4 + t, 1 - t >$
    L_2: $\mathbf{r} = < 2 - t, t, 2 + 3t >$
18. L_1: $\mathbf{r} = \mathbf{j} + t\mathbf{k}$
    L_2: $\mathbf{r} = \mathbf{k} + t(\mathbf{j} + \mathbf{k})$
19. Find the distance from the point (4, 2, 5) to the line with parametric equations $x = 4 - 3t$, $y = 2t$, $z = 3 - t$.
20. Let L be a line with direction vector $\mathbf{v}$ and P a point not on L. Let $\mathbf{w}$ be a vector from some point on L to P. Show that the distance from P to L is given by $||\mathbf{w} \times \mathbf{v}||/||\mathbf{v}||$.

In Probs. 21–28, find an equation for the plane with the given properties.

21. Passes through the point $(1,2,3)$ and has normal vector $< 2,-1,4 >$.
22. Passes through the point $(2,0,-4)$ and has normal vector $\mathbf{i} + 2\mathbf{j} - 3\mathbf{k}$.
23. Passes through the point $(3,4,5)$ and is parallel to the *yz*–plane.
24. Passes through the point $(3,4,5)$ and is perpendicular to the line given by $\mathbf{r} = < 3 + 2t, 5t, -4 + t >$.

25. Passes through the point $(1, 0, 1)$ and is parallel to the plane $x - 2y + 3z = 0$.

26. Passes through the origin and the points $(1, -1, 1)$ and $(2, 3, -1)$.

27. Passes through the points $(1, 2, 1)$, $(3, 4, 5)$, $(2, 4, 6)$.

28. Contains the line $\mathbf{r} = < 2 - t, 3 + 2t, 4t >$ and is perpendicular to the plane $x + 2y - z = 3$.

29. The lines in Ex. 4 are parallel. Find an equation for the plane that contains both lines.

30. The lines in Ex. 5 intersect. Find an equation for the plane that contains both lines.

In Probs. 31–34, find the angle between the planes.

31. $-3x + 2y + 5z = 11, \quad 3x + 2y + z = 5$

32. $x + y - z = 7, \quad 2x + y + 3z = 4$

33. $x + y + z = 1$, xy–plane

34. $3x + 2y + z = 4, \quad x + 3y - 2z = -5$

35. Find vector and parametric equations for the line of intersection of the planes $x + 2y + 2z = 6$ and $-4x + 3y + z = 5$.

36. Find vector and parametric equations for the line of intersection of the planes $x + y + z = 1$ and $x - y - z = 1$.

37. Write symmetric equations for the line $\mathbf{r} = < 2t, 1 - 3t, 2t >$.

38. Suppose that a line has direction vector $\mathbf{v} = < a, b, c >$. If $a \neq 0$ and $b \neq 0$, but $c = 0$, show that the symmetric equations for the line become

$$\frac{x - x_0}{a} = \frac{y - y_0}{b}, \qquad z = z_0.$$

39. Write symmetric equations for the line $\mathbf{r} = < 2 + 2t, 5, 7 - 6t >$.

40. Find the distance from the point $(1, 2, 3)$ to the plane $x + y + z = -4$.

41. Find the distance from the origin to the plane $3x - 2y + 4z = 1$.

42. Show that the distance D from the origin to the plane $ax + by + cz = d$ is

$$D = \frac{|d|}{\sqrt{a^2 + b^2 + c^2}}.$$

Consequently, if $< a, b, c >$ is a unit vector, then $|d|$ is the distance from the origin to the plane.

43. Show that the distance D from the point $P_0 = (x_0, y_0, z_0)$ to the plane $ax + by + cz = d$ is

$$D = \frac{|ax_0 + by_0 + cz_0 - d|}{\sqrt{a^2 + b^2 + c^2}}.$$

44. The planes $ax + by + cz = d_1$ and $ax + by + cz = d_2$ are parallel. Why? Show that the (constant perpendicular) distance D between them is

$$D = \frac{|d_1 - d_2|}{\sqrt{a^2 + b^2 + c^2}}.$$

45. Show that any two lines L_1 and L_2 in space that do not intersect must lie in parallel planes.

46. Use Probs. 44 and 45 to find the distance between the line that contains the points $(0,1,0)$ and $(1,0,0)$ and the line that contains the points $(-1,0,1)$ and $(1,1,1)$.

47. Solve the preceding problem again as follows: Find a vector **n** perpendicular to both lines. Then calculate the component of the vector from $(0,1,0)$ to $(1,1,1)$ along **n**. Explain why this does the job.

48. Assume that the plane determined by the nonzero vectors **a, b** and the plane determined by the nonzero vectors **c**, **d** are not parallel. Explain why $(\mathbf{a} \times \mathbf{b}) \times (\mathbf{c} \times \mathbf{d})$ is parallel to the line of intersection of the planes.

There are two useful parametric representations of lines in the xy–plane. The first, $\mathbf{r} = \mathbf{r}_0 + t\mathbf{v}$, is based on a vector parallel to the line and was discussed in the text. The second uses a vector normal to the line. You are asked to develop that representation in the next problem.

49. Let L be a line in the xy–plane, $\mathbf{r}_0 = \langle x_0, y_0 \rangle$ a point on it, and $\mathbf{n} = \langle a, b \rangle$ a vector perpendicular to L. (a) Make a sketch of this situation. (b) Show that the position vector $\mathbf{r} = \langle x, y \rangle$ to a variable point $P = (x, y)$ lies on L if and only if

$$\mathbf{n} \cdot (\mathbf{r} - \mathbf{r}_0) = 0.$$

This is the **vector normal equation** of the line. (c) Express the vector equation in scalar form to obtain

$$a(x - x_0) + b(y - y_0) = 0.$$

(d) If **n** is a unit vector, show that the previous equation can be expressed as

$$ax + by = d,$$

where $|d|$ is the distance from the origin to the line.

Chapter Highlights

Vectors in 2–space and 3–space are expressed by $\mathbf{v} = \langle a, b \rangle$ and $\mathbf{v} = \langle a, b, c \rangle$. They are represented geometrically as directed line segments. Important examples are velocity vectors and force vectors.

The dot product of vectors $\mathbf{v} = \langle v_1, v_2, v_3 \rangle$ and $\mathbf{w} = \langle w_1, w_2, w_3 \rangle$ is defined by

$$\mathbf{v} \cdot \mathbf{w} = v_1 w_1 + v_2 w_2 + v_3 w_3.$$

The dot product is expressed geometrically by

$$\mathbf{v} \cdot \mathbf{w} = ||\mathbf{v}|| \; ||\mathbf{w}|| \cos\theta,$$

where θ is the angle between $\mathbf{v}$ and $\mathbf{w}$. The vectors $\mathbf{v}$ and $\mathbf{w}$ are orthogonal if $\mathbf{v} \cdot \mathbf{w} = 0$. The dot product is used to express components and projections of one vector along another. For example, we might be interested in the component of a force in a particular direction.

The cross product of $\mathbf{v}$ and $\mathbf{w}$ is expressed in determinant form by

$$\mathbf{v} \times \mathbf{w} = \begin{vmatrix} \mathbf{i} & \mathbf{j} & \mathbf{k} \\ v_1 & v_2 & v_3 \\ w_1 & w_2 & w_3 \end{vmatrix}.$$

The cross product $\mathbf{v} \times \mathbf{w}$ is orthogonal to both $\mathbf{v}$ and $\mathbf{w}$. The length of $\mathbf{v} \times \mathbf{w}$ is

$$||\mathbf{v} \times \mathbf{w}|| = ||\mathbf{v}|| \; ||\mathbf{w}|| \sin\theta,$$

which is the area of the parallelogram formed by $\mathbf{v}$ and $\mathbf{w}$. Here θ is the angle between $\mathbf{v}$ and $\mathbf{w}$.

The position vector of a point $P = (x, y, z)$ is the vector $\mathbf{r} = < x, y, z >$. Position vectors are convenient for expressing equations for lines and planes in vector forms. In the coming chapters, they will be used to represent other curves and surfaces in vector forms.

Let $\mathbf{r}_0 = < x_0, y_0, z_0 >$ be the position vector for a point $P_0 = (x_0, y_0, z_0)$. The line L through P_0 with direction vector $\mathbf{v} = < a, b, c >$ has the vector equation

$$\mathbf{r} = \mathbf{r}_0 + t\mathbf{v}, \qquad -\infty < t < \infty.$$

Equivalent parametric equations are

$$x = x_0 + at, \qquad y = y_0 + bt, \qquad z = z_0 + ct.$$

A vector equation for a plane through a point P_0 with normal vector $\mathbf{n}$ is

$$\mathbf{n} \cdot (\mathbf{r} - \mathbf{r}_0) = 0.$$

An equivalent scalar equation is

$$a(x - x_0) + b(y - y_0) + (z - z_0) = 0.$$

The project for this chapter uses basic vector algebra to study familiar and important forces that affect virtually every human endeavor – frictional forces. Frictional forces have both beneficial and detrimental effects. For example, frictional forces help to make the tires on a car roll and provide traction. At the same time, additional gasoline is required to overcome the same frictional forces.

Chapter Project: Friction

Friction occurs when one object slides along another. Properties of frictional forces can be observed and analyzed by means of experiments involving objects sliding down inclined planes. This project offers both mathematical and experimental challenges.

In the discussion of friction, use the following consequence of Newton's second law: A body will remain at rest or in a state of uniform motion along a line if and only if the sum of the forces acting on the body is zero.

Problem 1 Let $\mathbf{F}_1, \mathbf{F}_2, \ldots, \mathbf{F}_n$ be given forces that are in equilibrium, meaning their sum is $\mathbf{0}$. Show that the sum of the components of the forces along any direction $\mathbf{n}$ in space also is 0.

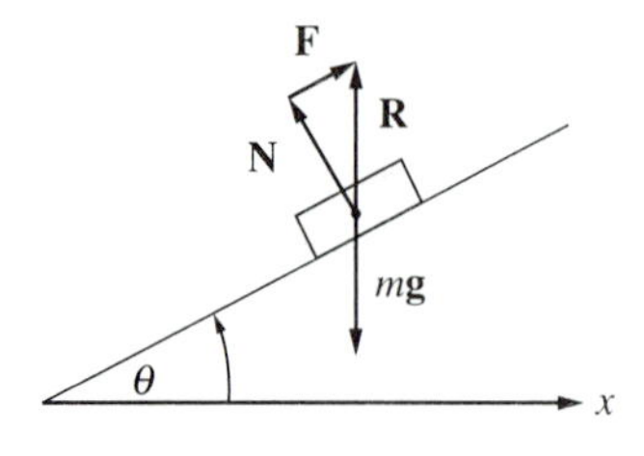

FIGURE 1

Figure 1 shows a block of mass m placed on an inclined plane. In the figure $\mathbf{R}$ is the reaction force of the inclined plane to the weight of the block. The projections of $\mathbf{R}$ perpendicular and parallel to the inclined plane are denoted by $\mathbf{N}$ and $\mathbf{F}$. We call $\mathbf{F}$ the frictional force. Imagine that the angle of inclination θ can vary. For example, when $\theta = 0$ the situation models a block on a horizontal table. Suppose the block is at rest on the table and that θ begins to increase gradually. It is an experimental fact (see Prob. 5) that at first the block will remain at rest as the frictional force $\mathbf{F}$ increases in magnitude so as to just balance the increasing component of the gravitational force down the inclined plane. At a critical angle $\theta = \theta^+$, the *angle of static friction* (or *repose),* the block begins to slide down the plane. At this critical angle, it is found by experiment that the magnitude of the frictional force satisfies

$$||\mathbf{F}|| = \mu^+ \, ||\mathbf{N}||,$$

where $\mathbf{F}$ and $\mathbf{N}$ are as in the figure when $\theta = \theta^+$ and the constant μ^+ is the *coefficient of static friction.*

Problem 2 Show that $\mu^+ = \tan\theta^+$.

Next, suppose the angle θ is large enough so the block is sliding down the inclined plane. Adjust the angle θ so that the block slides down the plane at constant speed. The angle in question, say θ^-, is the *angle of sliding friction.* In this context, there is a *coefficient of sliding friction* μ^- such that

$$||\mathbf{F}|| = \mu^- \, ||\mathbf{N}||,$$

with $\mathbf{F}$ and $\mathbf{N}$ as in the figure when $\theta = \theta^-$.

Problem 3 Show that $\mu^- = \tan\theta^-$.

Experiments show that $\theta^- < \theta^+$. So $\mu^- < \mu^+$.

Problem 4 Give a physical explanation for $\theta^- < \theta^+$.

Problem 5 Carry out the foregoing experiments for a particular block and an inclined plane of your choice. Measure θ^+ and θ^- and determine the coefficients of static and sliding friction.

Return to the inclined plane in Fig. 1. Assume again that the block is sitting at rest on the plane but that in addition to the forces shown in the figure a horizontal force $\mathbf{H}$ is applied to the block and that its line of action is through the center of the block. Let $\mathbf{H} = H\mathbf{i}$. If $H > 0$, $\mathbf{H}$ is directed to the right in Fig. 1. In the problems that follow you are asked to find the minimum and maximum values $H_{\min}$ and $H_{\max}$ of H that are consistent with static equilibrium. Observe that H will have its minimum value $H_{\min}$ when the block is just about to start slipping down the plane and that H will have its maximum value $H_{\max}$ when the block is just about to start moving up the plane.

Problem 6 After incorporating the new force **H** into Fig. 1, resolve the forces on the block into components parallel and perpendicular to the inclined plane and then show that $H_{\min}$ satisfies the equations

$$H_{\min} \sin\theta + mg\cos\theta = R_N,$$
$$H_{\min} \cos\theta + \mu^+ R_N = mg\sin\theta,$$

where R_N is the normal component of the reaction force **R** of the plane on the block and $\mu^+ R_N$ is the magnitude of the frictional force.

Problem 7 Now solve for $H_{\min}$ and after that show that

$$H_{\min} = mg\tan(\theta - \theta^+),$$

where θ^+ is the angle of static friction.

Problem 8 Does the result in Prob. 7 make good physical sense? Does it make sense for $\theta < \theta^+$, $\theta = \theta^+$, $\theta > \theta^+$? Discuss briefly.

Problem 9 Show that

$$H_{\max} = mg\tan(\theta + \theta^+).$$

Problem 10 Does the result in Prob. 9 make good physical sense? Discuss briefly.

Chapter Review Problems

1. Given $\mathbf{v} = <2,3>$ and $\mathbf{w} = <4,-1>$, find (a) $2\mathbf{v} - \mathbf{w}$ and (b) $3\mathbf{v} + 2\mathbf{w}$.
2. Given $\mathbf{v} = 3\mathbf{i} - 2\mathbf{j} + \mathbf{k}$ and $\mathbf{w} = -2\mathbf{i} + \mathbf{j} - 3\mathbf{k}$, find (a) $3\mathbf{v} + \mathbf{w}$ and (b) $2\mathbf{v} - 3\mathbf{w}$.
3. Let $\mathbf{v} = <3,-2,4>$ and $\mathbf{w} = <1,3,-2>$. Find (a) $\mathbf{v} \times \mathbf{w}$ and (b) $\mathbf{v} \cdot \mathbf{w}$.
4. Given $\mathbf{w} = <4,-2,4>$, find (a) $\|\mathbf{w}\|$ and (b) a unit vector **u** in the direction of **w**.
5. Given $P = (1,2,-3)$ and $Q = (4,4,3)$, find $\mathbf{v} = \overrightarrow{PQ}$ and $\|\mathbf{v}\|$.
6. Find the direction cosines and direction angles of $\mathbf{v} = <3,6,6>$.
7. Find the cosine of the angle between $\mathbf{v} = \mathbf{i} - 2\mathbf{j} + 2\mathbf{k}$ and $\mathbf{w} = 6\mathbf{i} + 3\mathbf{j} + 2\mathbf{k}$.
8. Find all unit vectors **u** perpendicular to $\mathbf{v} = <4,3>$. Illustrate with a sketch.
9. Find all unit vectors **u** in space perpendicular to $\mathbf{v} = <4,3,0>$. Illustrate with a sketch.
10. Let $\mathbf{v} = <1,-2,2>$ and $\mathbf{w} = <-1,0,t>$. Find a value of t so that $\mathbf{v} \perp \mathbf{w}$.
11. A ball is moving at speed 60 mph southwest over the *xy*–plane with angle of inclination 60°. (a) Find the velocity **v** of the ball. (b) Find the horizontal projection $\mathbf{v}_0$ of **v**. (c) Find the vertical projection $\mathbf{v}_1$ of **v**.

12. A sled is pulled 60 feet in a straight line across a frozen pond by a 10 pound force that acts at an angle inclined 30° to the direction of motion. (a) Find the effective force in the direction of motion. (b) Find the work done.

13. Find a vector perpendicular to both $\mathbf{v} = <2,1,3>$ and $\mathbf{w} = <1,-1,-1>$.

14. Use the dot product to show that an angle inscribed in a semicircle is a right angle. *Hint.* Choose rectangular coordinates so that the semicircle has center at $(0,0)$, diameter on the x–axis, and radius r. Let (x,y) be the vertex of the angle. Express the vectors extending from the vertex to the ends of the diameter of the semicircle in terms of x, y, and r.

Find vector parametric equations for each of the lines in Probs. 15–19.

15. The line through $P_0 = (3,0,2)$ parallel to the vector from $P_1 = (1,2,3)$ to $P_2 = (7,4,0)$.

16. The line through $P_0 = (2,-1,4)$ and $P_1 = (5,1,2)$.

17. The line through $P_0 = (4,1,-3)$ perpendicular to the xy–plane.

18. The line through $P_0 = (3,2,1)$ perpendicular to the *plane* $x + y = 0$. *Hint.* Draw a figure.

19. The line through $P_0 = (1,1,1)$ parallel to the line given by $\mathbf{r} = <2+t, 3-4t, 5+2t>$.

20. Find the minimum distance from the origin to the line given by $x = 3 + 2t$, $y = -3 - t$, $z = 2t$.

21. Find the symmetric equations for the line through $P_0 = (5,-3,1)$ and perpendicular to the plane $-4x + y - 2z = 7$.

22. Find the distance from the origin to the plane $3x + 4y - 5z = 10$.

23. Find the distance from the point $(2,1,-2)$ to the plane $3x + 4y - 5z = 10$.

24. Find the distance between the parallel planes $3x + 4y + 5z = 10$ and $3x + 4y + 5z = 20$.

Determine whether the vectors in Probs. 25 and 26 are coplanar.

25. $<1,2,6>$, $<3,5,9>$, and $<2,3,3>$.

26. $\mathbf{i} + 2\mathbf{j} - 3\mathbf{k}$, $3\mathbf{i} - \mathbf{j} + 4\mathbf{k}$, $2\mathbf{j} - \mathbf{k}$.

27. Where does the line

$$\frac{x+2}{3} = \frac{y-1}{2} = \frac{z+6}{1}$$

intersect the plane $2x + y - 3z = 20$?

28. Find a scalar equation for the plane consisting of all points which have the same distance from $P = (1,3,5)$ and $Q = (-1,5,3)$.

29. Show that the four points $P_1 = (2,2,3)$, $P_2 = (1,-3,-1)$, $P_3 = (1,-1,0)$, and $P_4 = (5,5,9)$ all lie on the same plane.

30. Find a point of intersection of the three planes $x + y = 3$, $x + z = 1$, and $y + z = 2$.

31. Given $\mathbf{v} = < 2, 2, 1 >$ and $\mathbf{w} = < 3, 4, 12 >$, find (a) the component of $\mathbf{v}$ along $\mathbf{w}$, (b) the component of $\mathbf{v}$ orthogonal to $\mathbf{w}$, (c) the projection of $\mathbf{v}$ along $\mathbf{w}$, (d) the projection of $\mathbf{v}$ orthogonal to $\mathbf{w}$.

32. For $\mathbf{v} \neq \mathbf{0}$ and $\mathbf{w} \neq \mathbf{0}$, (a) explain why the angle θ between $\mathbf{v}$ and $\mathbf{w}$ satisfies

$$\sin^2\theta = \frac{||\mathbf{v}||^2\, ||\mathbf{w}||^2 - (\mathbf{v} \cdot \mathbf{w})^2}{||\mathbf{v}||^2\, ||\mathbf{w}||^2},$$

and (b) explain why $\mathbf{v} \parallel \mathbf{w} \Leftrightarrow |\mathbf{v} \cdot \mathbf{w}| = ||\mathbf{v}||\, ||\mathbf{w}||$.

33. (a) Find the horizontal and vertical projections of $\mathbf{v} = < 6, 8, 5 >$. (b) Find the slope of $\mathbf{v}$ relative to the xy–plane. (c) Find the angle of inclination of $\mathbf{v}$.

34. Find a scalar equation for the plane containing the lines $\mathbf{r} = <3t, -4t, 5t>$ and $\mathbf{r} = < 1 + 3t, 1 - 4t, 1 + 5t >$.

35. Find the cosine of the angle between the planes $2x + 2y + z = 5$ and $3x + 4y - 12z = -5$.

36. Find the cosine of the angle between the planes $5x + 3y - 4z = 4$ and $y = x$.

37. Find vector and parametric equations for the line of intersection of the planes $y = x$ and $z = y$.

38. Find vector and parametric equations for the line of intersection of the planes $x - 2y + 3z = -2$ and $2x + 3y - 5z = 3$.

39. Show that the lines with parametric equations $x = 1 + t$, $y = 2t$, $z = 1 + 3t$ and $x = 3t$, $y = 2t$, $z = 2 + t$ intersect and find the point of intersection.

40. Find a scalar equation for the plane that contains the two lines in the previous problem.

41. Find a vector $\mathbf{n}$ that is perpendicular to the plane containing $P = (1, -1, 4)$, $Q = (2, 0, 1)$, $R = (0, 2, 3)$.

42. Find the area of the triangle with vertices the three points in the previous problem.

43. Find a scalar equation for the plane containing the point $P_0 = (2, -3, 5)$ and perpendicular to the line joining P_0 and $P_1 = (7, 1, 6)$.

CHAPTER 4
VECTOR CALCULUS

Vector calculus was developed, in large part, to solve motion problems in 2–space or 3–space. The flight of a space shuttle, a stunt pilot, or an acrobat can be analyzed using the techniques of vector calculus. A natural way to describe the motion of an object is to model the object as a moving point and to express the coordinates of the moving point as functions of time. In this context, the path of motion (trajectory of the object) is a parametric curve with time as parameter. We shall also meet parametric curves in other contexts. Typically, the parameter has a special geometric or physical significance.

In Sec. 4.1, parametric curves are treated in scalar form. We emphasize basic techniques for visualizing and graphing such curves. Vector functions are introduced in Sec. 4.2 and parametric curves are expressed in vector form. We learn that the graph of a vector function is a parametric curve. In addition, the concepts of limit, continuity, and differentiability are extended to vector functions. Then, in Sec. 4.3, velocity, speed, and acceleration are defined for motion in 2–space and 3–space and integrals of vector functions make their natural debut. Curvature of a curve, which describes bending, is introduced in Sec. 4.4. Curvature is related to acceleration in motion problems. Finally, the last section of the chapter is devoted to parametric curves expressed in polar coordinate form. This material has significant applications outside calculus but it is not necessary for later chapters of this text. Polar coordinates are particularly convenient for describing motions maintained by central forces, such as gravity. The section concludes with a modern version of Newton's derivation of Kepler's laws of planetary motion.

4.1 Parametric Curves

Parametric curves arise naturally as the trajectories (paths) of moving objects, with time as the parameter. For example, the location of a space shuttle is given by expressing its x–, y–, and z–coordinates as functions of time. Parametric curves also arise when curves are described in terms of natural geometric parameters. For example, a circle is conveniently described in terms of a central angle, which serves as a parameter. In this section, we concentrate on understanding the basic properties of parametric curves and we learn how to graph and visualize such curves. Physical applications will come later. We begin with parametric curves that lie in a plane. Along the way, we learn how to find slopes of plane parametric curves and how to use parametric equations to evaluate certain integrals. The section concludes with parametric curves in 3–space.

Parametric Curves in 2–Space

We met parametric equations for lines in Ch. 3. For example, $x = 1 + 2t$ and $y = 3 + 4t$ are parametric equations for the line L in the xy–plane, which passes through the point (1,3) when $t = 0$ and the point (3,7) when $t = 1$. As t increases, the moving point $P(t) = (1 + 2t, 3 + 4t)$ traces out the line L in the direction from (1,3) to (3,7), which is called the positive direction on L. Thus, L is a directed line.

In the same manner, **parametric equations**

$$x = x(t), \qquad y = y(t)$$

represent a directed curve C in the xy–plane under reasonable conditions on the functions $x(t)$ and $y(t)$. See Fig. 1. The variable t that helps us describe the curve is called a **parameter**. As t increases, the moving point $P(t) = (x(t), y(t))$ traces out C in a definite direction, called the *positive direction* on C. We call C a **parametric curve**. The curve is *continuous* (or *differentiable*) if $x(t)$ and $y(t)$ are continuous (or differentiable). In typical cases, the parameter is restricted to a given interval. Often, $-\infty < t < \infty$, as is the case for lines. If no domain for t is specified, we take the largest domain for which $x(t)$ and $y(t)$ make sense. Regardless of the meaning of the parameter t, you can always think of t as time and regard C as the path of a moving object with position $P(t) = (x(t), y(t))$ at time t.

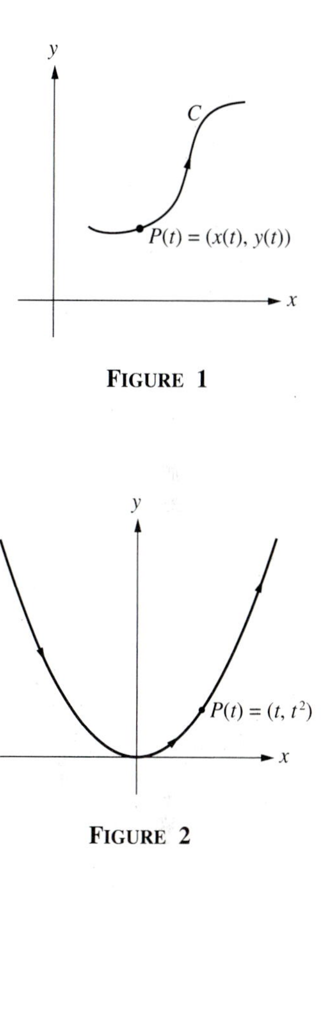

FIGURE 1

FIGURE 2

EXAMPLE 1. Identify the graph of the parametric curve C given by $x = t$, $y = t^2$.

Solution. Since $t = x$ and $y = t^2 = x^2$, the graph is the familiar parabola $y = x^2$ in Fig. 2. As t increases, the point $P(t) = (t, t^2)$ moves along C in the direction of the arrows, which is the positive direction on C. The preceding steps can be reversed. Starting with the parabola $y = x^2$, let $t = x$. Then $x = t$ and $y = t^2$ express the parabola in parametric form. □

Although t is traditional, any letter can be used as a parameter. For example, the parabola in Ex. 1 can be represented also by $x = s$, $y = s^2$.

A particular curve can be represented parametrically in many ways. For example, another parametric representation for the parabola $y = x^2$ is given by $x = t^3$, $y = t^6$. If t is time, the two parametric representations for the parabola, $x = t$, $y = t^2$ and $x = t^3$, $y = t^6$, can be interpreted as describing two different motions along the same parabola. In the first description, an object passes through the point (2,4) when $t = 2$, while in the second description, the object passes through the point (8,64) when $t = 2$. The two objects move along the parabola in the same direction but the actual motions are quite different. Yet another parameterization of the parabola $y = x^2$ is $x = -t$, $y = t^2$. Now, x decreases as t increases, and points on the parabola are swept out from right to left, rather than from left to right as in Fig. 2.

The graph of any function $y = f(x)$ can be recast in the parametric form $x = t$, $y = f(t)$. For example, $y = \sin x$ is expressed equivalently by $x = t$, $y = \sin t$. Thus, graphs of functions can be regarded as parametric curves, but not all parametric curves represent graphs of functions, as we shall see in a moment.

Figure 3 is a graph of the unit circle $x^2 + y^2 = 1$. A glance at the right triangle in the figure reveals that the circle has parametric equations

$$x = \cos\theta \quad \text{and} \quad y = \sin\theta \quad \text{for} \quad 0 \le \theta \le 2\pi.$$

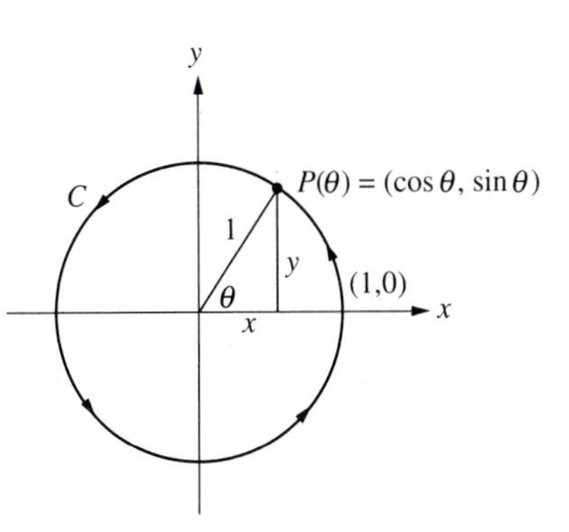

FIGURE 3

As θ increases from 0 to 2π, the point $P(\theta) = (\cos\theta, \sin\theta)$ moves once around the circle in the counterclockwise direction from (1,0) to (1,0). The parameter values $\theta = 0$ and $\theta = 2\pi$ determine the same point (1,0) on the circle. The unit circle is a parametric curve. But since it does not pass the vertical line test, it is not the graph of a function $y = f(x)$.

Graphs of polar equations $r = f(\theta)$ are also parametric curves. The next example makes it clear why this is true.

EXAMPLE 2. Express the polar curve $r = e^{\theta}$, which is a spiral, in parametric form.

Solution. Recall that the polar coordinates (r,θ) and the rectangular coordinates (x,y) of a point are related by $x = r\cos\theta$ and $y = r\sin\theta$. Since $r = e^{\theta}$ on the spiral, parametric equations for the spiral are

$$x = e^{\theta}\cos\theta, \qquad y = e^{\theta}\sin\theta. \ \square$$

Graphing Plane Parametric Curves

There are three common ways (often used in concert) to graph parametric equations: plot points, use calculus or other facts, eliminate the parameter and hope you will know the graph of the resulting equation. The next example illustrates all three approaches.

EXAMPLE 3. Sketch the graph of the parametric curve C given by $x = \sqrt{4t - t^2}$, $y = 2 - t$, for $0 \le t \le 4$.

Solution 1 (Plot Points). The following table gives a few representative points on C.

t	0	1	2	3	4
x	0	$\sqrt{3}$	2	$\sqrt{3}$	0
y	2	1	0	-1	-2

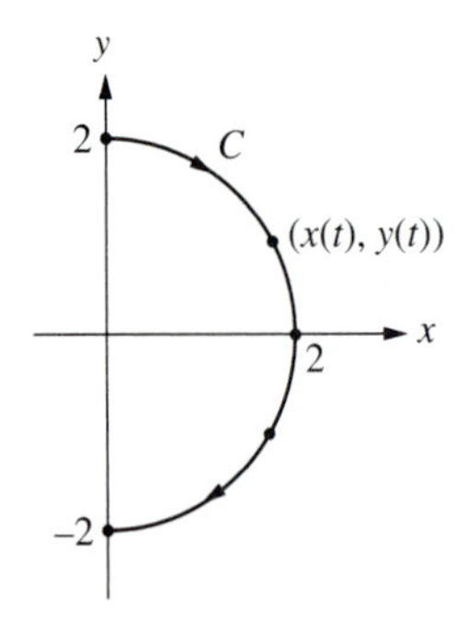

FIGURE 4

The points (x,y) in the table are plotted in Fig. 4 along with a curve that seems to fit. Plotting more points would give us more confidence in the graph. Such a plot can be obtained with the aid of a graphing calculator or graphics utility.

Solution 2 (Think). We can use calculus to determine how a point (x,y) moves on C from (0,2) when $t = 0$ to $(0, -2)$ when $t = 4$. Since

$$x = (4t - t^2)^{1/2} \quad \text{and} \quad x'(t) = \frac{2 - t}{\sqrt{4t - t^2}}$$

$x'(t) > 0$ for $0 \le t < 2$ and $x'(t) < 0$ for $2 < t \le 4$. Therefore,

$x(t)$ increases from 0 to 2 as t increases from 0 to 2,
$x(t)$ decreases from 2 to 0 as t increases from 2 to 4.

On the other hand,

$$y(t) = 2 - t \text{ decreases from 2 to } -2 \text{ as } t \text{ increases from 0 to 4.}$$

Thus, as t increases from 0 to 2, the point $(x(t), y(t))$ moves from (0,2) to (2,0) with $x(t)$ increasing and $y(t)$ decreasing, which gives a graph like the top half of Fig. 4. As t increases from 2 to 4, the point $(x(t), y(t))$ moves from (2,0) to $(0, -2)$ with $x(t)$ decreasing and $y(t)$ decreasing, which gives a graph like the bottom half of Fig. 4. Once again, we obtain something like Fig. 4.

Solution 3 (Eliminate the Parameter). We could solve $y = 2 - t$ for $t = 2 - y$ and substitute the result into $x = \sqrt{4t - t^2}$ to eliminate t. The elimination is easier if we notice instead that

$$x^2 + y^2 = 4t - t^2 + 4 - 4t + t^2 = 4.$$

Thus, the points (x,y) on C satisfy $x^2 + y^2 = 4$. Also, $0 \le x \le 2$ from the parametric equation for x. Consequently, the graph of C is the right half of the circle $x^2 + y^2 = 4$. □

Slopes of Plane Parametric Curves

Figure 5 shows a differentiable parametric curve C with parametric equations $x = x(t)$ and $y = y(t)$. Assume that $x'(t) > 0$ for t in some open interval. Then $x(t)$ increases as t increases, as indicated in Fig. 5. So there is an inverse function $t = t(x)$ and, by the inverse function rule,

$$\frac{dt}{dx} = \frac{1}{dx/dt}.$$

The composite function $y = y(t(x))$ gives y as a function of x. By the chain rule,

$$\frac{dy}{dx} = \frac{dy}{dt}\frac{dt}{dx} = \frac{dy/dt}{dx/dt},$$

which is the slope of the curve at $P(t)$ in Fig. 5. Virtually the same reasoning applies if $x'(t) < 0$ for t in some interval. In summary,

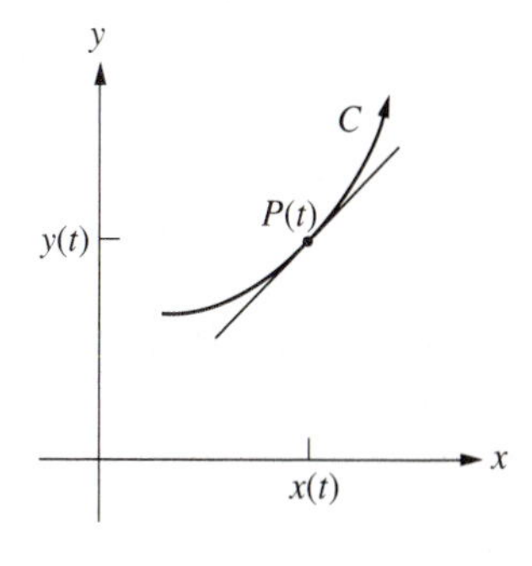

FIGURE 5

> the slope of a differentiable parametric curve C at a point $P = (x(t), y(t))$ is
>
> $$\frac{dy}{dx} = \frac{dy/dt}{dx/dt} \quad \text{if} \quad \frac{dx}{dt} \ne 0.$$

EXAMPLE 4. Identify the parametric curve with equations

$$x = 4\cos t, \qquad y = 3\sin t \quad \text{for} \quad 0 \le t \le 2\pi.$$

Then find the slope of the curve when $t = \pi/4$.

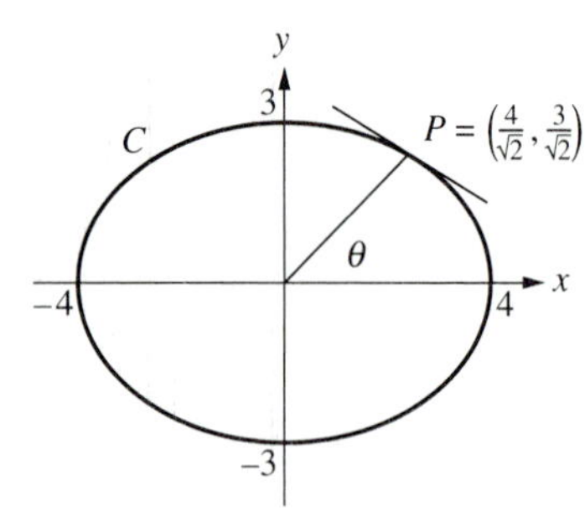

FIGURE 6

Solution. Since $x/4 = \cos t$, $y/3 = \sin t$, and $\cos^2 t + \sin^2 t = 1$,

$$\frac{x^2}{16} + \frac{y^2}{9} = 1.$$

The graph is the ellipse in Fig. 6. The point on the ellipse corresponding to $t = \pi/4$ is $P = (4/\sqrt{2}, 3/\sqrt{2})$. Since

$$\frac{dy}{dx} = \frac{dy/dt}{dx/dt} = \frac{3\cos t}{-4\sin t} \quad \text{and} \quad \left.\frac{dy}{dx}\right|_{t=\pi/4} = -\frac{3}{4},$$

the slope of the ellipse at the point P is $-3/4$. □

In Ex. 4 the parameter t is *not* the polar angle θ to a point $P(t)$ on the ellipse. In the example, t and θ are related by

$$\tan\theta = \frac{y}{x} = \frac{3\sin t}{4\cos t} = \frac{3}{4}\tan t.$$

For the point P in Fig. 6, $t = \pi/4$ and $\tan t = 1$, whereas the polar angle θ of P satisfies $\tan\theta = 3/4$ and $\theta \approx \pi/5$. See the problems for the geometric significance of the parameter t.

The line of reasoning in Ex. 4 establishes the following useful result.

> The ellipse $x^2/a^2 + y^2/b^2 = 1$ has parametric equations
>
> $$x = a\cos t, \quad y = b\sin t \quad \text{for} \quad 0 \le t \le 2\pi.$$

The ellipse is traversed in the counterclockwise direction with this parameterization. If $b = a$, we obtain the parametric equations $x = a\cos t$, $y = a\sin t$ for the circle $x^2 + y^2 = a^2$.

EXAMPLE 5. Identify the parametric curve C with

$$x = \sinh t, \qquad y = \cosh t.$$

Find the slope of the curve at any point.

Solution. Since $\cosh^2 t - \sinh^2 t = 1$ and $y = \cosh t > 0$ for all t,

$$y^2 - x^2 = 1, \qquad y = \sqrt{x^2 + 1}.$$

The graph is the upper branch of the hyperbola in Fig. 7. The slope at any point on C is

$$\frac{dy}{dx} = \frac{dy/dt}{dx/dt} = \frac{\sinh t}{\cosh t} = \tanh t$$

because $dx/dt = \cosh t \neq 0$. Alternatively, since $x = \sinh t$ and $y = \cosh t$,

$$\frac{dy}{dx} = \frac{x}{y} = \frac{x}{\sqrt{x^2+1}},$$

which also follows from implicit differentiation of $y^2 - x^2 = 1$. □

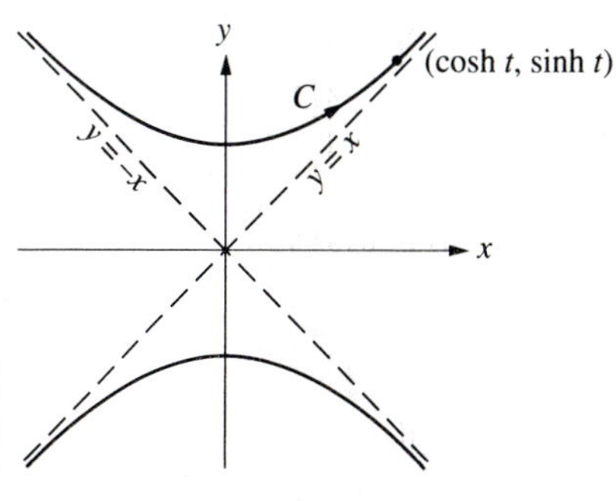

FIGURE 7

In all of the examples so far, we began with a parametric representation of a curve and then identified the curve. Now we reverse the procedure. We start with a curve described geometrically and then derive parametric equations for it. Figure 8 shows a curve C, called a **cycloid**, which is constructed as follows. Start with a circle with radius a and center $(0,a)$. Think of the circle as the tread of a bicycle tire and the x–axis as a road. The point labeled $P(0)$ is the point of contact of the tread with the road. Now let the circle roll along the x–axis without slipping. The point on the circle initially at $P(0)$ traces out the cycloid as the circle rolls along the x–axis. One arch of the cycloid is traced by the rolling circle as the central angle t in Fig. 8 increases from 0 to 2π. Congruent arches are traced by further revolutions of the circle.

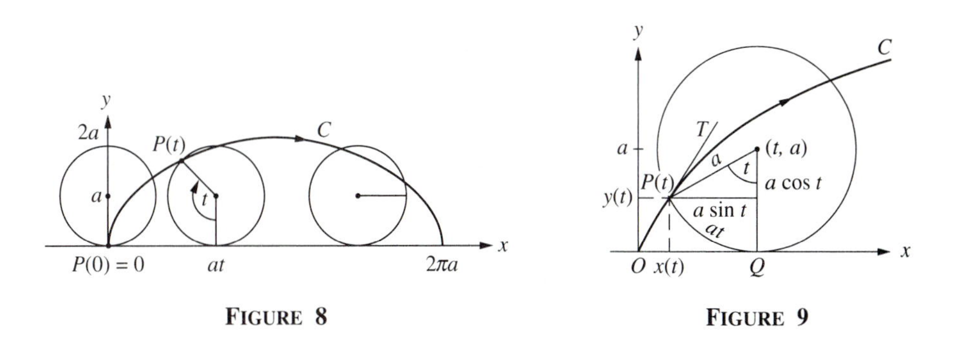

FIGURE 8

FIGURE 9

Points on the cycloid are conveniently located by means of the central angle t in Fig. 8. Figure 9 shows the same rolling circle and central angle t drawn with a larger scale. In Fig. 9, the circle has turned through an angle t with $0 < t < \pi/2$. The point $P(t) = (x(t), y(t))$ was originally at $P(0) = (0,0)$. Since there is no slippage as the circle rolls, the segment OQ has the same length at as the arc along the circle from Q to $P(t)$. From Fig. 9, the coordinates of $P(t) = (x(t), y(t))$ are $x(t) = at - a \sin t$ and $y(t) = a - a \cos t$.

> The cycloid has parametric equations
> $x = a(t - \sin t), \qquad y = a(1 - \cos t).$

Evidently, the cycloid is a differentiable parametric curve.

EXAMPLE 6. Find the slope of the cycloid when $t = \pi/3$.

Solution. The slope of the cycloid at any point where $dx/dt \neq 0$ is

$$\frac{dy}{dx} = \frac{dy/dt}{dx/dt} = \frac{a \sin t}{a(1-\cos t)} = \frac{\sin t}{1-\cos t}.$$

The slope of the cycloid at the point P corresponding to $t = \pi/3$ is

$$\frac{dy}{dx} = \frac{\sqrt{3}/2}{1 - 1/2} = \sqrt{3}.$$

Figure 9 is drawn to illustrate this case. Note that $x'(t) = 0$ for $t = 0$ and $t = 2\pi$. At the corresponding points $(x,y) = (0,0)$ and $(x,y) = (2\pi a,0)$, the derivative dy/dx fails to exist and the cycloid has vertical tangents. □

Integrals Related to Plane Parametric Curves

Parametric equations are useful in evaluating certain integrals. The parametric equations are used to change the variable of integration. Here is a typical illustration.

EXAMPLE 7. Find the area A under one arch of the cycloid $x = 2(t - \sin t)$ and $y = 2(1 - \cos t)$, with $0 \le t \le 2\pi$.

Solution. Let (x,y) be a variable point on the cycloid. From Fig. 8 with $a = 2$,

$$A = \int_0^{4\pi} y dx.$$

In order to evaluate this integral directly, we would have to know y explicitly as a function of x. However, a formula for y as a function of x is not readily available. (Try to find one.) So we take a different tack and make the change of variable suggested by the parametric representation of the cycloid. Let

$$x = 2(t - \sin t), \qquad 0 \le t \le 2\pi.$$

Then $y = 2(1 - \cos t)$ and $dx = 2(1 - \cos t)\,dt$. Since $x = 0$ when $t = 0$ and $x = 4\pi$ when $t = 2\pi$,

$$A = \int_0^{4\pi} y\,dx = \int_0^{2\pi} 4(1 - \cos t)^2\,dt.$$

Since $(1 - \cos t)^2 = 1 - 2\cos t + \cos^2 t$ and $\cos^2 t = \frac{1}{2}(1 + \cos 2t)$,

$$A = \int_0^{2\pi} 4\left(\frac{3}{2} - 2\cos t + \frac{1}{2}\cos 2t\right)dt = [\,6t - 8\sin t + \sin 2t\,]_0^{2\pi} = 12\pi. \square$$

In general, for parametric equations $x = x(t)$, $y = y(t)$ with $\alpha \le t \le \beta$, integration by substitution yields

$$\int_{x(\alpha)}^{x(\beta)} y dx = \int_\alpha^\beta y(t)x'(t)\,dt$$

if the standard conditions for changing variables in the definite integral are satisfied: $y(x)$ is continuous on the interval with endpoints $x(\alpha)$ and $x(\beta)$, and $y(t)$ and $x'(t)$ are continuous on the interval $[\,\alpha, \beta\,]$.

Parametric Curves in 3-Space

The essential distinction between parametric curves in 3–space and parametric curves in 2–space is the appearance of a third parametric equation to account for the third spatial dimension. We shall concentrate on visualizing and sketching graphs of parametric curves in space. Applications, especially to motions along curves, will come later in the chapter.

Once again, we start on familiar ground. The parametric equations

$$x = 1 + 2t, \qquad y = 3 + 4t, \qquad z = 5 + 6t,$$

represent a directed line in 3–space. Similarly, parametric equations

$$x = x(t), \qquad y = y(t), \qquad z = z(t),$$

represent a directed curve in 3–space under reasonable conditions on the functions $x(t)$, $y(t)$, and $z(t)$. Just as for parametric curves in the xy–plane, we call a parametric curve in 3–space *continuous* or *differentiable* if each of $x(t)$, $y(t)$, and $z(t)$ has the corresponding property.

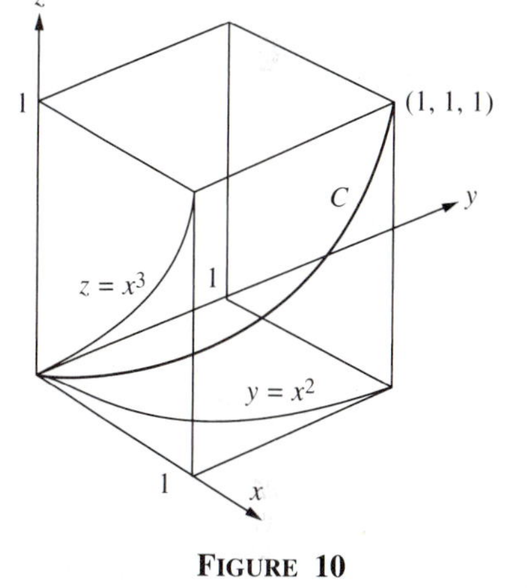

FIGURE 10

EXAMPLE 8. Sketch the graph of the curve C with parametric equations

$$x = t, \qquad y = t^2, \qquad z = t^3, \qquad 0 \leq t \leq 1.$$

Solution. As a first step, consider the projection of C onto the xy–plane given by $x = t$, $y = t^2$, $z = 0$, with $0 \leq t \leq 1$, or by $y = x^2$ and $z = 0$, with $0 \leq x \leq 1$, which is a piece of a parabola. This parabolic arc is what you would see if you looked straight down at C with your line of sight parallel to the z–axis. Similarly, the projection of C onto the xz–plane is given by $x = t$, $y = 0$, $z = t^3$, with $0 \leq t \leq 1$, or by $z = x^3$ and $y = 0$, with $0 \leq x \leq 1$, which is a piece of a cubic. This is what a side view of C, with sight line parallel to the y–axis, looks like. Both projections of C are shown in Fig. 10. With a little imagination, you should be able to visualize C from these projections. □

The unrestricted space curve represented by the parametric equations

$$x = t, \qquad y = t^2, \qquad z = t^3, \qquad -\infty < t < \infty,$$

is called a **twisted cubic**. Figure 10 should help you imagine what the full curve looks like.

We now describe another technique for visualizing parametric curves in 3–space that usually suits our purposes better. Let C be a typical parametric curve with moving point

$$P(t) = (x(t), y(t), z(t)).$$

The **projection of C onto the xy–plane** is the curve C_0 with moving point

$$P_0(t) = (x(t), y(t), 0).$$

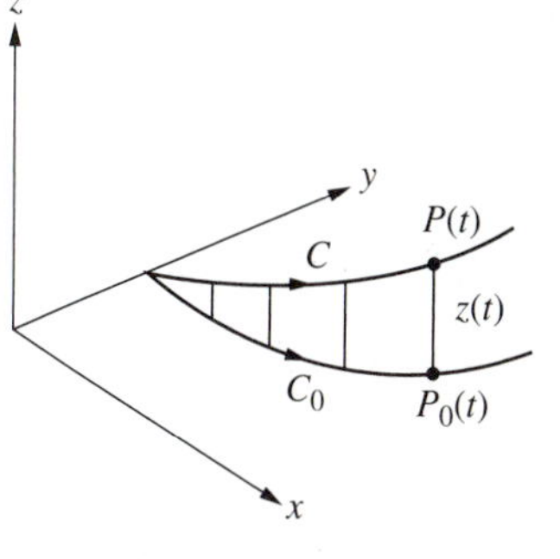

FIGURE 11

See Figure 11. The projection C_0 inherits its parameterization and positive direction from C. Whenever convenient, we write

$$P_0(t) = (x(t), y(t)).$$

The **elevation function** of $P(t)$ is $z(t)$. It is positive in Fig. 11. It could be negative or zero. We make a simple observation:

The curve C is determined by its projection C_0 onto the xy–plane and its elevation function $z(t)$.

Imagine that t is time and you walk along C_0 in the positive direction carrying a pole held vertically. The height of the pole is adjustable and at time t you adjust the height to be $z(t)$. Then the top of the pole traces out the curve C as you walk along C_0. Alternatively, think of C as a three–dimensional model of a trail in the mountains and C_0 as a map of the trail. With the map and the elevation function, you can reconstruct the three–dimensional model of the trail. Topographic maps provide the same sort of information, as we shall learn later when we study surfaces.

EXAMPLE 9. Describe the space curve C with parametric equations

$$x = \cos t, \qquad y = \sin t, \qquad z = 4 + 2\sin t \quad \text{for} \quad 0 \le t \le 2\pi.$$

Solution. The projection of C onto the xy–plane is the parametric curve C_0 with equations $x = \cos t,\ y = \sin t$ for $0 \le t \le 2\pi$. We recognize C_0 as the unit circle traversed from (1,0) around to (1,0) in the counterclockwise direction when viewed from above. Imagine walking around the circle C_0 in the counterclockwise sense carrying an adjustable pole with top at elevation $z = 4 + 2\sin t$ for $0 \le t \le 2\pi$. The following table indicates how the elevation varies as you walk around the circle C_0.

t	0	$\pi/2$	π	$3\pi/2$	2π
(x,y)	(1,0)	(0,1)	$(-1,0)$	$(0,-1)$	(1,0)
z	4	6	4	2	4

As you walk once around the circle from (1,0) to (1,0), the curve C which sits above the circle starts out at height 4, rises to height 6, falls to height 2, then finally rises back up to height 4. The curve looks much like a roller coaster. □

EXAMPLE 10. Sketch and describe the space curve C with parametric equations

$$x = \sinh t, \qquad y = \cosh t, \qquad z = \tanh t, \qquad t \ge 0.$$

Solution. The projection of C onto the xy–plane is the parametric curve C_0 determined by

$$x = \sinh t, \qquad y = \cosh t, \qquad z = 0, \qquad t \ge 0.$$

From Ex. 5 and the fact that $x = \sinh t \geq 0$ for $t \geq 0$, C_0 is the part of the hyperbola that lies in the first quadrant in Fig. 7. The elevation function for C is $z = \tanh t$. Standard properties of the hyperbolic tangent give

$$z = 0 \text{ for } t = 0,\ z \text{ increases as } t \text{ increases},\ z \to 1 \text{ as } t \to \infty.$$

We can visualize the graph of C as a trail rising over the hyperbolic arc C_0 in the xy–plane with elevation function $z = \tanh t$, as in Fig. 11. □

Graphics Utilities for Parametric Curves

Just as for graphs of functions, graphics calculators and graphics utilities provide means for graphing parametric curves. Depending on the sophistication of the particular utility and the problem at hand, you may need to manually set the graphing window in order to see the curve or to focus in on a particular portion of the curve. Usually, you can determine an appropriate graphing window either by using calculus to determine the maximum and minimum values of $x(t)$, $y(t)$, and $z(t)$ or by estimating these maxima and minima from graphs of the individual functions $x(t)$, $y(t)$, and $z(t)$ as t varies in the parameter interval.

We close this section with a simple Matlab m–file for graphing plane parametric curves. The program is written to graph the cycloid generated by rolling the unit circle along the x–axis. To run the program at the Matlab prompt, type "paragrf" and press the enter key.

```
% PARAGRF.M is a simple program for graphing a parametric
% curve C given by x = x(t), y = y(t) for
% a ≤ t ≤ b. The program asks for inputs a and b.
% You can edit the formulas for x(t) and y(t) to graph other
% curves.

xcoord = 't – sin(t)'
ycoord = '1 – cos(t)'

a = input('Enter smallest value of the parameter, a =  ? ');
b = input('Enter the largest value of the parameter, b =  ? ');

% Prepare to graph C at 201 points.

t = a:(b-a)/200:b;

% Evaluate x(t) and y(t) at the 201 equally spaced t–values.

x = eval(xcoord);
y = eval(ycoord);

% Plot the graph of C.

plot(x, y, x, zeros(x))
```

If you use the preceding m–file to graph the cycloid, say for $0 \leq t \leq 2\pi$, the graph Matlab displays will be distorted because the x– and y–axes are scaled differently, by default. To obtain a more accurate graph, use the Matlab command axis([0,2*pi,0,2*pi]) before executing the preceding m–file. An even more accurate graph is obtained by executing the foregoing axis command and the command axis('square') before running the m–file. Similar adjustments will be needed with other graphics utilities such as Maple and Mathematica.

PROBLEMS

In Probs. 1–14 parametric equations $x = x(t)$, $y = y(t)$ are given for a curve C in the xy–plane. (a) Eliminate the parameter and find a rectangular coordinate equation for C. (b) Sketch the curve and attach arrows indicating the direction that $P(t) = (x(t), y(t))$ moves along the curve as t increases.

1. $x(t) = t + 1, \quad y(t) = 2 - t$
2. $x(t) = t^3, \quad y(t) = 3 - 2t^3$
3. $x(t) = t^2 + 1, \quad y(t) = 2 - t$
4. $x(t) = t^4, \quad y(t) = 3 - 2t^4$
5. $x(t) = t - 1, \quad y(t) = t^2 + 1$
6. $x(t) = t - 4, \quad y(t) = t^2 - 6t + 10$
7. $x(t) = e^t, \quad y(t) = -2e^{2t}$
8. $x(t) = 2e^t, \quad y(t) = e^{-t}$
9. $x(t) = t^2, \quad y(t) = t^4$
10. $x(t) = t^2, \quad y(t) = -t$
11. $x(t) = 4\cos t, \quad y(t) = 4\sin t$
12. $x(t) = 4\cos t, \quad y((t) = 3\sin t$
13. $x(t) = 4\cos 2t, \quad y(t) = -4\sin 2t$
14. $x(t) = -4\cos 2t, \quad y(t) = 3\sin 2t$
15. $x(t) = \cos 2t, \quad y(t) = \sin t$
16. $x(t) = \cosh t, \quad y(t) = \sinh t$

In Probs. 17 – 22 find the slope of the given parametric curve at the indicated point.

17. $x(t) = t - 2, \quad y(t) = t^2 + 1$ at $(-2,1)$
18. $x(t) = t^2 - 2t + 3, \quad y(t) = t + 4$ when $t = 0$
19. $x(t) = \ln t, \quad y(t) = e^{-t}$ when $t = 1$
20. $x(t) = \cos 2t, \quad y(t) = \sin t$ at $(0, 1/\sqrt{2})$
21. $x(t) = 1/t, \quad y(t) = t$ at $(1,1)$
22. $x(t) = 3\sin t, \quad y(t) = 4\cos t$ when $t = 0$.
23. Find parametric equations for the circle $x^2 + y^2 = 9$ that sweep out the circle once in the clockwise sense starting from $(3,0)$.
24. Find parametric equations for the circle $x^2 + y^2 = 4$ that sweep out the circle once in the counterclockwise sense starting from $(-2,0)$.
25. Find parametric equations for the ellipse $x^2/9 + y^2/16 = 1$ that sweep out the ellipse once in the counterclockwise sense starting from $(3,0)$.
26. Find parametric equations for the ellipse $x^2/9 + y^2/16 = 1$ that sweep out the ellipse once in the clockwise sense starting from $(0, -4)$.
27. Describe how the points are swept out on the parametric curve $x = \sin t$, $y = \sin^2 t$ for $0 \leq t \leq \pi/2$ and sketch the curve.
28. Describe how the points are swept out on the parametric curve $x = \sin t$, $y = \sin^2 t$ for $0 \leq t \leq 2\pi$ and sketch the curve.

29. Let a and ω be positive constants. (a) Identify the parametric curve C given by $x = a\cos\omega t$ and $y = a\sin\omega t$ for $0 \le t \le 2\pi$. (b) Give a precise verbal description of the motion of an object if its position at time t is $P(t) = (a\cos\omega t, a\sin\omega t)$.

30. Let a and ω be positive constants. (a) Identify the parametric curve C given by $x = a\cosh\omega t$ and $y = a\sinh\omega t$ for $t \ge 0$. (b) Give a precise verbal description of the motion of an object if its position at time t is $P(t) = (a\cosh\omega t, a\sinh\omega t)$.

In Probs. 31–34: (a) Sketch the projection of the curve given by $x = x(t)$, $y = y(t)$, $z = z(t)$ on the xy–, yz–, and xz–planes. Use arrows to indicate the direction in which the projections are swept out as t increases. (b) Use this information to describe the curve.

31. $x(t)=t+1,\ y(t)=2-t,\ z(t)=2t+3$

32. $x(t) = t^2,\ y(t) = t,\ z(t) = t^3,\ t \ge 0$

33. $x(t) = 4\cos t,\quad y(t) = 4\sin t,\quad z(t) = 3\cos t$ for $0 \le t \le \pi/2$

34. $x(t) = 4\cos t,\quad y(t) = 2,\quad z(t) = -4\sin t$ for $0 \le t \le \pi/2$

In Probs. 35–38 a curve C is given by $x = x(t)$, $y = y(t)$, $z = z(t)$. (a) Describe its projection C_0 on the xy–plane. Use arrows to indicate the direction in which the projection is swept out as t increases. (b) Use (a) and the elevation function to sketch the graph of C.

35. $x(t) = \sin t,\quad y(t) = \cos t,\quad z(t) = -t$ for $0 \le t \le 2\pi$

36. $x(t) = 4\cos t,\quad y(t) = 3\sin t,\quad z(t) = 2\pi - t$ for $0 \le t \le 2\pi$

37. $x(t) = t,\ y(t) = \sqrt{t}, z(t)=4\cos t$ for $0 \le t \le \pi$

38. $x(t) = 2e^t,\quad y(t) = e^{-t},\quad z(t) = 3 - e^{-t}$ for $t \ge 0$

39. Figure 5 shows the graph of a curve C that can be expressed by either $x = x(t)$, $y = y(t)$ or by $y = f(x)$ near the point $P(t)$. We found that

$$\frac{dy}{dx} = \frac{y'(t)}{x'(t)} \quad \text{if} \quad x'(t) \ne 0$$

and the indicated derivatives exist. The curve in Fig. 5 also can be expressed as $x = g(y)$ near the point $P(t)$. (a) Show that

$$\frac{dx}{dy} = \frac{x'(t)}{y'(t)} \quad \text{if} \quad y'(t) \ne 0.$$

Then explain why C has a vertical tangent line at $P(t)$ if $x'(t) = 0$. (b) Show that the parametric curve $x = \cos t$, $y = \sin t$ has vertical tangent lines when $t = 0$ and π. Sketch the graph and display the vertical tangent lines.

40. Find the point(s), if any, where the parametric curve $x=4\cos 2t$ and $y = -4\sin 2t$ has vertical tangent lines.

41. Find the point(s), if any, where the parametric curve $x = \sin 2t$ and $y = \cos t$ has vertical tangent lines.

In Probs. 42–45 use the method in Ex. 7 to find the indicated area.

42. The area of the circle given by $x = x_0 + a\cos\theta$, $y = y_0 + a\sin\theta$.

43. The area under the curve C given by $x(t) = 2e^t$, $y(t) = e^{-t}$ for $0 \le t \le \ln 4$, and above the x–axis.

44. The area under the curve C given by $x(t) = t^3$, $y(t) = t^2$ for $-1 \le t \le 2$, and above the x–axis.

45. The area under the curve C given by $x(t) = \ln t$, $y(t) = 1/t$ for $1 \le t \le \infty$, and above the x–axis.

46. Thread is unwound from a spool with radius a, always keeping the unwound part straight. Take the spool to be the circle $x^2 + y^2 = a^2$. Initially the free end of the thread is at $(a,0)$. Let $P = (x,y)$ be the position of the free end as the thread unwinds counterclockwise. (a) Sketch this situation and the curve C that P sweeps out as the thread unwinds. (b) Let Q be the point on the spool where the thread leaves it. If θ is the angle between $\overline{OQ}$ and the positive x–axis, show that

$$x = a(\cos\theta + \theta\sin\theta), \qquad y = a(\sin\theta - \theta\cos\theta).$$

The curve with these equations is called the **involute** of the circle. (c) Find the slope of the involute when $\theta = \pi/2$ and $\theta = \pi$.

47. For the involute of the preceding circle, let θ_0 be the smallest positive angle such that P lies on the negative x–axis. (a) Find, in terms of θ_0 , the area below the involute and above the x–axis. (b) Use a root–finding method to approximate θ_0 and the area, say to two decimal places.

48. A circle in the xy–plane has center $(0,a)$, radius a, and tangent line L at $(0,2a)$. Let $\theta > 0$ be the angle between the positive x–axis and the ray from the origin that intersects the circle at Q and the tangent line L at R. The horizontal line through Q meets the vertical line through R at $P = (x,y)$. (a) Sketch the curve C swept out by P as θ varies from 0 to π. It is called the **Witch of Agnesi**. (b) Show that parametric equations for C are

$$x = 2a\cot\theta, \qquad y = 2a\sin^2\theta, \qquad 0 < \theta < \pi.$$

(c) Show that C has rectangular coordinate equation $x^2y = 4a^2\,(2a - y)$. (d) Find the area under the graph of C and above the x–axis.

49. Suppose a plane curve C can be described either by giving x as a function of y or by the parametric equations $x = x(t)$, $y = y(t)$ for $\alpha \le t \le \beta$. Show that

$$\int_{y(\alpha)}^{y(\beta)} x\,dy = \int_{\alpha}^{\beta} x(t)y'(t)\,dt.$$

What assumptions must you make in order to guarantee the formula is valid?

50. Use the preceding problem to find the area bounded by the curve C given by $x(t) = 4 - t^2$, $y(t) = t$ for $-2 \le t \le 2$ and the y–axis.

The next problem gives an interesting geometric construction for an ellipse and leads naturally to the parametric equations given in the text.

51. Assume that $0 < b \leq a$. Draw two circles with centers at the origin and radii a and b. Let the ray from the origin inclined at angle t radians to the positive x–axis cut the circle with radius b at Q and the circle with radius a at R. Let P be the intersection point of the vertical line through R and the horizontal line through Q. Show that P has coordinates $x = a\cos t$ and $y = b\sin t$. Conclude that P sweeps out the ellipse $x^2/a^2 + y^2/b^2 = 1$ as t increases from 0 to 2π. (The construction shows that t is not the polar angle of the point P, except for the special values $t = 0, \pi/2, \pi, 3\pi/2$, and 2π.)

4.2 Vector Functions and Curve Length

Parametric curves in 2–space or 3–space are expressed naturally as graphs of vector functions of the parameter. For example, the path swept out by a moving object is described by its position vector, which is a vector function of time. The vector point of view adds further insight into the properties of parametric curves and enables us to treat curves in 2–space and in 3–space simultaneously. Any vector function of a real independent variable can be interpreted as the position vector of a moving object if the independent variable is regarded as time. It is worthwhile to keep this point of view in mind.

We begin this section by representing parametric curves in vector form. Then we extend the notions of limits, continuity, and differentiability to vector functions. These concepts carry over from scalar calculus to vector calculus just by applying what we already know component by component. Later in the section, we shall find tangent vectors to curves and determine lengths of curves.

Parametric Curves and Vector Functions

Once again, we set the stage by returning to lines in 2–space. Recall that $x = 1 + 2t$ and $y = 3 + 4t$ are parametric equations for a line L in the xy–plane. A typical point on L is $P(t) = (1 + 2t, 3 + 4t)$. The corresponding position vector is $\mathbf{r}(t) = \, < 1 + 2t, 3 + 4t >$. We call $\mathbf{r}(t) = \, < 1 + 2t, 3 + 4t >$ a vector function of t. The graph of this vector function is the line L.

More generally, let $x = x(t)$ and $y = y(t)$ be parametric equations for a curve C in the xy–plane. In Fig. 1, $P(t) = (x(t), y(t))$ is a typical point on C. The corresponding position vector is $\mathbf{r}(t) = \, < x(t), y(t) >$. Then C is the graph of the vector function $r(t)$ and we say that C is represented by the vector equation

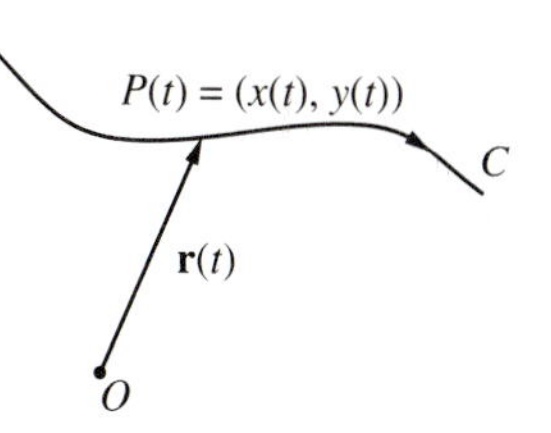

FIGURE 1

$$\mathbf{r} = \mathbf{r}(t) = \, < x(t), y(t) > .$$

For example, the parabola with parametric equations $x = t$, $y = t^2$ is represented by the vector equation $\mathbf{r} = \, < t, t^2 >$.

There is no significant difference for curves in 3–space. If a curve C has parametric equations $x = x(t)$, $y = y(t)$, $z = z(t)$, then a typical point on C is $P(t) = (x(t), y(t), z(t))$. Consequently, C is the graph of the vector function $\mathbf{r}(t) = \, < x(t), y(t), z(t) >$ and a vector equation for C is

$$\mathbf{r} = \mathbf{r}(t) = \, < x(t), y(t), z(t) > .$$

For example, the twisted cubic in Fig. 2 is represented in parametric form by $x = t, y = t^2, z = t^3$ and in vector form by $\mathbf{r} = < t, t^2, t^3 >$.

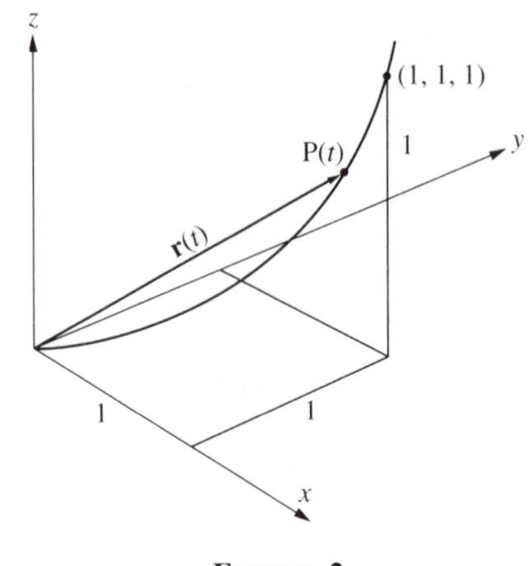

FIGURE 2

Vector functions can be expressed also in **i j k** notation. Thus, $\mathbf{r} = t\mathbf{i} + t^2\mathbf{j}$ represents the parabola $x = t$, $y = t^2$ and $\mathbf{r} = t\mathbf{i} + t^2\mathbf{j} + t^3\mathbf{k}$ represents the twisted cubic. Convenience and personal taste dictate which notation to use. Often the angle bracket notation saves writing.

EXAMPLE 1. Identify the graph C of the vector function $\mathbf{r}(\theta) = < \cos\theta, \sin\theta >$ for $\theta \geq 0$.

Solution. The points (x, y) on C are given by $x = \cos\theta$, $y = \sin\theta$ for $\theta \geq 0$. These parametric equations describe the unit circle $x^2 + y^2 = 1$ traversed repeatedly in the counterclockwise direction starting at (1,0). □

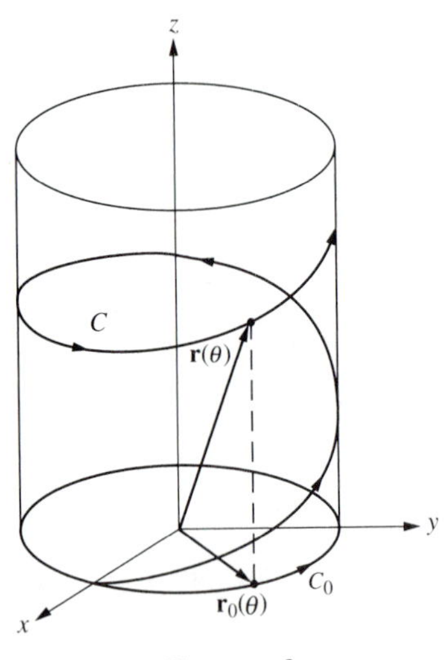

FIGURE 3

EXAMPLE 2. Describe the graph of the vector function $\mathbf{r}(\theta) = < \cos\theta, \sin\theta, \theta >$ for $\theta \geq 0$.

Solution. Points (x, y, z) on C are given by the parametric equations $x = \cos\theta$, $y = \sin\theta$, and $z = \theta$ for $\theta \geq 0$. The projection C_0 of C onto the *xy*–plane is the parametric curve with $x = \cos\theta$, $y = \sin\theta$, $z = 0$ or in vector form $\mathbf{r}_0(\theta) = < \cos\theta, \sin\theta, 0 >$ for $\theta \geq 0$. From Ex. 1, C_0 is the unit circle traversed repeatedly in the counterclockwise direction starting from (1,0,0). The elevation function for C is $z = \theta$. Consequently, the tip of the position vector $\mathbf{r}(\theta)$ lies vertically over the tip of the position vector $\mathbf{r}_0(\theta)$ and θ units above it. See Fig. 3. As θ increases, the tip of $\mathbf{r}(\theta)$ spirals upward around the right circular cylinder with base $x^2 + y^2 = 1$, $z = 0$. Thus, we obtain the graph of C in Fig. 3. □

The curve in Ex. 2 is called a **helix**. The plural of helix is helices. Other helices are given by the vector equation $\mathbf{r}(t) = < a\cos\omega t, a\sin\omega t, bt >$ where a, b, and ω are given constants. See the problems.

Terminology

Let C be a parametric curve given by $\mathbf{r} = \mathbf{r}(t)$ for $a \leq t \leq b$. We call $\mathbf{r}(a)$ the **initial point** of C and $\mathbf{r}(b)$ the **terminal point** of C. The parametric curve C

is **simple** if it has no self–intersections; that is, $\mathbf{r}(t_1) \neq \mathbf{r}(t_2)$ for $t_1 \neq t_2$ in the parameter interval. The curve in Fig. 4 is simple and the curve in Fig. 5 is not simple. We call C **closed** if its initial and terminal points are the same: $\mathbf{r}(a) = \mathbf{r}(b)$. We agree to call a closed curve *simple* if its only self–intersection is $\mathbf{r}(a) = \mathbf{r}(b)$. For example, the circle given by $\mathbf{r}(t) = < \cos t, \sin t >$ for $0 \leq t \leq 2\pi$ is a simple closed curve with $\mathbf{r}(0) = \mathbf{r}(2\pi) = (1,0)$.

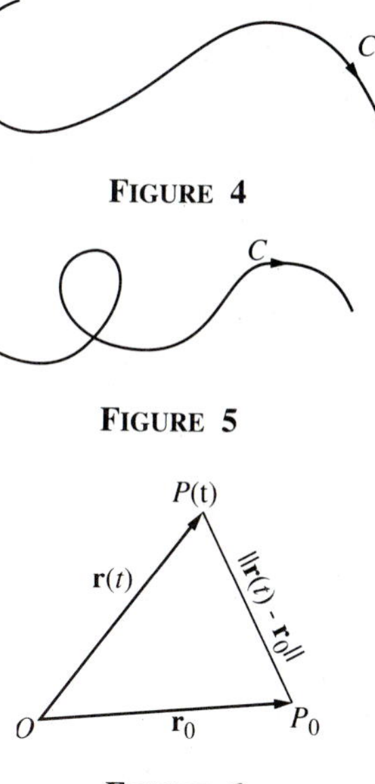

FIGURE 4

FIGURE 5

FIGURE 6

Limits of Vector Functions

Limits of vector functions are defined in much the same way as for real functions. Since there are no surprises, we shall proceed somewhat informally. As we mentioned before, any vector function $\mathbf{r}(t)$ can be regarded as the position vector of a point $P(t)$. We adopt that point of view in what follows.

The definition of a limit of a vector function is motivated and illustrated by Fig. 6, in which $\mathbf{r}(t)$ is the position vector for a variable point $P(t)$ and $\mathbf{r}_0$ is the position vector for a fixed point P_0. Observe that the distance from $P(t)$ to P_0 is $||\mathbf{r}(t) - \mathbf{r}_0||$. We call $\mathbf{r}_0$ the **limit** of the vector function $\mathbf{r}(t)$ as $t \to t_0$ if the distance $||\mathbf{r}(t) - \mathbf{r}_0||$ tends to zero as $t \to t_0$. In symbols,

$$\mathbf{r}(t) \to \mathbf{r}_0 \text{ as } t \to t_0 \text{ or } \lim_{t \to t_0} \mathbf{r}(t) = \mathbf{r}_0$$
$$\text{if } ||\mathbf{r}(t) - \mathbf{r}_0|| \to 0 \text{ as } t \to t_0.$$

The geometric meaning of the limit is that $P(t)$ can be made as near to P_0 as we like in Fig. 6 by taking t close enough to t_0 with $t \neq t_0$. Occasionally, we write $P(t) \to P_0$ instead of $\mathbf{r}(t) \to \mathbf{r}_0$. One–sided limits are defined in the obvious way, by letting $t \to t_0$ with $t > t_0$ or $t < t_0$.

As usual, let $\mathbf{r}(t) = < x(t), y(t), z(t) >$ and $\mathbf{r}_0 = < x_0, y_0, z_0 >$. The distance formula in 3–space gives

$$||\mathbf{r}(t) - \mathbf{r}_0|| = \sqrt{[x(t) - x_0]^2 + [y(t) - y_0]^2 + [z(t) - z_0]^2}.$$

By an easy argument,

$$\mathbf{r}(t) \to \mathbf{r}_0 \quad \text{as} \quad t \to t_0 \qquad \Leftrightarrow$$
$$x(t) \to x_0,\ y(t) \to y_0,\ z(t) \to z_0 \quad \text{as} \quad t \to t_0.$$

Thus, limits of vector functions are evaluated component by component.

EXAMPLE 3. Show that the vector function $\mathbf{r}(t) = < t, t^2, t^3 >$ has a limit as $t \to 2$ and find the limit.

Solution. From the preceding boxed result,

$$\mathbf{r}(t) = < t, t^2, t^3 > \to < 2,4,8 > \quad \text{as } t \to 2. \ \square$$

Continuity of Vector Functions

As you should expect, continuity is defined in terms of limits:

$\mathbf{r}(t)$ is **continuous at** t_0 if
$\mathbf{r}(t) \to \mathbf{r}(t_0)$ as $t \to t_0$.

It follows directly from the properties of limits given earlier that

$$\mathbf{r}(t) \text{ is continuous at } t_0 \iff x(t), y(t), z(t) \text{ are continuous at } t_0.$$

A vector function $\mathbf{r}(t)$ defined on an interval I is **continuous** if it is continuous at all t in I, with one–sided continuity at end points. Therefore,

a vector function is continuous $\iff$
each of its components is continuous.

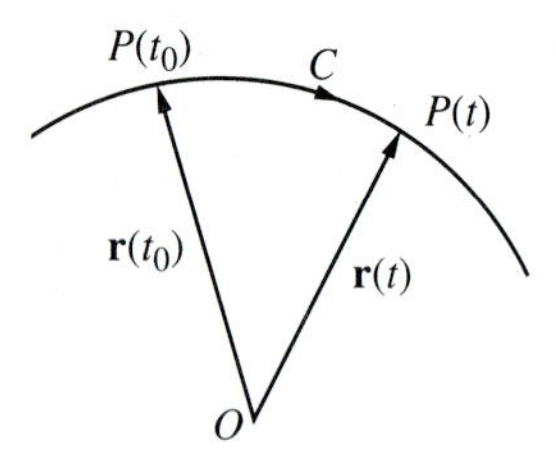

FIGURE 7

EXAMPLE 4. The vector function $\mathbf{r}(t) = \langle t, t^2, t^3 \rangle$ is continuous because the functions t, t^2, and t^3 are continuous.

The graph of a continuous vector function $\mathbf{r}(t)$ defined on an interval I is a directed curve with no gaps or breaks, such as in Fig. 7, where $P(t) \to P(t_0)$ as $t \to t_0$ for any t_0 in I.

Derivatives of Vector Functions

The derivative of a vector function $\mathbf{r}(t) = \langle x(t), y(t), z(t) \rangle$ is defined as the limit of a difference quotient, just as for real functions. The notation used in the vector case is borrowed from the scalar case and should seem quite reasonable.

Let $\Delta\mathbf{r} = \mathbf{r}(t + \Delta t) - \mathbf{r}(t)$. The **derivative** of $\mathbf{r}(t)$ is the vector function defined by

$$\mathbf{r}'(t) = \frac{d\mathbf{r}}{dt} = \lim_{\Delta t \to 0} \frac{\Delta\mathbf{r}}{\Delta t}$$

whenever the limit exists. One–sided derivatives of $\mathbf{r}(t)$ are defined in the same way, with the added provision that $\Delta t > 0$ or $\Delta t < 0$. The derivative of a vector function $\mathbf{r}(t)$ can be regarded as a rate of change, just as for real functions.

Next, we express $\mathbf{r}'(t)$ in terms of its components. Let

$$\Delta x = x(t + \Delta t) - x(t), \qquad \Delta y = y(t + \Delta t) - y(t), \qquad \Delta z = z(t + \Delta t) - z(t).$$

Then

$$\Delta\mathbf{r} = \mathbf{r}(t + \Delta t) - \mathbf{r}(t) = \langle \Delta x, \Delta y, \Delta z \rangle, \qquad \frac{\Delta\mathbf{r}}{\Delta t} = \left\langle \frac{\Delta x}{\Delta t}, \frac{\Delta y}{\Delta t}, \frac{\Delta z}{\Delta t} \right\rangle.$$

Let $\Delta t \to 0$ to reach the following important conclusion:

$$\mathbf{r}(t) \text{ is differentiable at } t \iff x(t), y(t), z(t) \text{ are differentiable at } t, \text{ in which case } \mathbf{r}'(t) = < x'(t), y'(t), z'(t) > .$$

Thus, vector functions are differentiated component by component.

EXAMPLE 5. Show that the vector function $\mathbf{r}(t) = < t^2, \sin 3t, e^{5t} >$ is differentiable and find its derivative.

Solution. Since each component of $\mathbf{r}(t)$ is differentiable, $\mathbf{r}(t)$ is differentiable and

$$\mathbf{r}'(t) = < 2t, 3\cos 3t, 5e^{5t} > . \quad \square$$

As expected, we say that a vector function $\mathbf{r}(t)$ defined on an interval I is **differentiable** if it is differentiable at each point in I, with one–sided derivatives at endpoints. Just as for real functions,

$$\mathbf{r}(t) \text{ differentiable} \quad \Rightarrow \quad \mathbf{r}(t) \text{ continuous.}$$

In Sec. 4.1 we defined a parametric curve C with parametric equations $x = x(t)$, $y = y(t)$, $z = z(t)$ to be continuous or differentiable if each of $x(t)$, $y(t)$, and $z(t)$ has the same property. Now we can express C by the vector equation $\mathbf{r} = \mathbf{r}(t)$ and with the definitions given earlier, C is continuous or differentiable if $\mathbf{r}(t)$ has the same property.

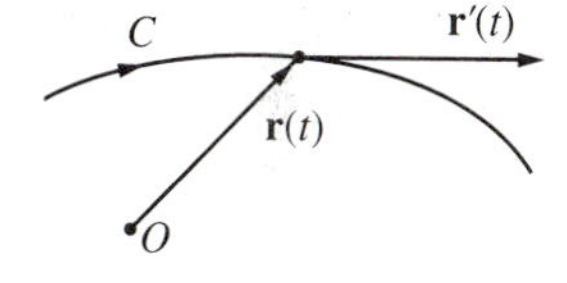

FIGURE 8

Tangent Vectors

A nonzero derivative $\mathbf{r}'(t)$ of a vector function $\mathbf{r}(t)$ is represented geometrically by a tangent vector, as illustrated in Fig. 8. The picture in Fig. 8 is justified by a geometric argument based on Fig. 9.

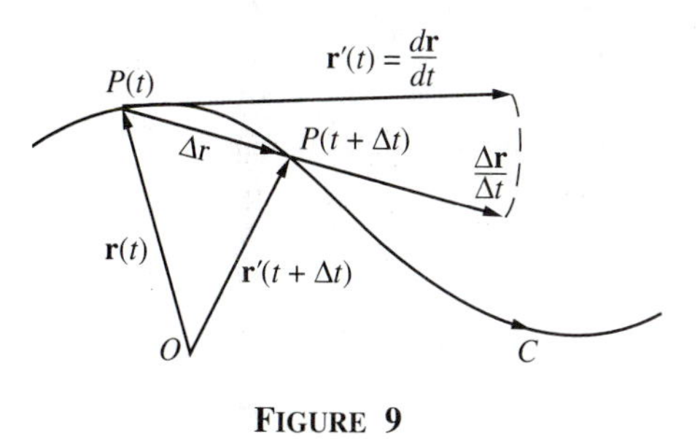

FIGURE 9

Assume that $\mathbf{r}(t)$ is differentiable and $\mathbf{r}'(t) \neq 0$ at a particular t. Then $\mathbf{r}(t)$ is continuous at t. By definition,

$$\mathbf{r}'(t) = \lim_{\Delta t \to 0} \frac{\Delta \mathbf{r}}{\Delta t}.$$

We are interested in the behavior of the vector $\Delta\mathbf{r}/\Delta t$ in Fig. 9 as $\Delta t \to 0$. For convenience, $0 < \Delta t < 1$. Observe that $\Delta\mathbf{r}$ is the vector from $P(t)$ to $P(t + \Delta t)$ and that $\Delta\mathbf{r}/\Delta t$ is an extension of $\Delta\mathbf{r}$ because $1/\Delta t > 1$. As $\Delta t \to 0$, the point $P(t + \Delta t)$ slides along the curve toward $P(t)$ and the vector $\Delta\mathbf{r}/\Delta t$ rotates into the position occupied by $\mathbf{r}'(t)$. We conclude that $\mathbf{r}'(t)$ is tangent to the curve at $P(t)$ in the sense of Fig. 8 and that $\mathbf{r}'(t)$ points in the positive direction along the curve. Actually, this discussion serves to explain what we mean by a tangent vector.

Definition *Tangent Vector to a Differentiable Curve*
Let C be a differentiable curve given by $\mathbf{r} = \mathbf{r}(t)$. If $\mathbf{r}'(t) \neq \mathbf{0}$, then $\mathbf{r}'(t)$ is the **tangent vector** to the curve C at the point $P(t)$ with position vector $\mathbf{r}(t)$.

The case with $\mathbf{r}'(t) = \mathbf{0}$ is exceptional because the zero vector has no direction.

A parametric curve C given by $\mathbf{r} = \mathbf{r}(t)$ is **smooth** if it has a nonzero tangent vector at each point and the tangent vectors vary continuously; that is, $\mathbf{r}'(t)$ is continuous and $\mathbf{r}'(t) \neq \mathbf{0}$ for all t. For example, the twisted cubic $\mathbf{r}(t) = < t, t^2, t^3 >$ is a smooth parametric curve because $\mathbf{r}'(t) = < 1, 2t, 3t^2 > \neq \mathbf{0}$ for all t.

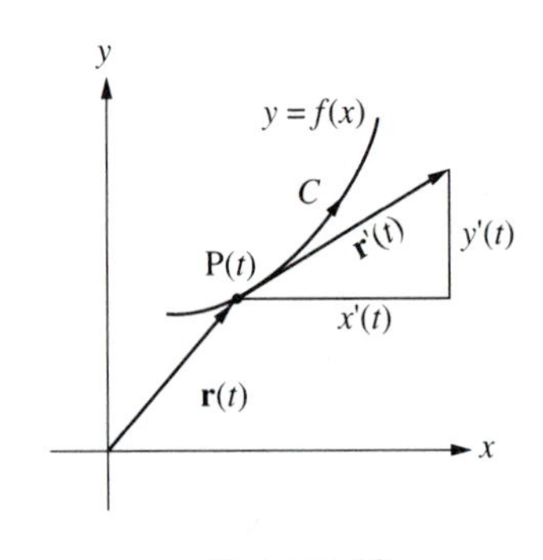

FIGURE 10

Figure 10 illustrates a tangent vector in the xy–plane. It shows a curve C that can be expressed either by $y = f(x)$ or in vector form by $\mathbf{r}(t) = < x(t), y(t) >$. The tangent vector at $P(t)$ is $\mathbf{r}'(t) = < x'(t), y'(t) >$. From the triangle in Fig. 10, the slope of C at $P(t)$ is

$$\frac{dy}{dx} = \frac{y'(t)}{x'(t)} = \frac{dy/dt}{dx/dt} \quad \text{if} \quad x'(t) \neq 0.$$

We reached the same conclusion by a chain rule argument in Sec. 4.1. In general, if $\mathbf{r}'(t_0) = < a,b >$ with $a \neq 0$, then the slope of C at $P(t_0)$ is $m = b/a$.

EXAMPLE 6. The graph of $\mathbf{r}(t) = < t, t^2 >$ is the parabola $y = x^2$ in Fig. 11. The tangent vector to the curve at $P(t)$ is $\mathbf{r}'(t) = < 1,2t >$. Since $\mathbf{r}'(1) = < 1,2 >$, the slope of the curve at $(1,1)$ is $m = 2/1 = 2$, which is obtained more directly by differentiating $y = x^2$. □

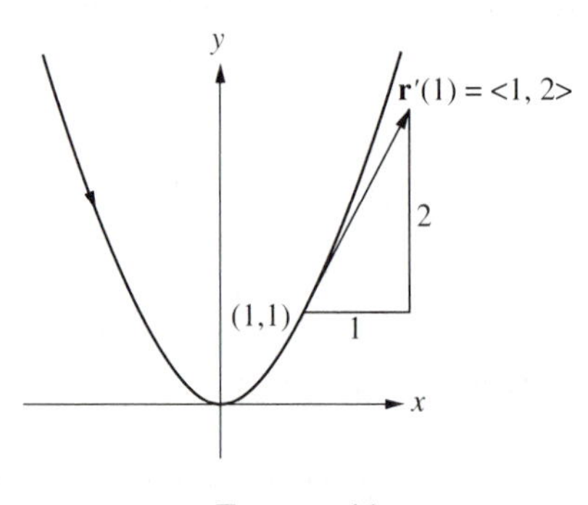

FIGURE 11

It should be apparent from Fig. 8 that a nonzero tangent vector at a point $P(t)$ on a curve is a direction vector for the tangent line to the curve at $P(t)$. For this reason, tangent vectors provide a convenient way to determine tangent lines to curves.

EXAMPLE 7. Find the tangent line to the twisted cubic $\mathbf{r}(t) = < t, t^2, t^3 >$ at $(1,1,1)$. See Fig. 2.

Solution. Since $\mathbf{r}'(t) = < 1, 2t, 3t^2 >$, the tangent vector at $(1,1,1)$ is $\mathbf{r}'(1) = < 1,2,3 >$, which is a direction vector for the tangent line at $(1,1,1)$. So a vector equation for the tangent line is

$$\mathbf{r} = \mathbf{r}(1) + t\mathbf{r}'(1) = < 1,1,1 > + t < 1,2,3 > .$$

Corresponding parametric equations for the tangent line are

$$x = 1 + t, \qquad y = 1 + 2t, \qquad z = 1 + 3t. \ \square$$

EXAMPLE 8. Show that the position vector $\mathbf{r}(\theta) = < \cos\theta, \sin\theta >$ to a point on the unit circle is always perpendicular to the tangent vector $\mathbf{r}'(\theta)$ to the circle at that point.

Solution. We must show that $\mathbf{r}(\theta) \cdot \mathbf{r}'(\theta) = 0$ for any θ. This is easy:

$$\mathbf{r}(\theta) \cdot \mathbf{r}'(\theta) = < \cos\theta, \sin\theta > \cdot < -\sin\theta, \cos\theta > = -\cos\theta\sin\theta + \sin\theta\cos\theta = 0. \ \square$$

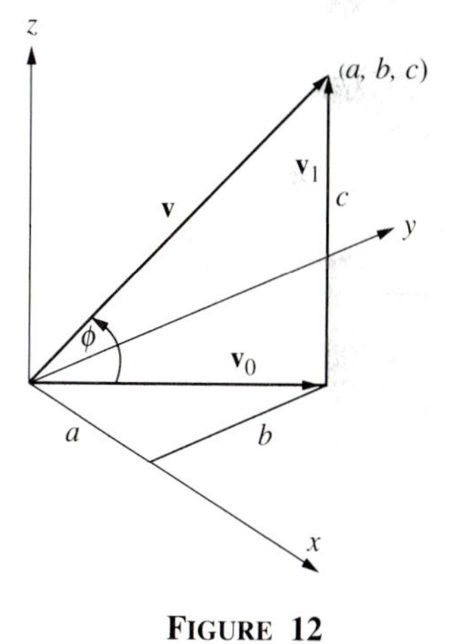

FIGURE 12

Visualizing Tangent Vectors in Space

Visualizing the direction of a tangent vector in 3–space is usually harder than in 2–space. To assist in this task, we recall a few ideas from Ch. 3. Let $\mathbf{v} = < a, b, c >$, as in Fig. 12. The horizontal and vertical projections of $\mathbf{v}$ are $\mathbf{v}_0 = < a, b, 0 >$ and $\mathbf{v}_1 = < 0, 0, c > = c\mathbf{k}$. The slope of $\mathbf{v}$ relative to the xy–plane is

$$m = \frac{c}{||\mathbf{v}_0||} = \frac{c}{\sqrt{a^2 + b^2}}.$$

The slope m and the angle of inclination φ of $\mathbf{v}$ are related by $m = \tan\varphi$. The direction of $\mathbf{v}$ is determined by $\mathbf{v}_0$ and either m or φ. We can also visualize the direction of $\mathbf{v}$ by recalling that $\mathbf{v} = \mathbf{v}_0 + \mathbf{v}_1$ and using the triangle in Fig. 12 to construct $\mathbf{v}$.

EXAMPLE 9. Describe the direction in space of the tangent vector $\mathbf{r}'(1)$ to the twisted cubic $\mathbf{r}(t) = < t, t^2, t^3 >$. Sketch $\mathbf{r}'(1)$.

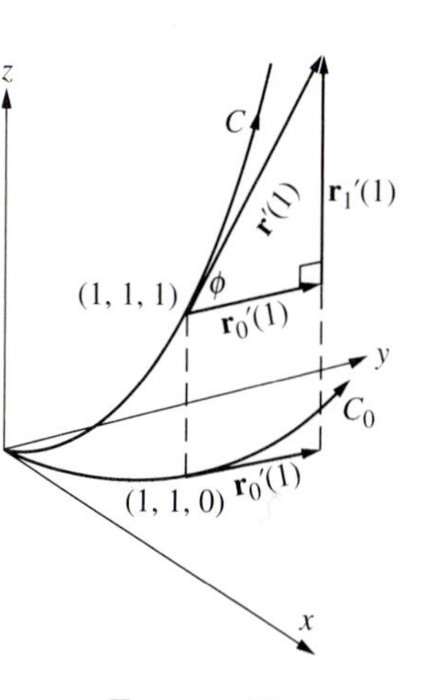

FIGURE 13

Solution. From Ex. 7, $\mathbf{r}'(1) = < 1,2,3 >$ is tangent to the curve C at $(1,1,1)$. Figure 13 shows the twisted cubic and the tangent vector $\mathbf{r}'(1)$. Projections will help us visualize $\mathbf{r}'(1)$. The horizontal projection of C is the parabola C_0 in the $xy-$plane determined by $\mathbf{r}_0(t) = < t, t^2, 0 >$. The horizontal projection of $\mathbf{r}'(1) = < 1,2,3 >$ is $\mathbf{r}_0'(1) = < 1,2,0 >$, which is tangent to C_0 at $(1,1,0)$. The vertical projection is $\mathbf{r}_1'(1) = < 0,0,3 > = 3\mathbf{k}$. The direction of $\mathbf{r}'(1)$ in space is determined by the direction of $\mathbf{r}_0'(1)$ in the xy–plane and the slope of $\mathbf{r}'(1)$ relative to the xy–plane, which is

$$m = \frac{3}{||\mathbf{r}_0'(1)||} = \frac{3}{\sqrt{5}} \cong 1.34.$$

Since $\tan\varphi = m$, the angle of inclination φ of $\mathbf{r}'(1)$ is about 53°. If we think of C as a three–dimensional model of a very steep trail and C_0 as a map of the trail, then the direction of $\mathbf{r}'(1)$ on the trail is determined by the direction of $\mathbf{r}_0'(1)$ on the map and either the slope m or the angle of inclination φ. Another way to visualize $\mathbf{r}'(1)$ is by means of $\mathbf{r}'(1) = \mathbf{r}_0'(1) + \mathbf{r}_1'(1)$ and the corresponding triangle in Fig. 13. □

EXAMPLE 10. Show that the tangent vector at any point on the circular helix $\mathbf{r}(\theta) = \langle \cos\theta, \sin\theta, \theta \rangle$ is inclined at 45° to the xy–plane.

Solution. The stated conclusion follows if we can show that the horizontal and vertical projections of the tangent vector to the helix at any point are of equal length. Then the right triangle in Fig. 14 with $\mathbf{r}'(\theta)$ as hypotenuse is a 45° right triangle. The tangent vector for the helix is

$$\mathbf{r}'(\theta) = \langle -\sin\theta, \cos\theta, 1 \rangle .$$

The horizontal projection of $\mathbf{r}'(\theta)$ is

$$\mathbf{r}_0'(\theta) = \langle -\sin\theta, \cos\theta, 0 \rangle ,$$

which is tangent to C_0 at $P_0(\theta)$ in Fig. 14. The vertical projection of $\mathbf{r}'(\theta)$ is $\mathbf{r}_1'(\theta) = \langle 0, 0, 1 \rangle = \mathbf{k}$. The triangle in Fig. 14 shows how to construct $\mathbf{r}'(\theta)$ from its projections. Since $||\mathbf{r}_0'(\theta)|| = 1$ and $||\mathbf{r}_1'(\theta)|| = 1$, the angle of inclination of $\mathbf{r}'(\theta)$ relative to the xy–plane is $\varphi = 45°$ at all points $P(\theta)$ on the helix. □

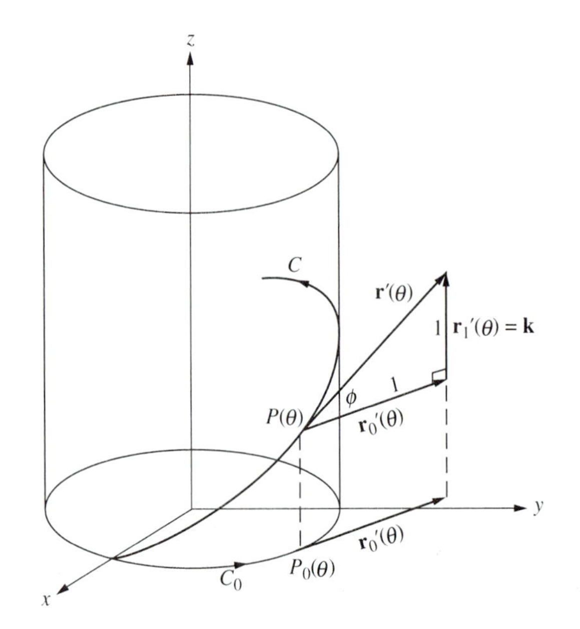

FIGURE 14

The Second Derivative

Just as for scalar functions, the **second derivative** of a vector function $\mathbf{r}(t) = <x(t), y(t), z(t)>$ is the derivative of the derivative. Since derivatives are evaluated component by component,

$$\mathbf{r}''(t) = <x''(t), y''(t), z''(t)> .$$

EXAMPLE 11. The first and second derivatives of the vector function $\mathbf{r}(t) = <t, t^2, t^3>$ are

$$\mathbf{r}'(t) = <1, 2t, 3t^2> \text{ and } \mathbf{r}''(t) = <0, 2, 6t> . \ \square$$

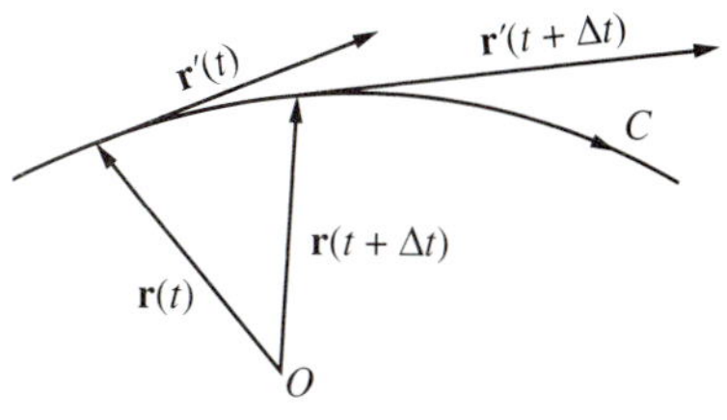

FIGURE 15

The second derivative $\mathbf{r}''(t)$ measures how the tangent vector $\mathbf{r}'(t)$ changes in length and direction as t changes. One way to illustrate this is based on

$$\mathbf{r}''(t) = \lim_{\Delta t \to 0} \frac{\mathbf{r}'(t + \Delta t) - \mathbf{r}'(t)}{\Delta t} = \lim_{\Delta t \to 0} \frac{\Delta \mathbf{r}'(t)}{\Delta t},$$

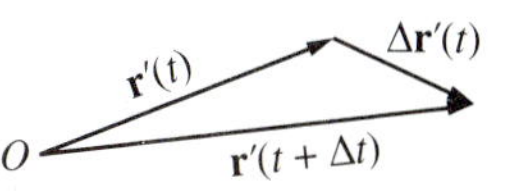

FIGURE 16

where $\Delta \mathbf{r}'(t) = \mathbf{r}'(t + \Delta t) - \mathbf{r}'(t)$. The vectors $\mathbf{r}'(t)$ and $\mathbf{r}'(t + \Delta t)$ in Fig. 15 are moved to the origin in Fig. 16 in order to display $\Delta \mathbf{r}'(t)$.

Another way to illustrate the effect of $\mathbf{r}''(t)$ on $\mathbf{r}'(t)$ makes use of a graph of $\mathbf{r}'(t)$. If $\mathbf{r}(t)$ is the position vector for a point moving along the curve in Fig. 15, then the tangent vector $\mathbf{r}'(t)$, moved to the origin, traces out the curve C' in Fig. 17 and $\mathbf{r}''(t)$ is tangent to C'. We see once again that $\mathbf{r}''(t)$ measures how $\mathbf{r}'(t)$ changes as t changes.

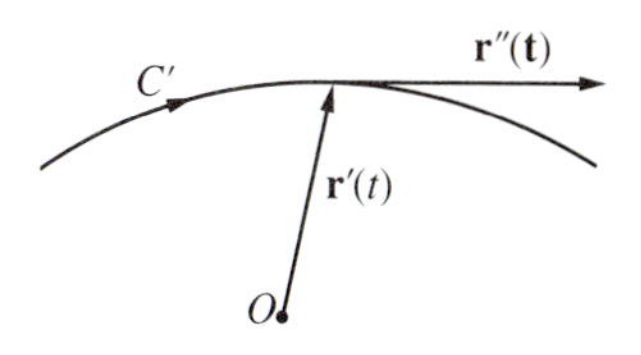

FIGURE 17

Curve Length

Lengths of parametric curves can be found by adapting the reasoning used in one variable calculus, where lengths of curves given explicitly by $y = f(x)$ or by $x = g(y)$ are determined. As in the one variable case, we approximate the curve by a polygonal line and use a Riemann sum and limit passage argument to express curve length as a definite integral. We shall merely sketch the derivation for parametric curves because it is nearly the same as for curves given explicitly by $y = f(x)$ or by $x = g(y)$.

Assume throughout that C is the graph of the differentiable vector function $\mathbf{r}(t)$ and that $\mathbf{r}'(t)$ is continuous for $a \le t \le b$. We also assume that C is simple or intersects itself a finite number of times. First, consider a curve C in the xy–plane. Then $\mathbf{r}(t) = <x(t), y(t)>$ and $\mathbf{r}'(t) = <x'(t), y'(t)>$, where $x'(t)$ and $y'(t)$ are continuous for $a \le t \le b$.

Partition the interval $a \le t \le b$:

$$a = t_0 < t_1 < \cdots t_{i-1} < t_i < \cdots < t_n = b.$$

In Fig. 18, C is the graph of $\mathbf{r}(t)$ and $P_i = (x_i, y_i) = (x(t_i), y(t_i))$. An approximation for the length of C is

FIGURE 18

$$L_n = \sum_{i=1}^{n} |P_{i-1}P_i|$$

Let $\Delta x_i = x_i - x_{i-1}$, $\Delta y_i = y_i - y_{i-1}$, and $\Delta t_i = t_i - t_{i-1}$. Then

$$|P_{i-1}P_i| = \sqrt{(\Delta x_i)^2 + (\Delta y_i)^2} = \sqrt{\left(\frac{\Delta x_i}{\Delta t_i}\right)^2 + \left(\frac{\Delta y_i}{\Delta t_i}\right)^2}\, \Delta t_i.$$

Two applications of the mean value theorem yield

$$\frac{\Delta x_i}{\Delta t_i} = x'(t_i^*), \qquad \frac{\Delta y_i}{\Delta t_i} = y'(t_i^{**})$$

for some t_i^* and t_i^{**} in (t_{i-1}, t_i). Therefore,

$$L_n = \sum_{i=1}^{n} \sqrt{[x'(t_i^*)]^2 + [y'(t_i^{**})]^2}\,\Delta t_i.$$

A limit passage yields an integral for the length of the curve:

$$L = \int_a^b \sqrt{\left(\frac{dx}{dt}\right)^2 + \left(\frac{dx}{dt}\right)^2}\, dt.$$

The corresponding integral for the length of a curve in 3–space is

$$L = \int_a^b \sqrt{\left(\frac{dx}{dt}\right)^2 + \left(\frac{dy}{dt}\right)^2 + \left(\frac{dz}{dt}\right)^2}\, dt.$$

The foregoing discussion serves to define what we mean by the length of a parametric curve and it provides an integral for calculating the length. The integrals for curve length in 2–space and 3–space are included in the single formula given next

Definition *Curve Length*
Let C be a curve given by $\mathbf{r} = \mathbf{r}(t)$ for $a \le t \le b$.
Assume that $\mathbf{r}'(t)$ exists and is continuous on $[a, b]$.
Then the length of C is

$$L = \int_a^b \|\mathbf{r}'(t)\|\, dt.$$

As in the discussion leading to the definition, we assume that C is simple or intersects itself a finite number of times. The definition includes as special cases the formulas for curve length of curves given by $y = f(x)$ or $x = g(y)$. See the problems.

EXAMPLE 12. Find the length L of the helix determined by

$$\mathbf{r}(\theta) = \langle 3\cos\theta, 3\sin\theta, 4\theta \rangle, \qquad 0 \le \theta \le 2\pi.$$

Solution. Since $\mathbf{r}'(\theta) = \langle -3\sin\theta, 3\cos\theta, 4 \rangle$ and $\|\mathbf{r}'(\theta)\| = 5$, the length is

$$L = \int_0^{2\pi} \|\mathbf{r}'(\theta)\|\, d\theta = 10\pi. \ \square$$

EXAMPLE 13. Find the length L of the cycloid given by $x = t - \sin t$, $y = 1 - \cos t$, for $0 \le t \le 2\pi$.

Solution. A vector formula for the cycloid is

$$\mathbf{r}(t) = \langle t - \sin t,\ 1 - \cos t \rangle, \qquad 0 \le t \le 2\pi.$$

Then $\mathbf{r}'(t) = \langle 1 - \cos t, \sin t \rangle$ and

$$\begin{aligned} ||\mathbf{r}'(t)||^2 &= 1 - 2\cos t + \cos^2 t + \sin^2 t \\ &= 2 - 2\cos t = 2(1 - \cos t) = 4\sin^2 \frac{t}{2}. \end{aligned}$$

Since $\sin t/2 \ge 0$ for $0 \le t \le 2\pi$, we conclude that $||r'(t)|| = 2\sin t/2$ and

$$L = \int_0^{2\pi} 2\sin\frac{t}{2}\,dt = \left[-4\cos\frac{t}{2}\right]_0^{2\pi} = 8. \ \square$$

EXAMPLE 14. Find the length L of the graph of

$$\mathbf{r}(t) = \langle 2t, t^2, \frac{1}{3}t^3 \rangle, \qquad 0 \le t \le 3.$$

This is another twisted cubic.

Solution. Since $\mathbf{r}'(t) = \langle 2, 2t, t^2 \rangle$,

$$||\mathbf{r}'(t)|| = \sqrt{4 + 4t^2 + t^4} = 2 + t^2.$$

Therefore,

$$L = \int_0^3 (2 + t^2)\,dt = \left[2t + \frac{1}{3}t^3\right]_0^3 = 15. \ \square$$

In the foregoing examples, the final expressions for $||\mathbf{r}'(t)||$ do not involve square roots. In real life, we are seldom so lucky. Often the integrals for curve length do not have elementary antiderivatives and must be evaluated numerically.

Algebraic Limit Laws and Differentiation Rules

Since limits and derivatives of vector functions are determined component by component, several limit laws and differentiation rules for real functions extend directly to vector functions. They are listed here for reference purposes. The proofs are omitted.

We begin with some useful algebraic limit laws:

> Let $\mathbf{r}_1(t) \to \mathbf{v}$, $\mathbf{r}_2(t) \to \mathbf{w}$ as $t \to t_0$. Then
> $\mathbf{r}_1(t) \pm \mathbf{r}_2(t) \to \mathbf{v} \pm \mathbf{w}$,
>
> $\lambda\mathbf{r}_1(t) \to \lambda\mathbf{v}$ for any scalar λ,
>
> $\mathbf{r}_1(t) \cdot \mathbf{r}_2(t) \to \mathbf{v} \cdot \mathbf{w}$,
>
> $\mathbf{r}_1(t) \times \mathbf{r}_2(t) \to \mathbf{v} \times \mathbf{w}$.

In the following differentiation rules, a, b are scalars, $f(t)$, $g(t)$ are scalar functions, $\mathbf{r}(t)$, $\mathbf{r}_1(t)$, $\mathbf{r}_2(t)$ are vector functions and all functions are differentiable at t.

$$[a\mathbf{r}_1(t) \pm b\mathbf{r}_2(t)]' = a\mathbf{r}_1'(t) \pm b\mathbf{r}_2'(t),$$
$$[f(t)\mathbf{r}(t)]' = f(t)\mathbf{r}'(t) + f'(t)\mathbf{r}(t),$$
$$\left[\frac{\mathbf{r}(t)}{g(t)}\right]' = \frac{g(t)\mathbf{r}'(t) - g'(t)\mathbf{r}(t)}{g(t)^2},$$
$$[\mathbf{r}_1(t) \cdot \mathbf{r}_2(t)]' = \mathbf{r}_1(t) \cdot \mathbf{r}_2'(t) + \mathbf{r}_1'(t) \cdot \mathbf{r}_2(t),$$
$$[\mathbf{r}_1(t) \times \mathbf{r}_2(t)]' = \mathbf{r}_1(t) \times \mathbf{r}_2'(t) + \mathbf{r}_1'(t) \times \mathbf{r}_2(t).$$

PROBLEMS

In Probs. 1–20 a curve C is given by $\mathbf{r} = \mathbf{r}(t)$. (a) Find $\mathbf{r}'(t)$ and $\mathbf{r}''(t)$ for the indicated value of t. (b) For a plane curve, find the slope of the curve at the same point. (c) For a space curve, find the horizontal and vertical projections of the tangent vector and its angle of inclination relative to the xy–plane.

1. $\mathbf{r} = \langle t-1, t^2+1 \rangle, \quad t=-2$
2. $\mathbf{r} = \langle t-4, t^2-6t+10 \rangle, \quad t=4$
3. $\mathbf{r} = e^t\mathbf{i} - 2e^{2t}\mathbf{j}, \quad t=1$
4. $\mathbf{r} = 2e^t\mathbf{i} + e^{-t}\mathbf{j}, \quad t=-1$
5. $\mathbf{r} = \langle t^2+1, 2-t^2 \rangle, \quad t=0$
6. $\mathbf{r} = \langle t^3, 3-2t^3 \rangle, \quad t=0$
7. $\mathbf{r} = \langle 4\cos t, 4\sin t \rangle, \quad t=0$
8. $\mathbf{r} = \langle 4\cos t, 3\sin t \rangle, \quad t=\pi/2$
9. $\mathbf{r} = \langle 4\cos t, -t, 4\sin t \rangle, \ t=0$
10. $\mathbf{r} = \langle 5t, 4\cos t, 3\sin t \rangle, \quad t=\pi/2$
11. $\mathbf{r} = (4\cos 2t)\mathbf{i} + (4\sin 2t)\mathbf{j} - 3t\mathbf{k}$, $t=\pi/2$
12. $\mathbf{r} = (4\cos 2t)\mathbf{i} + (3\sin 2t)\mathbf{j} + 4t\mathbf{k}$, $t=\pi$
13. $\mathbf{r} = \langle e^t, t^2+1 \rangle, \quad t=0$
14. $\mathbf{r} = \langle t, \sqrt{t} \rangle, \quad t=4$
15. $\mathbf{r} = (t^2+1)\mathbf{i} + (2t-1)\mathbf{j}, \quad t=2$
16. $\mathbf{r} = t^3\mathbf{i} - t^2\mathbf{j}, \quad t=2$
17. $\mathbf{r} = \langle 4\cos 2t, 3t, 4\sin 2t \rangle, \ t=\pi/4$
18. $\mathbf{r} = \langle \cos 2t, \sin t, \tan t \rangle, \ t=\pi/6$
19. $\mathbf{r} = (3-\cos t)\mathbf{i} + (2t-\sin t)\mathbf{j} + (4-e^{-t})\mathbf{k}, \quad t=0$
20. $\mathbf{r} = (t-\sin t)\mathbf{i} + (1-\cos t)\mathbf{j} + (\sec t)\mathbf{k}, \quad t=\pi/3$

In Probs. 21–28 find a vector equation for the tangent line to the given parametric curve at the indicated point.

21. $\mathbf{r} = e^t\mathbf{i} - 2e^{2t}\mathbf{j}$ at $(1,-2)$
22. $\mathbf{r} = 2e^t\mathbf{i} + e^{-t}\mathbf{j}$ at $(2,0)$
23. $\mathbf{r} = \langle 4\cos t, 4\sin t \rangle$ at $t=\pi/2$
24. $\mathbf{r} = \langle 4\cos t, 3\sin t \rangle$ at $t=\pi/4$
25. $\mathbf{r} = \langle \sin t, \cos t, -t \rangle$ at $(-1, 0, \pi/2)$
26. $\mathbf{r} = \langle 4\cos t, 3\sin t, 2\pi - t \rangle$ at $(-4, 0, \pi)$
27. $\mathbf{r} = \langle t, \sqrt{t}, 4\cos t \rangle$ at $t=\pi/2$
28. $\mathbf{r} = \langle 2e^t, e^{-t}, 3-e^{-t} \rangle$ at $t=0$
29. The curves given by $\mathbf{r} = \langle 1+t, t^2, t^3 \rangle$ and $\mathbf{r} = \langle \cos t, \sin t, t \rangle$ intersect at $(1,0,0)$. Find the angle between them. (That is, find the angle between their tangent vectors.)

30. Check that the curves $\mathbf{r} = (t - \sin t)\mathbf{i} + (1 - \cos t)\mathbf{j}$ and $\mathbf{r} = (t - \sqrt{3}/2)\mathbf{i} + (\cos t)\mathbf{j}$ intersect when $t = \pi/3$. Then find the angle between the curves at the point of intersection.

31. Let a and ω be positive constants and $0 \leq \theta \leq 2\pi$. Sketch and describe the curve C with vector equation (a) $\mathbf{r} = < a\cos\theta, a\sin\theta >$; (b) $\mathbf{r} = (a\cos\omega t)\mathbf{i} + (a\sin\omega t)\mathbf{j}$; and (c) $\mathbf{r} = < a\sin\theta, a\cos\theta >$.

32. Let a, b, and ω be positive constants and $\theta \geq 0$. Sketch and describe the curve C with vector equation (a) $\mathbf{r} = < a\cos\theta, a\sin\theta, b\theta >$; (b) $\mathbf{r} = < a\cos\omega\theta, a\sin\omega\theta, b\theta >$; and (c) $\mathbf{r} = < a\sin\theta, a\cos\theta, b\theta >$.

33. The involute of a circle has vector equation

$$\mathbf{r}(\theta) = < a(\cos\theta + \theta\sin\theta), a(\sin\theta - \theta\cos\theta) >.$$

Find the tangent vector to the involute when $\theta = \pi/2$ and $\theta = \pi$. (See Prob. 46 in Sec. 4.1.)

In Probs. 34–43 find the length of the given curve.

34. $\mathbf{r} = < \cos\theta, \sin\theta >, 0 \leq \theta \leq 2\pi$

35. $\mathbf{r} = < t^2, t^3 >, 0 \leq t \leq 1$

36. $x = t,\ \ y = t^2,\ \ 0 \leq t \leq 1$

37. $x = t,\ \ y = \sqrt{t},\ \ 0 \leq t \leq 4$

38. $\mathbf{r} = t\mathbf{i} + (\ln\cos t)\mathbf{j} - 4\mathbf{k}$, $0 \leq t \leq \pi/4$

39. $r = (\ln t)\mathbf{i} + 3\mathbf{j} - t\mathbf{k}$, $1 \leq t \leq 3$

40. $\mathbf{r} = (\cos\theta)\mathbf{i} + (\sin\theta)\mathbf{j} + \frac{1}{2}\theta^2\mathbf{k}$, $0 \leq t \leq \pi/2$

41. $\mathbf{r} = < e^t, e^{-t}, \sqrt{2}\,t >,\ \ 0 \leq t \leq 1$

42. $\mathbf{r} = (t - \ln t)\mathbf{i} + (t + \ln t)\mathbf{j} - 4\mathbf{k}$, $1 \leq t \leq e$

43. $\mathbf{r} = < \cosh t, \sinh t, t >$, $0 \leq t \leq \ln 4$

44. Show that

$$\frac{d}{dt}\|\mathbf{r}\|^2 = 2\mathbf{r} \cdot \frac{d\mathbf{r}}{dt}.$$

Conclude that $\mathbf{r}'(t)$ is orthogonal to $\mathbf{r}(t)$ if $\|\mathbf{r}(t)\|$ is constant for all t.

45. Assume that $\mathbf{r}$ is a differentiable vector function and that $\mathbf{r}(t) \neq \mathbf{0}$. Show that

$$\text{(a)}\ \frac{d}{dt}\|\mathbf{r}\| = \frac{\mathbf{r}\cdot\mathbf{r}'}{\|\mathbf{r}\|}, \qquad \text{(b)}\ \frac{d}{dt}\left(\frac{\mathbf{r}}{\|\mathbf{r}\|}\right) = \frac{\mathbf{r}'}{\|\mathbf{r}\|} - \frac{\mathbf{r}\cdot\mathbf{r}'}{\|\mathbf{r}\|^2}\,\mathbf{r}.$$

46. Let $\mathbf{u}$, $\mathbf{v}$, and $\mathbf{w}$ be differentiable vector functions. Show that

$$(\mathbf{u}\cdot\mathbf{v}\times\mathbf{w})' = \mathbf{u}'\cdot\mathbf{v}\times\mathbf{w} + \mathbf{u}\cdot\mathbf{v}'\times\mathbf{w} + \mathbf{u}\cdot\mathbf{v}\times\mathbf{w}'.$$

47. Express the result of the preceding problem as a rule for differentiating 3×3 determinants.

48. Explain why $\mathbf{r}(t)$ differentiable implies $\mathbf{r}(t)$ continuous follows directly from the corresponding result for scalar functions.

49. Assume that $\mathbf{r}_1(t) \to \mathbf{v}$ and $\mathbf{r}_2(t) \to \mathbf{w}$ as $t \to t_0$. Prove:
(a) $\mathbf{r}_1(t)\cdot\mathbf{r}_2(t) \to \mathbf{v}\cdot\mathbf{w}$ as $t \to t_0$,
(b) $\mathbf{r}_1(t)\times\mathbf{r}_2(t) \to \mathbf{v}\times\mathbf{w}$ as $t \to t_0$.

50. Assume that $\mathbf{r}_1(t)$ and $\mathbf{r}_2(t)$ are differentiable at t. Prove:
(a) $[\mathbf{r}_1(t) \cdot \mathbf{r}_2(t)]' = \mathbf{r}_1(t) \cdot \mathbf{r}_2'(t) + \mathbf{r}_1'(t) \cdot \mathbf{r}_2(t)$,
(b) $[\mathbf{r}_1(t) \times \mathbf{r}_2(t)]' = \mathbf{r}_1'(t) \times \mathbf{r}_2(t) + \mathbf{r}_1(t) \times \mathbf{r}_2'(t)$.

51. In one variable calculus, the length of a curve given by a continuously differentiable function $y = f(x)$ for $a \leq x \leq b$ is defined by $L = \int_a^b \sqrt{1 + f'(x)^2}\, dx$. Write the curve parametrically as $x = t$, $y = f(t)$ for $a \leq t \leq b$ and show that the formula for curve length of this parametric curve reduces to the preceding integral.

4.3 Velocity, Speed, and Acceleration

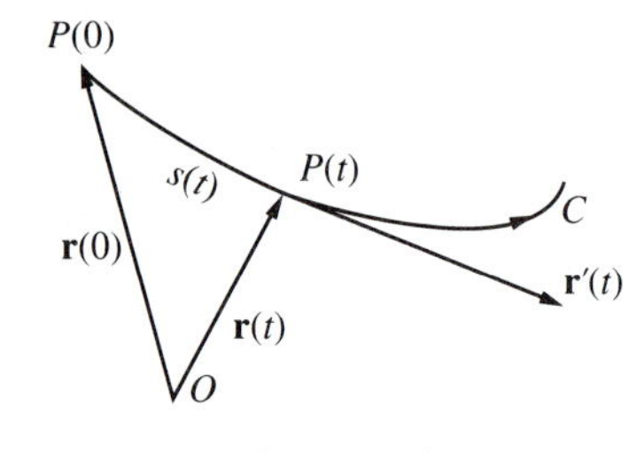

FIGURE 1

Vector functions are very effective tools for describing motions along curves. In this section, the concepts of velocity, speed, and acceleration are extended to two– and three–dimensional motions. Throughout the section, t is time in seconds or other units. For convenience, let t vary in an interval $I = [0, T]$ or $I = [0, \infty)$. Regard the curve in Fig. 1 as the path (or trajectory) of a moving point $P(t)$ in 2–space or 3–space.

Let $\mathbf{r}(t)$ be the position vector of $P(t)$. Thus,

$$\mathbf{r}(t) = < x(t), y(t) > \quad \text{or} \quad \mathbf{r}(t) = < x(t), y(t), z(t) >.$$

For brevity, we say that

$$\mathbf{r}(t) = \textbf{position}.$$

Then $\mathbf{r}'(t) =$ **velocity** and $\mathbf{r}''(t) =$ **acceleration**. Velocity is represented geometrically by the tangent vector $\mathbf{r}'(t)$ in Fig. 1, which points in the direction of motion. The length of the velocity vector is the speed. Acceleration measures change in velocity.

A force $\mathbf{F}(t)$ acting on a mass m causes an acceleration which satisfies Newton's second law of motion in the vector form $\mathbf{F}(t) = m\mathbf{r}''(t)$. If the force and the initial position and velocity are known, then two antidifferentiations yield the trajectory $\mathbf{r} = \mathbf{r}(t)$ of the mass m.

The length of the curve from $P(0)$ to $P(t)$ in Fig. 1 is $s = s(t)$. The speed at time t is $s'(t) = ||\mathbf{r}'(t)||$. For some purposes, it is convenient to determine position on a curve by distance s from the starting point. Then position is expressed by $\mathbf{r}(s)$ with s as parameter. As we shall see, $\mathbf{r}'(s)$ is a unit tangent vector.

Velocity and Speed

Just as for motion along a line, velocity in 2–space or 3–space is the rate of change of position with respect to time. Thus, by definition,

$$\textbf{velocity} = \mathbf{r}'(t).$$

We assume that $\mathbf{r}'(t)$ exists and is continuous. Then the velocity vector $\mathbf{r}'(t)$ in Fig. 1 varies continuously. The length of the velocity vector is the speed, which will turn out to be the same as the rate of change of distance with respect to time as the point $P(t)$ moves along the curve. In symbols,

$$\textbf{speed} = ||\mathbf{r}'(t)||.$$

Since each component of the position vector has units of length, the components of the velocity vector have units of length/time, such as ft/sec or meters/sec. The units for speed are also length/time.

EXAMPLE 1. The position of an object at time t is

$$\mathbf{r}(t) = \frac{1}{2}t^2\mathbf{i} - 2t\mathbf{j} + \frac{4}{3}t^{3/2}\mathbf{k}, \qquad t \geq 0.$$

Find the velocity and speed at any time t and at the particular time $t = 2$.

Solution. The velocity at time t is

$$\mathbf{r}'(t) = t\mathbf{i} - 2\mathbf{j} + 2t^{1/2}\mathbf{k},$$

and the speed is

$$\|\mathbf{r}'(t)\| = \sqrt{t^2 + 4 + 4t} = t + 2.$$

When $t = 2$, we have $\mathbf{r}'(2) = 2\mathbf{i} - 2\mathbf{j} + 2\sqrt{2}\mathbf{k}$ and $\|\mathbf{r}'(2)\| = 4$. □

We stated earlier that speed, defined as the length of the velocity vector, is also the rate of change of distance traveled with respect to time. This is easy to confirm. From Sec. 4.2, the distance along the curve in Fig. 1 from $P(0)$ to $P(t)$ is

$$s(t) = \int_0^t \|\mathbf{r}'(\tau)\|\,d\tau,$$

where τ is a dummy variable of integration. Occasionally we refer to $s(t)$ as the **arc length function** on C. Differentiate both sides of the equation for $s(t)$. By the fundamental theorem of calculus, the derivative of the right member is the integrand evaluated at the upper limit of integration. Thus,

$$\frac{ds}{dt} = \|\mathbf{r}'(t)\|,$$

which shows that the speed is the rate of change of distance traveled with respect to time, just as claimed. The formula $s(t) = \int_0^t \|\mathbf{r}'(\tau)\|\,d\tau$ tells us that distance traveled is the integral of the speed with respect to time, a result established for motion along a line in one variable calculus.

EXAMPLE 2. Find the distance traveled by the object in Ex. 1 during the time interval $0 \leq t \leq 2$.

Solution. From Ex. 1, the speed at any time is $s'(t) = \|\mathbf{r}'(t)\| = t + 2$. Since distance traveled during $0 \leq t \leq 2$ is the integral of the speed over this time interval, the distance traveled is

$$\int_0^2 \|\mathbf{r}'(t)\|\,dt = \int_0^2 (t + 2)\,dt = \left[\frac{1}{2}t^2 + 2t\right]_0^2 = 6. \ \square$$

EXAMPLE 3. Return to the cycloid in Fig. 8 of Sec. 4.1. Assume $a = 1$ and that the rolling circle that generates the cycloid turns at the constant rate of 1 rad/sec with the motion starting at time $t = 0$. Then at time t the point $P(t)$ has central angle t, as indicated in the figure. The cycloid has vector equation $\mathbf{r}(t) = < t - \sin t, 1 - \cos t >$ for $t \geq 0$. Find the speed of the point $P(t)$ at any time t and at time $t = \pi/3$.

Solution. We found in Ex. 12 of Sec. 4.2 that $||\mathbf{r}'(t)|| = 2 \sin t/2$. So the speed of the rolling point $P(t)$ at any time t is $s'(t) = ||\mathbf{r}'(t)|| = 2 \sin t/2$. In particular, $s'(t) = 1$ when $t = \pi/3$. □

If a curve C is not simple, then the distance traveled along C and the length of C may not be the same. Here is an example.

EXAMPLE 4. An object has position $\mathbf{r}(t) = < \cos t, \sin t >$ for $0 \leq t \leq 4\pi$. The object starts at (1,0) when $t = 0$, moves twice counterclockwise around the unit circle, and ends at (1,0) when $t = 4\pi$. The circle traversed twice is not a simple curve. The circle itself has length 2π whereas the distance (two full circles) traveled by the object is 4π, which is the integral of the speed $||\mathbf{r}'(t)|| = 1$ from 0 to 4π. □

Acceleration

Acceleration is the derivative of velocity; equivalently, acceleration is the second derivative of position with respect to time. Thus,

$$\mathbf{r}(t) = \text{position},$$
$$\mathbf{r}'(t) = \text{velocity},$$
$$\mathbf{r}''(t) = \text{acceleration}.$$

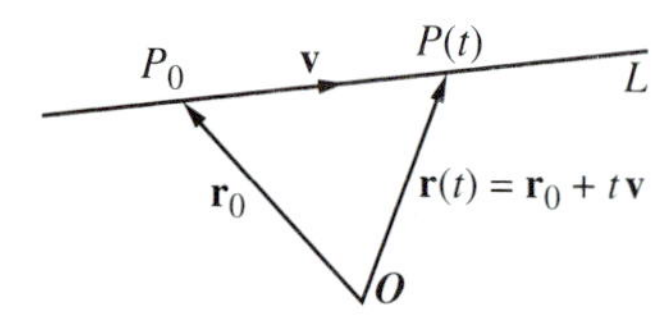

FIGURE 2

For the motion along the line L in Fig. 2 given by the familiar vector equation $\mathbf{r}(t) = \mathbf{r}_0 + t\mathbf{v}$ with $\mathbf{v} \neq \mathbf{0}$, we have

position	$\mathbf{r}(t) = \mathbf{r}_0 + t\mathbf{v}$,
velocity	$\mathbf{r}'(t) = \mathbf{v}$,
speed	$s'(t) = \lVert\mathbf{r}'(t)\rVert = \lVert\mathbf{v}\rVert$,
acceleration	$\mathbf{r}''(t) = \mathbf{0}$.

Since velocity is constant, acceleration is zero. Motion along a line is called **linear motion**. Linear motions with variable velocity appear in the problems.

Motion along a circle is called **circular motion**. We deal first with uniform circular motion in the *xy*–plane, which has many important applications. Suppose an object moves with position vector

$$\mathbf{r}(t) = \rho < \cos \omega t, \sin \omega t > = < \rho \cos \omega t, \rho \sin \omega t >$$

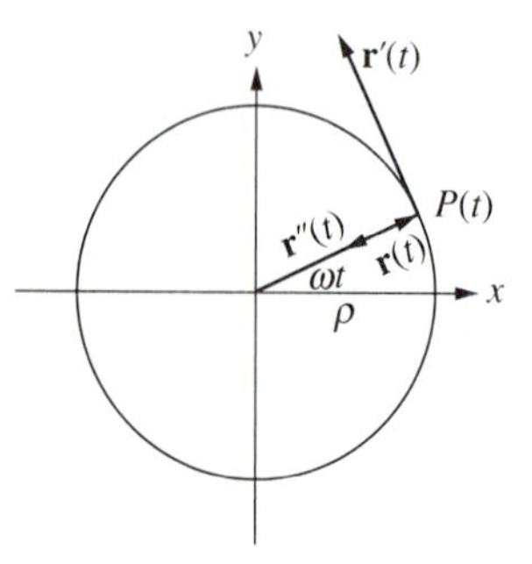

FIGURE 3

where ρ and ω are positive constants. The corresponding parametric equations are $x = \rho \cos \omega t$, $y = \rho \sin \omega t$. Then $x^2 + y^2 = \rho^2$ and $||\mathbf{r}(t)|| = \rho$. Thus, the object moves on the circle of radius ρ in Fig. 3. In the figure, $P(t) = (\rho \cos \omega t, \rho \sin \omega t)$ has the polar angle $\theta = \omega t$. Since $d\theta/dt = \omega$, the object moves with constant angular velocity ω. The motion is called *uniform* for this reason. For uniform circular motion:

$$\begin{array}{lll} \text{position} & \mathbf{r}(t) & = \rho < \cos \omega t, \sin \omega t >, \\ \text{velocity} & \mathbf{r}'(t) & = \rho\omega < -\sin \omega t, \cos \omega t >, \\ \text{speed} & s'(t) & = \|\mathbf{r}'(t)\| = \rho\omega, \\ \text{acceleration} & \mathbf{r}''(t) & = -\rho\omega^2 < \cos \omega t, \sin \omega t > = -\omega^2 \mathbf{r}(t). \end{array}$$

It is clear in Fig. 3 that $\mathbf{r}(t)$ and $\mathbf{r}'(t)$ are orthogonal, which can be confirmed also from $\mathbf{r}(t) \cdot \mathbf{r}'(t) = 0$. The acceleration $\mathbf{r}''(t) = -\omega^2 \mathbf{r}(t)$ experienced by an object in uniform circular motion always points toward the origin and is called **centripetal acceleration.**

The fact that the position vector and velocity vector are orthogonal for uniform circular motion is no accident. It happens for any circular motion, uniform or not. Indeed, suppose that an object moves on a circle. We choose the origin of coordinates at its center. The position vector of the object $\mathbf{r}(t)$ satisfies $\|\mathbf{r}(t)\| = \rho$, where ρ is the radius of the circle. Then

$$0 = \frac{d}{dt} \|\mathbf{r}(t)\|^2 = \frac{d}{dt} \mathbf{r}(t) \cdot \mathbf{r}(t) = 2\mathbf{r}(t) \cdot \mathbf{r}'(t),$$

$$\mathbf{r}(t) \cdot \mathbf{r}'(t) = 0, \qquad \mathbf{r}'(t) \perp \mathbf{r}(t).$$

Since the steps are reversible,

$$\boxed{\|\mathbf{r}(t)\| \text{ is constant} \quad \Leftrightarrow \quad \mathbf{r}'(t) \perp \mathbf{r}(t).}$$

By the same token,

$$\boxed{\|\mathbf{r}'(t)\| \text{ is constant} \quad \Leftrightarrow \quad \mathbf{r}''(t) \perp \mathbf{r}'(t).}$$

Since $\|\mathbf{r}'(t)\|$ is speed, the latter property tells us that, for motions with constant speed, the acceleration vector is orthogonal to the velocity vector. The acceleration acts to change the direction of the velocity vector but not its length. For example, an electron moving transverse to a uniform magnetic field can follow a path with variable velocity but constant speed. (See Prob. 30 for further details.)

EXAMPLE 5. An electron moves with position vector $\mathbf{r}(t) = < 4\cos t, 4\sin t, 3t >$ for $t \geq 0$. Identify the path of motion and find the velocity, speed, and acceleration. Then discuss the motion.

Solution. We recognize $\mathbf{r}(t)$ as the position vector for the helix in Fig. 4. The velocity, speed, and acceleration of the electron are

$$\begin{aligned} \mathbf{r}'(t) &= < -4\sin t, 4\cos t, 3 >, \\ \|\mathbf{r}'(t)\| &= 5, \\ \mathbf{r}''(t) &= < -4\cos t, -4\sin t, 0 >. \end{aligned}$$

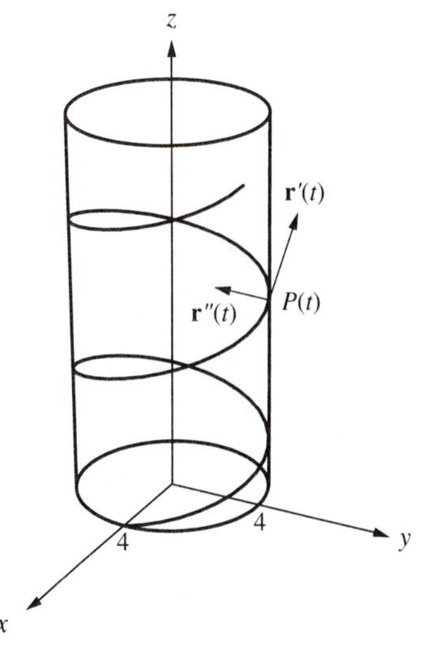

FIGURE 4

Since the speed $||\mathbf{r}'(t)||$ is constant, the velocity vector changes only in direction and not in length. It also follows from the constant speed that $\mathbf{r}''(t) \perp \mathbf{r}'(t)$, which can be verified directly in this case by using dot products. Notice too that the acceleration vector is horizontal because its z–component is zero. It always points toward the z–axis in Fig. 4. The velocity vector also has some special features. The horizontal component of velocity is $\mathbf{r}_0' = < -4\sin t,\ 4\cos t,\ 0 >$, which has constant length $||\mathbf{r}_0'(t)|| = 4$. The vertical component of velocity is $\mathbf{r}_1'(t) = < 0, 0, 3 >$, which has constant length 3. Consequently, the electron rises at the uniform rate of 3 units of length per sec. Furthermore, the velocity vector has constant slope $m = 3/4$ relative to the xy–plane and constant angle of inclination $\varphi = \arctan 3/4 \approx 37°$ as it changes direction along the helix. □

In the next example, the velocity vector changes both in length and in direction.

EXAMPLE 6. An object moves in 3–space with $\mathbf{r}(t) = < t\cos t,\ t\sin t,\ \sqrt{3}t >$ for $t \geq 0$. (a) Describe the path of motion and find the velocity, speed, and acceleration. (b) Find the distance traveled by the object during the time interval $0 \leq t \leq 2\pi$.

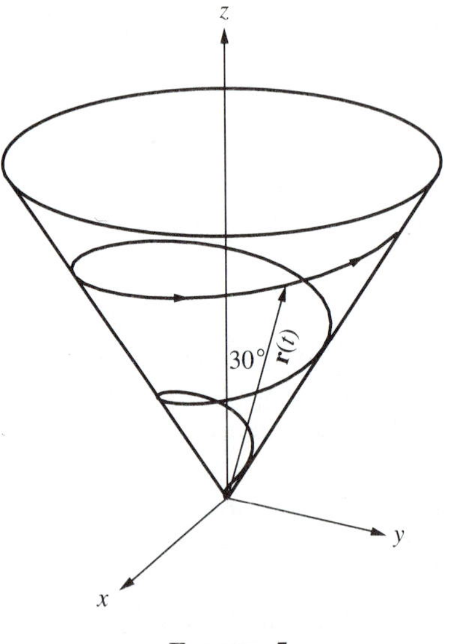

FIGURE 5

Solution. (a) We claim that the object ascends on a spiral that lies on a circular cone as shown in Fig. 5. To see why, notice that the direction angle γ between $\mathbf{r}(t)$ and the z–axis is given by

$$\cos\gamma = \frac{\mathbf{r}(t)\cdot\mathbf{k}}{||\mathbf{r}(t)||\cdot||\mathbf{k}||} = \frac{\sqrt{3}t}{2t} = \frac{\sqrt{3}}{2}, \qquad \gamma = 30°.$$

Thus, the position vector always is inclined at 30° relative to the z–axis and the object moves on the surface of the circular cone with apex angle 30°, as indicated in the figure. The velocity, speed, and acceleration are

$$\mathbf{r}'(t) = < \cos t - t\sin t,\ \sin t + t\cos t,\ \sqrt{3} >,$$
$$||\mathbf{r}'(t)|| = \sqrt{t^2+4},$$
$$\mathbf{r}''(t) = < -2\sin t - t\cos t,\ 2\cos t - t\sin t,\ 0 >.$$

Since the vertical component of velocity is constant, the object rises at a constant rate and the vertical component of acceleration is zero. (b) The distance traveled during the time interval $0 \le t \le 2\pi$ is

$$\int_0^{2\pi} ||\mathbf{r}'(t)||\, dt = \int_0^{2\pi} \sqrt{t^2+4}\, dt.$$

This integral can be evaluated by means of a trigonometric substitution and integration by parts. However, since this is not a lesson on techniques of integration, it is good enough to look the integral up in a table or use a CAS system. Either way,

$$\int_0^{2\pi} \sqrt{t^2+4}\, dt = \left[\frac{1}{2} t\sqrt{t^2+4} + 2\ln\left(\frac{\sqrt{t^2+4}+t}{2}\right)\right]_0^{2\pi}.$$

Consequently,

$$\int_0^{2\pi} ||\mathbf{r}'(t)||\, dt = 2\pi\sqrt{\pi^2+1} + 2\ln(\sqrt{\pi^2+1}+\pi) \approx 25.9. \ \square$$

Antiderivatives and Integrals

Antiderivatives, indefinite integrals, and definite integrals of vector functions are useful for much the same reasons that their scalar analogues are useful. For example, simple motion problems in space can be solved by antidifferentiation much as we did earlier for one–dimensional motion problems. Furthermore, solutions to more complicated motion problems are often conveniently represented by integrals.

Antiderivatives and integrals of vector functions and scalar functions are very similar. To heighten the similarity, we adopt the notation

$$\mathbf{f}(t) = < f_1(t), f_2(t), f_3(t) >, \qquad \mathbf{F}(t) = < F_1(t), F_2(t), F_3(t) >,$$

for t in some interval. As expected, $\mathbf{F}(t)$ is an **antiderivative** of $\mathbf{f}(t)$ if $\mathbf{F}'(t) = \mathbf{f}(t)$, and the **indefinite integral** of $\mathbf{f}(t)$ is given by

$$\int \mathbf{f}(t)\, dt = \mathbf{F}(t) + \mathbf{C} \quad \Leftrightarrow \quad \mathbf{F}'(t) = \mathbf{f}(t),$$

where $\mathbf{C}$ is any constant vector. Since each component of $\mathbf{F}$ is an antiderivative of the corresponding component of $\mathbf{f}$, it follows that

$$\int \mathbf{f}(t)\,dt = \left\langle \int f_1(t)\,dt,\ \int f_2(t)\,dt,\ \int f_3(t)\,dt \right\rangle.$$

Thus, vector functions are antidifferentiated (integrated) component by component.

EXAMPLE 7. Component–by–component integration gives

$$\int <2t, \cos t, e^{t/2}> dt = <t^2+c_1, \sin t + c_2, 2e^{t/2} + c_3> = <t^2, \sin t, 2e^{t/2}> + \mathbf{C},$$

where $\mathbf{C} = <c_1, c_2, c_3>$ is any constant vector. □

The definite integral of a continuous vector function $\mathbf{f}(t) = <f_1(t), f_2(t), f_3(t)>$, $a \leq t \leq b$, is defined as a limit of Riemann sums, just as in the scalar case. Since vectors are added componentwise and limits of vector functions are evaluated component by component, it follows easily that

$$\int_a^b \mathbf{f}(t)\,dt = \left\langle \int_a^b f_1(t)\,dt,\ \int_a^b f_2(t)\,dt,\ \int_a^b f_3(t)\,dt \right\rangle.$$

Both versions of the fundamental theorem of calculus carry over to vector functions; simply apply the corresponding scalar fundamental theorem to each component.

If $\mathbf{F}(t)$ is an antiderivative of $\mathbf{f}(t)$, then

$$\int_a^b \mathbf{f}(t)\,dt = [\mathbf{F}(t)]_a^b.$$

If $\mathbf{f}(t)$ is continuous on $[a,b]$, then

$$\frac{d}{dt}\int_a^t \mathbf{f}(\tau)\,d\tau = \mathbf{f}(t) \quad \text{for } t \text{ in } [a,b].$$

As in the scalar case, both indefinite and definite integrals are linear operations:

$$\int [\alpha\mathbf{f}(t) + \beta\mathbf{g}(t)]\,dt = \alpha\int \mathbf{f}(t)\,dt + \beta\int \mathbf{g}(t)\,dt,$$

$$\int_a^b [\alpha\mathbf{f}(t) + \beta\mathbf{g}(t)]\,dt = \alpha\int_a^b \mathbf{f}(t)\,dt + \beta\int_a^b \mathbf{g}(t)\,dt,$$

Newton's Second Law of Motion

Newton's second law of motion, force equals mass times acceleration, is used in scalar form in one variable calculus. The law is valid also in vector form: a force $\mathbf{F}(t)$ acting on a mass m produces an acceleration $\mathbf{r}''(t)$ in the same direction that satisfies

$$\boxed{\mathbf{F}(t) = m\mathbf{r}''(t).}$$

When no force acts on an object, Newton's law reduces to $\mathbf{r}''(t) = 0$ for all t. Then two antidifferentiations yield $\mathbf{r}(t) = \mathbf{r}_0 + t\mathbf{v}$ for some $\mathbf{r}_0$ and some $\mathbf{v}$; consequently, the object moves along a straight line with constant velocity and speed.

In Fig. 3, suppose that $\mathbf{r}(t) = \rho < \cos \omega t, \sin \omega t >$ describes uniform circular motion of an object with mass m. Then the force required to keep the mass in the circular orbit is given by

$$\mathbf{F}(t) = m\mathbf{r}''(t) = -m\omega^2\mathbf{r}(t).$$

The force always points toward the origin in Fig. 3. It is called a **centripetal force.** The equal and opposite force you would feel if you were at the center of the circle and twirling the object at the end of a string is called a **centrifugal force.**

EXAMPLE 8. A projectile with mass m is fired from the ground with angle of elevation 30° and initial velocity 800 feet per second. Find the flight path (trajectory) of the projectile. When and where does it strike the ground?

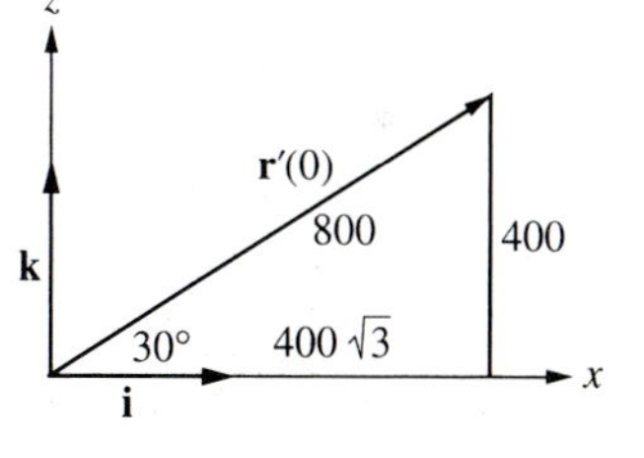

FIGURE 6

Solution. For convenience, suppose that the projectile is launched from the origin in a direction above the positive x–axis. Then $\mathbf{r}(0) = \mathbf{0}$ and, from Fig. 6, $\mathbf{r}'(0) = 400\sqrt{3}\,\mathbf{i} + 400\mathbf{k}$. Since English units are used and the motion takes place near the surface of the earth, we can approximate the gravitational force on the projectile by $-32m\mathbf{k}$ with the coordinate system in Fig. 6. By Newton's second law,

$$\mathbf{F}(t) = m\mathbf{r}''(t) = -32m\mathbf{k}, \qquad \mathbf{r}''(t) = -32\mathbf{k}.$$

One antidifferentiation yields $\mathbf{r}'(t) = -32t\mathbf{k} + \mathbf{C}$. Set $t = 0$ to find $\mathbf{C} = \mathbf{r}'(0)$. Therefore,

$$\mathbf{r}'(t) = -32t\mathbf{k} + \mathbf{r}'(0).$$

Another antidifferentiation gives $\mathbf{r}(t) = -16t^2\mathbf{k} + t\mathbf{r}'(0) + \mathbf{C}$. Once again set $t = 0$ to get $\mathbf{C} = \mathbf{r}(0) = \mathbf{0}$. Since $\mathbf{r}'(0) = 400\sqrt{3}\,\mathbf{i} + 400\mathbf{k}$, we obtain

$$\begin{aligned}\mathbf{r}(t) &= -16t^2\mathbf{k} + t\mathbf{r}'(0)\\ &= 400\sqrt{3}\,t\mathbf{i} + (400t - 16t^2)\mathbf{k}.\end{aligned}$$

Corresponding parametric equations for the flight path are

$$x = 400\sqrt{3}\,t, \qquad y = 0, \qquad z = 400t - 16t^2.$$

Notice that the entire flight takes place in the xz–plane because $y = 0$ for all t. The projectile strikes the ground at the time $t > 0$ when $z = 16t^2 - 400t = 0$ or $t = 25$ sec. Since $x = 10{,}000\sqrt{3} \approx 17{,}320$ feet when $t = 25$ seconds, the projectile lands about 3.3 miles down range from where it was fired. See the problems for more details about this motion. □

Newton actually stated his second law in a more general form:

$$\boxed{\frac{d}{dt}[m\mathbf{r}'(t)] = \mathbf{F}(t).}$$

The quantity $m\mathbf{r}'(t)$ is called the **momentum** of the object. If the mass is constant, as in the preceding examples, then the general form of Newton's second law reduces to $\mathbf{F} = m\mathbf{r}''(t)$. If mass is variable, as it is when a space shuttle rises into orbit, then the momentum form of Newton's second law must be used.

Arc Length Parameterization

As you travel along a road, you can determine your position either by time or by distance from the starting point. Linear motion illustrates this simple and important observation. Motion along the line L in Fig. 2 is expressed by $\mathbf{r}(t) = \mathbf{r}_0 + t\mathbf{v}$ with $\mathbf{v} \neq 0$. The velocity $\mathbf{r}'(t) = \mathbf{v}$ and speed $||\mathbf{v}||$ are constant. The distance from $P(0)$ to $P(t)$ is $s = ||\mathbf{v}||\,t$. So $t = s/||\mathbf{v}||$ and

$$r = \mathbf{r}_0 + \frac{s}{||\mathbf{v}||}\,\mathbf{v}.$$

Equivalently,

$$\mathbf{r} = \mathbf{r}_0 + s\mathbf{u}, \qquad \mathbf{u} = \mathbf{v}/||\mathbf{v}||,$$

where $\mathbf{u}$ is the unit vector in the direction of motion. The equation $\mathbf{r} = \mathbf{r}_0 + s\mathbf{u}$ expresses the position vector $\mathbf{r}$ for a point on the line as a function of arc length s.

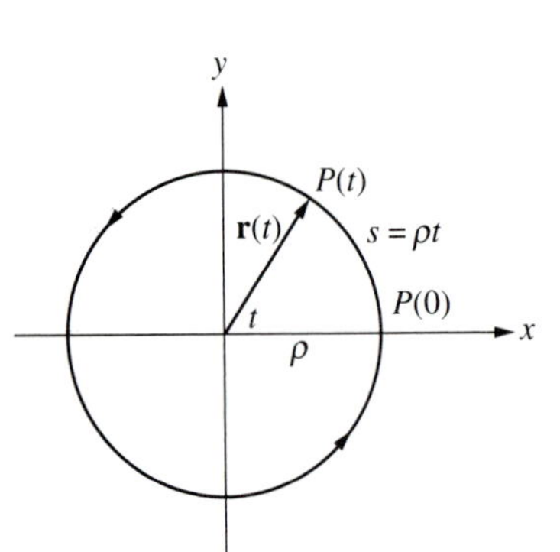

FIGURE 7

EXAMPLE 9. The position vector for the point $P(t)$ moving on the circle in Fig. 7 is $\mathbf{r}(t) = <\rho\cos t, \rho\sin t>$ for $0 \le t \le 2\pi$. Since the polar angle of $P(t)$ is equal to the time t, the circular motion is uniform with angular speed 1. Express the position vector as a function of arc length s on the circle measured counterclockwise from $(\rho, 0)$.

Solution. The arc length from $P(0) = (\rho, 0)$ to $P(t) = (\rho\cos t, \rho\sin t)$ is $s = \rho t$. Then $t = s/\rho$ and

$$\mathbf{r} = <\rho\cos(s/\rho), \rho\sin(s/\rho)>, \qquad 0 \le s \le 2\pi\rho.$$

This equation expresses the position vector $\mathbf{r}$ for a point on the circle as a function of arc length s. □

The change of parameter from t to arc length s can be made for any smooth curve. Recall that a curve $\mathbf{r} = \mathbf{r}(t)$ is smooth if $\mathbf{r}'(t)$ exists, is continuous, and is nonzero, so that $s'(t) = ||\mathbf{r}'(t)|| > 0$ for all t. These conditions are satisfied for the linear motion described above and for uniform circular motion. It follows that $s(t)$ increases as t increases. Therefore, $s(t)$ has an inverse function, which we express by $t(s)$. Substitute $t(s)$ into $\mathbf{r}(t)$ to obtain $\mathbf{r}(t(s))$, which is the **arc length parameterization** of C. To keep the notation simple, we write $\mathbf{r}(s)$ instead of $\mathbf{r}(t(s))$ when we regard $\mathbf{r}$ as a function of s. Thus, we can regard the position vector $\mathbf{r}$ as a function of t or s, according to the convenience of the moment.

It is usually difficult to find convenient formulas for $t(s)$ and $\mathbf{r}(s)$, but occasionally we get lucky, as in the next example.

EXAMPLE 10. Let $\mathbf{r}(t) = < t\cos t,\ t\sin t,\ (2\sqrt{2/3}\,)t^{3/2} >$ for $t \geq 0$. Find the arc length function $s = s(t)$ on C and its inverse function $t = t(s)$.

Solution. The graph of $\mathbf{r}(t)$ is an ascending spiral similar to the curve in Fig. 5. Straightforward calculations yield

$$\begin{aligned} \mathbf{r}'(t) &= < \cos t - t\sin t, \sin t + t\cos t, \sqrt{2}\, t^{1/2} >, \\ s'(t) &= ||\mathbf{r}'(t)|| = \sqrt{1 + t^2 + 2t} = t + 1, \\ s(t) &= \int_0^t ||\mathbf{r}'(\tau)||\, d\tau = \int_0^t (1 + \tau)\, d\tau = \frac{1}{2}t^2 + t. \end{aligned}$$

So the arc length function is given by $s = \frac{1}{2}t^2 + t$. We find the inverse function by solving for t as a function of s. The quickest way to do this is to multiply the equation for s by 2 and complete the square on t: $2s = t^2 + 2t$, $2s + 1 = t^2 + 2t + 1 = (t + 1)^2$ and, hence,

$$t = t(s) = \sqrt{2s + 1} - 1.$$

We could insert this expression for t into the formula for $\mathbf{r}(t)$ to find $\mathbf{r}(s)$; however, we do not bother to write out this messy formula. □

When a curve is reparameterized from t to s, the inverse function rule gives

$$\frac{dt}{ds} = \frac{1}{\frac{ds}{dt}}.$$

The chain rule for scalar functions extends easily to $\mathbf{r}(s)$ and $\mathbf{r}(t)$. Thus,

$$\frac{d\mathbf{r}}{ds} = \frac{d\mathbf{r}}{dt}\frac{dt}{ds}, \qquad \frac{d\mathbf{r}}{dt} = \frac{d\mathbf{r}}{ds}\frac{ds}{dt}.$$

To verify these formulas, just differentiate component by component. In prime notation, the latter formula is

$$\mathbf{r}'(t) = \mathbf{r}'(s)s'(t).$$

Some authors prefer to use a new letter to express the arc length parameterization of a curve C given originally by $\mathbf{r}(t)$. For example, the arc length parameterization may be expressed by $\mathbf{R}(s) = \mathbf{r}(t(s))$. Although this approach has some advantages, we believe that the notation $\mathbf{r}(t)$ and $\mathbf{r}(s)$ is better in the long run. However, you must be on your toes. For example, in the expressions $\mathbf{r}'(t)$ and $\mathbf{r}'(s)$, it is clear that prime means differentiation with respect to t in the first and differentiation with respect to s in the second. In order to interpret $\mathbf{r}'(2)$, it must be clear from the context whether $t = 2$ or $s = 2$ is meant.

The Unit Tangent Vector

Figure 8 shows a curve C with position vector $\mathbf{r}(t)$. The vector $\mathbf{r}'(t)$ is tangent to the curve at $P(t)$, as we know. The **unit tangent vector** or, more briefly, the **unit tangent** to C at $P(t)$ is defined by

$$\boxed{\mathbf{T} = \frac{\mathbf{r}'(t)}{||\mathbf{r}'(t)||}.}$$

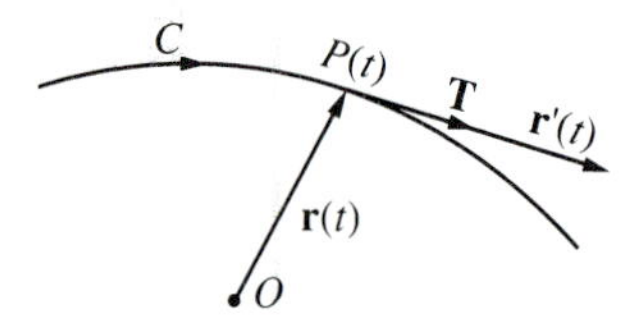

FIGURE 8

Clearly, $||\mathbf{T}|| = \mathbf{1}$ and $\mathbf{T}$ is tangent to the curve in Fig. 8 at $P(t)$. The unit tangent vector $\mathbf{T}$ tells us the direction of motion of the point $P(t)$ as it travels along the curve.

It is useful to express $\mathbf{T}$ in terms of arc length s. Since $\mathbf{r}'(t) = \mathbf{r}'(s)s'(t)$ by the chain rule, and $||\mathbf{r}'(t)|| = s'(t)$,

$$\mathbf{T} = \frac{\mathbf{r}'(t)}{||\mathbf{r}'(t)||} = \frac{\mathbf{r}'(s)s'(t)}{s'(t)} = \mathbf{r}'(s).$$

Thus, when C is parameterized by arc length, the unit tangent to C at $P(s)$ is given by

$$\boxed{\mathbf{T} = \mathbf{r}'(s).}$$

The unit tangent $\mathbf{T}$ can be regarded as a function of t or s. We write $\mathbf{T}(t)$ or $\mathbf{T}(s)$ when we want to indicate which independent variable is used. Since it is usually not possible to express $\mathbf{r}$ conveniently as a function of s, the pleasant looking formula $\mathbf{T} = \mathbf{r}'(s)$ primarily serves theoretical and conceptual purposes. The expression $\mathbf{T} = \mathbf{r}'(t)/||\mathbf{r}'(t)||$ is used in most practical calculations.

EXAMPLE 11. Let $\mathbf{r}(t) = < 2t, t^2, \frac{1}{3}t^3 >$. We met this vector function, which graphs as a twisted cubic, in Ex. 13 of Sec. 4.2. Since $\mathbf{r}'(t) = < 2, 2t, t^2 >$ and $||\mathbf{r}'(t)|| = 2 + t^2$, the unit tangent is

$$\mathbf{T}(t) = \frac{< 2, 2t, t^2 >}{2 + t^2}.$$

When $t = 2$ we have $\mathbf{T}(2) = \frac{1}{6} < 2,4,4 >$, which is tangent to the curve at $(4,4,8/3)$. Check directly that $||\mathbf{T}(2)|| = 1$. □

PROBLEMS

In Probs. 1–8: (a) Find the velocity, speed, and acceleration of the object whose position vector is given. (b) Sketch and describe the trajectory.

1. $\mathbf{r} = t\mathbf{i} + t^2\mathbf{j}$
2. $\mathbf{r} = (\sin t)\mathbf{i} + (4\sin t)\mathbf{j}$
3. $\mathbf{r} = < 2\cos t, 3t, 2\sin t >$
4. $\mathbf{r} = t^2(\mathbf{i} - 4\mathbf{j} + \mathbf{k})$
5. $\mathbf{r} = t\mathbf{i} + t^2\mathbf{j} + t^3\mathbf{k}$
6. $\mathbf{r} = (1 - \sin t)\mathbf{i} + (2 - \cos t)\mathbf{j} + \mathbf{k}$
7. $\mathbf{r} = < t, t^2, \ln t >$
8. $\mathbf{r} = < e^t\cos t, e^t\sin t, e^t >$

In Probs. 9–14 evaluate the integrals.

9. $\int (2t\mathbf{i} - e^{-t}\mathbf{j})\,dt$
10. $\int_0^1 (2t\mathbf{i} - e^{-t}\mathbf{j})\,dt$
11. $\int (e^t\mathbf{i} + (\cos^2 t)\mathbf{j} - (\sin 2t)\mathbf{k})\,dt$
12. $\int_0^\pi (e^t\mathbf{i} + (\cos^2 t)\mathbf{j} - (\sin 2t)\mathbf{k})\,dt$
13. $\int < 1/t, \ln t, \tan \pi t > dt$
14. $\int_1^2 < 1/t, \ln t, \tan \pi t > dt$

15. An object has position vector $\mathbf{r} = (a\cos\omega t)\mathbf{i} + (a\sin\omega t)\mathbf{j} + bt\mathbf{k}$, where a, b, and ω are positive constants and $t \geq 0$. Find (a) the velocity and acceleration, (b) the angle between the position vector and velocity vector, (c) the angle between the velocity vector and the acceleration vector, and (d) sketch the trajectory.

16. An object has position vector $\mathbf{r} = < a\sin\omega t, a\cos\omega t, bt >$, where a, b, and ω are positive constants and $t \geq 0$. Find (a) the velocity and acceleration, (b) the angle between the position vector and velocity vector, (c) the angle between the velocity vector and the acceleration vector, and (d) sketch the trajectory.

17. In Ex. 3 the point $P(t)$ moves on a cycloid. Where and when is the speed (a) a maximum and (b) a minimum?

18. A car with tires of radius a moves at constant speed v_0 along a straight road. Take the road to be the x–axis and assume the car moves to the right. (a) Explain briefly why the point on the tire which is at the origin at $t = 0$ is located by the position vector $\mathbf{r} = a(\theta - \sin\theta)\mathbf{i} + a(1 - \cos\theta)\mathbf{j}$ where $\theta = (v_0/a)t$. Assume that $0 \leq t \leq 2\pi a/v_0$ so the point on the tire makes one full revolution. (b) Find the velocity of the point. When is it moving most rapidly? (c) Show that the velocity points toward the highest point on the circle, except when the velocity is zero. (d) Find the acceleration vector and its magnitude. When is the magnitude greatest?

19. See Prob. 46 in Sec. 4.1. Assume the thread is unwound from the spool so that $\theta = \omega t$ with ω a positive constant. (a) Show that the position of the free end of the thread is $\mathbf{r} = < a(\cos\theta + \theta\sin\theta), a(\sin\theta - \theta\cos\theta) >$ where $\theta = \omega t$. (b) Find the velocity and speed at any time. (c) Find the acceleration and its magnitude. (d) Find the angle between the velocity vector and the acceleration vector.

20. An object moves in space with constant speed. Show that its velocity and acceleration are always orthogonal.

21. If the force on an object is always orthogonal to its velocity, show that the speed is constant.

22. An object moves on a sphere with center at P_0 and radius a. Show that the velocity vector is tangent to the sphere and that the acceleration vector is always directed toward P_0.

23. In Ex. 5, find the position, velocity, speed, and acceleration of the object when $t = 2\pi$.

24. In Ex. 8, (a) find the maximum elevation of the projectile and the time when it occurs, and (b) eliminate t to show that the trajectory is a parabolic arc in the xz–plane.

25. A fan in the outfield bleachers is 48 feet above the playing field and throws a baseball at 60 mph at an angle of elevation of 30° out into the playing field. Ignore air resistance. (a) Find the maximum height of the ball above the field. (b) Find the horizontal distance traveled before the ball hits the field.

26. A fairway is 360 yards long and nearly level. An oak tree 80 feet tall stands in the driving range 120 yards from the tee. If a golf ball is hit from the tee at a 30° angle, what minimum initial speed must it have in order to clear the tree? Ignore air resistance.

27. A projectile with mass m is fired from the ground with angle of elevation α and muzzle velocity (really speed) v_0. Ignore air resistance and assume the terrain is flat. Set up coordinates, as in Fig. 6, where the xz–plane is chosen to contain the initial velocity vector, the projectile is fired from the origin, and α is measured counterclockwise from the x–axis. (a) Show that the position of the projectile at time t is given by

$$\mathbf{r} = (v_0 \cos \alpha) t\mathbf{i} + \left[(v_0 \sin \alpha) t - \frac{1}{2} g t^2\right]\mathbf{k},$$

where g is the constant gravitational acceleration near the earth. (b) Find the maximum height of the projectile. (c) Find the range (that is, the horizontal distance traveled before impact) of the projectile. (d) What firing angle α gives the greatest range? (e) Eliminate the time and identify the trajectory.

28. Set up coordinates as in the preceding problem. Consider the motion of the projectile subject to both gravity and air resistance that is proportional to the velocity. Then $\mathbf{r}(0) = \mathbf{0}$ and $\mathbf{r}'(0) = \mathbf{v}(0) = v_0[(\cos \alpha)\mathbf{i} + (\sin \alpha)\mathbf{k}]$. (a) Show that the position vector satisfies the differential equation

$$m\mathbf{r}'' = -mg\mathbf{k} - c\mathbf{r}'$$

where $c > 0$ is the proportionality constant for the air resistance. (b) Integrate to find

$$\mathbf{r}' + \frac{c}{m} r = \mathbf{v}(0) - gt\mathbf{k}.$$

(c) Check that

$$(e^{(c/m)t}\mathbf{r})' = e^{(c/m)t}\left(\mathbf{r}' + \frac{c}{m}\mathbf{r}\right).$$

Use this fact in (b) to find $\mathbf{r}$ at any time t.

29. Assume $m = c = 2$, $g = 32$, and $\mathbf{r}'(0) = 100(\mathbf{i} + \sqrt{3}\,\mathbf{k})$ in the previous problem. (a) Find the maximum height of the projectile. (b) Find the range. *Hint.* You will need to use Newton's method or some other root–finder.

A particle with charge q, position vector $\mathbf{r}(t)$, and velocity vector $\mathbf{v} = \mathbf{r}'(t)$ moves in a magnetic field characterized by the magnetic field vector $\mathbf{B}$. In suitable units, the force (due to the magnetic field) on the particle is given by the **Biot–Savat law**

$$\mathbf{F} = q\mathbf{v} \times \mathbf{B}.$$

30. Let $\mathbf{v}(t) = < v_1(t), v_2(t), v_3(t) >$ be the velocity of the charged particle with charge q and mass m, as described above. Assume that the magnetic field is uniform and parallel to the z–axis so that $\mathbf{B} = B\mathbf{k}$ for some constant B. (a) Use Newton's second law and the Biot–Savat law to show that

$$v_1' = \omega v_2, \qquad v_2' = -\omega v_1, \qquad v_3' = 0, \qquad \omega = \frac{qB}{m}.$$

(b) Deduce that $v_3 = b$, a constant, and

$$v_1'' = -\omega^2 v_1, \qquad v_2 = \frac{1}{\omega} v_1'.$$

Check that $v_1 = a\cos\omega(t - \varphi)$, where a and φ are any constants, satisfies the first differential equation. It is a fact that all solutions of the first differential equation are of this form. Then show that $v_2 = a\sin\omega(t - \varphi)$. (c) Show that

$$\mathbf{r}'(t) = < a\cos\omega(t - \varphi), a\sin\omega(t - \varphi), b >$$

and integrate to obtain $\mathbf{r}(t)$. (d) What choices for the constants give the motion in Ex. 5?

31. The **impulse** of a force $\mathbf{F} = \mathbf{F}(t)$ acting over a time interval $t_0 \le t \le t_1$ is $\int_{t_0}^{t_1} \mathbf{F}(t)\,dt$. Show that the change in momentum of an object over the time interval $t_0 \le t \le t_1$ is equal to the impulse of the force acting on the object.

32. Let $\mathbf{f}(t) = < f_1(t), f_2(t), f_3(t) >$ be continuous for $a \le t \le b$. Use a Riemann sums and limit passage argument to show that

$$\int_a^b \mathbf{f}(t)\,dt = \left\langle \int_a^b f_1(t)\,dt, \int_a^b f_2(t)\,dt, \int_a^b f_3(t)\,dt \right\rangle.$$

In Probs. 33–38 a curve C is given parametrically by a position vector $\mathbf{r}(t)$. Find the function $t = t(s)$ required to express C in terms of the arc length parameter s. Assume $t \ge 0$ and in Prob. 36 also that $t \le \pi/4$.

33. $\mathbf{r} = < \cos t, \sin t, t >$

34. $\mathbf{r} = < 3\sin t, 4t, 3\cos t >$

35. $\mathbf{r} = (e^t\cos 2t)\mathbf{i} + (e^t\sin 2t)\mathbf{j} + 2e^t\mathbf{k}$

36. $\mathbf{r} = (\cos^3 t)\mathbf{i} + (\sin^3 t)\mathbf{k}$

37. $\mathbf{r} = < \cos t + t\sin t, \sin t - t\cos t >$

38. $\mathbf{r} = 3(\sin t + t\cos t)\mathbf{i} + 3(\cos t - t\sin t)\mathbf{j} + 4t\mathbf{k}$

4.4 Angular Velocity and Curvature

Two interrelated topics are studied in this section. Both have to do with turning of vectors. We begin with angular velocity for a plane motion. In Sec. 4.3 we observed that uniform circular motion has constant angular velocity. For other plane motions, variable angular velocity commonly occurs. Angular velocity plays a very important role in physics and engineering. Here we only touch on the subject. Further aspects and applications of angular velocity will be discussed in later chapters.

Curvature measures the bending of a curve. A circle with a large radius bends slowly while a circle with a small radius bends rapidly. Curvature is defined in terms of the turning of the unit tangent vector. There are important connections between curvature and acceleration which will be explored at the end of the section.

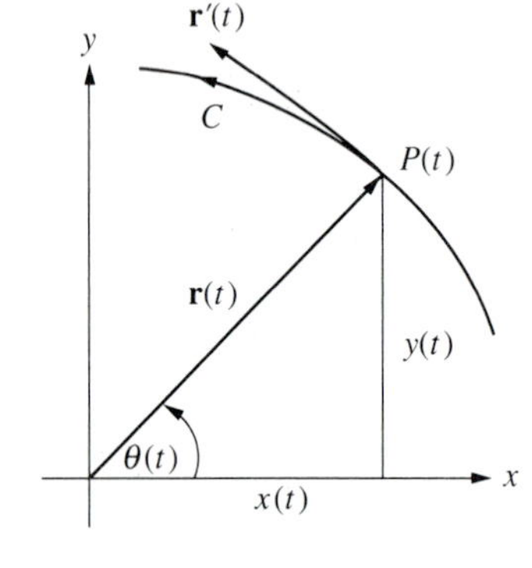

FIGURE 1

Angular Velocity

The **angular velocity** of a moving point $P(t) = (x(t), y(t))$ or of the corresponding position vector $\mathbf{r}(t) = < x(t), y(t) >$ in Fig. 1 is the rate of change of the angle $\theta(t)$ with respect to time. Thus,

$$\frac{d\theta}{dt} = \text{angular velocity.}$$

The angular velocity is positive if the motion is counterclockwise and it is negative if the motion is clockwise. Figure 1 illustrates the case with $d\theta/dt > 0$. The **angular speed** is the absolute value of the angular velocity:

$$\left| \frac{d\theta}{dt} \right| = \text{angular speed.}$$

A formula for angular velocity is easy to derive. In Fig. 1,

$$\tan\theta(t) = \frac{y(t)}{x(t)}, \qquad \theta(t) = \arctan\frac{y(t)}{x(t)}.$$

By the chain rule,

$$\frac{d\theta}{dt} = \frac{1}{1 + y^2/x^2} \cdot \frac{xy' - x'y}{x^2}.$$

Therefore,

$$\boxed{\frac{d\theta}{dt} = \frac{xy' - x'y}{x^2 + y^2}.}$$

Recall that the parametric equations $x(t) = \rho\cos\omega t$, $y(t) = \rho\sin\omega t$ with $\rho > 0$ and $\omega > 0$ describe counterclockwise motion around the circle with center (0,0) and radius ρ. Use the displayed formula to verify that the angular velocity and speed are equal to ω.

There are useful formulas for angular velocity and angular speed expressed in vector form. As usual, let $\mathbf{r}(t) = \langle x(t), y(t) \rangle$. Assume that $\mathbf{r}(t) \neq 0$ and $\mathbf{r}(t)$ is differentiable. Routine calculations give

$$\mathbf{r} \times \mathbf{r}' = (xy' - x'y) \cdot \mathbf{k}, \qquad \mathbf{r} \times \mathbf{r}' \cdot \mathbf{k} = (xy' - x'y)$$

and, consequently,

$$\boxed{\frac{d\theta}{dt} = \frac{\mathbf{r}(t) \times \mathbf{r}'(t) \cdot \mathbf{k}}{||\mathbf{r}(t)||^2}, \qquad \left|\frac{d\theta}{dt}\right| = \frac{||\mathbf{r}(t) \times \mathbf{r}'(t)||}{||\mathbf{r}(t)||^2}.}$$

EXAMPLE 1. Let $\mathbf{r}(t) = \langle t^2, t^3 \rangle$. Elementary calculations yield $\mathbf{r}(t) \times \mathbf{r}'(t) = t^4\mathbf{k}$ and $||\mathbf{r}(t)||^2 = t^4 + t^6$. So the angular velocity is

$$\frac{d\theta}{dt} = \frac{t^4\mathbf{k} \cdot \mathbf{k}}{t^4 + t^6} = \frac{1}{1 + t^2}. \quad \square$$

It is instructive to give a geometric derivation of the cross-product formulas for angular velocity and angular speed. A similar argument will be used to determine curvature. Figure 2 will help us explain why and how $\mathbf{r}(t) \times \mathbf{r}'(t)$ and $d\theta/dt$ are related. We begin with

$$\mathbf{r}(t) \times \mathbf{r}'(t) = \lim_{\Delta t \to 0} \mathbf{r}(t) \times \frac{\mathbf{r}(t + \Delta t) - \mathbf{r}(t)}{\Delta t}.$$

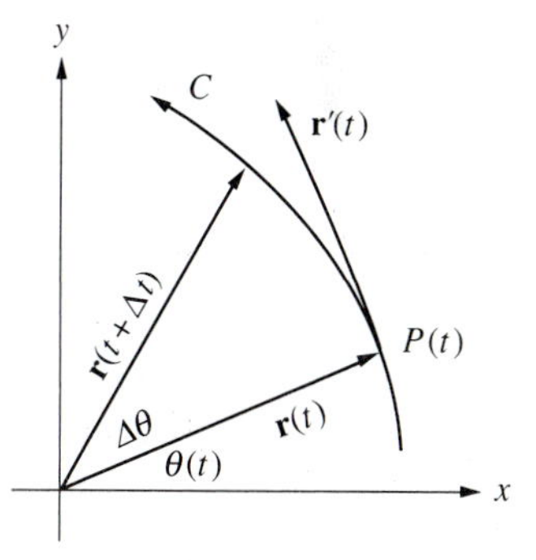

FIGURE 2

Since $\mathbf{r}(t) \times \mathbf{r}(t) = \mathbf{0}$,

$$\mathbf{r}(t) \times \mathbf{r}'(t) = \lim_{\Delta t \to 0} \frac{\mathbf{r}(t) \times \mathbf{r}(t + \Delta t)}{\Delta t}.$$

Recall that the cross product of vectors $\mathbf{v}$ and $\mathbf{w}$ is perpendicular to $\mathbf{v}$ and $\mathbf{w}$ and that $||\mathbf{v} \times \mathbf{w}|| = ||\mathbf{v}|| \; ||\mathbf{w}|| \sin\theta$, where θ is the angle between $\mathbf{v}$ and $\mathbf{w}$. Therefore, from Fig. 2,

$$\mathbf{r}(t) \times \mathbf{r}'(t + \Delta t) = ||\mathbf{r}(t)|| \;\; ||\mathbf{r}(t + \Delta t)|| \sin\Delta\theta \; \mathbf{k},$$

$$\frac{\mathbf{r}(t) \times \mathbf{r}'(t + \Delta t)}{\Delta t} = ||\mathbf{r}(t)|| \;\; ||\mathbf{r}(t + \Delta t)|| \frac{\sin\Delta\theta}{\Delta\theta} \frac{\Delta\theta}{\Delta t} \mathbf{k}.$$

Let $\Delta t \to 0$. Then $\mathbf{r}(t + \Delta t) \to \mathbf{r}(t)$, $\Delta\theta \to 0$,

$$\frac{\sin\Delta\theta}{\Delta\theta} \to 1 \text{ and } \frac{\Delta\theta}{\Delta t} \to \frac{d\theta}{dt}.$$

Hence,

$$\mathbf{r}(t) \times \mathbf{r}'(t) = \lim_{\Delta t \to 0} \frac{\mathbf{r}(t) \times \mathbf{r}(t + \Delta t)}{\Delta t} = ||\mathbf{r}(t)||^2 \frac{d\theta}{dt} \mathbf{k}.$$

The cross-product formulas for angular velocity and angular speed follow immediately.

Curvature

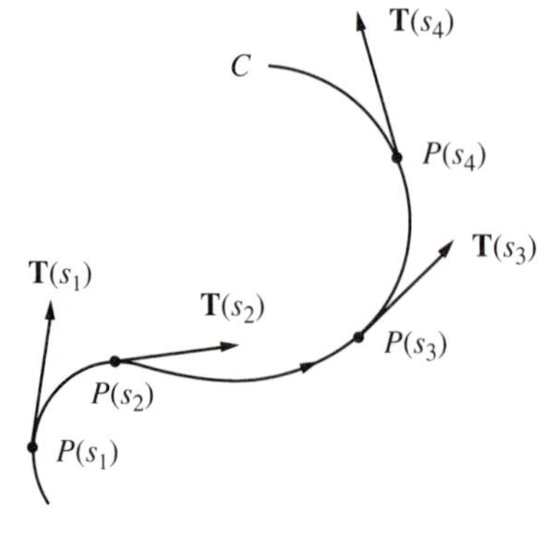

FIGURE 3

As we mentioned at the outset, curvature measures the bending of a curve. We first define curvature in geometric terms and then go on to develop several formulas for curvature that are convenient for calculation.

Figure 3 shows a curve C given by a vector function of arc length $\mathbf{r}(s)$ and several unit tangent vectors $\mathbf{T}(s) = \mathbf{r}'(s)$. As s varies, the unit tangent vectors change in direction but not in length. Evidently, the more rapidly $\mathbf{T}(s)$ changes direction as s changes, the more rapidly the curve bends. Moreover, the variations in $\mathbf{T}(s)$ indicate the direction of bending along the curve. These observations suggest that we define the curvature (bending) of a curve as $\mathbf{T}'(s) = \mathbf{r}''(s)$, which is a vector measure of bending. Since our interest is mainly in how rapidly or slowly the curve bends, the magnitude of $\mathbf{T}'(s)$ adequately describes bending for us.

Definition *Curvature*
The **curvature** of a curve C at a point $P(s)$ with position vector $\mathbf{r}(s)$ is

$$\kappa(s) = ||\mathbf{T}'(s)|| = ||\mathbf{r}''(s)||.$$

For a straight line, $\mathbf{T}(s)$ is a constant unit vector that points in the positive direction along the line; hence, $\mathbf{T}'(s) = \mathbf{0}$ and $\kappa = ||\mathbf{T}'(s)|| = 0$. If a curve is nearly straight, then κ is near zero. On the other hand, if a curve bends rapidly, then κ is correspondingly large.

EXAMPLE 2. Show that the circle $\mathbf{r}(t) = \rho < \cos t, \sin t >$ with radius ρ has constant curvature $\kappa = 1/\rho$. In words, the curvature is the reciprocal of the radius.

Solution. From Ex. 9 of Sec. 4.3, $s = \rho t$ and, hence, the circle has arc length parameterization $\mathbf{r}(s) = \rho < \cos(s/\rho), \sin(s/\rho) >$. Then

$$\mathbf{T}(s) = \mathbf{r}'(s) = < -\sin(s/\rho), \cos(s/\rho) >,$$

$$\mathbf{T}'(s) = \mathbf{r}''(s) = -\frac{1}{\rho} < \cos(s/\rho), \sin(s/\rho) >,$$

$$\kappa = ||\mathbf{T}'(s)|| = \frac{1}{\rho}.$$

The conclusion that $\kappa = 1/\rho$ should seem reasonable because a circle with a large radius bends slowly and a circle with a small radius bends rapidly. □

Curvature Formulas for Plane Curves

We shall develop several useful formulas for curvature, beginning with plane curves because the geometry is simpler. A curve parameterized by arc length is shown in Fig. 4. The angle between $\mathbf{r}'(s)$ and the horizontal unit vector

$\mathbf{i}$ is denoted by $\theta(s)$. The angle $\theta(s)$ is positive in Fig. 4, but it could be negative or zero. Curvature is expressed in terms of $\theta(s)$ by

$$\kappa(s) = \left|\frac{d\theta}{ds}\right|.$$

The derivation is easy. Since $||\mathbf{r}'(s)|| = 1$, Fig. 4 gives

$$\mathbf{r}'(s) = \cos\theta(s)\mathbf{i} + \sin\theta(s)\mathbf{j},$$

$$\mathbf{r}''(s) = -\sin\theta(s)\frac{d\theta}{ds}\mathbf{i} + \cos\theta(s)\frac{d\theta}{ds}\mathbf{j},$$

$$\kappa = ||\mathbf{r}''(s)|| = \left|\frac{d\theta}{ds}\right|.$$

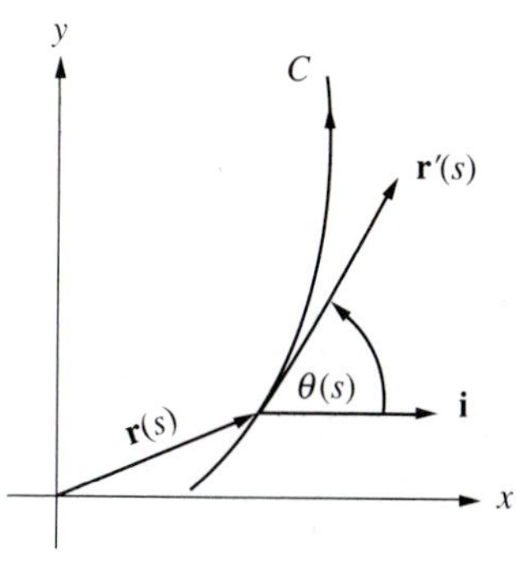

FIGURE 4

This formula adds to our understanding of curvature, but it is not very useful for calculation because it is generally difficult to represent curves in terms of arc length. It is more useful to express curvature in terms of the original curve parameter t. By the chain rule,

$$\frac{d\theta}{ds} = \frac{d\theta}{dt}\frac{dt}{ds} = \frac{d\theta}{dt}\Big/\frac{ds}{dt} = \frac{d\theta}{dt}\Big/||\mathbf{r}'(t)||,$$

where $d\theta/dt$ is the angular velocity of the tangent vector $\mathbf{r}'(t)$ shown in Fig. 5 and moved to the origin in Fig. 6. Hence,

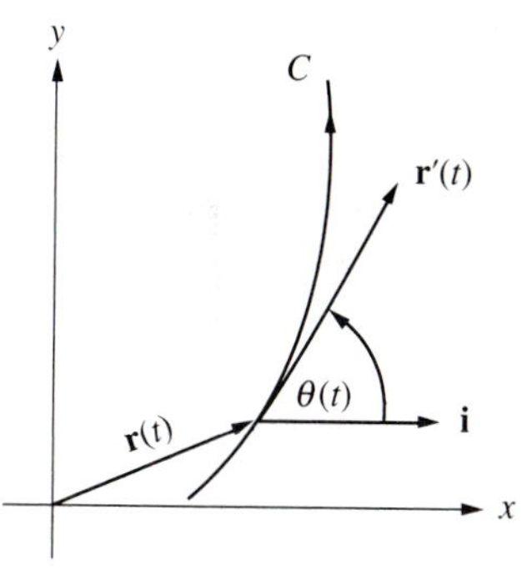

FIGURE 5

$$\kappa = \left|\frac{d\theta}{ds}\right| = \left|\frac{d\theta}{dt}\right|\Big/||\mathbf{r}'(t)||$$

Since $d\theta/dt$ is the angular velocity of $\mathbf{r}'(t)$, replace $\mathbf{r}(t)$ by $\mathbf{r}'(t)$ in the cross-product formula for angular speed to obtain

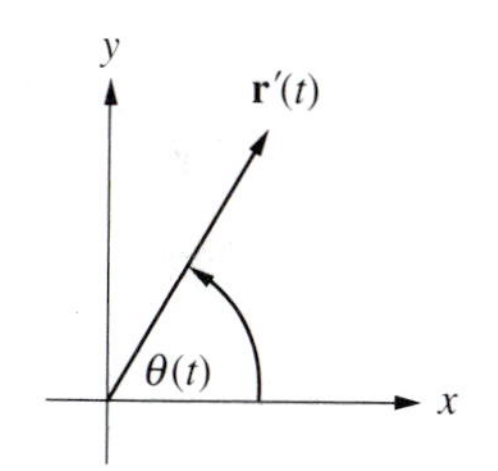

FIGURE 6

$$\left|\frac{d\theta}{dt}\right| = \frac{||\mathbf{r}'(t) \times \mathbf{r}''(t)||}{||\mathbf{r}'(t)||^2}.$$

Therefore, if $\mathbf{r}(t)$ is twice differentiable and $\mathbf{r}'(t) \neq 0$,

$$\boxed{\kappa(t) = \frac{||\mathbf{r}'(t) \times \mathbf{r}''(t)||}{||\mathbf{r}'(t)||^3}.}$$

Now consider a plane curve given parametrically by $x = x(t)$, $y = y(t)$. In the formula for $\kappa(t)$ let $\mathbf{r}(t) = \langle x(t), y(t) \rangle$ and carry out the cross product to obtain

$$\boxed{\kappa(t) = \frac{|x'(t)y''(t) - x''(t)y'(t)|}{[x'(t)^2 + y'(t)^2]^{3/2}}.}$$

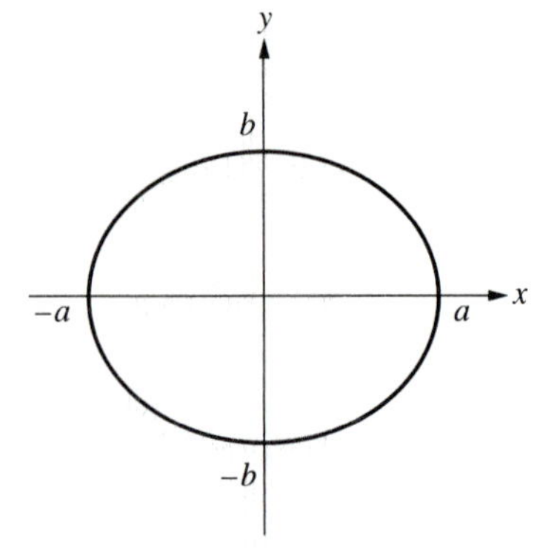

FIGURE 7

EXAMPLE 3. The ellipse

$$\frac{x^2}{a^2} + \frac{y^2}{b^2} = 1 \quad \text{with } 0 < b \le a,$$

which is graphed in Fig. 7, seems to bend most rapidly at the points $(\pm a, 0)$ and least rapidly at the points $(0, \pm b)$. Confirm these observations.

Solution. It is convenient to express the ellipse in parametric form:

$$x = a\cos t, \qquad y = b\sin t, \qquad 0 \le t \le 2\pi.$$

Elementary calculations yield

$$x'y'' - x''y' = ab,$$

$$(x')^2 + (y')^2 = a^2\sin^2 t + b^2\cos^2 t,$$

$$\kappa = \frac{ab}{[a^2\sin^2 t + b^2\cos^2 t]^{3/2}} = \frac{ab}{[b^2 + (a^2 - b^2)\sin^2 t]^{3/2}}.$$

Evidently, the curvature is largest when $\sin t = 0$ and $\cos t = \pm 1$. Then $(x,y) = (\pm a, 0)$ and $\kappa_{max} = a/b^2$. Likewise the curvature is smallest when $\sin t = \pm 1$ and $\cos t = 0$. Then $(x, y) = (0, \pm b)$ and $\kappa_{min} = b/a^2$. The ellipse becomes a circle if $b = a$, in which case the curvature formula reduces to $\kappa = 1/a$, the known curvature for a circle with radius a. □

Now consider a curve given explicitly by $y = y(x)$. Let $t = x$ in the parametric formula for curvature. Then $x' = 1$, $x'' = 0$, and

$$\boxed{\kappa(x) = \frac{|y''(x)|}{[1 + y'(x)^2]^{3/2}}.}$$

EXAMPLE 4. Find the curvature of the curve $y = 1/x$ for $x > 0$ in Fig. 8. Discuss the bending of the curve as $x \to 0$ and as $x \to \infty$. Also find the point of the curve where the curvature is greatest.

Solution. Since $y' = -x^{-2}$ and $y'' = 2x^{-3}$,

$$\kappa = \kappa(x) = \frac{|2x^{-3}|}{[1 + x^{-4}]^{3/2}} = \frac{2x^3}{[x^4 + 1]^{3/2}}.$$

Let $x \to 0$ or $x \to \infty$. Then $\kappa(x) \to 0$ and the curve becomes straighter and straighter, which is apparent in Fig. 8. The figure suggests that the maximum curvature occurs at (1,1), where $\kappa = 1/\sqrt{2}$. To confirm this, differentiate $\kappa(x)$ with respect to x and set the derivative equal to zero. □

FIGURE 8

Curvature Formulas for Space Curves

With relatively minor modifications, the principal formulas for curvature in the plane extend to space curves. The geometry is a little more complicated than for plane curves because there is no logical candidate to play the role of the angle $\theta(s)$. We begin with Figs. 9 and 10. In Fig. 9, s is fixed and $\Delta s > 0$. The unit tangent vectors $\mathbf{T}(s)$ and $\mathbf{T}(s + \Delta s)$ in Fig. 9 are moved to the origin in Fig. 10. The angle between $\mathbf{T}(s)$ and $\mathbf{T}(s + \Delta s)$ is denoted by $\theta(\Delta s)$. For a plane curve, $\theta(\Delta s) = \Delta\theta$. Since $\|\mathbf{T}(s)\| = 1$ and $\|\mathbf{T}(s + \Delta s)\| = 1$, $\theta(\Delta s)$ is also the arc length in Fig. 10. Curvature is expressed in terms of $\theta(\Delta s)$ by

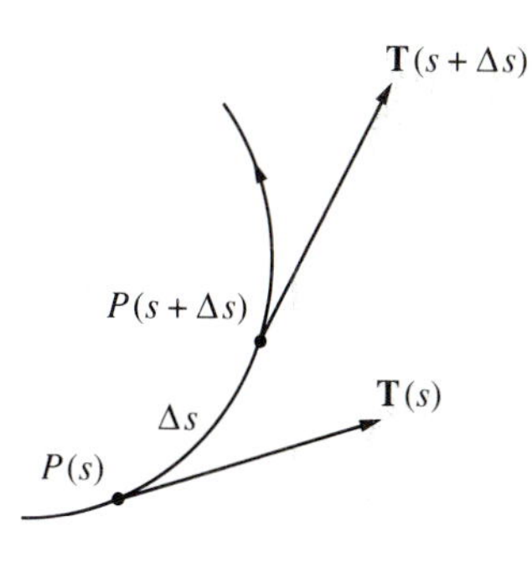

FIGURE 9

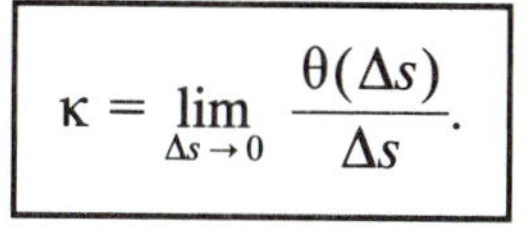

$$\boxed{\kappa = \lim_{\Delta s \to 0} \frac{\theta(\Delta s)}{\Delta s}.}$$

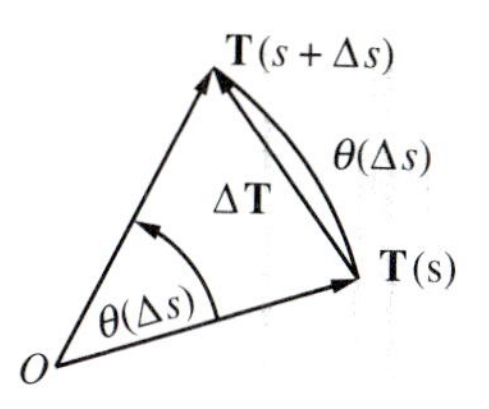

FIGURE 10

For plane curves, this reduces to our previous formula for curvature, $\kappa = |d\theta/ds|$.

To derive the displayed formula for curvature, we begin with

$$\kappa = \left\| \frac{d\mathbf{T}}{ds} \right\| = \lim_{\Delta s \to 0} \frac{\|\Delta \mathbf{T}\|}{\Delta s} = \lim_{\Delta s \to 0} \frac{\|\Delta \mathbf{T}\|}{\theta(\Delta s)} \frac{\theta(\Delta s)}{\Delta s}.$$

Either from the geometry in Fig. 10 or from $\|\Delta \mathbf{T}\| = 2 \sin\frac{1}{2}\theta(\Delta s)$,

$$\lim_{\Delta s \to 0} \frac{\|\Delta \mathbf{T}\|}{\theta(\Delta s)} = 1.$$

Therefore, as announced,

$$\kappa = \lim_{\Delta s \to 0} \frac{\theta(\Delta s)}{\Delta s}.$$

The cross-product formula for curvature in the plane is valid without change for space curves. If $\mathbf{r}(t)$ is twice differentiable and $\mathbf{r}'(t) \neq 0$,

$$\boxed{\kappa = \frac{\|\mathbf{r}'(t) \times \mathbf{r}''(t)\|}{\|\mathbf{r}'(t)\|^3}.}$$

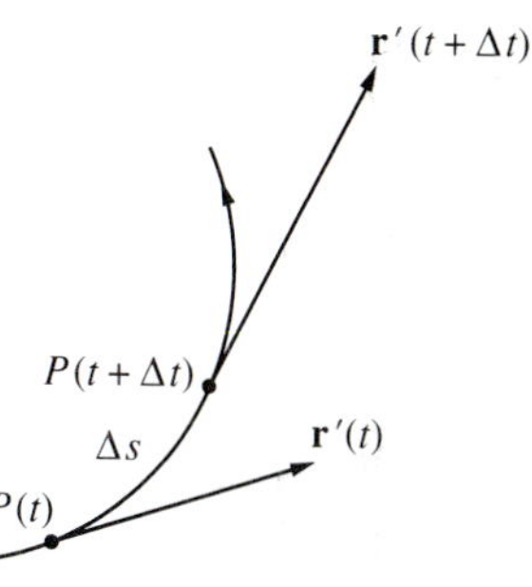

FIGURE 11

The derivation is based on Figs. 11 and 12, which are similar to Figs. 9 and 10. The major difference is that the unit tangent vectors in the earlier figures are replaced by the tangent vectors $\mathbf{r}'(t)$ and $\mathbf{r}'(t + \Delta t)$. The following argument should look familiar. It is adapted from the derivation of the cross-product formula for angular velocity. In Fig. 12,

$$\|\mathbf{r}'(t) \times \mathbf{r}'(t + \Delta t)\| = \|\mathbf{r}'(t)\| \quad \|\mathbf{r}'(t + \Delta t)\| \sin\theta(\Delta s).$$

Since $\mathbf{r}'(t) \times \mathbf{r}'(t) = \mathbf{0}$,

$$\left\| \mathbf{r}'(t) \times \frac{\mathbf{r}'(t + \Delta t) - \mathbf{r}'(t)}{\Delta t} \right\| = \|\mathbf{r}'(t)\| \; \|\mathbf{r}'(t + \Delta t)\| \frac{\sin\theta(\Delta s)}{\theta(\Delta s)} \frac{\theta(\Delta s)}{\Delta s} \frac{\Delta s}{\Delta t}.$$

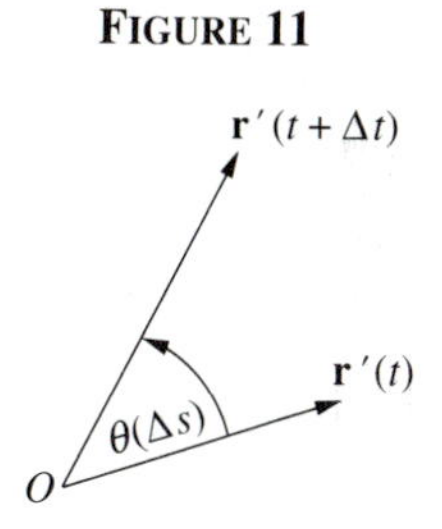

FIGURE 12

Let $\Delta t \to 0$. Then $\Delta s \to 0$, $\theta(\Delta s) \to 0$, and

$$\frac{\sin\theta(\Delta s)}{\theta(\Delta s)} \to 1, \qquad \frac{\theta(\Delta s)}{\Delta s} \to \kappa, \qquad \frac{\Delta s}{\Delta t} \to \frac{ds}{dt} = \|\mathbf{r}'(t)\|.$$

Therefore,

$$\|\mathbf{r}'(t) \times \mathbf{r}''(t)\| = \|\mathbf{r}'(t)\|^3\kappa.$$

The cross-product formula for κ follows immediately.

EXAMPLE 5. Let $\mathbf{r}(t) = \langle 2t, t^2, \frac{1}{3}t^3 \rangle$. This is a twisted cubic we have met twice before. Find its maximum curvature.

Solution. First we find the curvature at any point on the twisted cubic. We have $\mathbf{r}'(t) = \langle 2, 2t, t^2 \rangle$, $\mathbf{r}''(t) = \langle 0, 2, 2t \rangle$, $\|\mathbf{r}'(t)\| = t^2 + 2$, and

$$\mathbf{r}' \times \mathbf{r}'' = \begin{vmatrix} \mathbf{i} & \mathbf{j} & \mathbf{k} \\ 2 & 2t & t^2 \\ 0 & 2 & 2t \end{vmatrix} = 2t^2\,\mathbf{i} - 4t\mathbf{j} + 4\mathbf{k}.$$

Therefore, $\|\mathbf{r}' \times \mathbf{r}''\| = \sqrt{4t^4 + 16t^2 + 16} = 2(t^2 + 2)$ and

$$\kappa = \frac{\|\mathbf{r}' \times \mathbf{r}''\|}{\|\mathbf{r}'\|^3} = \frac{2(t^2+2)}{(t^2+2)^3} = \frac{2}{(t^2+2)^2}.$$

By inspection, the maximum curvature is $\kappa = 1/2$ when $t = 0$ and $P(0) = (0, 0, 0)$. □

Tangential and Normal Components of Acceleration*

As usual, starred material is not essential for material covered later in the book. We mentioned earlier that curvature and acceleration are closely related. To explore the connection, imagine that an object with mass m is moving in space with position vector $\mathbf{r}(t)$. As before, we assume that $\mathbf{r}(t)$ is twice differentiable and $\mathbf{r}'(t) \neq 0$. Our focus is on the acceleration vector

$$\mathbf{a}(t) = \mathbf{r}''(t).$$

There are two principal components of acceleration, both of physical significance. The tangential component gives the acceleration along the path. The normal component gives the acceleration perpendicular to the path.

To gain physical insight, we may suppose that the motion is caused by an external force $\mathbf{F}(t)$ related to the acceleration by Newton's second law, $\mathbf{F}(t) = m\mathbf{a}(t)$. The tangential and normal components of $\mathbf{F}(t)$ are just m times the corresponding components of $\mathbf{a}(t)$. The tangential component of $\mathbf{F}(t)$ tends to make the object speed up or slow down as it moves along its path. The normal component of $\mathbf{F}(t)$ tends to make the path curve away from the tangential direction. So it should be expected that the normal components of force and acceleration are closely related to curvature.

A piece of the trajectory is shown in Fig. 13. The **tangential component of acceleration** at a point P on the curve is the component of $\mathbf{a}(t)$ in the direction of the tangent vector $\mathbf{r}'(t)$. Thus,

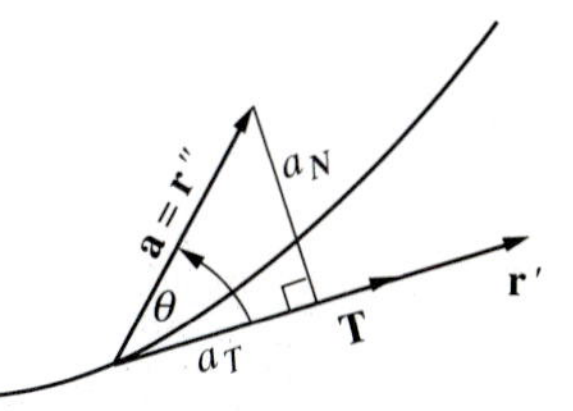

FIGURE 13

$$a_T = ||\mathbf{a}|| \cos\theta = \frac{\mathbf{r}''(t) \cdot \mathbf{r}'(t)}{||\mathbf{r}'(t)||}.$$

We also call a_T the **acceleration along the curve** because another formula for a_T is

$$a_T = s''(t).$$

It is easy to show that $a_T = s''(t)$. Since $s'(t) = ||\mathbf{r}'(t)||$,

$$s'(t)^2 = ||\mathbf{r}'(t)||^2 = \mathbf{r}'(t) \cdot \mathbf{r}'(t)$$

Differentiate to obtain

$$2s'(t)s''(t) = 2\mathbf{r}''(t) \cdot \mathbf{r}'(t),$$

$$a_T = \frac{\mathbf{r}''(t) \cdot \mathbf{r}'(t)}{||\mathbf{r}'(t)||} = s''(t).$$

The **normal component of acceleration** a_N is the component of $\mathbf{a}(t)$ orthogonal to $\mathbf{r}'(t)$. In Fig. 13,

$$a_N = ||\mathbf{r}''|| \sin\theta = \frac{||\mathbf{r}'||\ ||\mathbf{r}''||\ \sin\theta}{||\mathbf{r}'||} = \frac{||\mathbf{r}' \times \mathbf{r}''||}{||\mathbf{r}'||}.$$

Since $\kappa = ||\mathbf{r}' \times \mathbf{r}''|| / ||\mathbf{r}'||^3$ and $s'(t) = ||\mathbf{r}'(t)||$,

$$a_N = s'(t)^2\kappa.$$

This is the relation between curvature and the normal component of acceleration that was mentioned earlier. If a_N is calculated first, then the curvature is available from $\kappa = a_N/||\mathbf{r}'||^2$.

There is another useful formula for a_N. Since $a_N = s'(t)^2\kappa \geq 0$ and $a_T^2 + a_N^2 = ||\mathbf{a}||^2$, it follows that

$$a_N = \sqrt{||\mathbf{a}||^2 - a_T^2}.$$

EXAMPLE 6. Let

$$\mathbf{r}(t) = \frac{1}{2}t^2\mathbf{i} - 2t\mathbf{j} + \frac{4}{3}t^{3/2}\mathbf{k}, \qquad t \geq = 0.$$

This vector function appeared in Ex. 1 of Sec 4.3. Find a_T, a_N, and κ.

Solution. Routine calculations give

$$\mathbf{r}'(t) = t\mathbf{i} - 2\mathbf{j} + 2t^{1/2}\mathbf{k}, \qquad s'(t) = \|\mathbf{r}'(t)\| = t + 2,$$
$$\mathbf{r}''(t) = \mathbf{i} + t^{-1/2}\mathbf{k}, \qquad \|\mathbf{r}''(t)\|^2 = 1 + t^{-1}.$$

Since $\mathbf{a} = \mathbf{r}''$,

$$a_T = s''(t) = 1,$$
$$a_N = \sqrt{\|\mathbf{a}\|^2 - a_T^2} = 1/\sqrt{t},$$
$$\kappa = \frac{a_N}{s'(t)^2} = \frac{1}{(t+2)^2\sqrt{t}}.$$

As a check, find κ from the cross-product formula for curvature and then find $a_N = s'(t)^2\kappa$. If you are careful, you should get the same answers, however with more work. □

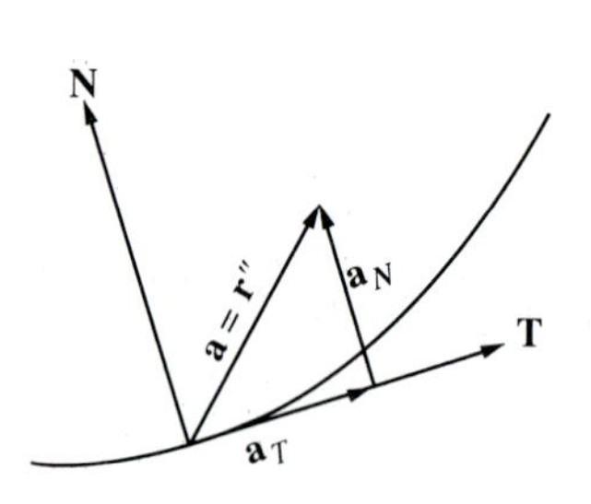

FIGURE 14

We close this section with a brief discussion of the tangential and normal projections of the acceleration vector. Figure 14 will serve as a guide. The unit tangent vector at a point P on the curve is $\mathbf{T} = \mathbf{r}'(s)$. The **tangential** and **normal projections** of $\mathbf{a}(t)$ are the vectors

$$\mathbf{a}_T = a_T\mathbf{T}, \qquad \mathbf{a}_N = \mathbf{a} - \mathbf{a}_T.$$

They satisfy

$$\mathbf{a} = \mathbf{a}_T + \mathbf{a}_N, \quad \mathbf{a}_T \perp \mathbf{a}_N, \quad \|\mathbf{a}_T\| = a_T = s''(t), \quad \|\mathbf{a}_N\| = a_N = s'(t)^2\kappa.$$

There are significant differences between the cases with zero and nonzero curvature. If $\mathbf{r}''(s) = \mathbf{0}$, then $\kappa = \|\mathbf{r}''(s)\| = 0$, $a_N = 0$, $\mathbf{a}_N = \mathbf{0}$, and $\mathbf{a} = \mathbf{a}_T$. Thus, if the curvature is zero at P, then the acceleration vector points in the tangential direction.

Figure 14 illustrates the case with $\mathbf{r}''(s) \neq 0$, $\kappa = \|\mathbf{r}''(s)\| > 0$, and $\|\mathbf{a}_N\| = a_N > 0$. Since $\|\mathbf{r}'(s)\|$ is constant, $\mathbf{r}''(s) \perp \mathbf{r}'(s)$. The **principal unit normal** or, more briefly, the **unit normal** at P is

$$\mathbf{N} = \frac{\mathbf{r}''(s)}{\|\mathbf{r}''(s)\|}.$$

Clearly, $\|\mathbf{T}\| = \|\mathbf{N}\| = 1$ and $\mathbf{T} \perp \mathbf{N}$. Two applications of the chain rule give $\mathbf{r}'(t) = \mathbf{r}'(s)s'(t)$ and

$$\mathbf{r}''(t) = \mathbf{r}'(s)s''(t) + r''(s)s'(t)^2,$$
$$= s''(t)\mathbf{T} + s'(t)^2\|\mathbf{r}''(s)\|\mathbf{N}.$$

Since $\mathbf{a}(t) = \mathbf{r}''(t)$ and $\kappa = ||\mathbf{r}''(s)||$,

$$\boxed{\mathbf{a}(t) = s''(t)\mathbf{T} + s'(t)^2\kappa\mathbf{N}.}$$

Since $\mathbf{a}_N = \mathbf{a} - \mathbf{a}_T$ and $\mathbf{a}_T = s''(t)$,

$$\mathbf{a}_N = s'(t)^2\kappa\mathbf{N} = a_N\mathbf{N}.$$

The boxed decomposition of $\mathbf{a}(t)$ enables us to interpret any motion as an instant-by-instant superposition of linear motion and circular motion. See the problems.

PROBLEMS

In Probs. 1-6 a curve C is given explicitly or by a vector function $\mathbf{r}(t)$. Find the angular velocity and angular speed.

1. $x = \cos 3t, \quad y = \sin 3t$
2. $x = \cos 2t, \quad y = -\sin 2t$
3. $\mathbf{r}(t) = < t\cos t, t\sin t >$
4. $\mathbf{r} = < e^t \cos t, e^t \sin t >$
5. $\mathbf{r} = < \cos^3 t, \sin^3 t >$
6. $\mathbf{r} = < t - \sin t, 1 - \cos t >$

In Probs. 7–16 a curve C is given explicitly or by a vector function $\mathbf{r}(t)$. Find the curvature.

7. $\mathbf{r} = < \cos 3t, -\sin 3t >$
8. $x = \cos^3 t, \quad y = \sin^3 t$
9. $x = 4t, \quad y = -3\cos t, \quad z = 3\sin t$
10. $\mathbf{r} = (e^t \cos 2t)\mathbf{i} + (e^t \sin 2t)\mathbf{j} + e^t\mathbf{k}$
11. $\mathbf{r} = < \theta - \cos\theta, 1 - \sin\theta >$
12. $< \cos\theta, \sin\theta, \cosh\theta >$
13. $y = \sin x$
14. $y = \sin 2x$
15. $y = \ln x$
16. $y = e^{-x^2/2}$
17. Find the curvature at any point on the cycloid $x = a(t - \sin t)$, $y = a(1 - \cos t)$, $0 < t < 2\pi$. Locate any maxima and minima of $\kappa(t)$. Sketch the graph of $\kappa(t)$.
18. Find the points on the curve given by $\mathbf{r} = < \sin t, \cos t, \sin t >$, $0 \le t \le 2\pi$, where the curvature is largest and smallest. Sketch the graph of $\kappa(t)$.
19. Find the points on the graph of $y = e^{-x}$ where the curvature is largest. Examine $\kappa(x)$ as $x \to \pm\infty$ and sketch the graph of $\kappa(x)$.
20. Find the points on the graph of $y = \arctan x$ where the curvature is greatest. Examine $\kappa(x)$ as $x \to \pm\infty$ and sketch the graph of $\kappa(x)$.

In Probs. 21–26 find the tangential and normal components of acceleration.

21. $\mathbf{r} = < t^2/2, t^3/3 >$
22. $\mathbf{r} = < t, t^2 >$
23. $\mathbf{r} = (\sin 2t)\mathbf{i} + (\cos 2t)\mathbf{j} + 3t\mathbf{k}$
24. $\mathbf{r} = (t\cos t)\mathbf{i} + (t\sin t)\mathbf{j} - 3t\mathbf{k}$
25. $\mathbf{r} = (\cosh t)\mathbf{i} + (\sinh t)\mathbf{j} + t\mathbf{k}$
26. $\mathbf{r} = (\ln\sec t)\mathbf{i} + t\mathbf{j} - \mathbf{k}$

27. The motion of an object is given by $\mathbf{r} = (\cos t)\mathbf{i} + (\sin 2t)\mathbf{j} + (\sin t)\mathbf{k}$. Find the tangential and normal components of acceleration.

28. An object moves with constant angular velocity $\omega > 0$ on a circle with center at the origin and radius ρ so that $\mathbf{r} = (\rho\cos\omega t)\mathbf{i} + (\rho\sin\omega t)\mathbf{j}$ describes the motion. Find a_T, a_N, κ, $\mathbf{T}$, and $\mathbf{N}$ for this motion.

29. An object has position vector $\mathbf{r} = (a\cos t)\mathbf{i} + (b\sin t)\mathbf{j}$ for $t \geq 0$. Identify and sketch the trajectory. Find a_T, a_N, κ, $\mathbf{T}$, and $\mathbf{N}$ when $t = \pi/4$.

30. An object moves on the hyperbola $x^2 - y^2 = 1$ so that its position at time t is $\mathbf{r} = < \cosh t, \sinh t >$. Find a_T, a_N, κ, $\mathbf{T}$, and $\mathbf{N}$ when $t = \ln 2$.

31. An object moves on the circular helix so that its position at time t is $\mathbf{r} = (\rho\cos\omega t)\mathbf{i} + (\rho\sin\omega t)\mathbf{j} + bt\mathbf{k}$. Find a_T, a_N, κ, $\mathbf{T}$, and $\mathbf{N}$ for this motion.

32. (a) Sketch the trajectory of an object with position vector $\mathbf{r} = < a\cos\omega t, b\sin\omega t, ct >$, where a, b, c, and ω are positive constants. (b) If $a = 4$, $b = 3$, $c = 5$, and $\omega = \pi/4$ find a_T, a_N, κ, $\mathbf{T}$, and $\mathbf{N}$ when $t = 1$.

33. The motion of an object is given by $\mathbf{r} = < e^t, e^{-t}, \sqrt{2t} >$. Fin a_T, a_N, κ, $\mathbf{T}$, and $\mathbf{N}$ for this motion.

34. If the thread in Prob. 46 of Sec. 4.1 unwinds at a constant angular velocity of 1 radian/sec and the spool has radius 1, then the free end has position given by $\mathbf{r} = (\cos t + t\sin t)\mathbf{i} + (\sin t - t\cos t)\mathbf{j}$. Find a_T, a_N, κ, $\mathbf{T}$, and $\mathbf{N}$ for this motion.

35. If the acceleration vector always points in the tangential direction, show that the motion takes place along a straight line. *Hint.* Show that $\mathbf{r}'' = s''(t)\,\mathbf{T}$. Use this and $\mathbf{T} = \mathbf{r}'/s'(t)$ to show that $d\mathbf{T}/dt = 0$.

36. Show that the principal unit normal vector can also be expressed by

$$\mathbf{N} = \frac{\mathbf{T}'(t)}{\|\mathbf{T}'(t)\|}.$$

This formula is often used to define the unit normal. *Hint.* Recall that $\mathbf{T} = \mathbf{r}'(t)/s'(t)$. Use the quotient rule to find $d\mathbf{T}/dt$. Then show that the formula above is the same as the formula for $\mathbf{N}$ given in the text.

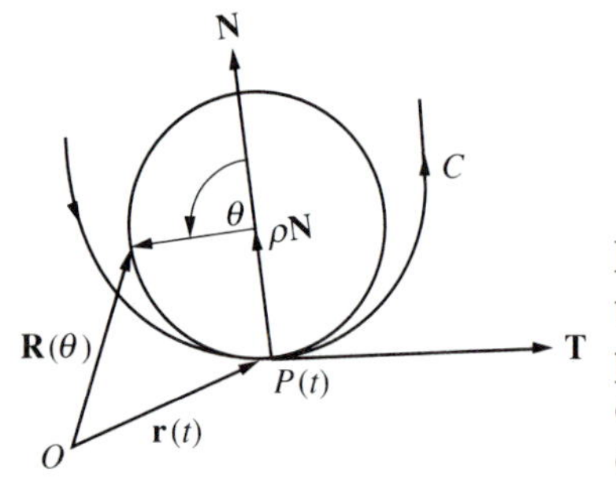

FIGURE 15

Figure 15 shows a curve C given by $\mathbf{r} = \mathbf{r}(t) = < x(t), y(t), z(t) >$. If $\kappa = \kappa(t) \neq 0$ at $P(t)$, then $\rho = 1/\kappa$ is called the **radius of curvature** of C at the point $P(t)$. The circle in the figure in the plane of $\mathbf{T}$ and $\mathbf{N}$ with radius ρ and center $\mathbf{r}(t) + \rho\mathbf{N}(t)$ is called the **circle of curvature** at the point $P(t)$. These concepts are most useful for plane curves.

37. Let C be a circle with radius a. (a) Show that the radius of curvature ρ is constant and equal to the radius a of the given circle. (b) Show that the circle of curvature at any point P coincides with the given circle.

38. Fix t and the point $P(t)$ in Fig. 15. (a) Show that the circle of curvature in Fig. 15 has vector equation

$$\mathbf{R}(\theta) = \mathbf{r}(t) + \rho\mathbf{N}(t) + (\rho\cos\theta)\mathbf{T}(t) + (\rho\sin\theta)\mathbf{N}(t), \quad 0 \leq \theta \leq 2\pi.$$

(b) Then show that C and the circle of curvature have the same unit tangent, unit normal, and curvature at the point $P(t)$.

39. Find (a) the radius of curvature and (b) the circle of curvature of the curve $y = x^2$ at the point (1,1). *Hint.* Write the curve in parametric form $\mathbf{r} = \langle t, t^2 \rangle$.

40. Find (a) the radius of curvature and (b) the circle of curvature of the curve $y = e^x$ at the point (0,1).

41. Find (a) the radius of curvature and (b) the circle of curvature of the curve $y = \ln x$ at the point (1,0).

42. Find (a) the radius of curvature and (b) the circle of curvature of the curve $y = \cosh x$ at the point (0,1).

43. Fix an instant in time t. Explain how the formula $\mathbf{a} = a_T\mathbf{T} + a_N\mathbf{N}$ permits you to regard any motion as an instant–by–instant superposition of rectilinear motion along $\mathbf{T}$ with constant acceleration $s''(t)$ and uniform circular motion on the circle of curvature with constant angular velocity $\omega = ||\mathbf{r}'(t)||/\rho = \kappa||\mathbf{r}'(t)||$.

4.5 Motion in Polar Coordinates*

Polar coordinates are particularly convenient for the study of rotary motion and other plane motions about a center of action. Important examples are motions maintained by central forces, such as gravitational and electrical forces. Any motion maintained by a central force lies in a plane, as we shall demonstrate. For this reason, plane motions have a much broader range of applications than you might have thought. As usual, the asterisk at the end of the title signifies that this material is not essential for topics covered later in the book.

After a brief discussion of central forces, we study polar parametric equations $r = r(t)$ and $\theta = \theta(t)$ for plane curves. Then we go on to express position, velocity, and acceleration vectors in convenient polar coordinate forms that reveal their radial and transverse components. We shall learn that the position vector of an object moving in a plane sweeps out equal areas in equal times if and only if a central force maintains the motion. The section and chapter conclude with a derivation of Kepler's laws of planetary motion from Newton's laws of motion and gravitation.

Throughout the section t denotes time and $\mathbf{r}(t)$ is the position vector of a point $P(t)$, which represents an object in motion. As before,

$$\mathbf{r}(t) = \text{position}, \qquad \mathbf{r}'(t) = \text{velocity}, \qquad \mathbf{r}''(t) = \text{acceleration}.$$

Primes always denote differentiation with respect to t. For brevity, the independent variable t is often omitted.

Central Forces

A **central force** is always directed toward or away from a particular point in space that may be taken as the origin. The force is **attractive** if it points toward the origin and **repulsive** if it points away from the origin. Gravitational and electrical forces are central.

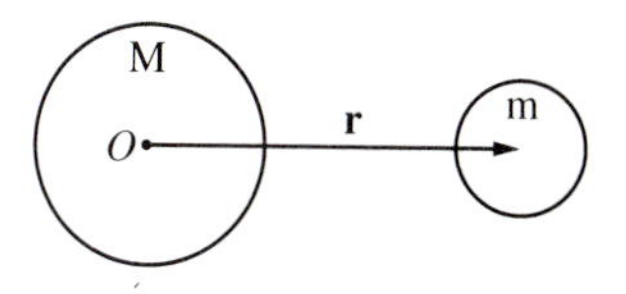

FIGURE 1

For example, consider the two masses M and m in Fig. 1. We regard both as point masses concentrated at their centers. In the figure, $\mathbf{r}$ is the vector extending from M to m. The force $\mathbf{F}$ exerted by M on m is given by Newton's law of gravitation:

$$\mathbf{F} = -\frac{GMm}{||\mathbf{r}||^3}\mathbf{r}.$$

The force $\mathbf{F}$ is directed toward the center of M and has magnitude $||\mathbf{F}|| = GMm/||\mathbf{r}||^2$. The gravitational force is an attractive central force.

In typical applications of Newton's law of gravitation, M is much greater than m. For example, M may be the mass of the sun and m the mass of a planet. Then the mass M remains virtually at rest as the mass m moves in the gravitational field of M. In such cases, it is convenient to choose the origin of coordinates at the center of M. Then $\mathbf{r} = \mathbf{r}(t)$ in Fig. 1 is the position vector of the mass m at time t. It follows from Newton's second law of motion $\mathbf{F} = m\mathbf{r}''$ and his law of gravitation that

$$\mathbf{r}'' = -\frac{GM}{||\mathbf{r}||^3}\mathbf{r}.$$

Evidently, a force $\mathbf{F}$ is central precisely when there is an origin of coordinates so that $\mathbf{F}$ and the position vector $\mathbf{r}$ are always parallel; in symbols, $\mathbf{F}||\mathbf{r}$. Since $\mathbf{F} = m\mathbf{r}''$,

the force $\mathbf{F}$ is central
$\Leftrightarrow \quad \mathbf{r}''||\mathbf{r}$.

For a central force, it follows from $\mathbf{r}''||\mathbf{r}$ that $\mathbf{r}'' \times \mathbf{r} = \mathbf{0}$. Differentiate $\mathbf{r}' \times \mathbf{r}$ and use $\mathbf{r}' \times \mathbf{r}' = \mathbf{0}$ to obtain

$$\frac{d}{dt}(\mathbf{r}' \times \mathbf{r}) = \mathbf{r}'' \times \mathbf{r} + \mathbf{r}' \times \mathbf{r}' = \mathbf{0} \quad \text{and} \quad \mathbf{r}'(t) \times \mathbf{r}(t) = \mathbf{C}$$

for some constant vector $\mathbf{C}$. Since $\mathbf{r}'(t) \times \mathbf{r}(t) \perp \mathbf{r}(t)$, the position vector $\mathbf{r}(t)$ is always perpendicular to $\mathbf{C}$. Consequently, $\mathbf{r}(t)$ lies in the plane through the origin that is perpendicular to $\mathbf{C}$. We have reached an important conclusion.

Motion under a central force takes place in a plane.

We shall return to motions maintained by central forces a little later.

Polar Parametric Equations

In Fig. 2, $r(t)$ and $\theta(t)$ are the polar coordinates at time t of a point $P(t)$ moving along a curve C in a plane. Then

$$r = r(t), \qquad \theta = \theta(t)$$

are **polar parametric equations** for the trajectory C.

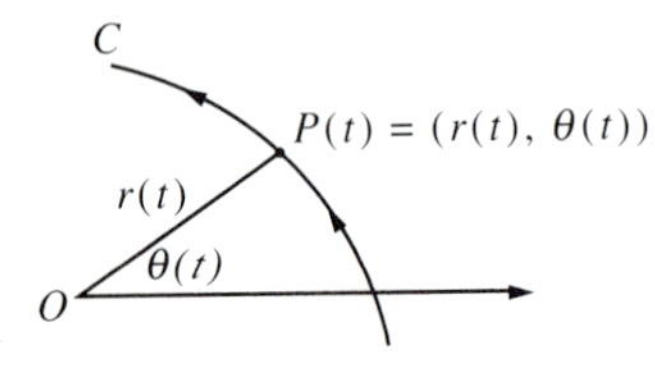

FIGURE 2

EXAMPLE 1. *Circular Motion.* Fix $\rho > 0$. Let $r(t) = \rho$ and $\theta'(t) > 0$ for all t. The point $(r(t), \theta(t))$ runs counterclockwise around the circle in Fig. 3 with angular velocity $\theta'(t)$ and angular acceleration $\theta''(t)$. Since $s = \rho\,\theta(t)$ in Fig. 3, the speed along the curve is $s'(t) = \rho\,\theta'(t)$. Uniform circular motion, which was treated in Sec. 4.3, is the special case given by $\theta(t) = \omega t$, where ω is constant. Then $\theta'(t) = \omega$ and $s'(t) = \rho\omega$, so the angular velocity and the speed along the curve are constant and the angular acceleration is zero. □

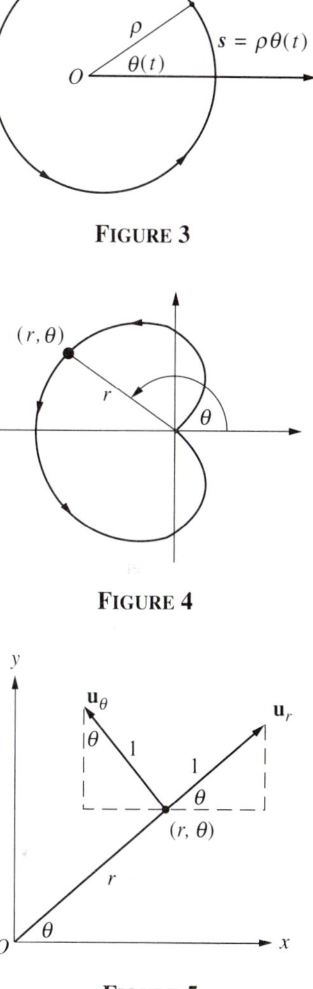

FIGURE 3

FIGURE 4

FIGURE 5

Any motion along a polar curve $r = r(\theta)$ can be put in polar parametric form. If θ is expressed as a function of t by $\theta = \theta(t)$ then $r = r(\theta(t))$ or, more briefly, $r = r(t)$. For example, consider motion along the cardioid $r = 1 - \cos\theta$ in Fig. 4. If the polar angle at time t is $\theta(t) = t^2$, then $r(t) = 1 - \cos(t^2)$. The polar parametric equations $\theta = t^2$ and $r = 1 - \cos(t^2)$ describe motion along the cardioid with variable angular velocity $\theta'(t) = 2t$ and constant angular acceleration $\theta''(t) = 2$.

Position, Velocity, and Acceleration in Polar Form

In order to express position, velocity, and acceleration vectors in polar form, we introduce polar analogues of the basic unit vectors **i** and **j** used in rectangular coordinates. Figure 5 shows the **radial unit vector $\mathbf{u}_r$** and the **transverse unit vector $\mathbf{u}_\theta$** at a point (r,θ) other than the origin. From the figure,

$$\mathbf{u}_r = \cos\theta\mathbf{i} + \sin\theta\mathbf{j},$$
$$\mathbf{u}_\theta = -\sin\theta\mathbf{i} + \cos\theta\mathbf{j}.$$

From these formulas or from the figure, $\|\mathbf{u}_1\| = 1$, $\|\mathbf{u}_\theta\| = 1$ and $\mathbf{u}_r \perp \mathbf{u}_\theta$. The vectors $\mathbf{u}_r$ and $\mathbf{u}_\theta$ depend only on θ and not on r. Thus, $\mathbf{u}_r$ and $\mathbf{u}_\theta$ are constant along any ray from the origin. On the positive x–axis, $\mathbf{u}_r = \mathbf{i}$ and $\mathbf{u}_\theta = \mathbf{j}$. On the positive y–axis, $\mathbf{u}_r = \mathbf{j}$ and $\mathbf{u}_\theta = -\mathbf{i}$. We shall need to know how $\mathbf{u}_r$ and $\mathbf{u}_\theta$ vary with θ. Differentiate the formulas for $\mathbf{u}_r$ and $\mathbf{u}_\theta$ to obtain

$$\frac{d\mathbf{u}_r}{d\theta} = \mathbf{u}_\theta, \qquad \frac{d\mathbf{u}_\theta}{d\theta} = -\mathbf{u}_r.$$

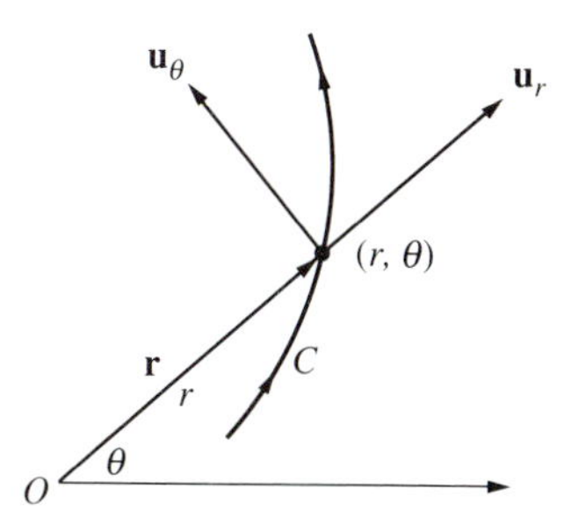

FIGURE 6

Consider a point moving along the curve C in Fig. 6 with polar parametric equations $r = r(t)$ and $\theta = \theta(t)$. Now the unit vectors $\mathbf{u}_r$ and $\mathbf{u}_\theta$ become functions of t along the trajectory:

$$\mathbf{u}_r(t) = \cos\theta(t)\mathbf{i} + \sin\theta(t)\mathbf{j}, \qquad \mathbf{u}_\theta(t) = -\sin\theta(t)\mathbf{i} + \cos\theta(t)\mathbf{j}.$$

By the chain rule,

$$\frac{d\mathbf{u}_r}{dt} = \frac{d\mathbf{u}_r}{d\theta}\frac{d\theta}{dt} = \mathbf{u}_\theta\frac{d\theta}{dt}, \qquad \frac{d\mathbf{u}_\theta}{dt} = \frac{d\mathbf{u}_\theta}{d\theta}\frac{d\theta}{dt} = -\mathbf{u}_r\frac{d\theta}{dt}.$$

Therefore

$$\mathbf{u}_r' = \theta'\mathbf{u}_\theta, \qquad \mathbf{u}_\theta' = -\theta'\mathbf{u}_r.$$

The position vector $\mathbf{r}$ for the point (r,θ) in Fig. 6 is also called the **radius vector** for the point. Note that

$$\|\mathbf{r}\| = r.$$

The position vector is represented in polar form by

$$\mathbf{r} = r\mathbf{u}_r.$$

Differentiate $\mathbf{r} = r\mathbf{u}_r$ with respect to t with the aid of the product rule and the foregoing displayed formulas to obtain the polar representation of velocity:

$$\mathbf{r}' = r'\mathbf{u}_r + r\theta'\mathbf{u}_\theta.$$

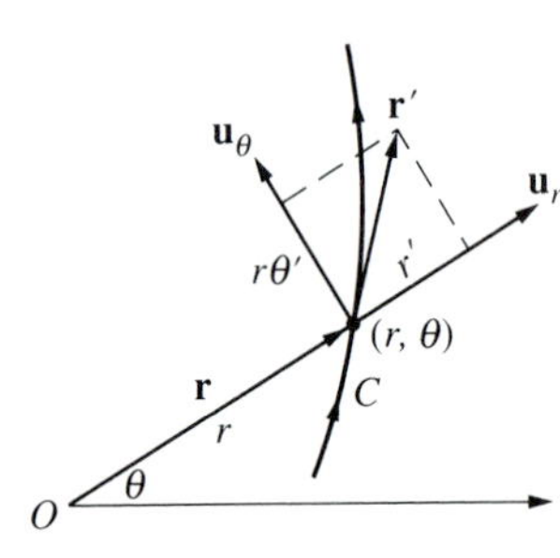

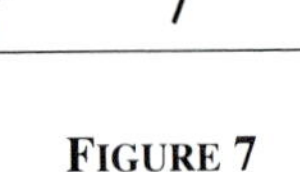

FIGURE 7

The **radial** and **transverse components of velocity** are

$$\mathbf{r}' \cdot \mathbf{u}_r = r', \qquad \mathbf{r}' \cdot \mathbf{u}_\theta = r\theta'.$$

These components are illustrated in Fig. 7. The speed of an object along its trajectory is given in polar form by

$$s'(t) = \|\mathbf{r}'\| = \sqrt{(r')^2 + (r\theta')^2} = \sqrt{\left(\frac{dr}{dt}\right)^2 + r^2\left(\frac{d\theta}{dt}\right)^2}.$$

Differentiate the velocity $\mathbf{r}' = r'\mathbf{u}_r + r\theta'\mathbf{u}_\theta$ with the aid of the product rule and the relations $\mathbf{u}_r' = \theta'\mathbf{u}_\theta$ and $\mathbf{u}_\theta' = -\theta'\mathbf{u}_r$ (see the problems) to obtain the polar representation of acceleration:

$$\mathbf{r}'' = [r'' - r(\theta')^2]\mathbf{u}_r + (2r'\theta' + r\theta'')\mathbf{u}_\theta.$$

The **radial** and **transverse components of acceleration** are

$$\mathbf{r}'' \cdot \mathbf{u}_r = r'' - r(\theta')^2, \qquad \mathbf{r}'' \cdot \mathbf{u}_\theta = 2r'\theta' + r\,\theta''.$$

EXAMPLE 2. An object moves along the polar curve $r = 1/(1 + \cos\theta)$ for $-\pi < \theta < \pi$ with $\theta(t) = 2t$. Find the velocity, speed, and acceleration at time $t = \pi/4$ when $\theta = \pi/2$ and $r = 1$.

Solution. Since $x = r\cos\theta$ and $r = 1 - r\cos\theta$ on the polar curve, it follows that $x^2 + y^2 = r^2 = (1-x)^2$ or $x = (1-y^2)/2$. The curve is a parabola, as shown in Fig. 8. In order to use the general formulas for velocity and acceleration in polar coordinates, we need to find r', r'', θ', and θ'' when $t = \pi/4$. Since $\theta(t) = 2t$, the angular velocity is $\theta'(t) = 2$ and the angular acceleration is $\theta''(t) = 0$. Since $r = (1+\cos\theta)^{-1}$ and $\theta'(t) = 2$, the chain rule gives

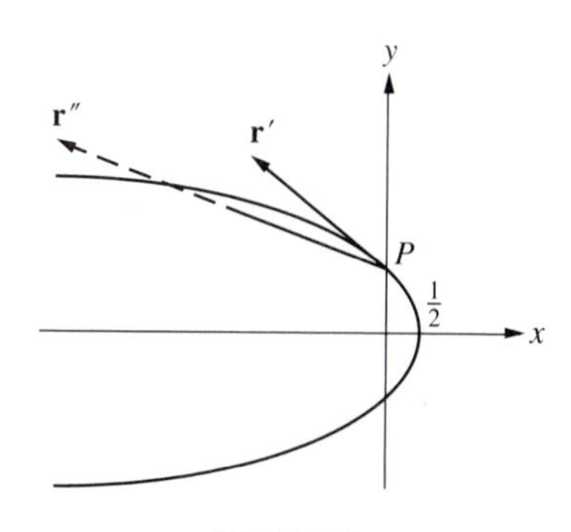

FIGURE 8

$$r' = \frac{dr}{dt} = \frac{dr}{d\theta}\frac{d\theta}{dt} = \frac{\sin\theta}{(1+\cos\theta)^2}\cdot 2 = 2\sin\theta(1+\cos\theta)^{-2},$$

$$r'' = [2\cos\theta(1+\cos\theta)^{-2} + 4\sin^2\theta(1+\cos\theta)^{-3}]\cdot 2.$$

Let $t = \pi/4$. Then $\theta = \pi/2$ and $r = 1$. The object is at the point $P = (1,\pi/2)$ in Fig. 8 and

$$\theta' = 2, \qquad \theta'' = 0, \qquad r' = 2, \qquad r'' = 8.$$

It is clear from the figure (or from the formulas for $\mathbf{u}_r$ and $\mathbf{u}_\theta$) that

$$\mathbf{u}_r = \mathbf{j}, \qquad \mathbf{u}_\theta = -\mathbf{i} \qquad \text{at } P.$$

Substitute these values into the general formulas for velocity, speed, and acceleration to obtain

$$\mathbf{r}' = 2\mathbf{j} - 2\mathbf{i}, \qquad \|\mathbf{r}'\| = 2\sqrt{2}, \qquad \mathbf{r}'' = 4\mathbf{j} - 8\mathbf{i} \quad \text{when } t = \pi/4.$$

The velocity and acceleration vectors are shown in Fig. 8. The actual length of $\mathbf{r}''$ is $4\sqrt{5}$. □

Area Swept Out by the Radius Vector

Newton discovered a surprising connection between the rate at which a radius vector sweeps out area and the nature of a force that causes a plane motion. We describe that discovery next. We are concerned particularly with the area swept out by the radius vector of a moving object. In this setting, the area becomes a function of time.

In Fig. 9, a curve C is represented by a polar equation $r = r(\theta)$. To avoid complications, assume that $r(\theta) > 0$ for all θ. Let $A(\theta)$ be the area inside the curve between the horizontal axis and any ray θ with $\theta > 0$. Then $A(0) = 0$, and

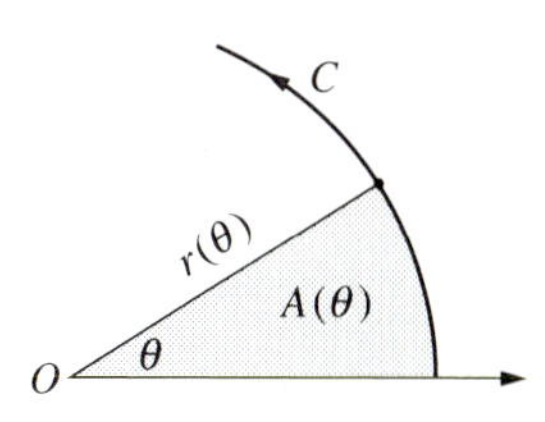

FIGURE 9

$$A(\theta) = \frac{1}{2}\int_0^\theta r(\varphi)^2 d\varphi, \qquad \frac{dA}{d\theta} = \frac{1}{2}r(\theta)^2,$$

where φ is a dummy variable of integration. The integral formula for the area is familiar from one variable calculus and differentiation using the fundamental theorem of calculus yields the expression for $dA/d\theta$.

Now let $\theta = \theta(t)$ and $r = r(t)$. Then $A = A(\theta(t))$ or, more briefly, $A = A(t)$. Then $A'(t)$ is the rate that area is swept out by the radius vector. By the chain rule:

$$\frac{dA}{dt} = \frac{dA}{d\theta}\frac{d\theta}{dt} = \frac{1}{2}r(\theta)^2\theta'(t).$$

Thus, $A' = \frac{1}{2}r^2\theta'$. Another differentiation yields $A'' = r\ r'\theta' + \frac{1}{2}\ r^2\ \theta''$. In summary,

$$\boxed{A' = \frac{1}{2}\,r^2\theta', \qquad A'' = \frac{1}{2}\,r(r\theta'' + 2r'\theta').}$$

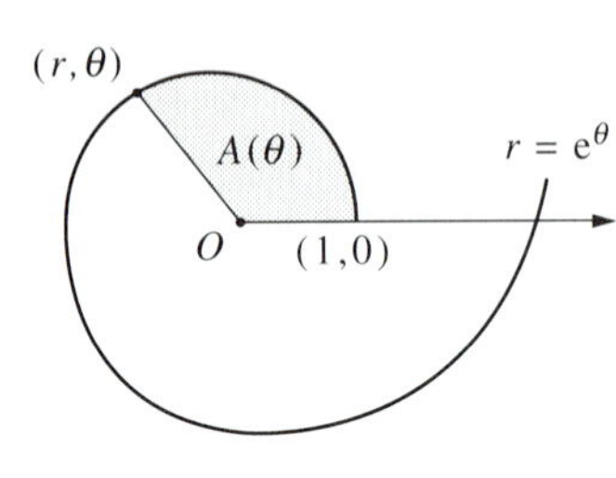

FIGURE 10

EXAMPLE 3. An object moves on the spiral $r = e^{\theta}$, $\theta \geq 0$, in Fig. 10 with $\theta = \ln(t + 1)$ for $t \geq 0$. Find $A'(t)$ and $A''(t)$.

Solution. First, $r = e^{\ln(t+1)} = t + 1$ and

$$r'(t) = 1, \qquad r'' = 0,$$

$$\theta'(t) = \frac{1}{t+1}, \qquad \theta'' = -\frac{1}{(t+1)^2}.$$

The general formulas for A' and A'' yield

$$A'(t) = \frac{1}{2}\,(t+1), \qquad A''(t) = \frac{1}{2}\,.\square$$

There is a close connection between $A(t)$ and the transverse components of velocity and acceleration, $\mathbf{r}' \cdot \mathbf{u}_\theta = r\theta'$ and $\mathbf{r}'' \cdot \mathbf{u}_\theta = 2r'\theta' + r\theta''$. The foregoing displayed formulas for A' and A'' are expressed equivalently by

$$A' = \frac{1}{2}\,r(\mathbf{r}' \cdot \mathbf{u}_\theta), \qquad A'' = \frac{1}{2}\,r(\mathbf{r}'' \cdot \mathbf{u}_\theta).$$

We could have used these formulas in Ex. 3 to find A' and A''.

Now we come to a great discovery of Newton concerning motion in a plane.

The radius vector of a moving object sweeps out equal areas in equal times $\Leftrightarrow$ a central force maintains the motion.

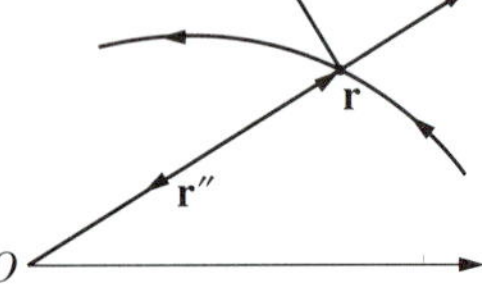

FIGURE 11

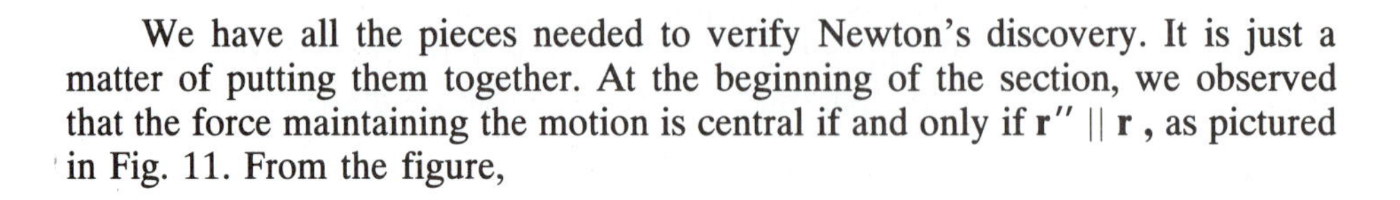

We have all the pieces needed to verify Newton's discovery. It is just a matter of putting them together. At the beginning of the section, we observed that the force maintaining the motion is central if and only if $\mathbf{r}'' \parallel \mathbf{r}$, as pictured in Fig. 11. From the figure,

$$\mathbf{r}'' \parallel \mathbf{r} \quad \Leftrightarrow \quad \mathbf{r}'' \perp \mathbf{u}_\theta \quad \Leftrightarrow \quad \mathbf{r}'' \cdot \mathbf{u}_\theta = 0.$$

Since $A'' = \frac{1}{2} r(\mathbf{r}'' \cdot \mathbf{u}_\theta)$ and $A(0) = 0$,

$$\begin{aligned} \mathbf{r}'' \| \mathbf{r} \quad &\Leftrightarrow \quad A''(t) = 0 \\ &\Leftrightarrow \quad A'(t) = c \\ &\Leftrightarrow \quad A(t) = ct \end{aligned}$$

for some constant c. A further equivalence is

$$\mathbf{r}'' \| \mathbf{r} \quad \Leftrightarrow \quad A(t_2) - A(t_1) = c(t_2 - t_1) \quad \text{for} \quad t_2 \geq t_1 \geq 0.$$

Thus, the area swept out during the time interval from t_1 to t_2 depends only on the constant c and the elapsed time $t_2 - t_1$. In other words, equal areas are swept out in equal times.

It is worth remembering that the radius vector sweeps out equal areas in equal times if and only if $A'(t) = c$ for some constant c, which is the rate of change of area with respect to time. In Ex. 3, $A'(t) = \frac{1}{2}(t+1)$ is not constant, so the radius vector does not sweep out equal areas in equal times and the force that causes the motion is not a central force.

EXAMPLE 4. We make an apparently small change in Ex. 3 and consider the motion along the spiral $r = e^\theta$ with $\theta = \frac{1}{2} \ln(t+1)$ instead of $\theta = \ln(t+1)$. (a) Show that the force that causes this motion is a central force. (b) Check directly that the radius vector sweeps out equal areas in equal times.

Solution. (a) Now $r(t) = e^{\ln\sqrt{t+1}} = \sqrt{t+1}$ and

$$r' = \frac{1}{2\sqrt{t+1}}, \qquad r'' = -\frac{1}{4(t+1)^{3/2}},$$

$$\theta' = \frac{1}{2(t+1)}, \qquad \theta'' = \frac{1}{2(t+1)^2}.$$

The general formulas for position, velocity, and acceleration in polar form yield

$$\mathbf{r} = \sqrt{t+1}\,\mathbf{u}_r, \quad \mathbf{r}' = \frac{1}{2\sqrt{t+1}}(\mathbf{u}_r + \mathbf{u}_\theta), \quad \mathbf{r}'' = -\frac{1}{2(t+1)^{3/2}}\mathbf{u}_r.$$

The first and last equations give

$$\mathbf{r}'' = -\frac{1}{2(t+1)^2}\mathbf{r}.$$

Therefore, the acceleration and force are central and area is swept out at a constant rate. (b) Let's check this area statement directly. The area $A(\theta)$ in Fig. 10 is

$$A(\theta) = \frac{1}{2}\int_0^\theta e^{2\varphi}\,d\varphi = \frac{1}{4}(e^{2\theta} - 1).$$

Since $\theta = \frac{1}{2}\ln(t+1)$, we find that $A(t) = \frac{1}{4}t$. Thus, area is swept out at the constant rate $A'(t) = \frac{1}{4}$, say, in ft²/sec. □

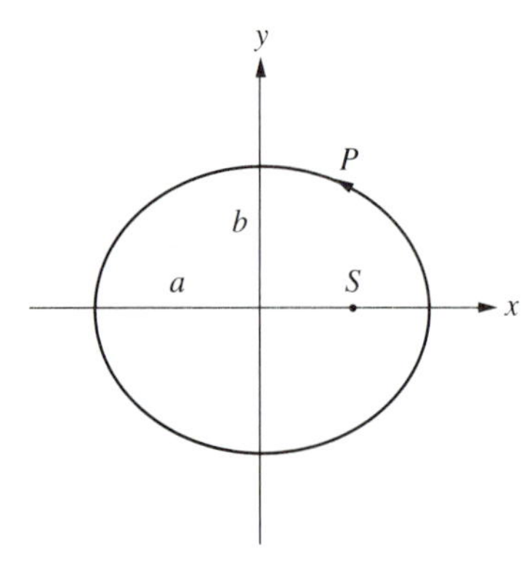

FIGURE 12

Kepler's Laws of Planetary Motion

The general results just developed for motion in polar coordinates have many applications. We illustrate their use in the context of planetary motion.

Preliminaries

Almost 400 years ago the German scientist Johannes Kepler (1571–1631) discovered three laws of planetary motion:

1. The orbit of each planet is an ellipse with the sun at one focus.
2. The radius vector of a planet sweeps out equal areas in equal times.
3. The square of the period of a planet is proportional to the cube of the semimajor axis of its elliptical orbit.

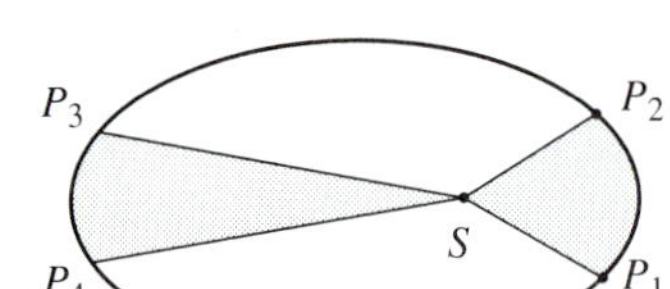

FIGURE 13

Kepler's first law is illustrated by Fig. 12. The orbit of a planet is an ellipse with semimajor axis a and semiminor axis b. The sun S is at the right-hand focus of the ellipse in Fig. 12. Figure 13 shows a planet at times t_1, t_2, t_3, t_4, with $t_2 - t_1 = t_4 - t_3$. By Kepler's second law, the shaded areas in Fig. 13 are equal. We infer that the planet moves more rapidly from P_1 to P_2, where it is nearer the sun, than from P_3 to P_4, when it is farther away. Let T be the period (the time for one complete revolution about the sun) of the planet in Fig. 12. Kepler's third law asserts that the ratio T^2/a^3 is the same for all the planets.

Kepler's laws are great landmarks in the history of science. Their discovery required decades of painstaking observations and calculations. Kepler had neither laws of motion nor tools of calculus to help him.

A half century later, Newton reasoned from Kepler's laws and experiments of Galileo to discover the universal law of gravitation. Then he derived Kepler's laws from the laws of motion and gravitation. These great achievements are all the more remarkable because Newton was only 22 or 23 years old at the time!

In the rest of this section, we use Newton's laws to derive Kepler's laws. In the problems, we turn the tables and ask you to derive Newton's law of gravitation from Kepler's laws and Newton's second law of motion.

To prepare for the derivation of Kepler's first law, we recall a few properties of ellipses. Each of these properties can be confirmed by elementary algebra or geometry. The ellipse

$$\frac{x^2}{a^2} + \frac{y^2}{b^2} = 1,$$

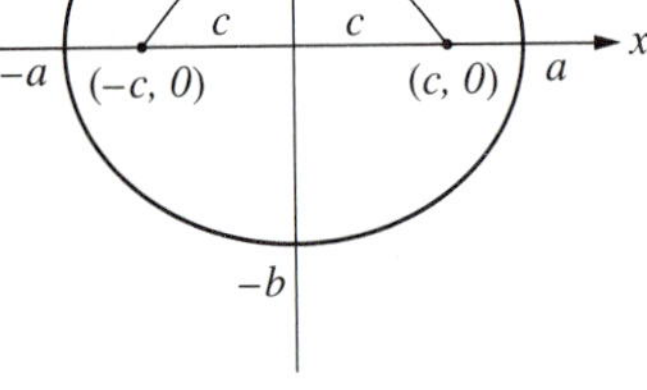

FIGURE 14

with $0 < b \le a$, is graphed in Fig. 14. The foci of the ellipse are the points $(c, 0)$ and $(-c, 0)$, where $c = \sqrt{a^2 - b^2}$ and $a^2 = b^2 + c^2$. If $a = b$, then the ellipse is a circle, $c = 0$, and there is only one focus, which is at the origin. The eccentricity of the ellipse is

$$e = \frac{c}{a} = \frac{\sqrt{a^2 - b^2}}{a} = \sqrt{1 - \left(\frac{b}{a}\right)^2},$$

where $0 \le e < 1$. The eccentricity measures how much the ellipse differs from a circle. The ellipse is a circle if $e = 0$. If a is very large compared with b, then e is near 1 and the ellipse is highly elongated.

Polar coordinates with the pole at the right–hand focus of the ellipse are introduced in Fig 15. The equation for the ellipse in terms of these polar coordinates is

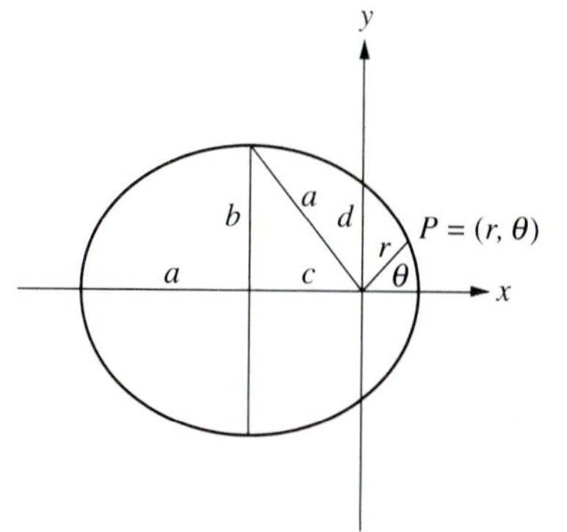

FIGURE 15

$$r = \frac{d}{1 + e\cos\theta}, \qquad d > 0, \qquad 0 \le e < 1,$$

where,

$$a = \frac{d}{1 - e^2}, \qquad b = \frac{d}{\sqrt{1 - e^2}}, \qquad c = \frac{de}{1 - e^2}, \qquad b^2 = ad.$$

Note that $r = d$ when $\theta = \pi/2$, as indicated in Fig. 15. The ellipse is a circle of radius d if $e = 0$. The graph of $r = d/(1 + e\cos\theta)$ is a parabola if $e = \pm 1$ and a hyperbola if $e > 1$ or $e < -1$.

Derivation of Kepler's Laws

Now we are ready to derive Kepler's laws from Newton's laws and the observation that the orbits of planets are bounded. Throughout, M is the mass of the sun and m the mass of a planet. It is an experimental fact that the sun remains virtually at rest as the planets revolve around it. Therefore, we choose the origin of coordinates at the center of the sun as in Fig. 1. Then $\mathbf{r}$ in Fig. 1 is the position vector or radius vector of a planet. We shall apply several of the general results for motion in polar coordinates to the present situation.

We begin by establishing Kepler's second law. Since the gravitational force exerted by the sun on each planet is a central force, we know that the motion takes place in a plane and the radius vector sweeps out equal areas in equal times. This proves Kepler's second law and is a first step toward proving Kepler's first law. We know the planet moves in a plane and this fact enables us to adopt a coordinate system as in Fig. 16.

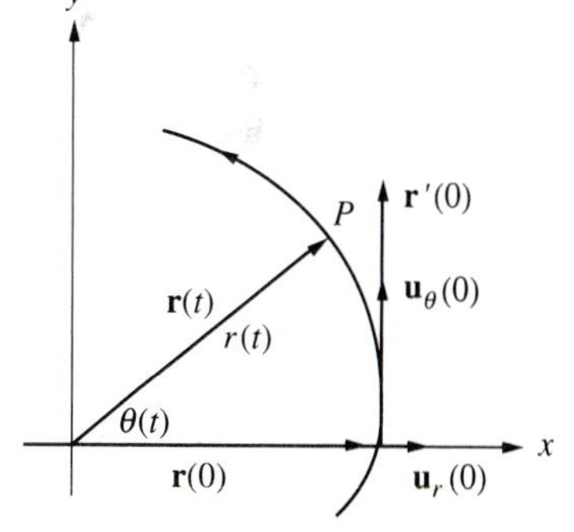

FIGURE 16

The coordinate system in Fig. 16 is chosen so that $\mathbf{r}(t)$ lies in the xy–plane, the sun is at the origin, the planet moves counterclockwise around the sun, and the positive x–axis passes through the planet at time $t = 0$ when it is nearest to the sun. This is called the *perihelion position* of the planet. Locate the planet by means of polar coordinates, as in the figure, where the position of the planet at time t is $(r,\theta) = (r(t),\theta(t))$. The corresponding position vector is given by

$$\mathbf{r}(t) = r(t)\,\mathbf{u}_r(t), \qquad r(t) = ||\mathbf{r}(t)||.$$

For brevity, let

$$r_0 = r(0), \qquad \theta_0{}' = \theta'(0), \qquad v_0 = ||\mathbf{r}'(0)||.$$

Since $r(t)$ takes its minimum at $t = 0$ when the planet is nearest to the sun, $r'(0) = 0$. Differentiate $r(t)^2 = \mathbf{r}(t) \cdot \mathbf{r}(t)$ to obtain $r(t)r'(t) = \mathbf{r}(t) \cdot \mathbf{r}'(t)$ and use $r'(0) = 0$ to see that $\mathbf{r}(0) \cdot \mathbf{r}'(0) = 0$. Hence,

$$\mathbf{r}'(0) \perp \mathbf{r}(0).$$

Consequently, since the motion is counterclockwise,

$$\mathbf{u}_r(0) = \mathbf{i}, \qquad \mathbf{u}_\theta(0) = \mathbf{j},$$
$$\mathbf{r}(0) = r_0\mathbf{i}, \qquad \mathbf{r}'(0) = v_0\mathbf{j}.$$

Recall that $\mathbf{r}' = r'\mathbf{u}_r + r\theta'\mathbf{u}_\theta$. Set $t = 0$ and use $r'(0) = 0$, to obtain

$$\mathbf{r}'(0) = r'(0)\mathbf{u}_r(0) + r(0)\theta'(0)\mathbf{u}_\theta(0) = r_0\theta_0'\mathbf{j}$$

and

$$v_0 = ||\mathbf{r}'(0)|| = r_0\,\theta_0'.$$

By Kepler's second law, the radius vector sweeps out area $A = A(t)$ at a constant rate. Since $A'(t) = \frac{1}{2}\,r^2\,\theta'$ and this rate is constant, we conclude that

$$A'(t) = \frac{1}{2}\,r^2\,\theta' = \frac{1}{2}\,{r_0}^2\,\theta_0' = \frac{1}{2}r_0v_0.$$

Antidifferentiate and use $A(0) = 0$ to obtain

$$A(t) = \frac{1}{2}r_0v_0t.$$

The equation $A'(t) = \frac{1}{2}\,r^2\theta' = \frac{1}{2}\,r_0v_0$ also yields

$$r(t)^2\theta'(t) = r_0v_0 \quad \text{for all } t.$$

Return now to Kepler's first law. We already know the orbit lies in a plane. A clever argument permits us to show that the orbit is an ellipse. First, express Newton's law of gravitation in the form

$$\mathbf{r}'' = -\frac{GM}{r^2}\,\mathbf{u}_r.$$

Then use $\mathbf{u}_\theta' = -\,\theta'\mathbf{u}_r$ and $r^2\theta' = r_0v_0$ to obtain

$$\mathbf{r}'' = \frac{GM}{r^2\theta'}\,\mathbf{u}_\theta' = \frac{GM}{r_0v_0}\,\mathbf{u}_\theta'$$

$$r'' = \alpha\mathbf{u}_\theta', \quad \alpha = \frac{GM}{r_0v_0}.$$

Antidifferentiation gives $\mathbf{r}' = \alpha\mathbf{u}_\theta + \mathbf{C}$ for some constant vector $\mathbf{C}$. Use $\mathbf{r}'(0) = v_0\mathbf{j}$ and $\mathbf{u}_\theta(0) = \mathbf{j}$ to find that $\mathbf{C} = (v_0 - \alpha)\mathbf{j}$ and

$$\mathbf{r}' = \alpha\mathbf{u}_\theta + \beta\mathbf{j}, \qquad \beta = v_0 - \alpha = \frac{r_0{v_0}^2 - GM}{r_0v_0}.$$

Since $\mathbf{u}_\theta = -\sin\theta\mathbf{i} + \cos\theta\mathbf{j}$, the transverse velocity of the planet is

$$\mathbf{r}' \cdot \mathbf{u}_\theta = \alpha + \beta\mathbf{j} \cdot \mathbf{u}_\theta = \alpha + \beta\cos\theta.$$

However, for any motion, $\mathbf{r}' \cdot \mathbf{u}_\theta = r\theta'$. Hence,

$$r\theta' = \alpha + \beta\cos\theta.$$

Multiply by r and use $r^2\theta' = r_0 v_0$ to obtain

$$r_0 v_0 = r(\alpha + \beta\cos\theta).$$

Since $r_0 v_0 > 0$, both factors on the right are positive and

$$r = \frac{r_0 v_0}{\alpha + \beta\cos\theta} = \frac{r_0 v_0/\alpha}{1 + (\beta/\alpha)\cos\theta}.$$

Let

$$d = \frac{r_0 v_0}{\alpha} = \frac{(r_0 v_0)^2}{GM}, \quad e = \frac{\beta}{\alpha} = \frac{r_0 v_0{}^2 - GM}{GM}.$$

Then

$$r = \frac{d}{1 + e\cos\theta}.$$

The graph of the polar equation $r = d/(1 + e\cos\theta)$ is an ellipse, parabola, or hyperbola with the sun at one focus. Since planetary orbits are bounded, the orbit must be an ellipse. This completes the derivation of Kepler's first law.

Finally, we establish Kepler's third law. Let T be the period of the planet in Fig. 15. Kepler's third law says that the ratio T^2/a^3 is the same for all the planets. The proof is based on two formulas for the area of the ellipse in Fig. 15. The standard formula for the area is πab. Since we showed previously that $A(t) = \frac{1}{2} r_0 v_0 t$, another formula for the area of the ellipse is $A(T) = \frac{1}{2} r_0 v_0 T$. Compare the two formulas for the area to find $\frac{1}{2} r_0 v_0 T = \pi ab$. Now solve for T, square, and use $b^2 = ad$ to obtain

$$T^2 = \frac{4\pi^2 a^2 b^2}{(r_0 v_0)^2} = \frac{4\pi^2 d}{(r_0 v_0)^2} a^3.$$

Finally, since we found earlier that $d = (r_0 v_0)^2/GM$,

$$\frac{T^2}{a^3} = \frac{4\pi^2}{GM}.$$

This is Kepler's third law. The constant $4\pi^2/GM$ is the same for all the planets because M is the mass of the sun and G is the universal gravitational constant.

This discussion completes the derivation of Kepler's laws from Newton's laws. We used vector notation and modern methods in our arguments but the line of reasoning we followed is essentially Newton's.

According to Kepler's third law, the periods T_1 and T_2 and semimajor axes a_1 and a_2 of two planets are related by

$$\frac{T_1^2}{a_1^3} = \frac{T_2^2}{a_2^3}, \quad \left(\frac{T_2}{T_1}\right)^2 = \left(\frac{a_2}{a_1}\right)^3, \quad \frac{a_2}{a_1} = \left(\frac{T_2}{T_1}\right)^{2/3}.$$

EXAMPLE 5. Earth is about 93 million miles from the sun and has a period of about 365 days. The period of Venus is about 225 days. Estimate the distance from Venus to the sun.

Solution. From Kepler's third law,

$$a_2 = a_1\left(\frac{T_2}{T_1}\right)^{2/3}.$$

We take $a_1 = 93$, $T_1 = 365$, and $T_2 = 225$ and find that $a_2 \approx 67$ million miles. Since both planets have nearly circular orbits, a_1 and a_2 may be regarded as approximate radii of the respective orbits. □

The foregoing derivation of Kepler's laws used Newton's laws, the boundedness of a planetary orbit, and the fact that the sun can be regarded as being at rest as a planet revolves about it. The same reasoning applies to any planet and its moons, or to the earth and its moon or any of its artificial satellites. Consequently, each of these systems obeys Kepler's three laws.

EXAMPLE 6. Communications satellites are placed in *geosynchronous orbits.* These are circular orbits in which the satellite remains directly above a particular point on Earth. Find the radius of such an orbit in the equatorial plane of Earth given that the moon orbits Earth with a period of 27.32 days and that the mean distance of the moon from Earth is about 238,900 miles.

Solution. We apply Kepler's third law to the earth, moon, satellite system to find the radius of a geosynchronous orbit. For the moon $a_1 = 238{,}900$ miles and $T_1 = 27.32$ days. For a geosynchronous orbit, $T_2 = 1$ day and, hence, as in Ex. 5,

$$a_2 = a_1\left(\frac{T_2}{T_1}\right)^{2/3} \approx (238{,}900)\left(\frac{1}{27.32}\right)^{2/3} \approx 26{,}340 \text{ miles. } \square$$

Other Orbits

In the derivation of Kepler's laws, we used the fact that planets move on bounded orbits to deduce that the eccentricity e in $r = d/(1 + e\cos\theta)$ satisfied $0 \le e < 1$. Without the boundedness assumption, it can happen that $e = 1$ and the orbit is a parabola or that $e > 1$ and the orbit is a hyperbola. This is the case for comets that pass through the solar system only once and for the *Voyager* spacecraft that is launched on an orbit that will take it past the outer planets and out of the solar system. Thus, when applied to general orbits of heavenly bodies, Kepler's first law is replaced by the statement that the orbits are conic sections.

PROBLEMS

In Probs. 1–8 an object moves on the given polar curve and the polar angle is the given function of time. (a) Express the velocity and acceleration in terms of $\mathbf{u}_r$ and $\mathbf{u}_\theta$. (b) Find $\mathbf{r}'(t)$ and $\mathbf{r}''(t)$ at the indicated time. (c) Identify the trajectory.

1. $r = 4, \quad \theta = t; \quad t = 0$
2. $r = 4, \quad \theta = t^2/2; \quad t = \sqrt{\pi}$
3. $r = 2\sin\theta, \quad \theta = \pi t; \quad t = 1$
4. $r = 2\cos\theta, \quad \theta = \pi(1 - e^{-t}); \quad t = 0$
5. $r = e^\theta, \quad \theta = t/4; \quad t = 0$
6. $r = 2\sin\theta - 4\cos\theta, \quad \theta = t/2; \quad t = \pi$
7. $r = 2/(1 + \sin\theta), \theta = 2t; t = \pi/4$
8. $r = 12/(9 + 7\cos^2\theta)^{1/2}, \theta = \sqrt{t}; t = \pi^2$
9. An object with mass m moves with constant angular velocity ω on the polar curve $r = a/(1 + \cos\theta)$, $a > 0$. Find the radial and transverse components of (a) the velocity and (b) the acceleration. (c) Identify the curve. (d) Is area swept out at a constant rate? Explain briefly.
10. An object with mass m moves with constant angular velocity ω on the polar curve $r = a(1 + \cos\theta)$, $a > 0$. Find the radial and transverse components of (a) the velocity and (b) the acceleration. (c) Identify the curve. (d) Is area swept out at a constant rate? Explain briefly.
11. An object with mass m moves counterclockwise with constant angular velocity ω on a circle with radius ρ. (a) Find its radial and transverse components of velocity and acceleration. (b) Sketch the circle and display the velocity and acceleration vectors at a typical point. (c) Is area swept out at a constant rate? Explain briefly.
12. An object with mass m moves counterclockwise with variable angular velocity $\omega = \omega(t)$ on a circle with radius ρ. (a) Find its radial and transverse components of velocity and acceleration. (b) Sketch the circle and display the velocity and acceleration vectors at a typical point. (c) Is area swept out at a constant rate? Explain briefly.
13. An object with mass m has position vector $\mathbf{r} = (a\cos\omega t)\mathbf{i} + (b\sin\omega t)\mathbf{j}$ where a, b, and ω are positive constants. (a) Show that an attractive, central force maintains the motion. (b) Is area swept out at a constant rate? Explain briefly. (c) Identify the orbit.
14. An object with mass m has position vector $\mathbf{r} = (a\cosh\omega t)\mathbf{i} + (a\sinh\omega t)\mathbf{j}$ where a and ω are positive constants. (a) Show that a repulsive, central force maintains the motion. (b) Is area swept out at a constant rate? Explain briefly. (c) Identify the orbit.
15. Differentiate $\mathbf{r}' = r'\mathbf{u}_r + r\theta'\mathbf{u}_\theta$ and use the relations $\mathbf{u}_r' = \theta'\mathbf{u}_\theta$ and $\mathbf{u}_\theta' = -\theta'\mathbf{u}_r$ to obtain $\mathbf{r}'' = (r'' - r(\theta')^2)\mathbf{u}_r + (2r'\theta' + r\theta'')\mathbf{u}_\theta$.
16. Let $\mathbf{r}(t)$ be the position vector of an object moving in a plane, as in Fig. 2. Show that

$$A = \frac{1}{2}\int_{t_1}^{t} \|\mathbf{r}(u) \times \mathbf{r}'(u)\|\, du$$

where $A = A(t)$ is the area swept out by the radius vector from a reference time t_1 to a time $t > t_1$. Conclude that

$$\frac{dA}{dt} = \frac{1}{2}||\mathbf{r}(t) \times \mathbf{r}'(t)||.$$

17. A silo, surrounded by pasture, has base a circle with radius a feet, say, $x^2 + y^2 = a^2$. A goat is tethered to the silo at $(a,0)$ by a rope of length πa feet. Find the area over which the goat can graze. *Hint.* Sketch the grazing area. Show that the outer boundary of the grazing area with $x \leq a$ and $y \geq 0$ is given by $\mathbf{r} = a\mathbf{u}_r + (a\pi - a\theta)\mathbf{u}_\theta$, $0 \leq \theta \leq \pi$. Use the area formula from the previous problem.

18. Let $\mathbf{r}(t)$ be the position vector of an object moving in a plane. (a) Use the fact that $\mathbf{u}_r$ and $\mathbf{u}_\theta$ are perpendicular to show that the speed $s'(t)$ of the object is expressed in polar form by $s'(t) = ||\mathbf{r}'(t)|| = \sqrt{(r')^2 + (r\theta')^2}$. *Hint.* $||\mathbf{r}'||^2 = \mathbf{r}' \cdot \mathbf{r}'$. (b) Conclude that the distance traveled by the object from time t_1 to time t_2 is $\int_{t_1}^{t_2} \sqrt{(r')^2 + (r\theta')^2}\, dt$.

19. Find the distance traveled by an object that moves on the cardioid $r = a(1 - \cos\theta)$ with $\theta = 2t$ for $0 \leq t \leq \pi$.

20. Find the distance traveled by the object in Ex. 2 for $0 \leq t \leq \pi/4$.

21. Find the distance traveled by the object in Ex. 3 for $0 \leq t \leq 1$.

22. Find the distance traveled by the object in Ex. 4 for $0 \leq t \leq 1$.

The following problems concern planetary motion and Kepler's laws.

23. In Fig. 16 the planet is closest to the sun when $\theta = 0$ and the perihelion distance is r_0. The *aphelion* distance r_1 is the maximum distance between the sun and planet. (a) Show that

$$r_1 = \frac{1 + e}{1 - e} r_0.$$

(b) Find r_1 for Earth given that $e = 0.017$ and $r_0 = 91$ million miles. Use this information to find the semimajor and semiminor axes for the orbit of Earth. (c) Find r_1 for Halley's comet given that $e = 0.967$ and $r_0 = 55$ million miles. Use this information to find the semimajor and semiminor axes of the orbit of Halley's comet.

24. The period of Pluto is about 248.35 (Earth) years. (a) Find the semimajor axis of the orbit of Pluto. (b) Pluto is the planet with the most eccentric orbit. The eccentricity is about 0.248. Calculate its perihelion and aphelion distances.

25. Deimos, the smaller moon of Mars, has a mean distance of about 14,600 miles from the planet and a period of about 30 hours. A day on Mars is about 24 hours and 37 minutes. Find the geosynchronous orbit for Mars.

The following problems outline Newton's derivation of the law of gravitation from Kepler's laws, Newton's laws of motion, and Galileo's inclined plane experiments.

26. By Kepler's first law each planet moves on an ellipse with the sun at one focus. Set up polar coordinates as in Fig. 16 and locate the planet by $r = r(t)$

and $\theta = \theta(t)$. As in the text, Kepler's second law can be expressed as $A' = \frac{1}{2}r^2\theta' = \frac{1}{2}r_0v_0$. Differentiate $r^2\theta' = r_0\theta_0$ and use the fact that $r > 0$ for any planet to conclude that the transverse component of acceleration is zero and, hence, that

$$\mathbf{r}'' = (r'' - r(\theta')^2)\mathbf{u}_r.$$

Problem 26 shows that the force of the sun on a planet is a central force. The next problem reveals that the magnitude of the force varies like $1/r^2$.

27. Use Kepler's first law $r = d(1 + e\cos\theta)^{-1}$ and $\theta' = r_0v_0/r^2$ from Prob. 26 to show: (a) $r' = (r_0v_0 e\sin\theta)/d$. (b) $r'' = [(r_0v_0)^2 e\cos\theta]/dr^2$. (c) $r'' - r(\theta')^2 = -(r_0v_0)^2/dr^2$, and, hence,

$$\mathbf{r}'' = -\frac{(r_0v_0)^2}{dr^2}\mathbf{u}_r = -\frac{(r_0v_0)^2}{dr^2}\left(\frac{\mathbf{r}}{r}\right).$$

Problems 26 and 27 show that Kepler's first two laws and Newton's second law imply that the force that maintains the planets in orbit is an attractive, central force whose magnitude varies as the inverse of the distance between the planet and the sun. At this point, the proportionality constant $(r_0v_0)^2/d$ appears to depend on the planet through the three orbital parameters r_0, v_0, and d. Kepler's third law implies that $(r_0v_0)^2/d$ depends only upon the sun.

28. Calculate the area of the ellipse in Fig. 15 in two ways to obtain

$$\frac{(r_0v_0)^2}{d} = 4\pi^2\left[\frac{a^3}{T^2}\right].$$

Conclude, as Newton did, that

$$\mathbf{r}'' = -\frac{\Gamma}{r^2}\left(\frac{\mathbf{r}}{r}\right), \qquad \Gamma = 4\pi^2\left(\frac{a^3}{T^2}\right),$$

where, by Kepler's third law, Γ is the *same* for all planets; it depends only upon the sun. Now, show that the force $\mathbf{F}$ of the sun on the planet is

$$\mathbf{F} = -\frac{\Gamma m}{r^2}\left(\frac{\mathbf{r}}{r}\right).$$

Newton also confirmed from experimental data that Kepler's three laws hold for the moons of Saturn and the moons of Jupiter. Consequently, the foregoing reasoning shows there is a constant Γ associated with Saturn and another constant Γ associated with Jupiter so that the force exerted by these planets on their moons is given by $\mathbf{F}$ in Prob. 28. Newton's discovery in Prob. 28 was the principal step in establishing the universality of the law of gravitation. The following problems tie up a few loose ends and make an essential connection with Galileo's work.

29. The result of Prob. 28 led Newton to conclude that each mass M (such as the sun) has associated with it a constant Γ_M such that any object with mass m (such as a planet) at a distance r from M is attracted to M by a force with magnitude

$$F_{M \text{ on } m} = \frac{m\Gamma_M}{r^2}.$$

Likewise, m exerts a force on M with magnitude

$$F_{m \text{ on } M} = \frac{M\Gamma_m}{r^2}.$$

By Newton's third (action–reaction) law, the force of m on M is equal in magnitude and opposite in direction from the force of M on m. Use these observations to conclude that

$$\frac{\Gamma_M}{M} = \frac{\Gamma_m}{m}.$$

Finally, think of m as a fixed "test mass" and let M represent any massive body. Deduce that the ratio Γ_M/M is the same for all masses. This common value is $G = \Gamma_M/M$, the *universal gravitational constant.*

30. Return to Prob. 28. Let M be the mass of the sun and m be the mass of the planet. Show that the force of the sun on the planet is

$$\mathbf{F} = -\frac{GMm}{r^2}\left(\frac{\mathbf{r}}{r}\right).$$

The next problem connects the experimental work of Galileo with the astronomical observations summarized in Kepler's laws.

31. Assume (it's true) that Earth attracts objects outside it just as if Earth were a point mass concentrated at its center. Galileo showed that the gravitational acceleration near Earth is about 32 ft/sec^2 in English units. Newton reasoned that if the same force of attraction that maintains planets in their orbits also attracted objects near the surface of Earth, then the gravitational acceleration measured by Galileo must be given also by Γ_E/R_E^2, where Γ_E is the value of Γ associated with Earth and R_E is radius of Earth. (a) Explain briefly how Newton reached this conclusion. (b) Then show that

$$\frac{\Gamma_E}{R_E^2} = \frac{4\pi^2 R_M^3/T_M^2}{R_E^2},$$

where R_M is the semimajor axis of the moon's orbit about Earth and T_M its period of rotation.

32. Newton knew that the radius of Earth was about 4000 miles, that the moon has a nearly circular orbit with radius about 60 Earth radii, and that the period of the moon is about 27 1/3 days. Use these data to calculate Γ_E/R_E^2 and compare with Galileo's result of 32 ft/sec^2.

Incidentally, Newton did not regard it as obvious that Earth would attract objects outside it just as if Earth were a point mass concentrated at its center. He proved that too! But that is a story for Ch. 8.

Chapter Highlights

A curve in 3–space is represented in parametric form by

$$x = x(t), \qquad y = y(t), \qquad z = z(t),$$

with t in some interval. A vector equation for the curve is

$$\mathbf{r} = \mathbf{r}(t) = \langle x(t), y(t), z(t) \rangle .$$

For a curve in the xy–plane, eliminate the z–coordinate.

The derivative of $\mathbf{r}(t)$ is

$$\mathbf{r}'(t) = \langle x'(t), y'(t), z'(t) \rangle .$$

If $\mathbf{r}'(t) \neq 0$, then $\mathbf{r}'(t)$ is tangent to the curve at $\mathbf{r}(t)$.

Assume that $\mathbf{r}(t)$ is twice differentiable and t is time. Then $\mathbf{r}(t)$ is the position vector of a moving point, $\mathbf{r}'(t)$ is the velocity, $||\mathbf{r}'(t)||$ is the speed, and $\mathbf{r}''(t)$ is the acceleration. The length of the curve $\mathbf{r} = \mathbf{r}(t)$, $a \leq t \leq b$ is

$$L = \int_a^b ||\mathbf{r}'(t)||\, dt.$$

Variable distance along the curve is

$$s(t) = \int_a^t ||\mathbf{r}'(\tau)||\, d\tau.$$

If a force $\mathbf{F}(t)$ acts on a mass m, then the motion is governed by Newton's second law in vector form, $\mathbf{F}(t) = m\mathbf{r}''(t)$. In typical problems, we are given the initial position $\mathbf{r}(0)$ and the initial velocity $\mathbf{r}'(0)$. Two antidifferentiations of $\mathbf{F}(t) = m\mathbf{r}''(t)$ produce $\mathbf{r}(t)$.

Curvature measures the bending of a curve. A formula for curvature is

$$\kappa = \frac{||\mathbf{r}'(t) \times \mathbf{r}''(t)||}{||\mathbf{r}'(t)||^3}.$$

For a curve $y = f(x)$ in 2–space,

$$\kappa = \frac{|f''(x)|}{[1 + f'(x)^2]^{3/2}}.$$

The tangential and normal components of acceleration along the curve are given by

$$a_T = s''(t) = \frac{\mathbf{r}'(t) \cdot \mathbf{r}''(t)}{||\mathbf{r}'||}, \qquad a_N = \kappa s'(t)^2 = \frac{||\mathbf{r}'(t) \times \mathbf{r}''(t)||}{||\mathbf{r}(t)||}.$$

The last section in Ch. 4 concerns motion in polar coordinates. The radial and tangential unit vectors $\mathbf{u}_r$ and $\mathbf{u}_\theta$ are attached to the point (r, θ) and they point in the radial and tangential directions. Suppose that a particle is moving in the plane with polar coordinates $r = r(t)$ and $\theta = \theta(t)$. Then the position, velocity, and acceleration are given by

$$\mathbf{r} = r\mathbf{u}_r, \qquad \mathbf{r}' = r'\mathbf{u}_r + r\theta'\mathbf{u}_\theta,$$
$$\mathbf{r}'' = [r'' - r(\theta')^2]\mathbf{u}_r + (2r'\theta' + r\theta'')\mathbf{u}_\theta.$$

These formulas are very useful in analyzing many motion problems, especially motion under the action of a central force such as gravity. The derivation of Kepler's laws of planetary motion is a case in point.

The chapter project concerns problems of pursuit. Position vectors, tangent vectors, and the arc length parameterization of a curve make natural appearances in pursuit problems and the methods used in the project can be employed in other motion problems. This project involves both exact and approximate methods of solution.

Chapter Project: Pursuit Problems

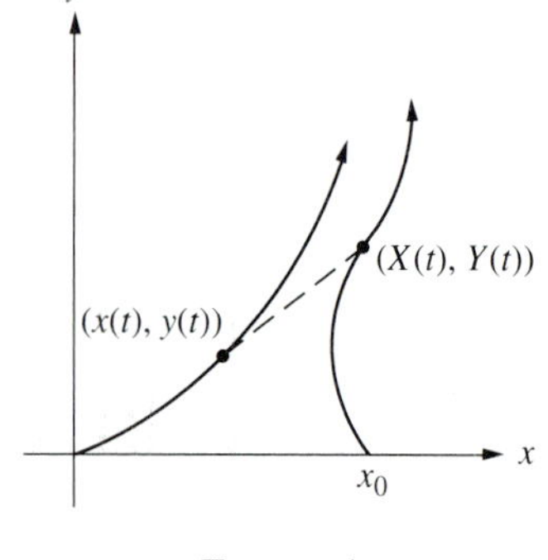

FIGURE 1

Pursuit problems arise in a variety of situations, for example, in military operations in the air and on the high seas and in the docking of one spacecraft with another. We shall describe pursuit problems in more down–to–earth terms: a dog spots a rabbit and chases it. Will the rabbit be caught or will it get away? This is a typical two–dimensional pursuit problem. It is illustrated in Fig. 1.

In the figure, the dog starts at the origin at time $t = 0$. The x–axis is chosen so that the rabbit is at $(x_0, 0)$ at time $t = 0$. Assume throughout that at each instant during the pursuit the dog adjusts its path to always head directly toward the rabbit; see Fig. 1. The speed of the dog at time t is $\sigma(t)$. The rabbit follows the path given by the position vector $\mathbf{R}(t) = \langle X(t), Y(t) \rangle$ for $t \geq 0$. The path of the dog is given by $\mathbf{r}(t) = \langle x(t), y(t) \rangle$.

Problem 1 Under the assumptions just stated, show that the curve of pursuit is determined by the initial value problem

$$\left\{ \begin{array}{l} \mathbf{r}'(t) = \sigma(t) \dfrac{\mathbf{R}(t) - \mathbf{r}(t)}{||\mathbf{R}(t) - \mathbf{r}(t)||}, \\ \mathbf{r}(0) = \langle x_0, 0 \rangle. \end{array} \right\}$$

Problem 2 Show that the vector initial value problem in Prob. 1 can be expressed in scalar form as follows.

$$\left\{ \begin{array}{ll} \mathbf{x}'(t) = \sigma(t) \dfrac{X(t) - x(t)}{\sqrt{[X(t) - x(t)]^2 + [Y(t) - y(t)]^2}}, & x(0) = x_0, \\ \mathbf{y}'(t) = \sigma(t) \dfrac{Y(t) - y(t)}{\sqrt{[X(t) - x(t)]^2 + [Y(t) - y(t)]^2}}, & y(0) = 0. \end{array} \right\}$$

In general, given the speed of the dog $\sigma(t)$ and the path of the rabbit $\mathbf{R}(t)$, the initial value problem in Probs. 1 and 2 is so complicated that it must be solved numerically. Further remarks about numerical solutions are given at the end of the project. For the moment, consider a special case that can be solved exactly.

Assume now that the rabbit runs with constant velocity $\mathbf{j}$ from (1,0) on the x–axis and that the dog pursues with constant speed 2, always heading directly toward the rabbit.

Problem 3 (a) Draw a figure, like Fig. 1, that illustrates this pursuit problem. Label the pursuit curve C. (b) Show that the pursuit path $\mathbf{r}(t) = < x(t), y(t) >$ is determined by the initial value problem

$$\left\{\begin{array}{ll} \mathbf{x}'(t) = \dfrac{2(1 - x(t))}{\sqrt{[1 - x(t)]^2 + [t - y(t)]^2}}, & x(0) = 1, \\ \mathbf{y}'(t) = \dfrac{2(t - y(t))}{\sqrt{[1 - x(t)]^2 + [t - y(t)]^2}}, & y(0) = 0. \end{array}\right\}$$

For the pursuit in Prob. 3, it is clear on physical grounds that $x(t) < 1$ before capture and, hence, $x'(t) > 0$ during the chase.

Problem 4 (a) Explain briefly why the curve of pursuit C with parametric equation $\mathbf{r}(t) = < x(t), y(t) >$ also can be expressed in the explicit form $y = y(x)$ for $0 \le x \le 1$. (b) Then use Prob. 3 to show that

$$\frac{dy}{dx} = \frac{t - y}{1 - x}, \qquad y(0) = 0, \qquad y'(0) = 0.$$

Problem 5 Let s be variable arc length on the pursuit curve measured from the origin to $(x, y(x))$. Explain why

$$\frac{ds}{dt} = 2 \text{ and } \frac{dt}{dx} = \frac{\sqrt{1 + y'(x)^2}}{2}.$$

Problem 6 Use Probs. 4 and 5 to obtain the following initial value problem for $y = y(x)$:

$$\left\{\begin{array}{ll} \dfrac{y''(x)}{\sqrt{1 + y'(x)^2}} = \dfrac{1}{2(1 - x)}, \\ y(0) = 0, \qquad y'(0) = 0. \end{array}\right\}$$

Problem 7 Verify that

$$\frac{d}{dx} \sinh^{-1} y'(x) = \frac{y''(x)}{\sqrt{1 + y'(x)^2}},$$

where $\sinh^{-1} x$ is the inverse hyperbolic sine. Then antidifferentiate and use $y'(0) = 0$ to obtain

$$\sinh^{-1} y'(x) = \ln (1 - x)^{-1/2}.$$

An identity for the inverse hyperbolic sine is useful at this point:

$$\sinh^{-1}(z) = \ln (z + \sqrt{1 + z^2}).$$

Problem 8 Use the preceding identity and Prob. 7 to show first that

$$y'(x) = \frac{1}{2}(1 - x)^{-1/2} - \frac{1}{2}(1 - x)^{1/2}.$$

Then integrate to find

$$y(x) = \frac{2}{3} - (1-x)^{1/2} + \frac{1}{3}(1-x)^{3/2}$$

for $0 \le x \le 1$.

Problem 9 Find the point where the dog catches the rabbit and find the total time of pursuit. Make a sketch (possibly using a graphics utility) of the curve of pursuit C and the path of flight of the rabbit.

As noted earlier, the general pursuit problem formulated in Probs. 2 and 3 normally resists exact solution and must be treated numerically. Such numerical solutions and accompanying graphics can be obtained using more powerful graphing calculators or more conveniently by use of high-level mathematics software such as Matlab, Maple, or Mathematica. If you are comfortable using one of these computing platforms, the last problem should prove interesting and informative.

Problem 10 The rabbit leaves (4,0) on the path $\mathbf{R}(t) = <3 + \cos t, t>$. The dog pursues the rabbit at constant speed 2, always heading directly toward the rabbit. (a) Use a numerical differential equation solver from your software package or calculator and its graphics utilities to make a plot of the pursuit curve. (b) Estimate the point of capture (if there is one) and the total time of pursuit.

Once you have solved the last problem you can easily modify your solution in order to try out other paths for the rabbit and other (possibly variable) speeds of pursuit for the dog.

Chapter Review Problems

In Probs. 1–4, express the parametric curves in terms of x and y. Identify the curves and sketch them.

1. $x = 2t + 3,\ y = 3t - 4$
2. $x = t - 1,\ y = 2t^2 + 1$
3. $x = 2\sin t,\ y = 3\cos t$
4. $x = 4\sin^4 t,\ y = 4\cos^4 t,\ 0 \le t \le \pi/2$
5. Find the slope of the parametric curve $x = 3\tan t + 1,\ y = 2\sec t + 3$ when $t = \pi/6$.
6. Find the slope of the parametric curve $x = \arcsin t,\ y = \ln(1 - t^2)$ when $t = 3/5$.
7. Find the slope of the polar curve $r = 3e^{\theta}$ when $\theta = \pi$.
8. Find all points where the graph of $x = t^2 + 4t,\ y = 4 - t^2$ has horizontal and vertical tangent lines.
9. Let $\mathbf{r}(t) = <3\ln t, t^{3/2}, t>$ for $t > 0$. When is $\mathbf{r}'(t) \,|\, \mathbf{r}''(t)$?
10. Let $\mathbf{r}'(t) = <t^2 + 2, -\cos t, 3e^t>$ and $\mathbf{R}(t) = <t + 1, -1, 3t^2>$. Find (a) the point of intersection of the curves and (b) the cosine of the angle between the curves at that point.
11. The path of a particle is marked by the position vector $\mathbf{r}(t) = 2\cos 2t\,\mathbf{i} + 3\cos t\,\mathbf{j}$ for $t \ge 0$. (a) Express the path of motion in terms

of x and y. (b) Identify the path and sketch it. (c) Describe how the particle moves along its path.

12. A particle of mass $m = 3$ with initial position vector $\mathbf{i} + 3\mathbf{j}$ and initial velocity $(1/2)\mathbf{j}$ moves along a curve C due to a force $\mathbf{F}(t) = 6\mathbf{k}$. (a) Find the position vector $\mathbf{r}(t)$ that traces out C. (b) Identify and sketch the curve C. (c) Describe how the particle moves along its path of motion.

13. A curve in 3–space is described by $x = e^t$, $y = e^{-t}$, $z = 2 - e^{-t}$ for $t \geq 0$. Describe (a) its projection on the xy–plane, (b) its elevation function, and (c) the curve itself.

14. Let $\mathbf{r}(t) = < 4t^3, 4t^4, 5t >$. (a) Find the horizontal and vertical projections of the tangent vector when $t = 1$. (b) Find the angle of inclination relative to the xy–plane of the tangent vector in (a).

15. Find the area between the x–axis and the curve $x = 3t + 2$, $y = t^2 - 4t + 3$ from $x = 2$ to $x = 5$.

16. Find a vector equation for the tangent line to the curve $\mathbf{r}(t) = < \sin 2t, e^{-3t}, t >$ when $t = 0$.

17. Let $\mathbf{r}(t) = < 2t, t^2, t^3/3 >$ for $t \geq 0$. When does the tangent vector make a $60°$ angle with the z–axis?

18. Find the position vector of a particle at time $t = \pi/2$ if $\mathbf{r}''(t) = -3\cos 2t\mathbf{i} - 4\sin 2t\mathbf{j}$, $\mathbf{r}(0) = (3/2)\mathbf{i}$, and $\mathbf{r}'(0) = 2\mathbf{j}$.

19. Let $\mathbf{r}(t) = e^t(\cos t\mathbf{i} + \sin t\mathbf{j})$. (a) Express the curve in polar form. (b) Find the angle between $\mathbf{r}(t)$ and $\mathbf{r}'(t)$.

20. A particle has position vector $\mathbf{r}(t) = < t\sin t, t\cos t, t^{1/2} >$ for $t \geq 0$. When is the speed a minimum?

21. A particle moves in the xy–plane with $x = t - \sin t$, $y = 1 - \cos t$ for $0 \leq t \leq 2\pi$, with t as time. Find its maximum speed.

22. The path of a particle moving in space is $\mathbf{r}(t) = t^2\mathbf{i} - 4t\mathbf{j} + (8/3)t^{3/2}\mathbf{k}$ for $t > 0$. Find its minimum speed.

23. The path of a particle at time t is given by $x = 3\cos t$ and $y = 4\sin t$ for $t \geq 0$. Find (a) the angular velocity when $t = \pi/4$, (b) the maximum angular velocity, and (c) the minimum angular velocity.

24. The path of a particle at time t is given by $x = 4\cos(1/t)$ and $y = 4\sin(1/t)$ for $t > 0$. (a) Find the angular velocity when $t = 6/\pi$. (b) Describe how the angular velocity varies as t increases. (c) Determine (if it exists) the angular velocity of the particle as $t \to 0^+$ and as $t \to \infty$.

25. Find the length of the curve $\mathbf{r}(t) = 2t^{1/2}\mathbf{i} + t\mathbf{j} + 2\mathbf{k}$ for $0 \leq t \leq 2$.

26. Find the length of the curve $\mathbf{r}(t) = e^t(\cos t\mathbf{j} + \sin t\mathbf{k})$ for $0 \leq t \leq 2$.

27. Find the length of the curve $\mathbf{r}(t) = 2t\mathbf{i} + t^2\mathbf{j} + (1/3)t^3\mathbf{k}$ for $0 \leq t \leq 3$.

28. Find the length of the curve $\mathbf{r}(t) = < t^3, 2t^2 >$ for $0 \leq t \leq 1$.

29. What is the total length of the curve $\mathbf{r}(t) = e^{-t} < \sin t, \cos t >$?

30. A curve has vector equation $\mathbf{r}(t) = < (3/2)t^2, 2t^2, t^3 >$ for $t \geq 0$. Find a formula for the arc length function $s(t)$.

31. A curve has vector equation $\mathbf{r}(t) = \langle (1/2)\cos 3t, (1/2)\sin 3t, t^{3/2} \rangle$ for $t \geq 0$. Find a formula for the arc length function $s(t)$.

32. Express the parametric curve $x = t^3$, $y = t^2$ for $t \geq 0$ in terms of arc length s as parameter.

33. Find the curvature of the graph of $y = \cosh x$.

34. Answer the following questions about the graph of $y = \ln x$. Find (a) the curvature $\kappa(x)$ and (b) the limiting values of the curvature as $x \to 0^+$ and as $x \to \infty$. (c) Explain briefly how the results in (b) might have been anticipated from the graph of $y = \ln x$. (d) Find the maximum of the curvature and the point where it occurs.

35. Let $\mathbf{r}(t) = \cos 2t\,\mathbf{i} + 2\sin^2 t\,\mathbf{j}$. (a) Find the curvature κ at any t. (b) Is the result in (a) surprising? Explain briefly.

36. Find the curvature when $t = \pi$ for the curve given by $\mathbf{r}(t) = \cos t\,\mathbf{i} + 2\sin t\,\mathbf{j} + t\,\mathbf{k}$.

37. A parametric curve is given by $\mathbf{r}(t) = \sin^3 t\,\mathbf{i} + \cos^3 t\,\mathbf{j}$ for $t \geq 0$. Find the unit tangent vector and the principal unit normal vector at any point.

38. A parametric curve is given by $x(t) = t^3 - 3t$, $y(t) = 3t^2$ for $t \geq 0$. Find the unit tangent vector and the principal unit normal vector at any point.

39. A particle has position vector $\mathbf{r}(t) = (t+1)(t-1)\mathbf{i} + t^2\mathbf{j}$ for $-2 \leq t \leq 2$. (a) Draw the path of motion and describe how the particle moves along the path. (b) Find $\mathbf{a}_T$ and $\mathbf{a}_N$. (Don't work too hard!)

40. A particle has position vector $\mathbf{r}(t) = \langle t, t^2, (2/3)t^3 \rangle$. Find (a) the curvature at any time t, (b) the maximum curvature, and (c) the tangential and normal components of acceleration at any time.

41. A particle moves with position vector $\mathbf{r}(t) = \langle e^t, e^{-t} \rangle$ for $t \geq 0$. Find (a) the speed at any time, (b) the minimum speed, (c) the tangential component of acceleration, and (d) the normal component of acceleration.

42. The position of a particle at time t is given by $\mathbf{r}(t) = \langle \cosh t, \sinh t \rangle$ for $t \geq 0$. (a) Identify the path of motion. Then find (b) the speed, (c) the curvature, and (d) a_T and a_N all when $t = \ln 2$.

43. Find the unit tangent vector and principal unit normal vector at $t = 4$ for the curve given by $\mathbf{r}(t) = \langle t^2, -4t, (8/3)t^{3/2} \rangle$.

44. Assume the curve in the previous problem is the path of motion of a particle. Calculate the tangential and normal components of acceleration by two different methods.

45. (a) Identify the polar curve $r = 2(1 + \sin\theta)$. (b) Suppose a particle moves on this polar curve so that $\theta = t$. Express $\mathbf{r}'(t)$ and $\mathbf{r}''(t)$ in terms of $\mathbf{u}_r$ and $\mathbf{u}_\theta$. (c) Find $\mathbf{r}'(t)$ and $\mathbf{r}''(t)$ when $t = \pi/2$.

46. Find the area $A(t)$ swept out by the position vector of a particle moving on the polar curve (a lemniscate) $r^2 = 4\cos 2\theta$ with $\theta = t/2$ from time 0 to any time t.

47. Find the radial and tranverse components of acceleration of an object that moves on the polar curve $r = \sin 2\theta$ with $\theta = t/2$.

48. In the context of the previous problem find (a) the speed of travel along the curve at any time and (b) the maximum and minimum speeds.

CHAPTER 5
DIFFERENTIAL CALCULUS FOR FUNCTIONS OF TWO AND THREE VARIABLES

In this chapter and the next, the principal results of differential calculus for functions of one variable are extended to functions of two and three variables. Applications to science and engineering are given as we go along. Many of the main ideas we shall meet apply as well to functions of more than three variables. Some of the extensions will be indicated as we proceed, but our focus will be on the two–and three–variable cases.

The chapter begins with a discussion of functions, equations, and graphs. The graphs are curves or surfaces, depending on the number of variables involved. A useful catalog of surfaces is developed in Sec. 5.1. Limits and continuity for functions of two or three variables are defined in much the same way as for functions of one variable. Derivatives are another matter. We shall meet partial derivatives, directional derivatives, and gradients. Partial derivatives and directional derivatives are particular ordinary derivatives associated with functions of several variables. They measure rates of change of functions in particular directions and have natural interpretations as slopes. The gradient vector, introduced in Sec. 5.5, is the analogue for functions of several variables of the derivative for functions of one variable. We explore the relationships among partial derivatives, directional derivatives, gradients, and chain rules as we go along. Typical applications to rate and motion problems are given.

5.1 Functions and Graphs

Functions of two, three, and even more variables occur frequently in calculus and its applications. The basic terminology and notation used for functions of several variables is essentially the same as for functions of one variable. We shall focus mainly on functions of two variables and deal to a lesser extent with functions of three variables. Functions of more than three variables will be mentioned only in passing.

The graph of a typical function of two variables or a typical equation involving three variables is a surface in 3–space. Planes and spheres are simple examples. As we go along, we shall meet several functions, equations, and graphs that will come up frequently in later sections and chapters. Often surfaces are not very easy to visualize. Traces, which are intersections of surfaces with various planes, are useful aids in visualization. Another device that helps us better understand functions of two or three variables is the concept of a level curve or surface. Graphing calculators and graphics utilities can be used to obtain graphs of functions, traces of surfaces, and level curves of surfaces.

Functions of Two Variables

Functions of two variables describe many common situations. In fact, we have been using them for quite some time. For example, the area A of a rectangle with sides x and y is the function of x and y given by $A = xy$. Here, x and y are *independent variables* and A is the *dependent variable.* Common notation for a function of two variables is $z = f(x,y)$ or $f(x,y)$ or f. In everyday use, a function $z = f(x,y)$ is given by a formula or a rule that tells us how z is determined by x and y. It is convenient to regard $f(x,y)$ either as a function of the two variables x and y or as a function of the variable point (x,y) in the plane. The **domain** of f is the set D_f of all prescribed or allowed points (x,y). The **range** of f is the set R_f of all values $z = f(x,y)$ with (x,y) in D_f. Just as for functions of one variable, if the domain is not given, then we take the largest domain for which the function makes sense. For example, the domain of the function $z = f(x,y) = x^2 + y^2$ is the entire xy–plane. The range consists of all $z \geq 0$.

EXAMPLE 1. Find the domain and range of the function

$$z = f(x,y) = \frac{1}{x^2 + y^2}.$$

Solution. Evidently the domain D_f consists of all $(x,y) \neq (0,0)$. The range R_f consists of all $z > 0$. To confirm this, first note that $z = f(x,y) > 0$ for all $(x,y) \neq 0$. So every number z in the range is positive. Now fix any $z > 0$. Then

$$z = f(x,y) = \frac{1}{x^2 + y^2} \quad \Leftrightarrow \quad x^2 + y^2 = \frac{1}{z}.$$

So $z = f(x,y)$ whenever (x,y) lies on the circle with center (0,0) and radius $1/\sqrt{z}$. Therefore, as claimed, the range R_f consists of all $z > 0$. □

Ranges are typically more difficult to determine than domains are. The reason is that z is in the range of a function f if and only if the equation $z = f(x,y)$ has a solution (x,y). For the function in Ex. 1, it was easy to solve $z = f(x,y)$ for (x,y). For other functions, it may be quite a challenge. In the next problem, we don't even try to find the range.

EXAMPLE 2. Find the domain of the function

$$f(x,y) = \sqrt{y - \sin x}\ \ln(4 - x^2 - y^2).$$

Solution. Since $\sqrt{y - \sin x}$ makes sense only for $y \geq \sin x$ and $\ln(4 - x^2 - y^2)$ makes sense only for $x^2 + y^2 < 4$,

$$(x,y) \text{ is in } D_f \quad \Leftrightarrow \quad y \geq \sin x \text{ and } x^2 + y^2 < 4.$$

The domain of f is the shaded region in Fig. 1. It consists of the points (x,y) lying on or above the graph of $y = \sin x$ and inside the circle with center $(0,0)$ and radius 2. □

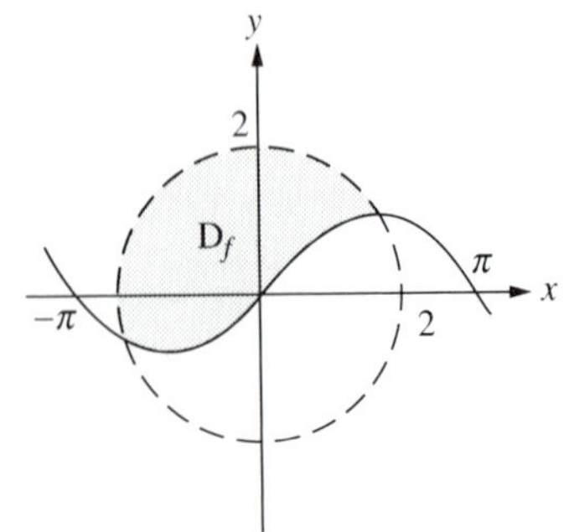

FIGURE 1

Functions of Three or More Variables

Just as for functions of two variables, we have used functions of three variables for quite a while. For example, the volume V of a box with sides x, y, and z is the function of three variables given by $V = xyz$. The notation and terminology used to describe functions of three or more variables are a natural extension of the notation used for functions of one and two variables. For example, $f(x,y,z) = e^{xyz}$ is a function of three variables. The domain of this function is all of 3–space and its range consists of all positive numbers. The same function can be expressed also by $w = e^{xyz}$.

Occasionally, we shall meet functions of more than three variables. The temperature T at a point (x,y,z) ordinarily varies with time t. Then T is a function of four variables. For example,

$$T = f(x,y,z,t) = \frac{e^{(x^2+y^2+z^2)/4t}}{(4\pi t)^{3/2}}$$

is the temperature distribution due to a point source of heat located at the origin at time $t = 0$. The independent variables x, y, z, t form a four–dimensional space called 4–space.

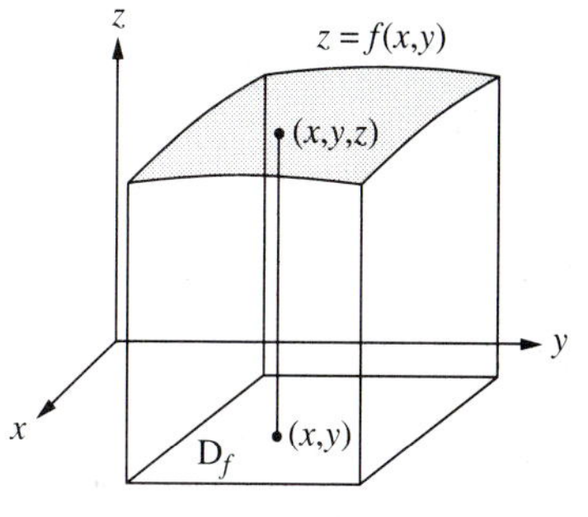

FIGURE 2

Graphs of Functions and Equations

We begin with functions of two variables. The **graph** of a function $z = f(x,y)$ is the set of points (x,y,z) in 3–space with (x,y) in D_f and $z = f(x,y)$. Typically, the graph is a surface, as in Fig. 2.

Aside from constant functions, the simplest functions of two variables are the linear functions. A *linear function* has the form $z = ax + by + c$ with constants a, b, and c. In most of mathematics beyond calculus a function $z = ax + by + c$ is called linear only if $c = 0$. However, in calculus, it is convenient to allow any value for c. The graph of a linear function is a nonvertical plane in 3–space. For example, consider $z = -3x - 2y + 6$. A quick way to graph this plane is to determine its intercepts with the three coordinate axes:

$$x = y = 0 \;\Rightarrow\; z = 6,$$
$$y = z = 0 \;\Rightarrow\; x = 2,$$
$$x = z = 0 \;\Rightarrow\; y = 3.$$

Figure 3 shows the part of the plane in the first octant.

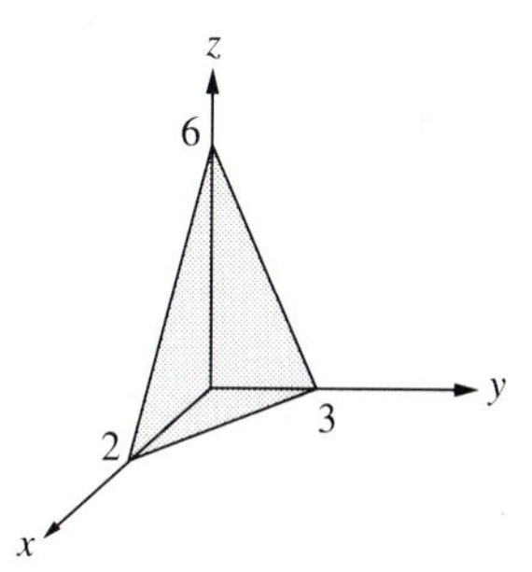

FIGURE 3

The **graph** of an equation $F(x,y,z) = c$ is the set of all points (x,y,z) that satisfies the equation. Graphs of functions and graphs of equations are closely related. For example, the function $z = -3x - 2y + 6$ and the equation $3x + 2y + z = 6$ have the same graph, the plane indicated in Fig. 3.

EXAMPLE 3. The equation $x^2 + y^2 + z^2 = a^2$ for the sphere with center $(0,0,0)$ and radius a can be expressed by two functions,

$$z = \pm\sqrt{a^2 - x^2 - y^2}.$$

The graphs of the two functions are the upper and lower hemispheres in Fig. 4. Both functions have the same domain, the *disk* $x^2 + y^2 \le a^2$ in the xy–plane. □

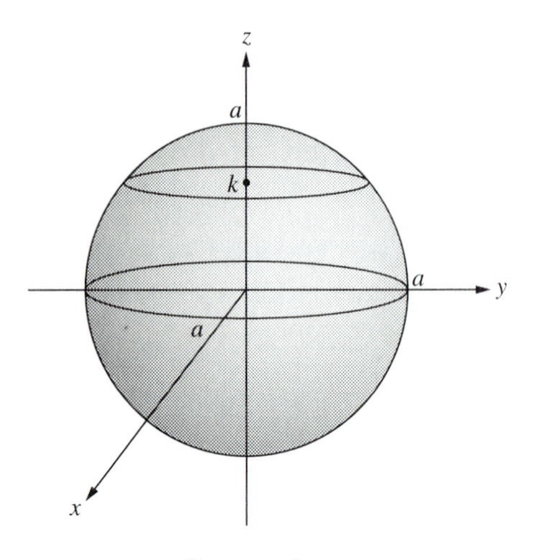

FIGURE 4

A **trace** of a surface is an intersection of the surface with a plane. Traces are usually curves. For example, the trace of the sphere $x^2 + y^2 + z^2 = a^2$ in the horizontal plane $z = k$ is the circle in Fig. 4 determined by

$$x^2 + y^2 = a^2 - k^2, \qquad z = k, \quad \text{if } -a < k < a.$$

The trace is a single point $(0, 0, k)$ if $k = \pm a$. The trace is the empty set if $k > a$ or $k < -a$.

An effective way to visualize or construct the graph of a function or equation is to examine its traces in well-chosen planes, particularly the planes $x = k$, $y = k$, $z = k$, which are parallel to the coordinate planes. Traces may reveal symmetries. If the traces in planes $z = k$ are circles with centers on the z–axis, or single points on the z–axis, or empty, then the surface is symmetric about the z–axis. For example, the sphere in Fig. 4 is symmetric about the z–axis. In fact, it is symmetric about all three axes. We shall make use of traces to develop a small catalog of functions, equations, and graphs that will reappear from time to time as we go along. With a little practice, you should be able to reconstruct the graphs from their traces whenever you need them.

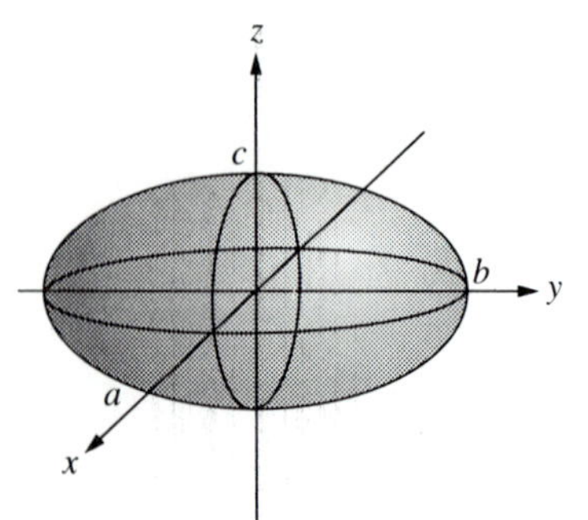

FIGURE 5

EXAMPLE 4. The equation

$$\boxed{\frac{x^2}{a^2} + \frac{y^2}{b^2} + \frac{z^2}{c^2} = 1,}$$

with a, b, c, positive, is graphed in Fig. 5. The surface is an **ellipsoid** with *center* $(0, 0, 0)$ and *semiaxes* a, b, c. It is a sphere if $a = b = c$. The name ellipsoid comes from the fact that the traces in the planes $x = k$, $y = k$, $z = k$, are ellipses (or single points or empty). For example, the trace in the xy–plane, where $z = 0$, is the ellipse

$$\frac{x^2}{a^2} + \frac{y^2}{b^2} = 1, \qquad z = 0.$$

The equation for the ellipsoid can be expressed by two functions,

$$z = \pm c\sqrt{1 - \frac{x^2}{a^2} - \frac{y^2}{b^2}},$$

which graph as the top and bottom halves of the ellipsoid. Both functions have the same domain. What is it? □

The following several surfaces are representative types of paraboloids, cones, and hyperboloids. Other types and examples are given in the problems.

EXAMPLE 5. The function

$$\boxed{z = x^2 + y^2}$$

is graphed in Fig. 6. The surface is a **circular paraboloid** with *vertex* $(0, 0, 0)$ and *axis* the z–axis. To see why the name fits, let's look at some traces. The trace in the plane $z = k$ is the circle $x^2 + y^2 = k$ if $k > 0$. The trace is the point $(0, 0, 0)$ if $k = 0$. The trace is empty if $k < 0$, which means that no points of the surface lie below the xy–plane. The trace in the xz–plane $(y = 0)$ is the parabola $z = x^2$, $y = 0$. The circular paraboloid is generated by rotating this parabola, or just the half with $x \geq 0$, around the z–axis. Therefore, the trace in any plane that contains the z–axis is a parabola and the circular paraboloid is a surface of revolution about the z–axis. □

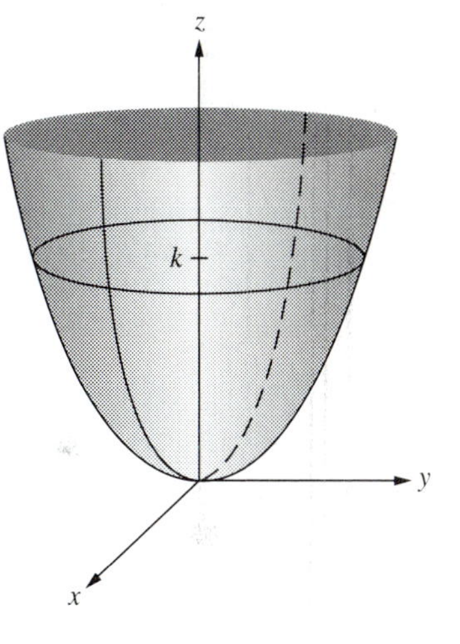

FIGURE 6

The graph of $z = c(x^2 + y^2)$ with any $c \neq 0$ is a circular paraboloid with axis the z–axis and which opens up if $c > 0$ and opens down if $c < 0$. How does the value of c affect the shape of the surface? Other circular paraboloids include

$$x = c(y^2 + z^2) \text{ and } y = c(x^2 + z^2).$$

Can you describe these paraboloids?

As noted in Ex. 5, circular paraboloids are surfaces of revolution. Any **surface of revolution** about the z–axis can be generated by rotating a curve $z = f(x)$ for $x \geq 0$ (or x in a smaller interval) about the z–axis. The surface of revolution is the graph of $z = f(\sqrt{x^2 + y^2})$. Thus, we have replaced x by the radial distance $\sqrt{x^2 + y^2}$ from the origin in the xy–plane. Surfaces of revolution about the z–axis can be recognized by observing that z depends on x and y only through the combination $x^2 + y^2$. Similarly, surfaces of revolution about the y–axis and x–axis involve only the combinations $x^2 + z^2$ and $y^2 + z^2$.

EXAMPLE 6. The function

$$z = \frac{x^2}{a^2} + \frac{y^2}{b^2},$$

with a and b positive, has a graph similar to Fig. 6. The surface is an **elliptic paraboloid** with *vertex* $(0,0,0)$ and *axis* the z–axis. It is a circular paraboloid if $a = b$. The traces in the planes $z = k$ with $k > 0$ are ellipses. The traces in the yz–plane $(x = 0)$ and the xz–plane $(y = 0)$ are parabolas. In fact, the trace in any plane that contains the z–axis is a parabola. See the problems. □

EXAMPLE 7. The equation

$$z^2 = x^2 + y^2$$

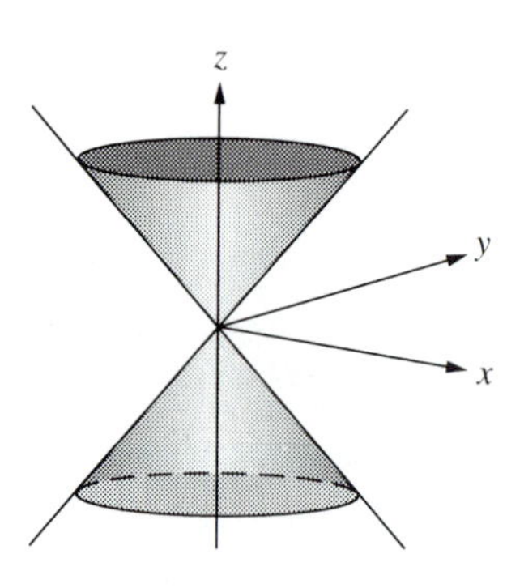

FIGURE 7

is graphed in Fig. 7. The surface is a **circular cone** with *vertex* (0,0,0) and *axis* the z–axis. The traces in the planes $z = k$ with $k \neq 0$ are circles. The trace in the yz–plane $(x = 0)$ consists of two lines, $z = \pm y$. The cone can be generated by rotating either line about the z–axis. You are asked to verify these observations in the problems. □

Other circular cones are

$$z^2 = c^2(x^2 + y^2), \qquad y^2 = c^2(x^2 + z^2), \qquad x^2 = c^2(y^2 + z^2),$$

with $c > 0$. Can you describe these cones?

EXAMPLE 8. The equation

$$z^2 = \frac{x^2}{a^2} + \frac{y^2}{b^2},$$

with a and b positive, has a graph similar to Fig. 7. The surface is an **elliptic cone** with *vertex* $(0,0,0)$ and *axis* the z–axis. It is a circular cone if $a = b$. The traces in the planes $z = k$ with $k \neq 0$ are ellipses. The trace in the yz–plane $(x = 0)$ consists of the two lines $z = \pm y/b$. In the problems you are asked to verify that the trace in any plane that contains the z–axis consists of two lines through the origin. □

The next three surfaces are more challenging to visualize and to sketch from their traces.

EXAMPLE 9. The function

$$\boxed{z = \frac{y^2}{b^2} - \frac{x^2}{a^2}}$$

is graphed in Fig. 8. The surface is a **hyperbolic paraboloid**. It is shaped rather like a saddle. The traces in the planes $z = k$ with $k \neq 0$ are hyperbolas. The traces in the xz–plane ($y = 0$) and the yz–plane ($x = 0$) are parabolas. Interesting features of this surface are explored further in the problems. □

FIGURE 8

EXAMPLE 10. The equation

$$\boxed{z^2 = \frac{x^2}{a^2} + \frac{y^2}{b^2} + 1}$$

is graphed in Fig. 9. This surface is a **hyperboloid of two sheets**. The traces in the xz–plane ($y = 0$) and the yz–plane ($x = 0$) are hyperbolas. Notice that $z^2 \geq 1$. So $z \geq 1$ or $z \leq -1$. No part of the surface lies between the planes $z = 1$ and $z = -1$. The traces in the planes $z = k$ with $k > 1$ and $k < -1$ are ellipses. The equation for the hyperboloid can be expressed by two functions,

$$z = \pm\sqrt{\frac{x^2}{a^2} + \frac{y^2}{b^2} + 1},$$

which give the upper and lower sheets in Fig. 9. □

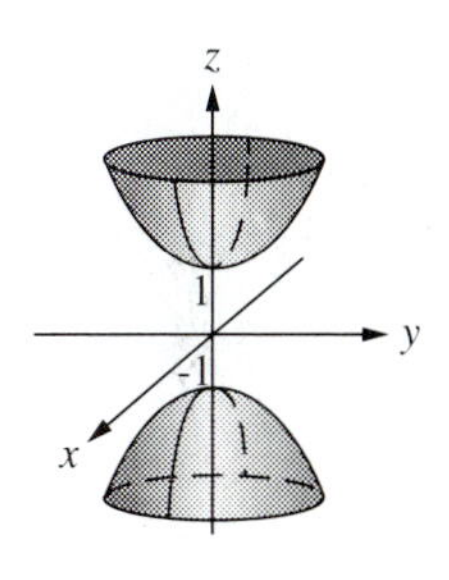

FIGURE 9

EXAMPLE 11. The equation

$$\boxed{z^2 = \frac{x^2}{a^2} + \frac{y^2}{b^2} - 1}$$

is graphed in Fig. 10. The surface is a **hyperboloid of one sheet**. The traces in the xz–plane ($y = 0$) and the yz–plane ($x = 0$) are hyperbolas. The equation can be expressed by two functions,

$$z = \pm\sqrt{\frac{x^2}{a^2} + \frac{y^2}{b^2} - 1}.$$

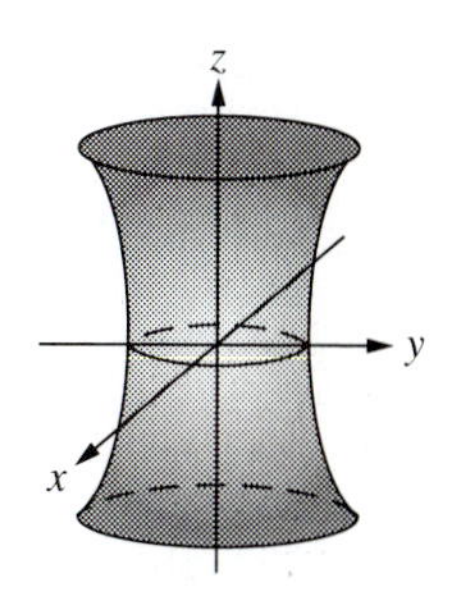

FIGURE 10

The common domain of these functions is the region in the xy–plane on or outside the ellipse

$$\frac{x^2}{a^2} + \frac{y^2}{b^2} = 1, \qquad z = 0,$$

which is the trace of the hyperboloid in the xy–plane. □

Examples 3–11 are special cases of quadric surfaces. A **quadric surface** is the graph of a quadratic equation in x, y, and z.The most general such equation is

$$Ax^2 + By^2 + Cz^2 + Dxy + Exz + Fyz + Gx + Hy + Iz + J = 0,$$

where $A, B, C, \ldots, J$ are constants. Any quadratic equation can be recast in one of the two forms

$$Ax^2 + By^2 + Cz^2 + J = 0, \qquad Ax^2 + By^2 + Cz^2 + Iz = 0$$

by translations and rotations of the coordinate axes. Examples 3–11 are of these simpler, but basic, types.

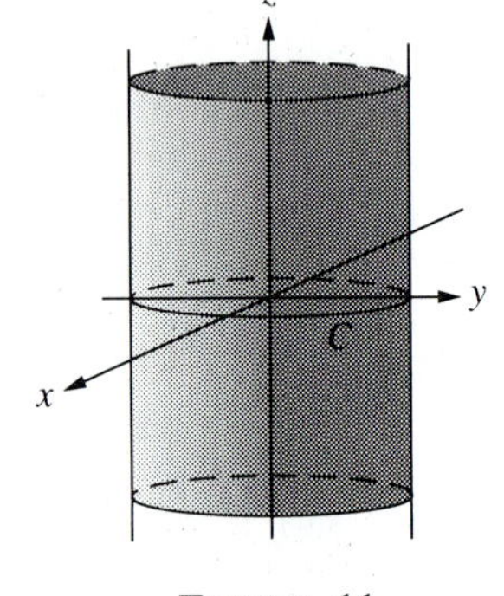

FIGURE 11

Cylinders and Their Equations

Cylindrical solids occurred earlier in calculus when volumes were found by the method of cross sections. Here we are interested in cylindrical surfaces, called cylinders for short. We begin with a familiar example.

EXAMPLE 12. The graph of $x^2 + y^2 = 4$ is a circle in the xy–plane. In 3–space, the graph consists of all (x, y, z) for which $x^2 + y^2 = 4$. Since z does not appear in the equation, it takes on all values. The surface shown in Fig. 11 is swept out by moving the circle C parallel to the z–axis. It is called a **(right) circular cylinder.** □

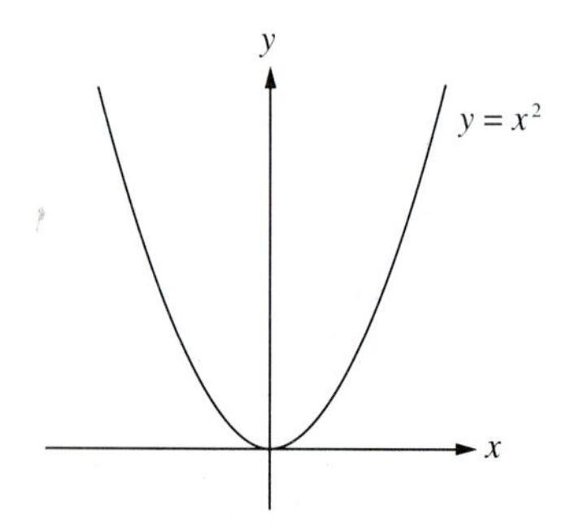

FIGURE 12

In general, a **cylinder** is a surface swept out when a plane curve is translated parallel to a line. The plane curve is a *directrix* of the cylinder and the line is an *axis* of the cylinder. A cylinder that is also a quadric surface, as in Fig. 11, is called a *quadric cylinder.*

The graph of a typical equation $y = f(x)$ or $F(x, y) = c$ in the xy–plane is a curve. The graph in 3–space consists of all points (x, y, z) which satisfy $y = f(x)$ or $F(x, y) = c$. Since any z satisfies either equation, the graph is a cylinder. For example, the graph of $y = x^2$ is a parabola in 2–space and a parabolic cylinder in 3–space. See Figs. 12 and 13.

The cylinder in Fig. 13 is a quadric cylinder. A cylinder that is not a quadric surface is the graph of $z = \sin x$ in 3–space. It looks like a corrugated roof.

Level Curves

It can be quite difficult to make a satisfactory sketch of the graph of a function $z = f(x, y)$ on a two–dimensional sheet of paper. The figures in the text required the skills of a professional draftsman. A way around this difficulty is provided by a different geometric description of a function $z = f(x, y)$. A simple example will illustrate the idea.

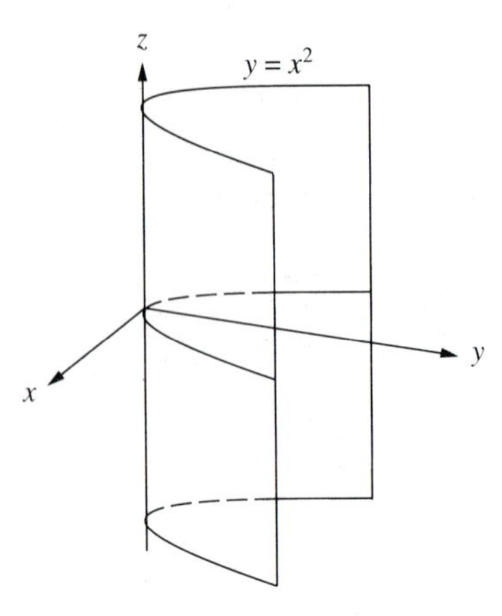

FIGURE 13

EXAMPLE 13. The function $z = f(x, y) = x^2 + y^2$ is graphed in Fig. 14, along with traces in the planes $z = 1$, $z = 4$, and $z = 9$, which are circles. The projections of these circles onto the xy–plane are shown in Fig. 15. For example, the trace $x^2 + y^2 = 4$, $z = 4$, has the projection $x^2 + y^2 = 4$, $z = 0$. In Fig. 15, the three circles are labeled $f = 1$, $f = 4$, $f = 9$ because these are the values of f on the circles. The three circles are called level curves of f because $f(x, y) = x^2 + y^2$ is constant on each of them. Here the word *level* means constant. Starting from Fig. 15, we can reconstruct the traces in Fig. 14 by lifting the circles in the xy–plane up to the right levels (or heights). If we had a great many level curves $f = k$ in Fig. 15, very close together, then we could construct a reasonable approximation of the surface in Fig. 14 by lifting all the level curves up to the right heights. □

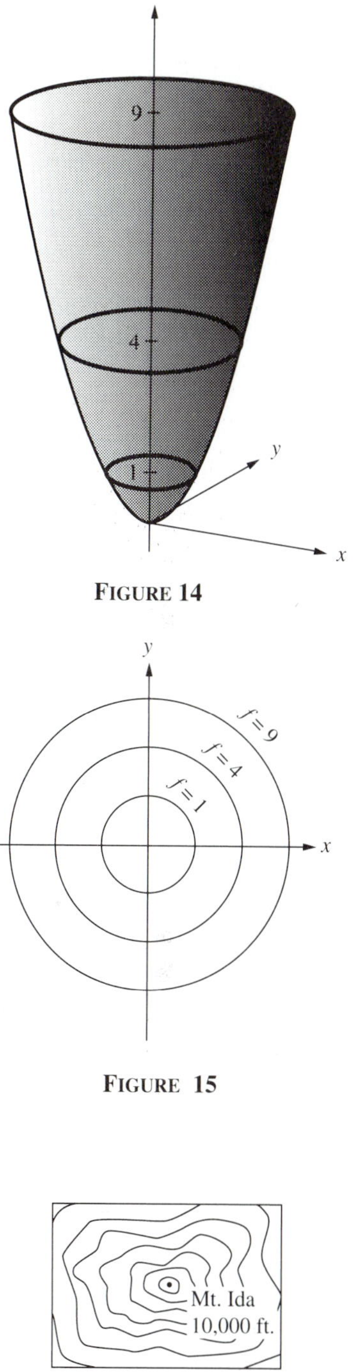

FIGURE 14

FIGURE 15

In general, a **level curve** of a function $f(x, y)$ is the set of points (x, y) where $f(x,y) = k$ for a constant k in the range of f.

Topographic maps use level curves to show equal elevations. The topographic map of Mt. Ida in Fig. 16 has level curves for elevations at every 1000 feet. The terrain is steeper where the level curves are closer together, and is flatter where the level curves are farther apart. Weather maps also use level curves. You have probably seen such maps with curves of constant temperature (called isotherms) or constant pressure (called isobars).

EXAMPLE 14. Use level curves to visualize the graph of

$$z = f(x,y) = x^2 - xy + y^2.$$

Solution. The graph of $f(x,y) = x^2 - xy + y^2 = 1$ in the xy–plane is an ellipse with center at the origin and axes along the lines obtained by rotating the x – and y–axes 45° counterclockwise. It is the level curve $f = 1$ in Fig. 17. The level curves $f = 4$ and $f = 9$, also shown in Fig. 17, are just larger copies of the ellipse $f = 1$. Raise the level curves up to their corresponding elevations to see that the surface $z = f(x,y)$ looks rather like an elliptic paraboloid. In fact, the surface is an elliptic paraboloid. It can be put into the form of Ex. 6 and Fig. 6 by rotating the x – and y–axes 45° counterclockwise. □

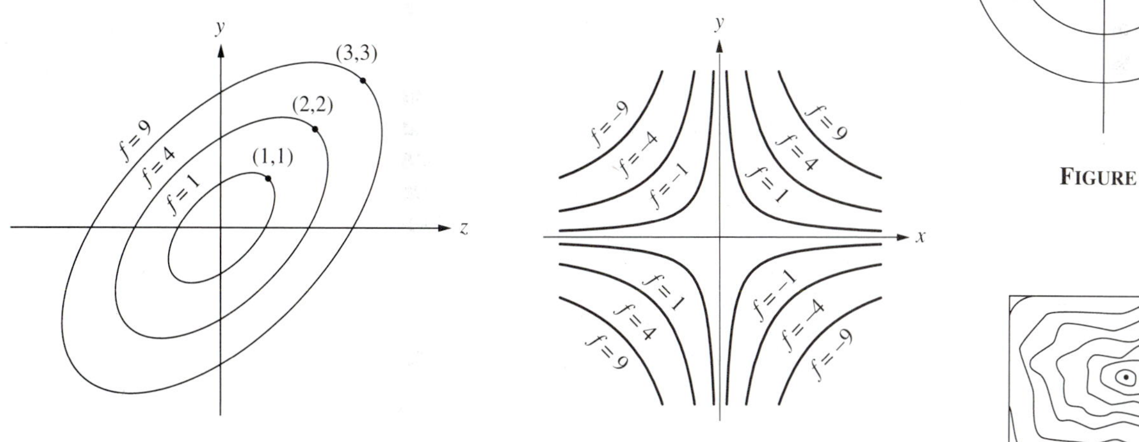

FIGURE 17

FIGURE 18

FIGURE 16

EXAMPLE 15. Use level curves to visualize the graph of $z = f(x,y) = xy$.

Solution. Figure 18 shows several level curves. Since $f(x,y) = 0$ for $x = 0$ or $y = 0$, the level curve $f = 0$ consists of two straight lines, the x–axis and the y–axis. The other level curves are hyperbolas. The symmetries in Fig. 18 indicate that the graph of f is symmetric through the origin and antisymmetric (symmetric except for sign change) across the x–axis and across the y–axis. In Fig. 18, $f(x,y)$ is positive and increases as (x,y) moves away from the origin in the first or third quadrant. In the second or fourth quadrant, $f(x,y)$ is negative and decreases as (x,y) moves away from the origin. It follows that the surface looks rather like a saddle. In fact, it is a 45° rotation of the surface in Ex. 9 and Fig. 8. □

Level Surfaces

To graph a function $w = f(x,y,z)$ of three variables would require a space of four variables. Obviously, we cannot sketch such a graph. However, we can sketch level surfaces in 3–space. A **level surface** of f is the set of points (x,y,z) which satisfies $f(x,y,z) = k$ for some k in the range of f. Level surfaces can tell us a good deal about how $f(x,y,z)$ varies.

EXAMPLE 16. Suppose that the temperature at (x,y,z) is given by

$$T = \frac{1}{x^2 + y^2 + z^2 + 1}.$$

Describe the level surfaces (isotherms) of T.

Solution. Clearly, $0 < T \leq 1$. The temperature is $T = 1$ at the origin and T decreases as (x,y,z) moves away from the origin. The level surfaces $T = k$ are the graphs of

$$x^2 + y^2 + z^2 = \frac{1}{k} - 1 = \frac{1-k}{k} \quad \text{for } 0 < k \leq 1.$$

For each k, the level surface $T = k$ is the sphere with center (0,0,0) and radius $c = \sqrt{(1-k)/k}$. □

EXAMPLE 17. Describe the level surfaces of

$$f(x,y,z) = z - \frac{1}{2}(x^2 + y^2).$$

Solution. The level surfaces $f = k$ are the graphs of

$$z = \frac{1}{2}(x^2 + y^2) + k$$

with k any constant. If $k = 0$, then $z = \frac{1}{2}(x^2 + y^2)$. This surface is the circular paraboloid labeled $f = 0$ in Fig. 19. The other level surfaces $f = k$ are vertical translates of this paraboloid, k units upward if $k > 0$, and k units downward if $k < 0$. □

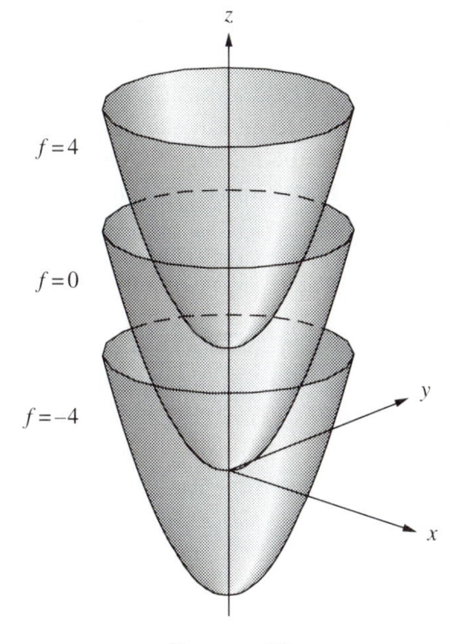

FIGURE 19

Computer Graphics

Computer graphics packages produce computer-generated graphs of functions of two variables, level curves of such functions, and level surfaces for functions of three variables. Some of the packages enable you to view a surface from various perspectives. For example, Matlab, Maple, and Mathematica include such packages. The better packages also use special projection techniques that help eliminate the distortion inherent in representing a three–dimensional object on a two–dimensional screen or printout. Figures 20 and 21 are typical examples of computer generated graphics.

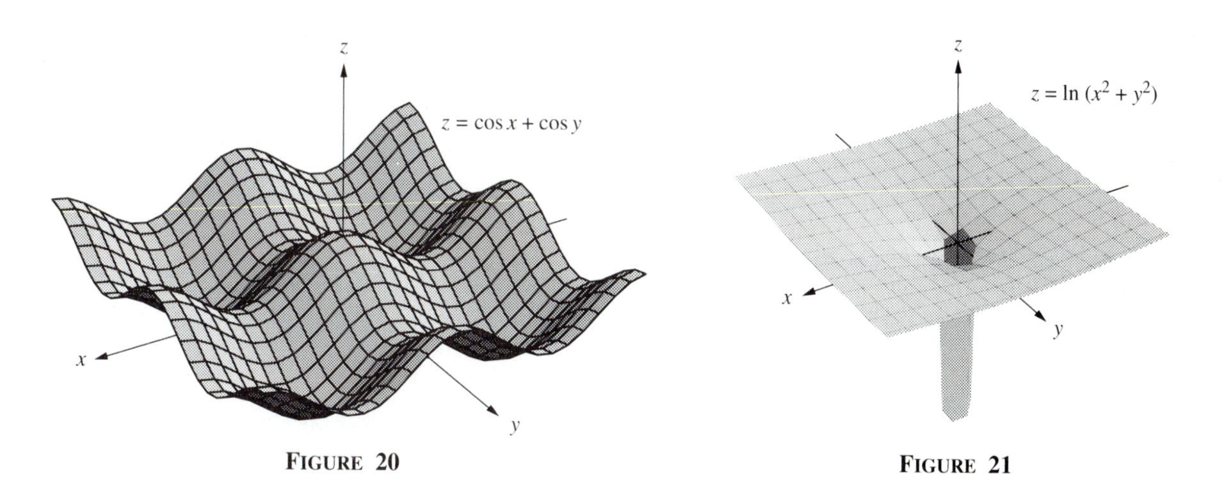

FIGURE 20 FIGURE 21

Effective use of three–dimensional graphics packages requires a good deal of expertise. Often the default presentation of a surface distorts and/or conceals important aspects of the surface. Both mathematical insight about what to expect and skill in the use of the software are needed to take full advantage of the available technology.

A simple Matlab m–file for graphing surfaces of the form $z = f(x,y)$ follows. The program is written to graph the function $f(x,y) = \cos x + \cos y$. Modest changes in the m–file will enable you to graph other surfaces. In the m–file, the spacing between grid points at which the function is evaluated is rather coarse so that the program will run on the student version of Matlab. If you have access to the complete version of Matlab, you may want to refine the grid somewhat and plot the graph on a larger rectangle.

```
% SPLOT.M is an m-file for plotting graphs of functions z  = f(x, y).
% The file is written to plot f(x, y) = cos(x) + cos(y). It can be modified
% easily to plot other functions. You may wish to type help meshdom
% and help mesh at a matlab prompt to get more information about these
Matlab
% surface plotting procedures.
 [x,y] = meshdom(-5:.4:5, -5:.4:5);   % Grid points at which the function is
                                      % evaluated.
 z = cos(x) + cos(y);     % Evaluate function at the grid points.
 M = [-37.5 60];          % Set the location from which the graph is viewed.
 mesh(z,M)                % Plot the graph of the function.
```

PROBLEMS

In Probs. 1–10, find (a) the domain and (b) the range of the given function.

1. $f(x,y) = x + y - 3$
2. $f(x,y) = \sin(x + y - 3)$
3. $f(x,y) = \sqrt{4 - x^2 - y^2}$
4. $f(x,y) = \ln(4 - x^2 - y^2)$
5. $f(x,y) = \arctan(y/x)$
6. $f(x,y) = \cos xy$
7. $f(x,y,z) = \arctan \dfrac{z}{\sqrt{x^2 + y^2}}$
8. $f(x,y,z) = \ln \dfrac{z}{\sqrt{x^2 + y^2}}$
9. $f(x,y,z) = e^{-1/(x^2 + y^2 + z^2)}$
10. $f(x,y,z) = \dfrac{1}{4x^2 + y^2 + 9z^2}$

In Probs. 11–34, (a) identify and (b) sketch the graph in 3–space of the given equation.

11. $x^2 + 4y^2 + 9z^2 = 36$
12. $9x^2 + 4y^2 + 9z^2 = 36$
13. $x^2 + 2y^2 - z = 0$
14. $x^2 + y^2 - 4z = 0$
15. $y = x^2 + z^2$
16. $x + 2y^2 = -z^2$
17. $z^2 = 4x^2 + 4y^2$
18. $x^2 = 4z^2 + 4y^2$
19. $4x^2 = z^2 - 9y^2$
20. $4z^2 = y^2 - 9x^2$
21. $z = x^2 - y^2$
22. $9y^2 - 16x^2 - 144z = 0$
23. $z^2 - x^2 - y^2 = 1$
24. $36z^2 - 1 = 4x^2 + 9y^2$

25. $x^2 + y^2 - z^2 = 1$

26. $x^2 + 16y^2 - 16z^2 = 16$

27. $x^2 + z^2 = 9$

28. $4z + y^2 = 0$

29. $xy = 4$

30. $4x^2 + 9y^2 = 36$

31. $z = \sin x$

32. $y = e^{-x^2}$

33. $x = \ln y$

34. $y = 4 - z^2$

In Probs. 35–38, identify the traces of the surface in planes of the indicated type. As usual, assume the xy–plane is horizontal. Note that vertical planes that contain the z–axis have equations of the form $y = mx$ for any constant m or $x = 0$.

35. $16x^2 + 4y^2 = 64$ in horizontal planes.

36. $16x^2 + 4y^2 = 64$ in (a) the yz–plane, (b) the xz–plane, (c) any vertical plane that contains the z–axis.

37. $z = xy$ in horizontal planes.

38. $z = xy$ in vertical planes that contain the z–axis.

In Probs. 39–48, identify and sketch typical level curves or level surfaces for the given function.

39. $f(x,y) = e^{x/y}$

40. $f(x,y) = e^{-(x+y-3)}$

41. $f(x,y) = \sin(x - y)$

42. $f(x,y) = \sin(x^2 + y^2)$

43. $f(x,y) = \ln(1 - x^2 + y)$

44. $f(x,y) = y/(x^2 + y^2)$

45. $f(x,y,z) = x^2 + y^2 - z$

46. $f(x,y,z) = x^2 + y^2 - z^2$

47. $f(x,y,z) = e^{4x^2 + 9y^2 + 25z^2}$

48. $f(x,y,z) = \ln[1/(4x^2 + y^2 + 9z^2)]$

All the quadric surfaces in Exs. 3–11 have center at the origin. Quadrics with centers at other points and without any cross-product terms xy, xz, or yz can be identified and sketched after completing the square to locate the center. We used the same idea earlier for circles and parabolas. In Probs. 49–52 find the center of the quadric surface. Then identify and sketch it.

49. $x^2 + y^2 + z^2 - 2x + 4y - 6z = 2$

50. $x^2 + y^2 - 4x - 16y - z + 73 = 0$

51. $x^2 + y^2 - z^2 - 6y - 2z + 5 = 0$

52. $9x^2 + 4y^2 - 36z^2 - 90x - 72z + 153 = 0$

53. The curve $z = x^2 + 1$ for $x \geq 0$ in the xz–plane is rotated about the z–axis. Identify the surface that is generated.

54. The curve $z = \sqrt{x^2 + 1}$ for $x \geq 0$ in the xz–plane is rotated about the z–axis. Identify the surface that is generated.

55. The curve $z = 1/y$ for $y \geq 0$ in the yz–plane is rotated about the z–axis. Find an equation for the surface generated and sketch it.

56. Find an equation for and sketch the surface generated when the curve $z = e^{-y^2}$ for $y \geq 0$ in the yz–plane is rotated about the z–axis.

57. None of the level curves in the examples in the text intersect. Is this always true? Explain why it is or give an example of a function that has two distinct level curves that intersect each other.

58. Answer the previous question for level surfaces instead of level curves.

Probs. 59–64 ask you to establish many of the facts stated in Exs. 4–11 about the basic quadric surfaces.

59. Show that the traces of the ellipsoid $x^2/a^2 + y^2/b^2 + z^2/c^2 = 1$ in the planes $z = k$ are ellipses, points, or empty depending on the value of k. Find the semiaxes of the trace ellipses.

60. Use polar coordinates to show that the traces of the circular paraboloid $z = x^2 + y^2$ in any plane that contains the z–axis is a parabola and that all such parabolas are congruent. *Hint.* The z–axis and the ray $\theta = \theta_0$ determine a plane with rectangular coordinates (r, z). Find the equation of the trace in this plane.

61. Sketch the paraboloids $x = c(y^2 + z^2)$ for (a) $c = 2$, (b) $c = 1/2$, and (c) $c = -2$.

62. Verify the following properties of an elliptic paraboloid $z = x^2/a^2 + y^2/b^2$. (a) The traces in the plane $z = k$ are ellipses, points, or empty. (b) The traces in any plane containing the z–axis is a parabola.

63. (a) Show that the traces of the hyperbolic paraboloid $z = y^2/b^2 - x^2/a^2$ in the planes $z = k$ for $k \neq 0$ are hyperbolas. (b) Find the trace for $k = 0$. (c) Find the traces in any plane that contains the z–axis.

64. Define $f(x, y) = y^2/b^2 - x^2/a^2$. Show that $f(x, 0)$ has a global minimum at $x = 0$ while $f(0, y)$ has a global maximum at $y = 0$. Can $f(0, 0)$ be a local or global maximum or minimum value of the function of two variables $f(x, y)$? Explain briefly.

65. Verify the following properties of the circular cone $z^2 = x^2 + y^2$. (a) The traces in the planes $z = k$ with $k \neq 0$ are circles. (b) The trace in the yz–plane consists of the two lines $z = \pm y$. (c) The trace in any plane containing the z–axis is a pair of lines inclined at 45° to the z–axis.

66. Verify the statements made about the traces of the elliptic cone in Ex. 8.

5.2 Limits and Continuity

Limits and continuity for functions of two or three variables are defined in much the same way as for functions of one variable. So the material in this section should seem quite natural. To prepare for the discussion of limits and continuity, we begin with some useful vocabulary about sets. As you will see, familiar words are given precise mathematical meanings that are akin to but are not the same as their meanings in colloquial use.

Properties of Sets in 2–space and 3–space

The domain of a function $f(x,y)$ or $f(x,y,z)$ is a set in 2–space or 3–space. In order to define limits and continuity for such a function, we need to know more about its domain. In particular, we need to understand the concepts open and closed, as well as bounded and unbounded, as they apply to sets in 2–space and 3–space.

Open and Closed Sets

For functions of one variable, it often made a difference whether a function was defined on an open interval or a closed interval. For example, a continuous function on a closed interval $[a,b]$ has maximum and minimum values, whereas a function on an open interval may have neither a maximum nor a minimum. The situation is similar for functions of two or three variables.

The two–dimensional analogue of an interval is a rectangle. Let $a < b$ and $c < d$. The set of points (x,y) with $a < x < b$ and $c < y < d$ forms an *open rectangle*. The inequalities $a \leq x \leq b$ and $c \leq y \leq d$ determine a *closed rectangle*. In Fig. 1, the open rectangle consists of the points inside the perimeter, while the closed rectangle consists of the points inside or on the perimeter.

We shall call a three–dimensional rectangular solid a **box**. The older name, rectangular parallelepiped, is just too long. An **open box** is determined by inequalities $a < x < b$, $c < y < d$, $e < z < f$, and a **closed box** by $a \leq x \leq b$, $c \leq y \leq d$, $e \leq z \leq f$.

Recall that the distance between the points $P = (x,y)$ and $P_0 = (x_0,y_0)$ in the plane is

$$|P\,P_0| = \sqrt{(x-x_0)^2 + (y-y_0)^2}.$$

The equation $(x-x_0)^2 + (y-y_0)^2 = a^2$ for the circle C with center (x_0,y_0) and radius $a > 0$ is expressed geometrically by $|PP_0| = a$. See Fig. 2. The points P inside C, which satisfy $|PP_0| < a$, form an **open disk**. The points P inside or on C, which satisfy $|P\,P_0| \leq a$, form a **closed disk**.

Very little changes in 3–space. The distance between $P = (x,y,z)$ and $P_0 = (x_0,y_0,z_0)$ is

$$|P\,P_0| = \sqrt{(x-x_0)^2 + (y-y_0)^2 + (z-z_0)^2}.$$

The equation $(x-x_0)^2 + (y-y_0)^2 + (z-z_0)^2 = a^2$ for the sphere with center (x_0,y_0,z_0) and radius $a > 0$ is expressed geometrically by $|P\,P_0| = a$. The points P inside the sphere, with $|P\,P_0| < a$, form an **open ball**. The points P inside or on the sphere, with $|P\,P_0| \leq a$, form a **closed ball**.

The notation P and P_0 for points will enable us to deal with sets in 2–space and 3–space at the same time. An open disk or ball $|P\,P_0| < a$ centered at P_0 also is called a **neighborhood** of P_0. In Fig. 3, the circle $|P\,P_0| = a$ is dashed to indicate that it does not belong to the neighborhood of P_0.

Let D be a nonempty set in 2–space or 3–space. For example, D may be the domain of a function. Figure 4 illustrates the two–dimensional case. A point P_0 is an **interior point** of D if P_0 has a neighborhood that lies entirely in D. The set of all interior points of D is the **interior** of D. The interior of a closed rectangle or box is the corresponding open rectangle or box. The interior of a closed disk or ball is the corresponding open disk or ball.

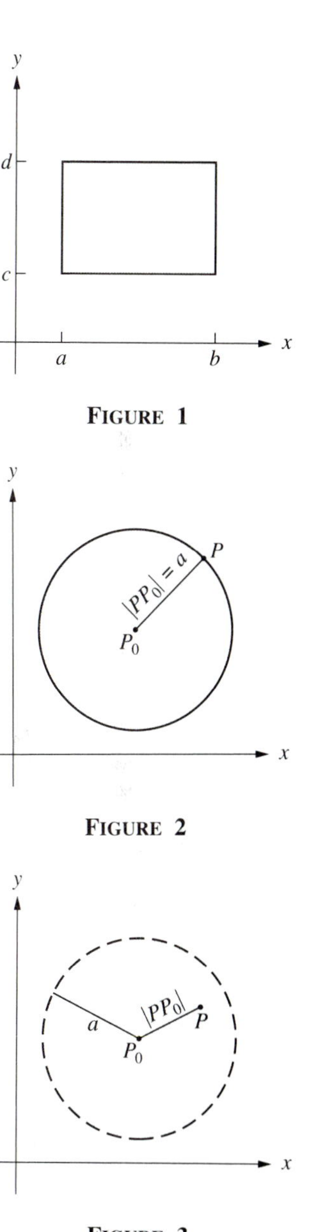

FIGURE 3

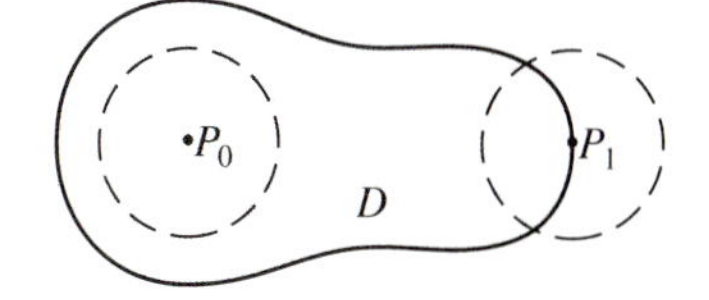

FIGURE 4

In general, a set is **open** if all its points are interior points. Open rectangles and open boxes are open sets. So are open disks and open balls.

EXAMPLE 1. The entire xy–plane is an open set. So is the set of all points $(x,y) \neq 0$. Do you see why every point of this set is an interior point of the set? Think about it. □

EXAMPLE 2. The entire 3–space is an open set. So is the set of all points $(x,y,z) \neq 0$. □

The boundary of a disk, as in Fig. 2, is the corresponding circle. Each point P on the circle is a boundary point of the disk. For any nonempty set D, as in Fig. 4, a point P_1 is a **boundary point** of D if every neighborhood of P_1 contains at least one point of D and at least one point which is not in D. The set of all boundary points of D is the **boundary** of D. For example, the boundary of a ball is the corresponding sphere. What is the boundary of a rectangle or a box?

The entire xy–plane has no boundary points. Neither does all of 3–space. The set in the xy–plane consisting of all $(x,y) \neq 0$ has the single boundary point $(0,0)$. Similarly, $(0,0,0)$ is the only boundary point of the set in 3–space consisting of all $(x,y,z) \neq 0$.

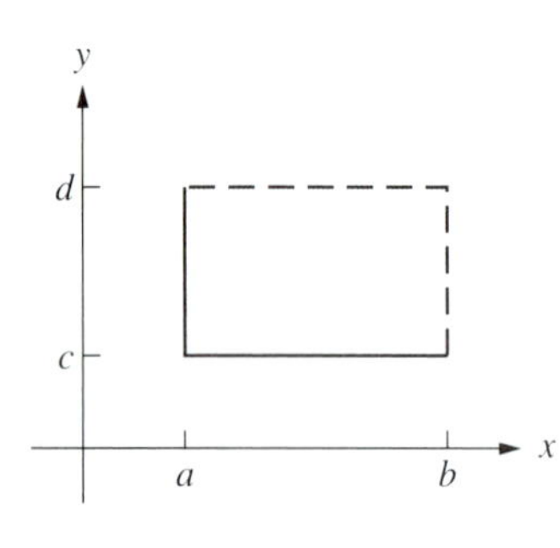

FIGURE 5

A set is **closed** if it contains all its boundary points. Closed disks and closed balls are closed sets. So are closed rectangles and closed boxes. In daily conversation, a door is either open or closed. But this is not true for sets. The rectangle in Fig. 5, determined by $a \leq x < b$ and $c \leq y < d$, contains part but not all of its boundary. It is neither open nor closed. This example should remind you of a half-open interval.

There is another departure from conversational usage of the terms *open* and *closed.* The entire xy–plane is open and, since it has no boundary points, it is also closed. In the same way, all of 3–space is both open and closed. These are the only nonempty sets in 2–space and 3–space that are both open and closed.

Our interest in open and closed sets is primarily for domains of functions. Here are a few examples.

EXAMPLE 3. The domain of $f(x,y) = \sqrt{4 - x^2 - y^2}$ is the closed disk $x^2 + y^2 \leq 4$. □

EXAMPLE 4. The domain of $f(x,y) = 1/\sqrt{4-x^2-y^2}$ is the open disk $x^2+y^2<4$. □

EXAMPLE 5. The domain of $f(x,y,z) = \ln(x^2 + y^2 + z^2)$ is the set of all $(x,y,z) \neq 0$ which, as we have seen, is open. □

EXAMPLE 6. The domain of $f(x,y) = \sqrt{4 - x^2 - y^2}/(x^2 + y^2)$ is the set of all (x,y) with $0 < x^2 + y^2 \leq 4$, which is neither open nor closed. This set may be called a *punctured disk.* □

Bounded Sets

A set on the real line is bounded if it lies in a finite interval $[a,b]$. Similarly, a set in 2–space or 3–space is **bounded** if it lies in a disk (or a ball). Obviously, any disk or ball is bounded. So is any rectangle or box. The domains of the functions in Exs. 3, 4, and 6 are bounded. The domain of the function in Ex. 5 is unbounded.

Connected Sets

There is another property of sets that plays a significant role from time to time. In everyday usage, the word *connected* conveys the idea of being joined together. A set is **arcwise connected** or, more briefly, **connected**, if any two points of the set can be joined by a continuous curve that lies entirely in the set. For example, any open or closed rectangle, box, disk, or ball is connected. The rectangle in Fig. 5, which is neither open nor closed, is connected. The set of points that comprise the earth and moon at any instant of time is disconnected (not connected).

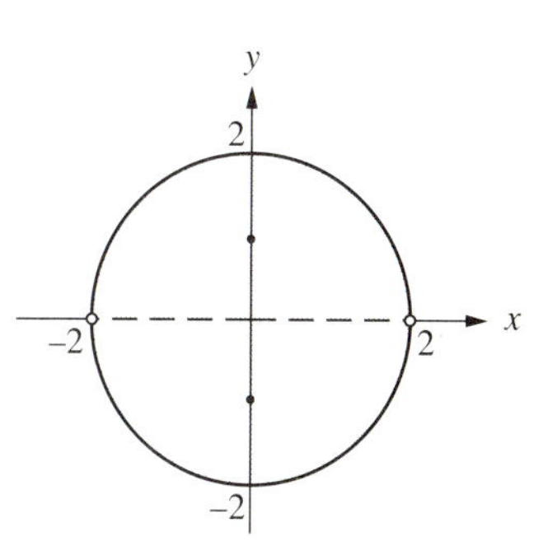

FIGURE 6

EXAMPLE 7. The domain D of the function $f(x,y) = \sqrt{4 - x^2 - y^2}/y$ is the set of points in the closed disk $x^2 + y^2 \leq 4$ that do not lie on the x–axis. See Fig. 6 where D is shaded and the dashed line indicates points not in D. Clearly there is no continuous curve that lies entirely in D and joins the points $(0,1)$ and $(0,-1)$ of D. So D is not connected. □

Limits

As we mentioned at the beginning of the section, limits of functions of two or more variables are defined very much like limits of functions of one variable. We discuss such limits and corresponding limit laws next.

Let's begin with an easy example, where it is obvious what the limit is. The limit statement

$$f(x,y) = x^3y^2 \to 2^3 3^2 = 24 \text{ as } (x,y) \to (2,3)$$

means that $f(x,y)$ can be made as near to 24 as we like by taking (x,y) near enough to $(2,3)$ with $(x,y) \neq (2,3)$. Likewise, since $(\sin u)/u \to 1$ as $u \to 0$, the function

$$f(x,y) = \frac{\sin\sqrt{x^2+y^2}}{\sqrt{x^2+y^2}} \quad \text{for} \quad (x,y) \neq (0,0)$$

has limit 1 as (x,y) tends to $(0,0)$. The statement,

$$f(x,y) \to 1 \text{ as } (x,y) \to (0,0),$$

means that $f(x,y)$ can be made as near to 1 as we like by taking $(x,y) \neq (0,0)$ near enough to $(0,0)$ or, equivalently, by making the distance $\sqrt{x^2+y^2}$ from (x,y) to $(0,0)$ small enough. The point (x,y) may approach $(0,0)$ in any manner: along the x–axis, along the y–axis, or even along an exotic curve such as $y = x \sin 1/x$ for $x > 0$, whose graph is suggested by Fig 7. The limit is 1 no matter how (x,y) approaches $(0,0)$. In other words, $f(x,y) \to 1$ as $x \to 0$ and $y \to 0$ *independently*.

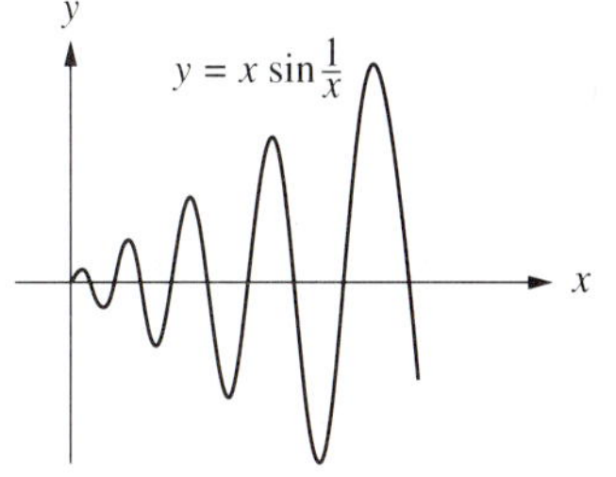

FIGURE 7

Now we turn to the general situation. In order to define limits of functions of two or three variables at the same time, we use again the notation

$$P = (x,y), \qquad P_0 = (x_0, y_0) \qquad \text{in 2–space,}$$
$$P = (x,y,z), \qquad P_0 = (x_0, y_0, z_0) \qquad \text{in 3–space.}$$

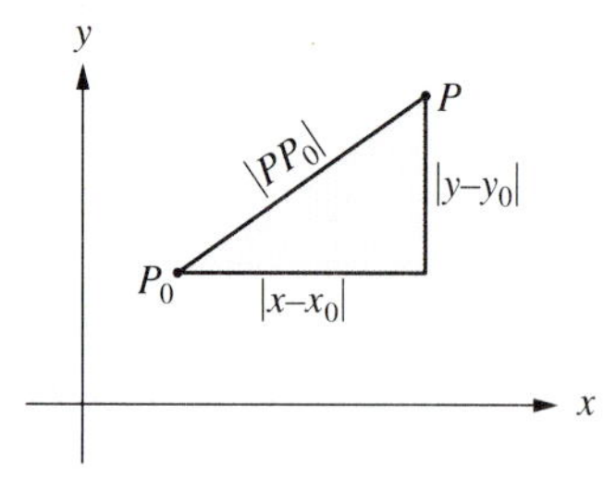

FIGURE 8

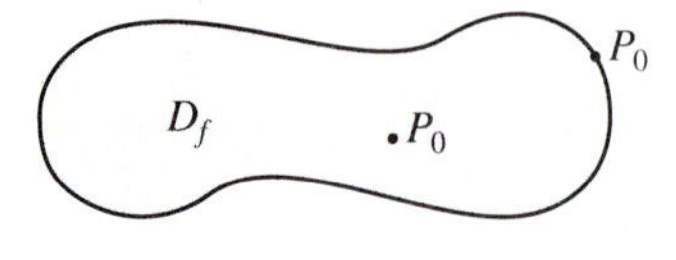

FIGURE 9

Before giving a careful definition of the limit relation $f(P) \to L$ as $P \to P_0$, we make a few observations. In 2–space,

$$P \to P_0 \quad \Leftrightarrow \quad |PP_0| \to 0 \quad \Leftrightarrow \quad x \to x_0, \quad y \to y_0 \text{ independently.}$$

In Fig. 8, imagine that P is moving nearer to P_0. In 3–space,

$$P \to P_0 \quad \Leftrightarrow \quad |PP_0| \to 0 \quad \Leftrightarrow \quad x \to x_0, \quad y \to y_0, \quad z \to z_0 \text{ independently.}$$

For $f(P) \to L$ as $P \to P_0$ to make any sense, there have to be points P in the domain of f that are arbitrarily near to P_0 with $P \neq P_0$. In other words, every neighborhood of P_0 must contain points P of D_f with $P \neq P_0$. Figure 9 illustrate two possibilities for P_0, one an interior point of D_f and the other a boundary point of D_f. Here is a verbal definition of a finite limit:

L is the limit of $f(P)$ as P approaches P_0 if
$f(P)$ is as near to L as we like
for all P in D_f near enough to P_0 with $P \neq P_0$.

A more concise version of this statement is

$$\boxed{f(P) \to L \text{ as } P \to P_0 \quad \Leftrightarrow \quad |f(P) - L| \to 0 \text{ as } |PP_0| \to 0.}$$

In particular, when $L = 0$,

$$f(P) \to 0 \text{ as } P \to P_0 \quad \Leftrightarrow \quad |f(P)| \to 0 \text{ as } |PP_0| \to 0.$$

The precise mathematical definition of a finite limit of a function of two or three variables will be given a little later. First, we make a few general comments and give some examples. If a limit exists it is unique, just as for functions of one variable. The algebraic limit laws and squeeze laws are virtually the same for functions of one, two, three, or more variables. They help us find many limits by inspection. We shall use limit laws and squeeze laws freely, without pausing to give formal statements.

EXAMPLE 8. Find the limit of $f(x,y,z) = x^2 e^y \sin z + \cos 2xz$ as $(x,y,z) \to (1,1,\pi/2)$.

Solution. This function is defined for all (x,y,z). So $(1,1,\pi/2)$ is an interior point of the domain of f. Since the trigonometric and exponential functions are continuous,

$$x^2 e^y \sin z \to 1 \cdot e \cdot 1 = e \text{ and } \cos 2xz \to \cos \pi = -1 \text{ as } (x,y,z) \to (1,1,\pi/2).$$

Hence, $f(x,y,z) \to e - 1$ as $(x,y,z) \to (1,1,\pi/2)$. □

EXAMPLE 9. Find the limit of $f(x,y) = \sqrt{4 - x^2 - y^2}$ as $(x,y) \to (0,2)$.

Solution. The graph of f is the hemisphere in Fig. 10. The domain D_f is the closed disk $x^2 + y^2 \leq 4$. Note that $(0,2)$ is a boundary point of D_f. By inspection,

$$f(x,y) = \sqrt{4 - x^2 - y^2} \to \sqrt{4 - 0 - 4} = 0 \text{ as } (x,y) \to (0,2).$$

This limit is suggested by Fig. 10. □

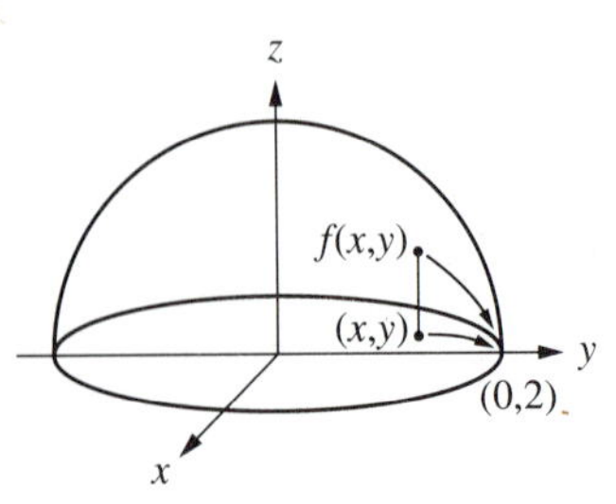

FIGURE 10

Infinite limits are useful for functions of two or more variables, just as they are for functions of one variable. We do not bother to give formal definitions because no essentially new ideas are involved. For example,

$$f(x,y) = \frac{1}{\sqrt{4 - x^2 - y^2}} \to \infty \text{ as } (x,y) \to (0,2),$$

where it is understood that (x,y) approaches $(0,2)$ through points in the domain of f, which is the open disk $x^2 + y^2 < 4$. Similarly,

$$f(x,y,z) = \ln(x^2 + y^2 + z^2) \to -\infty \text{ as } (x,y,z) \to (0,0,0).$$

The function in the next problem has a limiting behavior that may surprise you. Careful study of this problem will deepen your understanding of limits.

EXAMPLE 10. Does the function

$$f(x,y) = \frac{2xy}{x^2 + y^2}$$

have a limit as $(x,y) \to (0,0)$? If so, find it.

Solution 1. This function is defined for $(x,y) \neq (0,0)$ and is indeterminate of the form 0/0 as $(x,y) \to (0,0)$. Since $f(x,y) = 0$ if $x = 0$ or $y = 0$, it follows that $f(x,y) \to 0$ as $(x,y) \to (0,0)$ along either the x–axis or the y–axis. So you might suspect that f has limit 0. Not so fast! Let $y = \pm x$ in the formula for $f(x,y)$ to find that $f(x,y) = 1$ for $y = x$ and $f(x,y) = -1$ for $y = -x$, as shown by the level curves in Fig. 11. Therefore, $f(x,y) \to 1$ as $(x,y) \to (0,0)$ along the line $y = x$, and $f(x,y) \to -1$ as $(x,y) \to (0,0)$ along the line $y = -x$. Since a limit of $f(x,y)$ must be the same no matter how (x,y) approaches $(0,0)$, we are forced to conclude that $f(x,y)$ does not have a limit as $(x,y) \to (0,0)$.

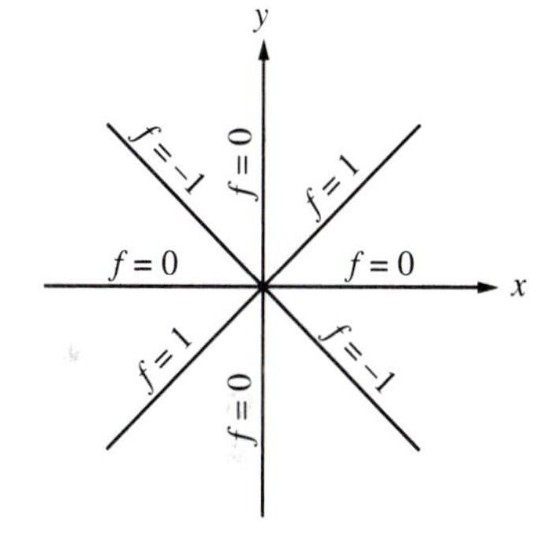

FIGURE 11

Solution 2. Polar coordinates reveal more about the behavior of $f(x,y)$ as (x,y) approaches $(0,0)$. Let $x = r\cos\theta$, $y = r\sin\theta$, and $x^2 + y^2 = r^2$ to obtain

$$f(x,y) = \frac{2xy}{x^2 + y^2} = \frac{2r^2\cos\theta\sin\theta}{r^2} = 2\cos\theta\sin\theta = \sin 2\theta.$$

Consequently, as $f(x,y)$ runs around the circle $x^2 + y^2 = r^2$ with any radius $r > 0$, $f(x,y)$ oscillates between 1 and -1. This is true no matter how small r is. So $z = f(x,y)$ takes on all values in the interval $-1 \leq z \leq 1$ as (x,y) varies in any neighborhood of $(0,0)$. We see again that $f(x,y)$ does not have a limit as $(x,y) \to (0,0)$. □

The function in the next example looks similar, but it behaves quite differently near $(0,0)$.

EXAMPLE 11. Does the function

$$f(x,y) = \frac{2xy}{\sqrt{x^2 + y^2}}$$

have a limit as $(x,y) \to (0,0)$? If so, find it.

Solution. As in Ex. 10, $f(x,y)$ is defined for $(x,y) \neq (0,0)$ and is indeterminate of the form 0/0 as $(x,y) \to (0,0)$. In the problems you are asked to show that $f(x,y) \to 0$ as $(x,y) \to (0,0)$ along any line through the origin. You may be amazed to learn that, by itself, this is not enough to guarantee that $f(x,y) \to 0$ as $(x,y) \to (0,0)$. (Again, see the problems.) The limit is 0 in this example, but we have to work a little harder to show it. Here are two proofs.

1. Since $|x| \leq \sqrt{x^2 + y^2}$ and $|y| \leq \sqrt{x^2 + y^2}$,

$$|f(x,y)| = \frac{2|x|\ |y|}{\sqrt{x^2 + y^2}} \leq 2\sqrt{x^2 + y^2}.$$

By a squeeze law, $|f(x,y)| \to 0$ and, hence, $f(x,y) \to 0$ as $(x,y) \to 0$.

2. Use polar coordinates to express f as

$$f(x,y) = \frac{2r^2 \cos\theta \sin\theta}{r} = r \sin 2\theta.$$

It follows that

$$|f(x,y)| \leq r = \sqrt{x^2 + y^2},$$

which implies, once again, that $f(x,y) \to 0$ as $(x,y) \to (0,0)$. □

It is informative to compare the functions in Exs. 10 and 11 as (x,y) runs around a circle $x^2 + y^2 = r^2$ with an arbitrarily small radius $r > 0$. In Ex. 10, f oscillates between 1 and -1 and there is no limit as $(x,y) \to (0,0)$. In Ex. 11, f oscillates between r and $-r$. As $r \to 0$ the oscillations "die out" and $f(x,y) \to 0$ as $(x,y) \to 0$.

We conclude this introduction to limits with the precise mathematical definition of a finite limit of a function $f(P)$ of two or three variables.

Definition *Finite Limit of a Function*
Assume that every neighborhood of P_0 contains points P in the domain of f with $P \neq P_0$. Then

$$f(P) \to L \text{ as } P \to P_0$$

if for any $\varepsilon > 0$, no matter how small,
there is a corresponding number $\delta > 0$ such that
$|f(P) - \mathrm{L}| < \varepsilon$ if $|P\,P_0| < \delta$, with P in D_f and $P \neq P_0$.

The precise mathematical definition of an infinite limit is similar to the corresponding definition for functions of one variable. We shall continue the practice of basing arguments and conclusions on the verbal definitions of limits, while keeping in mind that complete proofs must rest on the precise definitions.

Continuity

Continuity for functions of two or three variables is defined in terms of limits just as for functions of one variable.

Definition *Continuity*
A function $f(P)$ is **continuous at P_0** if
$f(P) \to f(P_0)$ as $P \to P_0$.
A function $f(P)$ is **continuous** if
it is continuous at every point P_0 in its domain.

EXAMPLE 12. As in Ex. 8, let $f(x, y, z) = x^2 e^y \sin z + \cos 2xz$. Then $f(1, 1, \pi/2) = e - 1$. By the reasoning in Ex. 8,

$$f(x, y, z) \to f(1, 1, \pi/2) \text{ as } (x, y, z) \to (1, 1, \pi/2).$$

Therefore, f is continuous at $(1, 1, \pi/2)$. By a similar argument, f is continuous at any (x, y, z). So f is a continuous function. □

The algebraic rules of continuity carry over without change to functions of two or three variables: sums, differences, products, quotients, scalar multiples, and compositions of continuous functions are continuous.

Most of the functions of two or three variables that we meet are continuous. They are combinations of familiar continuous functions of one variable: powers, roots, polynomials, and rational functions, as well as trigonometric, exponential, logarithmic, and hyperbolic functions. To show that such functions are continuous, it isn't necessary to take limits as in Ex. 12. For example, we know that $f(x) = x^2$ is a continuous function of x. It is true and should seem obvious that $g(x, y) = x^2$ is a continuous function of (x, y) and that $h(x, y, z) = x^2$ is a continuous function of (x, y, z). Similarly, $g(x, y) = e^y$ and $h(x, y, z) = \sin z$ are continuous. By the same token, all algebraic combinations and compositions of such continuous functions are continuous. Thus, we see by inspection that the function in Ex. 12 is continuous. Likewise, here are three more continuous functions:

$$f(x, y) = x^2 - xy + y^2,$$

$$g(x, y) = \sqrt{25 - x^2 - y^2},$$

$$h(x, y, z) = xy \ln (y - x) + \sin xyz.$$

Nothing much changes for functions of four or more variables. One example should suffice. Let

$$f(x, y, z, t) = e^{x^2 + y^2} \cos z \ln (t + 1).$$

By an easy argument,

$$f(x, y, z, t) \to f(0, 0, 0, 0) = 0 \text{ as } (x, y, z, t) \to (0, 0, 0, 0).$$

So f is continuous at $(0, 0, 0, 0)$. In fact, f is a continuous function because it is a combination of continuous functions of one variable.

PROBLEMS

In Probs. 1–10 do the following: (a) Describe the domain of the function with inequalities and or equations. (b) Sketch the domain, marking boundary points that are in the domain with a solid line and boundary points not in the domain with a dashed line. Then determine whether the domain is (c) open, closed, or neither (d) bounded or unbounded, and whether (e) it is connected or disconnected. (f) Explain briefly why $f(P) \to f(P_0)$ for any P and P_0 in the domain of f. What does this tell you about f?

1. $f(x,y) = \ln\sqrt{4 - x^2 - y^2}$
2. $f(x,y) = \ln\dfrac{\sqrt{4 - x^2 - y^2}}{y}$
3. $f(x,y) = \dfrac{\sqrt{36 - 4x^2 - 9y^2}}{\sqrt{x^2 + y^2 - 1}}$
4. $f(x,y) = \sqrt{x^2 + y^2 - 4}\,e^{-1/(x^2+y^2)}$
5. $f(x,y) = \ln x + \ln y$
6. $f(x,y) = \ln(xy)$
7. $f(x,y) = xy\ln(y/x)$
8. $f(x,y) = (x^2 + y^2)^{1/3}$
9. $f(x,y) = \sqrt{e^x - y}$
10. $f(x,y) = \arcsin(y/x)$

In Probs. 11–16 describe the domain of the function.

11. $f(x,y,z) = xy\ln(y - x) + \sin xyz$
12. $f(x,y,z) = e^{z/(x^2+y^2)}\ln z$
13. $f(x,y,z) = \dfrac{1}{z}\ln(xy)$
14. $f(x,y,z) = \ln\dfrac{xy}{z}$
15. $f(x,y,z,t) = e^{x^2+y^2}\sqrt{z}\ln(t + 1)$
16. $f(x,y,z,t) = \dfrac{1}{(4\pi t)^{3/2}}e^{-(x^2+y^2+z^2)/4t}$

In Probs. 17–26 use algebraic limit laws, squeeze laws, and continuity arguments to evaluate the given limit. (If you have a graphing calculator or graphics utility, you may find it informative to graph the function near the point in question.)

17. $\displaystyle\lim_{(x,y)\to(2,-3)} \frac{xy}{x^2 + y^2}$
18. $\displaystyle\lim_{(x,y)\to(0,-3)} e^{-x}\cos\frac{\pi y}{x - 4y}$
19. $\displaystyle\lim_{(x,y)\to(2,1)} \sqrt{\frac{x^2 + y^2}{x^3 - y^3}}$
20. $\displaystyle\lim_{(x,y)\to(2,1)} \sin\left(\frac{x^2 + y^2}{x + y + 1}\pi\right)$
21. $\displaystyle\lim_{(x,y)\to(0,1)} \frac{\ln(x + y)}{\cos xy}$
22. $\displaystyle\lim_{(x,y)\to(0,1)} \frac{\cos xy}{\ln(x + y)}$
23. $\displaystyle\lim_{(x,y)\to(1,1)} \frac{2x^2 - 3xy + y^2}{x - y}$
24. $\displaystyle\lim_{(x,y)\to(1,2)} \frac{2x^2 - 3xy + y^2}{2x - y}$
25. $\displaystyle\lim_{(x,y)\to(0,0)} \frac{1 - \cos(x^2 + y^2)}{x^2 + y^2}$
26. $\displaystyle\lim_{(x,y)\to(0,0)} \frac{\tan(x + y)}{x + y}$

Probs. 27–31 deal with issues raised in connection with Exs. 10 and 11.

27. The function $f(x,y) = 2xy/(x^2 + y^2)$ does not have a limit as $(x,y) \to (0,0)$; see Ex. 10. Nevertheless, show that the function does have a limit as $(x,y) \to (0,0)$ along any straight line through the origin. *Hint.* All such lines, except the y–axis, have an equation of the form $y = mx$.
28. Use the hint in the previous exercise to show that $f(x,y) = 2xy/\sqrt{x^2 + y^2}$, the function from Ex. 11, has limit 0 along every straight line through the origin.
29. (a) Show that the function $f(x,y) = x^2y/(x^4 + y^2)$ has limit 0 as $(x,y) \to (0,0)$ along any line through the origin. (b) Nevertheless, show that $f(x,y)$ does not have a limit as $(x,y) \to (0,0)$. *Hint.* For (b) try a parabolic approach to the origin.
30. Let $f(x,y) = (x^2 + y^2)\ln(x^2 + y^2)$ for $(x,y) \neq (0,0)$. Determine whether it is possible to define $f(0,0)$ so that f is continuous at the origin.
31. Let $f(x,y) = x^3/(x^2 + y^2)$ for $(x,y) \neq (0,0)$. Determine whether it is possible to define $f(0,0)$ so that f is continuous at the origin.

5.3 Partial Derivatives

Partial derivatives of a function of two or more variables are just ordinary derivatives obtained by differentiating with respect to one variable while the other variables are held constant. Virtually all of the properties of ordinary derivatives carry over to partial derivatives. In particular, partial derivatives have important interpretations as slopes and as rates of change.

Partial Derivatives of Functions of Two Variables

We begin with partial derivatives of functions of two variables. A simple example will tell the story. Let

$$f(x,y) = x^3y^2.$$

If $y = 1$, then $f(x,1) = x^3$ is a function of the single variable x. The partial derivative of f with respect to x when $y = 1$ is

$$\frac{\partial f}{\partial x}(x,1) = \frac{d}{dx}f(x,1) = \frac{d}{dx}x^3 = 3x^2.$$

The "curly d" distinguishes partial derivatives from ordinary derivatives. Similarly, if $x = 2$, then $f(2,y) = 8y^2$ is a function of y alone. The partial derivative of f with respect to y when $x = 2$ is

$$\frac{\partial f}{\partial y}(2,y) = \frac{d}{dy}f(2,y) = \frac{d}{dy}8y^2 = 16y.$$

At the point (2,1),

$$\frac{\partial f}{\partial x}(2,1) = 12, \qquad \frac{\partial f}{\partial y}(2,1) = 16.$$

Definition *Partial Derivatives*

The **partial derivative** of $f(x,y)$ **with respect to** x **when** $y = y_0$ is

$$\frac{\partial f}{\partial x}(x, y_0) = \frac{d}{dx} f(x, y_0)$$

whenever the ordinary derivative on the right exists.

The **partial derivative** of $f(x,y)$ **with respect to** y **when** $x = x_0$ **is**

$$\frac{\partial f}{\partial y}(x_0, y) = \frac{d}{dy} f(x_0, y)$$

whenever the ordinary derivative on the right exists.

When differentiating partially with respect to x or y, we generally regard y or x as fixed without giving it a particular value or changing the notation. Thus,

$$f(x,y) = x^3y^2 \quad \Rightarrow \quad \frac{\partial f}{\partial x}(x,y) = 3x^2y^2, \qquad \frac{\partial f}{\partial y}(x,y) = 2x^3y.$$

Partial derivatives of functions are new functions. The partial derivatives of $f(x,y) = x^3y^2$ are functions defined for all (x,y). In fact, they are continuous functions. Most of the functions we deal with have continuous partial derivatives.

Since partial derivatives are particular ordinary derivatives, all the properties of ordinary derivatives of functions of one variable, such as the algebraic differentiation rules and the chain rule, are valid for partial derivatives. We shall use these properties freely, without further comment.

EXAMPLE 1. Find the partial derivatives of $f(x,y) = \sqrt{x^2 + y^2}$.

Solution. First, $f(x,y) = (x^2 + y^2)^{1/2}$. Hold y fixed and differentiate with respect to x to obtain

$$\frac{\partial f}{\partial x} = \frac{1}{2}(x^2 + y^2)^{-1/2}(2x) = \frac{x}{\sqrt{x^2 + y^2}}.$$

Likewise, or by symmetry,

$$\frac{\partial f}{\partial y} = \frac{y}{\sqrt{x^2 + y^2}}.$$

Observe that the function $f(x,y) = \sqrt{x^2 + y^2}$ is defined for all (x,y), whereas the partial derivatives are defined only for $(x,y) \neq (0,0)$. Both of the partial derivatives are continuous functions. □

The prime notation used for ordinary derivatives is unsuitable for partial derivatives of $f(x,y)$ because there are two independent variables to differentiate with respect to. In place of a prime, we use subscripts:

$$f_x = \frac{\partial f}{\partial x}, \qquad f_y = \frac{\partial f}{\partial y}.$$

EXAMPLE 2. Let $f(x,y) = xy^2 + \sin(2x - 3y)$. Then

$$f_x(x,y) = y^2 + 2\cos(2x - 3y), \qquad f_y(x,y) = 2xy - 3\cos(2x - 3y)$$

for all (x,y). In particular,

$$f_x(\pi/2, \pi/3) = \frac{\pi^2}{9} + 2, \qquad f_y(\pi/2, \pi/3) = \frac{\pi^2}{3} - 3. \ \square$$

Partial derivatives of a function $z = f(x,y)$ are also denoted by

$$z_x = \frac{\partial z}{\partial x}, \qquad z_y = \frac{\partial z}{\partial y}.$$

EXAMPLE 3. Let $z = e^{x^2y}$. Then

$$z_x = 2xy\, e^{x^2y}, \qquad z_y = x^2 e^{x^2y}. \ \square$$

EXAMPLE 4. Let $z = x^3 \ln y$. Then

$$\frac{\partial z}{\partial x} = 3x^2 \ln y, \qquad \frac{\partial z}{\partial y} = \frac{x^3}{y}. \ \square$$

Some authors use a D–notation for partial derivatives. If $f(x,y) = x^3 \ln y$, then

$$D_x f = 3x^2 \ln y, \qquad D_y f = \frac{x^3}{y}. \ \square$$

Compare with Ex. 4.

Implicit Partial Differentiation

Implicit differentiation is just as useful for partial derivatives as it is for ordinary derivatives. There is nothing really new here because partial derivatives are particular ordinary derivatives.

EXAMPLE 5. The graph of the equation

$$9x^2 + 16y^2 + 4z^2 = 144$$

is the ellipsoid in Fig. 1. The equation implicitly determines two functions,

$$z = \pm \frac{1}{2}\sqrt{144 - 9x^2 - 16y^2},$$

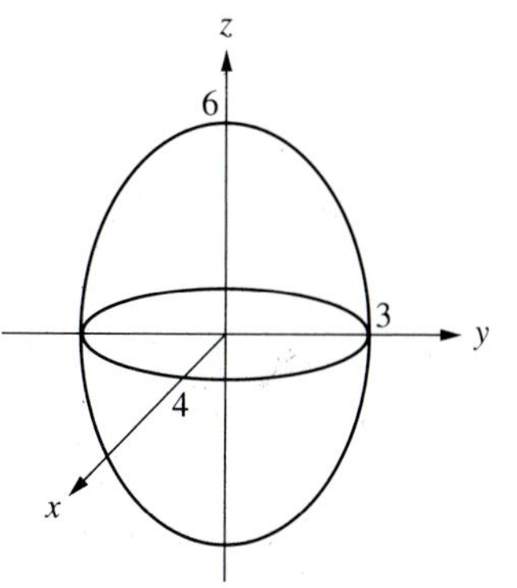

FIGURE 1

represented by the top and bottom halves of the ellipsoid. Find the partial derivatives of these two functions.

Solution. We could calculate the partial derivatives of z directly from the formulas for z in terms of x and y, but implicit differentiation of the original equation is easier. First, fix y and differentiate both sides of the equation for the ellipsoid with respect to x to obtain

$$18x + 8z\frac{\partial z}{\partial x} = 0, \qquad \frac{\partial z}{\partial x} = -\frac{9x}{4z}.$$

In the same way, fix x and differentiate with respect to y to obtain

$$32y + 8z\frac{\partial z}{\partial y} = 0, \qquad \frac{\partial z}{\partial y} = -\frac{4y}{z}.$$

The formulas for the partial derivatives are valid for both of the implicitly defined functions giving z in terms of x and y. The point $(x, y, z) = (8/3, 1, 4)$ is on the top half of the ellipsoid. At this point,

$$\frac{\partial z}{\partial x} = -\frac{3}{2}, \qquad \frac{\partial z}{\partial y} = -1. \ \square$$

Partial Derivatives as Slopes

Ordinary derivatives are slopes of tangent lines. So are partial derivatives, although the geometry is a little more involved. The story is simplest for linear functions. Let

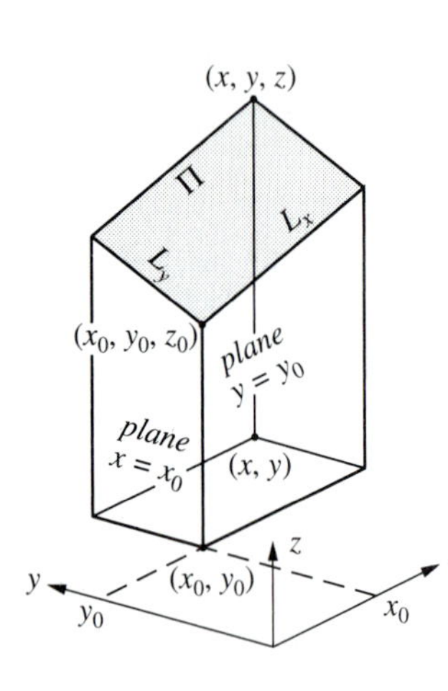

FIGURE 2

$$z = \alpha x + \beta y + \gamma$$

with any constants α, β, γ. As we know, the graph of a linear function in 3–space is a nonvertical plane Π. A piece of the plane is shown in Fig. 2, where (x_0, y_0) is arbitrary and

$$z_0 = \alpha x_0 + \beta y_0 + \gamma.$$

So (x_0, y_0, z_0) is any point in the plane Π. Subtract the equations for z and z_0 to obtain another equation for Π:

$$z - z_0 = \alpha(x - x_0) + \beta(y - y_0).$$

The line L_x in Fig. 2 is the intersection of the plane Π with the vertical plane $y = y_0$. Set $y = y_0$ in the preceding equation for Π to obtain the following equations for L_x:

$$z - z_0 = \alpha(x - x_0), \qquad y = y_0.$$

You should recognize that α is the slope of L_x defined in the usual manner as rise over run, and that $z - z_0 = \alpha(x - x_0)$ is a point-slope equation for L_x.

Similarly, the line L_y in Fig. 2 is the intersection of the plane Π with the vertical plane $x = x_0$. Equations for L_y are

$$z - z_0 = \beta(y - y_0), \qquad x = x_0.$$

The slope of L_y is β and $z - z_0 = \beta(y - y_0)$ is a point-slope equation for L_y.

It is natural to call

$$z - z_0 = \alpha(x - x_0) + \beta(y - y_0)$$

a **point-slope equation** for the plane Π. Since α and β are the slopes of the lines L_x and L_y, which lie in Π, we call α and β the *slopes of the plane* Π *in the x– and y–directions*. Note that

$$\frac{\partial z}{\partial x} = \alpha, \qquad \frac{\partial z}{\partial y} = \beta.$$

Thus, the partial derivatives of $z = \alpha x + \beta y + \gamma$ are the slopes of the plane Π in the x– and y–directions.

EXAMPLE 6. Find an equation for the plane Π through (2, 3, 4) with slope 3 in the x–direction and slope 2 in the y–direction.

Solution. Since $\alpha = 3$ and $\beta = 2$, the point-slope equation for Π is

$$z - 4 = 3(x - 2) + 2(y - 3),$$

which reduces to $z = 3x - 2y - 8$. □

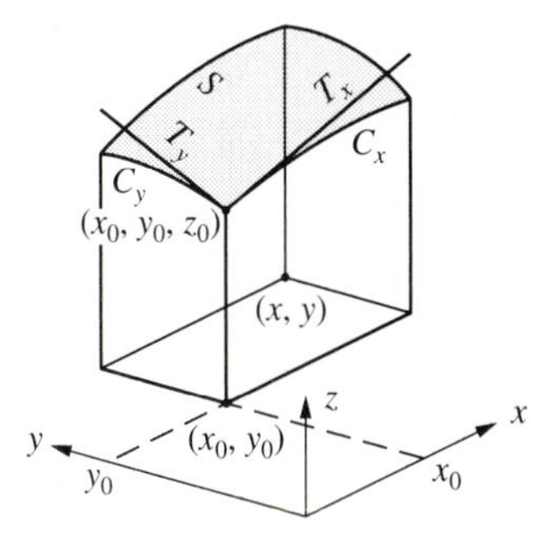

FIGURE 3

Now consider a function $z = f(x,y)$ which has partial derivatives $f_x(x_0,y_0)$ and $f_y(x_0,y_0)$ at an interior point (x_0,y_0) of its domain. Let $z_0 = f(x_0,y_0)$. The graph of f is a surface S containing the point (x_0,y_0,z_0). A piece of the surface is shown in Fig. 3.

The curve C_x in the figure is the intersection of S with the vertical plane $y = y_0$. Equations for C_x are

$$z = f(x,y_0), \qquad y = y_0.$$

The line T_x is tangent to C_x at (x_0,y_0,z_0). The slope of T_x and therefore of C_x at (x_0,y_0,z_0) is given by

$$\text{slope } T_x = \frac{d}{dx} f(x,y_0)\bigg|_{x_0} = f_x(x_0, y_0).$$

We say that $f_x(x_0,y_0)$ is the *slope of the surface S in the x–direction at* (x_0,y_0,z_0).

The curve C_y in Fig. 3 is the intersection of S with the vertical plane $x = x_0$. Equations for C_y are

$$z = f(x_0,y), \qquad x = x_0.$$

The line T_y is tangent to C_y at (x_0, y_0, z_0). The slope of T_y and therefore of C_y at (x_0, y_0, z_0) is given by

$$\text{slope } T_y = \frac{d}{dx} f(x_0, y)\bigg|_{y_0} = f_y(x_0, y_0).$$

We say that $f_y(x_0, y_0)$ is the *slope of S in the y–direction at* (x_0, y_0, z_0).

EXAMPLE 7. Let $z = f(x,y) = 8 - x^2 - y^2$. The graph of f is a circular paraboloid S opening down. (a) Find the slopes of S in the x– and y–directions at $(1,2,3)$. (b) Find equations for the curves C_x and C_y through $(1,2,3)$ and the corresponding tangent lines T_x and T_y.

Solution. (a) In this case,

$$f_x(x,y) = -2x, \qquad f_x(1,2) = -2,$$
$$f_y(x,y) = -2y, \qquad f_y(1,2) = -4.$$

The slopes of S in the x– and y–directions are $\alpha = -2$ and $\beta = -4$. (b) The intersections of S with the vertical planes $y = 2$ and $x = 1$ are given by

$$C_x: \quad z = 4 - x^2, \qquad y = 2,$$
$$C_y: \quad z = 7 - y^2, \qquad x = 1.$$

The tangent lines to C_x and C_y at $(1,2,3)$ have the slopes $\alpha = -2$ and $\beta = -4$. So the point-slope equations for these tangent lines are

$$T_x: \quad z - 3 = -2(x-1), \qquad y = 2,$$
$$T_y: \quad z - 3 = -4(y-2), \qquad x = 1. \ \square$$

In Ex. 7, the partial derivatives $f_x(x,y) = -2x$ and $f_y(x,y) = -2y$ are continuous functions of (x,y). This means that the slopes of the surface in the x– and y–directions vary gradually as the point (x,y) varies and the corresponding point (x,y,z) with $z = f(x,y)$ moves on the surface S.

Partial Derivatives as Rates of Change

Recall that the ordinary derivative $f'(x_0)$ of a function $f(x)$ is the rate of change of $f(x)$ with respect to x at $x = x_0$. Now consider a function $f(x,y)$ of two variables. Since $f_x(x_0, y_0)$ is the ordinary derivative of $f(x, y_0)$ at $x = x_0$,

> $f_x(x_0, y_0)$ is the rate of change of $f(x, y_0)$ with respect to x at $x = x_0$.

Likewise,

> $f_y(x_0, y_0)$ is the rate of change of $f(x_0, y)$ with respect to y at $y = y_0$.

EXAMPLE 8. Imagine that the xy–plane, with distances measured in feet, is a giant pancake griddle with temperature at (x, y) given by

$$T(x,y) = 375 + \frac{24}{\pi} \cos \frac{\pi x}{4} \sin \frac{\pi y}{4}$$

in degrees Fahrenheit. The trigonometric term models variations from a uniform temperature of 375° The partial derivatives of $T(x, y)$ are

$$\frac{\partial T}{\partial x} = -6 \sin \frac{\pi x}{4} \sin \frac{\pi y}{4}, \qquad \frac{\partial T}{\partial y} = 6 \cos \frac{\pi x}{4} \cos \frac{\pi y}{4}.$$

At the point (1,3),

$$\frac{\partial T}{\partial x}(1,3) = 3 \text{ deg/ft}, \qquad \frac{\partial T}{\partial y}(1,3) = -3 \text{ deg/ft}.$$

These are rates of change of temperature with respect to distance. As you walk through the point (1,3) in the positive x–direction (wearing asbestos shoes), the temperature at your feet increases 3 deg/ft. As you walk through (1,3) in the positive y–direction, the temperature decreases 3 deg/ft. □

Functions of Three or More Variables

Partial derivatives of functions of three or more variables involve nothing really new. A pair of examples will tell the story. First, let $f(x, y, z) = x^2 y^3 \sin 4z$. Then

$$f_x = 2x\, y^3 \sin 4z, \qquad f_y = 3x^2 y^2 \sin 4z, \qquad f_z = 4x^2 y^3 \cos 4z.$$

Now let

$$f(x, y, z, t) = e^{x^2 + y^2} \cos z \ln(t + 1).$$

You might think of $f(x, y, z, t)$ as the temperature at (x, y, z) at time t. Then the rate of change of temperature with respect to time at the point (x, y, z) in space is the partial derivative

$$f_t(x, y, z, t) = \frac{e^{x^2 + y^2} \cos z}{t + 1}.$$

PROBLEMS

In Probs. 1–8, (a) find $\partial f/\partial x$ and $\partial f/\partial y$, (b) evaluate the partials at the indicated point, and (c) find the slopes of the graph of f in the x– and y– directions at the point.

1. $f(x,y) = x^3 + 4x^2y - 8$, (1,2)
2. $f(x,y) = x/y + y/x$, $(-1,1)$
3. $f(x,y) = 1/(x^2 + y^2)$, (2,3)
4. $f(x,y) = \ln\sqrt{x^2 + y^2}$, (0,1)

5. $f(x,y) = \arctan(y/x)$, $(3,4)$
6. $f(x,y) = e^{-x/y}$, $(\ln 2, -2)$
7. $f(x,y) = x^2 e^{x \sin y}$, $(2,0)$
8. $f(x,y) = y^x$, $(3,-2)$

In Probs. 9–14 find z_x and z_y.

9. $z = e^{x-y}$
10. $z = \sin(x - 2y)$
11. $z = e^x \cos y$
12. $z = e^{-x} \ln y$
13. $z = \sinh(x + 3y)$
14. $z = \cosh xy^2$

In Probs. 15–20 find the partial derivatives of the given functions.

15. $f(x,y,z) = xy + yz + xz$
16. $f(x,y,z) = \ln(xyz)$
17. $w = e^{xyz}$
18. $w = 1/\sqrt{x^2 + y^2 + z^2}$
19. $w = (4\pi t)^{-3/2} e^{-(x^2+y^2+z^2)/4t}$
20. $w = \sqrt{t^2 - x^2 - y^2 - z^2}$
21. The equation $4x^2 + 3y^2 - z^2 = 0$ graphs as an elliptic cone and determines two functions z of x and y that have partial derivatives for $(x,y) \neq (0,0)$. (a) Use implicit partial differentiation to find the partial derivatives. (b) Evaluate them when $(x,y) = (2,0)$.
22. The point $(0, \pi, 1/2)$ satisfies the equation $e^{xz} - \sin xyz = 1$. Assume the equation determines z as a function of x and y near $(0, \pi, 1/2)$ and that z has partial derivatives. (a) Find these partial derivatives. (b) Evaluate them at $(0, \pi, 1/2)$.
23. The point $(1,1,2)$ satisfies the equation $z^5 - 4(x + y)z^2 = y^2 - x^2$ Assume the equation determines z as a function of x and y near $(1,1,2)$ and that z has partial derivatives. (a) Find these partial derivatives. (b) Find the slopes of the graph of z in the x–direction and y–direction at $(1,1,2)$.
24. The point $(4,1,0)$ satisfies the equation $e^z - y = \ln[(1 + xz)/y]$. Assume the equation determines z as a function of x and y near $(4,1,0)$ and that z has partial derivatives. (a) Find these partial derivatives. (b) Find the slopes of the graph of z in the x–direction and y–direction at $(4,1,0)$.
25. The change of variables from polar coordinates to rectangular coordinates $x = r\cos\theta$, $y = r\sin\theta$ expresses x and y as functions of r and θ. (a) Find the partial derivatives of x and y with respect to r and θ. (b) Use $r = \sqrt{x^2 + y^2}$ and $\theta = \arctan y/x$ to find the partials of r and θ with respect to x and y.
26. Let $x = e^r \cos\theta$, $y = e^r \sin\theta$. (a) Find the partial derivatives of x and y with respect to r and θ. (b) Solve for r and θ as functions of x and y. Then calculate the partial derivatives of r and θ with respect to x and y. (c) Compare $\partial x/\partial r$ and $1/(\partial r/\partial x)$; observe that they are *not* equal.
27. In (b) of the previous problem, you showed directly that the equations $x = e^r \cos\theta$, $y = e^r \sin\theta$ determine r and θ as functions of x and y. Use implicit partial differentiation, rather than the explicit formulas of the last problem, to find (a) $\partial r/\partial x$ and $\partial\theta/\partial x$ and (b) $\partial r/\partial y$ and $\partial\theta/\partial y$.
28. Verify that $z = (x^2 - y^2)/xy$ satisfies the partial differential equation $xz_x + yz_y = 0$.
29. Verify that $z = y^2/x$ satisfies the partial differential equation $xz_x + yz_y = z$.

30. Verify that $w = (xz + y^2)/yz$ satisfies the partial differential equation $xw_x + yw_y + zw_z = 0$.

31. A point moves on the trace of the ellipsoid $9x^2 + 16y^2 + 4z^2 = 144$ with the plane $y = 1$. Find the rate of change of z with respect to x when $x = 2$ and $z > 0$.

32. A point moves on the trace of the hyperboloid $9x^2 + 16y^2 - 4z^2 = 144$ with the plane $x = 2$. Find the rate of change of z with respect to y when $y = 3$.

33. The temperature in degrees Fahrenheit at position x at time t in a laterally insulated rod is $T(x,t) = e^{-t/4}\sin(x/2)$. Measure distance in inches and time in seconds. (a) Find the rate of change of temperature with respect to distance along the rod when $x = 4$ and $t = 1$. (b) Find the rate of change of temperature with respect to time when $x = 4$ and $t = 1$.

34. The temperature in degrees Fahrenheit at (x,y) in a laterally insulated plate is $T(x,y,t) = 375 + (24/\pi)e^{-\pi^2 t/8}\cos(\pi x/4)\,\sin(\pi y/4)$. Measure distances in inches and time in seconds. (a) Find the rate of change of temperature with respect to distance in the x–direction and in the y–direction at the point $(3,5)$ and at the times $t = 1$ and $t = 3$. (b) Find the rate of change of temperature with respect to time at the point $(3,5)$ and at the times $t = 1$ and $t = 3$.

For many purposes a gas can be adequately described by its pressure P, volume V, and temperature T. These variables are related by an *equation of state* of the form $f(P,V,T) = 0$. The form of the function f actually specifies a particular gas. For an *ideal gas* $f(P,V,T) = PV - nRT$, where n is the number of moles of the gas and $R > 0$ is the *ideal gas constant.* Real gases obey this law reasonably well when the density is low.

35. The ideal gas law determines each of P, V, and T as functions of the other two variables. Show that
$$\frac{\partial P}{\partial V}\cdot\frac{\partial V}{\partial T}\cdot\frac{\partial T}{\partial P} = -1.$$
Notice that partial derivatives don't "cancel" like ordinary derivatives do.

36. Suppose an equation of state $f(P,V,T) = 0$ determines each of the three variable as functions of the other two. Show that the partial derivative relation in the previous problem still holds.

37. The capacity of a gas to hold heat is expressed by its specific heat. Actually there are two useful specific heats, C_P and C_V. The subscripts refer to the heat capacity per mole of the gas at constant pressure and at constant volume, respectively. It is shown in thermodynamics that
$$C_P - C_V = T\frac{\partial P}{\partial T}\cdot\frac{\partial V}{\partial T}$$
for one mole of a gas. Show, for an ideal gas, that $C_P - C_V = R$, the ideal gas constant. Conclude that $C_P > C_V$ for an ideal gas.

Van der Waal's equation of state
$$\left(P + \frac{a}{V^2}\right)(V - b) = nRT$$

is derived to better account for the behavior of gases at higher densities. Here a, b are empirical constants determined by particular gases.

38. Find $C_P - C_V$ for a gas that satisfies Van der Waal's equation of state.

39. *Calandar's equation*,

$$PV = nRT + P(b - cT^{\gamma}),$$

where b, c, and γ are positive constants, is a reasonable equation of state for ordinary air. Find $C_P - C_V$ in this case.

The following problem shows that the existence of partial derivatives at a point is weaker than you might at first think. Compare with the one-variable result: If $f'(x_0)$ exists, then f is continuous at x_0. This fact and the next problem suggest that partial derivatives are not the analogues of "the derivative" for functions of several variables. Stay tuned.

40. Let

$$f(x,y) = \begin{Bmatrix} 2xy/(x^2 + y^2), & (x,y) \neq (0,0) \\ 0, & (x,y) = (0,0) \end{Bmatrix}.$$

Evidently f is continuous, except perhaps at the origin. (a) Explain why f is not continuous at the origin. (b) Nevertheless, show that f has partial derivatives at $(0,0)$; in fact, show that $f_x(0,0) = 0$ and $f_y(0,0) = 0$.

5.4 Tangent Planes, Linear Approximations, and Differentials

Tangent lines, linear approximations, and differentials for functions of one variable were studied in earlier chapters. Here we take up the corresponding ideas for functions of two and three variables.

Tangent Planes

The concept of a tangent plane to a sufficiently smooth surface is illustrated by an egg lying on a table. The plane of the table is tangent to the surface of the egg (the eggshell). You should be able to imagine the tangent plane to the eggshell at any other point of contact just as easily.

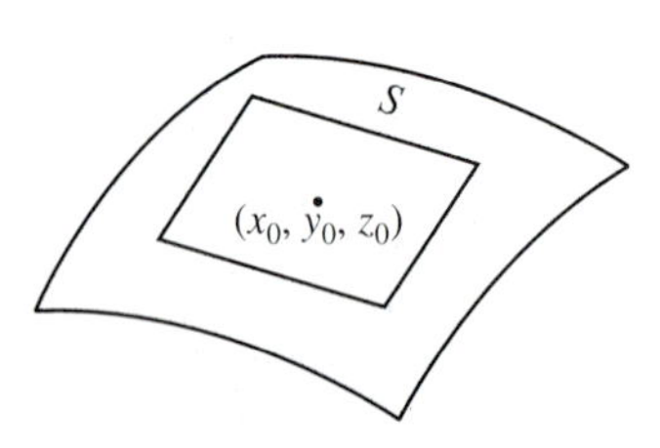

FIGURE 1

Let S be the graph of a function $z = f(x,y)$. Let $z_0 = f(x_0,y_0)$. Then (x_0,y_0,z_0) is a point on S. In order for S to have a tangent plane Π at (x_0,y_0,z_0), as illustrated in Fig. 1, the function $f(x,y)$ must be reasonably well behaved near the point of tangency. The following assumptions will enable us define the tangent plane. Let (x_0,y_0) be an interior point of the domain of f. Assume that the partial derivatives $f_x(x,y)$ and $f_y(x,y)$ exist for (x,y) near (x_0,y_0) (meaning for all (x,y) in some neighborhood of (x_0,y_0)) and that the partial derivatives are continuous at (x_0,y_0). Then

$$f_x(x,y) \to f_x(x_0,y_0) \text{ and } f_y(x,y) \to f_y(x_0,y_0) \text{ as } (x,y) \to (x_0,y_0).$$

Most of the functions we deal with have continuous partial derivatives at all points in their domains. So their graphs have tangent planes at all points.

A cutaway drawing of S is shown in Fig. 2. A similar figure appeared in Sec. 5.3. The curve C_x is the intersection of the surface S with the vertical plane

$y = y_0$ and T_x is the tangent line to C_x at (x_0,y_0,z_0). As we learned in Sec. 5.3, the slope of T_x is

$$\alpha = f_x(x_0,y_0).$$

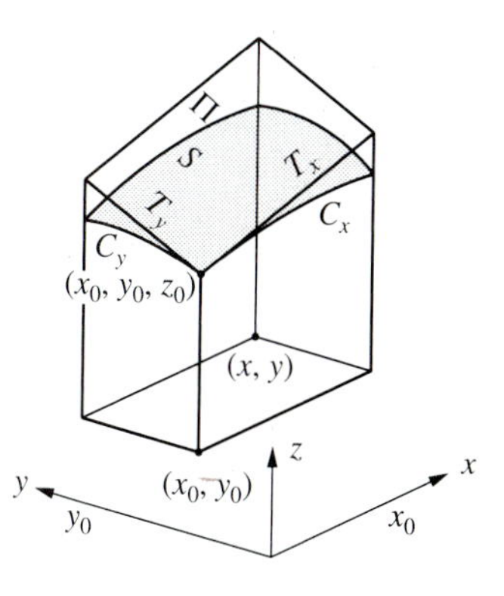

FIGURE 2

The curve C_y in Fig. 2 is the intersection of the surface S with the vertical plane $x = x_0$ and T_y is the tangent line to C_y at (x_0, y_0, z_0). The slope of T_y is

$$\beta = f_y(x_0,y_0).$$

There is a unique plane Π containing the two lines T_x and T_y. It has the point–slope equation

$$z - z_0 = \alpha(x - x_0) + \beta(y - y_0).$$

It will be shown later that Π contains *every* tangent line to a curve obtained by intersecting S with a vertical plane through (x_0,y_0,z_0). So it is reasonable to call Π the tangent plane to S at (x_0,y_0,z_0). The following definition summarizes our conclusions.

> **Definition** *Tangent Plane to a Surface $z = f(x,y)$*
> Assume that $f_x(x,y)$ and $f_y(x,y)$ exist for (x,y) near (x_0,y_0) and are continuous at (x_0,y_0). Let $z_0 = f(x_0,y_0)$. Then the **tangent plane** to the surface $z = f(x,y)$ at the point (x_0,y_0,z_0) is the plane with point-slope equation
> $$z - z_0 = \alpha(x - x_0) + \beta(y - y_0),$$
> where $\alpha = f_x(x_0,y_0)$ and $\beta = f_y(x_0,y_0)$.

Tangent planes to surfaces given by $x = g(y,z)$ and $y = h(x,z)$ are defined similarly.

A little later in this section and also in the next section we shall explore how well a tangent plane approximates a surface near the point of contact. But first we give examples of tangent planes to given surfaces.

EXAMPLE 1. Find an equation for the tangent plane to the paraboloid $z = x^2 + y^2$ at the point $(1,2,5)$.

Solution. In this case,

$$\alpha = \frac{\partial z}{\partial x} = 2x = 2 \quad \text{and} \quad \beta = \frac{\partial z}{\partial y} = 2y = 4 \quad \text{at} \quad (1,2).$$

So the point-slope equation for the tangent plane is

$$z - 5 = 2(x - 1) + 4(y - 2),$$

which reduces to $z = 2x + 4y - 5$. □

EXAMPLE 2. Find an equation for the tangent plane to the ellipsoid $4x^2 + 3y^2 + 2z^2 = 9$ at the point $(1, 1, 1)$.

Solution. We could solve for $z > 0$ in terms of x and y, and then find the partial derivatives. It is easier to find $\partial z/\partial x$ and $\partial z/\partial y$ by implicit differentiation:

$$8x + 4z \frac{\partial z}{\partial x} = 0, \qquad \frac{\partial z}{\partial x} = -\frac{2x}{z} = -2 \text{ at } (1,1,1),$$

$$6y + 4z \frac{\partial z}{\partial y} = 0, \qquad \frac{\partial z}{\partial y} = -\frac{3y}{2z} = -\frac{3}{2} \text{ at } (1,1,1).$$

Equations for the tangent plane are

$$z - 1 = -2(x-1) - \frac{3}{2}(y-1) \text{ and } 4x + 3y + 2z = 9. \ \square$$

Linear Approximations

Linear approximations for functions of two or three variables follow the same pattern as for functions of one variable. The graph of a linear approximation for a function $f(x)$ is a tangent line to the curve $y = f(x)$. Correspondingly, the graph of a linear approximation for a function $f(x, y)$ is a tangent plane to the surface $z = f(x, y)$. The situation for a function of three variables, $f(x, y, z)$, is similar. We'll come to it later.

Functions of One Variable

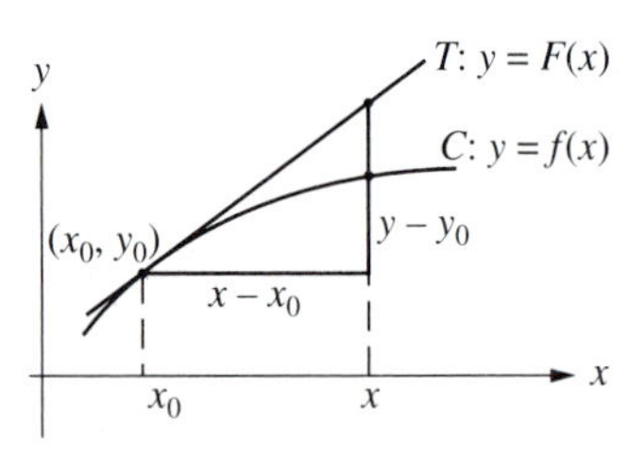

FIGURE 3

We begin with linear approximations for functions of one variable. Here we look at some ideas covered earlier in calculus from a slightly different point of view in order to tighten the correspondence between linear approximations for functions of one, two, and three variables. Essentially the same reasoning, based on the mean value theorem, is used in all three cases.

The curve C in Fig. 3 is the graph of a function $y = f(x)$. Let $y_0 = f(x_0)$. Assume that $f'(x)$ exists for x near x_0 and is continuous at x_0. The **linear approximation** for $f(x)$ at x_0 is the linear function

$$F(x) = f(x_0) + f'(x_0)(x - x_0).$$

It satisfies

$$F(x_0) = f(x_0), \qquad F'(x_0) = f'(x_0).$$

The graph of $y = F(x)$ is the tangent line T to the curve C in Fig. 3.

The **error of linear approximation** is

$$f(x) - F(x) = f(x) - f(x_0) - f'(x_0)(x - x_0).$$

In Fig. 3, $|f(x) - F(x)|$ is the vertical distance between the curve and the tangent line. The figure suggests that, for x near x_0, the vertical distance is small compared with the distance from x to x_0. This is easy to confirm. By the mean value theorem, $f(x) - f(x_0) = f'(c)(x - x_0)$ for some c between x_0 and x. It follows that

$$f(x) - F(x) = [f'(c) - f'(x_0)](x - x_0).$$

Let $x \to x_0$. Then, since $x_0 < c < x$ and f' is continuous at x_0, $c \to x_0$ and $f'(c) \to f'(x_0)$. Let $\varepsilon(x) = [f'(c) - f'(x_0)]$. Then

$$f(x) - F(x) = \varepsilon(x)(x - x_0), \text{ where } \varepsilon(x) \to 0 \text{ as } x \to x_0.$$

Consequently,

$$\frac{f(x) - F(x)}{x - x_0} \to 0 \quad \text{and} \quad \frac{|f(x) - F(x)|}{|x - x_0|} \to 0 \quad \text{as} \quad x \to x_0.$$

As expected, if x is near x_0, then the vertical distance $|f(x) - F(x)|$ between the graphs in Fig. 3 is small compared with the horizontal distance $|x - x_0|$ from x to x_0.

Functions of Two Variables

Again let the surface S be the graph of $z = f(x,y)$ in Fig. 2. Continue to assume that the partial derivatives $f_x(x,y)$ and $f_y(x,y)$ exist for (x,y) near (x_0,y_0), and are continuous at (x_0,y_0). Then the tangent plane Π to S at (x_0,y_0,z_0), where $z_0 = f(x_0,y_0)$, has the point-slope equation

$$z - z_0 = f_x(x_0,y_0)(x - x_0) + f_y(x_0,y_0)(y - y_0).$$

Another equation for Π is $z = F(x,y)$, where $F(x,y)$ is the linear function

$$\boxed{F(x,y) = f(x_0,y_0) + f_x(x_0,y_0)(x - x_0) + f_y(x_0,y_0)(y - y_0).}$$

We call $F(x,y)$ the **linear approximation** for $f(x,y)$ at (x_0,y_0). It satisfies

$$F(x_0,y_0) = f(x_0,y_0), \quad F_x(x_0,y_0) = f_x(x_0,y_0), \quad F_y(x_0,y_0) = f_y(x_0,y_0).$$

The **error of linear approximation** is $f(x,y) - F(x,y)$.

To reduce clutter and highlight the geometry, we shall use the familiar notation $P = (x,y)$ and $P_0 = (x_0,y_0)$ whenever it is advantageous to do so. Then the linear approximation for $f(P)$ is $F(P)$ and the error of linear approximation is $f(P) - F(P)$. The vertical distance between the surface S and the tangent plane Π in Fig. 2 is $|f(P) - F(P)|$. The figure suggests that, for P near P_0, the vertical distance is small compared with the distance from P to P_0, which is

$$|PP_0| = \sqrt{(x - x_0)^2 + (y - y_0)^2}.$$

Before making a general statement to that effect, let's look at an example.

EXAMPLE 3. Let $f(x,y) = 3x^2 + 2y^2$. The graph of f is an elliptic paraboloid opening up. Let $(x_0,y_0) = (1,2)$ and $z_0 = f(1,2) = 11$. Then

$$f_x(x,y) = 6x, \qquad f_y(x,y) = 4y,$$
$$f_x(1,2) = 6, \qquad f_y(1,2) = 8.$$

The linear approximation for $f(x,y)$ at $(1,2)$ is

$$F(x,y) = 11 + 6(x-1) + 8(y-2) = 6x + 8y - 11.$$

The error of linear approximation is

$$f(x,y) - F(x,y) = 3x^2 + 2y^2 - 6x - 8y + 11.$$

Complete the squares on the right to obtain

$$f(x,y) - F(x,y) = 3(x-1)^2 + 2(y-2)^2.$$

Now let $P=(x,y)$ and $P_0=(x_0,y_0)=(1,2)$. Then $|PP_0|=\sqrt{(x-1)^2+(y-2)^2}$ and

$$0 \leq f(P) - F(P) \leq 3|PP_0|^2,$$
$$\frac{f(P)-F(P)}{|PP_0|} \to 0 \text{ as } P \to P_0. \ \square$$

The conclusions of Ex. 3 are typical of the general situation.

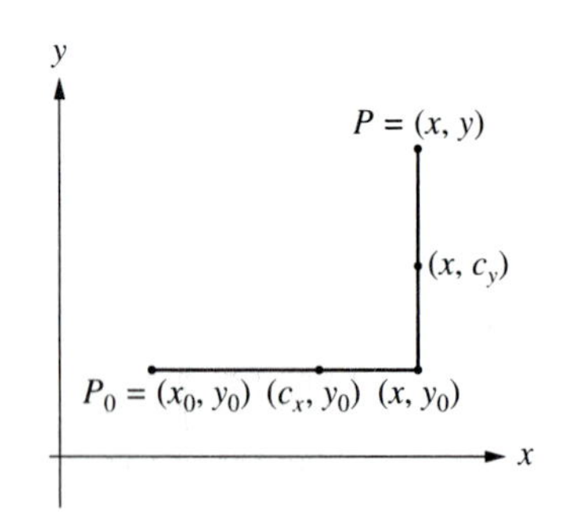

FIGURE 4

Theorem 1 *Error of Linear Approximation for $f(x,y)$*
Assume that $f(P)$ has partial derivatives for P near P_0, which are continuous at P_0. Let $F(P)$ be the linear approximation for f at P_0.
Then

$$f(P) - F(P) = \varepsilon_1(P)(x - x_0) + \varepsilon_2(P)(y - y_0),$$

where $\varepsilon_1(P) \to 0$ and $\varepsilon_2(P) \to 0$ as $P \to P_0$. Furthermore,

$$\frac{f(P)-F(P)}{|PP_0|} \to 0 \text{ as } P \to P_0.$$

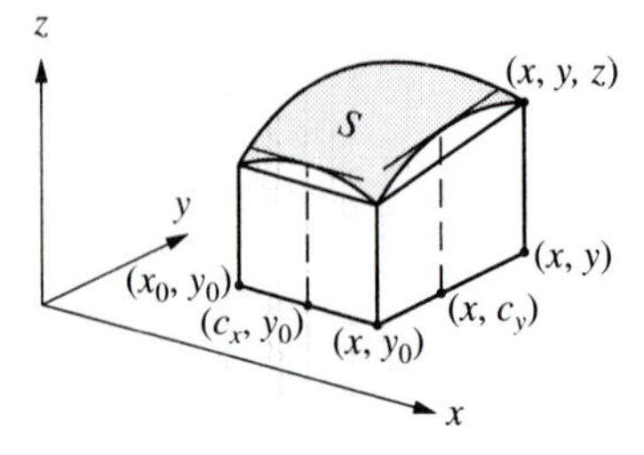

FIGURE 5

Proof. Our reasoning is guided by Figs. 4 and 5. Note that

$$f(x,y) = f(x_0,y_0) + [f(x,y_0) - f(x_0,y_0)] + [f(x,y) - f(x,y_0)].$$

On the right, $f(x_0,y_0)$ and $f(x,y_0)$ cancel out. Consider $f(x,y_0) - f(x_0,y_0)$, which is a function of x alone. Since $f_x(x,y_0)$ is the ordinary derivative of $f(x,y_0)$, the mean value theorem yields

$$f(x,y_0) - f(x_0,y_0) = f_x(c_x,y_0)(x - x_0)$$

for some c_x between x_0 and x. See Figs. 4 and 5. Similarly,

$$f(x,y) - f(x,y_0) = f_y(x,c_y)(y - y_0)$$

for some c_y between y_0 and y. It follows that

$$f(x,y) = f(x_0,y_0) + f_x(c_x,y_0)(x - x_0) + f_y(x,c_y)(y - y_0).$$

Since

$$F(x,y) = f(x_0,y_0) + f_x(x_0,y_0)(x - x_0) + f_y(x_0,y_0)(y - y_0),$$

subtraction yields

$$f(x,y) - F(x,y) = \varepsilon_1(x,y)(x - x_0) + \varepsilon_2(x,y)(y - y_0),$$

where

$$\varepsilon_1(x,y) = f_x(c_x,y_0) - f_x(x_0,y_0), \qquad \varepsilon_2(x,y) = f_y(x,c_y) - f_y(x_0,y_0).$$

Let $(x,y) \to (x_0,y_0)$. In view of Fig. 4, $(c_x,y_0) \to (x_0,y_0)$ and $(x,c_y) \to (x_0,y_0)$. Since f_x and f_y are continuous at (x_0,y_0),

$$\varepsilon_1(x,y) \to 0 \quad \text{and} \quad \varepsilon_2(x,y) \to 0 \quad \text{as} \quad (x,y) \to (x_0,y_0).$$

To express our findings more succinctly, let $P = (x,y)$ and $P_0 = (x_0,y_0)$. Then

$$f(P) - F(P) = \varepsilon_1(P)(x - x_0) + \varepsilon_2(P)(y - y_0),$$

where $\varepsilon_1(P) \to 0$ and $\varepsilon_2(P) \to 0$ as $P \to P_0$, which proves the first assertion in the theorem.

Either from Fig. 4 or from $|PP_0| = \sqrt{(x - x_0)^2 + (y - y_0)^2}$,

$$|x - x_0| \le |PP_0| \text{ and } |y - y_0| \le |PP_0|.$$

It follows that $|f(P) - F(P)| \le |PP_0|(|\varepsilon_1(P)| + |\varepsilon_2(P)|)$ and

$$\frac{|f(P) - F(P)|}{|PP_0|} \le |\varepsilon_1(P)| + |\varepsilon_2(P)| \to 0 \text{ as } P \to P_0,$$

which proves the second and final assertion of the theorem. □

EXAMPLE 4. Let $f(x,y) = e^{x^2+y^2}$. The graph of f is obtained by rotating the curve $z = e^{x^2}$ around the z–axis. The surface is similar to a paraboloid, but rises much more steeply. It should be evident that the tangent plane to the surface at $(0,0,1)$ is the horizontal plane $z = 1$ and, therefore, the linear approximation for $f(x,y)$ at $(0,0,1)$ is $F(x,y) = 1$. Verify this observation and the final conclusion of Th. 1.

Solution. In this case, $f(0,0) = 1$ and

$$f_x(x,y) = 2x\,e^{x^2}, \qquad f_y(x,y) = 2y\,e^{y^2},$$
$$f_x(0,0) = 0, \qquad f_y(0,0) = 0.$$

Therefore,

$$F(x,y) = f(0,0) + f_x(0,0)(x-0) + f_y(0,0)(y-0) = 1,$$

and the tangent plane has equation $z = 1$, just as expected. Let $P = (x,y)$, $P_0 = (0,0)$, and $r = |PP_0| = \sqrt{x^2+y^2}$. Then $f(P) = e^{r^2}$ and

$$\frac{f(P)-F(P)}{|PP_0|} = \frac{e^{r^2}-1}{r},$$

which is indeterminate of the form 0/0 as $r \to 0$. By l'Hôpital's rule,

$$\lim_{r\to 0}\frac{e^{r^2}-1}{r} = \lim_{r\to 0}\frac{2re^{r^2}}{1} = 0.$$

Therefore,

$$\frac{f(P)-F(P)}{|PP_0|} \to 0 \text{ as } |PP_0| \to 0,$$

in agreement with Th. 1. □

Functions of Three Variables

Linear approximations for functions of three variables are defined in basically the same way as for functions of two variables. So we shall be brief. Whenever convenient, the standard notation $P = (x,y,z)$ and $P_0 = (x_0,y_0,z_0)$ will be used.

A function $w = \alpha x + \beta y + \gamma z + \delta$, with any constants α, β, γ, and δ, will be called *linear*. The graph of such a linear function, which we cannot draw, is called a hyperplane. It is a four–dimensional counterpart of a plane in 3–space. A linear function which satisfies $w = w_0$ for $P = P_0$ is given by

$$w = w_0 + \alpha(x-x_0) + \beta(y-y_0) + \gamma(z-z_0).$$

This is a four–dimensional analogue of the point-slope equation for a line in 2–space or a plane in 3–space.

Given a function $f(P) = f(x,y,z)$ and an interior point $P_0 = (x_0,y_0,z_0)$ of the domain of f, assume that the partial derivatives $f_x(P)$, $f_y(P)$, and $f_z(P)$ exist for P near P_0 and are continuous at P_0. The **linear approximation** for $f(P)$ at P_0 is the linear function

$$F(P) = f(P_0) + f_x(P_0)(x-x_0) + f_y(P_0)(y-y_0) + f_z(P_0)(z-z_0).$$

It satisfies

$$F(P_0) = f(P_0), \qquad F_x(P_0) = f_x(P_0), \qquad F_y(P_0) = f_y(P_0), \qquad F_z(P_0) = f_z(P_0).$$

As before, the **error of linear approximation** is $f(P) - F(P)$.

EXAMPLE 5. Let $f(P) = \ln\sqrt{x^2+y^2+z^2}$. Find the linear approximation $F(P)$ for $f(P)$ at $P_0 = (1,2,2)$.

Solution. First, $f(P_0) = \ln 3$. Next, since $f(P) = \frac{1}{2}\ln(x^2+y^2+z^2)$, we obtain

$$f_x(P) = \frac{x}{x^2+y^2+z^2}, \qquad f_y(P) = \frac{y}{x^2+y^2+z^2}, \qquad f_z(P) = \frac{z}{x^2+y^2+z^2},$$

$$f_x(P_0) = \frac{1}{9}, \qquad f_y(P_0) = \frac{2}{9}, \qquad f_z(P_0) = \frac{2}{9}.$$

So the linear approximation for f at $P_0 = (1,2,2)$ is

$$F(P) = \ln 3 + \frac{1}{9}(x-1) + \frac{2}{9}(y-2) + \frac{2}{9}(z-2).$$

Since $F(P_0) = f(P_0) = \ln 3$ and both functions $F(P)$ and $f(P)$ are continuous at P_0, $f(P) - F(P) \to 0$ as $P \to P_0$. □

The error of linear approximation, $f(P) - F(P)$, is estimated in the next theorem.

Theorem 2 *Error of Linear Approximation for $f(x,y,z)$*
Assume that $f(P)$ has partial derivatives for P near P_0, which are continuous at P_0. Let $F(P)$ be the linear approximation for f at P_0. Then

$$f(P)-F(P)=\varepsilon_1(P)(x-x_0)+\varepsilon_2(P)(y-y_0)+\varepsilon_3(P)(z-z_0),$$

where $\varepsilon_1(P), \varepsilon_2(P), \varepsilon_3(P) \to 0$ as $P \to P_0$. Furthermore,

$$\frac{f(P)-F(P)}{|PP_0|} \to 0 \text{ as } P \to P_0.$$

The proof is essentially the same as for Th. 1. The key steps are outlined in the problems.

Differentials

Differentials describe linear approximations with different notation. Once again the case for a function of one variable serves as a model. So we begin with a brief review.

Functions of One Variable

In Fig. 6, which is the same as Fig. 3 but with new labels, the curve C is the graph of a differentiable function $y = f(x)$ and T is the tangent line to C at (x,y). The change in y corresponding to a change Δx in x is $\Delta y = f(x+\Delta x) - f(x)$. The differential of $y = f(x)$ is given by

$$dy = f'(x)\Delta x \text{ or } df = f'(x)\Delta x.$$

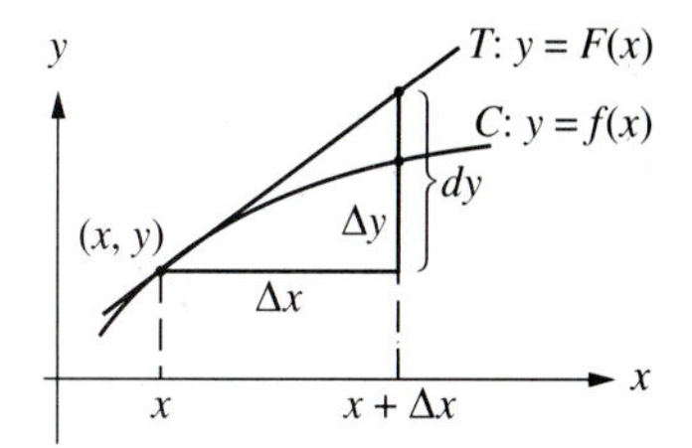

FIGURE 6

The differential dy is a good approximation for Δy when Δx is small. In Fig. 6, dy would be the change in y if the curve C were replaced by the tangent line T. The error of linear approximation is $\Delta y - dy$. It satisfies

$$\Delta y - dy = \varepsilon \Delta x, \qquad \text{where } \varepsilon \to 0 \text{ as } \Delta x \to 0.$$

This is just our earlier result, $f(x) - F(x) = \varepsilon(x)(x - x_0)$, expressed in differential notation.

When dealing with differentials, it is common practice to denote a change in x by dx instead of Δx. Then $dy = f'(x)\, dx$.

Functions of Two Variables

Now let $z = f(x,y)$ and assume that $f(x,y)$ has continuous partial derivatives. In Fig. 7, which is a copy of Fig. 2 with different labels, the surface S is the graph of f and Π is the tangent plane to S at (x,y,z). The change in z corresponding to changes Δx and Δy in x and y is $\Delta z = f(x + \Delta x, y + \Delta y) - f(x,y)$. The **differential** of $z = f(x,y)$ is given by

$$dz = \frac{\partial z}{\partial x}\Delta x + \frac{\partial z}{\partial y}\Delta y \text{ or } df = f_x(x,y)\,\Delta x + f_y(x,y)\,\Delta y.$$

The differential dz is a good approximation for Δz when Δx and Δy are small. Then

$$\Delta z \approx dz = \frac{\partial z}{\partial x}\Delta x + \frac{\partial z}{\partial y}\Delta y.$$

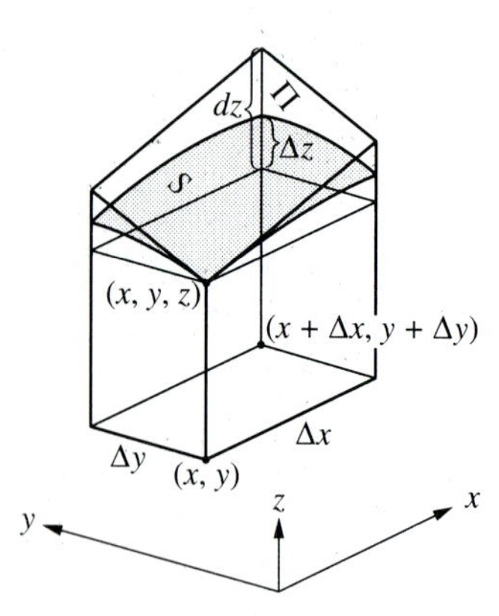

FIGURE 7

This is called the *differential approximation* for Δz.

In Fig. 7, dz would be the change in z if the surface S were replaced by the tangent plane Π. The error of linear approximation is $\Delta z - dz$ in differential notation. The vertical distance between the surface S and the tangent plane Π in Fig. 7 is $\Delta z - dz$. From Th. 1, with change to differential notation,

$$\Delta z - dz = \varepsilon_1 \Delta x + \varepsilon_2 \Delta y,$$

where $\varepsilon_1 \to 0$ and $\varepsilon_2 \to 0$ as $(\Delta x, \Delta y) \to (0,0)$. Hence, Th. 3 follows immediately from Th. 1.

Theorem 3 *Error of the Differential Approximation*
Assume that $f(x,y)$ has continuous partial derivatives for (x,y) in an open set. Then

$$\Delta z = \frac{\partial z}{\partial x}\Delta x + \frac{\partial z}{\partial y}\Delta y + \varepsilon_1 \Delta x + \varepsilon_2 \Delta y,$$

where $\varepsilon_1 \to 0$ and $\varepsilon_2 \to 0$ as $(\Delta x, \Delta y) \to (0,0)$.

If changes in x and y are denoted by dx and dy rather than Δx and Δy, then the differential dz is given by

$$\boxed{dz = \frac{\partial z}{\partial x} dx + \frac{\partial z}{\partial y} dy.}$$

Functions of Three Variables

Differentials for functions of three variables are defined as expected. The **differential** of $w = f(P)$ is given by

$$dw = \frac{\partial w}{\partial x}\Delta x + \frac{\partial w}{\partial y}\Delta y + \frac{\partial w}{\partial z}\Delta z \quad \text{or} \quad df = f_x(P)\,\Delta x + f_y(P)\,\Delta y + f_z(P)\,\Delta z.$$

The change in w from $P = (x, y, z)$ to $Q = (x + \Delta x, \Delta y + \Delta y, \Delta z + \Delta z)$ is $\Delta w = f(Q) - f(P)$. The differential approximation for Δw is

$$\Delta w \approx dw = \frac{\partial w}{\partial x}\Delta x + \frac{\partial w}{\partial y}\Delta y + \frac{\partial w}{\partial z}\Delta z.$$

The error of linear approximation is $\Delta w - dw$. It satisfies

$$\Delta w - dw = \varepsilon_1 \Delta x + \varepsilon_2 \Delta y + \varepsilon_3 \Delta z, \text{ where}$$
$$\varepsilon_1 \to 0, \quad \varepsilon_2 \to 0, \quad \varepsilon_3 \to 0 \text{ as } (\Delta x, \Delta y, \Delta z) \to (0,0,0).$$

If changes in x, y, and z are denoted by dx, dy, and dz, then the differential dw is given by

$$dw = \frac{\partial w}{\partial x} dx + \frac{\partial w}{\partial y} dy + \frac{\partial w}{\partial z} dz.$$

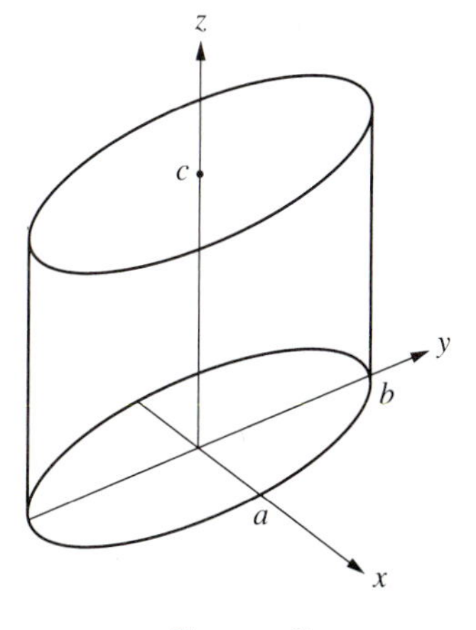

FIGURE 8

EXAMPLE 6. A solid metal cylinder with elliptical cross-sections is shown in Fig. 8. The ellipse in the xy–plane is given by

$$\frac{x^2}{a^2} + \frac{y^2}{b^2} = 1.$$

It has area πab. The volume of the cylinder is $V = \pi abc$. If $a = 2$, $b = 3$, and $c = 5$, then $V = 30\pi$. After being in the hot sun for a few hours, the dimensions of the cylinder increase by the amounts $da = 0.01$, $db = 0.015$, and $dc = 0.025$. The new volume is $V + \Delta V = 30.452\pi$, correct to three decimal places. So the change in volume is $\Delta V = 0.452\pi$. Let's compare this with the differential approximation

$$\begin{aligned} dV &= \frac{\partial V}{\partial a} da + \frac{\partial V}{\partial b} db + \frac{\partial V}{\partial c} dc \\ &= 15\pi(0.01) + 10\pi(0.015) + 6\pi(0.025) = 0.45\pi. \end{aligned}$$

Thus, $\Delta V - dV \approx 0.002\pi$, which is very small compared with the changes in a, b, c. □

PROBLEMS

In Probs. 1–10 find an equation for the tangent plane to the surface at the given point.

1. The hyperbolic paraboloid $z = xy$ at $(2, 3, 6)$
2. The surface $z = e^{-xy}$ at $(1/2,\ 1/3,\ e^{-1/6})$
3. The sphere $x^2 + y^2 + z^2 = 25$ at $(3, 0, 4)$
4. The surface $z = \arctan(y/x)$ at $(2, -2, -\pi/4)$
5. The ellipsoid $2x^2 + 3y^2 + 4z^2 = 48$ at $(4, 2, 1)$
6. The elliptic paraboloid $2x^2 + 3y^2 - 4z = 46$ at $(3, 4, 5)$
7. The elliptic cone $x^2 + 5y^2 - z^2 = 0$ at $(4, 2, 6)$
8. The cylinder $x^2 + z^2 = 25$ at $(3, -2, 4)$
9. The surface $x^{2/3} + y^{2/3} + z^{2/3} = 21$ at $(-1, -8, 64)$
10. The surface $z = x^y$ at $(2, -1, 1/2)$
11. Let $P_0 = (x_0, y_0, z_0)$ be a point on the ellipsoid $x^2/a^2 + y^2/b^2 + z^2/c^2 = 1$. Show that the tangent plane at P_0 has equation $x_0x/a^2 + y_0y/b^2 + z_0z/c^2 = 1$.
12. Let $a > 0$ and $P_0 = (x_0, y_0, z_0)$ be a point with all positive coordinates that lies on the surface $xyz = a$. The tangent plane to the surface at P_0 cuts a tetrahedron from the first octant. Show that this tetrahedron has the same volume for all choices of P_0 that lie on the surface and find the common volume.
13. Let $a > 0$. Show that the sum of the squares of the intercepts of any tangent plane to the surface $x^{2/3} + y^{2/3} + z^{2/3} = a^{2/3}$ is a constant and find the constant sum.
14. Let $a > 0$. Find the analogue of the previous exercise for the surface $x^{1/2} + y^{1/2} + z^{1/2} = a^{1/2}$.

Two surfaces are tangent at a point if they have the same tangent plane at the point.

15. Show that the cone $z^2 = x^2 + y^2$ and the paraboloid $20z = 2x^2 + 2y^2 + 50$ are tangent at $(3, 4, 5)$.
16. Show that the paraboloid $z = x^2 + y^2$ and the sphere $x^2 + y^2 + z^2 - 6x + 4 = 0$ are tangent at $(1, 0, 1)$.

In Probs. 17–22 find the linear approximation to the function at the given point.

17. $f(x, y) = x^2 - xy + y^2$ at $(1, -1)$
18. $f(x, y) = x/y + y/x$ at $(1, -1)$
19. $f(x, y) = \arctan(y/x)$ at $(1, \sqrt{3})$

20. $f(x,y) = e^{-(x^2+y^2)/8}$ at $(2,3)$

21. $f(x,y,z) = xy^2 + yz^2 + zx^2$ at $(1,2,3)$

22. $f(x,y,z) = e^{x^2-y^2-2z}$ at $(3,-1,4)$

In Probs. 23–25 use reasoning like that in Ex. 3 to verify directly that

$$\frac{f(x,y) - F(x,y)}{\sqrt{(x-x_0)^2 + (y-y_0)^2}} \to 0 \text{ as } (x,y) \to (x_0 y_0)$$

for the given function and at the given point.

23. $f(x,y) = x^2 - 2y^2$ at $(1,-1)$

24. $f(x,y) = 2x^2 + 3y^2 + 4x - 2y + 5$ at $(2,1)$

25. $f(x,y) = x^2 - xy + y^2$ at $(1,-1)$

In Probs. 26–31 find the differential of the function.

26. $f(x,y) = \sqrt{x^2+y^2}$

27. $f(x,y) = x/y + y/x$

28. $z = e^{-xy}$

29. $z = \sin xy$

30. $f(x,y,z) = e^{-z}(x^2+y^2)^{-1}$

31. $w = e^{x^2+y^2-2z}$

32. If a triangle has sides a and b that include an angle θ, then the area A of the triangle is given by $A = \frac{1}{2}ab\sin\theta$. So A is a function of a, b, and θ. (a) Find dA. (b) Measurements of a triangular garden give $a = 20$ feet, $b = 15$ feet, and $\theta = 30°$. Suppose the length and angle measurements may be in error by at most 1 inch and 1/2° respectively. Use differentials to estimate the error (uncertainty) in the area of the garden. (c) Estimate the relative error in the area.

33. All the interior faces of a jewelry box with dimensions 5 by 4 by 3 inches are covered with 1/8–inch felt. Use differentials to estimate (a) the reduction in the volume of the box and (b) the relative reduction in volume.

34. The total resistance R of two resistors R_1 and R_2 connected in parallel is given by $1/R = 1/R_1 + 1/R_2$. Suppose R_1 and R_2 are rated as 100 ohms and 200 ohms, respectively, with an error of at most 1.5%. Use differentials to estimate (a) the error (uncertainty) in the total resistance R and (b) the relative uncertainty.

35. The period T of a simple pendulum executing small oscillations is $T = 2\pi\sqrt{L/g}$ where L is the length of the pendulum and g is the (roughly) constant acceleration of gravity near the surface of the earth. Actually, g varies slightly over the surface of the earth; in fact, $|\Delta g/g| \le 0.0053$ from experimental data. Also, L expands and contracts as the temperature changes. For a steel rod $|\Delta L/L| \le 0.0002$ as the temperature changes over the 36° range from freezing to room temperature. (a) Find the differential of T as a function of L and g. (b) Use differentials to estimate the relative uncertainty in T due to the relative changes in g and L above.

36. For n moles of an ideal gas, $PV = nRT$, where R is the ideal gas constant. (See the problems in Sec. 5.2.) If the volume of the gas increases by 4% and the temperature decreases by 2%, use differentials to estimate the percent change in pressure.

37. Newton's law of gravitation is $F = GmM/r^2$. (a) Use differentials to estimate the relative change in F that results from a 1% increase in each mass and a 1% decrease in the distance. (b) If m is the mass of the earth and M the mass of the sun, which change in mass in (a) is most significant or are they equally significant for the relative change in the gravitational force? Explain briefly.

38. A ship is anchored offshore at B. You are onshore at A. To estimate the distance to the ship, you lay off a baseline of $b = 500$ feet and measure $\angle CAB = \alpha = 45°$ and $\angle ACB = \gamma = 60°$. According to the law of sines $(\sin B)/b = (\sin C)/c$, where $c = |AB|$ and $b = |AC|$. Find the distance c from you to the ship and estimate the greatest error in c if distances and angles are measured accurately to the nearest 1/8 foot and 1/2°, respectively.

39. Figures 2 and 3 suggest that $f(x) \approx F(x)$ and $f(x,y) \approx F(x,y)$ for x near x_0 and (x,y) near (x_0,y_0) even before a careful error analysis is made. Justify these approximations on continuity grounds.

40. Prove Th. 2. *Hint.* Pattern your argument on the proof of Th. 1. A box with opposite vertices $P_0 = (x_0,y_0,z_0)$ and $P = (x,y,z)$ replaces Fig. 4. Let $P_1 = (x,y_0,z_0)$ and $P_2 = (x,y,z_0)$. Then

$$f(P) = f(P_0) + [f(P_1) - f(P_0)] + [f(P_2) - f(P_1)] + [f(P) - f(P_2)].$$

41. In the context of Th. 2, prove that

$$f(P) = F(P) + \varepsilon(P)|PP_0| \text{ where } \varepsilon(P) \to 0 \text{ as } P \to P_0.$$

This is the analogue of $f(x) = F(x) + \varepsilon(x)(x - x_0)$ where $\varepsilon(x) \to 0$ as $x \to x_0$ for functions of one variable.

5.5 Chain Rules and Directional Derivatives

The chain rule for functions of one variable gives a recipe for differentiating composite functions such as $f(x(t))$. Chain rules for functions of two or more variables serve the same purpose. They enable us to differentiate composite functions such as $f(x(t), y(t))$. Directional derivatives are illustrated by variable temperature $f(x,y)$ in a plane. The directional derivative of f at a particular point (x_0,y_0) is the rate of change of temperature with respect to distance away from (x_0,y_0) in that direction. As we shall see, chain rules and directional derivatives are closely related.

The Basic Chain Rule

Let's begin by recalling the earlier chain rule. Let $y = f(x)$ and $x = x(t)$. The composite function $y = f(x(t))$ expresses y in terms of t. Its derivative is given by the chain rule

$$\frac{dy}{dt} = \frac{dy}{dx}\frac{dx}{dt}$$

whenever the derivatives on the right exist. As an example, let $y = \ln x$ and $x = t^2$. Then

$$\frac{dy}{dt} = \frac{dy}{dx}\frac{dx}{dt} = \frac{1}{x}\cdot 2t = \frac{1}{t^2}\cdot 2t = \frac{2}{t}.$$

A direct evaluation gives the same result:

$$y = \ln x = \ln t^2 = 2\ln t \;\Rightarrow\; \frac{dy}{dt} = \frac{2}{t}.$$

There are several chain rules for functions of two or more variables. We focus on one of them that will serve as a model for all the others. Let

$$z = f(x,y), \qquad x = x(t), \qquad y = y(t).$$

The composite function $z = f(x(t), y(t))$ gives z in terms of t. Its derivative is given by the **basic chain rule**

$$\frac{dz}{dt} = \frac{\partial z}{\partial x}\frac{dx}{dt} + \frac{\partial z}{\partial y}\frac{dy}{dt}.$$

We shall derive the basic chain rule under reasonable hypotheses shortly. But first, let's explore what it means and how it works. The first thing to notice is that the basic chain rule is made up of two simpler chain rules, which are the rates of change of z with respect to t "through x" and "through y."

The basic chain rule gives a general framework for dealing with related rates problems, which were solved by more elementary means earlier in calculus.

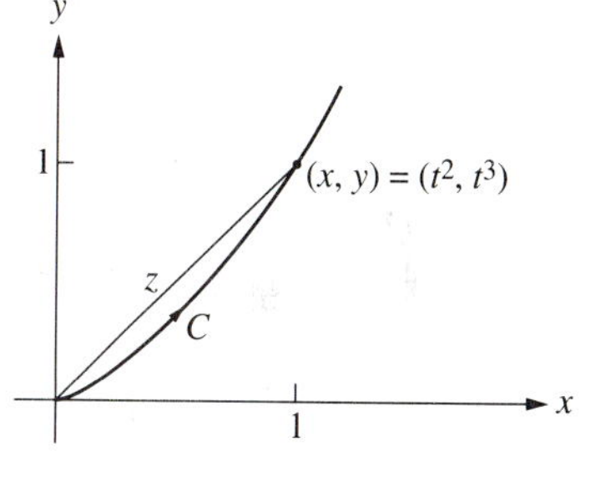

FIGURE 1

EXAMPLE 1. An object with coordinates (x,y) moves in the xy–plane so that $x = t^2$ and $y = t^3$ at time $t \geq 0$. It traces out the parametric curve in Fig. 1. How fast is the object moving away from the origin at time $t = 1$? Measure distance in feet and time in seconds.

Solution. The distance from $(0,0)$ to (x,y) is $z = \sqrt{x^2+y^2}$. The rate of change of z with respect to t is

$$\frac{dz}{dt} = \frac{\partial z}{\partial x}\frac{dx}{dt} + \frac{\partial z}{\partial y}\frac{dy}{dt} = \frac{x}{\sqrt{x^2+y^2}}\cdot 2t + \frac{y}{\sqrt{x^2+y^2}}\cdot 3t^2.$$

At time $t = 1$, we have $x = 1$, $y = 1$, $z = \sqrt{2}$, and

$$\frac{dz}{dt} = \frac{1}{\sqrt{2}}\cdot 2 + \frac{1}{\sqrt{2}}\cdot 3 = \frac{5}{2}\sqrt{2} \approx 3.54 \text{ ft/sec.} \;\square$$

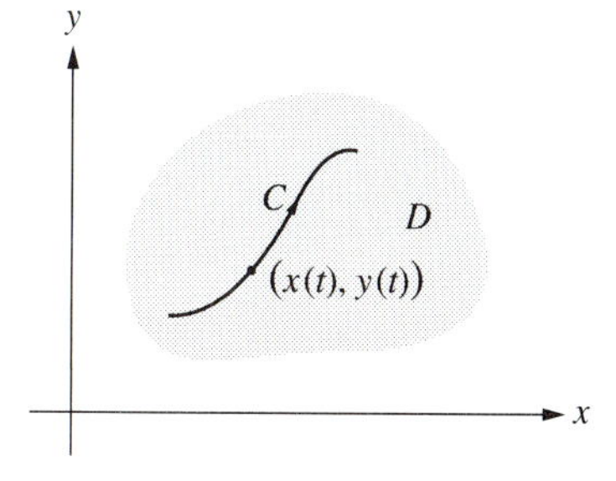

FIGURE 2

The general situation, with $z = f(x,y)$ and $x = x(t)$, $y = y(t)$, is illustrated in Fig. 2. The point $(x,y) = (x(t), y(t))$ traces out the curve C as t traverses an interval I. We assume that $z = f(x,y)$ is defined for (x,y) in an open set D containing C. Then the *composite function* $z = f(x(t), y(t))$ is defined for t in I. For example, if $z = f(x,y)$ is temperature at (x,y) and t is time, then $z = f(x(t), y(t))$ gives the temperature along the curve C and the basic chain rule gives the rate of change of temperature with respect to time as the point $(x,y) = (x(t), y(t))$ moves along C.

The following theorem justifies the basic chain rule in the setting illustrated by Fig. 2.

Theorem 1 *The Basic Chain Rule*
Let $z = f(x,y)$ and $x = x(t)$, $y = y(t)$.
Assume that f_x and f_y exist and are continuous on D and that $x(t)$ and $y(t)$ are differentiable on I.
Then $z = f(x(t), y(t))$ is differentiable on I and

$$\frac{dz}{dt} = \frac{\partial z}{\partial x}\frac{dx}{dt} + \frac{\partial z}{\partial y}\frac{dy}{dt}.$$

Proof. Fix t in I. A change Δt in t produces changes

$$\Delta x = x(t + \Delta t) - x(t), \qquad \Delta y = y(t + \Delta t) - y(t),$$
$$\Delta z = f(x + \Delta x, y + \Delta y) - f(x,y).$$

From Th. 3 of Sec. 5.4,

$$\Delta z = \frac{\partial z}{\partial x}\Delta x + \frac{\partial z}{\partial y}\Delta y + \varepsilon_1 \Delta x + \varepsilon_2 \Delta y,$$

where $\varepsilon_1 \to 0$ and $\varepsilon_2 \to 0$ as $(\Delta x, \Delta y) \to (0,0)$. Then

$$\frac{\Delta z}{\Delta t} = \frac{\partial z}{\partial x}\frac{\Delta x}{\Delta t} + \frac{\partial z}{\partial y}\frac{\Delta y}{\Delta t} + \varepsilon_1 \frac{\Delta x}{\Delta t} + \varepsilon_2 \frac{\Delta y}{\Delta t}.$$

Let $\Delta t \to 0$. Since $x(t)$ and $y(t)$ are differentiable and hence continuous, $\Delta x \to 0$ and $\Delta y \to 0$, so that $\varepsilon_1 \to 0$ and $\varepsilon_2 \to 0$. Therefore, passing to the limit as $\Delta t \to 0$ in the previous equation gives

$$\frac{dz}{dt} = \frac{\partial z}{\partial x}\frac{dx}{dt} + \frac{\partial z}{\partial y}\frac{dy}{dt}. \quad \square$$

Other forms of the basic chain rule in Th. 1 are

$$\frac{df}{dt} = \frac{\partial f}{\partial x}\frac{dx}{dt} + \frac{\partial f}{\partial y}\frac{dy}{dt} \quad \text{and} \quad \frac{df}{dt} = f_x\frac{dx}{dt} + f_y\frac{dy}{dt}.$$

When working with chain rules, it is essential to keep track of the independent and dependent variables. To help, we shall record chain rules in the abbreviated format:

$$z = f(x,y), \qquad x = x(t), \qquad y = y(t) \qquad \Rightarrow$$
$$\frac{dz}{dt} = \frac{\partial z}{\partial x}\frac{dx}{dt} + \frac{\partial z}{\partial y}\frac{dy}{dt}.$$

It is understood that the functions satisfy the appropriate hypotheses; in this case they are given in Th. 1.

EXAMPLE 2. At a particular moment, a solid cone has height $h = 6$ cm, radius of the base $r = 4$ cm, the height is increasing 3 cm/sec, and the radius is increasing 1 cm/sec. How fast is the volume increasing?

Solution. The volume of the cone is $V = \frac{1}{3}\pi r^2 h$, where r and h are functions of time. They increase at the rates $dr/dt = 1$ and $dh/dt = 3$ at the time in question. By the basic chain rule,

$$\frac{dV}{dt} = \frac{\partial V}{\partial r}\frac{dr}{dt} + \frac{\partial V}{\partial h}\frac{dh}{dt} = \frac{2}{3}\pi rh \cdot 1 + \frac{1}{3}\pi r^2 \cdot 3.$$

Let $r = 4$ and $h = 6$ to obtain $dV/dt = 32\pi\, cm^3/sec$ at the indicated moment. □

EXAMPLE 3. Figure 3 models a circular wire with radius 1 foot. There are heating elements in the wire. The temperature at (x, y) is maintained at $z = 64 + x^2 + 4y^2$ degrees Fahrenheit. Describe the variation in temperature recorded by a thermometer that starts at $(1, 0)$ and moves counterclockwise around the wire with speed 1 ft/sec.

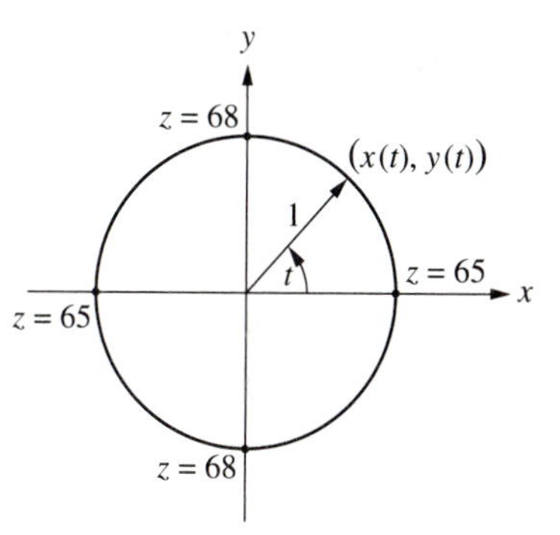

FIGURE 3

Solution. The thermometer is located at the position $(x, y) = (\cos t, \sin t)$ at time $t \geq 0$. (Why?) The rate of change of temperature with respect to time is dz/dt. By the basic chain rule,

$$\frac{dz}{dt} = \frac{\partial z}{\partial x}\frac{dx}{dt} + \frac{\partial z}{\partial y}\frac{dy}{dt} = 2x(-\sin t) + 8y(\cos t),$$

$$\frac{dz}{dt} = 6 \sin t \cos t = 3 \sin 2t.$$

Note that $dz/dt = 0$ for $t = 0, \pi/2, \pi, 3\pi/2, 2\pi$. This information is recorded in the table:

t	0	$\pi/2$	π	$3\pi/2$	2π
(x,y)	(1,0)	(0,1)	$(-1,0)$	$(0,-1)$	(1,0)
z	65	68	65	68	65
dz/dt	0	0	0	0	0

From the table and the critical point methods of differential calculus,

$$z \text{ increases for } 0 \leq t \leq \pi/2 \text{ and } \pi \leq t \leq 3\pi/2,$$

$$z \text{ decreases for } \pi/2 \leq t \leq \pi \text{ and } 3\,\pi/2 \leq t \leq 2\pi.$$

In summary, the temperature alternately increases and decreases between 65 and 68 degrees in each quadrant as the thermometer runs counterclockwise around the circular wire. □

Variants of the Basic Chain Rule

There are several variants of the basic chain rule. We discuss three of them that are used quite often.

First, let $z = f(x,y)$ and $y = y(x)$. Then $z = f(x,y(x))$ is a function of x. The chain rule for this case is

$$z = f(x,y), \qquad y = y(x) \qquad \Rightarrow \qquad \frac{dz}{dx} = \frac{\partial z}{\partial x} + \frac{\partial z}{\partial y}\frac{dy}{dx}.$$

The hypotheses are similar to those in Th. 1. Assume that $y'(x)$ exists for x in an interval I. Then the graph of $y(x)$ is a smooth curve like C in Fig. 2. Assume also that f_x and f_y exist and are continuous in an open set D containing C. The chain rule gives the rate of change of z with respect to x as (x,y) moves along the curve C.

Although dz/dx and $\partial z/\partial x$ look similar, they are not the same. The variable y is held fixed in the evaluation of $\partial z/\partial x$, whereas y is a function of x in the evaluation of dz/dx.

This new chain rule can be derived in the same way as for Th. 1. However, a separate proof is unnecessary because it a special case of the basic chain rule. In Th. 1 let $t = x$. Then $dx/dt = 1$ and $dy/dt = dy/dx$, so Th. 1 yields the chain rule displayed previously.

A companion chain rule is

$$z = f(x,y), \qquad x = x(y) \qquad \Rightarrow \qquad \frac{dz}{dy} = \frac{\partial z}{\partial x}\frac{dx}{dy} + \frac{\partial z}{\partial y},$$

with similar hypotheses. This, too, is a special case of the basic chain rule.

From now on, we shall refer to any of the chain rules we cover as *the chain rule.* It will be clear from the context which particular chain rule is meant.

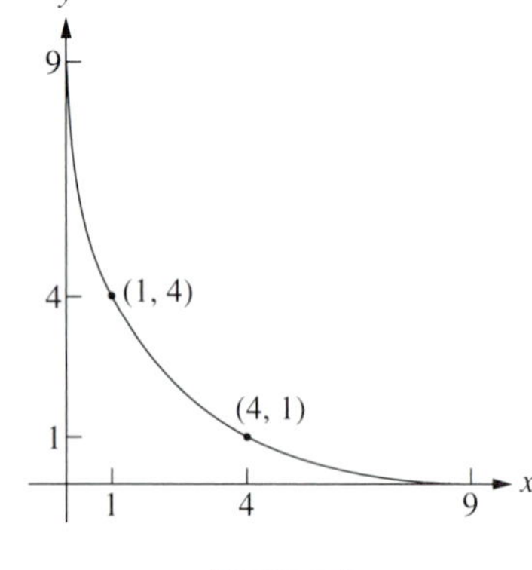

FIGURE 4

EXAMPLE 4. Let $z = x^2 - xy + y^2$. Define y as a function of x implicitly by $x^{1/2} + y^{1/2} = 3$, as illustrated in Fig. 4. Find dz/dx for $x = 1$ and $x = 4$.

Solution. By the chain rule,

$$\frac{dz}{dx} = \frac{\partial z}{\partial x} + \frac{\partial z}{\partial y}\frac{dy}{dx} = (2x - y) + (2y - x)\frac{dy}{dx}.$$

We could find dy/dx by first solving $x^{1/2} + y^{1/2} = 3$ for y and then differentiating. But it is more efficient to use implicit differentiation:

$$\frac{1}{2}x^{-1/2} + \frac{1}{2}y^{-1/2}\frac{dy}{dx} = 0, \qquad \frac{dy}{dx} = -\sqrt{\frac{y}{x}}.$$

Then

$$\frac{dz}{dx} = (2x - y) - (2y - x)\sqrt{\frac{y}{x}}.$$

Now it is easy to find dz/dx when $x = 1$ and $x = 4$:

$$x = 1 \quad \Rightarrow \quad y = 4 \text{ and } \frac{dz}{dx} = -16,$$

$$x = 4 \quad \Rightarrow \quad y = 1 \text{ and } \frac{dz}{dx} = 8.$$

As (x,y) slides down the curve in Fig. 4, z decreases as (x,y) passes through the point $(1,4)$ and z increases as (x,y) passes through the point $(4,1)$. If z represents elevation in feet on a mountain and the curve in Fig. 4 is a climbing route on a map with the x–axis pointing east, then dz/dx is the rate of change of elevation with respect to eastward movement along the trail. The route descends very steeply at $(1,4)$ and climbs very steeply at $(4,1)$. This is a technical climb! □

Now we come to the third variant of the basic chain rule. Suppose that $z = f(x,y)$ and $x = x(s,t)$, $y = y(s,t)$. Then $z = f(x(s,t), y(s,t))$ is a function of s and t. There are two chain rules for this situation:

$$z = f(x,y), \qquad x = x(s,t), \qquad y = y(s,t) \quad \Rightarrow$$

$$\frac{\partial z}{\partial s} = \frac{\partial z}{\partial x}\frac{\partial x}{\partial s} + \frac{\partial z}{\partial y}\frac{\partial y}{\partial s}, \qquad \frac{\partial z}{\partial t} = \frac{\partial z}{\partial x}\frac{\partial x}{\partial t} + \frac{\partial z}{\partial y}\frac{\partial y}{\partial t}.$$

Since the partial derivatives with respect to s and t are just ordinary derivatives with respect to one of these variables with the other variable held fixed, these chain rules are special cases of the basic chain rule. The hypotheses are transcriptions of the conditions in Th. 1.

EXAMPLE 5. Let $z = e^{xy}$ and $x = s^2 + t^2$, $y = s^2 - t^2$. Find $\partial z/\partial s$ and $\partial z/\partial t$. Evaluate the partial derivatives when $s = 1$ and $t = 1$.

Solution. By the chain rule,

$$\frac{\partial z}{\partial s} = \frac{\partial z}{\partial x}\frac{\partial x}{\partial s} + \frac{\partial z}{\partial y}\frac{\partial y}{\partial s} = ye^{xy}(2s) + xe^{xy}(2s) = 2s(x+y)e^{xy},$$

$$\frac{\partial z}{\partial t} = \frac{\partial z}{\partial x}\frac{\partial x}{\partial t} + \frac{\partial z}{\partial y}\frac{\partial y}{\partial t} = ye^{xy}(2t) + xe^{xy}\cdot(-2t) = -2t(x-y)e^{xy}.$$

Since $x = s^2 + t^2$ and $y = s^2 - t^2$,

$$\frac{\partial z}{\partial s} = 4s^3 e^{s^4 - t^4} \quad \text{and} \quad \frac{\partial z}{\partial t} = -4t^3 e^{s^4 - t^4}.$$

As a check, note that the same results follow from $z = e^{xy} = e^{s^4 - t^4}$. Finally, let $s = 1$ and $t = 1$. Then $x = 2$ and $y = 0$. Either pair of formulas for the partial derivatives gives

$$\frac{\partial z}{\partial s} = 4 \quad \text{and} \quad \frac{\partial z}{\partial t} = -4. \ \square$$

Changes of Variable

The chain rule plays a key role when changes of variable are made. For example, we may start with a problem described in rectangular coordinates x, y, and decide that it would be better to work in polar coordinates r, θ. Say we start with a function $z = f(x,y)$ and wish to change to polar coordinates. Then z becomes a function of r and θ through the relations $x = r\cos\theta$ and $y = r\sin\theta$. By the chain rule,

$$\frac{\partial z}{\partial r} = \frac{\partial z}{\partial x}\frac{\partial x}{\partial r} + \frac{\partial z}{\partial y}\frac{\partial y}{\partial r} = \frac{\partial z}{\partial x}\cos\theta + \frac{\partial z}{\partial y}\sin\theta,$$

$$\frac{\partial z}{\partial \theta} = \frac{\partial z}{\partial x}\frac{\partial x}{\partial \theta} + \frac{\partial z}{\partial y}\frac{\partial y}{\partial \theta} = \frac{\partial z}{\partial x}(-r\sin\theta) + \frac{\partial z}{\partial y}(r\cos\theta).$$

These special cases of the chain rule do not occur often enough for them to be worth memorizing. It is better to derive them on the spot when they are needed, as in the next example.

EXAMPLE 6. Let $z = xy$. Find $\partial z/\partial r$ and $\partial z/\partial \theta$.

Solution. In this case,

$$\frac{\partial z}{\partial r} = \frac{\partial z}{\partial x}\cos\theta + \frac{\partial z}{\partial y}\sin\theta = y\cos\theta + x\sin\theta,$$

$$\frac{\partial z}{\partial \theta} = \frac{\partial z}{\partial x}(-r\sin\theta) + \frac{\partial z}{\partial y}(r\cos\theta) = y(-r\sin\theta) + x(r\cos\theta).$$

Since $x = r\cos\theta$ and $y = r\sin\theta$,

$$\frac{\partial z}{\partial r} = 2r\sin\theta\cos\theta = r\sin 2\theta \quad \text{and} \quad \frac{\partial z}{\partial \theta} = r^2(\cos^2\theta - \sin^2\theta) = r^2\cos 2\theta.$$

As a check, verify that these results follow directly from $z = xy = r^2\sin\theta\cos\theta$. $\square$

Chain Rules for Functions of Three Variables

The direct analogue of the basic chain rule for a function of three variables is

$$w = f(x,y,z), \quad x = x(t), \quad y = y(t), \quad z = z(t) \quad \Rightarrow$$
$$\frac{dw}{dt} = \frac{\partial w}{\partial x}\frac{dx}{dt} + \frac{\partial w}{\partial y}\frac{dy}{dt} + \frac{\partial w}{\partial z}\frac{dz}{dt},$$

where f has continuous partial derivatives and $x(t)$, $y(t)$, $z(t)$ are differentiable on appropriate domains. The proof, which we omit, is almost the same as for Th. 1. It is based on Th. 2 of Sec. 5.4. See the problems.

EXAMPLE 7. Suppose that a wire has the shape of the twisted cubic $x(t) = t$, $y(t) = t^2$, $z(t) = t^3$ for $t \geq 0$. The temperature at $(x,y,z) = (t,t^2,t^3)$ is $w = \ln(1 + x^2 + y^2 + z^2)$. Find the rate of change of temperature when $t = 1$.

Solution. By the chain rule,

$$\frac{dw}{dt} = \frac{2x}{1 + x^2 + y^2 + z^2}(1) + \frac{2y}{1 + x^2 + y^2 + z^2}(2t) + \frac{2z}{1 + x^2 + y^2 + z^2}(3t^2).$$

Let $t = 1$. Then $(x,y,z) = (1,1,1)$ and $dw/dt = 3$. □

The variants of the basic chain rule for functions of two variables have obvious counterparts for functions of three variables. We merely mention them for reference. First,

$$w = f(x,y,z), \quad y = y(x), \quad z = z(x), \quad \Rightarrow$$
$$\frac{dw}{dx} = \frac{\partial w}{\partial x} + \frac{\partial w}{\partial y}\frac{dy}{dx} + \frac{\partial w}{\partial z}\frac{dz}{dx}.$$

The corresponding formulas for dw/dy and dw/dz, when y or z are the independent variables, are very similar. What are they? Finally, two applications of the basic chain rule for functions of three variables give the following pair of chain rules:

$$w = f(x,y,z), \quad x = x(s,t), \quad y = y(s,t), \quad z = z(s,t) \quad \Rightarrow$$
$$\frac{\partial w}{\partial s} = \frac{\partial w}{\partial x}\frac{\partial x}{\partial s} + \frac{\partial w}{\partial y}\frac{\partial y}{\partial s} + \frac{\partial w}{\partial z}\frac{\partial z}{\partial s},$$
$$\frac{\partial w}{\partial t} = \frac{\partial w}{\partial x}\frac{\partial x}{\partial t} + \frac{\partial w}{\partial y}\frac{\partial y}{\partial t} + \frac{\partial w}{\partial z}\frac{\partial z}{\partial t}.$$

For illustrations of these chain rules, see the problems. Chain rules for functions of more than three variables follow the same patterns.

Directional Derivatives for Functions of Two Variables

Recall that partial derivatives of a function $f(x,y)$ are slopes of the surface $z = f(x,y)$ in the x– and y–directions or, equivalently, rates of change of f in the x– and y–directions. Partial derivatives are special cases of directional derivatives, which are slopes of the surface $z = f(x,y)$ in any direction. Equivalently, directional derivatives are rates of change of f in any direction. Imagine that the surface $z = f(x,y)$ models a hillside. Then directional derivatives tell us how steep the terrain is in various directions, such as east, north, or southwest. At first glance, directional derivatives may seem to have nothing to do with chain rules. However, as we shall learn, directional derivatives are evaluated with the aid of chain rules.

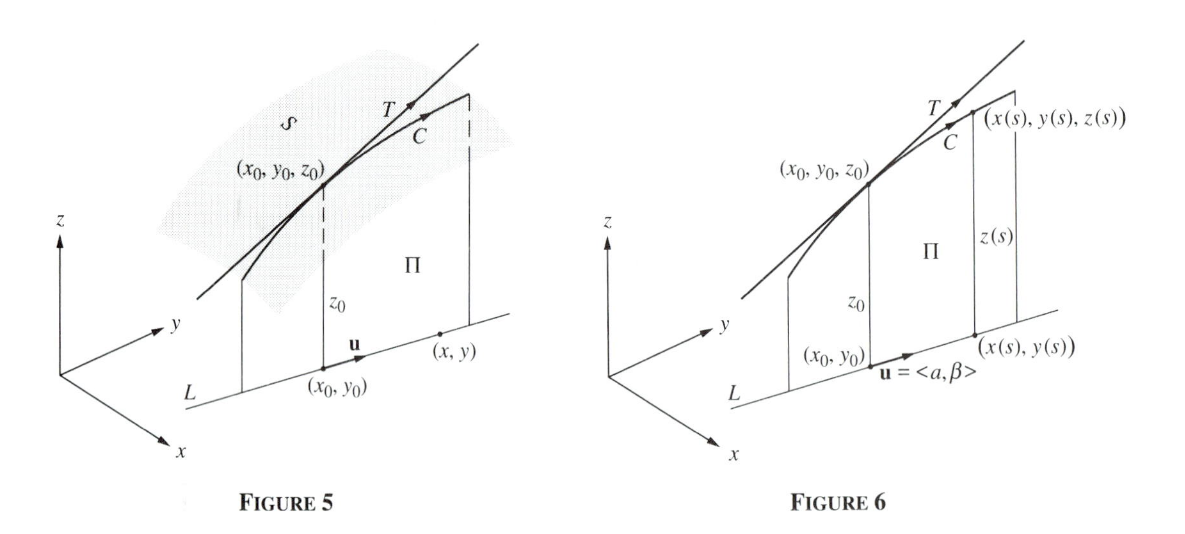

FIGURE 5

FIGURE 6

Figure 5 will help us explain directional derivatives. In the figure, S is the graph of $z = f(x,y)$. Assume throughout that f has continuous partial derivatives $f_x(x,y)$ and $f_y(x,y)$ for (x,y) in a neighborhood of (x_0,y_0). The unit vector $\mathbf{u}$ in Fig. 5 determines an arbitrary direction in the xy–plane. The line L passes through (x_0,y_0) with direction $\mathbf{u}$. The vertical plane Π containing L cuts the surface S in a curve C, which is the graph of $z = f(x,y)$ for (x,y) on L. The directed line T in Fig. 5 is tangent to C at (x_0,y_0,z_0). In other words, T is tangent to the surface S in the $\mathbf{u}$–direction at (x_0,y_0,z_0).

The directional derivative of f at (x_0,y_0,z_0) in the $\mathbf{u}$–direction is denoted by $D_{\mathbf{u}}f(x_0,y_0)$ and is defined geometrically by

$$\boxed{D_{\mathbf{u}}f(x_0,y_0) = \text{slope } T.}$$

It follows from our earlier geometric interpretations of partial derivatives that the directional derivatives in the directions $\mathbf{u} = \mathbf{i}$ and $\mathbf{u} = \mathbf{j}$ are

$$D_{\mathbf{i}}\, f(x_0,y_0) = f_x(x_0,y_0) \quad \text{and} \quad D_{\mathbf{j}}\, f(x_0,y_0) = f_y(x_0,y_0).$$

A sharpening of the foregoing discussion will lead to an analytic definition of the directional derivative. The reasoning is based on Fig. 6, which displays the vertical plane Π of Fig. 5 with additional structure. In Fig. 6, $\mathbf{u} = < a, b >$. The line L is labeled the s–axis, with $s = 0$ at (x_0, y_0) and the positive direction determined by $\mathbf{u}$. Then s measures (signed) distance from (x_0, y_0) along L. A vector equation for L is

$$\mathbf{r} = \mathbf{r}_0 + s\mathbf{u}, \text{ where } \mathbf{r} = < x, y >,\ \mathbf{r}_0 = < x_0, y_0 > .$$

Parametric equations for L are

$$x(s) = x_0 + as, \qquad y(s) = y_0 + bs.$$

Corresponding parametric equations for C are

$$x(s) = x_0 + as, \qquad y(s) = y_0 + bs, \qquad z(s) = f(x(s), y(s)).$$

The curve C lies in the vertical plane Π, which we can regard as the sz–plane. The derivative of f at (x_0, y_0) in the $\mathbf{u}$–direction is given by

$$D_{\mathbf{u}}f(x_0, y_0) = \text{slope } T = \left.\frac{df}{ds}\right|_{s=0}.$$

Since $f(x(s), y(s)) = f(x_0 + as, y_0 + bs)$,

$$D_{\mathbf{u}}f(x_0, y_0) = \left.\frac{d}{ds} f(x_0 + as, y_0 + bs)\right|_{s=0}.$$

By the chain rule,

$$D_{\mathbf{u}}f(x_0, y_0) = af_x(x_0, y_0) + bf_y(x_0, y_0).$$

In summary, we have arrived at the following analytical definition of a directional derivative.

Definition *Directional Derivative of $f(x, y)$*
Assume that $f(x, y)$ has continuous partial derivatives $f_x(x, y)$ and $f_y(x, y)$ for (x, y) near (x_0, y_0).
The **directional derivative** of $f(x, y)$ at (x_0, y_0) in the direction of a unit vector $\mathbf{u} = < a, b >$ is

$$D_{\mathbf{u}}\, f(x_0, y_0) = af_x(x_0\, ,\, y_0) + bf_y(x_0, y_0).$$

EXAMPLE 8. Let $f(x, y) = 3x^2 + 2y^2$ and $\mathbf{u} = < 3/5, 4/5 >$. Then $f_x = 6x$, $f_y = 4y$, and

$$D_{\mathbf{u}}\, f(1, 2) = \frac{3}{5} \cdot 6 + \frac{4}{5} \cdot 8 = 10.$$

This directional derivative has an informative geometric interpretation. The graph of $z = f(x,y) = 3x^2 + 2y^2$ is an elliptic paraboloid which opens up. The slope of the paraboloid at $(1,2,11)$ in the **u**–direction is $D_{\mathbf{u}}f(1,2) = 10$. □

In Sec. 5.4 we claimed that the tangent plane to a surface $z = f(x,y)$ at a point (x_0, y_0, z_0) contains every tangent line to the surface through (x_0, y_0, z_0). A proof is based on directional derivatives. We give the main steps. In Figs. 5 and 6, T is any tangent line to the surface S at (x_0, y_0, z_0). The point–slope equation for T in the sz–plane Π is

$$z - z_0 = D_{\mathbf{u}}f(x_0, y_0)s = asf_x(x_0, y_0) + bsf_y(x_0, y_0).$$

Since $x = x_0 + as$ and $y = y_0 + bs$ on T,

$$z - z_0 = f_x(x_0, y_0)(x - x_0) + f_y(x_0, y_0)(y - y_0) \text{ on } T.$$

This is the point–slope equation for the tangent plane to S at (x_0, y_0, z_0). So T lies in the tangent plane.

Since directional derivatives determine slopes and slopes are rates of change,

$$D_{\mathbf{u}}f(x_0, y_0) = \text{the rate of change of } f$$
$$\text{with respect to distance from } (x_0, y_0)$$
$$\text{in the } \mathbf{u}\text{–direction.}$$

EXAMPLE 9. The temperature in degrees centigrade at the point (x,y) is $f(x,y) = \sqrt{2x^2 + y^2 + 19}$. How rapidly does the temperature change as you move away from $(2,3)$ in the directions $\mathbf{u} = < 3/5, 4/5 >$ and $\mathbf{u} = < 3/5, -4/5 >$? Measure distances in meters.

Solution. We must find

$$D_{\mathbf{u}}f(2,3) = af_x(2,3) + bf_y(2,3)$$

for the two given directions $\mathbf{u} = < a, b >$. First,

$$f_x(x,y) = \frac{2x}{\sqrt{2x^2 + y^2 + 19}}, \qquad f_x(2,3) = \frac{2}{3},$$

$$f_y(x,y) = \frac{y}{\sqrt{2x^2 + y^2 + 19}}, \qquad f_y(2,3) = \frac{1}{2}.$$

For $\mathbf{u} = < 3/5, 4/5 >$,

$$D_{\mathbf{u}}f(2,3) = \left(\frac{3}{5}\right)\left(\frac{2}{3}\right) + \left(\frac{4}{5}\right)\left(\frac{1}{2}\right) = \frac{4}{5} \text{ deg/meter.}$$

The temperature is increasing 4/5 degrees per meter as you leave $(2,3)$ in the direction $\mathbf{u} = < 3/5, 4/5 >$. For $\mathbf{u} = < 3/5, -4/5 >$,

$$D_{\mathbf{u}} f(2,3) = \left(\frac{3}{5}\right)\left(\frac{2}{3}\right) - \left(\frac{4}{5}\right)\left(\frac{1}{2}\right) = 0 \text{ deg/meter.}$$

The temperature is neither increasing nor decreasing as you leave $(2,3)$ in the direction $\mathbf{u} = < 3/5, -4/5 >$. □

Bear in mind that $\mathbf{u}$ is a unit vector in $D_{\mathbf{u}} f$. However, directions are often described by vectors of various lengths or by angles. In Ex. 9, we could ask for the rate of change of temperature leaving $(2,3)$ in the direction of the vector $\mathbf{v} = < 4,3 >$. A unit vector in this direction is $\mathbf{u} = \mathbf{v}/||\mathbf{v}|| = < 4/5, 3/5 >$, and the requested rate is

$$D_{\mathbf{u}} f(2,3) = \left(\frac{4}{5}\right)\left(\frac{2}{3}\right) + \left(\frac{3}{5}\right)\left(\frac{1}{2}\right) = \frac{5}{6} \text{ deg/meter.}$$

How rapidly is the temperature changing from $(2,3)$ in the direction of the ray $\theta = 30°$?

Directional Derivatives for Functions of Three Variables

Directional derivatives for functions of three variables follow the same pattern as for functions of two variables. We simply summarize the key results.

Assume that $f(x,y,z)$ has continuous partial derivatives for (x,y,z) near (x_0, y_0, z_0). Specify a direction in space by means of a unit vector $\mathbf{u} = < a, b, c >$. The **directional derivative** of f at (x_0, y_0, z_0) in the direction $\mathbf{u}$ is

$$\boxed{D_{\mathbf{u}} f(x_0, y_0, z_0) = a f_x(x_0, y_0, z_0) + b f_y(x_0, y_0, z_0) + c f_z(x_0, y_0, z_0).}$$

Directional derivatives are rates of change:

$$D_{\mathbf{u}} f(x_0, y_0, z_0) = \text{the rate of change of } f$$
$$\text{with respect to distance from } (x_0, y_0, z_0)$$
$$\text{in the } \mathbf{u}\text{–direction.}$$

EXAMPLE 10. Let $f(x,y,z) = 2x \sin y - 3y \cos z + 6ze^x$. Find the directional derivative of f at $(0,0,0)$ in the direction of $\mathbf{v} = < 1,2,2 >$.

Solution. Since $||\mathbf{v}|| = 3$, the unit vector in the direction of $\mathbf{v}$ is $\mathbf{u} = \frac{1}{3} < 1,2,2 > = \langle\frac{1}{3}, \frac{2}{3}, \frac{2}{3}\rangle$. The partial derivatives of f are

$$f_x = 2 \sin y + 6z\, e^x, \qquad f_y = 2x \cos y - 3 \cos z, \qquad f_z = 3y \sin z + 6\, e^x.$$

At $(0,0,0)$,

$$f_x(0,0,0) = 0, \qquad f_y(0,0,0) = -3, \qquad f_z(0,0,0) = 6.$$

Therefore,

$$D_{\mathbf{u}} f(0,0,0) = \left(\frac{1}{3}\right)(0) + \left(\frac{2}{3}\right)(-3) + \left(\frac{2}{3}\right)(6) = 2.$$

If f is temperature in degrees Fahrenheit and distance is measured in feet, then the temperature increases 2 deg/ft at $(0,0,0)$ in the direction of $\mathbf{u}$ (or $\mathbf{v}$). □

Directional derivatives for functions of two or three variables have convenient algebraic properties which follow easily from the definitions:

$$D_{\mathbf{u}}(a_1 f_1 + a_2 f_2) = a_1 D_{\mathbf{u}} f_1 + a_2 D_{\mathbf{u}} f_2,$$
$$D_{-\mathbf{u}} f = -D_{\mathbf{u}} f.$$

In view of the second property, the temperature in Ex. 10 decreases 2 deg/ft at $(0,0,0)$ in the direction of $-\mathbf{u}$ (or $-\mathbf{v}$).

Implicit Function Theorems and Implicit Differentiation*

Implicit function theorems and the chain rule will help us clear up some unfinished business about implicit differentiation. To set the scene, we look at a standard implicit differentiation problem.

EXAMPLE 11. Let $3x^3 - 2x^2y^2 + y^3 = 3$. Note that $(x,y) = (1,2)$ satisfies the equation. Find dy/dx at $(1,2)$.

Solution. Assume, as is customary in one–variable calculus, that the given equation determines y as a differentiable function of x, at least for (x,y) near $(1,2)$. Differentiate the given equation implicitly to obtain

$$9x^2 - 4xy^2 - 4x^2y\,\frac{dy}{dx} + 3y^2\,\frac{dy}{dx} = 0,$$

$$\frac{dy}{dx} = \frac{4xy^2 - 9x^2}{3y^2 - 4x^2y}, \qquad \frac{dy}{dx} = \frac{7}{4} \quad \text{at} \quad (x,y) = (1,2).$$

But how do we know that the equation $3x^3 - 2x^2y^2 + y^3 = 3$ determines y as a differentiable function of x for (x,y) near $(1,2)$? The validity of the foregoing solution depends on an affirmative answer to this question. □

The same question comes up whenever implicit differentiation is used. Thus, consider an equation $f(x,y) = c$. Let $f(x_0,y_0) = c$. Suppose we know that $f(x,y) = c$ determines y as a differentiable function of x for (x,y) near (x_0,y_0). To find dy/dx at (x_0,y_0) by means of implicit differentiation, we differentiate both sides of the equation $f(x,y) = c$, with $y = y(x)$, and then solve for dy/dx. The steps are usually tailored to the particular problem, as in Ex. 11. The chain rule gives a general formula for dy/dx when $f(x,y) = c$ and $y = y(x)$:

$$\frac{d}{dx} f(x,y) = f_x(x,y) + f_y(x,y)\frac{dy}{dx} = 0,$$

$$\frac{dy}{dx} = -\frac{f_x(x_0,y_0)}{f_y(x_0,y_0)},$$

provided $f_y(x_0,y_0) \neq 0$ and the hypotheses for the chain rule are satisfied.

The validity of the implicit differentiation procedure for finding dy/dx is based on the following theorem. Proofs are given in more advanced courses.

Theorem 2 *The Implicit Function Theorem for $f(x,y)$*
Assume that $f(x,y)$ has continuous partial derivatives for (x,y) near (x_0,y_0). Let $f(x_0, y_0) = c$.
If $f_y(x_0,y_0) \neq 0$, then the equation $f(x,y) = c$ determines y as a differentiable function of x near (x_0,y_0) and

$$\frac{dy}{dx} = -\frac{f_x}{f_y}.$$

If $f_x(x_0,y_0) \neq 0$, then the equation $f(x,y) = c$ determines x as a differentiable function of y near (x_0,y_0) and

$$\frac{dx}{dy} = -\frac{f_y}{f_x}.$$

The formulas for dy/dx and dx/dy come from implicit differentiation of $f(x,y) = c$, using the chain rule. The conclusion of the theorem can be stated informally as follows. You can solve the equation $f(x,y) = c$ for x or y near a point on the graph of $f(x,y) = c$ if the partial derivative of f with respect to that variable is nonzero.

Now, consider $f(x,y,z) = c$. Such an equation often implicitly determines z as a function of x and y. The partial derivatives of such an implicitly defined function $z = z(x,y)$ can be found by implicit partial differentiation. Here is a typical example.

EXAMPLE 12. Let $f(x,y,z) = ye^{x-z} + z$ and consider $ye^{x-z} + z = 3$. Note that $(x,y,z) = (1,2,1)$ satisfies the equation. Find z_x and z_y at $(1,2,1)$.

Solution. Assume (pending verification) that the equation determines z as a function of x and y with continuous partial derivatives z_x and z_y for (x,y,z) near $(1,2,1)$. Differentiate $ye^{x-z} + z = 3$ partially with respect to x and y to obtain

$$ye^{x-z}(1 - z_x) + z_x = 0, \qquad z_x = \frac{ye^{x-z}}{ye^{x-z} - 1},$$

$$e^{x-z}(1 - yz_y) + z_y = 0, \qquad z_y = \frac{e^{x-z}}{ye^{x-z} - 1}.$$

Finally, let $(x,y,z)=(1,2,1)$ to get $z_x=2$ and $z_y=1$. □

Now consider an equation $f(x,y,z)=c$ in 3–space. Let $f(x_0,y_0,z_0)=c$. Suppose we know that $f(x,y,z)=c$ determines z as a function of x and y with continuous partial derivatives z_x and z_y for (x,y,z) near (x_0,y_0,z_0). To find z_x and z_y by implicit differentiation, we differentiate $f(x,y,z)=c$ partially with respect to x and y, then solve for z_x and z_y. This is usually done ad hoc, as in Ex. 12. The chain rule yields the general formulas

$$z_x=-\frac{f_x}{f_z},\qquad z_y=-\frac{f_y}{f_z}\quad \text{at}\quad (x_0,y_0,z_0),$$

provided $f_z(x_0,y_0,z_0)\neq 0$ and the hypotheses for the chain rule are satisfied. See the problems.

The use of implicit differentiation to find z_x and z_y is justified by the implicit function theorem for a function of three variables.

Theorem 3 *The Implicit Function Theorem for $f(x,y,z)$*
Assume that $f(x,y,z)$ has continuous partial derivatives for (x,y,z) near (x_0,y_0,z_0). Let $f(x_0,y_0,z_0)=c$. If $f_z(x_0,y_0,z_0)\neq 0$, then the equation $f(x,y,z)=c$ determines z as a function of x and y near (x_0,y_0,z_0) with continuous partial derivatives given by

$$\frac{\partial z}{\partial x}=-\frac{f_x}{f_z},\qquad \frac{\partial z}{\partial y}=-\frac{f_y}{f_z}.$$

Similar results hold if $f_x(x_0,y_0,z_0)\neq 0$ or $f_y(x_0,y_0,z_0)\neq 0$. What are they?

An informal statement of Th. 3 is: $f(x,y,z)=c$ can be solved for z as a function of x and y near any point where $f_z\neq 0$, and the partial derivatives of $z(x,y)$ can be found by implicit partial differentiation of $f(x,y,z)=c$.

PROBLEMS

In Probs. 1–6 find dz/dt or dw/dt in two ways: (a) Use a basic chain rule and then express the final result in terms of t. (b) Express z directly as a function of t and then differentiate.

1. $z=1/\sqrt{x^2+y^2}$, $\quad x=\sin t$, $\quad y=\cos t$
2. $z=\arctan(y/x)$, $\quad x=t^2$, $\quad y=t^3$
3. $z=e^{x^2+y^2}$, $\quad x=t+1/t$, $\quad y=t-1/t$
4. $z=3xy/(x^2-y^2)$, $\quad x=\cosh t$, $\quad y=\sinh t$
5. $w=\ln(x^2+y^2+z^2)$, $\quad x=\sin 2t$, $\quad y=\cos 2t$, $\quad z=3t$
6. $w=e^{-xy/z}$, $\quad x=t$, $\quad y=t^2$, $\quad z=t^3$

In Probs. 7–12 use chain rules, much as in Exs. 1, 2 and 3.

7. The position of an object at time $t \geq 0$ is $x = 3\cos t$, $y = 4\sin t$. Sketch the path of motion and determine how fast the object is moving away from (or approaching) the origin when (a) $t = \pi/6$, (b) $t = 3\pi/4$.

8. An object moves in the xy–plane with position vector $\mathbf{r}(t)$. How fast is the object approaching or moving away from the origin at the time $t = 3$ when $\mathbf{r}(3) = < 4, -5 >$ and $\mathbf{r}'(3) = < -1, 3 >$?

9. The portion of usable lumber in a Douglas fir tree with diameter 2 feet and height 80 feet is approximately a right circular cylinder with these dimensions. If the height is increasing at 6 inches per year and the diameter at 3/8 inch per year, find the rate at which usable lumber is being added to the tree.

10. At a particular moment, a rectangular box has dimensions $a = 12$, $b = 16$, and $c = 20$ inches and a, b are increasing at 2 in./sec while c is decreasing at 4 in./sec. Find the rate at which (a) the volume and (b) the surface area of the box are changing at the instant in question.

11. The temperature at (x, y) in the xy–plane is maintained at $T = 32 + x^2 - xy + 4y^2$ degrees Fahrenheit. A thermometer moves along the ellipse $x = 2\cos t$, $y = \sin t$ for $t \geq 0$. Describe the variation in temperature recorded by the thermometer as it moves once around the ellipse. In particular, find the maximum and minimum temperatures.

12. Let $T = T(x, y, z)$ be the temperature at (x, y, z) in space. At the point $(1, 2, 3)$ the temperature is increasing at 2 deg/ft in the x–direction, at 4 deg/ft in the y–direction, and at -3 deg/ft in the z–direction. How fast is the temperature changing with respect to distance traveled as you pass through $(1, 2, 3)$ with velocity $< 2, -3, 1 >$?

In Probs. 13–16 find $\partial z/\partial s$ and $\partial z/\partial t$ at the given point in two ways. (a) Use the chain rule for $z = f(x, y)$ where x and y are function of s and t. (b) Express z directly in terms of s and t and then differentiate.

13. $z = x^2 - xy + y^2$, $\quad x = s\cos t$, $\quad y = s\sin t$, $\quad (s, t) = (2, \pi/6)$

14. $z = x/(x^2 + y^2)$, $\quad x = s^2 - t^2$, $\quad y = 2st$, $\quad (s, t) = (1, 1)$

15. $z = \arctan(x^2 - y^2)$, $\quad x = e^s \cos t$, $\quad y = e^s \sin t$, $\quad (s, t) = (0, \pi)$

16. $z = e^{x/y}$, $\quad x = s^2 + t^2$, $\quad y = s^2 - t^2$, $\quad (s, t) = (2, -1)$

In Probs. 17–18, z is a function of x and y. Assume the second equation determines y as a differentiable function of x and find dz/dx at the given point.

17. $z = x^2 - xy + y^2$, $\quad 2xy + e^{x-y} = 3$, $\quad$ at $x = 1$, $y = 1$

18. $z = \ln(x^2 + y^2)$, $\quad x^2y^3 - 2xy^2 + 3xy = 6$, $\quad$ at $x = 2$

In Probs. 19–26 assume that f has continuous partial derivatives.

19. If $z = f(x, y)$ and you decide to change to polar coordinates via $x = r\cos\theta$, $y = r\sin\theta$, show that

$$\left(\frac{\partial z}{\partial x}\right)^2 + \left(\frac{\partial z}{\partial y}\right)^2 = \left(\frac{\partial z}{\partial r}\right)^2 + \frac{1}{r^2}\left(\frac{\partial z}{\partial \theta}\right)^2.$$

20. If $z = f(r,\theta)$ and you decide to change to rectangular coordinates via $r = \sqrt{x^2 + y^2}$, $\theta = \arctan(y/x)$, show that

$$\frac{\partial z}{\partial x} = \frac{\partial z}{\partial r}\cos\theta - \frac{\partial z}{\partial \theta}\frac{\sin\theta}{r},$$

$$\frac{\partial z}{\partial y} = \frac{\partial z}{\partial r}\sin\theta + \frac{\partial z}{\partial \theta}\frac{\cos\theta}{r}.$$

21. Obtain the formulas for $\partial z/\partial x$ and $\partial z/\partial y$ in Prob. 20 by regarding the pair of equations preceding Ex. 6 as simultaneous linear equations for these partial derivatives.

22. If $z = f(x,y)$ and $x = e^u \cos v$, $y = e^u \sin v$, show that

$$(z_x)^2 + (z_y)^2 = e^{-2u}[(z_u)^2 + (z_v)^2].$$

23. If $z = f(x - y, x + y)$ show that $z_x + z_y = 0$. *Hint.* Let $z = f(u,v)$ where $u = x - y$ and $v = x + y$.

24. If $z = f(x/y, y/x)$ show that $xz_x + yz_y = 0$.

25. If $w = f(x/z, y/z)$ show that $xw_x + yw_y + zw_z = 0$.

26. If $w = f((y - x)/xy, (z - y)/yz)$ show that $x^2w_x + y^2w_y + z^2w_z = 0$.

In Probs. 27–29 use the following chain rule, whose proof is left for Prob. 30.

$$z = f(u), \qquad u = u(x,y) \qquad \Rightarrow$$

$$\frac{\partial z}{\partial x} = f'(u)\frac{\partial u}{\partial x}, \qquad \frac{\partial z}{\partial y} = f'(u)\frac{\partial u}{\partial y}$$

27. If $z = f(y^2 - x^2)$, then $yz_x + xz_y = 0$.

28. If c is a constant and $z = f(x + ct)$, then $z_t - cz_x = 0$.

29. If $z = f([x^2 - y^2]/[x^2 + y^2])$, then $xz_x + yz_y = 0$.

30. Prove the chain rule highlighted above. *Hint.* Remember that partial derivatives are particular ordinary derivatives.

31. Two functions $u(x,y)$ and $v(x,y)$ satisfy the **Cauchy–Riemann equations** if $u_x = v_y$ and $u_y = -v_x$. These equations occur naturally in calculus of functions of a complex variable and in problems of two–dimensional fluid flow and electromagnetics. Show that the Cauchy–Riemann equations become

$$u_r = \frac{1}{r}v_\theta \text{ and } v_r = -\frac{1}{r}u_\theta$$

when u and v are expressed in terms of the polar coordinates r and θ.

32. The temperature at (x,y) is $T(x,y) = 72/(x^2 + 4y^2 + 1)$ degrees Fahrenheit. Find the rate of change of temperature at $(2,1)$ in the direction of the vector $< 3,4 >$.

33. The density of a thin rectangular plate is given by $\sigma(x,y) = 9/\sqrt{4 + x^2 + y^2}$ in units of mass per area. How fast is the density at $(1,2)$ changing in the direction inclined at (a) 30° and (b) 210° to the positive x–axis?

34. The electric potential (voltage) at a point (x,y) in the xy–plane set up by a line of charge along the z–axis is given by $V = \ln(x^2 + y^2)^{-1/2}$. (a) In what directions from the point $(3,4)$ does V increase and decrease most rapidly? (b) In what direction from $(3,4)$ does V have rate of change zero? (c) What is the relationship between the directions in (a) and (b)? *Hint.* For (a), $\mathbf{u} = \langle \cos\theta, \sin\theta \rangle$ is a unit vector inclined at angle θ to the positive x–axis.

35. Find the directional derivative of $f(x,y,z) = (x/y) - (y/x) + 5z$ at $(1,2,3)$ in the direction of $\langle 2,2,1 \rangle$.

36. Fix (x_0,y_0). Suppose $f(x,y)$ has continuous partial derivatives. Let $\mathbf{u}_1$ be a direction from (x_0,y_0) in which the directional derivative of f is 0 and let $\mathbf{u}_2$ be a direction in which the directional derivative is either a maximum or a minimum. Show that $\mathbf{u}_1$ is perpendicular to $\mathbf{u}_2$. *Hint.* See the hint for Prob. 34.

37. If $z = f(x,y)$ with x and y as independent variables, then $dz = z_x dx + z_y dy$. Suppose now that x and y are functions of s and t so that $x = x(s,t)$ and $y = y(s,t)$ are no longer independent. Show that the formula for dz still remains valid if dx and dy are interpreted as the differentials of the functions $x = x(s,t)$ and $y = y(s,t)$. *Hint.* Since z is a function of the independent variables s and t, $dz = z_s ds + z_t dt$. Use the chain rule to express z_s and z_t in terms of z_x and z_y.

38. Suppose that $z = f(x,y)$ and $x = x(s,t,u)$ and $y = y(s,t,u)$. What is the chain rule for $\partial z/\partial u$? Justify briefly.

39. Establish the formulas

    (a) $D_{\mathbf{u}}(a_1 f_1 + a_2 f_2) = a_1 D_{\mathbf{u}} f_1 + a_2 D_{\mathbf{u}} f_2$,

    (b) $D_{-\mathbf{u}} f = -D_{\mathbf{u}} f$,

    at any point (x_0,y_0) where f, f_1, and f_2 have directional derivatives.

40. (a) Use the implicit function theorem to show that the equation $3x^3 - 2x^2y^2 + y^3 = 3$ from Ex. 11 does determine y as a differentiable function of x near $(x,y) = (1,2)$. (b) Show that the same equation determines x as a differentiable function of y near $(x,y) = (1,2)$ and find dx/dy when $y = 2$.

41. Use the implicit function theorem to show that the equation

$$x^2y^2 + xz^3 - x^4 + zy^3 = 6$$

determines z as a function of x and y with continuous partial derivatives near $(1,-1,2)$. Then find $\partial z/\partial x$ and $\partial z/\partial y$ when $x = 1$ and $y = -1$.

42. Use the implicit function theorem to show that the equation

$$\sin(x + y) + \cos(y + z) + \tan(x + z) = 1$$

determines z as a function of x and y with continuous partial derivatives near $(0,0,0)$. Then find $\partial z/\partial x$ and $\partial z/\partial y$ when $x = 0$ and $y = 0$.

43. Use implicit partial differentiation to verify the formulas $z_x = -f_x/f_z$ and $z_y = -f_y/f_z$ in Th. 3.

5.6 Gradients

The gradient of a function is a vector whose components are the partial derivatives of the function. Gradients play the same role for functions of two or more variables that derivatives play for functions of one variable. Chain rules, directional derivatives, and tangent planes have vector forms involving the gradient which give a great deal of geometric insight, as we shall see.

As usual, we restrict the discussion to functions of two and three variables.

Definition *The Gradient*
The **gradient** (or gradient vector) of a function f is

$$\nabla f = < f_x, f_y > \quad \text{for} \quad f(x,y),$$

$$\nabla f = < f_x, f_y, f_z > \quad \text{for} \quad f(x,y,z),$$

provided the indicated partial derivatives exist.

The symbol ∇ is read "gradient" or "grad" or "del." In order to deal with functions $f(x,y)$ and $f(x,y,z)$ in a unified manner, let $P = (x,y)$ or $P = (x,y,z)$. The gradient of f is a vector function of position. This fact is expressed by $\nabla f(P) = < f_x(P), f_y(P) >$ or $\nabla f(P) = < f_x(P), f_y(P), f_z(P) >$. The gradient vector $\nabla f(P)$ is usually attached to the point P, as illustrated in our first example.

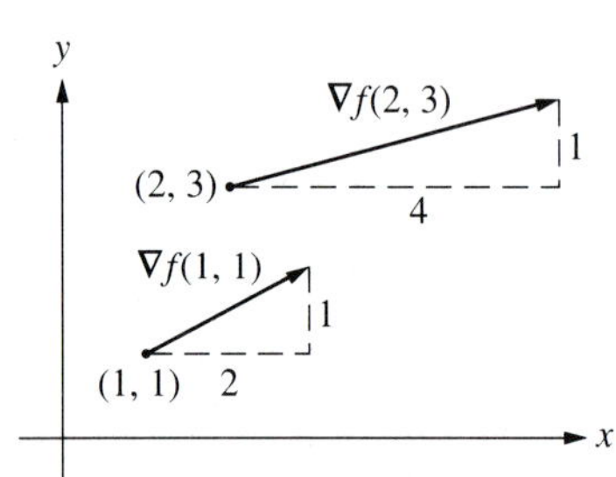

FIGURE 1

EXAMPLE 1. Let $f(x,y) = x^2 + y$. Then $\nabla f(x,y) = < f_x, f_y > = < 2x, 1 >$. Figure 1 displays the vector $\nabla f(1,1) = < 2,1 >$ attached to the point $(1,1)$ and the vector $\nabla f(2,3) = < 4,1 >$ attached to the point $(2,3)$. □

For convenience, we often use the position vectors $\mathbf{r} = < x,y >$ and $\mathbf{r} = < x,y,z >$ in place of the points $P = (x,y)$ and $= (x,y,z)$. Then we write $f(\mathbf{r})$ and $\nabla f(\mathbf{r})$ instead of $f(P)$ and $\nabla f(P)$.

EXAMPLE 2. Find the gradient of

$$f(\mathbf{r}) = \frac{1}{||\mathbf{r}||} = \frac{1}{\sqrt{x^2 + y^2 + z^2}} \qquad \text{for } \mathbf{r} \neq \mathbf{0}.$$

Then evaluate the gradient at $\mathbf{r} = < 1,2,2 >$.

Solution. Routine calculations give

$$\frac{\partial f}{\partial x} = \frac{-x}{(x^2 + y^2 + z^2)^{3/2}}, \quad \frac{\partial f}{\partial y} = \frac{-y}{(x^2 + y^2 + z^2)^{3/2}}, \quad \frac{\partial f}{\partial z} = \frac{-z}{(x^2 + y^2 + z^2)^{3/2}}.$$

Since $\mathbf{r} = < x,y,z >$ and $(x^2 + y^2 + z^2)^{3/2} = ||\mathbf{r}||^3$,

$$\nabla f(\mathbf{r}) = \left\langle \frac{-x}{||\mathbf{r}||^3}, \frac{-y}{||\mathbf{r}||^3}, \frac{-z}{||\mathbf{r}||^3} \right\rangle = \frac{-\mathbf{r}}{||\mathbf{r}||^3},$$

$$||\nabla f(\mathbf{r})|| = \frac{1}{||\mathbf{r}||^2}.$$

Let $r = ||\mathbf{r}|| = \sqrt{x^2 + y^2 + z^2}$. Then the foregoing results can be expressed as

$$\nabla\left(\frac{1}{r}\right) = \frac{-\mathbf{r}}{r^3} \quad \text{and} \quad \left\|\nabla\left(\frac{1}{r}\right)\right\| = \frac{1}{r^2}.$$

Apart from a constant factor, $\nabla(1/r)$ models the inverse square law for gravitational attraction. As a special case, let $\mathbf{r} = <1,2,2>$. Then $r = ||\mathbf{r}|| = 3$ and

$$\nabla\left(\frac{1}{r}\right) = -\frac{1}{27} < 1,2,2 >, \left\|\nabla\left(\frac{1}{r}\right)\right\| = \frac{1}{9}.$$

See Fig. 2. (The gradient $\nabla(1/r)$ is not drawn to scale.) □

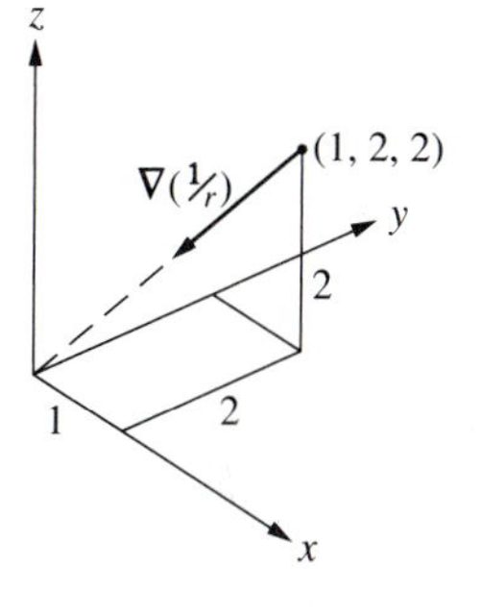

FIGURE 2

Gradient vectors often represent forces or velocities. Such applications, including gravitational attraction, will be explored as we go along (especially in Chs. 7 and 8).

Gradients and Derivatives

There is a close analogy between gradients of functions of two or three variables and derivatives of functions of one variable. For example, gradients obey differentiation rules such as $\nabla(f+g) = \nabla f + \nabla g$ and $\nabla(fg) = f\nabla g + g\nabla f$. Additional differentiation rules are given in Prob. 40.

For a function $f(x)$ defined on an open interval I,

$$f \text{ is constant on } I \quad \Leftrightarrow \quad f' = 0 \text{ on } I.$$

There is a corresponding result for a function $f(x,y)$ defined on an open rectangle R:

$$\boxed{f \text{ is constant on } R \quad \Leftrightarrow \quad \nabla f = \mathbf{0} \text{ on } R.}$$

To establish this useful property of the gradient, use its one-variable counterpart twice:

$$\nabla f = \mathbf{0} \text{ on } R \quad \Leftrightarrow \quad f_x = 0 \quad \text{and} \quad f_y = 0 \text{ on } R \quad \Leftrightarrow$$

$$f \text{ is constant on horizontal and vertical lines in } R \quad \Leftrightarrow$$

$$f \text{ is constant on } R.$$

Later in the section, we shall establish the foregoing result with the rectangle replaced by a much more general region.

Another aspect of the analogy between gradients and derivatives concerns differentiability. If $f(x)$ is differentiable at x_0, a slight reformulation of the definition of a derivative gives

$$\frac{f(x) - f(x_0) - f'(x_0)(x - x_0)}{x - x_0} \to 0 \quad \text{as} \quad x \to x_0.$$

The gradients of $f(x,y)$ and $f(x,y,z)$ satisfy the similar relation

$$\frac{f(\mathbf{r}) - f(\mathbf{r}_0) - \nabla f(\mathbf{r}_0) \cdot (\mathbf{r} - \mathbf{r}_0)}{||\mathbf{r} - \mathbf{r}_0||} \to 0 \text{ as } \mathbf{r} \to \mathbf{r}_0,$$

when the partial derivatives of f are continuous at $\mathbf{r}_0$. This limit follows directly from Ths. 1 and 2 of Sec. 5.4. In more advanced treatments of calculus, a function $f(\mathbf{r})$ is said to be **differentiable at $\mathbf{r}_0$** if there is a vector $\mathbf{v}_0$ that satisfies

$$\frac{f(\mathbf{r}) - f(\mathbf{r}_0) - \mathbf{v}_0 \cdot (\mathbf{r} - \mathbf{r}_0)}{||\mathbf{r} - \mathbf{r}_0||} \to 0 \text{ as } \mathbf{r} \to \mathbf{r}_0.$$

Then it follows quite easily that $\mathbf{v}_0 = \nabla f(\mathbf{r}_0)$, as you are asked to verify in the problems. Now Ths. 1 and 2 in Sec. 5.4 tell us that a function $f(\mathbf{r})$ is differentiable at $\mathbf{r}_0$ if it has partial derivatives near $\mathbf{r}_0$ that are continuous at $\mathbf{r}_0$. Consequently, for a function f defined on an open set D,

$$f \text{ has continuous partials on } D \quad \Rightarrow \quad f \text{ is differentiable on } D.$$

The converse of this displayed statement is false.

Throughout the rest of this section, we assume that f has continuous partial derivatives in the region of interest. However, several of the general results that follow are valid under the weaker hypothesis that f is merely differentiable, but that is a story primarily for the problems and for advanced calculus.

Gradients and Directional Derivatives

Directional derivatives can be expressed conveniently using gradients. Moreover, the resulting formula leads us to a deeper understanding of directional derivatives and gradients.

In Sec. 5.4 we established the following formulas for directional derivatives. If f has continuous partial derivatives at P and $\mathbf{u}$ is a unit vector, then

$$D_{\mathbf{u}} f(P) = af_x(P) + bf_y(P) \quad \text{for } P = (x,y), \quad \mathbf{u} = \langle a,b \rangle,$$
$$D_{\mathbf{u}} f(P) = af_x(P) + bf_y(P) + cf_z(P) \quad \text{for } P = (x,y,z), \quad \mathbf{u} = \langle a,b,c \rangle.$$

Both formulas are expressed in vector form by

$$\boxed{D_{\mathbf{u}} f(P) = \nabla f(P) \cdot \mathbf{u}.}$$

EXAMPLE 3. Let $f(x,y) = 2x^2 - xy - 2y^2$. Find the directional derivative of f in the direction from $(2,1)$ to $(6,4)$.

Solution. First, $\nabla f = < 4x - y, -x - 4y >$ and $\nabla f(2,1) = < 7, -6 >$. The vector from $(2,1)$ to $(6,4)$ is $\mathbf{v} = < 4,3 >$. The unit vector in the same direction is $\mathbf{u} = \frac{1}{5} < 4,3 >$. Therefore,

$$D_{\mathbf{u}} f(2,1) = < 7, -6 > \cdot \frac{1}{5} < 4,3 > = \frac{1}{5}(28 - 18) = 2. \square$$

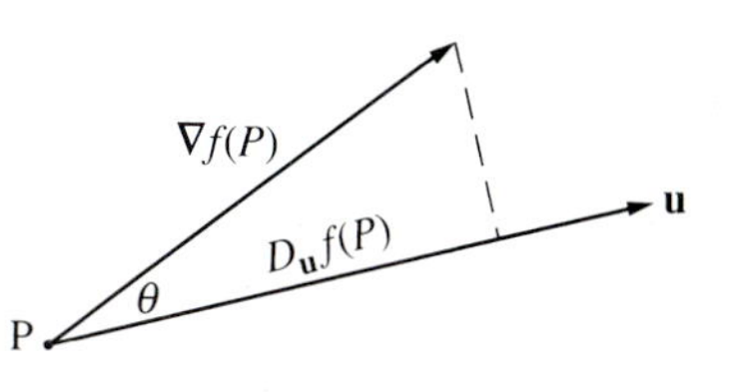

FIGURE 3

The formula $D_{\mathbf{u}} f(P) = \nabla f(P) \cdot \mathbf{u}$ sheds important new light on gradients and directional derivatives. In Fig. 3, the point P is fixed, $\nabla f(P) \neq \mathbf{0}$, $||\mathbf{u}|| = 1$ and

$$D_{\mathbf{u}} f(P) = \nabla f(P) \cdot \mathbf{u} = ||\nabla f(P)|| \cos \theta.$$

The directional derivative $D_{\mathbf{u}} f(P)$ is the component of $\nabla f(P)$ in the direction $\mathbf{u}$. Consider $D_{\mathbf{u}} f(P)$ as $\mathbf{u}$ varies over all directions. From the figure or from $D_{\mathbf{u}} f(P) = ||\nabla f(P)|| \cos \theta$, the maximum value of $D_{\mathbf{u}} f(P)$ over all directions $\mathbf{u}$ is attained when $\theta = 0$, $\cos \theta = 1$, and $\mathbf{u}$ points in the same direction as $\nabla f(P)$. Therefore,

$$\max_{\mathbf{u}} D_{\mathbf{u}} f(P) = ||\nabla f(P)||.$$

Since the directional derivative $D_{\mathbf{u}} f(P)$ is the rate of change of f with respect to distance from P in the **u**–direction,

> $f(P)$ increases most rapidly in the direction of $\nabla f(P)$ and the maximum rate of increase is $||\nabla f(P)||$.

Also from Fig. 3 or $D_{\mathbf{u}} f(P) = ||\nabla f(P)|| \cos \theta$, the minimum value of $D_{\mathbf{u}} f(P)$ over all directions $\mathbf{u}$ is attained when $\theta = \pi$, $\cos \theta = -1$, and $\mathbf{u}$ points in the direction opposite to $\nabla f(P)$. Therefore,

$$\min_{\mathbf{u}} D_{\mathbf{u}} f(P) = -||\nabla f(P)||.$$

Since a negative rate of increase is a rate of decrease,

> $f(P)$ decreases most rapidly in the direction of $-\nabla f(P)$ and the maximum rate of decrease is $||\nabla f(P)||$.

Thus, the maximum rates of increase and decrease are the same, but occur in opposite directions. This should seem obvious. For example, if the elevation on a hillside increases most rapidly if you walk eastward from a given point, then it will decrease most rapidly if you turn around and walk westward from that point.

EXAMPLE 4. Return to the pancake griddle of Ex. 8 in Sec. 5.3. The temperature at (x, y) is

$$T(x,y) = 375 + \frac{24}{\pi}\cos\frac{\pi x}{4}\sin\frac{\pi y}{4}$$

in degrees Fahrenheit. Find the maximum rate of increase and decrease of temperature with respect to distance from $(1,3)$ and the directions in which they occur.

Solution. First,

$$\nabla T = < T_x, T_y > = \left\langle -6\sin\frac{\pi x}{4}\sin\frac{\pi y}{4}, 6\cos\frac{\pi x}{4}\cos\frac{\pi y}{4}\right\rangle,$$

$$\nabla T(1,3) = < -3, 3 > = 3 < -1, 1 >.$$

The maximum rate of increase of temperature at $(1,3)$ is $||\nabla T(1,3)|| = 3\sqrt{2}$ deg/ft. It occurs in the direction of $\nabla T(1,3) = < -3, 3 >$ or $< -1, 1 >$. This direction may be called northwest with the usual orientation of the coordinate axes. The maximum rate of decrease of temperature at $(1,3)$ also is $||\nabla T(1,3)|| = 3\sqrt{2}$ deg/ft. It occurs in the direction of $-\nabla T(1,3) = < 3, -3 >$ or of $< 1, -1 >$, which may be called southeast. □

The gradient of a function $z = f(x,y)$ is denoted either by $\nabla f = < f_x, f_y >$ or by $\nabla z = < z_x, z_y >$. The latter notation often saves writing.

EXAMPLE 5. The equation $4x^2 + y^2 + 4z^2 = 24$ with $z \geq 0$ defines z implicitly as a function of x and y. Find the maximum directional derivative of $z = z(x,y)$ at $(x,y) = (1,2)$ and its direction. Interpret the results geometrically.

Solution. First, note that $z = 2$ when $(x,y) = (1,2)$. We shall need $\nabla z = < z_x, z_y >$. To find z_x and z_y, differentiate $4x^2 + y^2 + 4z^2 = 24$ implicitly with respect to x and y:

$$8x + 8z\, z_x = 0, \qquad z_x = -\frac{x}{z},$$

$$2y + 8z\, z_y = 0, \qquad z_y = -\frac{y}{4z}.$$

Then

$$z_x(1,2) = -\frac{1}{2}, \qquad z_y(1,2) = -\frac{1}{4},$$

$$\nabla z(1,2) = \left\langle -\frac{1}{2}, -\frac{1}{4}\right\rangle = -\frac{1}{4} < 2, 1 >.$$

So the maximum directional derivative at $(1,2)$ is

$$||\nabla z(1,2)|| = \frac{1}{4}\sqrt{5} \approx 0.56,$$

which occurs in the direction of $\nabla z(1,2)$. See Fig. 4.

The graph of $4x^2 + y^2 + 4z^2 = 24$ with $z \geq 0$ is the top half of an ellipsoid in Fig. 5. Think of it as a hill with elevation function $z = z(x, y)$. The maximum directional derivative is the maximum slope of the surface in any direction at $(1,2,2)$. Therefore, $\nabla z(1,2)$ gives the direction of steepest ascent at $(1,2,2)$ and $||\nabla z(1,2)|| \approx 0.56$ is the maximum rate of ascent leaving $(1,2,2)$. □

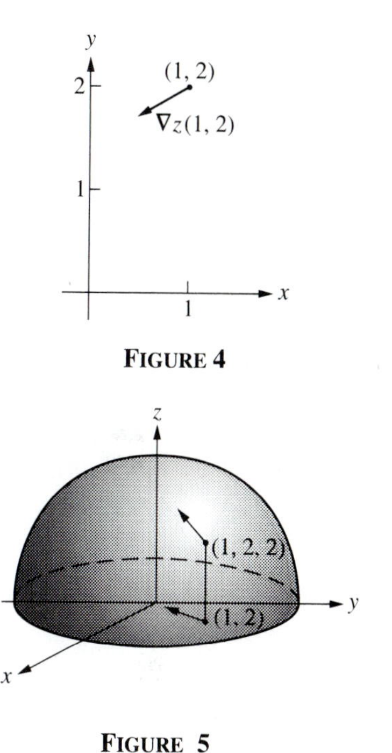

FIGURE 4

FIGURE 5

Chain Rules in Vector Form

The basic chain rule looks like a dot product of two vectors. This observation leads to a vector form for the chain rule that gives more insight into the rule, especially in applications to motion problems.

The basic chain rule,

$$\frac{d}{dt} f(x(t), y(t)) = \frac{\partial f}{\partial x}\frac{dx}{dt} + \frac{\partial f}{\partial y}\frac{dy}{dt},$$

tells us how $f(x(t), y(t))$ changes as $(x(t), y(t))$ moves along a curve C, as in Fig. 6. The curve is expressed in vector form by $\mathbf{r}(t) = < x(t), y(t) >$. The vector

$$\mathbf{r}'(t) = \frac{d\mathbf{r}}{dt} = \left\langle \frac{dx}{dt}, \frac{dy}{dt} \right\rangle$$

is tangent to C at $\mathbf{r}(t)$. Since $\nabla f = < \partial f/\partial x, \partial f/\partial y >$, the chain rule can be expressed in vector form by

$$\frac{df}{dt} = \nabla f \cdot \frac{d\mathbf{r}}{dt}.$$

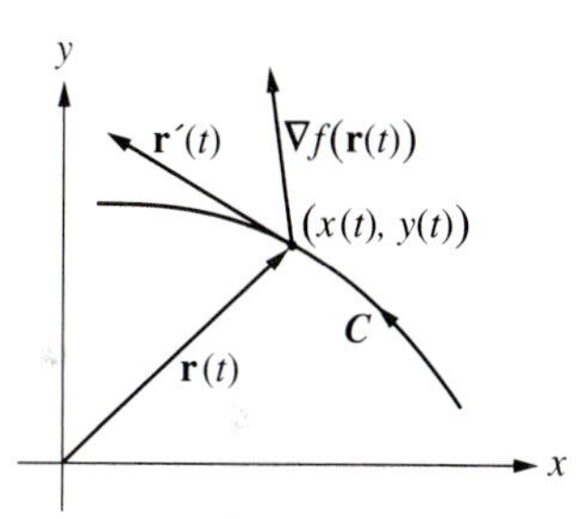

FIGURE 6

It is the same story for the basic chain rule in 3-space,

$$\frac{d}{dt} f(x(t), y(t), z(t)) = \frac{\partial f}{\partial x}\frac{dx}{dt} + \frac{\partial f}{\partial y}\frac{dy}{dt} + \frac{\partial f}{\partial z}\frac{dz}{dt}.$$

Now $\mathbf{r}(t) = < x(t), y(t), z(t) >$ describes a curve C in 3-space and the vector

$$\mathbf{r}'(t) = \frac{d\mathbf{r}}{dt} = \left\langle \frac{dx}{dt}, \frac{dy}{dt}, \frac{dz}{dt} \right\rangle$$

is tangent to C at $\mathbf{r}(t)$. Once again, $df/dt = \nabla f \cdot d\mathbf{r}/dt$.

In summary, both chain rules are expressed by the single vector formula

$$\boxed{\frac{df}{dt} = \nabla f \cdot \frac{d\mathbf{r}}{dt}.}$$

In more detail,

$$\frac{d}{dt}\, f(\mathbf{r}(t)) = \nabla f(\mathbf{r}(t)) \cdot \mathbf{r}'(t).$$

EXAMPLE 6. Let $f(x,y) = e^{xy}$, where $x = \cos t$, $y = \sin t$ for $0 \le t \le \pi/2$. Use the vector form of the chain rule to find df/dt when $t = \pi/6$.

Solution. In this case, $\nabla f(x,y) = e^{xy} < y, x >$ and

$$\nabla f(\sqrt{3}/2,\ 1/2) = e^{\sqrt{3}/4} < 1/2,\ \sqrt{3}/2 > = \frac{1}{2}\, e^{\sqrt{3}/4} < 1,\ \sqrt{3} >.$$

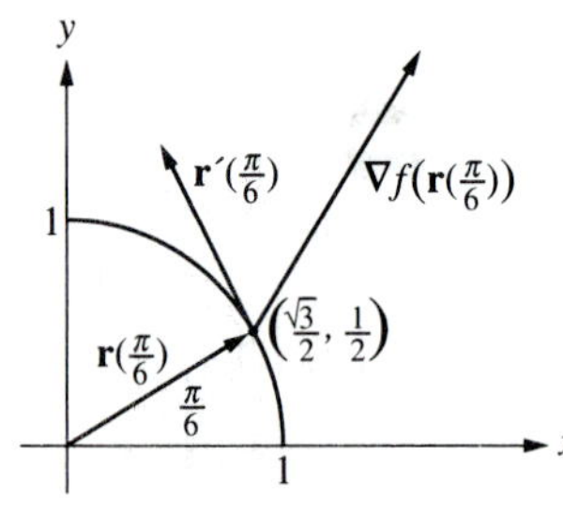

FIGURE 7

The graph of $x = \cos t$, $y = \sin t$ for $0 \le t \le \pi/2$ is the quarter circle in Fig. 7. It is expressed in vector form by $\mathbf{r}(t) = < \cos t, \sin t >$. Then $\mathbf{r}'(t) = < -\sin t, \cos t >$. So $\mathbf{r}(\pi/6) = < \sqrt{3}/2,\ 1/2 >$ and

$$\mathbf{r}'(\pi/6) = < -1/2,\ \sqrt{3}/2 > = \frac{1}{2} < -1,\ \sqrt{3} >.$$

Put the pieces together to obtain

$$\frac{df}{dt} = \nabla f \cdot \frac{d\mathbf{r}}{dt} = \frac{1}{2}\, e^{\sqrt{3}/4} \approx 0.77 \text{ when } t = \pi/6. \ \square$$

EXAMPLE 7. The position vector of a moving object is $\mathbf{r}(t) = <x(t), y(t)>$ at time t. When $t = 2$, the position is $\mathbf{r}(2) = <3, 4>$ and the velocity is $\mathbf{r}'(2) = < -2, -1 >$. A quick sketch of the velocity vector shows that the object is moving closer to the origin when $t = 2$. How fast is its distance from the origin decreasing at this time? Measure distance in feet and time in seconds.

Solution. The distance of the object from the origin is $f(x,y) = \sqrt{x^2 + y^2}$, where $x = x(t)$ and $y = y(t)$. When $t = 2$ and $< x, y > = <3, 4>$,

$$\nabla f = \left\langle \frac{x}{\sqrt{x^2+y^2}}, \frac{y}{\sqrt{x^2+y^2}} \right\rangle = \left\langle \frac{3}{5}, \frac{4}{5} \right\rangle.$$

By the chain rule,

$$\frac{df}{dt} = \nabla f \cdot \mathbf{r}'(t) = \left\langle \frac{3}{5}, \frac{4}{5} \right\rangle \cdot < -2, -1 > = -2 \text{ ft/sec when } t = 2.$$

The minus sign indicates that the object is moving closer to the origin at time $t = 2$, which we already observed. The distance from the origin is decreasing 2 ft/sec at that time. $\square$

Zero Gradients and Constant Functions

Earlier we found that a function $f(x,y)$ is constant on an open rectangle R if and only if $\nabla f = 0$ on R. Now we use the vector form of the chain rule to extend this result to functions defined on more general domains. Recall that a set is connected if any two of its points can be joined by a continuous curve that lies entirely in the set. Furthermore, if the set is open, we can choose the connecting curve to have a continuously varying tangent. (We shall need this geometrically plausible fact in the proof of the next theorem.)

> **Theorem 1** *Zero Gradients Characterize Constant Functions*
> Let f have continuous partials in a connected open set D in 2–space or 3–space. Then
> $$\nabla f = \mathbf{0} \text{ on } D \quad \Leftrightarrow \quad f \text{ is constant on } D.$$

Proof. Of course, if f is constant on D, then $\nabla f = \mathbf{0}$ on D. Conversely, assume that $\nabla f = \mathbf{0}$ on D. We must prove that f is constant on D. Fix a point $P_0 = (x_0, y_0)$ or (x_0, y_0, z_0) in D and let $P = (x, y)$ or (x, y, z) be any point in D. Since D is open and connected, there is a differentiable curve C, say given by $\mathbf{r}(t)$ for $a \le t \le b$, that begins at P_0, ends at P, and has a continuously varying tangent $\mathbf{r}'(t)$. The two–dimensional case is illustrated in Fig. 8. Since $\nabla f = \mathbf{0}$ on D, the chain rule gives

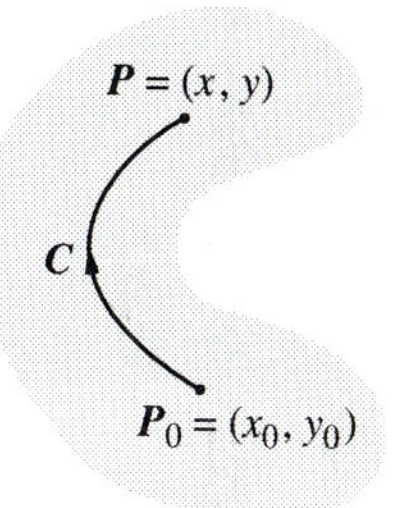

FIGURE 8

$$\frac{d}{dt} f(\mathbf{r}(t)) = \nabla f(\mathbf{r}(t)) \cdot \mathbf{r}'(t) = 0.$$

Consequently, the function $f(\mathbf{r}(t))$ of the single variable t is constant for $a \le t \le b$, so that

$$f(P) = f(\mathbf{r}(t)) = f(\mathbf{r}(a)) = f(P_0)$$

for any point P in D. Thus, the function f is constant on D. □

Theorem 1 is valid also under the weaker hypothesis that f is differentiable in D. See the problems.

An immediate and important consequence of Th. 1 follows. Suppose that f and g are differentiable on a connected open set D. Then

$$\boxed{\nabla f = \nabla g \text{ on } D \quad \Leftrightarrow \quad f - g = c \text{ for some constant } c.}$$

In words, two functions on D with the same gradient differ by a constant.

Gradients and Level Curves

Level curves were introduced in Sec. 5.1. A level curve of a function $f(x, y)$ is the graph of an equation $f(x, y) = c$ with any c in the range of f. Here we are interested in the interplay between gradients and level curves. An example will show what to expect.

EXAMPLE 8. Let $f(x, y) = x^2 + 2y^2$. The level curves $f = c$ with $c > 0$ are ellipses. The point $(2, 1)$ lies on the level curve $f = 6$, which is shown in Fig.

9. The gradient of f is $\nabla f(x,y) = < 2x, 4y >$. In particular, $\nabla f(2,1) = <4,4>$. We claim that $\nabla f(2,1)$ is perpendicular to the level curve $f = 6$, which means that $\nabla f(2,1)$ is perpendicular to the tangent line T at $(2,1)$. Equivalently, $\nabla f(2,1)$ is perpendicular to a tangent vector to the curve at $(2,1)$. To find a tangent vector, express the top half of the level curve $x^2 + 2y^2 = 6$ in the vector form $\mathbf{r}(x) = < x, y >$ with x as parameter and y an implicitly defined function of x. Then $\mathbf{r}'(x) = < 1, y' >$ and we can find y' by implicit differentiation of $x^2 + 2y^2 = 6$:

$$2x + 4yy' = 0, \qquad y' = -\frac{x}{2y}, \qquad y' = -1 \text{ at } (2,1).$$

Therefore, $\mathbf{r}'(2) = < 1, -1 >$ is a tangent vector to the level curve at $(2,1)$ as shown in Fig. 9. Since

$$\nabla f(2,1) \cdot \mathbf{r}'(2) = < 4, 4 > \cdot < 1, -1 > = 0,$$

$\nabla f(2,1) \perp \mathbf{r}'(2)$ and the gradient is perpendicular to the level curve at (2,1). In the same way, the gradient of $f(x,y) = x^2 + 2y^2$ at any point $(x,y) \neq (0,0)$ is perpendicular to the level curve of f through (x,y). □

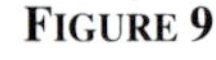

FIGURE 9

For the function $f(x,y) = x^2 + 2y^2$ in Ex. 8, the level curve $f(x,y) = 0$ is exceptional. It consists of the single point $(0,0)$. At this point, $\nabla f(0,0) = \mathbf{0}$, which has no direction. Thus, a level curve of a differentiable function f is not necessarily a curve in the usual sense. However, most of the level curves $f(x,y) = c$ we shall deal with are authentic curves. If $\nabla f(x,y) \neq \mathbf{0}$ for (x,y) in an open set D, then the level curves $f(x,y) = c$ in D are differentiable curves or pieces of differential curves, as illustrated in Fig. 10. This will not be proved. It is a consequence of the implicit function theorem for $f(x,y)$ given in Sec. 5.5.

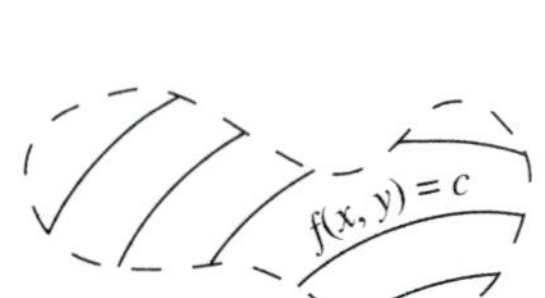
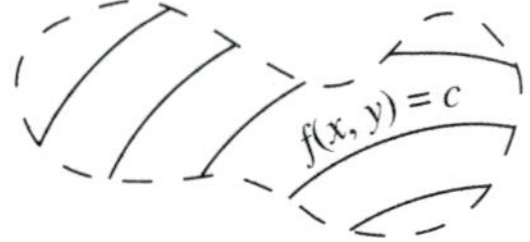

FIGURE 10

We observed in Ex. 8 that the gradient of $f(x,y) = x^2 + 2y^2$ at any point $(x,y) \neq (0,0)$ is perpendicular to the level curve of f through (x,y). This behavior is typical.

Theorem 2 *Gradients Are Perpendicular to Level Curves*
Let C be a level curve $f = c$ of a function $f(x,y)$ with continuous partial derivatives in an open set containing C. Assume that $\nabla f(x,y) \neq \mathbf{0}$ along C. Then $\nabla f(x,y)$ is perpendicular to C at any point (x,y) on C.

Proof. Express C in vector form $\mathbf{r} = \mathbf{r}(t)$, as shown in Fig. 11. Then $\mathbf{r}'(t)$ is a tangent vector to C. Since $f = c$ on C, $f(\mathbf{r}(t)) = c$ and, by the vector form of the chain rule,

$$\nabla f(\mathbf{r}(t)) \cdot \mathbf{r}'(t) = \frac{d}{dt} f(\mathbf{r}(t)) = 0.$$

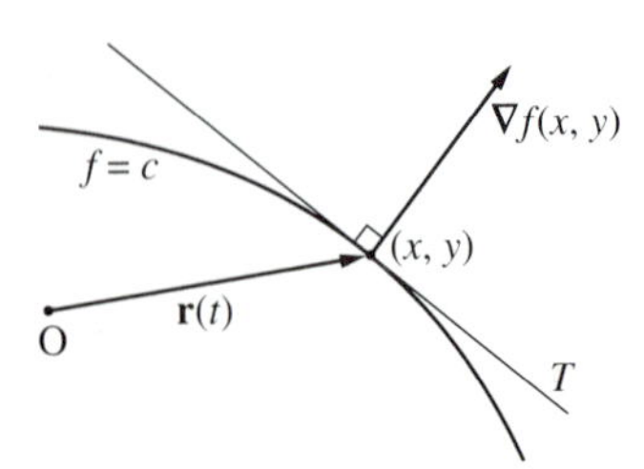

FIGURE 11

So $\nabla f(\mathbf{r}(t)) \perp \mathbf{r}'(t)$ and $\nabla f(x,y)$ is perpendicular to C at (x,y). □

EXAMPLE 9. Let $f(x,y) = xy$ for $x > 0$, $y > 0$. The level curves, $f(x,y) = c$ for $c > 0$, are branches of hyperbolas. Three level curves are shown in Fig. 12. The gradient of f is $\nabla f(x,y) = <y,x>$. Clearly, $\nabla f \neq \mathbf{0}$ at any point (x,y) on a level curve $f = c$ in Fig. 12. Hence, by Th. 2, $\nabla f = <y,x>$ is perpendicular at (x,y) to the level curve through (x,y). □

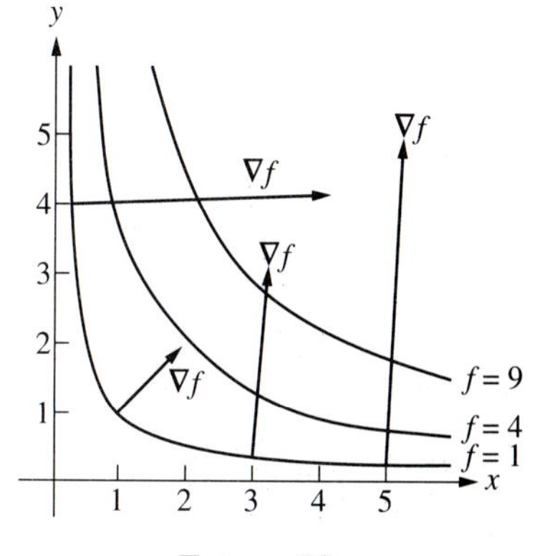

FIGURE 12

Often we are interested in the behavior of a function $f(x,y)$ only for (x,y) near a point (x_0,y_0) where $\nabla f(x_0,y_0) \neq \mathbf{0}$. Let $f(x_0,y_0) = c$ and assume that f has continuous partials near (x_0,y_0). Since $\nabla f(x,y)$ is continuous, $\nabla f(x,y) \neq \mathbf{0}$ for (x,y) near (x_0,y_0) and the level curve $f(x,y) = c$ is a differentiable curve for (x,y) near (x_0, y_0). By Th. 2, $\nabla f(x_0,y_0)$ is perpendicular to the level curve at (x_0,y_0).

EXAMPLE 10. Let $f(x,y) = x^3 + y^3 - 2xy$. Then $f(1,2) = 5$. Show that $\nabla f(1,2)$ is perpendicular to the level curve $f(x,y) = x^3 + y^3 - 2xy = 5$ at $(1,2)$.

Solution. In this case, $\nabla f(x,y) = <3x^2 - 2y, 3y^2 - 2x>$. Since f has continuous partials, f is differentiable. Since $\nabla f(1,2) = <-1,10> \neq 0$, the level curve $f(x,y) = 5$ is a differentiable curve for (x,y) near $(1,2)$ and $\nabla f(1,2)$ is perpendicular to the level curve at $(1,2)$. Notice that we have obtained all this information without a picture to rely on. □

Gradients and Level Surfaces

Recall that a level surface of a function $f(x,y,z)$ is the graph of $f(x,y,z) = c$ with any c in the range of f. The relationship between level surfaces and gradients is a good deal like the story for level curves and gradients. A simple example will lead the way.

EXAMPLE 11. Let $f(x,y,z) = x^2 + y^2 + z^2$. The level surfaces $f = c$ with $c > 0$ are spheres. The gradient of f is $\nabla f(x,y,z) = 2 <x,y,z>$, which points in the radial direction if $(x,y,z) \neq (0,0,0)$. Hence, for $(x,y,z) \neq (0,0,0)$, $\nabla f(x,y,z)$ is perpendicular to the spherical level surface through (x,y,z), which means that $\nabla f(x,y,z)$ is perpendicular to the tangent plane to the sphere at (x,y,z). The level surface $f = 0$ is exceptional; it consists of the single point $(0,0,0)$, where $\nabla f(0,0,0) = \mathbf{0}$. □

In Ex. 11, the gradient vector at any nonzero point is perpendicular to the level surface through that point. Theorem 3 that follows tells us that this behavior is typical. To put Th. 3 in proper perspective, we begin with a few preliminary remarks. Suppose that $f(x,y,z)$ has continuous partial derivatives and $\nabla f(x,y,z) \neq 0$ for (x,y,z) in an open set D. Then the level surfaces $f(x,y,z) = c$ in D are genuine surfaces or pieces of surfaces. In a small enough neighborhood of any point in D, the level surface through that point can be expressed in one of the standard forms $z = z(x,y)$ or $x = x(y,z)$ or $y = y(x,z)$. At any point (x,y,z) on a level surface of f, the surface has a tangent plane. These assertions will not be proved. They are consequences of the implicit function theorem for $f(x,y,z)$ given in Sec. 5.5.

Theorem 3 *Gradients Are Perpendicular to Level Surfaces*
Let S be a level surface $f = c$ for a function $f(P)$ that has continuous partial derivatives in an open set containing S. Assume that $\nabla f(P) \neq \mathbf{0}$ on S. Then $\nabla f(P)$ is perpendicular to S at any point P on S.

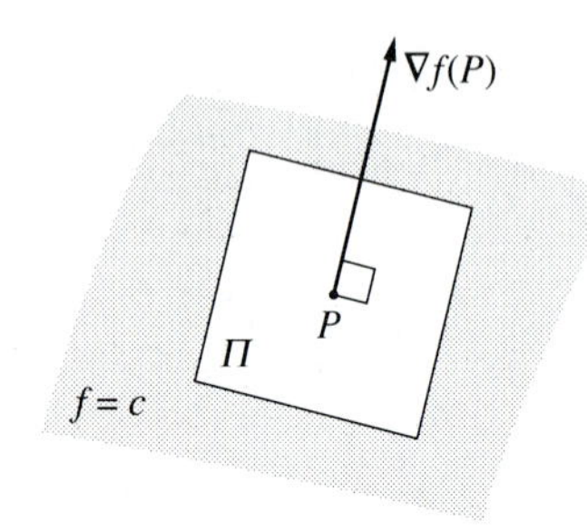

FIGURE 13

Proof. Fix a point P on S. We must show that $\nabla f(P)$ is perpendicular to the tangent plane Π to S at P, as indicated in Fig. 13. As noted earlier, near P the surface S can be expressed as the graph of an equation of the form $z = z(x,y)$ or $x = x(y,z)$ or $y = y(x,z)$. To be definite, assume that the level surface S is given by $z = z(x,y)$. As we have done before, intersect the surface S with any vertical plane through P to obtain a curve C. It is a consequence of the implicit function theorem that C is a differentiable curve near P. Express C in vector form $\mathbf{r} = \mathbf{r}(t)$ with $\mathbf{r}(0)$ the position vector of P. Then $\mathbf{r}'(0)$ is a tangent vector to C at P. Since $f = c$ on C, $f(\mathbf{r}(t)) = c$ and, by the vector form of the chain rule,

$$\nabla f(\mathbf{r}(t)) \cdot \mathbf{r}'(t) = \frac{d}{dt} f(\mathbf{r}(t)) = 0.$$

Set $t = 0$ to find $\nabla f(\mathbf{r}(0)) \perp \mathbf{r}'(0)$. Since $\mathbf{r}(0)$ is the position vector of P, $\nabla f(P)$ is perpendicular to the tangent line to C through P. In fact, we have shown that $\nabla f(P)$ is perpendicular to every tangent line to S through P, because the foregoing argument applies to the curve C obtained by intersecting S with any vertical plane through P. Therefore, $\nabla f(P)$ is perpendicular to the tangent plane Π at P. □

A consequence of Th. 3 is that a vector equation for the tangent plane Π at P_0 is

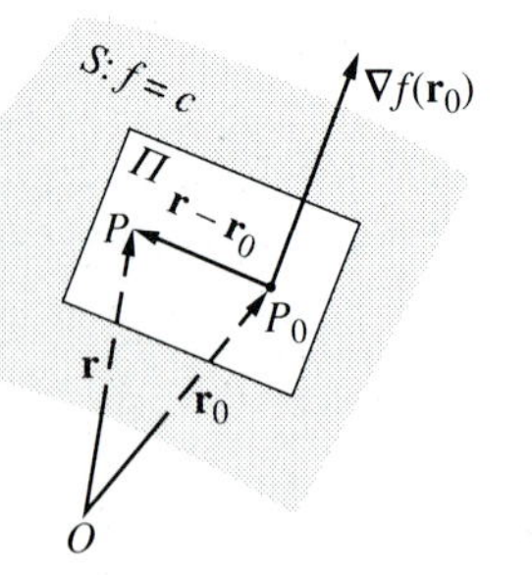

FIGURE 14

$$\boxed{\nabla f(\mathbf{r}_0) \cdot (\mathbf{r} - \mathbf{r}_0) = \mathbf{0},}$$

where $\mathbf{r} = < x, y, z >$ and $\mathbf{r}_0 = < x_0, y_0, z_0 >$. See Fig. 14.

EXAMPLE 12. Find vector and scalar equations for the tangent plane Π to the level surface $f(x,y,z) = xyz + e^{x+y+z} = 7$ at the point $(3, -2, -1)$.

Solution. First, notice that $(x,y,z) = (3, -2, -1)$ satisfies the given equation, so it is a point on the surface. The gradient of f is

$$\nabla f(x,y,z) = < yz + e^{x+y+z}, xz + e^{x+y+z}, xy + e^{x+y+z} >.$$

So $\nabla f(3, -2, -1) = < 3, -2, -5 >$, which is perpendicular to the tangent plane Π. Consequently, a vector equation for Π is

$$< 3, -2, -5 > \cdot < x - 3, y + 2, z + 1 > = 0.$$

Scalar equations for Π are

$$3(x-3)-2(y+2)-5(z+1)=0 \text{ and } 3x-2y-5z=18. \ \square$$

Although the equations given previously for tangent planes presumed that we were dealing with level surfaces, they apply to other surfaces as well. For example, if the surface in Ex. 12 is given in the form $xyz = 7 - e^{x+y+z}$, we can put it in level surface form merely by transposing the exponential term to the left-hand side. In general, any equation for a surface can be put in level surface form by transposing all nonconstant terms (or all terms) to the left-hand side.

EXAMPLE 13. Find an equation for the tangent plane to the surface

$$\frac{1}{x}+\frac{1}{y}+\frac{1}{z}=2+e^{(x-y)/z}$$

at the point $(1,1,1)$.

Solution. The given surface is the level surface $f=2$ for the function $f(x,y,z)=x^{-1}+y^{-1}+z^{-1}-e^{(x-y)/z}$. Since

$$\nabla f = \left\langle -x^{-2}-\frac{1}{z}\,e^{(x-y)/z},\ -y^{-2}+\frac{1}{z}\,e^{(x-y)/z},\ -z^{-2}+\frac{x-y}{z^2}\,e^{(x-y)/z}\right\rangle$$

and $\nabla f(1,1,1) = <-2,0,-1>$, a vector equation for the tangent plane is

$$<-2,0,-1> \cdot <x-1, y-1, z-1> = 0.$$

A scalar equation is

$$2x+z-3=0.$$

Since y is missing, the tangent plane is parallel to the y–axis. $\square$

PROBLEMS

In Probs. 1–10 find ∇f at the indicated point.

1. $f(x,y)=x^2+1/y^2,\ (2,-3)$
2. $f(x,y)=x/y-y/x,\ (2,-1)$
3. $f(x,y)=e^{2x}\sin \pi y,\ (1,1/2)$
4. $f(x,y)=\cos xy,\ (3,\pi)$
5. $f(x,y)=\arctan(x/y),\ (1,-1)$
6. $f(x,y)=\arcsin(x/y),\ (2,4)$
7. $f(x,y,z)=xe^{yz},\ (1,0,1)$
8. $f(x,y,z)=\sin xyz,\ (1/2,1/3,\pi)$
9. $f(x,y,z)=3xy/(x^2-z^2),\ (3,2,1)$
10. $f(x,y,z)=e^{xyz}\cos z,\ (2,-2,\pi/2)$

In Probs. 11–14 let $\mathbf{r} = <x,y>$ or $<x,y,z>$ and $r = \|\mathbf{r}\|$.

11. Show that

$$\nabla r = \frac{\mathbf{r}}{r}, \qquad \mathbf{r} \neq \mathbf{0}.$$

12. Let $f(x,y) = \ln\sqrt{x^2+y^2}$. Show that

$$\nabla f = \frac{\mathbf{r}}{r^2}, \qquad \mathbf{r} \neq \mathbf{0}.$$

13. Let $f(x,y,z) = (x^2+y^2+z^2)^{-n/2} = r^{-n}$, where n is a constant. Show that

$$\nabla r^{-n} = -\frac{n}{r^{n+2}}\mathbf{r}, \qquad \mathbf{r} \neq \mathbf{0}.$$

14. Let f be a differentiable function of one variable. Show that

$$\nabla f(r) = f'(r)\nabla \mathbf{r} = \frac{f'(r)}{r}\mathbf{r}.$$

In Probs. 15–20 find the directional derivative of the function at the given point and in the indicated direction.

15. $f(x,y) = x^2 + 1/y^2$, $(2,3)$, $< 1, -1 >$
16. $f(x,y) = x/y - y/x$, $(2,-1)$, $< 2, -1 >$
17. $f(x,y) = e^{2x}\sin \pi y$, $(1,1/2)$, $\theta = \pi/6$
18. $f(x,y) = \arctan(x/y)$, $(1,-1)$, $\theta = -\pi/4$
19. $f(x,y,z) = xe^{yz}$, $(1,0,1)$, $< 1,2,2 >$
20. $f(x,y,z) = \sin xyz$, $(1/2, 1/3, \pi)$, $< 1,1,1 >$
21. The temperature in degrees centigrade at (x,y,z) is $T = 2x^2 - y^2 + 4z^2$. Distances are in meters. (a) Find the rate of change of temperature at $(1,-2,1)$ in the direction of the vector $< 1,2,3 >$. (b) Find the direction from $(1,-2,1)$ in which the temperature increases most rapidly and the maximum rate of increase.
22. The terrain on a mountain above the reference level $z = 0$ has the shape given by $9x^2 + 4y^2 + 36z^2 = 33$, $z \geq 0$, where distance is in miles. So the elevation on the mountain at (x,y,z) is z. (a) Find the rate of change of elevation with respect to distance moved horizontally when departing $(1,2,2/3)$ in the direction of steepest ascent.
23. The electrical potential (voltage) at a point in the xy–plane is $V = 120e^{-x/5}$ $\sin(\pi y/24)$. The electric field strength at (x,y), which is the force per unit charge that acts on a charged particle at (x,y), is given by $\mathbf{E} = -\nabla V$. Experimental measurements show that V decreases most rapidly in the direction of $\mathbf{E}$. What is the mathematical reason for this experimental observation? Explain briefly.
24. The directional derivative of $f(x,y,z)$ is greatest at $(1,2,3)$ in the direction of $< 2,-1,2 >$ and the maximum rate of increase is 12. (a) Find ∇f at $(1,2,3)$. (b) Find the rate of change of f in the direction $< 1,1,1 >$ at $(1,2,3)$.
25. Find the rate of change of temperature $T = e^{xy/z}$ with respect to time t as a thermometer moves along the circular helix $\mathbf{r} = < \cos 3t, \sin 3t, 4t >$ at the instant when $t = \pi/2$.

26. Find the rate of change of elevation $z = 10{,}000 - 2x^2 - 3y^2$ with respect to time t at time $t = 3$ if you follow a path on a map given by $\mathbf{r} = \langle x, y \rangle = \langle t^2 - 4t, t^2 \rangle$.

27. Let $V = e^{-2x} \sin 2y$, as in Prob. 23. A charged particle, of unit charge, moves in the xy–plane with position vector $\mathbf{r}(t) = \langle x(t), y(t) \rangle$. At time $t = 4$, $\mathbf{r}(4) = \langle 5, 12 \rangle$ and $\mathbf{r}'(4) = \langle 3, -4 \rangle$. (a) How fast is the potential at the particle increasing or decreasing when $t = 4$? (b) What is the force on the particle at $t = 4$?

28. If $f(x,y) = e^{-xy}$ and $g(x,y) = 2x^2 - 3y^2$, find the rate of change of f at $(1,0)$ in the direction of most rapid increase of g.

29. If $f(x,y,z) = x^2y + y^2z + z^2x$ and $g(x,y,z) = x^2 + y^2 + z^2$, find the rate of change of f at $(1,1,1)$ in the direction of most rapid decrease of g.

30. Find a unit vector normal to the ellipse $4x^2 + 5y^2 = 96$ at (a) $(-2,4)$ and (b) $(2,-4)$.

31. The graph of the equation $x^2 + y^2 - \arctan xy = 1$ is a differentiable curve near $(0,1)$. Find a normal vector at $(0,1)$ and an equation for the tangent line at this point.

32. Find a unit vector normal to the paraboloid $x^2 + y^2 + z = 9$ at the point $(2,-1,4)$. Then find an equation for the tangent plane at this point.

33. The equation $21 + \ln(x + y + z) = x^2 + y^2 + z^2$ determines z as a function of x and y with continuous partial derivatives near the point $(-1,-2,4)$. Find the tangent plane to the surface at that point.

34. Find any tangent planes to the paraboloid $z = 3x^2 + 2y^2$ that are parallel to the plane $x + y + z = 1$.

35. Find any tangent planes to the ellipsoid $3x^2 + 2y^2 + z^2 = 66$ that are parallel to the plane $x + y + z = 1$.

36. Let a, b be constants and f and g be functions of two or three variables that have partial derivatives. Show that

    (a) $\nabla(af + bg) = a\nabla f + b\nabla g$,

    (b) $\nabla(fg) = f\nabla g + g\nabla f$,

    (c) $\nabla\left(\dfrac{f}{g}\right) = \dfrac{g\nabla f - f\nabla g}{g^2}, \qquad g \neq 0,$

    (d) $\nabla(f^p) = pf^{p-1}\nabla f$ whenever f^{p-1} is defined.

37. Assume that $f(\mathbf{r})$ with $\mathbf{r} = \langle x, y, z \rangle$ and $\mathbf{r}_0 = \langle x_0, y_0, z_0 \rangle$ satisfies

$$\frac{f(\mathbf{r}) - f(\mathbf{r}_0) - \mathbf{v}_0 \cdot (\mathbf{r} - \mathbf{r}_0)}{\|\mathbf{r} - \mathbf{r}_0\|} \to 0 \text{ as } \mathbf{r} \to \mathbf{r}_0$$

for some vector $\mathbf{v}_0 = \langle a, b, c \rangle$. Prove that $\mathbf{v}_0 = \nabla f(\mathbf{r}_0)$. *Hint.* Let $\varepsilon(\mathbf{r})$ be the preceding quotient, so that

$$f(\mathbf{r}) - f(\mathbf{r}_0) = \mathbf{v}_0 \cdot (\mathbf{r} - \mathbf{r}_0) + \varepsilon(\mathbf{r})\|\mathbf{r} - \mathbf{r}_0\|.$$

Choose $\mathbf{r} = \langle x_0 + \Delta x, y_0, z_0 \rangle$ and let $\Delta x \to 0$ to show that $a = f_x(\mathbf{r}_0)$.

38. If $f(\mathbf{r})$ is differentiable at $\mathbf{r}_0$, prove that f has a directional derivative at $\mathbf{r}_0$ in every direction $\mathbf{u}$ and $D_{\mathbf{u}} f(\mathbf{r}_0) = \nabla f(\mathbf{r}_0) \cdot \mathbf{u}$.

39. Let $\mathbf{r}(s)$ be the arc length parameterization of a curve C. Recall that $\mathbf{u} = \mathbf{r}'(s)$ is the unit tangent to C at $\mathbf{r}(s)$. Assume that f has continuous partial derivatives in an open set that contains C. Show that

$$\frac{d}{ds} f(\mathbf{r}(s)) = D_{\mathbf{u}} f(\mathbf{r}(s)) \quad \text{for} \quad \mathbf{u} = \mathbf{r}'(s).$$

In words, the rate of change of f with respect to arc length along C is equal to the directional derivative of f in the tangential direction to C at any point.

40. *(The mean value theorem for functions of several variables)* Let **a**, **b** be fixed vectors in 2– or 3–space. Let $f(\mathbf{r})$ be defined at points on and near the line L extending from **a** and **b**. Let $F(t) = f(\mathbf{a} + t\mathbf{b})$ for $0 \le t \le 1$.
(a) Apply the one-variable mean value theorem to F to obtain

$$\boxed{f(\mathbf{b}) - f(\mathbf{a}) = \nabla f(\mathbf{c}) \cdot (\mathbf{b} - \mathbf{a}),}$$

where **c** is some point on the line segment joining **a** and **b**.

(b) What hypotheses on f are sufficient to justify the steps needed to obtain the foregoing formula?

41. Suppose the surfaces $f(x, y, z) = c_1$ and $g(x, y, z) = c_2$ intersect in a differentiable curve C. Let $P_0 = (x_0, y_0, z_0)$ lie on C. Assume that $\nabla f(P_0) \times \nabla g(P_0) \neq \mathbf{0}$. Show that $\nabla f(P_0) \times \nabla g(P_0)$ is tangent to C at P_0. (*Remark.* If f and g have continuous partial derivatives near P_0 and $\nabla f(P_0) \times \nabla g(P_0) \neq \mathbf{0}$, then the surfaces do indeed intersect in a differentiable curve near P_0.)

42. Check that the point $(1, 3, 2)$ lies on the curve of intersection of the ellipsoid $4x^2 + y^2 + 3z^2 = 25$ and the plane $4x + 2y - z = 8$. Find a tangent vector to the curve of intersection of these surfaces at $(1, 3, 2)$.

43. Let $\mathbf{u}_r$ and $\mathbf{u}_\theta$ be the polar coordinate unit vectors from Sec. 4.5. If a function $f(x, y)$ with continuous partials is expressed in terms of r and θ through $x = r\cos\theta$, $y = r\sin\theta$, show that

$$\boxed{\nabla f = \frac{\partial f}{\partial r}\mathbf{u}_r + \frac{1}{r}\frac{\partial f}{\partial \theta}\mathbf{u}_\theta.}$$

44. Suppose that f is differentiable at $\mathbf{r}_0$ and that $D_{\mathbf{u}} f(\mathbf{r}_0) = 0$ in two nonparallel directions $\mathbf{u} = \mathbf{u}_1$ and $\mathbf{u} = \mathbf{u}_2$. Show that $D_{\mathbf{u}} f(\mathbf{r}_0) = 0$ for all directions $\mathbf{u}$.

45. Use the mean value theorem in Prob. 44 to give another proof of Th. 1 for any open convex set. A set D is **convex** if the line segment joining any two of its points lies entirely in the set.

46. Show that Th. 1 is valid if f is only assumed to be differentiable. *Hint.* Use the fact that any two points in an open set can be joined by a polygonal line

with sides parallel to the coordinate axes. Also use the mean value theorem of one-variable calculus.

47. Here is a useful extension of Th. 1 that will be needed in Ch. 8. In Th. 1 let $D = U \cup B$ consist of an open set U and part of its boundary B. Assume that f has continuous partial derivatives in U and is continuous on D. Prove:

$$\nabla f = \mathbf{0} \text{ in } U \quad \Leftrightarrow \quad f \text{ is constant on } D.$$

Hint. Apply Th. 1 to f restricted to the open set U. Then use a continuity argument.

Chapter Highlights

These highlights will focus on a function $f(x,y)$ of two variables. Similar remarks apply to functions of three or more variables.

Partial derivatives of $f(x,y)$ are just ordinary derivatives with one of the variables held constant. Thus,

$$\frac{\partial f}{\partial x}(x,y) = \frac{d}{dx} f(x,y),$$

where y is held constant. Just as for ordinary derivatives, partial derivatives have interpretations in terms of slopes and rates of change. To avoid repetition in what follows, assume that f has continuous first partials in a neighborhood of a point (x_0,y_0). Let $z_0 = f(x_0,y_0)$.

The tangent plane to the surface $z = f(x,y)$ at (x_0,y_0,z_0) is represented in point-slope form by

$$z - z_0 = \alpha(x - x_0) + \beta(y - y_0), \qquad \alpha = f_x(x_0,y_0),\ \beta = f_y(x_0,y_0).$$

The tangent plane is the graph of

$$F(x,y) = z_0 + \alpha(x - x_0) + \beta(y - y_0),$$

which is the linear approximation for $f(x,y)$ at (x_0,y_0).

The basic chain rule for a composite function $z = f(x,y)$, $x = x(t)$, $y = y(t)$, is

$$\frac{dz}{dt} = \frac{\partial z}{\partial x}\frac{dx}{dt} + \frac{\partial z}{\partial y}\frac{dy}{dt}.$$

The other chain rules are similar in appearance.

The directional derivative of $f(x,y)$ at (x_0,y_0) in the direction of a unit vector $\mathbf{u} = \langle a, b \rangle$ is given by

$$D_{\mathbf{u}} f(x_0,y_0) = \frac{d}{ds} f(x_0 + as, y_0 + bs)\bigg|_{s=0}.$$

The gradient of f is the vector $\nabla f = < f_x, f_y >$. Directional derivatives and gradients are related by

$$D_{\mathbf{u}} f = \nabla f \cdot \mathbf{u}.$$

The gradient points in the direction of maximum increase of f and the length of the gradient vector is the maximum rate of increase. Suppose that $\nabla f(x_0, y_0) \neq \mathbf{0}$. Then $\nabla f(x_0, y_0)$ is perpendicular to the level curve of f through (x_0, y_0).

Chain rules have vector forms. Thus, let f be a function defined along a curve $\mathbf{r} = \mathbf{r}(t)$. Then

$$f'(t) = \nabla f(\mathbf{r}(t)) \cdot \mathbf{r}'(t).$$

The project for this chapter introduces important practical and theoretical optimization procedures based on properties of the gradient. We shall describe the ideas in the context of finding the steepest trail up a mountain or the path along which temperature increases or decreases most rapidly. More sophisticated applications occur in the underground exploration for resources, efficient operation of chemical manufacturing plants, models of the U.S. economy, and lens design. The methods introduced here are especially useful when many independent variables are involved; however, the basic ideas can be explained and illustrated in the case of two independent variables. So we stick to two independent variables in the project.

Chapter Project: Curves of Steepest Descent and Ascent

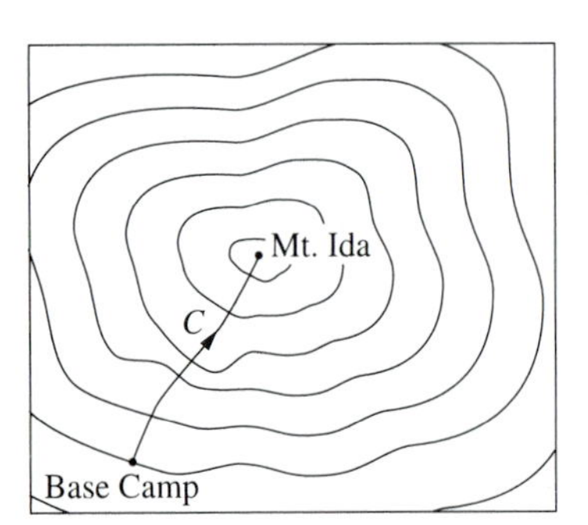

FIGURE 1

Figure 1 shows a climbing route C on a topographic map of Mt. Ida from base camp at 5000 feet to the summit at 10,000 feet. Let $z = f(x, y)$ be the elevation above sea level at the point (x, y) on the map. The curves of constant elevation shown on the map are the level curves of the elevation function f.

Problem 1 The climbing route C in Fig. 1 starts at the base camp, extends to the summit, and crosses every level curve at right angles. (a) Why is $\nabla f(x, y)$ tangent to C at each point on C and why does the gradient point in the positive direction along C? (b) Explain briefly but clearly why it is reasonable to call C a curve of steepest ascent up Mt. Ida.

Problem 2 Now you are at the top of Mt. Ida. Reverse the direction of C in Fig. 1. You are ready to ski down to base camp and you like a challenge. Explain why it is reasonable to call C (with its new direction) a curve of steepest descent down Mt. Ida.

The same idea carries over to level curves of other functions. If a directed curve C always heads in the direction of the gradient of a function f, it is called a **curve of steepest ascent** for the function. If the curve heads in the opposite direction, it is a **curve of steepest descent**. The discussion that follows pertains to curves of steepest descent. A parallel discussion applies to curves of steepest ascent.

For functions of two variables, curves of steepest descent can be approximated graphically as indicated by the following example. The temperature at a point (x, y) in the desert is $f(x, y) = x^2 + 2y^2$ in suitable units. It is very hot at

(2, 1). What path do you take from (2, 1) in order to cool down as rapidly as possible? Of course, the desired path is the curve of steepest descent from (2, 1) for the temperature $f(x, y) = 2^2 + 2y^2$. (Here steepest descent means most rapid temperature decrease.)

Problem 3 (a) Sketch (perhaps using a graphics utility) the isotherms (level curves of constant temperature) $f = 6, 4, 2, 1$ where $f(x, y) = x^2 + 2y^2$. (b) Then sketch the curve C of steepest descent that starts at (2, 1). (c) Where will C end? Why?

The graphical method in Prob. 3 gives a useful approximation of C. Next a general method is developed that enables you to find a curve of steepest descent by solving an initial value problem. To this end, let C be a curve of steepest descent and assume that C has the parametric representation

$$\mathbf{r} = \mathbf{r}(t) = < x(t), y(t) >$$

for $t \geq 0$ in some interval. Then

$$\mathbf{r}(0) = < x(0), y(0) > = < x_0, y_0 > = \mathbf{r_0}$$

marks the initial point on C. Finally, assume that C is a smooth parametric curve so that $\mathbf{r}'(t)$ is continuous and never zero.

Problem 4 Explain why $\mathbf{r}'(t)$ and $\nabla f(\mathbf{r}(t))$ must point in opposite directions at each point $\mathbf{r}(t)$ on the curve C of steepest descent. Conclude that

$$\mathbf{r}'(t) = -p(\mathbf{r}(t))\nabla f(\mathbf{r}(t)), \qquad \mathbf{r}(0) = \mathbf{r}_0,$$

where p is a positive continuous function on C.

The positive continuous function $p = p(\mathbf{r}(t))$ in Prob. 4 arises because the condition of steepest descent specifies only the direction of the tangent vector $\mathbf{r}'(t)$ and not its length. Since only the direction of descent is relevant in this context, you are free to choose p in any convenient way. Often the choice $p(\mathbf{r}(t)) \equiv 1$ is appropriate. In general, you seek a choice for p that simplifies the expression $-p(\mathbf{r}(t))\nabla f(\mathbf{r}(t))$ and hopefully makes it possible (or even easy) to integrate.

Problem 5 Return to Prob. 3. Show that the curve of steepest descent C is determined by the scalar initial value problem

$$x'(t) = -\,2x(t), \qquad x(0) = 2,$$

$$y'(t) = -\,4y(t), \qquad y(0) = 1.$$

Solve for $x(t)$ and $y(t)$. Eliminate the parameter and express y directly as a function of x. Next graph the steepest descent curve and the constant temperature curves $f = 6, 4, 2, 1$. Compare with your sketch in Prob. 3.

Problem 6 The temperature at a point in the plane is $T = 68 + 32e^{-(4x^2 + 3y^2)}$ degrees Fahrenheit. You are located at (0, 0) where the temperature is 100°. You prefer room temperature of 68°. What path would you follow to cool off as rapidly as possible? At what point will you first reach room temperature? Make a graph of the path of steepest descent.

In most applications of steepest descent techniques, especially when many independent variables are involved, the initial value problems cannot be solved in closed form. Then numerical differential equation solvers that come with all the major scientific software libraries can be used to generate approximate solutions.

Now it's time for a change in perspective. The path of steepest descent for the temperature in Prob. 3 ends at the point where the temperature is a minimum. Steepest descent methods provide effective tools for approximating minimum values of functions, especially functions of several variables. Effective numerical implementation of such methods involves some real challenges but the underlying idea is simple and natural. Let $\mathbf{r}_0$ be an initial guess at the point where a given function f assumes its minimum value and let $f(\mathbf{r}_0)$ be the initial estimate of the minimum. Since $-\nabla f(\mathbf{r}_0)$ points in the direction of most rapid decrease of f, a natural choice for an "improved" estimate of the minimum is

$$f(\mathbf{r}_1) \text{ where } \mathbf{r}_1 = \mathbf{r}_0 - h\nabla f(\mathbf{r}_0) \text{ for some } h > 0.$$

The idea is to choose h so that $f(\mathbf{r}_1) < f(\mathbf{r}_0)$. Now repeat the process. Use $\mathbf{r}_1$ as the new initial guess at the point where the minimum occurs, find a new $h > 0$ such that $f(\mathbf{r}_2) < f(\mathbf{r}_1)$ where $\mathbf{r}_2 = \mathbf{r}_1 - h\nabla f(\mathbf{r}_1)$, and so on. The expectation is that after a reasonable number of steps an acceptable approximation for the minimum value of f is obtained. As noted earlier, the development of this basic idea into a robust numerical method is quite challenging; we simply mention that several good steepest descent algorithms are available in the standard software libraries.

Problem 7 Let $f(x,y) = x^4 + y^4 + x^2 + y^2 - 2x + 6y + 6$. This function has a unique global minimum value at some point (a,b). That is, $f(a,b) < f(x,y)$ for all $(x,y) \neq (a,b)$. Experiment with the ideas described previously and use a CAS or scientific calculator to approximate the minimum of f and the point where it occurs.

Chapter Review Problems

1. (a) Find the domain of $f(x,y) = \ln(y/x)$. (b) Express the level curve $f = -1$ with y as a function of x.

2. Identify the traces of the surface $x^2 - y^2 + z^2 = 1$ in the following planes: (a) the xy–plane, (b) the plane perpendicular to the y–axis through $(0,1,0)$, and (c) the plane $y = x$.

3. Identify the level curve of $f(x,y) = x^2/(x^2 + y^2)$ which passes through the point $(3,4)$.

4. Let S be the set of points in the xy–plane with $(x + y)/2x \leq 1$. (a) Identify and sketch S. (b) Is S open? (c) Is S closed?

5. Identify the surface $9x^2 - y^2 + 16z^2 = 0$. Find the traces in the planes: xy–plane, yz–plane, plane $y = 1$.

6. Show that

$$\lim_{(x,y)\to(0,0)} \frac{y - x^2}{x - y^2}$$

does not exist.

7. Let $f(x,y) = (10x^2y^2)/(x^4 + y^4)$. Find the limit of $f(x,y)$ as $(x,y) \to (0,0)$ along the ray $\theta = \pi/3$.

8. (a) Find the limit, if it exists, of $f(x,y) = 3x^2y/(x^2 + y^2)$ as $(x,y) \to (0,0)$. (b) Is f continuous at $(0,0)$?

In Probs. 9–10, find the linear approximation of the given function at the indicated point.

9. $f(x,y,z) = x^{2/3} + y^{2/3} + z^{2/3}$ at $(8,8,8)$.

10. $f(x,y) = (2 + x - y)^2$ at $(3, -1)$.

11. Given that $xy + yz + z^2x = 5$ determines z as a function of (x,y) with continuous first partials, find $\partial z/\partial x$ and $\partial z/\partial y$.

12. Find the slopes of the surface $x^2 + 2y^2 + 4z^2 = 24$ at $(2, 2, \sqrt{3})$ in the x– and y–directions.

13. Find equations for the tangent line to the curve of intersection of the surface $z = x^2 + y^2$ and the plane $y = 1$ at $(2,1,5)$.

14. Find equations for the line that lies in the plane $x = 2$ and is tangent to the cone $z^2 = x^2 + y^2$ at $(2, 2\sqrt{3}, 4)$.

15. You are climbing a hill with altitude given by $z = 6x - x^2 - y^2$. As you pass through $(1,1,4)$ heading northeast, how fast are you climbing with respect to horizontal distance?

16. Find the point–slope equation for the tangent plane to the surface $z = \ln(x^2 + y^2)$ at (a) $(-1,0,0)$, (b) $(-1,1,\ln 2)$.

17. Find the differential of $f(x,y) = x \tan y + y \sec x$.

18. Use differentials to estimate $f(0.9, 1.1, 1.2)$ if $w = f(x,y,z) = xy + xz + yz$.

19. Let the lengths of the sides of a triangle be x, z, 3, and let θ be the angle opposite the side with length z. Find the differential dz when $x = 4$, $\theta = \pi/2$, $\Delta x = 0.05$, and $\Delta\theta = 0.1$.

20. Let $f(x,y,z) = ye^x + \sin z$, $y = \cos x$, $z = 2x$. Find df/dx when $x = \pi/4$.

21. Find the point on the elliptic paraboloid $z = 9 - x^2 - 4y^2$ where the tangent plane is parallel to the plane $z = 4x$.

22. Find an equation for the tangent plane to the level surface of $f(x,y,z) = xyz$ through $(1/2, -2, -1)$.

23. Find the tangent plane to the ellipsoid $x^2 + 4y^2 + 9z^2 = 22$ at $(3, 1, -1)$.

24. Let $f(x,y,z) = xy + yz + xz$. Let S be the level surface containing the point $(1, -1, 7)$. Find the tangent plane to S at $(1, -1, 7)$.

25. Find the tangent plane to the surface $e^z = \cos x/\cos y$ at $(\pi/6, \pi/6, 0)$.

26. Let $z = e^{x^2 - y^2}$. Find $\partial z/\partial r$ and $\partial z/\partial \theta$ when the polar coordinates satisfy $(r, \theta) = (2, \pi/6)$.

27. Find the tangent plane to the level surface of $f(x, y, z) = 3x^{1/3} + 3y^{1/3} + z$ through $(8, 1, 8)$.

28. Let $z = y \ln(x^2 + y^2)$, $x = 2s + 2t$, $y = 3t - s$. Find $\partial z/\partial t$ and $\partial z/\partial s$ when $s = 1$ and $t = 1$.

29. Let $z = 9x^2 + 4y^2$. Evaluate $\partial^2 z/\partial r^2$ when the polar coordinates satisfy $(r, \theta) = (1, \pi/3)$.

30. Find a unit vector perpendicular to the curve $3x^2 - xy + y^2 = 5$ at $(1, -1)$.

31. Find the unit normal vectors to the surface $xyz = -4$ at the point $(2, -2, 1)$.

32. A skier is at the point $(2, 1, 1)$ on a slope with elevation $z = 6y - x^2 - y^2$. What is the direction of steepest descent and how steep is the slope (relative to the horizontal xy–plane) in that direction?

33. Let $z = x^2 + 2xy + y^2$, $x = t \cos t$, and $y = t \sin t$. Find dz/dt when $t = \pi/2$.

34. Let $f(x, y, z) = 2x^2y + y^2z - x^2 z$. Find the maximum rate of change of f with respect to distance away from $(1, 0, 1)$ in any direction.

35. Let $f(x, y) = 25/(x^2 + y^2)$. Find (a) $\nabla f(3, 4)$, (b) the directional derivative of f at $(3, 4)$ in the direction from $(3,4)$ to $(7,7)$, and (c) the maximum value that a directional derivative of f at $(3, 4)$ can have.

36. Let $f(x, y, z) = \ln \sqrt{x^2 + y^2 + z^2}$. Find the rate of change of f with respect to distance away from $(1, 2, 3)$ in the direction toward $(5, 5, 3)$.

37. Find the directional derivative of $f(x, y) = \arctan(y/x)$ at $(3,4)$ in the direction from $(0, 0)$ to $(3, 4)$. Justify your answer on geometric grounds.

38. Find the directional derivative of $f(x, y, z) = \ln(x^2 + y^2 + z^2)$ at $(3, 4, 5)$ in the direction from $(3, 4, 5)$ to $(5, 5, 7)$.

39. Find the directional derivative of $f(x, y, z) = (x - y - z)/(x + y + z)$ at $(2, -1, 1)$ in the direction of $\mathbf{v} = 2\mathbf{i} + \mathbf{j} + \mathbf{k}$.

40. Find parametric equations for the normal line to the surface $x^2/4 + y^2/4 + z^2/2 = 1$ at $(1, 1, 1)$.

41. The temperature at $(x, y, z) \neq (0, 0, 0)$ is given by $T = 1/(x^2 + y^2 + z^2)$. How fast is the temperature changing at the time $t = 1$ as you move along the curve $\mathbf{r}(t) = < t, t^2/2, t^4/16 >$?

42. The position of an object at time $t \geq 0$ is $(x, y) = (16 \cosh t, 20 \sinh t)$. How fast is the object moving away from the origin at time $t = \ln 2$?

43. Let $f(x, y) = \arctan(y/x)$ and $x = \cos 2t$, $y = \sin 2t$. Find df/dt by two different methods.

44. Let $f(x, y) = e^{x^2 - y^2}$ with $x = \cos t$, $y = \sin t$. Find df/dt when $t = \pi/4$.

45. Let $f(x, y) = \sqrt{16x^2 + 9y^2}$, $x = \sqrt{t}$, $y = t/2$. Find df/dt when $t = 4$.

46. Let $f(x,y) = 3xy - x^2$ and $\mathbf{r}(t) = < t^2, 2t >$. Find:
(a) ∇f and $\mathbf{r}'(t)$ when $t = 1$; (b) $df(\mathbf{r}(t))/dt$ when $t = 1$;
(c) the directional derivative of f in the direction of $\mathbf{r}'(t)$ when $t = 1$;
(d) the cosine of the angle between ∇f and $\mathbf{r}'(t)$ when $t = 1$.

47. (a) Show that the function

$$u = u(x,t) = \frac{1}{\sqrt{4\pi c}} e^{-x^2/4ct}$$

satisfies the *diffusion (or heat) equation* $u_t = cu_{xx}$. (b) Let $c = 1$. Use a graphics utility to obtain portraits of $u(x,t)$ for $t = 1/2$, 1, 2, and 8. (c) Explain briefly why the portraits in (b) are consistent with $u(x,t)$ describing the diffusion of a substance in solution which is initially concentrated at $x = 0$.

CHAPTER 6
MAX–MIN PROBLEMS FOR FUNCTIONS OF TWO AND THREE VARIABLES

This chapter extends the treatment of max–min problems for functions of one independent variable to problems that involve functions of two and three independent variables. Such max–min problems are treated in a fashion that builds squarely upon the corresponding treatment of max–min problems in the one-variable case. Furthermore, the methods developed here for functions of two and three variables extend readily to functions of more than three independent variables. Such functions come up frequently in applications. For example, models for the national economy involve literally hundreds (even thousands) of variables and these variables are to be chosen so that the economy operates in a certain optimal way.

Recall that the max–min problems for functions of one variable typically involve constraints or side conditions. In some cases the constraints can be eliminated to obtain an unconstrained max–min problem. In other cases, methods based on implicit differentiation enable us to work directly with the constraint. Such methods often involve less work. These same methods are available for functions of two and three variables. The chapter concludes with an approach to constrained max–min problems that does not come up in the one-variable case. It is called the method of Lagrange multipliers and is closely related to methods based on implicit differentiation.

6.1 Maximum and Minimum Values

Max–min problems are just as important for functions of two or three variables as they are for functions of a single variable. As usual, let $P = (x, y)$ or (x, y, z) and $P_0 = (x_0, y_0)$ or (x_0, y_0, z_0). The definitions of local and global extreme values of a function $f(P)$ are essentially the same as for $f(x)$.

Definition *Local Extreme Values*

Let P_0 be an interior point of the domain of f.

Then $f(P_0)$ is a **local maximum** of f if

$f(P_0) \geq f(P)$ for all P in some neighborhood of P_0.

Likewise, $f(P_0)$ is a **local minimum** of f if

$f(P_0) \leq f(P)$ for all P in some neighborhood of P_0.

In either case, $f(P_0)$ is a **local extremum** of f.

Definition *Global Extreme Values*
Let f be defined on a set D and P_0 be a point in D.
Then $f(P_0)$ is the **(global) maximum** of f on D if

$$f(P_0) \geq f(P) \quad \text{for all } P \text{ in } D,$$

and $f(P_0)$ is the **(global) minimum** of f on D if

$$f(P_0) \leq f(P) \quad \text{for all } P \text{ in } D,$$

In either case, $f(P_0)$ is a **global extremum** of f.

Notice that local extrema can occur only at interior points of the domain of f. In contrast, a global extremum can occur at an interior point or a boundary point in the domain of f. Of course, if a global max or min occurs at an interior point, it is also a local max or min.

The max–min theorem for functions of one variable states that a continuous function $f(x)$ on a closed interval $[a, b]$ attains maximum and minimum values at certain points in $[a, b]$. The corresponding result for functions of two or three variables follows.

Theorem 1 *Max–Min Theorem*
A continuous function f on a closed and bounded set D attains maximum and minimum values at certain points in D.

All of the examples of extreme values given in the text involve functions of two variables. Functions of three variables will appear in the problems. Although our first example is rather simple, it does have enough structure to illustrate important features of extreme values for typical functions.

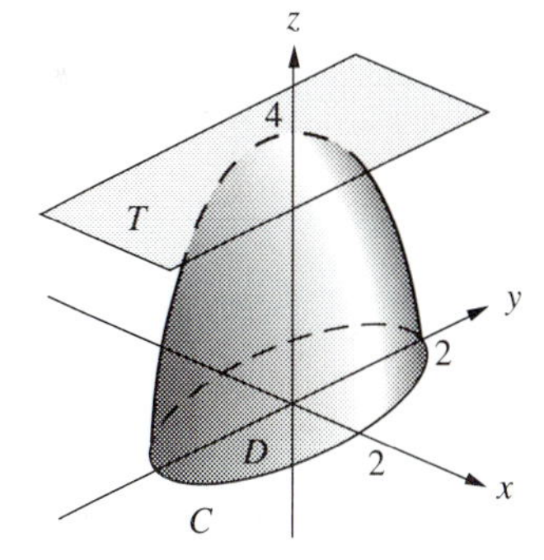

FIGURE 1

EXAMPLE 1. Let $f(x, y) = 4 - x^2 - y^2$ for $x^2 + y^2 \leq 4$. The graph of f in Fig. 1 is the part of the circular paraboloid $z = 4 - x^2 - y^2$ with $z \geq 0$. The domain of f is the closed disk D in the xy–plane bounded by the circle C with equation $x^2 + y^2 = 4$. The max–min theorem guarantees that f attains maximum and minimum values at certain points in D. It is apparent from the figure and the formula for f that

$$\max_D f = 4 \quad \text{and} \quad \min_D f = 0.$$

The maximum value 4 is attained at $(0, 0)$ and the minimum value 0 is attained at all points on the boundary C of the domain D. Since $(0, 0)$ is an interior point of D, $f(0, 0) = 4$ is also a local max of f. Figure 1 suggests that the graph of f has a horizontal tangent plane T at $(0, 0, 4)$, by analogy with a horizontal tangent line for a local max of a function of one variable. This is easy to confirm. The point-slope equation for the tangent plane at $(0, 0, 4)$ is

$$z - 4 = f_x(0, 0)(x - 0) + f_y(0, 0)(y - 0).$$

Since $f_x(x,y) = -2x$ and $f_y(x,y) = -2y$, we have $f_x(0,0) = 0$ and $f_y(0,0) = 0$. Equivalently, $\nabla f(0,0) = \mathbf{0}$. So the point-slope equation reduces to $z = 4$ and the tangent plane is horizontal. □

Suppose that the domain D of f in Ex. 1 is enlarged to the entire xy–plane. Then D is unbounded and the graph of f is the entire circular paraboloid. Now f has no minimum value, but it still has its local and global maximum value at $(0,0)$, where $\nabla f = \mathbf{0}$.

Our next example illustrates a local maximum that is not a global maximum.

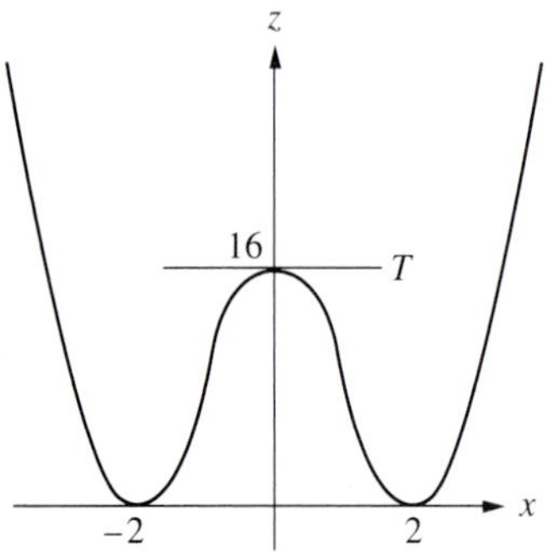

FIGURE 2

EXAMPLE 2. The curve $z = (x^2 - 4)^2$ is graphed in Fig. 2. Note that this function has a local maximum at $(0, 16)$, where the curve has a horizontal tangent line T, but there is no global maximum. Rotate the curve about the z–axis to obtain the surface represented by

$$z = f(x,y) = (x^2 + y^2 - 4)^2.$$

It should be apparent from the geometry that f has no global max but does have the local max $z = 16$ at $(0,0)$, where the tangent plane is horizontal. We also see that the local and global minimum value of f is 0, which is attained at every point (x,y) on the circle $x^2 + y^2 = 4$. The xy–plane, which is horizontal, is tangent to the surface at every point on this circle. □

If a function $f(x)$ is differentiable at x_0 and $f(x_0)$ is a local maximum or minimum, then $f'(x_0) = 0$. The corresponding result for a function of two or three variables is given next.

Theorem 2 *The Gradient Is Zero at a Local Extreme Value*
If $f(P_0)$ is a local max or min of f and $\nabla f(P_0)$ exists, then $\nabla f(P_0) = \mathbf{0}$. Thus,

$$f_x(P_0) = 0, \text{ and } f_y(P_0) = 0 \qquad \text{for } f(x,y),$$

$$f_x(P_0) = 0, \; f_y(P_0) = 0, \; f_z(P_0) = 0 \qquad \text{for } f(x,y,z).$$

Proof. There are several cases with proofs that are all alike. For example, suppose that $f(x,y)$ has a local max at (x_0, y_0). Then $f(x, y_0)$, which is a function of x alone, has a local max at $x = x_0$. Since $f_x(x, y_0)$ is the ordinary derivative of $f(x, y_0)$, we must have $f_x(x_0, y_0) = 0$. Similarly, $f(x_0, y)$ has a local max at (x_0, y_0) and $f_y(x_0, y_0) = 0$. So $\nabla f(x_0, y_0) = \mathbf{0}$. The proofs for a local min and for $f(x,y,z)$ are virtually the same. □

Suppose that $f(x,y)$ has continuous partial derivatives for (x,y) near (x_0, y_0). Let $z_0 = f(x_0, y_0)$. Then the surface $z = f(x,y)$ has a tangent plane at (x_0, y_0, z_0). The point-slope equation for the tangent plane is

$$z - z_0 = f_x(x_0, y_0)(x - x_0) + f_y(x_0, y_0)(y - y_0).$$

The equation reduces to $z = z_0$ if $f_x(x_0, y_0) = 0$ and $f_y(x_0, y_0) = 0$. Therefore,

$$\nabla f(x_0, y_0) = \mathbf{0} \quad \Leftrightarrow \quad f_x(x_0, y_0) = f_y(x_0, y_0) = 0 \quad \Leftrightarrow$$
the tangent plane at (x_0, y_0, z_0) is horizontal.

In view of Th. 2, the tangent plane to the graph of f at (x_0, y_0) is horizontal if $f(x_0, y_0)$ is a local max or min. Be careful not to read too much into this statement. Zero partial derivatives and a horizontal tangent plane do not always signal a local max or min, as the next example shows.

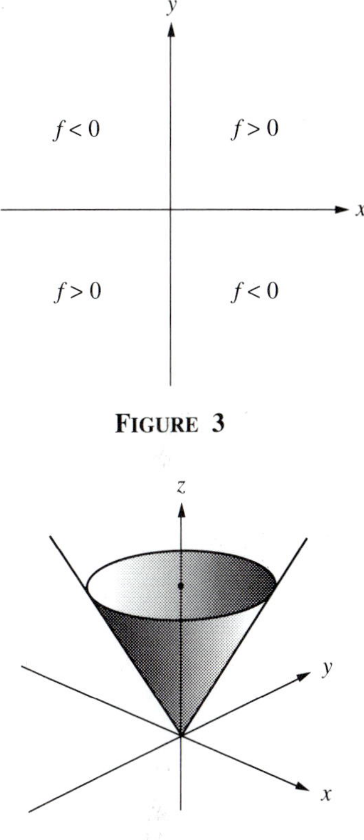

FIGURE 3

FIGURE 4

EXAMPLE 3. Let $f(x, y) = xy$. In Sec. 5.1 we learned that the surface is shaped like a saddle, with $f(x, y) > 0$ for (x, y) in the first and third quadrants, and $f(x, y) < 0$ for (x, y) in the second and fourth quadrants, as indicated in Fig. 3. For this reason, $(0, 0, 0)$ is called a saddle point. Clearly, $f(0, 0) = 0$ is not a local max or min. Since f has the partial derivatives $f_x = y$ and $f_y = x$, which are continuous everywhere, and $f_x(0, 0) = f_y(0, 0) = 0$, the surface has the horizontal tangent plane $z = 0$ (the xy–plane) at $(0, 0, 0)$. □

It is convenient to define a saddle point of a function more broadly than in Ex. 3. Let (x_0, y_0) be an interior point of the domain of f and let $z_0 = f(x_0, y_0)$. Then (x_0, y_0, z_0) is a **saddle point** of f or its graph if $\nabla f(x_0, y_0) = \mathbf{0}$ but (x_0, y_0) is not a local max or min of f.

A function of one variable can have a local extreme value at a point where it is not differentiable. A familiar example is $f(x) = |x|$ at $x = 0$. Similar behavior occurs for functions of more than one variable.

EXAMPLE 4. The graph of $f(x, y) = \sqrt{x^2 + y^2}$ in Fig. 4 is the upper half of a cone. Apparently, $f(0, 0) = 0$ is both a local and global minimum of f. However, neither $f_x(0, 0)$ nor $f_y(0, 0)$ exists (see the problems), so that $\nabla f(x_0, y_0)$ does not exist. □

The foregoing examples lead to the definition of a critical point of a function of two or three variables.

Definition *Critical Point*
Let P_0 be an interior point of the domain of f.
Then P_0 is a **critical point** of f if
$\nabla f(P_0) = \mathbf{0}$ or $\nabla f(P_0)$ does not exist.

The definition of a critical point has the immediate consequence:

Theorem 3 *Local Extreme Values*
Local extreme values can occur only at critical points.

The same theorem occurs earlier in calculus for functions of one variable. Most of the critical points we shall deal with are for functions of two variables $f(x, y)$ with $\nabla f(x_0, y_0) = \mathbf{0}$. Then the graph of f has a horizontal tangent plane at (x_0, y_0, z_0), where $z_0 = f(x_0, y_0)$. Theorem 3 yields the following strategy for finding the local extreme values of a function f.

1. Find the critical points of f.
2. At each critical point, determine whether f has a local max, a local min, or neither.

In this section, we use geometric, algebraic, or analytic means based on the particular problem for carrying out step 2. A more systematic test, based on second–order partial derivatives, will be given in the next section. Although the direct methods in this section usually require more work, they give considerably more information and insight. Often they give a very good idea of what a surface looks like and reveal global extreme values. Although the second derivative test of the next section does help identify local extreme values, it does not tell whether a local extreme value is a global extreme value.

EXAMPLE 5. Find all local maxima and minima of the quadratic function $f(x,y) = x^2 + y^2 - 2x - 4y + 8$.

Solution. Since $\nabla f(x,y) = \, < 2x - 2, \; 2y - 4 >$ exists for all (x,y), any critical points of f occur where $\nabla f = \, < f_x, f_y > \, = \mathbf{0}$:

$$f_x = 2x - 2 = 0, \qquad f_y = 2y - 4 = 0.$$

The only critical point is $(x,y) = (1,2)$. To determine whether f has a local max or min at (1, 2), we need to know something about the surface $z = f(x,y)$ for (x,y) near (1, 2). A useful procedure for studying quadratics is to complete the squares on x and y:

$$f(x,y) = (x^2 - 2x + 1) + (y^2 - 4y + 4) + 8 - 1 - 4 = (x-1)^2 + (y-2)^2 + 3.$$

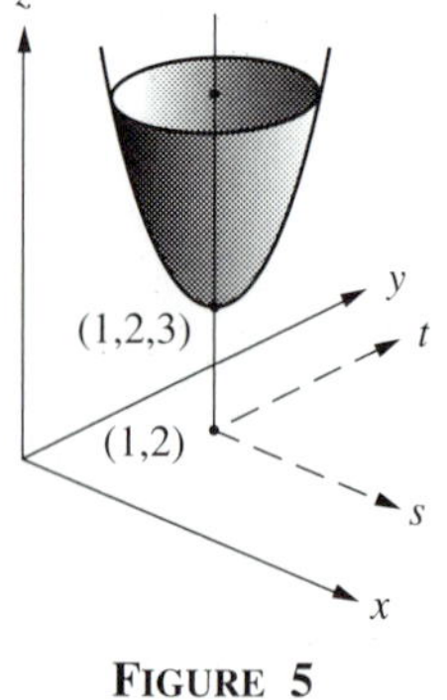

FIGURE 5

It follows that $f(x,y) > f(1,2) = 3$ for all $(x,y) \neq (1,2)$. So $f(1,2) = 3$ is a local minimum and the global minimum of f. The graph of f is the circular paraboloid in Fig. 5, which has vertex at $(1,2)$ and opens up. □

The minimum in Ex. 5 could have been found without calculus, just by completing the square. However, the use of calculus focuses our attention on the critical point, where any local extreme value must occur. The technique used in Ex. 5 is closely related to the change of variables $s = x - 1$ and $t = y - 2$, followed by a translation of axes that places the critical point at $(s,t) = (0,0)$, as suggested in Fig. 5.

Locating Global Extreme Values

The importance of critical points and local extreme values comes mainly from their roles in narrowing the search for global extreme values. The following theorem summarizes what may happen in typical situations.

Theorem 4 *Locating Global Extreme Values*
Global extreme values of a continuous function f on a set D can occur only at critical points or boundary points in D. If D is closed and bounded, then the global maximum and minimum values of f are the largest and smallest values of $f(P)$ as P runs through all critical points and boundary points in D.

The analogue of Th. 4 occurs earlier in calculus for a function $f(x)$ defined on a closed interval $[a, b]$. In the problems you are asked to give the easy proof of Th. 4.

The remaining examples in this section are more challenging than those we have addressed so far. They illustrate several very useful techniques for finding global extreme values.

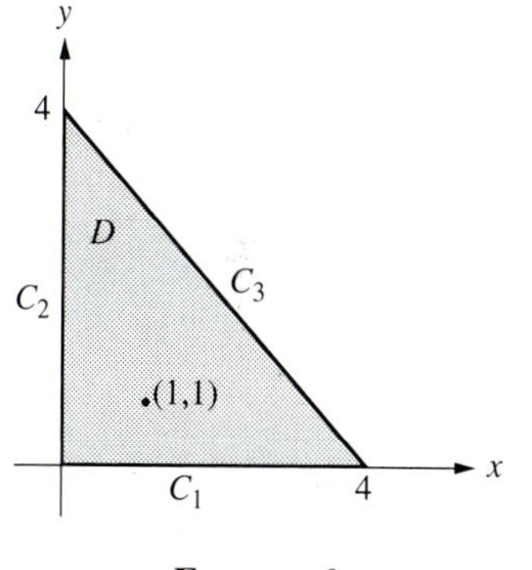

FIGURE 6

EXAMPLE 6. Let $f(x,y) = x^3 - 3xy + y^3 + 6$ for $x \geq 0$, $y \geq 0$, and $x + y \leq 4$. Find all local and global extreme values of f.

Solution. The domain of f is the shaded triangle D in Fig. 6. Since D is closed and bounded, we can use Th. 4 to help find the extreme values of f. The critical points of f are any interior points (x,y) of D where

$$f_x(x,y) = 3x^2 - 3y = 0, \qquad f_y(x,y) = 3y^2 - 3x = 0.$$

These equations give $y = x^2$ and $x = y^2 = x^4$. The solutions are $(x,y) = (0,0)$ and $(x,y) = (1,1)$. Since $(1,1)$ is an interior point of D, but $(0,0)$ is not, the only critical point of f is $(1,1)$, where $f(1,1) = 5$. According to Th. 4, the global max and min of f occur either at $(1,1)$ or somewhere on the boundary of D.

The boundary of D consists of the three line segments C_1, C_2, C_3 in Fig. 6. On C_1, we have $y = 0$, $0 \leq x \leq 4$, and $f(x,0) = x^3 + 6$. So

$$f \text{ increases from 6 to 70 on } C_1.$$

Likewise,

$$f \text{ increases from 6 to 70 on } C_2.$$

Next consider f on C_3:

$$f(x,y) = x^3 - 3xy + y^3 + 6 \text{ with } y = 4 - x \text{ and } 0 \leq x \leq 4.$$

First, note that $f(0,4) = f(4,0) = 70$ at the end points of C_3. To find any local extrema of f on C_3, we could let $y = 4 - x$ in the formula for f, differentiate, and then solve $df/dx = 0$ for x. However, it is less work to use implicit differentiation. Thus, consider $f(x,y)$ as a function of x for $0 \leq x \leq 4$, with $y = 4 - x$. Then $dy/dx = -1$ and

$$\frac{df}{dx} = 3x^2 - 3x\frac{dy}{dx} - 3y + 3y^2\frac{dy}{dx} = 3(x^2 - y^2 + x - y) = 3(x - y)(x + y + 1).$$

Since $y = 4 - x$, a little algebra yields

$$\frac{df}{dx} = 30(x-2), \qquad \frac{df}{dx} = 0 \text{ when } x = 2.$$

Then $y = 2$ and $f(2,2) = 10$. In summary,

$$f = 10 \text{ at } (2,2), \text{ where } df/dx = 0 \text{ on } C_3,$$

$$f = 70 \text{ at the end points } (0,4) \text{ and } (4,0) \text{ of } C_3.$$

Consequently, the maximum of f on C_3 is 70 and the minimum is 10. Combine the results for C_1, C_2, and C_3 to find that the maximum of f on the boundary of D is 70 and the minimum is 6. Finally, since $f = 5$ at the critical point $(1,1)$, we conclude that

$$\min_D f = f(1,1) = 5 \text{ and } \max_D f = f(0,4) = f(4,0) = 70. \ \square$$

EXAMPLE 7. As in Ex. 6, let $f(x,y) = x^3 - 3xy + y^3 + 6$, but now with domain D the entire first quadrant $x \geq 0$, $y \geq 0$. Find the global extreme values of f.

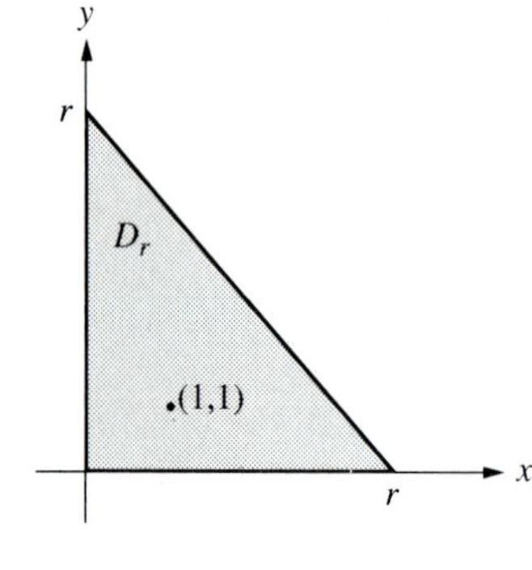

FIGURE 7

Solution. Just as in Ex. 6, the only critical point of f in D is $(1,1)$, where $f(1,1) = 5$. Let $r \geq 4$. In Fig. 7, D_r is the set of points (x,y) in D with $x + y \leq r$. In Ex. 6 we considered f on D_4. Virtually the same reasoning used there (see the problems) shows that

$$\min_{D_r} f = f(1,1) = 5 \text{ and } \max_{D_r} f = f(r,0) = f(0,r) = r^3 + 6.$$

Let $r \to \infty$ to see that

$$\min_D f = f(1,1) = 5$$

and f has no maximum on D. $\square$

EXAMPLE 8. Find all local and global extreme values of

$$f(x,y) = xy + \frac{1}{x} + \frac{1}{y} \text{ for } x > 0, y > 0.$$

Solution. The domain D is the open first quadrant. Observe that

$$f(x,y) \to \infty \text{ as } x \to 0^+ \text{ or } x \to \infty,$$

$$f(x,y) \to \infty \text{ as } y \to 0^+ \text{ or } y \to \infty.$$

Therefore, f has no maximum value on D. Let's look for local minimum values. The gradient of f is

$$\nabla f(x,y) = \left\langle y - \frac{1}{x^2},\ x - \frac{1}{y^2} \right\rangle \text{ for } (x,y) \text{ in } D.$$

Any critical points in D occur at points where $\nabla f = \mathbf{0}$:

$$y = 1/x^2, \qquad x = 1/y^2.$$

It follows easily that the only critical point of f in D is $(1,1)$, where $f(1,1) = 3$. We shall describe two ways to verify $f(1,1) = 3$ is the local and global minimum of f. Each method offers valuable ideas and insights that can be used in other problems.

Method 1. A similar strategy was used in Ex. 7. For $r > 3$, let D_r be the closed region bounded by the square C_r in Fig. 8 with opposite corners $(1/r, 1/r)$ and (r^2, r^2). The critical point $(1,1)$ lies inside C_r. The minimum of f on D_r must occur at $(1,1)$ or on C_r. An examination of the formula for $f(x,y)$ on each of the four sides of the square C_r shows that $f(x,y) > r > 3$ for all (x,y) on C_r. No calculus is necessary. (You are asked to carry out the details in the problems.) It follows that $f(1,1) = 3$ is the minimum of f on D_r. Let $r \to \infty$ to conclude that $f(1,1)$ is the minimum of f on D.

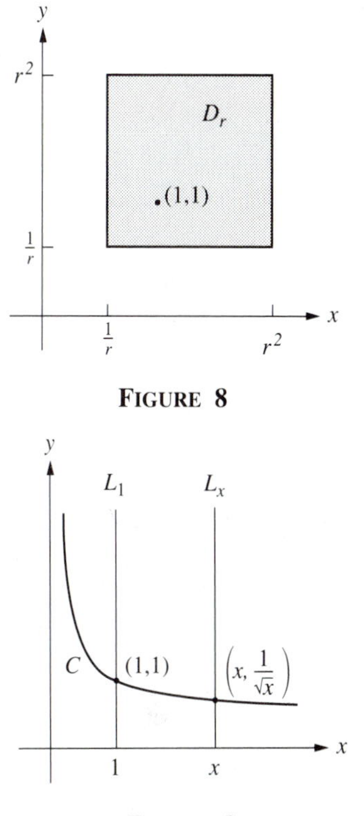

FIGURE 8

FIGURE 9

Method 2. Regard the open first quadrant as being made up of all the vertical half–lines L_x in Fig. 9. For each $x > 0$, L_x consists of the points (x,y) with $0 < y < \infty$. The game plan is to find the minimum value of f on each half-line L_x. The smallest of these minimum values will be the minimum of f on D. Let's begin. Fix $x > 0$ and minimize $f(x,y)$ for $0 < y < \infty$. Since

$$f_y(x,y) = x - \frac{1}{y^2} = 0 \text{ only for } y = 1/\sqrt{x}$$

and $f(x,y) \to \infty$ as $y \to 0^+$ or $y \to \infty$, the minimum of f on L_x is

$$f(x,\ 1/\sqrt{x}) = 2\sqrt{x} + 1/x.$$

Now minimize $f(x,\ 1/\sqrt{x}) = 2\sqrt{x} + 1/x$ for $0 < x < \infty$. Since

$$\frac{d}{dx} f(x,\ 1/\sqrt{x}) = \frac{1}{\sqrt{x}} - \frac{1}{x^2} = 0 \text{ only for } x = 1$$

and $f(x,\ 1/\sqrt{x}) \to \infty$ as $x \to 0^+$ or $x \to \infty$, the minimum of $f(x,\ 1/\sqrt{x})$ is $f(1,1) = 3$. This is the global minimum of f on D. Figure 9 gives a picture of what we have done. The minimum of f on L_x occurs at the point (x,y) with $y = 1/\sqrt{x}$. These points form the curve C in Fig. 9. The global minimum of f on D is the minimum of f on C. □

EXAMPLE 9. A rectangular shipping crate must have a 24 ft^3 capacity. It costs \$2/ft^2 to make the top and bottom, \$3/ft^2 to make the front and back, and \$4/ft^2 to make the other two sides. Find the minimum cost to make such a crate.

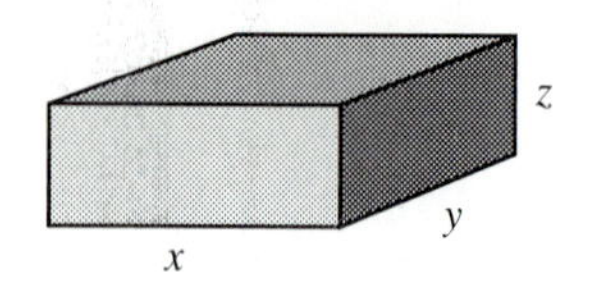

FIGURE 10

Solution. In Fig. 10, the dimensions of a crate are $x > 0$, $y > 0$, $z > 0$. The cost of construction in dollars is

$$C = 4xy + 6xz + 8yz.$$

Since $xyz = 24$, we have $z = 24/xy$ and, after a little algebra,

$$C = 4\left(xy + \frac{48}{x} + \frac{36}{y}\right).$$

The cost function is very similar to the function in Ex. 8. The reasoning used there can be employed again to show that C attains a global minimum where the partial derivatives of C with respect to x and y are zero. This procedure yields $x = 4$, $y = 3$, and $C = \$144$.

Instead of going through the procedure in Ex. 8 all over again, we can make a change of variables that reduces this problem to Ex. 8. Let $x = as$ and $y = bt$, where a and b are positive constants to be determined shortly. Then $s > 0$, $t > 0$, and

$$C = 4\left(abst + \frac{48}{as} + \frac{36}{bt}\right).$$

The cost function C will have virtually the same form as in Ex. 8 if we can choose a and b such that

$$ab = \frac{48}{a} = \frac{36}{b}.$$

From the second equation, $b = (3/4)a$. Then the first equation gives $a = 4$, $b = 3$, and the formula for C becomes

$$C = 4\left(12st + \frac{12}{s} + \frac{12}{t}\right) = 48\left(st + \frac{1}{s} + \frac{1}{t}\right).$$

From Ex. 8, the minimum cost occurs when $s = 1$, $t = 1$ and, hence, $x = 4$, $y = 3$. Then $z = 24/xy = 2$. These are the dimensions of the cheapest crate. The minimum cost is $C = \$144$. □

One of the messages of Ex. 9 is that problems often can be simplified by a change of variables that brings us to a more familiar setting. Max–min problems are challenging and important. Their solutions often involve a blend of techniques from calculus, algebra, and numerical analysis. We have only scratched the surface of the techniques that are available.

PROBLEMS

In Probs. 1–10 (a) find all critical points of the function and (b) determine whether each critical point corresponds to a local or global maximum, minimum, or neither one.

1. $f(x,y) = 4x^2 + 3y^2 - 8x + 12y$
2. $f(x,y) = 25 + 8y - 3x^2 - y^2$

3. $f(x,y) = x^2 - y^2$
4. $f(x,y) = x^2 - y^2 - 2x - 6y - 5$
5. $f(x,y) = (x^2 + y^2)^{-1/2}$
6. $f(x,y) = (x^2 + y^2)^{1/3}$
7. $f(x,y) = e^{-x^2-y^2}$
8. $f(x,y) = \cos\pi(x^2 + y^2)$
9. $f(x,y,z) = 4x^2 + 3y^2 + z^2 - 8x + 12y - 6z + 25$
10. $f(x,y,z) = (x^2 + y^2)e^{-z^2}$

In Probs. 11–20 utilize Th. 4 to find the global maximum and global minimum of the function.

11. $f(x,y) = x^2 + 2y^2 - 2x - 8y$ for $x \geq 0, y \geq 0, x + y \leq 4$
12. $f(x,y) = x^3 + 3xy + y^3$ for $|x| \leq 2, |y| \leq 2$
13. $f(x,y) = x^2 + 2y^2 - 2x - 8y$ for $x \geq 0, y \geq 0, y \leq 4 - x^2$
14. $f(x,y) = x^3 + 3xy + y^3$ for $0 \leq x, y \leq 1$
15. $f(x,y) = 4x^3y - 4xy^3$ for $x^2 + y^2 \leq 1$
16. $f(x,y) = x^4 - 6x^2y^2 + y^4$ for $x^2 + y^2 \leq 4$
17. $f(x,y) = \sin x + \sin y + \sin(x + y)$ for $0 \leq x, y \leq 2\pi$
18. $f(x,y) = e^{(x-y)/(x^2+y)}$ for $1 \leq x^2 + y^2 \leq 9$
19. $f(x,y) = (x + y + 1)/(x^2 + y^2 + 1)$ for $x^2 + y^2 \leq 4$
20. $f(x,y) = xy$ for $9x^2 + 16y^2 \leq 144$

In Probs. 21–26, Th. 4 does not apply directly. (a) Assume that the asserted max or min exists and find it. (b) Then use reasoning such as in Exs. 7–9 to establish that the max or min exists.

21. $f(x,y) = 3xy - x^2y - xy^2$ for $x,\ y \geq 0$ has a maximum.
22. $f(x,y) = (x + y + 1)/(x^2 + y^2 + 1)$ has a maximum.
23. $f(x,y) = 2x^2 + y^2 + 1/(2x^2 + y^2)$ for $x,\ y \geq 0$ and $(x,y) \neq (0,0)$ has a minimum.
24. $f(x,y) = xye^{-(x^2+y^2)}$ has a maximum and a minimum.
25. $f(x,y) = (2x^2 + y^2)e^{1-x^2-y^2}$ has a maximum.
26. $f(x,y) = (x^2 + y^2)^{-1}\ln\sqrt{xy}$ for $x,\ y > 0$ has a maximum.

In Probs. 27–44 find the required quantities. Also give either a mathematical reason or a geometric or physical plausibility argument that explains why the asserted max or min exists.

27. Find the maximum volume of a rectangular box (with top) that has surface area 24 ft^2
28. Find the maximum volume of a rectangular box (with top) that has fixed surface area S square units.
29. Find the maximum volume of a rectangular package that can be sent from a U.S. Post Office. Postal regulations limit the sum of the length and girth (the distance around the package perpendicular to the length) to 108 inches.

30. Find the dimensions and maximum volume of a rectangular package if the sum of its length and girth is fixed at L units.

31. The base of a rectangular crate costs twice as much per square foot as its top and other sides. Find the most economical proportions if the box must have a 36 ft^3 capacity.

32. A rectangular box with no top has fixed volume. Find the shape that makes the surface area a minimum.

33. A rectangular box with no top has fixed surface area. Find the shape that makes the volume a maximum.

34. A long metal sheet 2 feet wide is molded into a trough by bending up two sides of equal length that are inclined at the same angle to the bottom. (So the cross sections of the trough are isosceles trapezoids.) Find the length of the side and the angle that maximize the carrying capacity of the trough.

35. A rectangular box has three faces in the coordinate planes and one vertex in the plane $x/a + y/b + z/c = 1$, where a, b, $c > 0$. Find the largest volume of such a box.

36. Find the volume of the largest rectangular box with each face parallel to one of the coordinate planes and with vertices on the ellipse $x^2/a^2 + y^2/b^2 + z^2/c^2 = 1$.

37. The volume of an ellipsoid with semiaxes a, b, and c is $V = (4/3)\pi abc$. If the sum of the semiaxes is fixed, find the ellipsoid that has the largest volume.

38. A triangle has vertices (x_1, y_1), (x_2, y_2), (x_3, y_3). Find the point (x, y) in the plane of the triangle such that the sum of the squares of the distances from the vertices of the triangle is a minimum.

39. If α, β, and γ are the angles of a triangle, find the maximum value of $\sin\alpha + \sin\beta + \sin\gamma$ Is there a minimum value? Explain briefly.

40. A triangle is inscribed in a fixed circle. Among all such triangles, show that the one with greatest area is an equilateral triangle. *Hint.* Express the area in terms of the central angles subtended by the sides of the triangle.

41. Find the equation of the plane through $(1, 2, 2)$ that cuts off the smallest volume from the first octant.

42. When a steady electric current (charge/sec) I flows through a wire with constant resistance R, the heat/sec generated is proportional to RI^2. Three wires, with resistances R_1, R_2, and R_3, connect two terminals A and B. A steady current I entering A will produce currents I_1, I_2, and I_3 in the wires leading to B such that the total heat/sec generated in the three wires is minimized. Show that $I_1R_1 = I_2R_2 = I_3R_3$.

43. A rectangular box is inscribed in a hemisphere with radius a. The base of the box lies in the planar base of the hemisphere. Find the maximum volume of such a box.

44. The three points (x_1, y_1, z_1), (x_2, y_2, z_2), and (x_3, y_3, z_3) do not lie in a plane. Find the plane such that the sum of the squares of the distances of the given points to the plane is a small as possible.

45. In connection with Ex. 4, show that neither $f_x(0,0)$ nor $f_y(0,0)$ exists for $f(x,y) = \sqrt{x^2 + y^2}$. *Hint.* $f_x(0,0)$ is the ordinary derivative of what function of x at $x = 0$?

46. Prove Theorem 4.

47. Verify the statements made in Ex. 7 that $\min f = 5$ and $\max f = r^3 + 6$ on the domain D_r.

48. Assume $a, b, c > 0$. Find the global extrema of $f(x,y) = axy + b/x + c/y$ for $x, y > 0$.

49. Solve Prob. 48 if $a > 0$, $b < 0$ and $c < 0$.

50. Solve Prob. 48 if $a > 0$, $b > 0$ and $c < 0$.

6.2 Higher-Order Partial Derivatives and The Second Partials Test

Second- and higher-order partial derivatives are obtained by repeated partial differentiation. In calculus we are interested primarily in second-order partial derivatives. They arise in the second partials test, which is used to detect whether a function of several variables has a local maximum or minimum at a critical point. Second-order partial derivatives also appear in fundamental partial differential equations of science and engineering, such as the heat equation and the wave equation. Third- and fourth-order partial derivatives occur in physical applications too, but that is a topic for more advanced courses.

Higher-Order Partial Derivatives

Higher-order partial derivatives are just partial derivatives of partial derivatives, just as higher-order ordinary derivatives are derivatives of ordinary derivatives. All that is really new here is the notation used to denote higher order-partial derivatives. We shall illustrate that notation primarily by means of examples. As usual, we restrict our attention to functions of two and three variables.

EXAMPLE 1. Let $f(x,y) = x^3y^5$. The partial derivatives of f are

$$f_x = 3x^2y^5, \qquad f_y = 5x^3y^4.$$

Partial differentiation of f_x and f_y yields the four partial derivatives of f of second order:

$$f_{xx} = 6xy^5, \qquad f_{yy} = 20x^3y^3,$$
$$f_{xy} = 15x^2y^4, \qquad f_{yx} = 15x^2y^4. \ \square$$

In general, the **second-order partial derivatives** of a function $f(x,y)$ are defined by

$$f_{xx} = (f_x)_x, \qquad f_{yy} = (f_y)_y,$$
$$f_{xy} = (f_x)_y, \qquad f_{yx} = (f_y)_x.$$

For brevity they are called the **second partials** of f. Similarly, f_x and f_y are called the **first partials** of f.

You may have noticed that the **mixed partials** f_{xy} and f_{yx} in Ex. 1 are equal. Although this pleasant state of affairs does not always happen, it is typical of functions you are likely to meet.

Theorem 1 *Equality of Mixed Partials*
Assume that the mixed partials $f_{xy}(x, y)$ and $f_{yx}(x, y)$ exist for (x, y) in a neighborhood of (x_0, y_0) and are continuous at (x_0, y_0). Then

$$f_{xy}(x_0, y_0) = f_{yx}(x_0, y_0).$$

The proof, which is not difficult but would distract us at the moment, is given at the end of the section. We shall confirm particular cases of the theorem and related results in the next two examples.

EXAMPLE 2. Let $f(x, y) = \sin(x^2y)$. Then

$$f_x = 2xy\cos(x^2y), \qquad f_y = x^2\cos(x^2y),$$

$$f_{xx} = 2y\cos(x^2y) - 4x^2y^2\sin(x^2y), \qquad f_{yy} = -x^4\sin(x^2y)$$

$$f_{xy} = 2x\cos(x^2y) - 2x^3y\sin(x^2y), \qquad f_{yx} = 2x\cos(x^2y) - 2x^3y\sin(x^2y).$$

Thus, $f_{xy} = f_{yx}$, in accord with Th. 1. □

Partial derivatives of the third and higher orders are defined in the expected way. For $f(x, y) = x^3y^5$ in Ex. 1,

$$f_{xxx} = 6y^5, \qquad f_{yyy} = 60x^3y^2,$$

$$f_{xxy} = 30xy^4, \quad f_{yyx} = 60x^2y^3,$$

and so on. An extension of Th. 1 establishes the equality of mixed partials such as $f_{xxy} = f_{xyx} = f_{yxx}$. The story is virtually the same for functions of three or more variables. One example should suffice.

EXAMPLE 3. Let $f(x, y, z) = x^2y^3z^5$. Then

$$f_x = 2xy^3z^5, \qquad f_y = 3x^2y^2z^5, \qquad f_z = 5x^2y^3z^4,$$

$$f_{xy} = 6xy^2z^5, \qquad f_{yz} = 15x^2y^2z^4, \qquad f_{zx} = 10xy^3z^4,$$

$$f_{xyz} = 30xy^2z^4, \qquad f_{yzx} = 30xy^2z^4, \qquad f_{zxy} = 30xy^2z^4.$$

By the extension of Th. 1, $f_{xyz} = f_{xzy} = f_{yzx} = f_{yxz} = f_{zxy} = f_{zyx}$, and so forth, as you can easily confirm for this particular function. □

There are several notations for partial derivatives of higher order. For example, the second-order partials of $z = f(x,y)$ are expressed by

$$f_{xx} = z_{xx} = \frac{\partial}{\partial x}\left(\frac{\partial f}{\partial x}\right) = \frac{\partial^2 f}{\partial x^2}, \qquad f_{yy} = z_{yy} = \frac{\partial}{\partial y}\left(\frac{\partial f}{\partial y}\right) = \frac{\partial^2 f}{\partial y^2},$$

$$f_{xy} = z_{xy} = \frac{\partial}{\partial y}\left(\frac{\partial f}{\partial x}\right) = \frac{\partial^2 f}{\partial y \partial x}, \qquad f_{yx} = z_{yx} = \frac{\partial}{\partial x}\left(\frac{\partial f}{\partial y}\right) = \frac{\partial^2 f}{\partial x \partial y}.$$

EXAMPLE 4. The *partial differential equation*

$$\frac{\partial u}{\partial t} = a\,\frac{\partial^2 u}{\partial x^2},$$

with a a positive constant, is called the *one–dimensional heat equation*. It models the variation in temperature $u = u(x,t)$ of a thin, laterally insulated rod at position x and time t. Show that $u(x,t) = e^{-4at}\cos 2x$ satisfies the heat equation.

Solution. By inspection,

$$\frac{\partial u}{\partial x} = -2e^{-4at}\sin 2x,$$

$$\frac{\partial^2 u}{\partial x^2} = -4e^{-4at}\cos 2x, \qquad \frac{\partial u}{\partial t} = -4a\,e^{-4at}\cos 2x.$$

Compare the last two partial derivatives to see that $u(x,t)$ satisfies the heat equation. □

The remainder of this section is devoted to second-order partial derivatives for functions $z = f(x,y)$. We shall use the notation $f_{xx}, f_{yy}, f_{xy}, f_{yx}$ for the second partials.

Recall that the first partial $f_x(x_0,y_0)$ is the ordinary first derivative of $f(x,y_0)$ with respect to x at (x_0,y_0). In the same manner, the second partial $f_{xx}(x_0,y_0)$ is the ordinary second derivative of $f(x,y_0)$ with respect to x at (x_0,y_0). If $f_{xx}(x,y_0)$ is positive (or negative) for x in an interval I, then the graph of $f(x,y_0)$ is concave up (or down) for x in I. Similar remarks pertain to the second partial f_{yy}. There are more complicated interpretations of mixed partials, but we shall not need them.

The Second Partials Test

The second partials test for a function of two variables $f(x,y)$ is analogous to the second derivative test for $f(x)$. Assume that $f''(x)$ exists for x near x_0 and is continuous at x_0. A convenient form of the second derivative test is

$f'(x_0) = 0, \quad f''(x_0) > 0 \quad \Rightarrow$ f is concave up near x_0 and $f(x_0)$ is a local min,

$f'(x_0) = 0, \quad f''(x_0) < 0 \quad \Rightarrow$ f is concave down near x_0 and $f(x_0)$ is a local max.

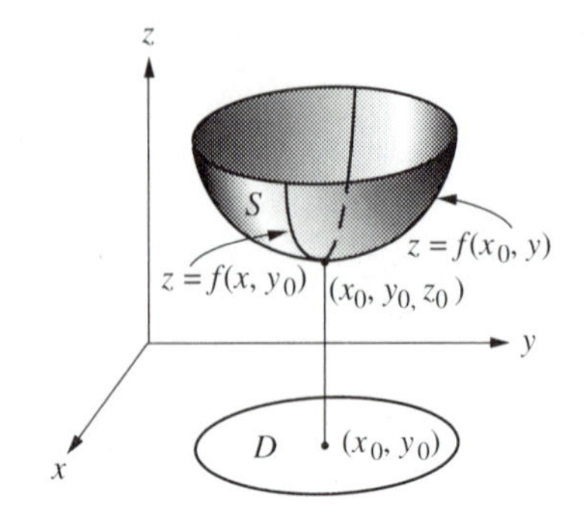

FIGURE 1

Let's take a closer look at the case of a local min. It follows from $f''(x_0) > 0$ and the continuity of $f''(x)$ at x_0 that $f''(x) > 0$ for x near x_0. Therefore, f is concave up near x_0 and since $f'(x_0) = 0$, $f(x_0)$ is a local min.

Similar reasoning leads to the second partials test for $f(x,y)$. Figure 1 will serve as a guide. It illustrates the case of a local minimum. Similar figures would illustrate a local maximum and a saddle point. We have met examples of all three types before. The circular paraboloid $z = x^2 + y^2$ has a local min at (0,0). The circular paraboloid $z = 4 - x^2 - y^2$ has a local max at (0,0). The hyperbolic paraboloid $z = xy$ has a saddle point at (0,0).

Assume that $\nabla f(x_0,y_0) = \mathbf{0}$ and that the second partials of $f(x,y)$ exist for (x,y) in a neighborhood D of (x_0,y_0) and are continuous at (x_0,y_0). Let $z_0 = f(x_0,y_0)$. In Fig. 1, S is the graph of $z = f(x,y)$ for (x,y) in D. The tangent plane to S at (x_0,y_0,z_0) is horizontal. The traces (intersections) of S in the vertical planes $y = y_0$ and $x = x_0$ are the curves $z = f(x,y_0)$ and $z = f(x_0,y)$. Since $f_x(x_0,y_0) = 0$ and $f_y(x_0,y_0) = 0$, both of these curves have horizontal tangent lines at (x_0,y_0). Suppose that $f_{xx}(x_0,y_0) > 0$ and $f_{yy}(x_0,y_0) > 0$. Since these are the ordinary second derivatives of $f(x, y_0)$ and $f(x_0,y)$ at (x_0,y_0), the second derivative test tells us that each of the curves $z = f(x,y_0)$ and $z = f(x_0,y)$ is concave up near (x_0,y_0) and each has a local min at (x_0,y_0).

The foregoing discussion and Fig. 1 might tempt you to believe that for $f(x,y)$ to have a local min at the critical point (x_0,y_0), it is enough that $f_{xx}(x_0,y_0) > 0$ and $f_{yy}(x_0,y_0) > 0$. However, that is not true. Here is a counter-example.

EXAMPLE 5. Let $f(x,y) = x^2 - 3xy + y^2$. Any critical points satisfy

$$f_x = 2x - 3y = 0, \qquad f_y = 2y - 3x = 0.$$

By simple algebra, the only critical point is $(0,0)$. The traces $z = f(x,0) = x^2$ and $z = f(0,y) = y^2$ are obviously concave up, which follows from the fact that the second partials $f_{xx}(x,y) = 2$ and $f_{yy}(x,y) = 2$ are positive for all (x,y). However, the trace with $y = x$ is $f(x,x) = -x^2$, which is concave down. So the function f has a saddle point at $(0,0)$. □

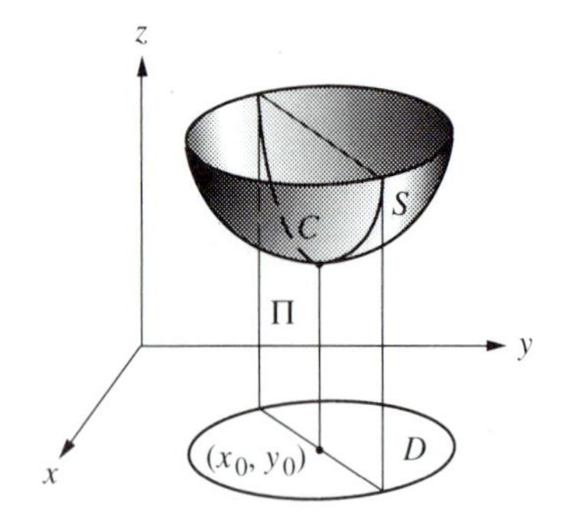

FIGURE 2

Example 5 shows that the graph of a function can be concave up in the x–direction and in the y–direction from a critical point, but may not be concave up in all directions from the critical point.

Now return to the general case. In Fig. 2, D is a neighborhood of a point (x_0,y_0) where $\nabla f(x_0,y_0) = \mathbf{0}$, Π is any vertical plane through (x_0,y_0), and C is the trace of the surface $z = f(x,y)$ in Π. Imagine that Π varies over all such vertical planes. If all the traces C are concave up over D, then f has a local min at (x_0,y_0). If they are all concave down over D, then f has a local max at (x_0,y_0). If some are concave up and some are concave down, then f has a saddle point at (x_0, y_0). The second partials test gives conditions on $f_{xx}(x_0,y_0)$, $f_{yy}(x_0,y_0)$ and $f_{xy}(x_0,y_0)$, which guarantee that one of these three concavity properties actually occurs near (x_0,y_0) and, consequently, it tells us whether f has a local min, a local max, or a saddle point at (x_0,y_0).

Suppose that $\nabla f(x_0,y_0) = \mathbf{0}$. Equivalently, $f_x(x_0,y_0) = 0$ and $f_y(x_0,y_0) = 0$. The second partials test will be expressed in terms of the constants

$$\boxed{A = f_{xx}(x_0,y_0), \quad B = f_{xy}(x_0,y_0), \quad C = f_{yy}(x_0,y_0), \quad \Delta = AC - B^2.}$$

Theorem 2 *The Second Partials Test*
Let $\nabla f(x_0, y_0) = \mathbf{0}$. Assume that the second partials of f exist for (x, y) near (x_0, y_0) and are continuous at (x_0, y_0).
If $A > 0$, $C > 0$ and $\Delta > 0$, then $f(x_0, y_0)$ is a local min.
If $A < 0$, $C < 0$ and $\Delta > 0$, then $f(x_0, y_0)$ is a local max.
If $\Delta < 0$, then f has a saddle point at (x_0, y_0).

The proof is given at the end of the section. It reveals that the three cases of the theorem guarantee that the traces of f in a neighborhood of (x_0, y_0) are, respectively, all concave up, all concave down, some concave up and some concave down. The second partials test gives us no information if $\Delta = 0$.

To better understand the sign conditions in Th. 2, we turn to four functions with familiar graphs. For each of the four functions in the following table, $\nabla f(0,0) = \mathbf{0}$ and the second partials are evaluated at $(0,0)$.

$f(x,y)$	A	B	C	Δ	Type
$x^2 + y^2$	2	0	2	4	local min
$4 - x^2 + y^2$	-2	0	-2	4	local max
$x^2 - y^2$	2	0	-2	-4	saddle point
xy	0	1	0	-1	saddle point

The conclusions in the table agree with our knowledge of the four graphs, which are, respectively, a circular paraboloid opening up, a circular paraboloid opening down, and two hyperbolic paraboloids.

EXAMPLE 6. Find the critical points of $f(x, y) = x^2 - xy + y^2$ and classify them; that is, determine whether f has a local max, a local min, or a saddle point at each critical point.

Solution. Any critical points satisfy

$$f_x = 2x - y = 0, \qquad f_y = 2y - x = 0.$$

The only critical point is $(0,0)$, where

$$A = f_{xx}(0,0) = 2, \qquad B = f_{xy}(0,0) = -1, \qquad C = f_{yy}(0,0) = 2.$$

Since $A > 0$, $C > 0$, and $\Delta = AC - B^2 = 3 > 0$, $f(0,0) = 0$ is a local minimum by the second partials test. In fact, the graph of f is an elliptic paraboloid with vertex at $(0,0,0)$ which opens up. We met this paraboloid earlier in Ex. 14 of Sec. 5.1. □

EXAMPLE 7. We found in Ex. 5 that the graph of $f(x, y) = x^2 - 3xy + y^2$ has a saddle point at $(0,0)$. Derive the same conclusion from the second partials test.

Solution. In this case, by inspection,

$$A = f_{xx}(0,0) = 2, \qquad B = f_{xy}(0,0) = -3, \qquad C = f_{yy}(0,0) = 2.$$

Since $\Delta = AC - B^2 = -5 < 0$, the second partials test tells us that f has a saddle point at $(0,0)$. □

The functions in Exs. 6 and 7 look a lot alike, but $f(x,y) = x^2 - xy + y^2$ has a local min at the critical point $(0,0)$ and $f(x,y) = x^2 - 3xy + y^2$ has a saddle point at $(0,0)$. The crucial difference lies in the size of B relative to A and C.

EXAMPLE 8. Find and classify the critical points of $f(x,y) = x^3 - 12xy + 12y^2 + 3$.

Solution. Any critical points satisfy

$$f_x = 3x^2 - 12y = 0, \qquad f_y = 24y - 12x = 0.$$

Then $x^2 = 4y$ and $4y = 2x$, so $x^2 = 2x$ and $x = 0$ or $x = 2$. The corresponding values of y are $y = 0$ and $y = 1$. So the critical points of f are $(0,0)$ and $(2,1)$. Also,

$$f_{xx} = 6x, \qquad f_{xy} = -12, \qquad f_{yy} = 24.$$

The following table is convenient for carrying out the second partials test.

(x,y)	$A = f_{xx}(x,y)$	$B = f_{xy}(x,y)$	$C = f_{yy}(x,y)$	$\Delta = AC - B^2$
$(0,0)$	0	-12	24	-144
$(2,1)$	12	-12	24	144

From the table and the second partials test, f has a saddle point at $(0,0)$ and a local min at $(2,1)$, where $f(2,1) = -1$. □

EXAMPLE 9. Find and classify the critical points of

$$f(x,y) = x^4 + 2x^2y^2 + y^4 - 8x^2 - 8y^2 + 16.$$

Solution. Any critical points satisfy

$$f_x = 4x^3 + 4xy^2 - 16x = 4x(x^2 + y^2 - 4) = 0,$$

$$f_y = 4y^3 + 4x^2y - 16y = 4y(x^2 + y^2 - 4) = 0.$$

The critical points of f are $(0,0)$ and any point (x,y) on the circle $x^2 + y^2 = 4$. This function has infinitely many critical points! Also,

$$f_{xx} = 12x^2 + 4y^2 - 16, \qquad f_{xy} = 8xy, \qquad f_{yy} = 12y^2 + 4x^2 - 16.$$

As in Ex. 7, we make a table.

(x,y)	$A = f_{xx}(x,y)$	$B = f_{xy}(x,y)$	$C = f_{yy}(x,y)$	$\Delta = AC - B^2$
$(0,0)$	-16	0	-16	256
$x^2 + y^2 = 4$	$8x^2$	$8xy$	$8y^2$	0

By the second partials test, $f(0,0) = 16$ is a local maximum of f. But the test gives us no information about the critical points on the circle $x^2 + y^2 = 4$ because $\Delta = 0$ at each of these points. Actually, we know what happens because $f(x,y)$ is the same function, $f(x,y) = (x^2 + y^2 - 4)^2$, that we met in Ex. 2 of Sec. 6.1. We learned there that $f(0,0) = 16$ is a local maximum but not a global maximum and that f attains the local and global minimum 0 at every point of the circle $x^2 + y^2 = 4$. The facts that $f(0,0) = 16$ is not the global max and that $f(x,y) = 0$ is the global min of f cannot be deduced from the second partials test. □

The function in our final example behaves much like $x^2 - y^2$ near the origin but, otherwise, has a much richer structure.

EXAMPLE 10. Classify the critical points of $f(x,y) = \cos y - \cos x$.

Solution. First, observe that $f(x,y)$ is periodic with period 2π in both x and y. So the surface $z = f(x,y)$ with (x,y) in the basic square $-\pi \leq x \leq \pi$, $-\pi \leq y \leq \pi$ is repeated over and over again in adjacent squares. You might think of it as the surface of a very orderly ocean. Now let's look for the critical points of f. Since $f_x = \sin x$ and $f_y = -\sin y$, $(0,0)$ is a critical point of f. Since $f_{xx} = \cos x$, $f_{xy} = 0$, and $f_{yy} = -\cos y$, we have

$$A = f_{xx}(0,0) = 1, \qquad B = f_{xy}(0,0) = 0, \qquad C = f_{yy}(0,0) = -1.$$

Then $\Delta = AC - B^2 = -1$ so f has a saddle point at (0,0). The function f has many more critical points. Since

$$f_x = \sin x = 0 \iff x = n\pi, \quad n \text{ any integer},$$

$$f_y = -\sin y = 0 \iff y = m\pi, \quad m \text{ any integer},$$

f has the critical points $(n\pi, m\pi)$, where m and n are any integers. In the problems you are asked to verify the entries in the following table.

n	m	$(n\pi, m\pi)$
even	odd	local min
odd	even	local max
even	even	saddle point
odd	odd	saddle point

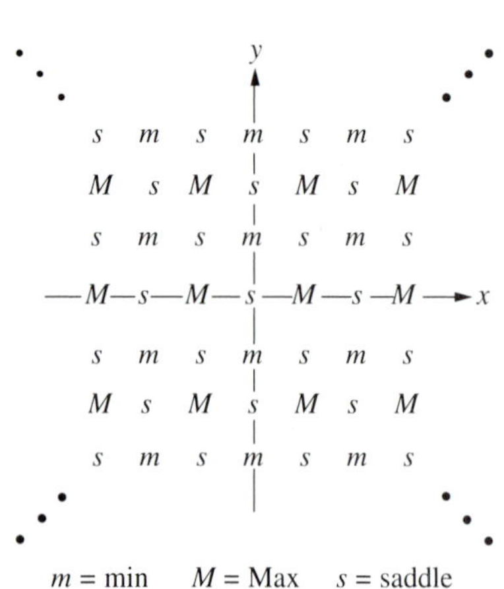

FIGURE 3

Figure 3 shows the pattern of these critical points near the origin. (The table should enable you to imagine the wave pattern in the ocean.) □

Proofs of Theorems 1 and 2*

As promised earlier, we close this section by proving Ths. 1 and 2. We begin with the second partials test. The proof is based on first- and second-order directional derivatives. The arguments add considerably to our geometric insight about the bending of surfaces and to our understanding of the second partials test. The proof of Th. 2 is short and fairly easy but less illuminating. We give it because the equality of mixed partials is used routinely over and over again in calculations of higher-order partials.

Proof of the Second Partials Test

First, we develop a convenient test that enables us to recognize when the traces of a surface in a vertical plane are concave up or down near a particular point on the surface. For this purpose, we use the derivative of f along a vertical trace C and, more particularly, the second derivative of f along C. The first derivative along C is a directional derivative. The second derivative along C is a second-order directional derivative. Most of the notation and structure for carrying out this program is borrowed from Sec. 5.5. Once the test for concavity is established, there is not much more to the proof of the second partials test. In the figures that follow, (x_0, y_0) could be any interior point of the domain of the function, but the figures illustrate the case when (x_0, y_0) is a critical point because that is the case of primary interest to us now.

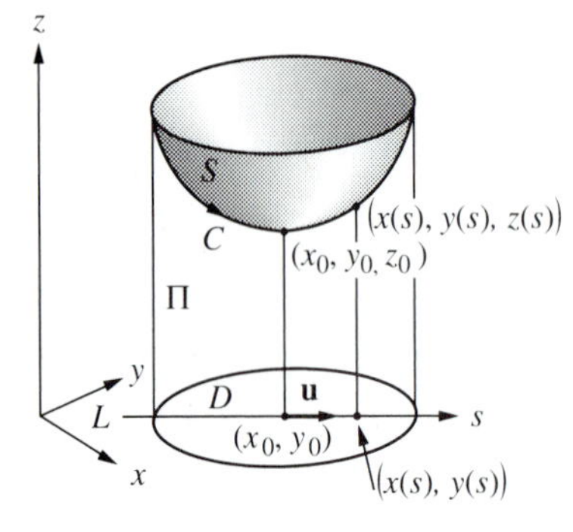

FIGURE 4

In Fig. 4, $\mathbf{u} = \langle h, k \rangle$ is any unit vector in the *xy*–plane attached to the point (x_0, y_0), Π is the vertical plane through **u**, and C is the trace of the surface S in the plane Π. The line L containing the unit vector **u** is labeled the *s*–axis with $s = 0$ at (x_0, y_0) and the positive direction determined by **u.** Just as in Sec. 5.5, s measures (signed) distance from (x_0, y_0) along L and Π can be regarded as the *sz*–plane. Figure 4 has been drawn to make it appear that the *sz*–plane is the plane of the paper. Parametric equations for L are

$$x(s) = x_0 + hs, \qquad y(s) = y_0 + ks.$$

Corresponding parametric equations for C are

$$x(s) = x_0 + hs, \qquad y(s) = y_0 + ks, \qquad z(s) = f(x(s), y(s)).$$

The directional derivative of f at $(x(s), y(s))$ in direction **u** is

$$D_{\mathbf{u}} f(x(s), y(s)) = \frac{d}{ds} f(x(s), y(s)).$$

This derivative is the slope of C at $(x(s), y(s), z(s))$.

The **second-order directional derivative** of f at $(x(s), y(s))$ in the direction **u** is

$$D_{\mathbf{u}}^2 f(x(s), y(s)) = \frac{d^2}{ds^2} f(x(s), y(s)).$$

Evidently, if $D_{\mathbf{u}}^2 f(x(s), y(s))$ is positive (or negative) for s in an interval I, then C is concave up (or down) for s in I.

Useful expressions for $D_{\mathbf{u}} f$ and $D_{\mathbf{u}}^2 f$ are obtained with the aid of the chain rule. Since $x(s) = x_0 + hs$ and $y(s) = y_0 + ks$, the chain rule gives

$$D_{\mathbf{u}} f(x(s), y(s)) = \frac{d}{ds} f(x(s), y(s)) = f_x(x(s), y(s))h + f_y(x(s), y(s))k.$$

We have just rederived $D_{\mathbf{u}} f = \nabla f \cdot \mathbf{u} = hf_x + kf_y$. Another application of the chain rule gives

$$D_{\mathbf{u}}^2 f = \frac{d}{ds}\left(\frac{df}{ds}\right) = \frac{d}{ds}(f_x h + f_y k) = (f_{xx} h + f_{yx} k)h + (f_{xy} h + f_{yy} k)k,$$

$$\boxed{D_{\mathbf{u}}^2 f = f_{xx} h^2 + 2f_{xy} hk + f_{yy} k^2,}$$

where all derivatives are evaluated at $(x(s), y(s))$ and we have used $f_{xy} = f_{yx}$. In particular, $D_{\mathbf{i}}^2 f = f_{xx}$ and $D_{\mathbf{j}}^2 f = f_{yy}$, as you might have expected.

Now we are ready to complete the proof of Th. 2. Let

$$A = f_{xx}(x_0, y_0), \qquad B = f_{xy}(x_0, y_0), \qquad C = f_{yy}(x_0, y_0).$$

For convenience, also let

$$\alpha = f_{xx}(x, y), \qquad \beta = f_{xy}(x, y), \qquad \gamma = f_{yy}(x, y).$$

Then

$$D_{\mathbf{u}}^2 f(x, y) = \alpha h^2 + 2\,\beta hk + \gamma k^2.$$

As in the first part of the theorem, assume that $A > 0$, $C > 0$, and $\Delta = AC - B^2 > 0$. Since the second partials are continuous at (x_0, y_0), there is a neighborhood D of (x_0, y_0) such that

$$\alpha > 0, \qquad \gamma > 0, \qquad \alpha\gamma - \beta^2 > 0 \qquad \text{for } (x, y) \text{ in } D.$$

Let D be this particular neighborhood of (x_0, y_0) in the rest of the proof and in Fig. 4. For (x, y) in D, complete the square on h in the preceding formula for $D_{\mathbf{u}}^2 f(x, y)$ to obtain

$$\begin{aligned} D_{\mathbf{u}}^2 f(x, y) &= \alpha\left(h^2 + \frac{2\beta hk}{\alpha} + \frac{\beta^2 k^2}{\alpha^2}\right) + \gamma k^2 - \frac{\beta^2 k^2}{\alpha} \\ &= \alpha\left(h + \frac{\beta k}{\alpha}\right)^2 + \frac{(\alpha\gamma - \beta^2)k^2}{\alpha} \end{aligned}$$

where $\alpha > 0$ and $\alpha\gamma - \beta^2 > 0$. Clearly, if $k \neq 0$, then $D_{\mathbf{u}}^2 f(x, y) > 0$. If $k = 0$, then $h = \pm 1$ and, again, $D_{\mathbf{u}}^2 f(x, y) > 0$. Thus, $D_{\mathbf{u}}^2 f(x, y) > 0$ for all directions $\mathbf{u}$ and all (x, y) on the part of L that lies in D. It follows that every trace C of the surface S in a vertical plane through (x_0, y_0) is concave up and $f(x,y) > f(x_0, y_0)$ for all $(x,y) \neq (x_0, y_0)$ in D, as indicated in Fig. 4. Therefore, $f(x_0, y_0)$ is a local minimum of f. This proves the first part of Th. 2. Virtually the same argument proves the second part. The proof of the third part, which is similar, is outlined in the problems.

A Proof of $f_{xy} = f_{yx}$

Finally, we prove Th. 1. Assume that $f_{xy}(x, y)$ and $f_{yx}(x,y)$ exist for (x,y) in a neighborhood of (x_0, y_0) and are continuous at (x_0, y_0). We shall prove that

$$f_{xy}(x_0, y_0) = f_{yx}(x_0, y_0) = \lim_{(x,y)\to(0,0)} \frac{f(x,y) - f(x,y_0) - f(x_0,y) + f(x_0,y_0)}{(x - x_0)(y - y_0)}.$$

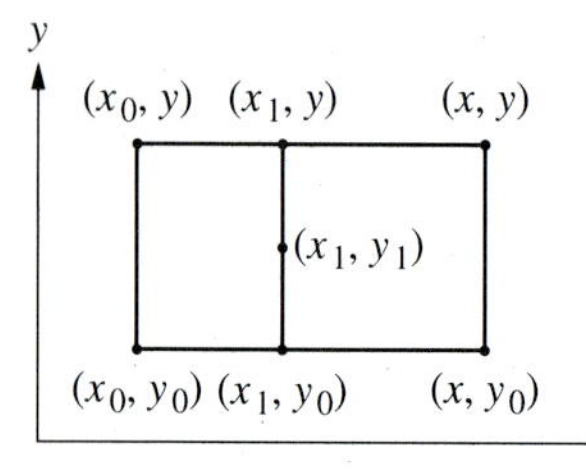

FIGURE 5

Figure 5 will serve as a guide. Let

$$g(x,y) = f(x,y) - f(x,y_0).$$

Then $g(x_0,y) = f(x_0,y) - f(x_0,y_0)$ and

$$f(x,y) - f(x,y_0) - f(x_0,y) + f(x_0,y_0) = g(x,y) - g(x_0,y).$$

By the mean value theorem, applied to the first variable of g,

$$g(x,y) - g(x_0,y) = g_x(x_1,y)(x - x_0)$$
for some x_1 between x_0 and x.

Since $g(x,y) = f(x,y) - f(x,y_0)$,

$$g_x(x_1,y) = f_x(x_1,y) - f_x(x_1,y_0).$$

The points (x_1,y) and (x_1,y_0) are shown in Fig. 5. Another application of the mean value theorem, this time on the second variable of f_x, yields

$$g_x(x_1,y) = f_x(x_1,y) - f_x(x_1,y_0) = f_{xy}(x_1,y_1)(y - y_0)$$
for some y_1 between y_0 and y.

Again refer to Fig. 5. Put the pieces together to obtain

$$f(x,y) - f(x,y_0) - f(x_0,y) + f(x_0,y_0) = g(x,y) - g(x_0,y),$$

$$g(x,y) - g(x_0,y) = g_x(x_1,y)(x - x_0) = f_{xy}(x_1,y_1)(x - x_0)(y - y_0),$$

$$f_{xy}(x_1,y_1) = \frac{f(x,y) - f(x,y_0) - f(x_0,y) + f(x_0,y_0)}{(x - x_0)(y - y_0)}.$$

Let $(x,y) \to (x_0,y_0)$. In view of Fig. 5, $(x_1,y_1) \to (x_0,y_0)$. Since f_{xy} is continuous at (x_0,y_0), $f_{xy}(x_1,y_1) \to f_{xy}(x_0,y_0)$ and

$$f_{xy}(x_0,y_0) = \lim_{(x,y)\to(x_0,y_0)} \frac{f(x,y) - f(x,y_0) - f(x_0,y) + f(x_0,y_0)}{(x - x_0)(y - y_0)}.$$

Essentially the same argument with the roles of x and y reversed shows that $f_{yx}(x_0, y_0)$ is equal to the same limit. How does Fig. 5 change? This completes the proof of Th. 1.

PROBLEMS

In Probs. 1–4 show that the given function satisfies the one–dimensional heat equation $u_t = au_{xx}$.

1. $e^{-4at} \cos 2x$
2. $e^{-n^2 at} \sin nx$, n any integer
3. $e^{-at} \sin x + 3e^{-9at} \cos 3x$
4. $(4\pi at)^{-1/2} e^{-x^2/4at}$

In Probs. 5–8 verify that the given function satisfies the *two–dimensional heat equation*

$$u_t = a(u_{xx} + u_{yy}),$$

where a is a positive constant and $u = u(x, y, t)$ is the temperature in a laterally insulated plate.

5. $e^{-2at} \sin x \cos y$
6. $e^{-5at} \cos 2x \sin y$
7. $e^{-(m^2+n^2)at} \sin mx \cos v$,
8. $\dfrac{1}{4\pi at} e^{-(x^2+y^2)/4at}$

A solution $u = u(x, y)$ of the two–dimensional heat equation that is independent of the time is called a *steady–state temperature distribution.* Since $u_t = 0$ for such a solution, it satisfies

$$u_{xx} + u_{yy} = 0,$$

which is called *Laplace's equation.* In Probs. 9–12 show that the given function satisfies the Laplace equation.

9. $\ln \sqrt{x^2 + y^2}$
10. $\arctan y/x$
11. $e^{x^2 - y^2} \sin 2xy$
12. $e^{-2x} \cos y$

The *one–dimensional wave equation*

$$u_{tt} = c^2 u_{xx}$$

models a variety of one–dimensional wave motions. Here $c > 0$ is the speed of wave propagation and $u = u(x, t)$ is the displacement at position x at time t from an equilibrium position. In Probs. 13–16 show that the given function satisfies the wave equation.

13. $\sin(x + ct)$
14. $3\cos(x - ct)$
15. $\sin 2x \cos 2ct$
16. $\cos h4(x - ct)$
17. If f and g are twice differentiable functions of one variable, show that $u(x, t) = f(x - ct) + g(x + ct)$ satisfies the wave equation.
18. Let $u = u(x, t)$ be a solution to the wave equation $u_{tt} = c^2 u_{xx}$. Show that under the change of variables

$$\xi = x - ct, \quad \eta = x + ct \quad \Leftrightarrow \quad x = \frac{\xi + \eta}{2}, \quad t = \frac{\eta - \xi}{2c}$$

the wave equation becomes

$$u_{\xi\eta} = 0.$$

19. Use Prob. 18 to show that all solutions $u = u(x,t)$, with continuous second partials, of the wave equation $u_{tt} = c^2 u_{xx}$ have the form given in Prob. 17. *Hint.* Antidifferentiate partially in $u_{\xi\eta} = 0$ to obtain $u = f(\xi) + g(\eta)$.

20. Let $u = u(x,y,t)$ be a solution to the two–dimensional heat equation $u_t = a(u_{xx} + u_{yy})$. To study heat conduction in a circular plate it is advantageous to change the space variables to polar coordinates via $x = r\cos\theta$, $y = r\sin\theta$. Then u becomes a function of r, θ, and t. Show that

$$u_{xx} + u_{yy} = u_{rr} + \frac{1}{r}u_r + \frac{1}{r^2}u_{\theta\theta}.$$

Now write down the heat equation and Laplace's equation when position in the plane is marked by polar coordinates.

In Probs. 21–34 find all the critical points of the given function and classify them as local maxima, minima, or saddle points.

21. $f(x,y) = x^2 + xy + y^2$
22. $f(x,y) = x^2 + 3xy + y^2$
23. $f(x,y) = xy + 1/x + 1/y$
24. $f(x,y) = xy(3 - x - y)$
25. $f(x,y) = x^4 + y^4 - 4xy$
26. $f(x,y) = xy^2 - x^2y - y + x$
27. $f(x,y) = 3xy - x^3 - y^3$
28. $f(x,y) = xy^2 + 3xy + x$
29. $f(x,y) = 4x^2e^y - 2x^4 - e^{4y}$
30. $f(x,y) = 3xe^y - x^3 - e^{3y}$
31. $f(x,y) = 8xye^{-(2x^2+y^2)}$
32. $f(x,y) = (2x^2 + y^2)e^{1-x^2-y^2}$
33. $f(x,y) = \cos y - 2\cos x$
34. $f(x,y) = \sin x + \sin y + \sin(x+y)$

35. As in Th. 2, assume that $\nabla f(x_0,y_0) = \mathbf{0}$ and let $A = f_{xx}(x_0,y_0)$, $B = f_{xy}(x_0,y_0)$, $C = f_{yy}(x_0,y_0)$. Verify the following statements:
    1. If A, $C > 0$ and $B = 0$, then $z_0 = f(x_0,y_0)$ is a local min.
    2. If A, $C < 0$ and $B = 0$, then $z_0 = f(x_0,y_0)$ is a local max.
    3. If $A > 0 > C$ or $A < 0 < C$, then $z_0 = f(x_0,y_0)$ is a saddle point.
    4. If $A = C = 0$ and $B \neq 0$, then $z_0 = f(x_0,y_0)$ is a saddle point.

    These four cases are illustrated by the functions in the table following Th. 2.

36. The only critical points not covered by the previous problem have A, C with the same sign and $B \neq 0$. For this case, explain briefly and make precise the following statement: The type of critical point depends on the size of B relative to A and C.

37. Let $f(x,y) = \frac{1}{2}Ax^2 + Bxy + \frac{1}{2}Cy^2$, where A, B, and C are constants. Complete the square to give an algebraic proof of Th. 2 for this quadratic function.

38. For each function in (a), (b), (c) explain why the second partials test does not give a conclusion at the critical point $(0,0)$. Then explain why the function has a global min in (a), a global max in (b), and a saddle in (c):

$$\text{(a)}\, x^4 + y^4 \qquad \text{(b)}\, -x^4 - y^4 \qquad \text{(c)}\, xy^2.$$

The next problem shows that all vertical traces of the graph of a function can be concave up on an interval through the critical point without a minimum occurring there.

39. Let $f(x,y) = (y - x^2)(y - 2x^2)$.
(a) Show that $(0,0)$ is the only critical point of f.
(b) Show that Th. 2 gives no information about the critical point.
(c) Show that f has a saddle point at the origin.
(d) Show that the trace of the graph of f in every vertical plane through the origin is concave up on some open interval containing $(0,0)$.
(e) Why doesn't the result in (d) contradict the reasoning used in the text in the context of Fig. 2?

40. Verify the entries in the table in Ex. 10 by applying the second partials test to each critical point (m,n).

41. Classify all the critical points of $f(x,y) = \cos x + \cos y$. Then use this information to describe the surface $z = \cos x + \cos y$. More particularly, describe the surface as a very regular mountain range. Indicate peaks, valleys, and saddles with a sketch like Fig. 3.

42. Prove part (2) of the second partials test.

43. Prove part (3) of the second partials test. *Hint.* If $A = C = 0$ show that $D^2_{\mathbf{u}} f(x_0, y_0)$ is positive in some directions and negative in others. If $A \neq 0$, use the expression for $D^2_{\mathbf{u}} f(x,y)$ obtained by completing the square in the text. Show that $D^2_{\mathbf{u}} f(x,y)$ is positive in some directions and negative in others.

44. Complete the proof of Th. 1 by showing that $f_{yx}(x_0, y_0)$ can be expressed by the same limit as $f_{xy}(x_0, y_0)$. Show how Fig. 5 changes in your argument.

6.3 Constrained Max–Min Problems and Lagrange Multipliers

Constrained max–min problems for functions of two variables occur early in differential calculus. Such problems are solved either by using the constraint equation to eliminate one variable or by using implicit differentiation. The use of implicit differentiation is closely related to the method of Lagrange multipliers, which is the principal new idea in this section. After dealing with constrained max–min problems for functions of two variables, we shall move on to similar problems for functions of three or more variables.

Constrained Max–Min Problems for Functions of Two Variables

You have met constrained max–min problem earlier in calculus. Here is a typical example. Find the dimensions and area of the largest rectangular field that has perimeter 160 yards. This is a constrained maximum problem. Let x and y be adjacent sides of the field. We seek to maximize the area $A = xy$ subject to the constraint $2x + 2y = 160$ or $x + y = 80$. Since lengths are never negative, $0 \leq x \leq 80$ and $0 \leq y \leq 80$. You can easily check that the maximum area, $A = 1600$ square yards, is attained for $x = y = 40$ yards.

A typical constrained max–min problem for a function of two variables has the form:

$$\text{maximize or minimize } f(x,y)$$
$$\text{for } (x,y) \text{ on a } \textit{constraint curve } C.$$

In the area problem, the constraint curve is the line segment $x + y = 80$ for $0 \leq x \leq 80$.

As we learned in Sec. 6.1, a max–min problem for a function $f(x,y)$ on a closed and bounded set D breaks naturally into two problems, one in the interior of D and the other on the boundary C of D. The problem on the boundary is a constrained max–min problem.

EXAMPLE 1. Find the global maximum and minimum of

$$f(x,y) = x^2 - xy + y^2 \quad \text{for} \quad x^2 + y^2 \leq 8.$$

Solution. First, f does have global maximum and minimum values because it is continuous on the closed and bounded disk D with $x^2 + y^2 \leq 8$. The max and min occur either at critical points in the interior of D where $\nabla f = \langle 2x - y, 2y - x \rangle = \mathbf{0}$ or on the boundary of D, which is the circle C with $x^2 + y^2 = 8$. The only critical point is $(0,0)$, where $f(0,0) = 0$. This value must be compared with the largest and smallest values of f on C. Thus, we are led to the constrained extremum problem: Find the maximum and minimum values of $f(x,y) = x^2 - xy + y^2$ for (x,y) on the circle C. We shall solve this problem two ways.

Method 1. Direct Approach. On the circle C with $x^2 + y^2 = 8$, we have $f(x,y) = 8 - xy$. Since $f(-x, -y) = f(x,y)$, the function f takes on the same values on the upper and lower semicircles of C. So we can restrict (x,y) to the upper semicircle C^+ given by $x^2 + y^2 = 8$ with $y \geq 0$. On C^+, $y = y(x) = \sqrt{8 - x^2}$ and

$$f(x, \sqrt{8 - x^2}) = 8 - x\sqrt{8 - x^2} \quad \text{for} \quad -2\sqrt{2} \leq x \leq 2\sqrt{2}.$$

Now seek the critical points of f on C^+:

$$\frac{df}{dx} = -x \cdot \frac{-x}{\sqrt{8 - x^2}} - \sqrt{8 - x^2} = \frac{2x^2 - 8}{\sqrt{8 - x^2}}$$

$$\frac{df}{dx} = 0 \text{ for } x = \pm 2.$$

The information we need is recorded in the table:

x	$-2\sqrt{2}$	-2	2	$2\sqrt{2}$
y	0	2	2	0
f	8	12	4	8
df/dx		0	0	

The table and $f(-x, -y) = f(x,y)$ reveal that

$$\max_C f = f(-2,2) = f(2,-2) = 12,$$

$$\min_C f = f(2,2) = f(-2,-2) = 4.$$

Method 2. Implicit Differentiation. The critical points of $f(x,y)$ with (x,y) on C can be found by implicit differentiation:

$$x^2 + y^2 = 8 \quad \Rightarrow \quad 2x + 2y\frac{dy}{dx} = 0 \quad \Rightarrow \quad \frac{dy}{dx} = -\frac{x}{y},$$

$$f(x,y) = 8 - xy \quad \Rightarrow \quad \frac{df}{dx} = -x\frac{dy}{dx} - y = \frac{x^2}{y} - y = \frac{x^2 - y^2}{y},$$

$$\frac{df}{dx} = 0 \text{ for } y^2 = x^2, \quad y = \pm x.$$

Let $y = \pm x$ in the constraint equation $x^2 + y^2 = 8$ to find

$$\frac{df}{dx} = 0 \text{ for } x = \pm 2, \quad y = \pm 2.$$

We conclude, as in Method 1, that $\max_C f = 12$ and $\min_C f = 4$.

Finally, return to the original problem. Compare the extreme values of f on C with $f(0,0) = 0$ to conclude that

$$\max_D f = f(-2,2) = f(2,-2) = 12, \qquad \min_D f = f(0,0) = 0.$$

Figure 1 illustrates what we have found. □

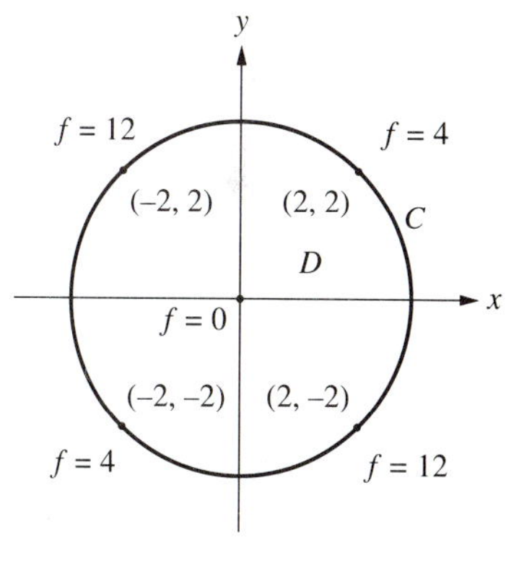

FIGURE 1

If a constraint curve C is given by parametric equations $x = x(t)$ and $y = y(t)$, then $f(x,y) = f(x(t), y(t))$ on C and the extreme values of $f(x(t), y(t))$ can be found by the usual one–variable techniques. In Ex. 1 we could have represented the circle C in terms of the polar angle θ (or t). In the problems you are asked to find the extreme values of $f(x,y)$ on the circle C by this method.

Again consider the general constrained max–min problem:

maximize or minimize $f(x,y)$
for (x,y) on a constraint curve C.

The constraint curve C may or may not have end points. For example, the curve $g(x,y) = x + y = 80$ for $0 \leq x \leq 80$ has the end points $(0,80)$ and $(80,0)$, whereas the curves $g(x,y) = x^2 + y^2 = 4$ and $g(x,y) = xy = 1$ have no end points. The general strategy for solving constrained max–min problems should seem familiar: We seek local maximum and minimum values of f at interior points of C (i.e., not endpoints of C) and evaluate f at any end points of C. If C, considered as a set in the xy–plane, is closed and bounded, then f attains global maximum and minimum values on C, which can be obtained by comparing the local extreme values of f and the values of f at any end points of C.

Lagrange Multipliers

The method of Lagrange multipliers is designed especially for constrained max–min problems of the form

$$\begin{gathered}\text{maximize or minimize } f(x,y)\\ \text{subject to a constraint } g(x,y)=c.\end{gathered}$$

Thus, the constraint curve C is expressed as a level curve of $g(x,y)$. The method of Lagrange multipliers exploits the fact that gradients are perpendicular to level curves. Assume that $f(x,y)$ and $g(x,y)$ have continuous first partials in an open set containing the constraint curve. We learned in Sec. 5.6 that if $\nabla f(x_0,y_0)\neq\mathbf{0}$ and $f(x_0,y_0)=c$, then the level curve $f(x,y)$ is a differentiable curve for (x,y) near (x_0,y_0) and $\nabla f(x_0,y_0)$ is perpendicular to the level curve.

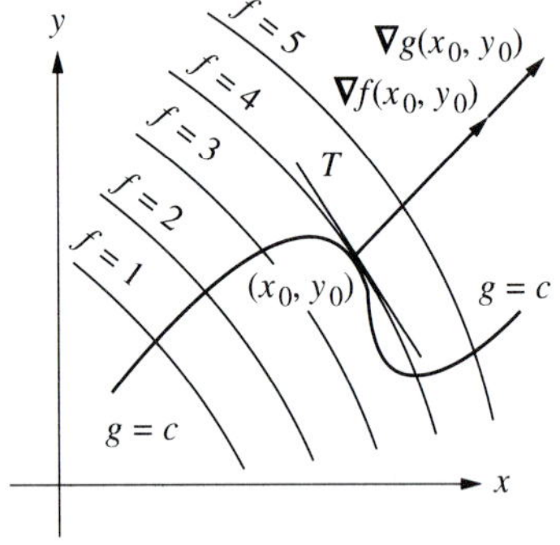

FIGURE 2

The method of Lagrange multipliers will be explained with the aid of Fig. 2. The level curves of f indicate that $f(x,y)$ increases as (x,y) moves in a "northeasterly" direction. In the figure, a local maximum value of f on the constraint curve $g=c$ occurs at (x_0,y_0), where a level curve of f and the constraint curve have a common tangent line T. Assume $\nabla g(x_0,y_0)\neq\mathbf{0}$. Since both $\nabla f(x_0,y_0)$ and $\nabla g(x_0,y_0)$ are perpendicular to T, they are parallel. Therefore, $\nabla f(x_0,y_0)=\lambda\nabla g(x_0,y_0)$ for some λ. The number λ is called a **Lagrange multiplier**. Theorem 1 summarizes the situation.

> **Theorem 1** *Lagrange Multipliers*
> Let $f(x,y)$ and $g(x,y)$ have continuous first partials in an open set D, which contains a curve $g(x,y)=c$. Assume that a local maximum or minimum of $f(x,y)$ subject to the constraint $g(x,y)=c$ occurs at a point (x_0,y_0) where $\nabla g(x_0,y_0)\neq\mathbf{0}$. Then
> $$\nabla f(x_0,y_0)=\lambda\nabla g(x_0,y_0) \text{ for some } \lambda.$$

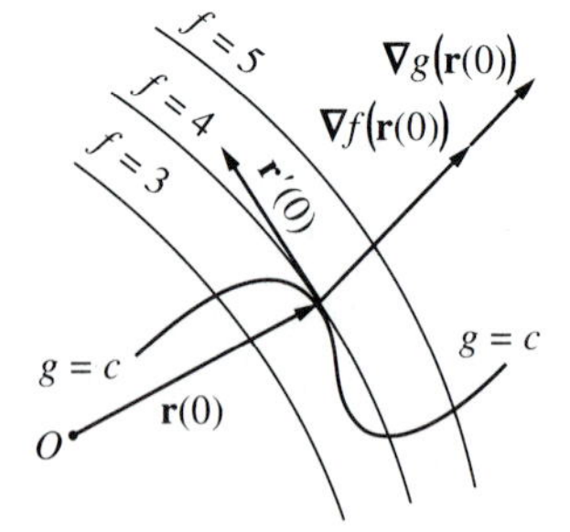

FIGURE 3

Proof. The proof is short and instructive. See Fig. 3 for the setting. Near (x_0,y_0) the constraint curve $g(x,y)=c$ can be represented in vector form $\mathbf{r}=\mathbf{r}(t)$ with $\mathbf{r}(0)=\ <x_0,y_0>$ and $\mathbf{r}'(0)\neq\mathbf{0}$. Since $g(\mathbf{r}(t))=c$ for t near 0,

$$\left.\frac{d}{dt}g(\mathbf{r}(t))\right|_{t=0}=\nabla g(\mathbf{r}(0))\cdot\mathbf{r}'(0)=0,$$
$$\nabla g(\mathbf{r}(0))\perp\mathbf{r}'(0).$$

On the other hand, since $f(\mathbf{r}(t))$ has a local max or min at 0,

$$\left.\frac{d}{dt}f(\mathbf{r}(t))\right|_{t=0}=\nabla f(\mathbf{r}(0))\cdot\mathbf{r}'(0)=0,$$
$$\nabla f(\mathbf{r}(0))\perp\mathbf{r}'(0).$$

Since both $\nabla f(\mathbf{r}(0))$ and $\nabla g(\mathbf{r}(0))$ are perpendicular to $\mathbf{r}'(0)$ in Fig. 3, they must be parallel. Therefore, $\nabla f(x_0, y_0) = \lambda \nabla g(x_0, y_0)$ for some λ. □

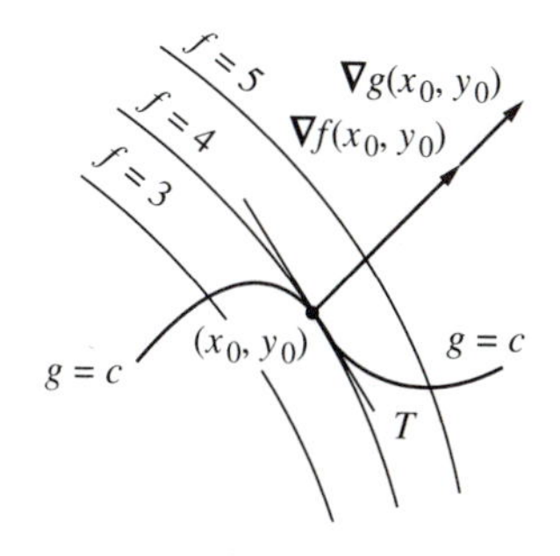

FIGURE 4

The converse of Th. 1 is false. At the point (x_0, y_0) in Fig. 4, $\nabla f(x_0, y_0)$ and $\nabla g(x_0, y_0)$ are parallel, but $f(x_0, y_0)$ is not a local max or min of f on C.

In the setting of Th. 1, the vector equation $\nabla f(x_0, y_0) = \lambda \nabla g(x_0, y_0)$ and the fact that (x_0, y_0) lies on the constraint curve provide three scalar equations for x_0, y_0, and λ:

$$f_x(x_0, y_0) = \lambda g_x(x_0, y_0),$$
$$f_y(x_0, y_0) = \lambda g_y(x_0, y_0),$$
$$g(x_0, y_0) = c.$$

In practical applications of Lagrange's method, these three equations are solved for x_0, y_0, and λ . In the process, we often eliminate λ because we are primarily interested in x_0 and y_0. The equations displayed previously can be expressed briefly as

$$\boxed{f_x = \lambda g_x, \qquad f_y = \lambda g_y, \qquad g = c.}$$

Let's call these equations the **Lagrange equations**. If the constraint curve C is a closed and bounded set, then the extreme values of f on C are the largest and smallest values of $f(x,y)$ as (x,y) runs through the solutions of the Lagrange equations and any end points of C. We give several applications of Lagrange multipliers, which demonstrate the advantages of this technique. You will see that the method of Lagrange multipliers is often shorter and easier than other methods.

EXAMPLE 2. In Ex. 1, the extreme values of $f(x,y) = 8 - xy$ for (x,y) on the circle C with $g(x,y) = x^2 + y^2 = 4$ were found by a direct method and by implicit differentiation. Find these extreme values by using Lagrange multipliers.

Solution. In this case, the Lagrange equations become

$$-y = \lambda \cdot 2x, \qquad -x = \lambda \cdot 2y, \qquad x^2 + y^2 = 8.$$

To eliminate λ, multiply the first equation by y, the second by x, and subtract to obtain $x^2 - y^2 = 0$. Then $y = \pm x$ and, by the constraint equation, $2x^2 = 8$, $x = \pm 2$ and, hence, $y = \pm 2$. The corresponding values of $f(x,y)$ are recorded in the table:

(x,y)	$(2,2)$	$(-2,2)$	$(-2,-2)$	$(2,-2)$
$f(x,y)$	4	12	4	12

Since C is a closed and bounded set and C has no endpoints, the extreme values of f are the largest and smallest values in the table:

$$\max_C f = 12, \qquad \min_C f = 4,$$

which agree with our previous results. □

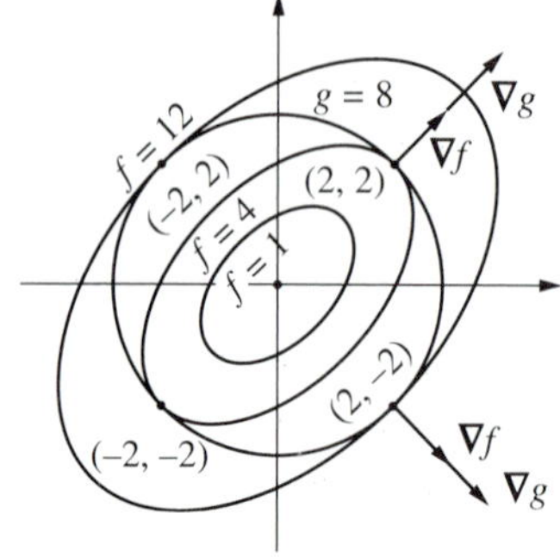
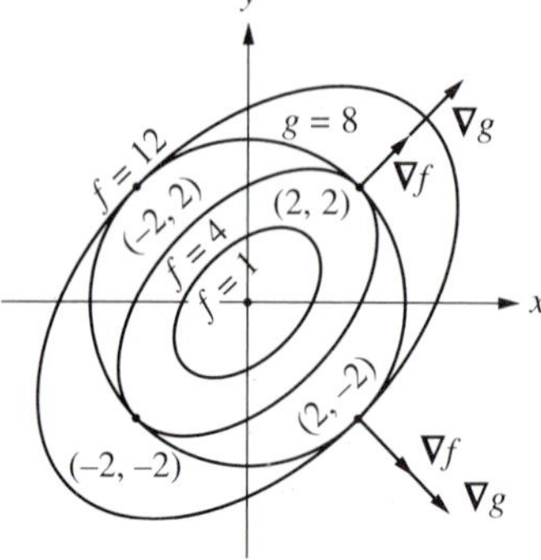

FIGURE 5

Although it is not needed for the solution, Fig. 5 shows the geometry behind the method of Lagrange multipliers in Ex. 2. Notice how the value of $f(x,y)$ increases and decreases between 4 and 12 as (x,y) runs around the circle $g = 8$.

EXAMPLE 3. Find the minimum distance from the point $(1,1)$ to the curve $xy = 4$ with $x > 0$, $y > 0$.

Solution. Figure 6 illustrates the situation. We seek to minimize

$$d(x,y) = \sqrt{(x-1)^2 + (y-1)^2} \quad \text{with } g(x,y) = xy = 4.$$

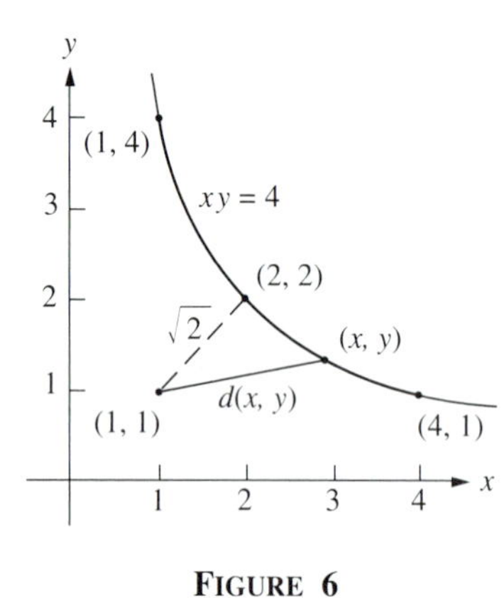

FIGURE 6

It is evident from the figure that $d(x,y)$ does have a minimum (and you can probably guess what it is). A more careful argument (that we omit) is based on the continuity of the function $d(x,y)$ and the fact that $d(x,y) \to \infty$ as $x \to 0^+$ or $x \to \infty$.

An equivalent and simpler problem is to minimize the square of the distance

$$f(x,y) = d(x,y)^2 = (x-1)^2 + (y-1)^2 \text{ with } g(x,y) = xy = 4.$$

The Lagrange equations $f_x = \lambda g_x$, $f_y = \lambda g_y$, and $g = 4$ give

$$2(x-1) = \lambda y, \qquad 2(y-1) = \lambda x, \qquad xy = 4.$$

Multiply the first equation by x, the second equation by y, and subtract to eliminate λ and obtain

$$2x(x-1) = 2y(y-1), \quad x^2 - y^2 - x + y = 0, \quad (x-y)(x+y-1) = 0.$$

So either $y = x$ or $y = 1 - x$. Try both in the constraint equation $xy = 4$. If $y = x$, we obtain $x = 2$ and $y = 2$. If $y = 1 - x$, we obtain $x^2 - x + 4 = 0$, which has no real solutions. Why? Only the point $(2,2)$ satisfies the Lagrange equations (with $\lambda = 1$). Since we know that f does have a minimum on C, it must be at $(2,2)$. Thus, $\min f = f(2,2) = 2$ and the minimum distance from $(1,1)$ to the curve $xy = 4$ is $d(1,1) = \sqrt{2}$, as indicated in Fig. 6. □

Often the hardest part of a constrained max–min problem is to show that a max or min exists. In practice, we usually rely on geometric or physical intuition, as we did in Ex. 3, to convince ourselves that there really is a maximum or minimum.

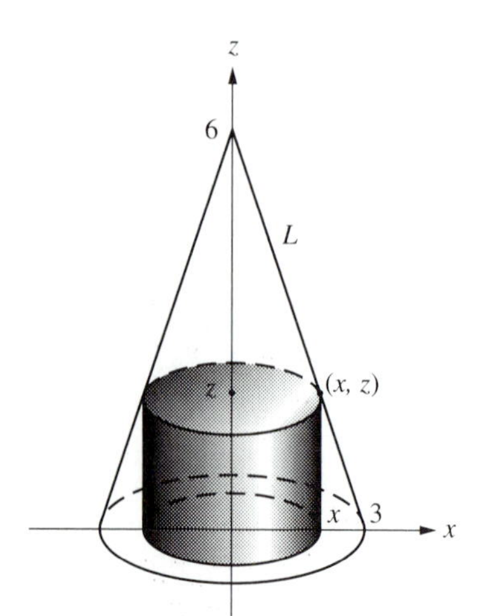

FIGURE 7

EXAMPLE 4. Find the maximum volume that a cylinder inscribed in the cone in Fig. 7 can have.

Solution. The line segment L in the figure has equation

$$\frac{x}{3} + \frac{y}{6} = 1 \text{ or } 2x + y = 6.$$

The volume of the inscribed cylinder is $V = \pi x^2 y$. We are asked to maximize

$$V = \pi x^2 y \quad \text{with } g(x, y) = 2x + y = 6.$$

Apparently, there is a maximum volume V for some (x, y) with $x > 0$, $y > 0$. We use Lagrange multipliers to find (x, y). The Lagrange equations are

$$2\pi xy = \lambda \cdot 2, \qquad \pi x^2 = \lambda \cdot 1, \qquad 2x + y = 6.$$

Then $2\pi xy = 2\lambda = 2\pi x^2$ and $y = x$. Since $2x + y = 6$, we find $x = 2$, $y = 2$. The maximum volume is $V = 8\pi$. □

Constrained Max–Min Problems for Functions of Three Variables

We shall deal briefly with Lagrange multipliers for constrained max–min problems of the form:

$$\begin{gathered}\text{maximize or minimize } f(x, y, z)\\ \text{subject to a constraint } g(x, y, z) = c.\end{gathered}$$

Assume that $f(x, y, z)$ and $g(x, y, z)$ have continuous first partials in an open set which contains the level surface $g(x, y, z) = c$. Just as in the case of two variables, if f has a local max or min at a point (x_0, y_0, z_0) on the constraint surface $g = c$ where $\nabla g(x_0, y_0, z_0) \neq 0$, then

$$\nabla f(x_0, y_0, z_0) = \lambda \nabla g(x_0, y_0, z_0)$$

for some λ called a Lagrange multiplier. This statement is proved by essentially the same argument used for functions of two variables; see the problems. The vector equation and the constraint equation give us four Lagrange equations in the four unknowns x_0, y_0, z_0 and λ. More briefly, the Lagrange equations for functions of three variables are

$$\boxed{f_x = \lambda g_x, \qquad f_y = \lambda g_y, \qquad f_z = \lambda g_z, \qquad g = c.}$$

These equations are used in the same way as before to solve constrained max–min problems. Similarly, if f and g are functions of four variables, x, y, z, and t, then there is another Lagrange equation, $f_t = \lambda g_t$.

EXAMPLE 5. Find the largest volume of a rectangular box with faces parallel to the coordinate planes that can be inscribed in the ellipsoid

$$\frac{x^2}{4} + \frac{y^2}{9} + \frac{z^2}{9} = 1.$$

Solution. By symmetry of the ellipsoid and box, we can confine our attention to the first octant. In Fig. 8, the point (x, y, z) lies on the ellipsoid. We seek to maximize the volume

$$V = 8xyz \quad \text{with} \quad g(x, y, z) = \frac{x^2}{4} + \frac{y^2}{9} + \frac{z^2}{9} = 1.$$

Since $V = 8xyz$ is continuous and the part of the ellipsoid with $x \geq 0$, $y \geq 0$, and $z \geq 0$ is closed and bounded, V has maximum and minimum values. The minimum value $V = 0$ occurs at every point on the ellipsoid with $x = 0$, $y = 0$, or $z = 0$. The maximum occurs at some point on the ellipsoid with $x > 0$, $y > 0$, $z > 0$. For comparison purposes, and for the different insights they give, we discuss three methods for finding the maximum of V.

Method 1. Direct Method. Solve the ellipsoid equation for

$$z = 3\sqrt{1 - \frac{x^2}{4} - \frac{y^2}{9}}$$

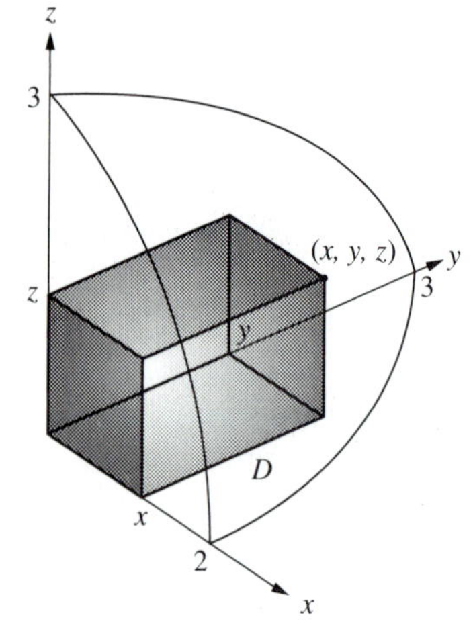

FIGURE 8

and substitute into $V = 8xyz$ to obtain

$$V = 24xy\sqrt{1 - \frac{x^2}{4} - \frac{y^2}{9}} \quad \text{for } (x, y) \text{ in } D,$$

where D is the region in the xy–plane in Fig. 8 with

$$\frac{x^2}{4} + \frac{y^2}{9} \leq 1 \text{ and } x,\, y \geq 0.$$

Now we have a standard max–min problem for a function of two variables: Find the maximum of $V = V(x, y)$ on the closed and bounded set D. We do not proceed further with this solution because there are more attractive approaches coming.

Method 2. Implicit Differentiation. Rather than eliminating z as in Method 1, we regard z as a function of (x, y) defined implicitly by the ellipsoid equation. Then two implicit differentiations of the ellipsoid equation yield

$$\frac{2x}{4} + \frac{2z}{9} z_x = 0, \qquad z_x = -\frac{9x}{4z},$$

$$\frac{2y}{9} + \frac{2z}{9} z_y = 0, \qquad z_y = -\frac{y}{z}.$$

Now we can find the criti[cal points of $V = 8xyz$ with V and z regarded as a functions of (x, y) with $x > 0$ and $y > 0$. By the chain rule and a little algebra,

$$V_x = 8y(z + xz_x) = 8y\left(z - \frac{9x^2}{4z}\right) = \frac{2y}{z}(4z^2 - 9x^2),$$

$$V_y = 8x(z + yz_y) = 8x\left(z - \frac{y^2}{z}\right) = \frac{8x}{z}(z^2 - y^2).$$

Evidently, $V_x = 0$ and $V_y = 0$ with $x, y, z > 0$ only if $x = 2z/3$ and $y = z$. Substitute these relations into the ellipsoid equation to obtain

$$\frac{z^2}{9} + \frac{z^2}{9} + \frac{z^2}{9} = 1, \qquad z^2 = 3,$$

$$z = \sqrt{3}, \qquad x = \frac{2}{3}\sqrt{3}, \qquad y = \sqrt{3}.$$

The maximum volume of the box is $V_{\max} = 16\sqrt{3}$.

Method 3. Lagrange Multipliers. The Lagrange equations $V_x = \lambda g_x$, $V_y = \lambda gy$, $V_z = \lambda g_z$, and $g = 1$ give

$$8yz = \lambda \cdot \frac{x}{2}, \qquad 8xz = \lambda \cdot \frac{2y}{9}, \qquad 8xy = \lambda \cdot \frac{2z}{9}, \qquad g(x,y,z) = 1.$$

Since we know that the maximum occurs at a point with $x, y, z > 0$, we also have $\lambda > 0$. Then

$$8xyz = \frac{1}{2}\lambda x^2 = \frac{2}{9}\lambda y^2 = \frac{2}{9}\lambda z^2.$$

It follows that $y = z$ and $x = 2z/3$. As in Method 2, put these results into the ellipsoid equation to obtain $z = \sqrt{3}$, $x = 2\sqrt{3}/3$, $y = \sqrt{3}$, and $V_{\max} = 16\sqrt{3}$. Clearly, the method of Lagrange multipliers is quickest. It immediately yields relations among the variables x, y, z that determine[the location of the maximum. □

We have just scratched the surface of the method of Lagrange multipliers for solving constrained max–min problems. In other problems, constraint inequalities, such as $g \le c$ or $g > c$ come into play, and problems with more than one constraint occur, such as to maximize or minimize $f(x,y,z)$ subject to $g(x,y,z) = c$ and $h(x,y,z) = k$.

PROBLEMS

In Probs. 1–8 use the method of Lagrange multipliers to determine the points at which local extrema of f may occur on the given constraint curve or surface. Then determine the global max and min, if they exist.

1. $f(x,y) = xy$ on the ellipse $x^2 + 4y^2 = 8$.
2. $f(x,y) = x^2 + 4y^2$ on the hyperbola $xy = 8$.
3. $f(x,y) = x^2 - xy + y^2$ on the circle $x^2 + y^2 = 1$.
4. $f(x,y) = x^2 + y^2$ on the ellipse $x^2 - xy + y^2 = 1$.
5. $f(x,y,z) = x + y + z$ on the surface $x^4 + y^4 + z^4 = 1$.
6. $f(x,y,z) = x^4 + y^4 + z^4$ on the plane $x + y + z = 1$.
7. $f(x,y,z) = xyz$ on the surface $xy + 2xz + 3yz = 18$.

8. $f(x,y,z) = 2x + y + 2z$ on the sphere $x^2 + y^2 + z^2 = 12$.

In Probs. 9–12 find the global max and min of f. Use Lagrange multipliers to find the extrema on the boundary of the domain of f.

9. $f(x,y) = x^2 + xy + y^2$ for $x^2 + y^2 \leq 1$.
10. $f(x,y) = x^4 - y^4$ for $x^2 + 4y^2 \leq 4$.
11. $f(x,y) = x^2 + y^2$ for the solid ellipse $x^2 + xy + y^2 \leq 3$.
12. $f(x,y) = x^3 - 3xy + y^3 + 6$ for $x^2 + y^2 \leq 8$.

In Probs. 13–41, use the method of Lagrange multipliers to find the indicated max or min.

13. Find the rectangle with the largest area among all rectangles with a fixed perimeter.
14. Find the rectangle with smallest perimeter among all rectangles with a fixed area.
15. Find the rectangular box with largest volume among all boxes with a fixed surface area.
16. Find the rectangular box with smallest surface area among all boxes with a fixed volume.
17. The top and bottom of a cylindrical can are aluminum and cost twice as much to produce as the curved side. If the can must have a capacity of 256 cubic inches, find the most economical proportions.
18. The top and bottom of a cylindrical can are aluminum and cost twice as much to produce as the curved side. If the can must have a fixed volume, find the most economical proportions.
19. Problem 31 in Sec. 6.1
20. Problem 32 in Sec. 6.1
21. Problem 34 in Sec. 6.1
22. Problem 35 in Sec. 6.1
23. Problem 37 in Sec. 6.1
24. Problem in 38 Sec. 6.1
25. Problem 39 in Sec. 6.1
26. Problem 40 in Sec. 6.1
27. Problem 41 in Sec. 6.1
28. Problem 42 in Sec. 6.1
29. A beam with rectangular cross sections is to be cut from a circular log with diameter 2 feet. The strength of a beam is proportional to the product of width times the square of its height. Find the shape of the strongest beam that can be cut from the log.
30. In the previous exercise four additional beams with rectangular cross sections are cut from the four "waste" pieces left from the initial cuts. Find the dimensions of these additional beams if each is as strong as possible.
31. Suppose a beam is cut from a log with elliptical cross sections $x^2/a^2 + y^2/b^2 = 1$. Find the shape of the strongest beam. See Prob. 29.
32. A trough, with a fixed cross–sectional area of 96 square inches, is to be made from a long piece of sheet metal by bending up equal lengths from each side along the length of the trough. Find the length of one of the equal

sides and the angle between the bottom of the trough and a side if the width of the sheet is to be as small as possible.

33. (a) Let $a > 0$. Maximize $(xyz)^{1/3}$ subject to $x + y + z = a$, over all positive numbers x, y, and z. (b) Use (a) to show that $(xyz)^{1/3} \le \frac{1}{3}(x + y + z)$.

34. Let a be a nonzero constant. Find the point(s) on the surface $xyz = a$ that are closest to the origin.

35. Maximize the sum of the squares of three numbers if their product is fixed, say equal to $a \neq 0$.

36. Find the extrema of f in Ex. 1 on the boundary curve C by using parametric equations for the circle C.

37. State and prove the analogue of Th. 1 for functions of three variables. *Hint.* Let $P_0 = (x_0, y_0, z_0)$. Recall that $\nabla g(P_0)$ is perpendicular to the tangent plane at P_0. Show that $\nabla f(P_0)$ is perpendicular to every vector that lies in the tangent plane.

38. *Herron's formula* $A = \sqrt{s(s-a)(s-b)(s-c)}$ expresses the area A of a triangle in terms of its sides a, b, c, and the semiperimeter $s = \frac{1}{2}(a + b + c)$. Find the triangle of largest area among all triangles with a fixed perimeter.

Chapter Highlights

An interior point P_0 of the domain of f is a critical point of f if $\nabla f(P_0) = \mathbf{0}$ or $\nabla f(P_0)$ does not exist. Local extreme values of f can occur only at critical points. In most cases, the second derivative test tells whether f has a local max, a local min, or a saddle point at a critical point. Global extreme values of f on a set D can occur only at critical points or boundary points in D. To find global extrema, evaluate f at the critical points in D and compare those values with the extreme values of f on the boundary of D.

A typical constrained max-min problem in 2–space or 3–space has the form:

$$\text{maximize or minimize } f(P)$$
$$\text{subject to a constraint } g(P) = c.$$

One way to solve such a problem is to eliminate the constraint. This direct approach often leads to unpleasant calculations or may not be possible to carry out in practice. Another more practical approach uses implicit differentiation. The related method of Lagrange multipliers, based on $\nabla f = \lambda \nabla g$, very often is quicker.

The chapter project is about optimal location. Problems of this kind occur often, particularly in civil engineering. For example, suppose a new electric power plant is to be built to serve several factories. Where should the plant be located? In principle many factors must be considered to answer this question such as what are the zoning laws, what environmental issues must be considered, where are the raw materials located that the plant needs, and how far should the plant be located relative to the factories. To keep the project manageable and still informative, we restrict the siting analysis to one criterion–locate the plant so the sum of its distances from all the factories is as small as possible. This criterion will help minimize various costs such as the expense of building and maintaining power lines, the loss of electricity during transmission, and certain transportation expenses. The project involves a mathematical model with an informative minimization analysis, a physical construction for estimating the optimal location, and a numerical method that can approximate the optimal location as accurately as desired.

Chapter Project: Optimal Location

Imagine that you are an engineer in charge of locating an electric power plant to serve three factories. Straight power lines will connect the plant to each factory. To keep cost down, you decide to locate the plant so as to minimize the total length of the three power lines. This project is longer than most. First, the optimal location problem is attacked with calculus, algebra, and geometry. Then a method of successive approximations is used to estimate the coordinates of the optimal location.

The factories are located at points P_i by rectangular coordinates $P_i = (x_i, y_i)$ for $i = 1, 2, 3$. The (as yet unknown) position of the power plant is $P = (x, y)$. Let

$$d_i(x, y) = \text{ the distance of factory } i \text{ from the power plant at } P = (x, y).$$

For convenience, let T be the solid triangle with vertices P_1, P_2, and P_3.

Problem 1 Write out a formula for $d_i(x, y)$ and explain why your goal is to locate the power plant at the point $P = (x, y)$ that minimizes

$$D(P) = D(x, y) = d_1(x, y) + d_2(x, y) + d_3(x, y).$$

Problem 2 If P' is outside the triangle T, show that there is another point P on one of the sides of the triangle such that $D(P) < D(P')$. *Hint.* Remember the triangle inequality from geometry.

Problem 3 (a) Explain why $D(x, y)$ has a global minimum as (x, y) varies over the xy–plane. (b) If the global minimum occurs at (x, y), explain why (x, y) lies in T.

Problem 3 Show that the global minimum of D must occur either at P_1, P_2, P_3, or a point $P = (x, y)$ with $P \neq P_i$ for which

$$(*) \qquad \left\{ \begin{array}{l} \dfrac{x - x_1}{d_1} + \dfrac{x - x_2}{d_2} + \dfrac{x - x_3}{d_3} = 0, \\[2ex] \dfrac{y - y_1}{d_1} + \dfrac{y - y_2}{d_2} + \dfrac{y - y_3}{d_3} = 0. \end{array} \right\}$$

The equations $(*)$ in Prob. 4 are not as simple to solve for x and y as they may at first appear to be because d_1, d_2, and d_3 depend on x and y.

If there is a point $P = (x, y)$ that satisfies $(*)$, then the global minimum of D occurs at (x, y), as the next three problems reveal.

Problem 5 Suppose there is a point $P = (x, y)$ whose coordinates satisfy (*). Transpose the terms in (*) containing x_3 and y_3 to the right side, square, and add the equations to obtain

$$\frac{(x - x_1)(x - x_2) + (y - y_1)(y - y_2)}{d_1 d_2} = -\frac{1}{2}.$$

Let $\mathbf{r} = \langle x, y \rangle$ and $\mathbf{r}_i = \langle x_i, y_i \rangle$ be the position vectors for P and P_i.

Problem 6 (a) In the context of Prob. 5, show that

$$\frac{(\mathbf{r} - \mathbf{r}_1) \cdot (\mathbf{r} - \mathbf{r}_2)}{\|\mathbf{r} - \mathbf{r}_1\|} \|\mathbf{r} - \mathbf{r}_2\| = -\frac{1}{2}.$$

(b) Conclude that the angle θ between $\mathbf{r}_1 - \mathbf{r}$ and $\mathbf{r}_2 - \mathbf{r}$ is equal to 120°.
(c) Explain why the angles between $\mathbf{r}_1 - \mathbf{r}$ and $\mathbf{r}_3 - \mathbf{r}$ and between $\mathbf{r}_2 - \mathbf{r}$ and $\mathbf{r}_3 - \mathbf{r}$ also are equal to 120°.

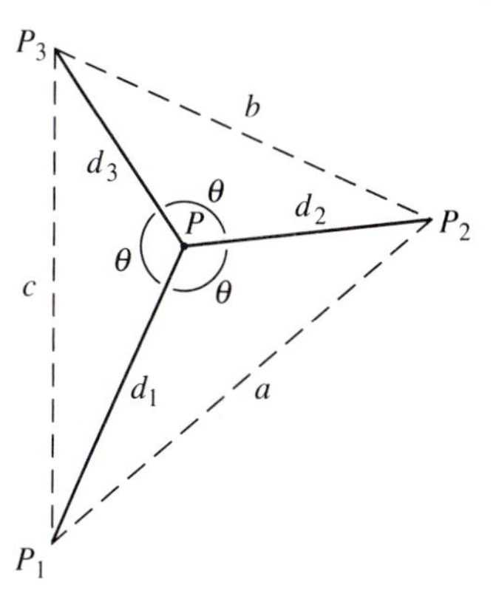

FIGURE 1

Figure 1 illustrates the conclusions of the last problem. The coordinates of P satisfy (*). It should be obvious from Fig. 1 that P must lie inside the triangle T as indicated and that there can be at most one point P with the property that the three angles marked θ are all equal to 120°. These facts will be assumed henceforth. However, it is important to realize that the equations (*) need not have a solution and, consequently, there may be no such point P. See Prob. 8.

Problem 7 In the context of Probs. 5–6, show that $D(P) < D(P_i)$ for $i = 1, 2, 3$. *Hint.* Refer to Fig. 1. Use the law of cosines to show that $a^2 = d_1^2 + d_2^2 + d_1 d_2$. Conclude that $\frac{1}{2}\,d_1 + d_2 < a$. Likewise, $\frac{1}{2}\,d_1 + d_3 < c$. Add the inequalities.

In summary, you have learned that if there is a point $P = (x, y)$ whose coordinates satisfy (*), then P lies inside T and is the unique point at which D assumes its global minimum. Thus, P is the site for the power plant.

The next problem shows that not all factory configurations admit a point P whose coordinates satisfy (*).

Problem 8 If there is a point $P = (x, y)$ that satisfies (*), then show that no angle in the triangle T can equal or exceed 120°. *Hint.* The sum of the angles in any triangle is 180°. Refer to Fig. 1.

Problem 9 The factory locations could be such that one of the angles in triangle T is 120° or greater. Suppose the obtuse angle is at vertex P_1. Where should the factory be located in this case? Explain.

The next problem gives a physical construction of the point $P = (x, y)$ that satisfies equations (*) in case no angle in the triangle T is 120° or greater. Use of this construction with a scale model of the factory locations would enable you to find the approximate location of the power plant.

Problem 10 Get out your tool box. On a flat board, drive in nails at the points P_1, P_2, and P_3. Join three rather long, thin rigid rods together to form a capital Y with 120° angles at the "center" of the Y. Slide two of the rods against the nails at P_1 and P_2 and adjust the Y until its third side rests against the nail at P_3. The center of the Y is the point P where the power plant should be built. (You may want to drive in the nails at locations as in Prob. 12 for comparison purposes.)

Next, the method of successive approximations will be used to find approximate values for the solutions to the system of equations (*), when the angles in triangle T are all less than 120°. Recall that $\mathbf{r} = \langle x, y \rangle$ and $\mathbf{r}_i = \langle x_i, y_i \rangle$.

Problem 11 (a) Show that (*) can be expressed as the vector equation

$$\mathbf{r} = \mathbf{F}(\mathbf{r}),$$

where

$$\mathbf{F}(\mathbf{r}) = \alpha_1 \mathbf{r}_1 + \alpha_2 \mathbf{r}_2 + \alpha_3 \mathbf{r}_3.$$

(b) Determine $\alpha_i = \alpha_i(\mathbf{r}) = \alpha_i(x, y)$ in terms of $d_i(x, y)$.
(c) Verify that $\alpha_i > 0$ and $\alpha_1 + \alpha_2 + \alpha_3 = 1$.

We mention in passing that the equation $\mathbf{r} = \alpha_1 \mathbf{r}_1 + \alpha_2 \mathbf{r}_2 + \alpha_3 \mathbf{r}_3$, which is a vector form of (*), and a little algebra can be used to show that any solution to (*) lies inside the triangle T, a fact we already deduced on geometric grounds earlier.

Choose an initial guess $\mathbf{r}^0 = \langle x^0, y^0 \rangle$ inside T for the optimal location of the power plant. Then calculate successive approximations

$$\mathbf{r}^{n+1} = \mathbf{F}(\mathbf{r}^n) = \alpha_1(\mathbf{r}^n)\mathbf{r}_1 + \alpha_2(\mathbf{r}^n)\mathbf{r}_2 + \alpha_3(\mathbf{r}^n)\mathbf{r}_3, \quad n = 0, 1, 2, \ldots.$$

Superscripts are used for the successive approximations because subscripts are already in use to mark the factories. It is shown in numerical analysis courses that the sequence of successive approximations converges to the solution of (*), which is the optimal location for the power plant. The procedure just described is easily programmed on a scientific calculator or with appropriate software.

Problem 12 Let $P_1 = (0, 5)$, $P_2 = (6, 7)$, and $P_3 = (7, 0)$. Take $\mathbf{r}^0 = \langle 4, 4 \rangle$ and calculate several successive approximations. Use the results to approximate the optimal location for the power plant.

The techniques developed previously enable you to treat more general optimal location problems. Suppose the three factories have substantially different power needs, so that the three power lines have different costs. Let c_i be the cost per mile of the ith power line, $i = 1, 2, 3$. Alternatively, the power plant could be replaced by a source of raw materials for the three factories. In either case, you could determine an optimal location by minimizing the total cost

$$C(x, y) = c_1 d_1(x, y) + c_2 d_2(x, y) + c_3 d_3(x, y).$$

If $c_1 = c_2 = c_3$, then $C(x, y) = D(x, y)$, the case treated earlier.

Problem 13 What system of equation replaces (*) in the minimum cost problem?

Problem 14 (a) Repeat the line of reasoning used in Probs. 5 and 6 to establish that

$$\cos\theta_1 = \frac{c_3^2 - c_1^2 - c_2^2}{2c_1c_2},$$

where θ_1 is the angle between $\overline{PP_2}$ and $\overline{PP_3}$ in Fig. 1.
(b) Find corresponding formulas that determine the other two angles at P.

The physical construction in Prob. 10 with a new Y modified as required by Prob. 14 can be used to approximate the optimal location.

Problem 15 The iterative procedure for approximating the optimal location applies virtually without change to the new situation. Only the values of the α_i in Prob. 11 must be changed slightly. (a) Find the new α_i. (b) Let the factories be located at the points in Prob. 12 and the cost per mile of the power lines be $c_1 = \$10{,}000$, $c_2 = \$30{,}000$, and $c_3 = \$40{,}000$. Start with the initial guess $\mathbf{r}^0 = \langle 4, 4 \rangle$ and carry out several steps in the successive approximation procedure to estimate the optimal location.

The foregoing intriguing developments are just the beginning of the story. Similar methods can be used to solve optimal location problems with many sites (factories, cities, work stations, or whatever) that are to be connected to a server.

Chapter Review Problems

In Probs. 1–5, find all the critical points of the function and classify them as local maxima, local minima, or saddle points.

1. $f(x,y) = xy(1 - x^2 - y^2)$
2. $z = x^2y^2 - x^2 - y^2$
3. $z = x^3 - 12xy + 12y^2 + 3$
4. $f(x,y) = xy + 4/x + 2/y$ for $x > 0$, $y > 0$
5. $f(x,y) = x^3 + 3x^2y + \frac{9}{2}x^2 + \frac{9}{2}y^2 + 3xy$
6. $f(x,y) = \left(\frac{1}{x} - 1\right)\left(\frac{2}{y} + 1\right)$
7. (a) Find all the critical points of $f(x,y) = 4x^2 - 2x^2y + 4xy + y^2$. (b) What does the second partials test say about them?
8. (a) Find and classify all critical points of $f(x,y) = 4x + xy - x^2 - y^2$. (b) Determine any global extrema.
9. (a) Find and classify all critical points of $f(x,y) = 4x^2y^2 - x^2 - y^2$. (b) Find the maximum and minimum of f for $x, y \geq 0$, $x^2 + y^2 \leq 1$.
10. A rectangular box with no top has volume 18 ft^3. The bottom costs 20 cents per ft^2. Each side costs 15 cents per ft^2. Find the minimum cost to make such a box.
11. Find the maximum of $f(x,y) = xy$ on the ellipse $9x^2 + 4y^2 = 72$.
12. Find the global maximum of $f(x,y) = 4y + xy - x^2 - y^2$.
13. Find the global maximum of $f(x,y) = x^4 + 2y^2 - 4y + 5$.

14. Find the maximum and minimum values of $f(x,y) = x^2 + 2y^2 - x$ on the unit disk with center $(0,0)$.

15. Find the minimum distance between the line L_1 given by $x = 3t$, $y = 2t$, $z = t$ and the line L_2 given by $x = t$, $y = t - 3$, $z = t$.

16. Find the extreme values of the function $f(x,y) = xy - 3x - 2y$ on the closed triangular region in the xy–plane with vertices $(0,0)$, $(0,4)$, and $(8,4)$.

17. A rectangular box has three faces in the coordinate planes and one vertex (x,y,z) on the plane through the points $(3,0,0)$, $(0,9,0)$, and $(0,0,4)$. (a) Find the largest volume the box can have. (b) Why is there a maximum but no minimum volume?

18. Find the maximum volume of a rectangular box with sides x, y, z such that $2x + y + 3z = 6$.

19. Let S be the surface $z = \sqrt{x^2 + y^2}$ with $z \leq 1$. Find the point on S nearest the point $(3,0,0)$ and the shortest distance from this point to S.

20. A triangular plate D is bounded by the x–axis, the y–axis, and the line from $(1,0)$ to $(0,2)$. The temperature at (x,y) in D is $T(x,y) = x^2 + xy + 2y^2 - 3x + 2y$. Find the maximum and minimum temperatures on the plate.

21. Find the minimum distance between the curves $y = x^2$ and $y = x - 1$.

22. Find the minimum distance from the surface $xy - z^2 - 3y + 12 = 0$ to the origin.

23. Find the minimum value of $f(x,y) = 2/x + 3/y$ with $x > 0$, $y > 0$ for (x,y) on the ellipse $9x^2 + 4y^2 = 72$.

24. Find the maximum value of a cylinder that can be fit inside a right circular cone with radius of base 3 inches and height 4 inches.

25. Find the minimum distance from the origin to the plane $x + 2y + 2z = 18$ in two ways: (a) minimize the square of the distance; (b) use elementary geometric methods – the shortest line to the origin is perpendicular to the plane.

26. Find the maximum volume of an ellipsoid $x^2/a^2 + y^2/b^2 + z^2/c^2 = 1$ that passes through the point $(1,2,2)$.

27. Find the maximum volume of a rectangular box with the sum of its edges equal to 120 inches.

28. Minimize the function $f(x,y,z) = 1/x^2 + 4/y^2 + 9/z^2$ subject to the constraint $xyz = 12$.

29. Let $f(\alpha, \beta, \gamma) = \cos\alpha + \cos\beta + \cos\gamma$, where α, β, and γ are angles in a triangle. (a) Find the maximum value of f. (b) Show that f has no minimum.

Show that each of the functions in Probs. 30–33 satisfies Laplace's equation.

30. $\ln(x^2 + y^2)$

31. $e^x \sin y$

32. $e^y \sin x$

33. $\arctan(y/x)$

CHAPTER 7
INTEGRAL CALCULUS FOR FUNCTIONS OF TWO AND THREE VARIABLES

In this chapter we learn how to integrate functions of two and three variables. The new integrals, called double and triple integrals, are defined as limits of Riemann sums in a manner strictly analogous to the way ordinary (single) Riemann integrals are defined in one–variable calculus. Single integrals are integrals over line segments. Double integrals are integrals over plane regions, and triple integrals are integrals over solid regions in space.

Not surprisingly, single, double, and triple integrals share many common properties, such as linearity. These properties and basic geometric and physical interpretations are covered in Secs. 7.1 and 7.2.

Double and triple integrals are usually evaluated by performing two or three single integrals. The ease with which the single integrals can be evaluated depends crucially on the coordinate system used to describe the integrals. A good choice of coordinates takes advantage of any special symmetry of the region of integration and of the integrand. We illustrate the use of rectangular, polar, cylindrical, and spherical coordinates for this purpose.

Applications of double and triple integrals abound. For example, they can be used to find areas, volumes, masses, charges, centers of mass, moments of inertia, and the total force exerted by one solid body on another. In fact, multiple integrals are essential for the formulation of the basic physical laws and concepts that enable us to describe the interactions among solid bodies, as distinguished from interactions of point masses.

7.1 Double Integrals in Rectangular Coordinates

Double integrals are natural two–dimensional analogues of the definite integrals (also called single integrals) that you met earlier in calculus. Double integrals are defined as limits of double sums. However, the definition is rarely used to evaluate particular integrals. Instead, two single integrals are performed. In one–variable calculus, area interpretations help you discover and understand basic properties of single integrals. Volume interpretations play the same role for double integrals.

Overview

The story of double integrals begins with the problem of finding the volume under a surface. For the time being we avoid technical details in favor of the big picture. Let $f(x,y)$ be a positive, continuous function defined on a rectangle R with $a \le x \le b$, $c \le y \le d$. In Fig. 1, S is the surface $z = f(x,y)$ for (x,y) in R. We seek the volume of the solid region D between the surface S and the rectangle R.

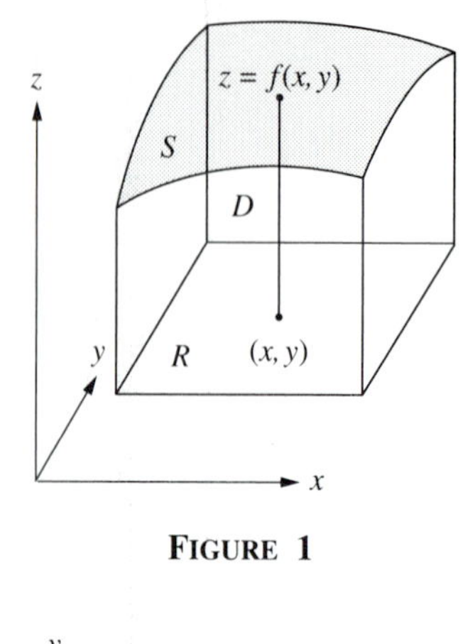

FIGURE 1

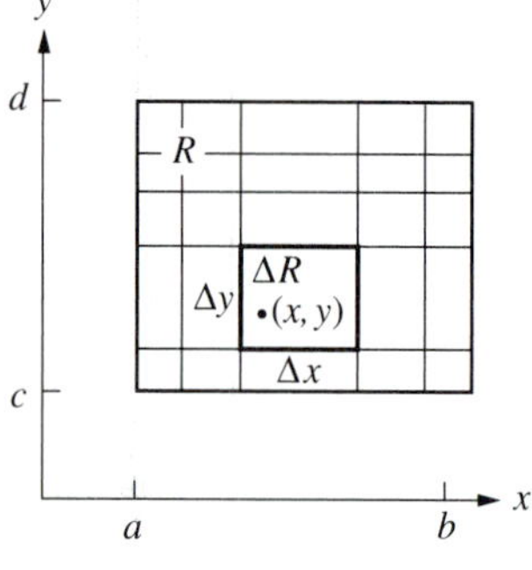

FIGURE 2

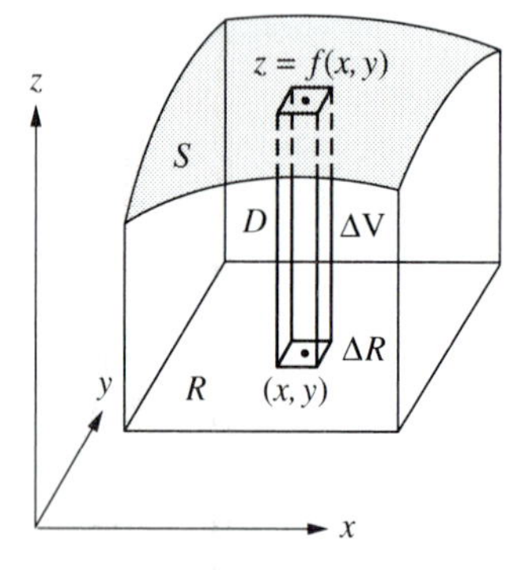

FIGURE 3

In Fig. 2, R is partitioned into smaller rectangles ΔR with areas $\Delta A = \Delta x\, \Delta y$. A typical rectangle ΔR is shown both in Fig. 2 and in Fig. 3. The volume ΔV of the "tower" with base ΔR and reaching up to the surface S is given approximately by $\Delta V \approx f(x,y)\,\Delta A$ for any choice of (x,y) in ΔR. The total volume V of D is given approximately by

$$V \approx \Sigma\Sigma f(x,y)\,\Delta A,$$

where the summation extends over all the rectangles ΔR in a partition of R, as in Fig. 2. The two summation signs indicate that we might sum first in one direction and then in the other direction in Fig. 2, perhaps first horizontally and then vertically. Better approximations for V should be expected if the rectangles ΔR in the partition of R are made smaller and their number is increased.

The volume V of D is given by

$$V = \lim \Sigma\Sigma f(x,y)\,\Delta A,$$

where the limit is taken as the number of rectangles ΔR in a partition of R increases and the maximum length and width of a rectangle approach zero. Then the largest area ΔA also approaches zero.

Now let $f(x,y)$ be any continuous function on R, which is not necessarily positive. The double integral of f over R is defined by

$$\iint_R f dA = \lim \Sigma\Sigma f(x,y)\,\Delta A,$$

where $\Sigma\Sigma f(x,y)\Delta A$ is a Riemann sum in shorthand. A more precise definition, for a larger class of functions $f(x,y)$, will be given a little later. The symbol dA in the integral, called an **area element**, is a reminder of ΔA in the Riemann sum. If $f(x,y) \geq 0$ on R, the volume between the surface $z = f(x,y)$ and the rectangle R is defined by $V = \int\int_R f dA$. If f varies in sign, then $\int\int_R f dA$ is net volume, with volume above the xy–plane counted positive and volume below the xy–plane counted negative.

Now it's time to flesh out this overview of double integrals. We begin with a few remarks about double sums. The expression

$$\sum_{i=1}^{m} \sum_{j=1}^{n} a_{ij}$$

is called a *double sum*. The numbers a_{ij} for $i = 1, 2, \ldots, m$ and $j = 1, 2, \ldots, n$ may be added in any order. For example, we may sum first on j and then on i or first on i and then on j. Thus,

$$\sum_{i=1}^{m} \sum_{j=1}^{n} a_{ij} = \sum_{i=1}^{m} \left(\sum_{j=1}^{n} a_{ij} \right) = \sum_{j=1}^{n} \left(\sum_{i=1}^{m} a_{ij} \right).$$

The latter two sums are called *iterated sums*.

Double sums are often expressed in abbreviated forms such as

$$\sum_i \sum_j a_{ij} \quad \text{and} \quad \sum \sum a_{ij}$$

if the limits of summation are understood from the context. Double sums have algebraic properties like those of single sums. For example,

$$\sum \sum c a_{ij} = c \sum \sum a_{ij}, \qquad \sum \sum (a_{ij} + b_{ij}) = \sum \sum a_{ij} + \sum \sum b_{ij}.$$

Double Integrals Over Rectangles

The double integral of a function of two variables over a rectangle is defined as a limit of Riemann sums in a manner strictly analogous to the definition of a single integral. Let R be a rectangle with $a \le x \le b$ and $c \le y \le d$. Partition each of the intervals $[a,b]$ and $[c,d]$:

$$a = x_0 < x_1 < \cdots < x_i < \cdots < x_m = b,$$

$$c = y_0 < y_1 < \cdots < y_j < \cdots < y_n = d,$$

$$\Delta x_i = x_i - x_{i-1}, \qquad \Delta y_j = y_j - y_{j-1}.$$

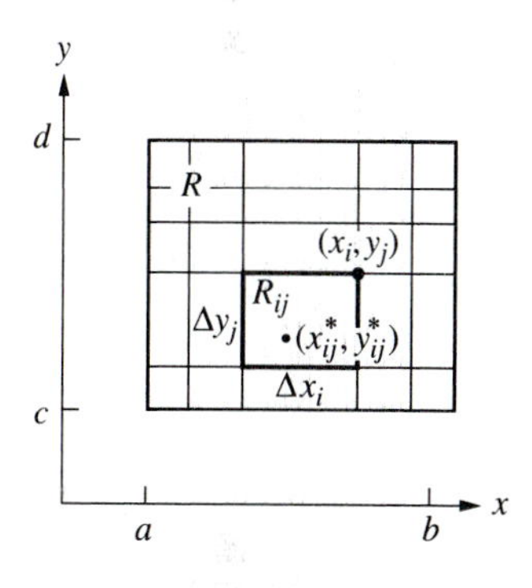

FIGURE 4

The partitions of $[a,b]$ and $[c,d]$ determine a partition P_{mn} of R into smaller rectangles R_{ij} with $x_{i-1} \le x \le x_i$ and $y_{j-1} \le y \le y_j$ as shown in Fig. 4. The area of ΔR_{ij} is $\Delta A_{ij} = \Delta x_i \, \Delta y_j$. The **mesh** $|P_{mn}|$ of P_{mn} is the maximum length of a diagonal of a rectangle in the partition. Pick any point (x_{ij}^*, y_{ij}^*) in each rectangle ΔR_{ij}. The **Riemann sum** for f corresponding to the partition P_{mn} of R and the choice of the points (x_{ij}^*, y_{ij}^*) is

$$\sum_{i=1}^{m} \sum_{j=1}^{n} f(x_{ij}^*, y_{ij}^*) \, \Delta A_{ij}.$$

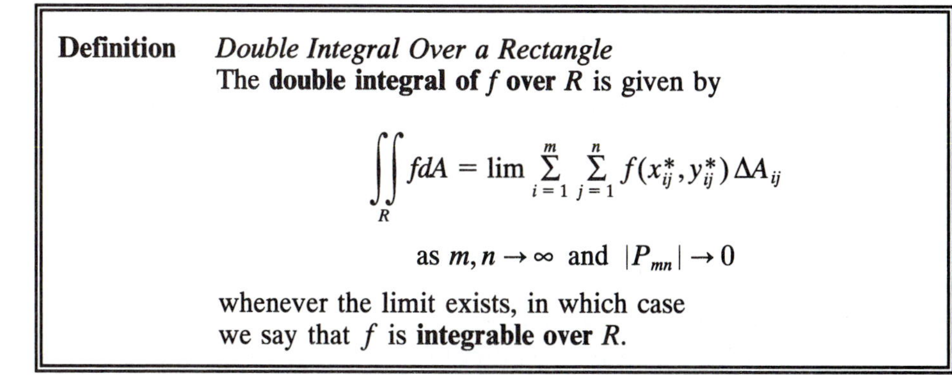

Definition *Double Integral Over a Rectangle*
The **double integral of** f **over** R is given by

$$\iint_R f\,dA = \lim \sum_{i=1}^{m} \sum_{j=1}^{n} f(x_{ij}^*, y_{ij}^*) \, \Delta A_{ij}$$

as $m, n \to \infty$ and $|P_{mn}| \to 0$

whenever the limit exists, in which case we say that f is **integrable over** R.

In more detail, f is integrable over R if for any $\varepsilon > 0$, no matter how small, there is a number J such that

$$\left| J - \sum_{i=1}^{m} \sum_{j=1}^{n} f(x_{ij}^*, y_{ij}^*) \, \Delta A_{ij} \right| < \varepsilon$$

for all m and n sufficiently large, for all partitions P_{mn} with sufficiently small mesh $|P_{mn}|$, and for all choices of the points (x_{ij}^*, y_{ij}^*) in ΔR_{ij}. Then $\iint_R f dA = J$. The shorthand notation

$$\iint_R f dA = \lim \Sigma\Sigma f(x,y)\,\Delta A$$

reminds us of the definition of the double integral. We shall use this abbreviated notation frequently to describe properties of integrable functions and, later on, to set up applications of definite integrals.

Motivated by our discussion in the overview, we define the **volume** V between the graph of $z = f(x,y) \geq 0$ and the rectangle R by $V = \iint_R f dA$, whenever the integral exists. For example, if $f(x,y) = h$, a constant, for (x,y) in R, then the solid region D in Figs. 1 and 3 becomes a rectangular box with length $l = b - a$, width $w = d - c$, and height h. The area of the base is $A = lw$ and the volume of the box is

$$V = \iint_R h dA = \lim \Sigma\Sigma\, h\Delta A = h \lim \Sigma\Sigma\,\Delta A = hA = lwh,$$

as we should expect.

Most of the functions you are likely to meet are integrable. For example,

$$f \text{ continuous on } R \;\Rightarrow\; f \text{ integrable over } R.$$

We omit the proof, which is given in advanced calculus.

Iterated Integrals Over Rectangles

As we mentioned earlier, the double integral of a particular function is usually evaluated by performing two single integrations. An example will show how it goes. Let $f(x,y) = xy$ for (x,y) in the rectangle R with $0 \leq x \leq 2$, $0 \leq y \leq 1$. Then

$$\iint_R f dA = \int_0^1 \left[\int_0^2 xy dx\right] dy = \int_0^1 \left[\frac{yx^2}{2}\right]_{x=0}^{x=2} dy = \int_0^1 2y\,dy = 1.$$

We integrate first with respect to x, holding y fixed and then we integrate with respect to y. Thus, y is regarded as a constant in the integral on x. The same result is obtained if we integrate first with respect to y, holding x fixed, and then we integrate with respect to x. Try it.

Now let $f(x,y)$ be a continuous function defined on a rectangle R with $a \leq x \leq b$, $c \leq y \leq d$. The two **iterated integrals** of f are

$$\int_a^b \int_c^d f(x,y)\,dydx = \int_a^b \left(\int_c^d f(x,y)\,dy\right) dx,$$

$$\int_c^d \int_a^b f(x,y)\,dxdy = \int_c^d \left(\int_a^b f(x,y) dx\right) dy.$$

The parentheses, which indicate the order of integration, ordinarily are omitted. In the first iterated integral, $f(x,y)$ is integrated with respect to y while x is held fixed and then the resulting function of x is integrated. The second iterated integral is evaluated in the reverse order. Thus, iterated integrals are evaluated "inside out."

The basic relationship between a double integral and the corresponding iterated integrals is spelled out in the following theorem.

> **Theorem 1** *Double Integrals Expressed as Iterated Integrals*
> If f is continuous on R, then
>
> $$\iint_R f dA = \int_a^b \int_c^d f(x,y)\,dydx = \int_c^d \int_a^b f(x,y)dxdy.$$

A proof of Th. 1 can be based on a careful limit passage argument using Riemann sums expressed in the iterated forms:

$$\Sigma\Sigma f(x,y)\Delta A = \Sigma[\Sigma f(x,y)\Delta y]\Delta x = \Sigma[\Sigma f(x,y)\Delta x]\Delta y.$$

The details of the argument are given in advanced calculus.

EXAMPLE 1. Integrate $f(x,y) = xy^2$ over the rectangle R with $2 \le x \le 3$ and $0 \le y \le 1$.

Solution. The double integral is given by either of the iterated integrals:

$$\int_2^3 \int_0^1 xy^2 dydx = \int_2^3 \left[\frac{1}{3}xy^3\right]_{y=0}^{y=1} dx = \int_2^3 \frac{1}{3}xdx = \left[\frac{1}{6}x^2\right]_2^3 = \frac{5}{6},$$

$$\int_0^1 \int_2^3 xy^2 dxdy = \int_0^1 \left[\frac{1}{2}x^2y^2\right]_{x=2}^{x=3} dy = \int_0^1 \frac{5}{2}y^2dy = \left[\frac{5}{6}y^3\right]_0^1 = \frac{5}{6}.$$

The two iterated integrals give the same result, as they must by Th. 1. □

Iterated integrals are closely related to the method of cross sections for finding volumes, which was introduced in one-variable calculus. To see the connection, let f be a positive, continuous function defined on a rectangle R, as in Fig. 5. For each x in $[a,b]$, the cross section at x is the plane region under the graph of $z = f(x,y)$, $c \le y \le d$. The cross section at x has area

$$A(x) = \int_c^d f(x,y)\,dy.$$

By the method of cross sections, the volume under S in Fig. 5 is

$$V = \int_a^b A(x)\,dx = \int_a^b \int_c^d f(x,y)\,dydx.$$

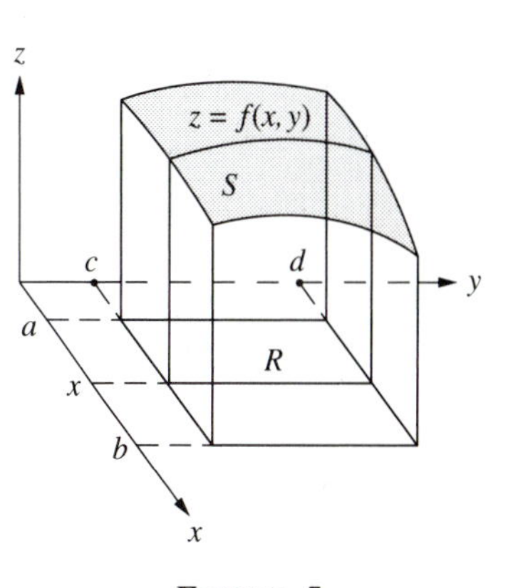

FIGURE 5

So V is expressed as an iterated integral. Similarly,

$$V = \int_c^d A(y)\,dy = \int_c^d \int_a^b f(x,y)\,dxdy.$$

Thus, for a positive, continuous function f on a rectangle R, the double integral over R and the method of cross sections give the same value for the volume under the graph of f. At an intuitive level, the fact that volume is given either by a double integral or by iterated integrals (via the method of cross sections) serves as a geometric argument in support of Th. 1.

EXAMPLE 2. Integrate $f(x,y) = 3x^2 - 4xy + 3y^2$ over the rectangle with $0 \le x \le 1$ and $1 \le y \le 2$.

Solution. Let's integrate first with respect to y:

$$\int_0^1 \int_1^2 (3x^2 - 4xy + 3y^2)\,dydx = \int_0^1 [3x^2y - 2xy^2 + y^3]_{y=1}^{y=2}\,dx$$

$$= \int_0^1 (3x^2 - 6x + 7)\,dx = [x^3 - 3x^2 + 7x]_0^1 = 5.$$

As a check, verify that the value of the other iterated integral is 5. □

EXAMPLE 3. Integrate $f(x,y) = xye^{x^2y}$ over $0 \le x \le 1,\ 0 \le y \le 2$.

Solution. This time, let's integrate first with respect to x:

$$\int_0^2 \int_0^1 xye^{x^2y}\,dxdy = \int_0^2 \left[\frac{1}{2}e^{x^2y}\right]_{x=0}^{x=1} dy$$

$$= \frac{1}{2}\int_0^2 (e^y - 1)\,dy = \frac{1}{2}[e^y - y]_0^2 = \frac{1}{2}(e^2 - 3).$$

If we had tried to integrate first with respect to y, we would have been faced with an unpleasant integration by parts. In this example, one of the iterated integrals is easier to evaluate than the other. More about this situation later. □

Sometimes the integrand in an iterated integral is a product $g(x)h(y)$. Then (see the problems) either iterated integral is a product of single integrals:

$$\boxed{\int_a^b \int_c^d g(x)h(y)\,dydx = \int_a^b f(x)\,dx \cdot \int_c^d h(y)\,dy.}$$

For example,

$$\int_0^1 \int_1^2 xe^{x^2+y}\,dydx = \int_0^1 xe^{x^2}\,dx \cdot \int_1^2 e^y\,dy = \frac{1}{2}(e-1)\cdot(e^2-e) = \frac{1}{2}e(e-1)^2.$$

Double Integrals Over More General Regions

Typical applications lead to double integrals over bounded regions R that are more general than rectangles. We define such double integrals next and give volume interpretations.

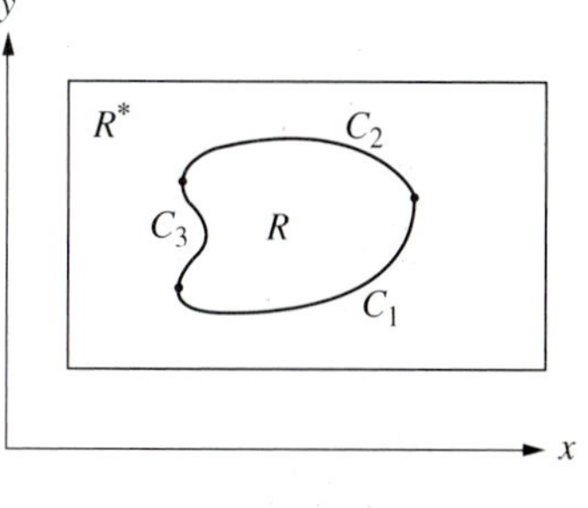

FIGURE 6

We deal only with regions R in the xy–plane of the type illustrated in Fig. 6. The boundary of R consists of a finite number of curves $C_1, C_2, C_3, \ldots$ that are graphs of continuous functions $y = y(x)$ or $x = x(y)$ defined on closed intervals. Let $f(x,y)$ be a function defined on such a region R. Enclose R in a rectangle R^*, as shown in Fig. 6, and define the auxiliary function $F(x,y)$ on R^* by

$$F(x,y) = \begin{cases} f(x,y) & \text{for } (x,y) \text{ in } R, \\ 0 & \text{for } (x,y) \text{ in } R^* \text{ but not in } R. \end{cases}$$

Since R^* is a rectangle, we know what it means for F to be integrable over the rectangle R^*. If F is integrable over R^*, then we say that f is **integrable over** R, in which case the **double integral of** f **over** R is defined by

$$\iint_R f dA = \iint_{R^*} F dA.$$

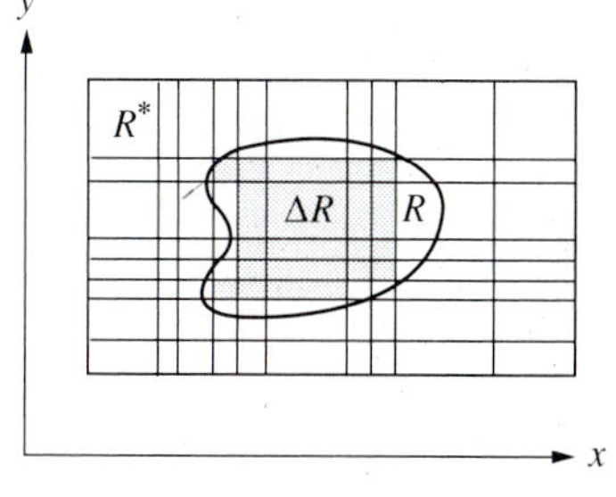

FIGURE 7

To gain a better understanding of the definition just given, we express $\iint_{R^*} F dA$ as a limit of Riemann sums. Figure 7 shows a typical partition of R^* into smaller rectangles ΔR. For any choice of (x,y) in each ΔR,

$$\iint_R f dA = \iint_{R^*} F dA = \lim \Sigma\Sigma F(x,y)\,\Delta A,$$

where ΔA is the area of ΔR. Let's call the rectangles ΔR in Fig. 7 that lie entirely in R **inner rectangles**. The inner rectangles are shaded in Fig. 7. Since $F(x,y) = f(x,y)$ for (x,y) in R, $F(x,y)\,\Delta A = f(x,y)\,\Delta A$, no matter how (x,y) is chosen in an inner rectangle ΔR. The other rectangles ΔR in Fig. 7 contain points outside of R. For these rectangles, choose (x,y) in ΔR and outside of R. Then $F(x,y) = 0$ and $F(x,y)\,\Delta A = 0$. Hence,

$$\Sigma\Sigma F(x,y)\,\Delta A = \Sigma'\Sigma' f(x,y)\,\Delta A,$$

where the primes indicate that the double sum on the right is over the inner rectangles of R. Thus, finally,

$$\iint_R f dA = \lim \Sigma'\Sigma' f(x,y)\,\Delta A.$$

We call $\Sigma'\Sigma' f(x,y)\,\Delta A$ an **inner Riemann sum** for f over R. Since the inner Riemann sums do not involve points outside of R, the values of these sums and the value of their limit $\iint_R f dA$ do not depend on the choice of the auxiliary rectangle R^* enclosing R.

Earlier we learned that continuous functions on closed rectangles are integrable. The same is true for the more general regions R described previously:

$$\boxed{f \text{ continuous on } R \quad \Rightarrow \quad f \text{ integrable over } R.}$$

We deal only with integrable functions in the rest of this section.

For the special case of $\int\int_R f dA$ with $f(x,y) = 1$ for all (x,y) in R,

$$\iint\limits_R 1 dA = \lim \Sigma'\Sigma' \Delta A.$$

Imagine that the mesh of the partition in Fig. 7 tends to zero. The figure suggests that the inner Riemann sums $\Sigma'\Sigma' \Delta A$ approach the area of R. Thus, we are led to define the **area of *R*** by

$$\boxed{A = \iint\limits_R 1 dA.}$$

If $f \geq 0$ on R, then, as before, the volume under the surface $z = f(x,y)$ and above the xy–plane is defined by

$$\boxed{V = \iint\limits_R f dA.}$$

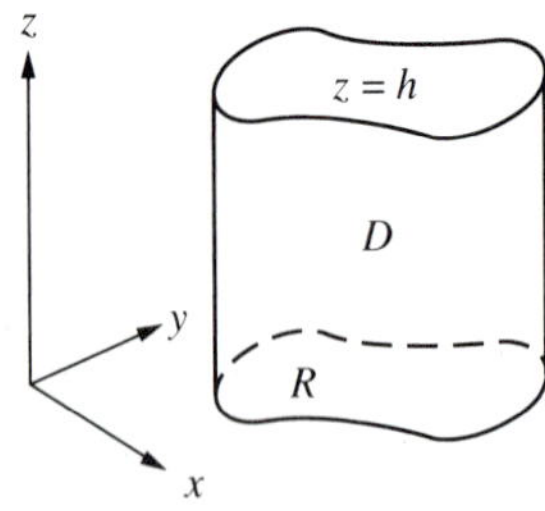

FIGURE 8

In particular, let $f(x,y) = h > 0$ for all (x,y) in R. The region under the graph of f is the solid cylinder D in Fig. 8. The volume of the cylinder is

$$V = \iint\limits_R f dA = \lim \Sigma'\Sigma' h \Delta A = h \lim \Sigma'\Sigma' \Delta A = hA.$$

That is, the volume of the cylinder is the area of the base times the height. This fact, generally used without proof, plays a fundamental role in the method of cross sections for finding volumes earlier in calculus.

If $f \leq 0$ on R, then $\int\int_R f dA$ is the negative of the volume between the surface $z = f(x,y)$ and the xy–plane. If f takes both positive and negative values, then $\int\int_R f dA$ is net volume.

Algebraic Properties of Double Integrals

The algebraic properties of double integrals are faithful analogues of corresponding properties for single integrals. Since there are no surprises, we merely summarize the results.

Linearity Properties

$$\iint_R cf\,dA = c\iint_R f\,dA.$$

$$\iint_R (f+g)\,dA = \iint_R f\,dA + \iint_R g\,dA.$$

Repeated applications of the linearity properties give

$$\iint_R (c_1f_1 + c_2f_2 + \cdots c_nf_n)\,dA = c_1\iint_R f_1\,dA + c_2\iint_R f_2\,dA + \cdots + c_n\iint_R f_n\,dA.$$

Domain Additivity Property
Let R be the union of two nonoverlapping regions R_1 and R_2. Then

$$\iint_R f\,dA = \iint_{R_1} f\,dA + \iint_{R_2} f\,dA.$$

Nonoverlapping regions can share boundary points but not interior points, as illustrated in Fig. 9. This property enables us to break a complicated region of integration into simpler regions.

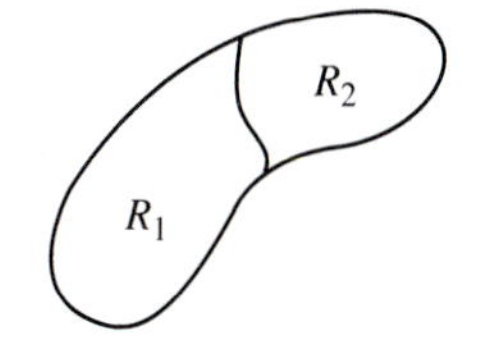

FIGURE 9

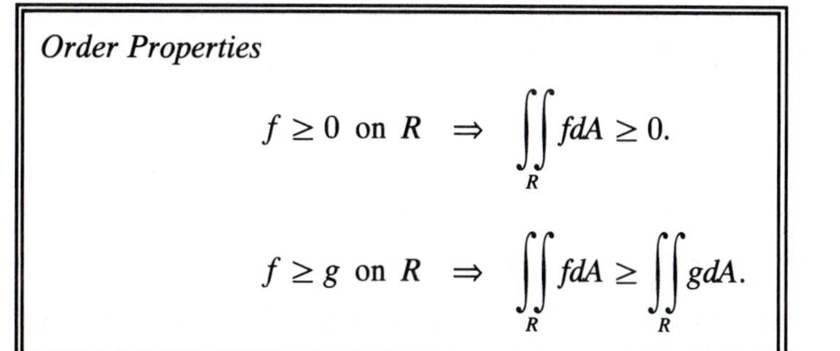

Order Properties

$$f \geq 0 \text{ on } R \;\Rightarrow\; \iint_R f\,dA \geq 0.$$

$$f \geq g \text{ on } R \;\Rightarrow\; \iint_R f\,dA \geq \iint_R g\,dA.$$

An important consequence of the order properties is

$$\left|\iint_R f dA\right| \le \iint_R |f|\, dA.$$

Comparison Properties
Let A be the area of R. Then

$$m \le f \le M \text{ on } R \quad \Rightarrow \quad mA \le \iint_R f dA \le MA.$$

These algebraic properties of double integrals have obvious interpretations in terms of volumes when the integrands are nonnegative.

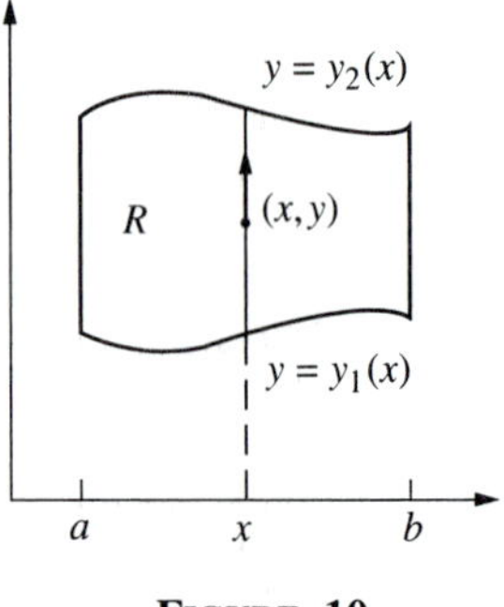

FIGURE 10

Evaluation of Double Integrals as Iterated Integrals

We showed earlier that double integrals over rectangles can be expressed as iterated integrals. There are similar results for double integrals over more general regions.

A region R in the xy–plane is **y–simple** (or **vertically simple**) if it lies between the graphs of two continuous functions $y = y_1(x)$ and $y = y_2(x)$ for $a \le x \le b$, as shown in Fig. 10. Thus, (x, y) lies in R if

$$a \le x \le b \text{ and } y_1(x) \le y \le y_2(x).$$

The double integral of a continuous function f over R can be expressed as an iterated integral:

$$\iint_R f dA = \int_a^b \int_{y_1(x)}^{y_2(x)} f(x, y)\, dy dx.$$

This equality can be proved by using Riemann sums. In the inner integral, x is fixed and the point (x, y) traces the vertical line segment across R in Fig. 10. In the outer integral, x varies from a to b and the vertical line in Fig. 10 sweeps across the region R. The case with $f(x, y) = 1$ gives the area of R:

$$A = \iint_R 1 dA = \int_a^b \int_{y_1(x)}^{y_2(x)} 1\, dy dx = \int_a^b [y_2(x) \; - y_1(x)] dx.$$

This formula for the area between two curves was found by other means in one–variable calculus.

EXAMPLE 4. Let R be the region in Fig. 11 between the parabolas $y = x^2$ and $x = y^2$. Find the area of R and the volume of the solid region D under the graph of $z = xy$ for (x,y) in R.

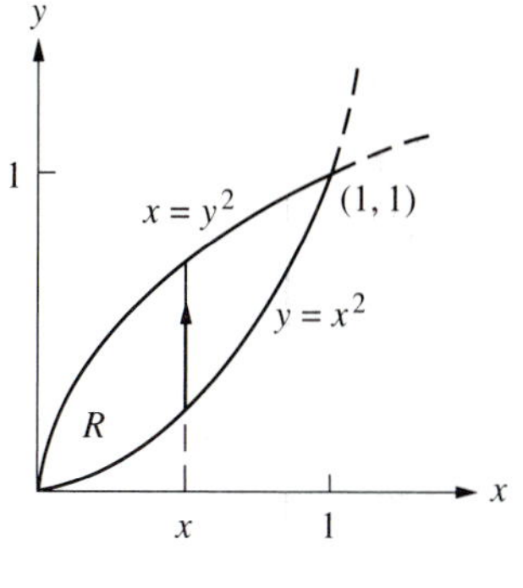

FIGURE 11

Solution. The area and volume we seek are

$$A = \iint_R 1 dA \quad \text{and} \quad V = \iint_R xy dA.$$

The two parabolas intersect at (1,1). Since $y = x^2$ on the lower boundary of R and $y = \sqrt{x}$ on the upper boundary of R, the region R is y–simple with

$$0 \le x \le 1 \quad \text{and} \quad x^2 \le y \le \sqrt{x}.$$

Consequently, the area of R is

$$A = \iint_R 1 dA = \int_0^1 \int_{x^2}^{\sqrt{x}} 1 dy dx = \int_0^1 (x^{1/2} - x^2)\, dx = \frac{2}{3} - \frac{1}{3} = \frac{1}{3},$$

and the volume of D is

$$V = \iint_R xy dA = \int_0^1 \int_{x^2}^{\sqrt{x}} xy dy dx = \int_0^1 \left[\frac{1}{2} xy^2\right]_{y = x^2}^{y = \sqrt{x}} dx,$$

$$V = \frac{1}{2}\int_0^1 (x^2 - x^5)\, dx = \frac{1}{2}\left(\frac{1}{3} - \frac{1}{6}\right) = \frac{1}{12}. \quad \square$$

Volumes between two surfaces can be found by subtraction. We give just one example here. Such volumes are considered further in the next section.

EXAMPLE 5. Again let R be the region between the parabolas $y = x^2$ and $x = y^2$ in Fig. 11. Find the volume V between the surfaces $z = xy$ and $z = 1$ for (x,y) in R.

Solution. For any (x,y) in R, Fig. 11 reveals that $0 \le x \le 1$, $0 \le y \le 1$ and, hence, $xy \le 1$. So the surface $z = xy$ lies below the horizontal plane $z = 1$ for (x,y) in R. Consequently, the volume we seek can be expressed as $V = V_2 - V_1$ where V_2 is the volume under the graph of $z = 1$ and V_1 is the volume under the graph of $z = xy$. Now,

$$V_1 = \iint_R xy dA, \qquad V_2 = \iint_R 1 dA = A,$$

where A is the area of R. From Ex. 4, $V_1 = 1/12$ and $V_2 = 1/3$. Therefore,

$$V = \frac{1}{3} - \frac{1}{12} = \frac{1}{4}. \quad \square$$

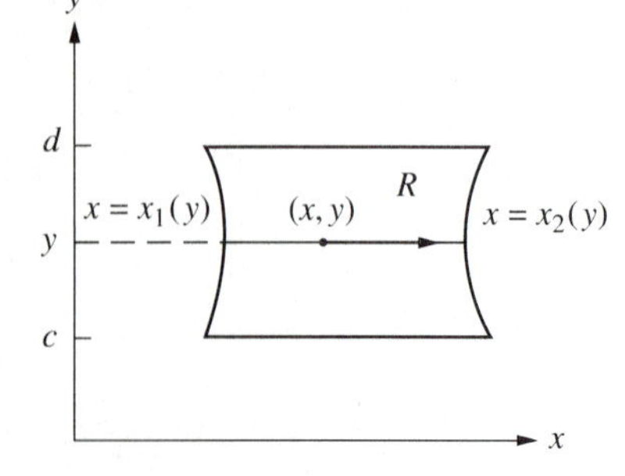

FIGURE 12

A region R in the xy–plane is **x–simple** (or **horizontally simple**) if it lies between the graphs of two continuous functions $x = x_1(y)$ and $x = x_2(y)$ for $c \leq y \leq d$, as shown in Fig. 12. Thus, R consists of the points (x, y) with

$$c \leq y \leq d \quad \text{and} \quad x_1(y) \leq x \leq x_2(y).$$

The double integral of a continuous function f over R can be expressed as an iterated integral:

$$\iint_R f dA = \int_c^d \int_{x_1(y)}^{x_2(y)} f(x, y)\, dxdy.$$

In the inner integral, y is fixed and the point (x, y) traces the horizontal line segment across R in Fig. 12. In the outer integral, y varies from c to d and the horizontal line in Fig. 11 sweeps across the region R. The region R in Ex. 4 and Fig. 11 is both x–simple and y–simple. So there are two ways to calculate the integrals for area and volume. Both methods give the same results.

EXAMPLE 6. Let R be the region in the xy–plane bounded by the line $y = x$ and the parabola $x = 2 - y^2$. Evaluate $\iint_R y dA$. Interpret your answer as a net volume.

Solution. The line and the parabola intersect where

$$y = x = 2 - y^2, \quad y^2 + y - 2 = 0, \quad (y - 1)(y + 2) = 0, \quad y = 1, -2.$$

So the points of intersection are

$$(x, y) = (1,1), \quad (x, y) = (-2, -2).$$

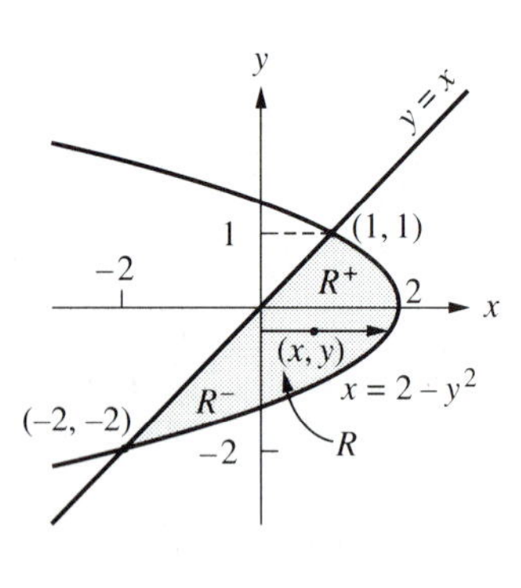

FIGURE 13

See Fig. 13. The region R is horizontally simple with $-2 \leq y \leq 1$ and $y \leq x \leq 2 - y^2$. Therefore,

$$\iint_R y dA = \int_{-2}^{1} \int_{y}^{2-y^2} y dxdy = \int_{-2}^{1} [xy]_{x=y}^{x=2-y^2} dy = \int_{-2}^{1} (2y - y^3 - y^2)\, dy = -\frac{33}{4}.$$

This is the net volume between the plane $z = y$ and the region R in Fig. 13. The plane $z = y$ lies above the xy–plane for (x, y) in R^+ and below the xy–plane for (x, y) in R^-. It should be clear from the geometry that the net volume is negative. In fact, it is $-33/4$ cubic units. □

The region R in Fig. 13 is also y–simple, but it is less efficient to evaluate the y–integral first because the upper boundary of R is given by different formulas for y in the intervals $-2 \leq x \leq 1$ and $1 \leq x \leq 2$. Many of the regions that we encounter are both x–simple and y–simple. Then we have two iterated integrals to choose between. Quite often, one choice is much better than the other. If the going gets tough with one iterated integral, try the other.

EXAMPLE 7. Let R be the triangular region in the xy–plane bounded by $y = 0$, $x = 1$, and $y = x$. Find the volume V between R and the surface $z = e^{-x^2}$.

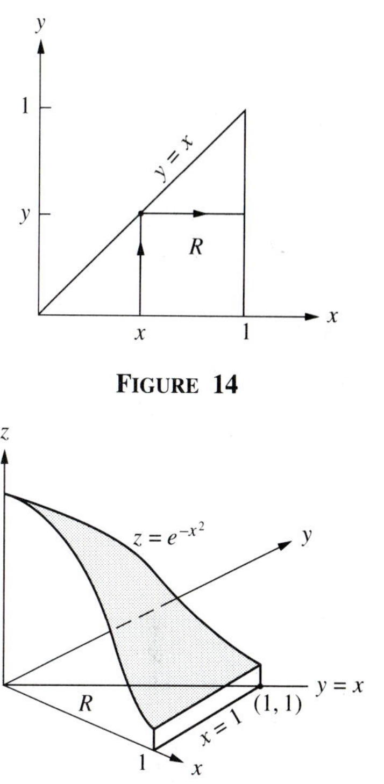

FIGURE 14

FIGURE 15

Solution. See Figs. 14 and 15. The region R in Fig. 14 is both x–simple and y–simple. The x–simple description of R is

$$0 \le y \le 1, \qquad y \le x \le 1.$$

It follows that the volume V is given by

$$V = \iint_R f dA = \int_0^1 \int_y^1 e^{-x^2} dx dy.$$

Now we are stuck! The function e^{-x^2} has no elementary antiderivative. Let's try the y–simple description of R:

$$0 \le x \le 1, \qquad 0 \le y \le x.$$

Then

$$V = \int_0^1 \int_0^x e^{-x^2} dy dx = \int_0^1 x e^{-x^2} dx = \left[-\frac{1}{2} e^{-x^2} \right]_0^1 = \frac{1}{2}(1 - e^{-1}).$$

In this case, one of the iterated integrals was easy and the other was impossible! □

Suppose that we had been asked in the first place to find the value of the iterated integral

$$\int_0^1 \int_y^1 e^{-x^2} dx dy.$$

The region of integration is revealed by the limits of integration. For each *fixed* y with $0 \le y \le 1$, x varies between y and 1. With y fixed in $0 \le y \le 1$, the point (x, y) runs across the horizontal line segment in Fig. 14. As y runs from 0 to 1, these horizontal line segments sweep out the triangle R in the same figure. Thus, R is the region of integration and

$$\int_0^1 \int_y^1 e^{-x^2} dx dy = \iint_R e^{-x^2} dA.$$

Now we can express the double integral as an iterated integral in the opposite order and evaluate it as we did in Ex. 7. This method of reversing the order of integration consists of two steps. Given an iterated integral, first identify and graph the region of integration R. Then use the graph of R to express the given iterated integral as a new iterated integral with the order of integration reversed.

EXAMPLE 8. (a) Identify and graph the region R of integration corresponding to the iterated integral

$$\int_1^e \int_0^{\ln x} f(x, y) \, dy dx.$$

(b) Express the iterated integral as an iterated integral with the order of integration reversed. (c) Evaluate both integrals with $f(x,y)=1$ to find the area A of R in two ways.

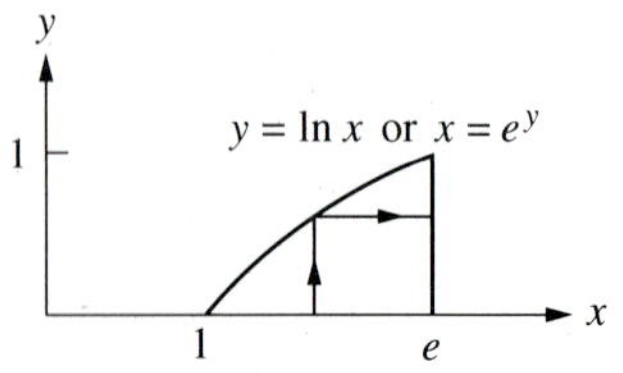

FIGURE 16

Solution. (a) From the given integral, R consists of the points (x,y) with $1 \le x \le e$ and $0 \le y \le \ln x$. See Fig. 16. (b) From the figure, R is given also by $0 \le y \le 1$ and $e^y \le x \le e$. Therefore,

$$\int_1^e \int_0^{\ln x} f(x,y)\,dydx = \int_0^1 \int_{e^y}^{e} f(x,y)\,dxdy.$$

(c) The evaluation of these integrals with $f(x,y)=1$ is left to you. You should find $A = e-2$ both ways. The second evaluation is easier. □

Regions that are neither horizontally nor vertically simple often can be decomposed into a finite number of nonoverlapping regions, each of which is either horizontally or vertically simple. Double integrals over such regions can be evaluated using this fact and the domain additivity property.

Symmetry

Taking advantage of symmetry can save a lot of work in evaluating double integrals. A few examples will illustrate what we mean.

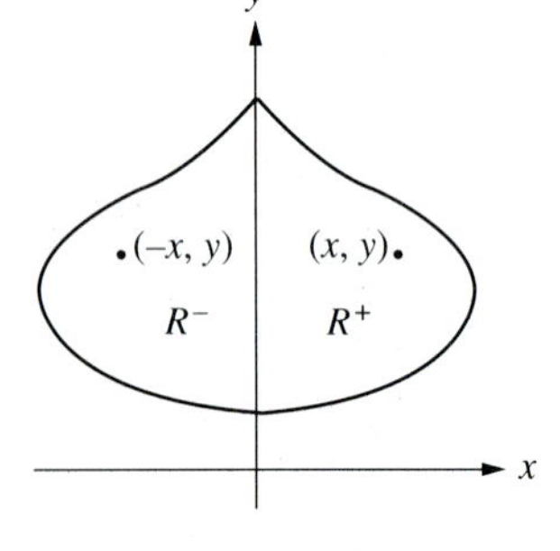

FIGURE 17

A region R is **symmetric in x** if, as in Fig. 17,

$$(-x,y) \text{ is in } R \quad \Leftrightarrow \quad (x,y) \text{ is in } R.$$

Let R^+ and R^- be the parts of R with $x \ge 0$ and $x \le 0$, respectively. Then R^+ and R^- are mirror images of each other across the y–axis. Suppose that f is defined on R and that $f(-x,y) = f(x,y)$. In words, f takes on the same values at symmetric points of R. Then, by a Riemann sum and limit passage argument,

$$\iint_{R^-} f dA = \iint_{R^+} f dA \quad \text{and} \quad \iint_{R} f dA = 2\iint_{R^+} f dA.$$

If $f \ge 0$ on R, these results have obvious volume interpretations.

EXAMPLE 9. Let R be the semicircular disk with $x^2+y^2 \le 4$, $y \ge 0$. Find the volume V of the solid wedge in Fig. 18 lying between R and the slanting plane $z = f(x,y) = y$.

Solution. Clearly, R is symmetric in x and $f(-x,y) = f(x,y)$. Therefore,

$$V = \iint_R y dA = 2\iint_{R^+} y dA.$$

The region R^+ in Fig. 18 is y–simple with

$$0 \le x \le 2 \text{ and } 0 \le y \le \sqrt{4 - x^2}.$$

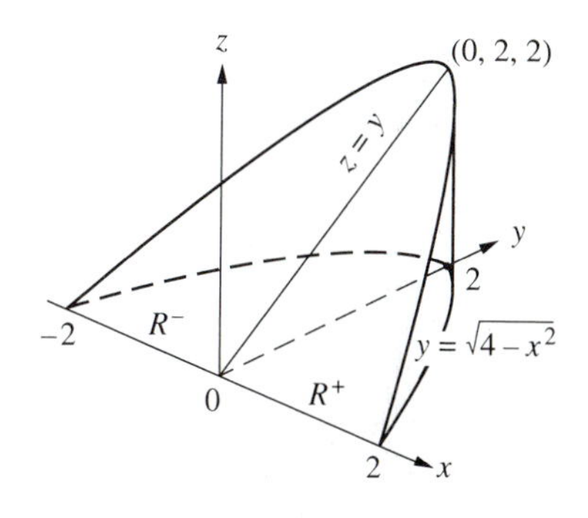

FIGURE 18

So the volume of the wedge is

$$V = 2\int_0^2\int_0^{\sqrt{4-x^2}} y\,dy\,dx = \int_0^2 [y^2]_{y=0}^{y=\sqrt{4-x^2}}\,dx,$$

$$V = \int_0^2 (4 - x^2)\,dx = \frac{16}{3}.$$

The region R^+ is also x–simple, but the integration that way is harder. □

Suppose that R is symmetric in x and $f(-x,y) = -f(x,y)$. Thus, f takes opposite values at symmetric points. Riemann sum and limit passage arguments or volume comparisons reveal that

$$\iint_{R^-} f\,dA = -\iint_{R^+} f\,dA \quad \text{and} \quad \iint_R f\,dA = 0.$$

For example, let $f(x,y) = x$ for (x,y) in the region R of Ex. 9 and Fig. 18. Then $f(-x,y) = -f(x,y)$ and, by symmetry, $\iint_R x\,dA = 0$. There is no need to carry out the integration to reach this conclusion.

Similar results pertain to regions R in the xy–plane that are symmetric in y or are symmetric about the origin. See the problems.

PROBLEMS

Evaluate the double integral in Probs. 1–6.

1. $\iint_R x^2y\,dA$
   R: $0 \le x \le 1$, $2 \le y \le 3$

2. $\iint_R y\sqrt{x}\,dA$
   R: $0 \le x \le 4$, $0 \le y \le 2$

3. $\iint_R e^{x+2y}\,dA$
   R: $-1 \le x \le 1$, $0 \le y \le 2$

4. $\iint_R \ln y^x\,dA$
   R: $-1 \le x \le 2$, $1 \le y \le e$

5. $\iint_R x \sin y\,dA$
   R: $2 \le x \le 4$, $0 \le y \le \pi$

6. $\iint_R \sin(x + y)\,dA$
   R: $0 \le x \le \pi$, $0 \le y \le \pi/2$

In Probs. 7–12, find the volume or net volume between the xy–plane and the graph of the function f over the domain R.

7. $f(x,y) = 20 - 2x^2 - y^2$
   R: $0 \le x \le 3, \quad 0 \le y \le 1$

8. $f(x,y) = 20 - 2x^2 - y^2$
   R: $0 \le x \le 3, \quad 0 \le y \le 2$

9. $f(x,y) = xye^{xy^2}$
   R: $-1 \le x \le 2, \quad 0 \le y \le 1$

10. $f(x,y) = y \cos xy$
    R: $0 \le x \le 1, 0 \le y \le \pi/2$

11. $f(x,y) = x\sqrt{x^2 + y}$
    R: $0 \le x \le 1, \quad 0 \le y \le 2$

12. $f(x,y) = 2xy/(x^2 + y^2)$
    R: $1 \le x \le 2, \quad 0 \le y \le 1$

In Probs. 13–29, (a) make a sketch of the solid and then (b) find the volume of the solid. Your sketch in (a) need only be accurate enough to find the volume.

13. The solid under the plane $z = 6 - 2x - 3y$ and above the region in the xy–plane bounded by the curves $y = x^2$ and $x = y^2$.

14. The solid under the plane $z = 6 - 2x - 3y$, above the xy–plane, and inside the cylinder $x^2 + y^2 = 1$.

15. The solid under the graph of $z = (\sin x)/x$ and above the region in the xy–plane bounded by the lines $y = x$, $x = 1$, and the x–axis. (Interpret $\sin x/x$ at $x = 0$ as its limiting value as $x \to 0$.)

16. The solid under the graph of $z = e^{y/x}$ and above the region in the xy–plane bounded by $y = x^2$, $x = 1$, and $y = 0$. (If $(x,y) \to (0,0)$ in this region, $e^{y/x} \to e^0 = 1$. Interpret $e^{y/x}$ as this limiting value at the origin.)

17. The solid under the graph of the parabolic cylinder $z = 3x^2$ and above the region in the xy–plane bounded by $y = 2x^2$ and $y = 1 + x^2$.

18. The solid under the graph of the hyperbolic paraboloid $z = 2xy$ and above the region in the xy–plane bounded by $x = 2y^2$ and $x = 1 + y^2$.

19. The solid under the graph of $z = xe^{y^2}$ and above the region in the xy–plane bounded by $y = x^2$, $x = 0$, and $y = 1$.

20. The solid under the graph of $z = \sin \pi y^2$ and above the region in the xy–plane bounded by $y = x$, $x = 0$, and $y = 1$.

21. The solid bounded by the paraboloid $z = x^2 + y^2$ and the plane $z = 4$.

22. The solid under the paraboloid $4z = 16 - x^2 - y^2$, above the xy–plane, and inside the cylinder $x^2 + y^2 = 4$.

23. The solid bounded by the surfaces $2z = y^2 + 2x$, $y = x$, $x + y = 4$, $y = 0$ and $z = 0$.

24. The solid in the first octant bounded by the planes $x + y + z = 9$, $3x + 2y = 18$, and $3x + y = 9$.

25. The solid bounded by the cylinder $x^2 + y^2 = 9$ and the planes $y + z = 4$ and $z = 0$.

26. The solid that lies under the plane $z = 2y$ and above the region in the xy–plane bounded by the x–axis and the parabola $y = 4 - x^2$.

27. The solid in the first octant between the cylinders $y^2 = ax$ and $x^2 = ay$ and under the plane $z = x + y$, where a is a positive constant.

28. The solid in the first octant bounded by the cylinders $x^2 + y^2 = a^2$ and $y^2 + z^2 = a^2$, where a is a positive constant.

29. The solid below the parabola $z = 7 - x^2 - y^2$, above the paraboloid $z = x^2 + y^2 - 1$, and over the region $0 \leq x, y \leq 1$ in the xy–plane.

In Probs. 30–37 do the following: (a) Identify and graph the region of integration. (b) Express the iterated integral as an iterated integral in the opposite order and evaluate it.

30. $\int_0^1 \int_x^1 e^{-y^2} dy\, dx$

31. $\int_0^1 \int_y^1 \frac{\sin x}{x} dx\, dy$

32. $\int_0^1 \int_{\sqrt{y}}^1 e^{y/x} dx\, dy$

33. $\int_0^1 \int_{y^2}^1 ye^{x^2} dx\, dy$

34. $\int_0^1 \int_y^1 \sin \pi x^2 dx\, dy$

35. $\int_0^2 \int_{x/2}^1 e^{y^2} dy\, dx$

36. $\int_0^2 \int_{x^2}^4 x(1 + y^2)^{-1/2} dy\, dx$

37. $\int_1^2 \int_0^{\pi} y \sin xy\, dy\, dx$

38. If f varies in sign over R, explain briefly, using sketches and shorthand Riemann sum notation, why $\iint_R f dA$ gives the net volume between the xy–plane and the graph of f.

39. If $f(x,y) = g(x)h(y)$, with g and h continuous on $a \leq x \leq b$ and $c \leq y \leq d$, show that

$$\iint_R f dA = \int_a^b g(x)\, dx \cdot \int_c^d h(y)\, dy.$$

Hence, the two iterated integrals of f also are equal to the product on the right.

40. Use (shorthand) Riemann sum, limit passage reasoning to support each of the following properties boxed in the text: (a) linearity properties; (b) domain additivity property; (c) order properties; and (d) comparison properties.

41. (a) Show that $|\iint_R f dA| \leq \iint_R |f| dA$.
(b) Interpret (a) with volumes.

42. Establish the symmetry property $\iint_R f dA = 2\iint_{R^+} f dA$, where R is symmetric in x and $f(-x,y) = f(x,y)$ for all points (x,y) in R. For simplicity, assume that R is vertically simple. *Hint.* Express the double integral over R as an iterated integral and use symmetry results for ordinary integrals.

43. Establish the symmetry property $\iint_R f dA = 0$, where R is symmetric in y and $f(x,-y) = -f(x,y)$ for all points (x,y) in R. For simplicity, assume that R is horizontally simple.

44. Suppose the region R in the xy–plane is *symmetric about the origin;* that is,

(x, y) is in $R \Leftrightarrow (-x, -y)$ is in R. Let R^+ consist of the points (x, y) in R with $x \geq 0$. Explain briefly, using shorthand Riemann sum notation, why

$$(a)\, f(-x, -y) = f(x, y) \quad \Rightarrow \quad \iint_R f dA = 2\iint_{R^+} f dA,$$

$$(b)\, f(-x, -y) = -f(x, y) \quad \Rightarrow \quad \iint_R f dA = 0.$$

In Probs. 45–50 evaluate the double integral where R is the region inside the square with vertices $(-1, -1)$, $(1, -1)$, $(1, 1)$, and $(-1, 1)$ and outside the square with vertices $(-1, 0)$, $(0, -1)$, $(1, 0)$, and $(0, 1)$. *Hint.* Take advantage of symmetry.

45. $\iint_R x^2 dA$

46. $\iint_R x^2 y dA$

47. $\iint_R x \sin xy\, dA$

48. $\iint_R \sin xy\, dA$

49. $\iint_R (x^2 + y^2)\, dA$

50. $\iint_R |xy|\, dA$

51. Let R be the region that is simultaneously beneath the parabola $y = x^2 + 1$, to the left of the parabola $x = y^2$, and inside the rectangle with sides parallel to the coordinate axes and opposite vertices $(-1, -2)$ and $(1, 2)$. Evaluate $\iint_R xy\, dA$.

52. Evaluate $\iint_R xy\, dA$ where R is the region inside the rectangle with sides parallel to the coordinate axes and opposite vertices $(-1, -1)$ and $(1, 2)$ and outside the rectangle with vertices $(-1, 0)$, $(0, -1)$, $(1, 1)$, and $(-1, 0)$.

7.2 Triple Integrals in Rectangular Coordinates

Triple integrals are natural three–dimensional generalizations of single and double integrals. Since the definitions and main properties involve no really new ideas, we shall be brief. This will give us more time to spend on interesting applications of triple integrals involving mass and charge.

Triple Integrals Over Boxes

We deal first with triple integrals over rectangular boxes in 3–space. Let $f(x, y, z)$ be a function defined on a box B with

$$a \leq x \leq b, \qquad c \leq y \leq d, \qquad p \leq z \leq q.$$

Partition B into smaller boxes ΔB, with volumes $\Delta V = \Delta x\, \Delta y\, \Delta z$, as indicated in

Fig. 1. Choose any point (x, y, z) in each box ΔB. A Riemann sum for f over B is expressed in shorthand notation by

$$\sum\sum\sum f(x,y,z)\,\Delta V.$$

The **triple integral of** f **over** B is defined by

$$\iiint_B f\,dV = \lim \sum\sum\sum f(x,y,z)\,\Delta V$$

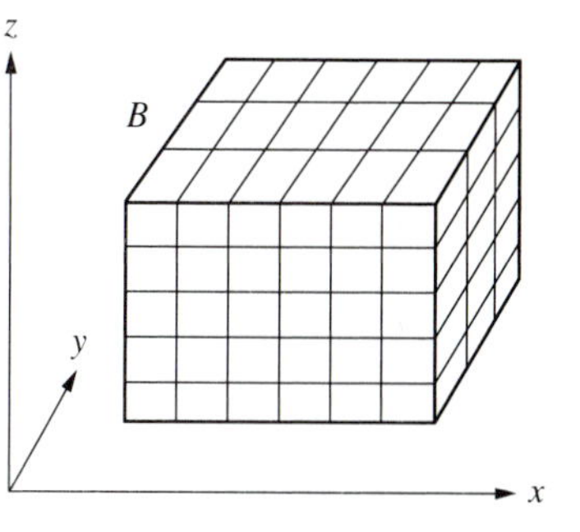

FIGURE 1

whenever the limit exists, in which case f is **integrable over** B. The limit is taken as the number of boxes ΔB in a partition of B increases and the maximum length of the diagonals of the boxes ΔB approaches zero. Then the largest volume ΔV of a box ΔB also approaches zero.

The triple integral can be expressed as an iterated integral in six ways, with six orders of integration. In particular,

$$\iiint_B f\,dV = \int_a^b \int_c^d \int_p^q f(x,y,z)\,dz\,dy\,dx.$$

As in Sec. 7.1, iterated integrals are evaluated "inside out." Here we integrate first with respect to z, then with respect to y, and finally with respect to x. Briefly, the order of integration is zyx.

EXAMPLE 1. Let B be the box with $3 \le x \le 4$, $1 \le y \le 2$, $0 \le z \le 1$. Evaluate $\iiint_B 24xy^2z^3\,dV$.

Solution. Let's integrate in the order zxy. Then

$$\iiint_B f\,dV = \int_1^2 \int_3^4 \int_0^1 24xy^2z^3\,dz\,dx\,dy = \int_1^2 \int_3^4 [6xy^2z^4]_{z=0}^{z=1}\,dx\,dy$$

$$= \int_1^2 \int_3^4 6xy^2\,dx\,dy = \int_1^2 [3x^2y^2]_{x=3}^{x=4}\,dy = \int_1^2 21y^2\,dy = 49.$$

Calculate one of the other five iterated integrals to see that it has the same value. □

The integrand in Ex. 1 is a product of the form $g(x)h(y)k(z)$, where g, h, k are continuous functions. In this case (see the problems),

$$\iiint_B g(x)h(y)k(z)\,dV = \int_a^b g(x)\,dx \cdot \int_c^d h(y)\,dy \cdot \int_p^q k(z)\,dz.$$

The integral in Ex. 1 is evaluated more quickly by means of this product formula:

$$\iiint_B 24xy^2z^3dV = 24\int_3^4 xdx \cdot \int_1^2 y^2dy \cdot \int_0^1 z^3dz = 24 \cdot \frac{7}{2} \cdot \frac{7}{3} \cdot \frac{1}{4} = 49.$$

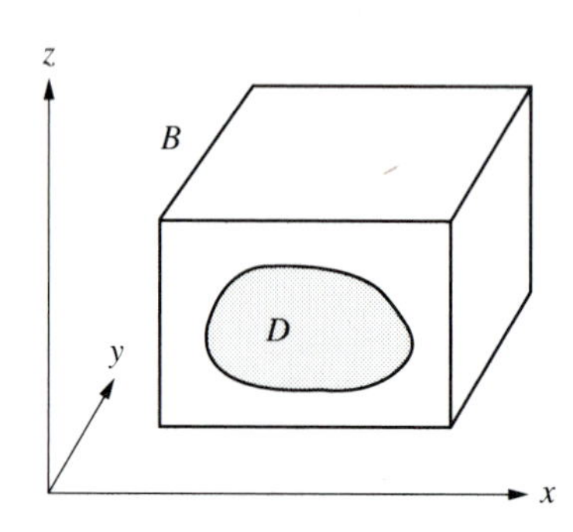

FIGURE 2

Triple Integrals Over More General Regions

Triple integrals over regions more general than boxes are defined and evaluated by natural analogues of corresponding results for double integrals.

Let D be a bounded region in 3–space. For example, D might be a solid ellipsoid. Enclose D in a box B, as illustrated in Fig. 2. Let $f(x,y,z)$ be a function on D. Define an auxiliary function $F(x,y,z)$ on B by

$$F(x,y,z) = \begin{cases} f(x,y,z) & \text{for } (x,y,z) \text{ in } D, \\ 0 & \text{for } (x,y,z) \text{ in } B \text{ but not in } D. \end{cases}$$

Then f is **integrable over** D if F is integrable over B, in which case

$$\iiint_D fdV = \iiint_B FdV.$$

Imagine that the box B in Fig. 2 is partitioned into smaller boxes ΔB, as illustrated in Fig. 1, and let (x,y,z) be a point in each box ΔB. If ΔB contains a point outside of D, choose any point (x,y,z) in ΔB which lies outside of D. Then the triple integral of f over D is a limit of **inner Riemann sums,**

$$\iiint_D fdV = \lim \Sigma'\Sigma'\Sigma'\ f(x,y,z)\,\Delta V,$$

where the summation is over the *inner boxes* ΔB, which lie entirely inside D. It follows that $\iiint_D fdV$ is independent of the choice of the auxiliary box B containing D.

The type of region D we have in mind is bounded by a surface, such as an ellipsoid, made up of the graphs of a finite number of continuous functions of two variables. It is not worth the trouble to be more specific. For all the regions D we shall deal with,

$$f \text{ continuous on } D \quad \Rightarrow \quad f \text{ integrable over } D.$$

All the particular functions we shall integrate in this section are continuous.

Let $f = 1$ on D. Then f is integrable over D and

$$\iiint_D 1dV = \lim \Sigma'\Sigma'\Sigma'\,\Delta V.$$

Intuition suggests that this limit should be the volume of D. So we define the **volume** of D as

$$V = \iiint_D 1 dV.$$

The four algebraic properties of double integrals displayed in Sec. 7.1 carry over almost verbatim to triple integrals. Only the comparison properties change slightly. Let V be the volume of D. Then

$$m \leq f \leq M \text{ on } D \quad \Rightarrow \quad mV \leq \iiint_D f dV \leq MV.$$

We shall not bother to reformulate the other properties.

Triple Integrals by Iterated Integration

As you should expect, triple integrals usually are evaluated by iterated integration. We explain how this is done next for rather general regions.

A solid region D is ***z*–simple** if it lies between the graphs of two continuous functions $z = z_1(x,y)$ and $z = z_2(x,y)$ for (x,y) in a plane region R. Thus, D consists of the points (x,y,z) with

$$(x,y) \text{ in } R \text{ and } z_1(x,y) \leq z \leq z_2(x,y).$$

In Fig. 3, R is the projection of D onto the xy–plane. For convenience, we assume also that R is either x–simple or y–simple. A special case of a z–simple solid is a region between a surface $z = z_2(x,y) = 0$ and the xy–plane. Then $z_1(x,y) = 0$.

For a z–simple region, as in Fig. 3, the triple integral of a function $f(x,y,z)$ over D can be expressed as the iterated integral

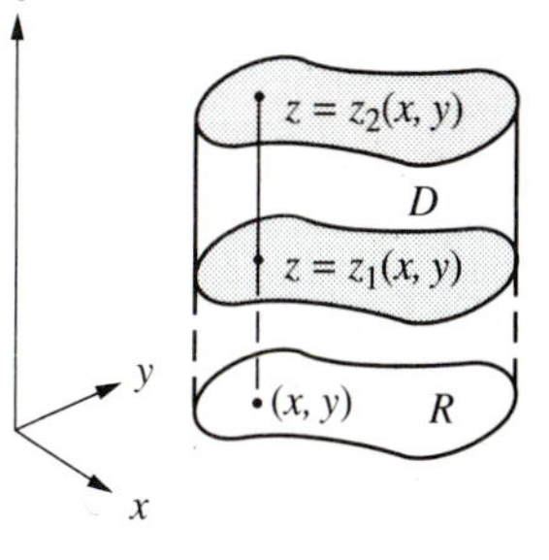

FIGURE 3

$$\iiint_D f dV = \iint_R \left[\int_{z_1(x,y)}^{z_2(x,y)} f(x,y,z)\, dz \right] dA.$$

The volume of D is

$$V = \iiint_D 1 dV = \iint_R \left[\int_{z_1(x,y)}^{z_2(x,y)} 1 dz \right] dA = \iint_R [z_2(x,y) - z_1(x,y)] dA.$$

The special case with $z_1(x,y) = 0$ and $z_2(x,y) = f(x,y) \geq 0$ gives the volume

$$V = \iint_R f(x,y)\,dA$$

between the surface $z = f(x,y)$ and the region R, in agreement with the volume formula in Sec. 7.1. We usually use the foregoing double integral formulas given to calculate volumes of z–simple solids. However, for illustrative purposes, the triple integral formula is used in the next example.

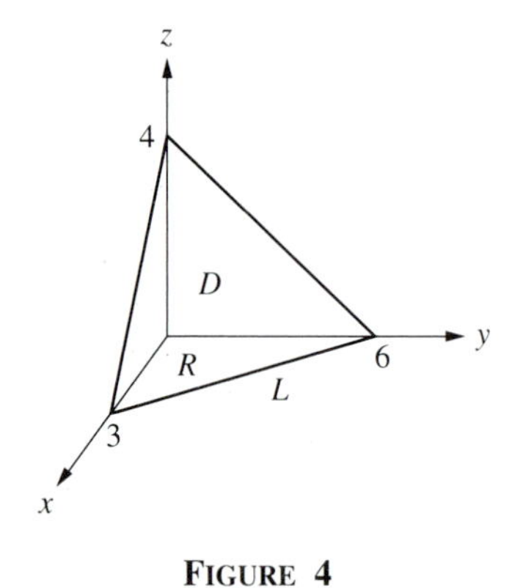

FIGURE 4

EXAMPLE 2. Let D be the tetrahedron in Fig. 4. Find the volume of D and evaluate $\iiint_D x\,dV$.

Solution. The base of the tetrahedron is the region R shown in Figs. 4 and 5. The line segment L is the graph of

$$\frac{x}{3} + \frac{y}{6} = 1 \quad \text{or} \quad y = 6 - 2x \quad \text{for} \quad 0 \leq x \leq 3.$$

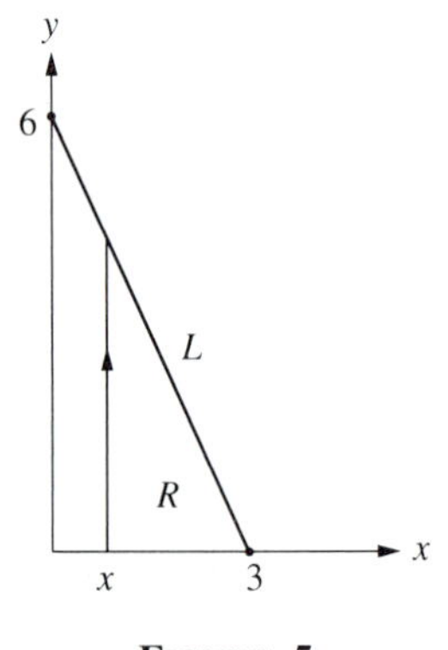

FIGURE 5

Observe that R is y–simple with

$$0 \leq x \leq 3 \quad \text{and} \quad 0 \leq y \leq 6 - 2x.$$

The slanting face of the tetrahedron in Fig. 4 is part of the plane given by

$$\frac{x}{3} + \frac{y}{6} + \frac{z}{4} = 1 \quad \text{or} \quad z = \frac{2}{3}(6 - 2x - y).$$

Therefore,

$$V = \iint_R \left[\int_0^{\frac{2}{3}(6-2x-y)} 1\,dz \right] dydx = \int_0^3 \int_0^{6-2x} \frac{2}{3}(6 - 2x - y)\,dydx$$

$$= \frac{2}{3}\int_0^3 \left[6y - 2xy - \frac{1}{2}y^2 \right]_{y=0}^{y=6-2x} dx = \frac{2}{3}\int_0^3 (18 - 12x + 2x^2)\,dx = 12.$$

The calculation of $\iiint_D x\,dV$ is very similar:

$$\iiint_D x\,dV = \iint_R \left[\int_0^{\frac{2}{3}(6-2x-y)} x\,dz \right] dydx = \int_0^3 \int_0^{6-2x} \frac{2}{3}x(6 - 2x - y)\,dydx$$

$$= \frac{2}{3}\int_0^3 \left[x\left(6y - 2xy - \frac{1}{2}y^2\right) \right]_{y=0}^{y=6-2x} dx = \frac{2}{3}\int_0^3 (18x - 12x^2 + 2x^3)\,dx = 9. \square$$

EXAMPLE 3. Let D be the solid region between the parabolic cylinders $z = y^2$ and $z = 2 - y^2$ for $0 \leq x \leq 3$. Find the volume of D and evaluate $\iiint_D x\,dV$.

Solution. The first step is to visualize the solid region D. In Fig. 6, imagine that the x–axis comes straight out of the page. The shaded region B in the yz–plane, where $x = 0$, lies between the parabolas $z = y^2$ and $z = 2 - y^2$. The two parabolas intersect where $y = \pm 1$. The solid D is generated by translating B parallel to the x–axis from $x = 0$ to $x = 3$. This should enable you to see that D is a right cylinder with base B and height 3. So the volume of D is $V = 3A$, where A is the area of B. By symmetry in y,

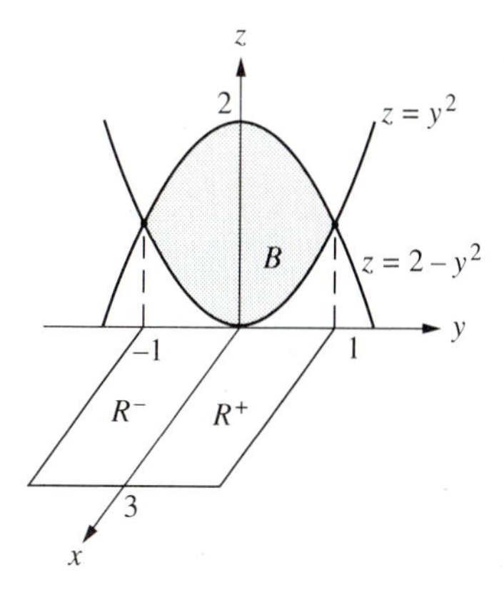

FIGURE 6

$$A = 2\int_0^1 [(2 - y^2) - y^2]\,dy = \frac{8}{3}, \qquad V = 3A = 8.$$

It remains to evaluate $\iiint_D x\,dV$. From Fig. 6, D is z–simple and its projection on the xy–plane is the rectangle R with $0 \le x \le 3$ and $-1 \le y \le 1$. Therefore,

$$\iiint_D x\,dV = \iint_R \left[\int_{z=y^2}^{z=2-y^2} x\,dz\right] dA = \iint_R x(2 - 2y^2)\,dA$$

$$= 2\int_0^1\int_0^3 x(2 - 2y^2)\,dx\,dy$$

$$= 4\int_0^3 x\,dx \cdot \int_0^1 (1 - y^2)\,dy = 4 \cdot \frac{9}{2} \cdot \frac{2}{3} = 12. \ \square$$

A solid region D is **x–simple** if it lies between the graphs of two continuous functions $x = x_1(y,z)$ and $x = x_2(y,z)$. Similarly, D is **y–simple** if it lies between the graphs of two continuous functions $y = y_1(x,z)$ and $y = y_2(x,z)$. The formulas for triple integrals over x–simple and y–simple regions are very similar to those given previously for triple integrals over z–simple regions. The variables x, y, and z just trade places. The region D in Ex. 3 is simple in all three coordinates. The triple integral $\iiint_D x\,dV$ could be evaluated in any order.

Mass and Mass Density

Suppose that the solid D in Fig. 2 is made of homogeneous material. This is a mathematical idealization that ignores atomic and subatomic structure. The word *homogeneous* means that the mass density, which is mass per unit volume, is the same for every piece of D that is regular enough to have a volume. Commonly used units for density are slugs/ft^3 in English units and g/cm^3 in the metric system.

Now suppose that the solid D in Fig. 2 is possibly inhomogeneous. Then the mass density $\sigma(x,y,z)$ at a point (x,y,z) in D is defined by

$$\sigma(x,y,z) = \lim_{\Delta V \to 0} \frac{\Delta M}{\Delta V},$$

where ΔM is the mass and ΔV is the volume in a small cube ΔB containing the point (x,y,z). (The shape of ΔB is unimportant if the largest dimension of ΔB tends to 0.) By definition, $\sigma(x,y,z) \ge 0$. We shall deal only with continuous mass density functions $\sigma(x,y,z)$. If there is no danger of misunderstanding, we call $\sigma(x,y,z)$ the density rather than the mass density.

Partition the box B in Fig. 2 into small cubes ΔB. Then the inner Riemann sum

$$\Sigma'\Sigma'\Sigma'\sigma(x,y,z)\,\Delta V$$

is an approximation for the mass of D. The **mass** of D is defined by

$$M = \iiint_D \sigma dV = \lim \Sigma'\Sigma'\Sigma'\sigma(x,y,z)\,\Delta V.$$

(The limit of the Riemann sums is taken in the usual manner.) The algebraic properties of triple integrals have obvious interpretations in terms of mass. For example, the domain additivity property,

$$\iiint_D \sigma dV = \iiint_{D_1} \sigma dV + \iiint_{D_2} \sigma dV,$$

expresses the fact that the mass of D is the sum of the masses of two nonoverlapping parts of D.

EXAMPLE 4. Suppose that the tetrahedron in Fig. 4 is a solid with density $\sigma(x,y,z) = x$. Then, from Ex. 2, the mass is $M = \iiint_D x dV = 9$. □

In Ex. 4 there is a small problem. The density $\sigma(x,y,z) = x$ seems to have units of length rather than mass per unit volume. We shall assume that all expressions for mass densities have been multiplied by an appropriate factor with numerical value 1 to make the units correct. To save writing, we do not specify the units of mass. In Ex. 4, $M = 9$ means 9 units of mass.

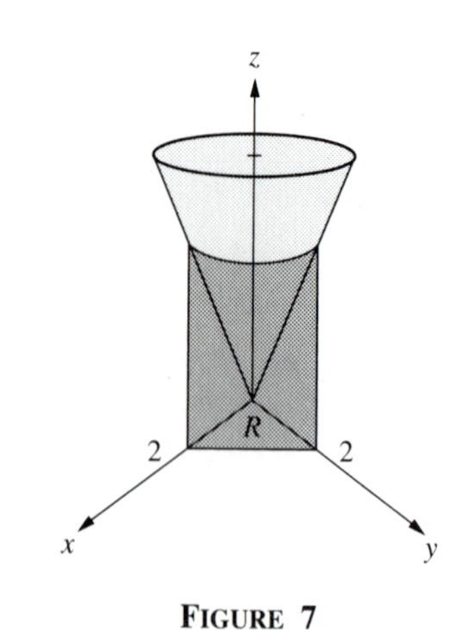

FIGURE 7

EXAMPLE 5. Let D be the solid that lies between the triangular region R in Fig. 7 and the upper half of the circular cone $z^2 = 9(x^2 + y^2)$. Let $\sigma(x,y,z) = z$ be the density of D. Find the mass of D.

Solution. The region R is y–simple with

$$0 \le x \le 2 \text{ and } 0 \le y \le 2 - x.$$

The solid D is z–simple with

$$(x,y) \text{ in } R \text{ and } 0 \le z \le 3\sqrt{x^2 + y^2}$$

Consequently, the mass of D is given by

$$M = \iiint_D \sigma dV = \iint_R \left[\int_0^{3\sqrt{x^2+y^2}} z dz \right] dA = \frac{9}{2} \iint_R (x^2 + y^2) \, dA,$$

$$M = \frac{9}{2} \int_0^2 \int_0^{2-x} (x^2 + y^2) \, dy dx = \frac{9}{2} \int_0^2 \left[x^2 y + \frac{y^3}{3} \right]_{y=0}^{y=2-x} dx,$$

$$M = \frac{9}{2} \int_0^2 \left[2x^2 - x^3 + \frac{(2-x)^3}{3} \right] dx = \frac{9}{2} \left[\frac{2x^3}{3} - \frac{x^4}{4} - \frac{(2-x)^4}{12} \right]_0^2 = 12. \ \square$$

Charge and Charge Density

Charge densities, also commonly denoted by σ, are useful in electromagnetic applications. The charge density in a solid D as in Fig. 2 is defined by

$$\sigma(x, y, z) = \lim_{\Delta V \to 0} \frac{\Delta Q}{\Delta V},$$

where ΔQ is the electric charge in a small cube ΔB with volume ΔV containing (x, y, z). Since electric charges can be positive or negative, $\sigma(x, y, z)$ can vary in sign. The **net charge** on D is defined by

$$\boxed{Q = \iiint_D \sigma dV.}$$

In Ex. 5, if $\sigma(x, y, z) = z$ is charge density, then the net charge on D is $Q = 12$. If $\sigma(x, y, z) = -z$, then $Q = -12$. As for mass, we assume that the formulas for charge are adjusted for correct units.

Symmetry

Just as for double integrals, taking advantage of symmetry can save a lot of work in evaluating triple integrals. A solid region D is **symmetric in z** (or symmetric across the xy–plane) if

$$(x, y, -z) \text{ is in } D \quad \Leftrightarrow \quad (x, y, z) \text{ is in } D.$$

Let D^+ and D^- be the parts of D with $z \geq 0$ and $z \leq 0$, respectively. Then D^+ and D^- are mirror images of each other across the xy–plane. If f is defined on D and $f(x, y, -z) = f(x, y, z)$, so that f takes the same value at symmetric points of D, then

$$\iiint_{D^-} f dV = \iiint_{D^+} f dV, \qquad \iiint_D f dV = 2 \iiint_{D^+} f dV.$$

On the other hand, if $f(x,y,-z) = -f(x,y,z)$, so that f takes opposite values at symmetric points of D, then

$$\iiint_{D^-} f dV = -\iiint_{D^+} f dV, \qquad \iiint_D f dV = 0.$$

Masses and charges furnish physical examples of these results. If a solid body D is symmetric in z and the mass density of the body takes the same value at symmetric points, then the mass of D is twice the mass of the top half of D. If the charge on D takes on opposite values at symmetric points, then the net charge on D is zero.

There are strictly analogous definitions and results for regions that are **symmetric in** x and **symmetric in** y. The region in the next example is symmetric in x.

EXAMPLE 6. Let D be the solid wedge that lies between the semicircular disk $x^2 + y^2 \le 4$, $y \ge 0$ and the slanting plane $z = y$. Assume that the mass density is $\sigma(x,y,z) = |x|$. Find the mass M.

Solution. We met this wedge in Ex. 9 and Fig. 18 of Sec. 7.1. It will be helpful to look at the figure again. Since

$$(-x,y,z) \text{ is in } D \quad \Leftrightarrow \quad (x,y,z) \text{ is in } D,$$

D is symmetric in x. Since $\sigma(-x,y,z) = \sigma(x,y,z)$,

$$M = \iiint_D \sigma dV = 2\iiint_{D^+} \sigma dV,$$

where D^+ is the half of D with $x \ge 0$. The projection of D^+ on the xy–plane is the quarter circle R^+ in Fig. 18 with $x \ge 0$, $y \ge 0$, and $x^2 + y^2 \le 4$. A y–simple description of R^+ is

$$0 \le x \le 2 \text{ and } 0 \le y \le \sqrt{4 - x^2}.$$

The solid region D^+ is z–simple with

$$(x,y) \text{ in } R^+ \text{ and } 0 \le z \le y.$$

Since $\sigma(x,y,z) = |x| = x$ on R^+,

$$M = 2\iiint_{D^+} x dV = 2\iint_{R^+}\left[\int_0^y x dz\right] dA = 2\iint_{R^+} xy dA,$$

$$M = 2\int_0^2\int_0^{\sqrt{4-x^2}} xy dy dx = \int_0^2 [xy^2]_{y=0}^{y=\sqrt{4-x^2}} dx,$$

$$M = \int_0^2 (4x - x^3) dx = 4. \ \square$$

EXAMPLE 7. The ball D with center at the origin and radius 4 has density $\sigma(x,y,z) = x^2$. Since both D and σ are symmetric in x, y, z, three applications of symmetry yield

$$M = \iiint_D x^2 dV = 8 \iiint_{D_1} x^2 dV,$$

where D_1 is the part of the ball in the first octant. The integral over D_1 is hard to evaluate in rectangular coordinates. Stay tuned. □

EXAMPLE 8. Suppose that the ball D in Ex. 7 carries a charge density $\sigma(x,y,z) = x + y + z$. Find the net charge Q.

Solution. By symmetry of D in x and the fact that $f(x,y,z) = x$ takes opposite values at the symmetric points (x,y,z) and $(-x,y,z)$ in D

$$\iiint_D x dV = 0.$$

Likewise,

$$\iiint_D y dV = \iiint_D z dV = 0.$$

Add the three integrals to see that the net charge on D is

$$Q = \iiint_D (x + y + z) dV = 0. \quad \square$$

Mass for Plane Regions

Figure 8 shows a thin plate or *lamina D,* perhaps made of sheet metal. More precisely, D is a right cylinder with base R in the xy–plane and height h, which is small compared with the dimensions of R.

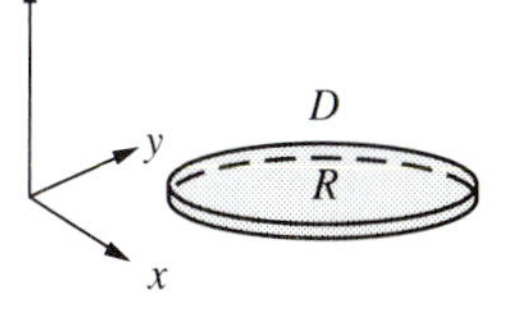

FIGURE 8

Assume that the density $\sigma(x,y,z)$ is constant on any vertical line segment in D. Then

$$\sigma(x,y,z) = \sigma(x,y,0) \text{ for } (x,y) \text{ in } R \text{ and } 0 \leq z \leq h.$$

The mass of the plate is

$$M = \iiint_D \sigma dV = \iint_R \left[\int_0^h \sigma(x,y,z) dz \right] dA = \iint_R h \, \sigma(x,y,0) dA.$$

It is convenient to define

$$\sigma(x,y) = h\,\sigma(x,y,0).$$

The units of $\sigma(x,y)$ are (length) $\times$ (mass/volume) $=$ mass/area. Then

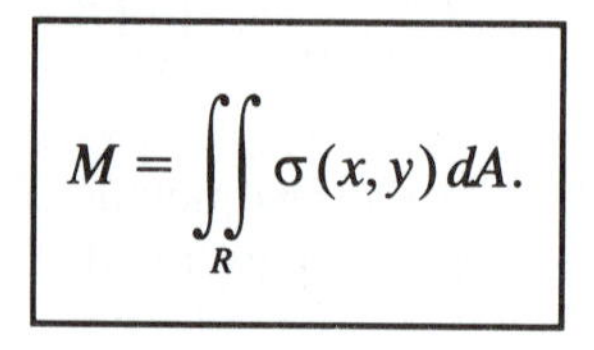

$$M = \iint_R \sigma(x,y)\,dA.$$

We can regard the plate in Fig. 8 as if it were the two–dimensional region R with mass density $\sigma(x,y)$ in units of mass/area. It is not necessary for D to be thin. The critical feature is that the mass per unit volume must be constant on each vertical line segment in D. The double integral for the mass is really a triple integral with the z–integration already performed. In the examples that follow, it may be helpful to think of the idealized two–dimensional plates as three–dimensional plates with constant thickness.

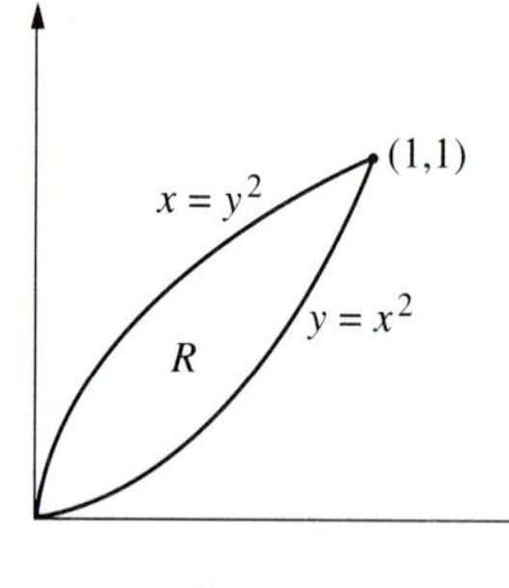

FIGURE 9

EXAMPLE 9. Assume that the plate R in Fig. 9, bounded by the parabolas $y = x^2$ and $x = y^2$, has the mass density $\sigma(x,y) = xy$. Find the mass of the plate.

Solution. The mass of the plate is

$$M = \iint_R xy\,dA.$$

Does the double integral look familiar? In Ex. 4 of Sec. 7.1 the same integral gave the volume of the solid between R and the surface $z = xy$. The value of the integral was found to be 1/12. So the mass of the plate is $M = 1/12$.□

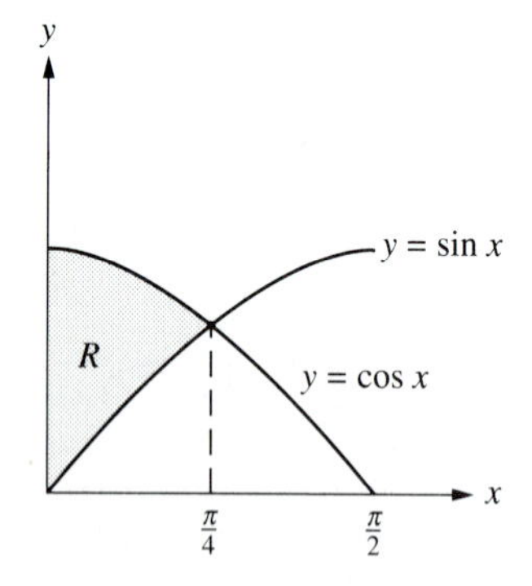

FIGURE 10

EXAMPLE 10. Assume that the plate R, bounded by the curves $y = \sin x$, $y = \cos x$, and $x = 0$, has density $\sigma(x,y) = e^x$. Find the mass of the plate.

Solution. The region R, shown in Fig. 10, is y–simple with $0 \le x \le \pi/4$ and $\sin x \le y \le \cos x$. The mass is

$$M = \iint_R \sigma\,dA = \int_0^{\pi/4}\int_{\sin x}^{\cos x} e^x\,dy dx,$$

$$M = \int_0^{\pi/4} e^x(\cos x - \sin x)\,dx.$$

Either by integration by parts, educated guessing, or by use of a CAS or integral table, an antiderivative of $e^x(\cos x - \sin x)$ is $e^x \cos x$. Hence,

$$M = [e^x \cos x]_0^{\pi/4} = \frac{1}{2}\sqrt{2}\,e^{\pi/4} \approx 0.55. \ \square$$

Charge for Plane Regions

Charges for plane regions come up frequently. It is an experimental fact that the charges on an insulated conductor are concentrated on its surface. If the plane region R in Fig. 8 represents a surface that carries an areal charge density $\sigma(x,y)$, then the net charge Q on R is

$$\boxed{Q = \iint_R \sigma \, dA.}$$

EXAMPLE 11. Let R be the region in the xy–plane bounded by the line $y = x$ and the parabola $x = 2 - y^2$. This region appeared in Ex. 6 and Fig. 13 of Sec. 7.1. We found that $\iint_R y\,dA = -33/4$. If R carries the charge density $\sigma(x,y) = y$, then the net charge on R is

$$Q = \iint_R y\,dA = -33/4.$$

The minus sign means that there are more electrons than protons on R. □

PROBLEMS

In Probs. 1–4, evaluate the triple integral over the given box.

1. $\iiint_B (x + y + z)\,dV$
   B: $0 \le x \le 1, \quad 0 \le y \le 1, \quad 0 \le z \le 1$

2. $\iiint_B xyz\,dV$
   B: $0 \le x \le 1, \quad 2 \le y \le 3, \quad 1 \le z \le 2$

3. $\iiint_B 4xz \sin y\,dV$
   B: $1 \le x \le 2, \quad 0 \le y \le \pi, \quad 0 \le z \le 4$

4. $\iiint_B xy \sin xz\,dV$
   B: $1/3 \le x \le 1/2, \ 0 \le y \le 1, \ 0 \le z \le \pi$

In Probs. 5–10, evaluate the iterated integral.

5. $\int_0^1 \int_0^{1-z} \int_0^2 dx\,dy\,dz$

6. $\int_0^2 \int_0^1 \int_0^{1-y} dz\,dx\,dy$

7. $\int_0^1 \int_0^{z^2} \int_0^{x+z} x\,dy\,dx\,dz$

8. $\int_0^2 \int_{y^2}^1 \int_0^{1-x} x\,dz\,dx\,dy$

9. $\int_0^1 \int_0^x \int_0^{\ln y} e^{x+y+z}\,dz\,dy\,dx$

10. $\int_1^3 \int_x^{x^2} \int_0^{\ln z} dy\,dz\,dx$

In Probs. 11–18, sketch the region D and evaluate $\iiint_D f\,dV$.

11. $f(x,y,z) = z$; D the tetrahedron in Ex. 2.

12. $f(x,y,z) = xy$; D the tetrahedron in Ex. 2.

13. $f(x,y,z) = yz$; D the solid in the first octant bounded by the coordinate planes, the plane $y + z = 1$, and the paraboloid $x = 5 - y^2 - z^2$.

14. $f(x,y,z) = xy \sin \pi z$; D the solid in the first octant bounded by the parabolic cylinder $z = 4 - x^2$ and the planes $z = 0$, $y = x$, and $y = 0$.

15. $f(x,y,z) = e^x$; D the solid between the yz plane and the graph of the function $x = \ln(y + z)$ defined on the region in the yz–plane bounded by the curves $z = y^2$, $z = y$, and the line $y = 2$.

16. $f(x,y,z) = e^z$; D the solid under the graph of the function $z = x + 2y$ defined on the region in the xy–plane bounded by the curves $y = x^2$, $y = 0$, and $x = 1$.

17. $f(x,y,z) = z$; D the solid in the first octant bounded by the cylinder $y^2 + z^2 = a^2$ and the planes $y = 0$, $x = 0$, and $x + y = a$. Here a is a positive constant.

18. $f(x,y,z) = x^2 y^2$; D the solid bounded above by the cylinder $y^2 + z = a^2$, below by the plane $y + z = a$, and on the sides by the planes $x = 0$ and $x = b$. Here $a \geq 1$ and b are positive constants.

19. The tetrahedron in Ex. 2 has density $\sigma(x,y,z) = \sqrt{x}$. Find its mass.

20. The tetrahedron cut from the first octant by the plane $x/a + y/b + z/c = 1$, where a, b, $c > 0$, is a solid with density $\sigma(x,y,z) = z$. Find its mass.

21. (a) The tetrahedron in the previous exercise has density $\sigma(x,y,z) = x^2$. Find its mass. (b) Find the mass if $\sigma(x,y,z) = x^2 + y^2$. *Hint.* Use symmetry and part (a).

22. The solid under the paraboloid $z = x^2 + y^2$ and above the square $0 \leq x, y \leq 2$ in the xy–plane has density $\sigma(x,y,z) = x^2 + y^2 + z^2$. Find its mass.

23. Find the mass of the solid in Ex. 5 if the density is $\sigma(x,y,z) = yz$.

24. The solid in Ex. 5 carries a charge density $\sigma(x,y,z) = z - 3$. Find the net charge on the solid.

25. The cube with $0 \leq x, y, z \leq 1$ has density $\sigma(x,y,z) = x$. Find its mass.

26. Find the mass of a cube of edge a if the density at any point is proportional to the distance from one face.

27. The cube with $0 \leq x, y, z \leq 1$ has density $\sigma(x,y,z) = x^2 + y^2$. Find its mass.

28. Find the mass of a cube of edge a if the density at any point is proportional to the square of the distance from one edge.

29. Find the mass of the solid cylinder bounded by the surfaces $x^2 + y^2 = 16$, $z = 0$, and $z = 8$ if the density $\sigma(x,y,z) = z$.

30. A solid right circular cylinder has base radius a, height h, and density proportional to the distance to one of its bases. Find its mass.

31. The solid bounded by the cylinders $x^2 + y^2 = 4$ and $x^2 + z^2 = 4$ has density $\sigma(x,y,z) = |x|$. Find the mass. *Hint.* Use symmetry. Sketch the part of the solid in the first quadrant.

32. The solid bounded by the cylinders $x^2 + y^2 = a^2$ and $x^2 + z^2 = a^2$ has density proportional to the square of the distance to the z–axis. Find the mass.

33. The solid under the paraboloid $z = x^2 + y^2$ and above the region in the xy–plane bounded by the lines $x = 0$, $y = 0$, and $x + y = 2$ has density $\sigma(x,y,z) = xy$. Find the mass.

34. The solid in the previous exercise has a charge density $\sigma(x,y,z) = z - 2$. Find the net charge on the solid.

35. The rectangular plate $0 \le x \le 2$, $0 \le y \le 1$ in the xy–plane has (areal) mass density $\sigma(x,y) = xy$. Find its mass.

36. The rectangular plate $0 \le x \le a$, $0 \le y \le b$ has density proportional to the product of the distances from two adjacent sides. Find the mass.

37. The region in the xy–plane bounded by the curves $y = 4x^2$ and $y = x^4$ has density $\sigma(x,y) = \sqrt{|xy|}$. Find the mass.

38. The region in the previous exercise carries a charge density $\sigma(x,y) = x(y - 8)$. Find the net charge.

39. The region in the xy–plane bounded by the curves $y = \cos x$, $y = \sin x$, and $x = 0$ has density $\sigma(x,y) = 2xy$. Find the mass.

40. Give a careful definition of an x–simple region D in space. Choose your notation so that the analogue of the formula accompanying Fig. 3 is

$$\iiint_D f(x,y,z)dV = \iint_R \left[\int_{x_1(y,z)}^{x_2(y,z)} f(x,y,z)\,dx \right] dA.$$

41. Return to the cylindrical solid D of Ex. 3. Assume the solid has density $\sigma(x,y,z) = |xy|$. The region is both z–simple and x–simple. Express the mass M of D by treating D as (a) z–simple and (b) x–simple. (c) Evaluate M using either (a) or (b), whichever you prefer.

42. If R is vertically simple with description $a \le x \le b$, $y_1(x) \le y \le y_2(x)$, then show that the formula accompanying Fig. 3 can be expressed as

$$\iiint_D f dV = \int_a^b \int_{y_1(x)}^{y_2(x)} \int_{z_1(x,y)}^{z_2(x,y)} f(x,y,z)\,dz\,dy\,dx.$$

43. Find the formula analogous to the one in the previous problem if R is horizontally simple with description $c \le y \le d$, $x_1(y) \le x \le x_2(y)$.

44. Let D be a z–simple region and use the notation from Fig. 3. Assume D is symmetric in y (symmetric in the xz–plane) with D^+ the part of D with $y \ge 0$. Show:

    (a) $z_1(x,-y) = z_1(x,y)$ and $z_2(x,-y) = z_2(x,y)$;

    (b) $f(x,-y,z) = f(x,y,z) \Rightarrow \iiint_D f dV = 2\iiint_{D^+} f dV$;

    (c) $f(x,-y,z) = -f(x,y,z) \Rightarrow \iiint_D f dV = 0$.

    Hint. Use corresponding results for double integrals.

45. If D is symmetric in the xz–plane use shorthand Riemann sum and limit passage reasoning to support the following conclusions:

    (a) $f(x,-y,z) = f(x,y,z) \Rightarrow \iiint_D f dV = 2\iiint_{D^+} f dV$;

    (b) $f(x,-y,z) = -f(x,y,z) \Rightarrow \iiint_D f dV = 0$.

46. Provide the missing symmetry arguments in Ex. 7 and, hence, establish that $\iiint_D x^2 dV = 8\iiint_{D_1} x^2 dV$.

7.3 Double and Triple Integrals in Polar and Cylindrical Coordinates

Double integrals over plane regions with some degree of central symmetry often are easier to evaluate in polar coordinates. Cylindrical coordinates in 3–space consist of polar coordinates in a plane and a coordinate axis perpendicular to the plane. Cylindrical coordinates are very convenient for the evaluation of triple integrals over solid regions that have an axis of symmetry or have some degree of symmetry about an axis. Integrals that may be very difficult or even impossible to evaluate in rectangular coordinates are often easy to evaluate in polar or cylindrical coordinates.

Integrals in polar and cylindrical coordinates are defined as limits of Riemann sums based on partitions that are especially adapted to these coordinate systems. We begin by introducing more general Riemann sums that will serve not only for integrals in polar and cylindrical coordinates, but also for integrals in spherical coordinates in the next section.

More General Riemann Sums

Integrals in rectangular coordinates were defined as limits of Riemann sums over partitions consisting of rectangles or boxes. In Sec. 7.1 the integral of a function $f(x,y)$ over a rectangle R was defined as the limit of Riemann sums,

$$\iint_R f dA = \lim \Sigma\Sigma f(x,y)\,\Delta A,$$

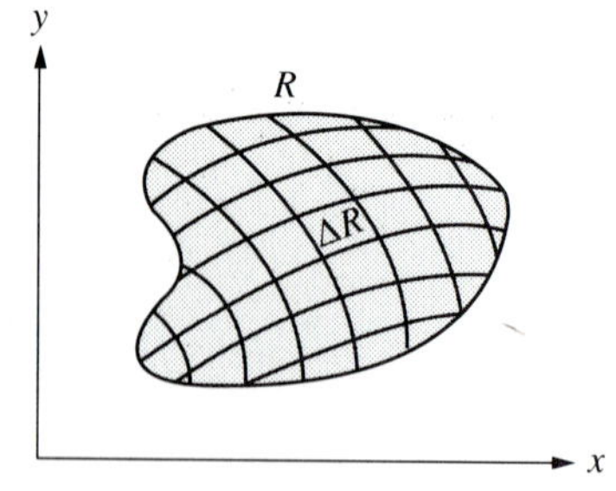

FIGURE 1

where the summation is over rectangles ΔR with areas ΔA in a partition of R. The actual shapes of the subregions ΔR in a partition are not very important. It is enough for each ΔR to be closed and regular enough in shape to have an area ΔA. Figure 1 illustrates the more general partitions that we have in mind. The *diameter* of ΔR is the greatest distance between any two points of ΔR. The *mesh* of a partition is the largest diameter of any subregion ΔR in the partition. Finally, as before, $\iint_R f dA$ is the limit of Riemann sums as the mesh of the partition tends to 0.

In the same way, the triple integral of a function $f(x,y,z)$ over a solid region D is a limit of Riemann sums,

$$\iiint_D f dV = \lim \Sigma\Sigma\Sigma f(x,y,z)\,\Delta V,$$

where ΔV is the volume of a closed subregion ΔD in a partition of D and (x,y,z)

is any point in ΔD. The limit is taken as the mesh of the partition, which is the largest diameter of a subregion ΔD, tends to zero.

It is implicit in the double and triple integral notation we have used that the limits of the more general Riemann sums introduced here are all the same and are the same as the limits of the Riemann sums based on rectangles and boxes in Secs. 7.1 and 7.2. The fact that the limits really are all the same should seem very plausible and it is true; however, the proof is in the realm of advanced calculus.

Polar and Cylindrical Coordinates

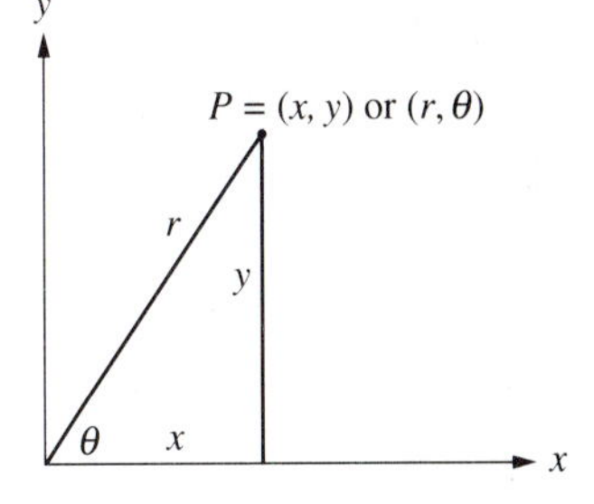

FIGURE 2

Polar coordinates (r,θ) are used in one–variable calculus to solve certain area problems and to aid in graphing certain equations. Figure 2 reminds us of the relationship between rectangular and polar coordinates:

$$x = r\cos\theta, \qquad r^2 = x^2 + y^2,$$
$$y = r\sin\theta, \qquad \tan\theta = y/x.$$

In one–variable calculus, r is sometimes allowed to be negative. Here it is convenient to allow only $r \geq 0$. This restriction is in force throughout the section.

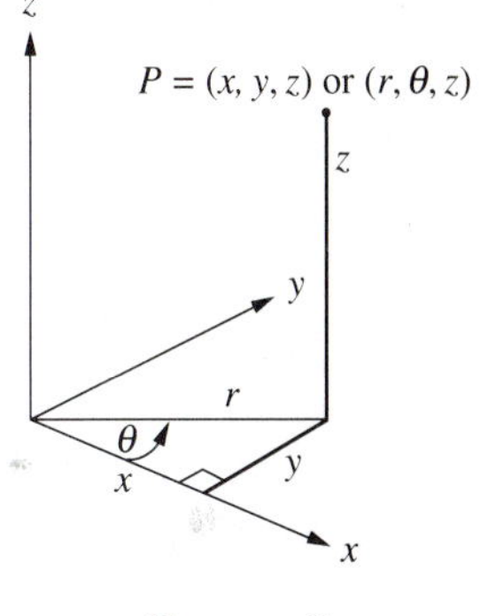

FIGURE 3

Cylindrical coordinates (r,θ,z) consist of polar coordinates (r,θ) and a rectangular coordinate z along an axis perpendicular to the $r\theta$–plane. A point P in 3-space can be located by its rectangular coordinates (x,y,z) or by its **cylindrical coordinates** (r,θ,z), as indicated in Fig. 3. The relationships between rectangular and cylindrical coordinates can be read from Fig. 3:

$$x = r\cos\theta, \qquad r^2 = x^2 + y^2$$
$$y = r\sin\theta, \qquad \tan\theta = y/x$$
$$z = z, \qquad z = z.$$

The graphs of equations $r = c$, $\theta = c$, and $z = c$ in cylindrical coordinates are of particular interest. To begin with, the graph of the polar coordinate equation $r = c$ with $c > 0$ is the circle in the xy–plane with radius c centered at the origin. In cylindrical coordinates, the equation $r = c$ places no restriction on z. So the graph of $r = c$ is the right circular cylinder obtained by translating the circle $r = c$, $z = 0$, parallel to the z–axis. In polar coordinates, the graph of $\theta = c$ is a ray from the origin. The angle between the ray and the positive x–axis is c radians. In cylindrical coordinates, the graph of $\theta = c$ is the half-plane that contains the ray just mentioned and has the z–axis for its edge. The special case with $\theta = 0$ is the half of the xz–plane with $x \geq 0$. Rotate this half-plane through the angle c to obtain the half-plane $\theta = c$. Finally, the graph of $z = c$ is a horizontal plane.

We shall be interested especially in surfaces that are symmetric about the z–axis or, in other words, surfaces of revolution about the z–axis. A simple example is the cylinder $r = c$ mentioned earlier. Note the absence of θ in the equation. The graph of any cylindrical coordinate equation $z = f(r)$ or $F(r,z) = 0$ is symmetric about the z–axis. The absence of θ means that the traces

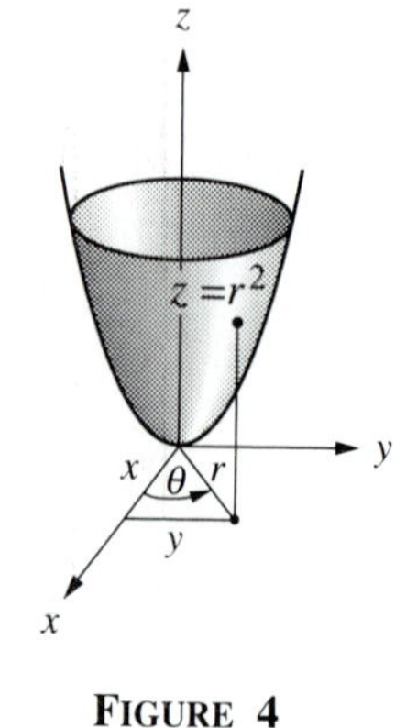

FIGURE 4

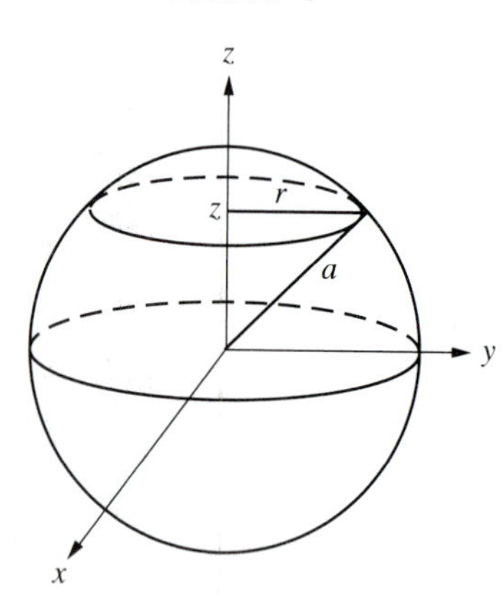

FIGURE 5

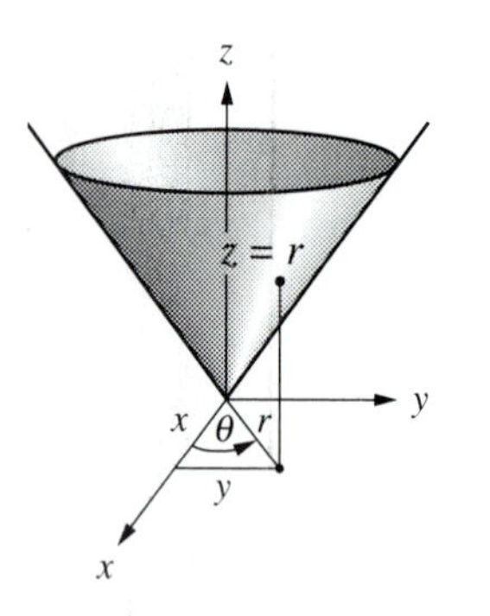

FIGURE 6

of such surfaces in all the half-planes $\theta = c$ look alike. Consequently, the entire surface can be generated by rotating the trace of the surface in the half-plane $\theta = 0$ about the z–axis. The following three familiar surfaces are symmetric about the z-axis. Their frequently used equations in cylindrical coordinates involve only r and z.

EXAMPLE 1. Circular paraboloid:

$$z = x^2 + y^2 \quad \Leftrightarrow \quad z = r^2.$$

See Fig. 4. □

EXAMPLE 2. Sphere:

$$x^2 + y^2 + z^2 = a^2 \quad \Leftrightarrow \quad r^2 + z^2 = a^2 \text{ or } z = \pm\sqrt{a^2 - r^2}$$

See Fig. 5. □

EXAMPLE 3. Half-cone:

$$z^2 = x^2 + y^2, \quad z \geq 0 \quad \Leftrightarrow \quad z = r.$$

See Fig. 6. The full cone $z^2 = x^2 + y^2$ is given by $z = \pm r$. □

Double Integrals Over Polar Rectangles

Polar rectangles play the same role in a polar coordinate system that ordinary rectangles play in a rectangular coordinate system. A **polar rectangle** R consists of points (r, θ) with $a \leq r \leq b$ and $c \leq \theta \leq d$, where $0 \leq a < b$ and $c < d \leq c + 2\pi$. See Fig. 7. Special cases of polar rectangles include:

R is a disk if $0 \leq r \leq b$ and $0 \leq \theta \leq 2\pi$,
R is a circular sector if $0 \leq r \leq b$ and $d < c + 2\pi$,
R is an annulus if $0 < a \leq r \leq b$ and $0 \leq \theta \leq 2\pi$.

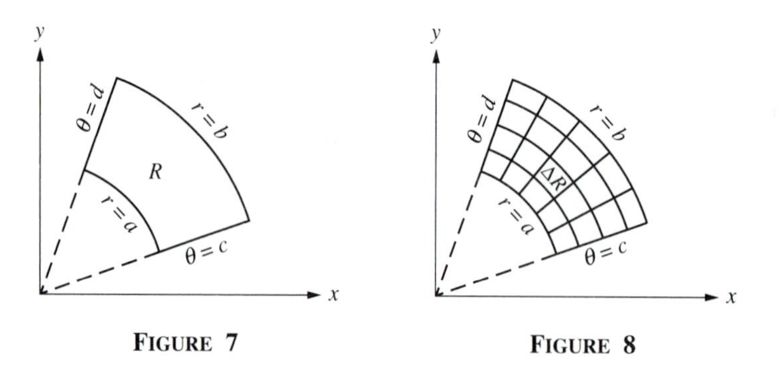

FIGURE 7 FIGURE 8

Double integrals over polar rectangles are the polar coordinate analogues of double integrals over ordinary rectangles. They are conveniently evaluated by iterated integration, as we show next.

Let $f(r,\theta)$ be a positive continuous function defined on the polar rectangle in Fig. 7. We shall express the integral $\iint_R f dA$ for the volume under the surface $z = f(r,\theta)$ in terms of polar coordinates. For this purpose, partition R into smaller polar rectangles ΔR, as in Fig. 8. A typical rectangle ΔR is enlarged in Fig. 9. The point (r,θ) lies at the center of ΔR. The area of ΔR, regarded as the difference of two circular sectors, is

$$\Delta A = \frac{1}{2}\left(r + \frac{1}{2}\Delta r\right)^2 \Delta\theta - \frac{1}{2}\left(r - \frac{1}{2}\Delta r\right)^2 \Delta\theta,$$

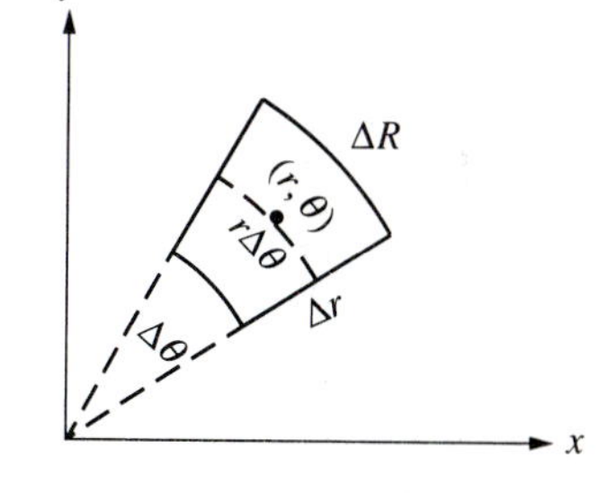

FIGURE 9

which simplifies to

$$\Delta A = r\,\Delta\theta\,\Delta r.$$

This is the product of the lengths $r\Delta\theta$ and Δr in Fig 9. We may regard ΔR as a distorted rectangle with sides $r\Delta\theta$ and Δr.

In Fig. 10, S is the graph of $z = f(r,\theta)$ for (r,θ) in R. The tower with base ΔR reaching up to S has volume ΔV given approximately by

$$\Delta V \approx f(r,\theta)\Delta A = f(r,\theta)\,r\Delta\theta\Delta r.$$

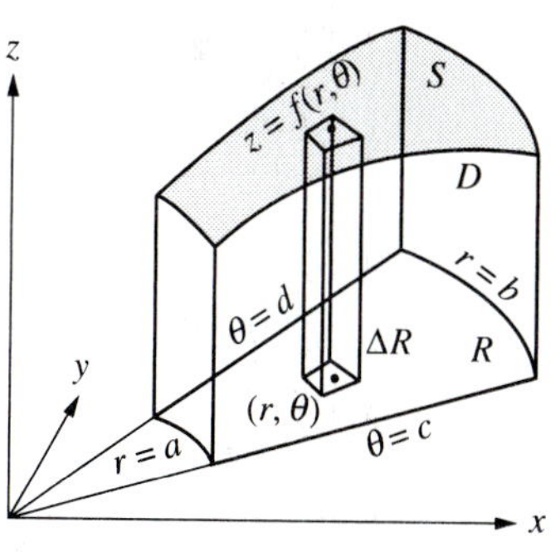

FIGURE 10

The total volume under S is

$$V = \iint_R f(r,\theta)\,dA = \lim \Sigma\Sigma f(r,\theta)\,r\Delta\theta\Delta r$$

in our usual shorthand for Riemann sums. The limit of the Riemann sums can be expressed as an iterated integral in r and θ in either order. This conclusion is valid for any continuous function f on R:

> **Theorem 1** *Double Integrals in Polar Coordinates*
> If $f(r,\theta)$ is continuous on the polar rectangle R with $a \le r \le b, \quad c \le \theta \le d$, then
>
> $$\iint_R f(r,\theta)\,dA = \int_a^b\int_c^d f(r,\theta)\,r\,d\theta\,dr = \int_c^d\int_a^b f(r,\theta)\,r\,dr\,d\theta.$$

The area element dA is expressed in polar coordinates by

$$\boxed{dA = r\,d\theta\,dr,}$$

which is a reminder of $\Delta A = r\Delta\theta\Delta r$ in the Riemann sums for the double integral.

Theorem 1 can be deduced from the corresponding theorem for double integrals in rectangular coordinates given in Sec. 7.1. The argument is informative and not particularly difficult. We start with

$$\iint_R f(r,\theta)\,dA = \lim \Sigma\Sigma f(r,\theta)\,\Delta A = \lim \Sigma\Sigma f(r,\theta)\,r\Delta\theta\Delta r.$$

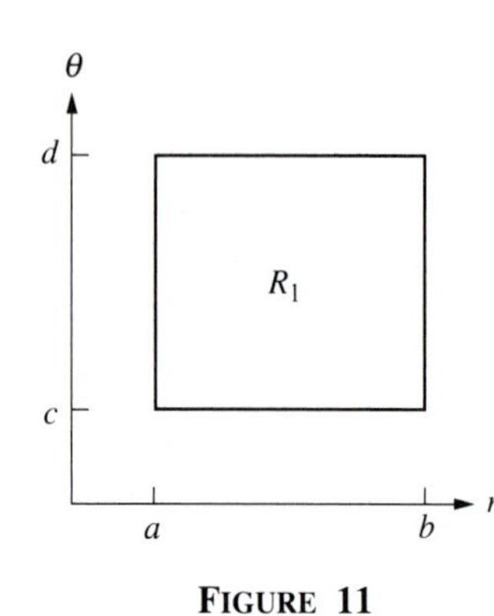

FIGURE 11

Although r and θ are polar coordinates, the numerical values of the Riemann sums do not depend on this fact. In the Riemann sums, r and θ are just numbers with $a \le r \le b$ and $c \le \theta \le d$. Let's make a new rectangular coordinate system with the r–axis horizontal and the θ–axis vertical, as in Fig. 11. Then the points (r,θ) form a rectangle R_1 in this $r\theta$–plane.

Since $f(r,\theta)r$ is continuous, it is integrable over the rectangle R_1. Thus,

$$\iint_{R_1} f(r,\theta)\,r\,d\theta\,dr = \lim \Sigma\Sigma[f(r,\theta)\,r]\,\Delta\theta\Delta r.$$

Compare the last two Riemann sums to obtain

$$\iint_R f(r,\theta)\,dA = \iint_{R_1} f(r,\theta)\,r\,d\theta\,dr.$$

By Th. 1 in Sec. 7.1, the double integral of $f(r,\theta)r$ over R_1 can be expressed as an iterated integral in either order. Consequently,

$$\iint_R f(r,\theta)\,dA = \int_a^b\int_c^d f(r,\theta)\,r\,d\theta\,dr = \int_c^d\int_a^b f(r,\theta)\,r\,dr\,d\theta,$$

which is the conclusion of the theorem.

EXAMPLE 4. Verify that the volume of the sphere $x^2 + y^2 + z^2 = a^2$ with radius a is $V = \frac{4}{3}\pi a^3$.

Solution. By symmetry in z, the volume of the sphere is twice the volume of its top half. From Ex. 2, the upper surface of the sphere is expressed in cylindrical coordinates by

$$z = \sqrt{a^2 - r^2} \quad \text{for} \quad 0 \le r \le a, \quad 0 \le \theta \le 2\pi.$$

Therefore,

$$V = 2\int_0^a\int_0^{2\pi}(a^2 - r^2)^{1/2}\,r\,d\theta\,dr = 2\int_0^{2\pi} 1\,d\theta \cdot \int_0^a (a^2 - r^2)^{1/2}\,r\,dr,$$

$$V = 4\pi\left[-\frac{1}{3}(a^2 - r^2)^{3/2}\right)_0^a = \frac{4}{3}\pi a^3. \;\square$$

EXAMPLE 5. Find the volume V of the solid region D below the paraboloid $z = 4 - x^2 - y^2$ and above the xy–plane.

Solution. The paraboloid opens down and the vertex is at $(x, y, z) = (0, 0, 4)$. The surface lies on or above the xy–plane for $x^2 + y^2 \le 4$. So the solid D is described in rectangular coordinates by

$$0 \le z \le 4 - x^2 - y^2 \quad \text{for} \quad x^2 + y^2 \le 4$$

and in cylindrical coordinates by

$$0 \le z \le 4 - r^2 \quad \text{for} \quad 0 \le r \le 2, 0 \le \theta \le 2\pi.$$

Therefore,

$$V = \int_0^2 \int_0^{2\pi} (4 - r^2)\, r\, d\theta\, dr = 2\pi \int_0^2 (4r - r^3)\, dr = 8\,\pi. \ \square$$

EXAMPLE 6. Find the volume of the solid region D inside the top half of the cone $z^2 = x^2 + y^2$ and the top half of the ellipsoid $2x^2 + 2y^2 + z^2 = 12$.

Solution. The two surfaces, shown in Fig. 12, are represented in cylindrical coordinates by $z = r$ and $z^2 = 12 - 2r^2$ with $z \ge 0$. The surfaces intersect where

$$r^2 = z^2 = 12 - 2r^2, \quad z^2 = r^2 = 4, \quad z = r = 2.$$

FIGURE 12

Hence, the surfaces intersect in the circle C with $r = 2$ and $z = 2$ shown in Fig. 12. The projection of D on the xy–plane is the disk R with $0 \le r \le 2$. So D is described by

$$r \le z \le (12 - 2r^2)^{1/2} \quad \text{for} \quad 0 \le r \le 2, \quad 0 \le \theta \le 2\pi.$$

The volume of D is

$$V = \int_0^2 \int_0^{2\pi} [(12 - 2r^2)^{1/2} - r]\, r\, d\theta\, dr = 2\pi \int_0^2 [(12 - 2r^2)^{1/2} r - r^2]\, dr,$$

$$V = 2\pi \left[-\frac{1}{6}(12 - 2r^2)^{3/2} - \frac{1}{3} r^3 \right]_0^2 = 8\pi(\sqrt{3} - 1) \approx 18.4. \ \square$$

EXAMPLE 7. Let $c > 0$. Find the volume V_c under the surface $z = e^{-(x^2 + y^2)}$ for $x^2 + y^2 \le c^2$.

Solution. In cylindrical coordinates, $z = e^{-r^2}$ for $0 \le r \le c$ and $0 \le \theta \le 2\pi$. The graph, which is symmetric about the z–axis, is shown in Fig. 13. The volume is

$$V_c = \int_0^c \int_0^{2\pi} e^{-r^2} r\, d\theta\, dr = 2\pi \int_0^c e^{-r^2} r\, dr = 2\pi \left[-\frac{1}{2} e^{-r^2} \right]_0^c = \pi(1 - e^{-c^2}). \ \square$$

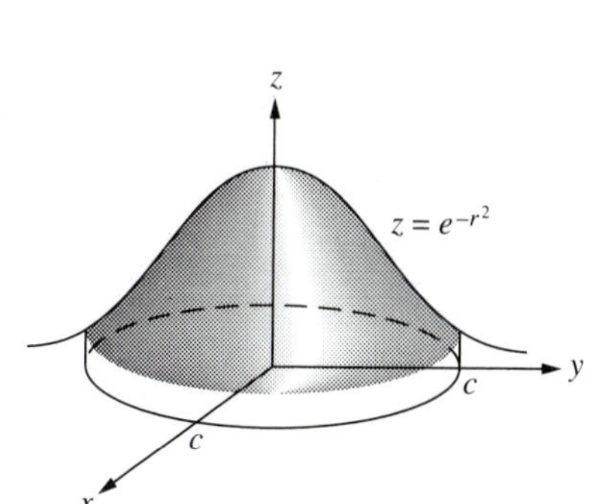

FIGURE 13

Let $c \to \infty$ in Ex. 7 to obtain the volume $V = \pi$ for the region under the entire surface $z = e^{-r^2}$ for $0 \leq r < \infty$. An argument based on $r^2 = x^2 + y^2$ (see the problems) yields

$$V = \int_{-\infty}^{\infty} e^{-x^2} dx \cdot \int_{-\infty}^{\infty} e^{-y^2} dy = \left(\int_{-\infty}^{\infty} e^{-x^2} dx \right)^2 .$$

Since $V = \pi$, it follows that

$$\boxed{\int_{-\infty}^{\infty} e^{-x^2} dx = \sqrt{\pi}.}$$

This exact evaluation is all the more remarkable because e^{-x^2} has no elementary antiderivative. This integral is very useful in the physical sciences and in probability and statistics.

Double Integrals Over Radially Simple Regions

Applications of double integrals often involve integrations over regions that are more general than polar rectangles. Many such regions are radially simple. Then the double integrals can be evaluated conveniently as iterated integrals, as we describe next.

A region R in the *xy*–plane is **radially simple** if it consists of points (r, θ) with

$$c \leq \theta \leq d \quad \text{and} \quad g(\theta) \leq r \leq h(\theta),$$

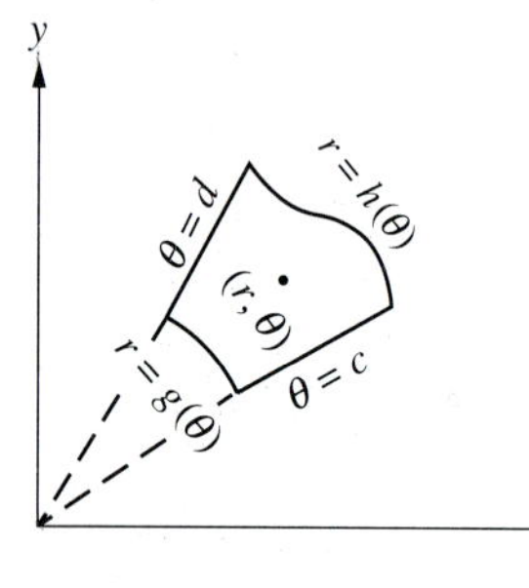

FIGURE 14

where g and h are continuous functions and $d \leq c + 2\pi$. See Fig. 14. The double integral of a continuous function $f(r, \theta)$ over R can be expressed as an iterated integral as follows:

$$\boxed{\iint_R f(r, \theta)\, dA = \int_c^d \int_{g(\theta)}^{h(\theta)} f(r, \theta)\, r dr d\theta.}$$

This iterated integration formula can be derived by an argument similar to that for Th. 1.

EXAMPLE 8. The area A of the radially simple region R in Fig. 14 is given by

$$A = \iint_R 1\, dA = \int_c^d \int_{g(\theta)}^{h(\theta)} r dr d\theta = \frac{1}{2} \int_c^d [h(\theta)^2 - g(\theta)^2]\, d\theta.$$

This formula is derived by other means in one–variable calculus. □

The next example looks hard but simplifies considerably when cylindrical coordinates are used.

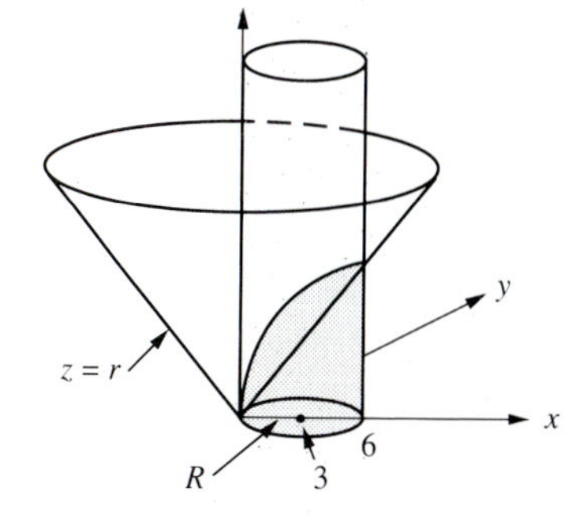

FIGURE 15

EXAMPLE 9. Let R be the disk $(x-3)^2 + y^2 \le 9$ in the xy–plane. Find the volume V between R and the upper half of the cone $z^2 = x^2 + y^2$. See Figs. 15 and 16.

Solution. We change to cylindrical coordinates to take advantage of the partial symmetry about the z–axis. The half-cone is represented by $z = r$. The circle $(x-3)^2 + y^2 = 9$ has the polar equation $r = 6\cos\theta$; see Fig. 16. So the disk R is radially simple with

$$-\frac{\pi}{2} \le \theta \le \frac{\pi}{2} \quad \text{and} \quad 0 \le r \le 6\cos\theta.$$

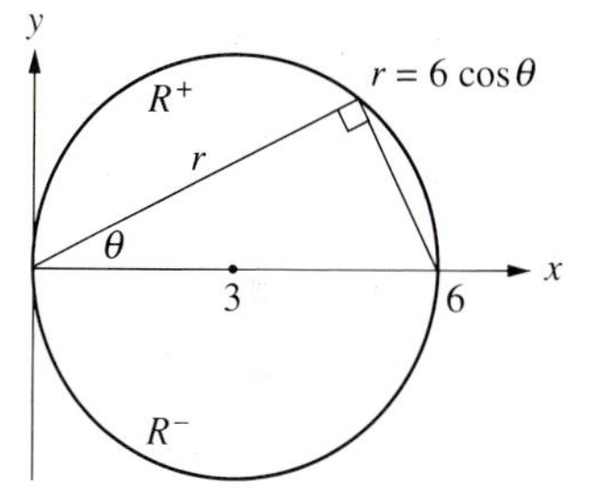

FIGURE 16

Let R^+ be the half of R with $0 \le \theta \le \pi/2$. By symmetry,

$$V = 2\iint_{R^+} r dA = 2\int_0^{\pi/2}\int_0^{6\cos\theta} r^2 dr d\theta = \frac{2}{3}\int_0^{\pi/2} [r^3]_{r=0}^{r=6\cos\theta} dr,$$

$$V = 144\int_0^{\pi/2} \cos^3\theta \, d\theta = \int_0^{\pi/2} (1-\sin^2\theta)\cos\theta d\theta,$$

$$V = 144\left[\sin\theta - \frac{1}{3}\sin^3\theta\right]_0^{\pi/2} = 96.$$

If the disk R carries a mass density $\sigma(r,\theta) = r$, then, again using symmetry, the total mass M of the disk is

$$M = 2\iint_{R^+} \sigma dA = 2\iint_{R^+} r dA = 96,$$

the same numerical value as the volume V. □

Triple Integrals in Cylindrical Coordinates

Triple integrals are expressed in cylindrical coordinates by adapting reasoning we have used earlier to represent integrals in rectangular coordinates and polar coordinates. Thus, we shall be rather brief and just record the essential facts. We begin with triple integrals over cylindrical blocks and move on to integrals over more general regions.

The role of a box in rectangular coordinates is played in cylindrical coordinates by a **cylindrical block** D, which consists of points (r,θ,z) with

$$a \le r \le b, \qquad c \le \theta \le d, \qquad p \le z \le q.$$

As usual, $a \ge 0$ and $d \le c + 2\pi$. See Fig. 17. The projection of D on the xy–plane is the polar rectangle R with $a \le r \le b$ and $c \le \theta \le d$. Note that D is a cylinder with base a copy of R and height $q - p$. In particular, if R is a circular

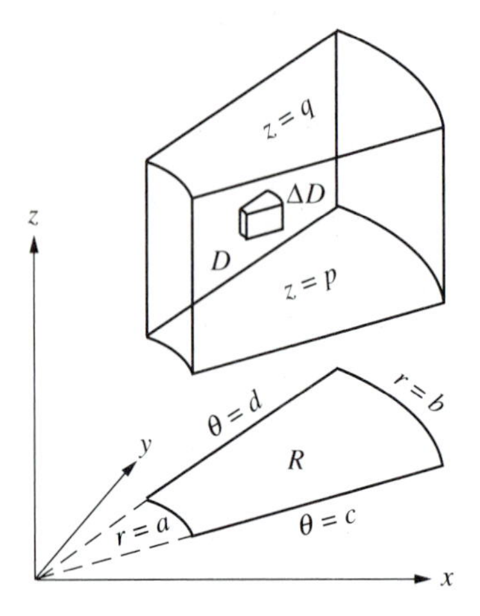

FIGURE 17

disk, then D is a right circular cylinder. What does D look like if R is a sector or an annulus?

Imagine that D is partitioned into many small cylindrical blocks ΔD. A typical block ΔD is shown in Fig. 17 and greatly enlarged in Fig. 18. The projection of ΔD on the xy–plane is the polar rectangle ΔR. The point (r, θ, z) in Fig. 18 is at the center of ΔD. So (r, θ) is at the center of ΔR and the area of ΔR is $\Delta A = r\Delta r\Delta\theta$. The volume of the cylinder ΔD is

$$\Delta V = \Delta A\, \Delta z = r\Delta\, r\, \Delta\,\theta\, \Delta z.$$

Let $f(r, \theta, z)$ be a continuous function defined on D. Then

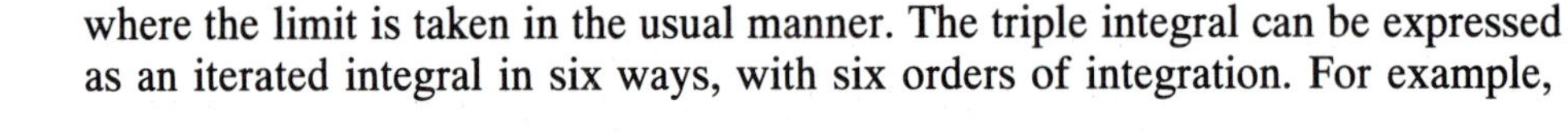

$$\iiint_D f dV = \lim \sum\sum\sum f(r,\theta,z)\,\Delta V = \lim \sum\sum\sum f(r,\theta,z)\, r\,\Delta r\,\Delta\,\theta\,\Delta z,$$

where the limit is taken in the usual manner. The triple integral can be expressed as an iterated integral in six ways, with six orders of integration. For example,

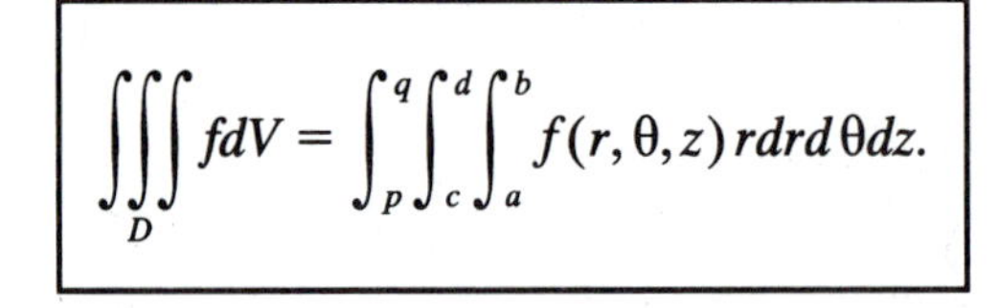

$$\boxed{\iiint_D f dV = \int_p^q \int_c^d \int_a^b f(r,\theta,z)\, r dr d\theta dz.}$$

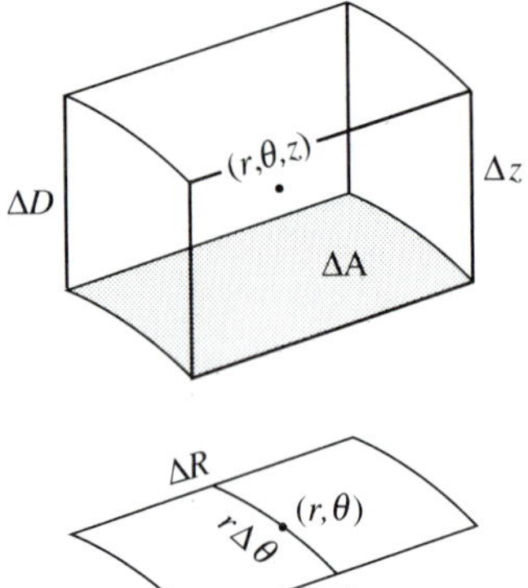

FIGURE 18

The volume element dV is expressed in cylindrical coordinates by

$$\boxed{dV = r dr d\theta\, dz,}$$

which reminds us of $\Delta V = r\,\Delta r\,\Delta\theta\Delta z$ in the Riemann sums for the triple integral.

EXAMPLE 10. Let D be the circular cylinder with $0 \le r \le 2$, $0 \le \theta \le 2\pi$, $0 \le z \le 1/2$. Evaluate $\iiint_D r^2 dV$.

Solution. Since the integrand does not involve θ or z, we integrate first with respect to these variables:

$$\iiint_D r^2 dV = \int_0^2 \int_0^{1/2} \int_0^{2\pi} r^2 r d\theta\, dz dr = \pi \int_0^2 r^3 dr = 4\pi.$$

This is the mass of D if the mass density is $\sigma(r, \theta, z) = r^2$. □

If $f(r, \theta, z) = g(r)h(\theta)k(z)$, then the triple integral of f over a cylindrical block can be expressed as a product of single integrals:

$$\boxed{\int_a^b \int_c^d \int_p^q g(r)h(\theta)k(z)\, r dz d\theta dr = \int_a^b g(r) r dr \cdot \int_a^b h(\theta)\, d\theta \cdot \int_a^b k(z)\, dz.}$$

We ask you to verify this formula in the problems. It could have been used in Ex. 10.

A solid region D is **cylindrically simple** if it consists of points (r, θ, z) with

$$(r, \theta) \text{ in } R \text{ and } z_1(r, \theta) \le z \le z_2(r, \theta),$$

where $z_1(r,\theta)$ and $z_2(r,\theta)$ are continuous and R is radially simple. See Fig. 19. The triple integral of a continuous function $f(r, \theta, z)$ over a cylindrically simple region D can be expressed as an iterated integral as follows:

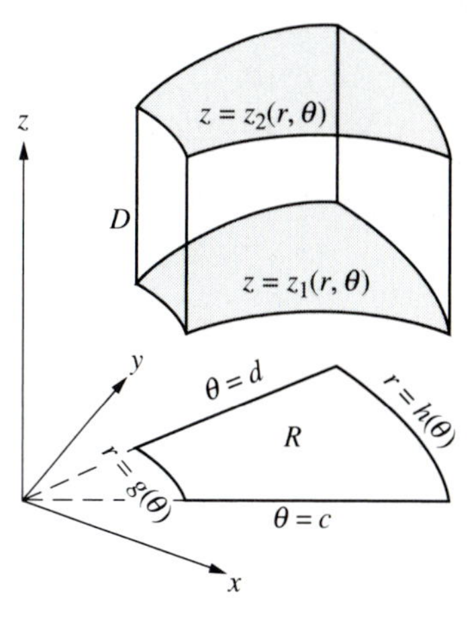

FIGURE 19

$$\iiint_D f dV = \iint_R \left[\int_{z_1(r,\theta)}^{z_2(r,\theta)} f(r, \theta, z)\, dz \right] dA.$$

Since R is radially simple, the double integral over R also can be expressed as an iterated integral, as described earlier. Doing so expresses the triple integral by three successive single integrals.

EXAMPLE 11. Let D be the solid region above the xy–plane and below the paraboloid $z = 4 - x^2 - y^2$. Given the density $\sigma(x, y, z) = x^2$, find the mass M of D.

Solution. The region D appeared in Ex. 5. It is described by $0 \le z \le 4 - r^2$ for (r, θ) in the disk R with $0 \le r \le 2$ and $0 \le \theta \le 2\pi$. Since $\sigma = x^2 = r^2 \cos^2 \theta$,

$$M = \iiint_D \sigma dV = \iint_R \left[\int_0^{4 - r^2} r^2 \cos^2 \theta dz \right] dA$$

$$= \iint_R r^2 \cos^2 \theta (4 - r^2)\, r dr d\theta$$

$$= \int_0^2 (4r^3 - r^5)\, dr \cdot \int_0^{2\pi} \frac{1}{2}(1 + \cos 2\theta)\, d\theta$$

$$= \left[r^4 - \frac{1}{6} r^6 \right]_0^2 \cdot \left[\frac{1}{2}\theta + \frac{1}{4}\sin 2\theta \right]_0^{2\pi} = \frac{16}{3}\pi. \ \square$$

EXAMPLE 12. Let D be the solid wedge with

$$x^2 + y^2 \le 4, \quad y \ge 0, \quad \text{and} \quad 0 \le z \le y.$$

The wedge is shown in Fig. 18 of Sec 7.1, which you should consult as we go along. Given that the wedge has density $\sigma(x, y, z) = \sqrt{x^2 + y^2}$, find its mass M.

Solution. By symmetry in x,

$$M = 2\iiint_{D^+} \sigma dV,$$

where D^+ is the half of D with $x \geq 0$. Let's express D^+ in cylindrical coordinates. Since $y = r\sin\theta$, the point (r, θ, z) is in D^+ if

$$0 \leq r \leq 2, \qquad 0 \leq \theta \leq \frac{\pi}{2}, \qquad 0 \leq z \leq r\sin\theta.$$

So D^+ is cylindrically simple. Since $\sigma = \sqrt{x^2 + y^2} = r$,

$$M = 2\int_0^2\int_0^{\pi/2}\int_0^{r\sin\theta} r^2 dz d\theta dr = 2\int_0^2\int_0^{\pi/2} r^3 \sin\theta \, d\theta dr,$$

$$M = 2\int_0^2 r^3 dr \cdot \int_0^{\pi/2} \sin\theta d\theta = 2 \cdot 4 \cdot 1 = 8. \ \square$$

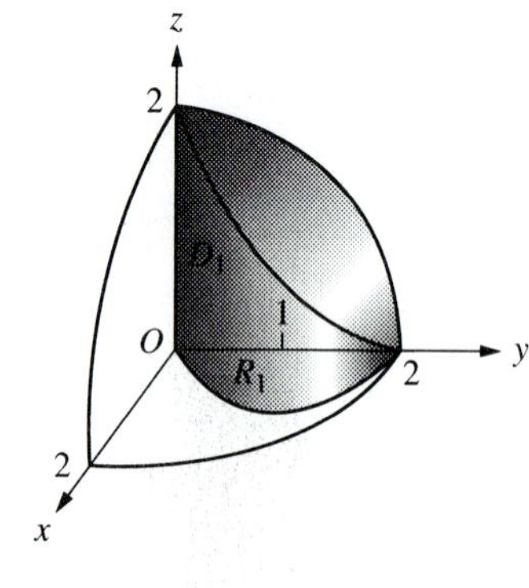

FIGURE 20

EXAMPLE 13. Let D be the solid inside both the sphere $x^2 + y^2 + z^2 = 4$ and the cylinder $x^2 + (y-1)^2 = 1$. Evaluate $\iiint_D |xz| dV$.

Solution. Figure 20 shows the part D_1 of D that lies in the first octant, where $x, y, z \geq 0$. By symmetry in x and z,

$$\iiint_D |xz|\, dV = 4\iiint_{D_1} xz\, dV.$$

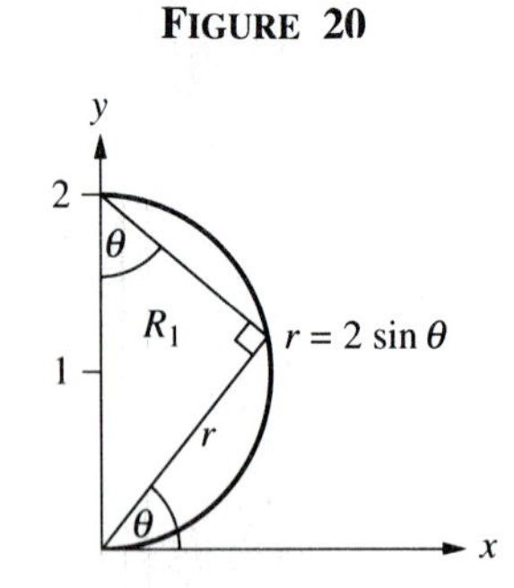

FIGURE 21

The upper half of the sphere is expressed in cylindrical coordinates by $r^2 + z^2 = 4$ with $z \geq 0$ or by $z = \sqrt{4 - r^2}$. The projection of the solid D_1 onto the xy–plane is the half-disk R_1 shown in Fig. 20 and also in Fig. 21. It is expressed in polar coordinates by

$$0 \leq r \leq 2\sin\theta \text{ and } 0 \leq \theta \leq \pi/2.$$

This is a radially simple description of R_1. The solid D_1 is described by

$$(r, \theta) \text{ in } R_1 \text{ and } 0 \leq z \leq \sqrt{4 - r^2}.$$

So D_1 is cylindrically simple and

$$\iiint_D |xz| dV = 4\iiint_{D_1} xz dV = 4\iint_{R_1}\left[\int_0^{\sqrt{4-r^2}} r\cos\theta\, z dz\right] dA$$

$$= 2\iint_{R_1} r\cos\theta (4 - r^2)\, r dr d\theta$$

$$= 2\int_0^{\pi/2}\int_0^{2\sin\theta} \cos\theta\,(4r^2 - r^4)\,dr\,d\theta$$

$$= 2\int_0^{\pi/2} \cos\theta \left[\frac{4}{3}r^3 - \frac{1}{5}r^5\right]_{r=0}^{r=\sin\theta} d\theta$$

$$= 64\int_0^{\pi/2} \left(\frac{1}{3}\sin^3\theta - \frac{1}{5}\sin^5\theta\right)\cos\theta\,d\theta$$

$$= 64\left[\frac{1}{12}\sin^4\theta - \frac{1}{30}\sin^6\theta\right]_0^{\pi/2} = \frac{16}{5} = 3.2. \ \square$$

PROBLEMS

In Probs. 1–10, sketch the graph of the cylindrical coordinate equation.

1. $r = 5$
2. $r = -3$
3. $\theta = \pi/4$
4. $\theta = 5\pi/6$
5. $r = 4\sec\theta$
6. $r = 4\csc\theta$
7. $z = \sqrt{3}\,r$
8. $z = \sqrt{3}\,r^2$
9. $r = 2\sin\theta$
10. $r = 2\cos\theta$

In Probs. 11–16, express the iterated integral as a double integral over an appropriate region R. Then use polar coordinates to evaluate the double integral.

11. $\displaystyle\int_{-1}^{1}\int_0^{\sqrt{1-y^2}} \sqrt{x^2+y^2}\,dx\,dy$
12. $\displaystyle\int_0^1\int_0^{\sqrt{1-x^2}} \frac{1}{1+x^2+y^2}\,dy\,dx$
13. $\displaystyle\int_0^2\int_0^{\sqrt{4-x^2}} \arctan\frac{y}{x}\,dy\,dx$
14. $\displaystyle\int_0^2\int_0^{\sqrt{4-y^2}} e^{x^2+y^2}\,dx\,dy$
15. $\displaystyle\int_0^1\int_0^{\sqrt{x}} \frac{1}{\sqrt{x^2+y^2}}\,dy\,dx$
16. $\displaystyle\int_0^1\int_x^1 \frac{1}{(1+x^2+y^2)^{3/2}}\,dx\,dy$

In Probs. 17–30, find the volume of the solid.

17. The region under the paraboloid $z = \frac{1}{2}(x^2+y^2)$, inside the cylinder $x^2+y^2=4$, and above the xy–plane.
18. The region under the top half of the cone $z^2 = x^2+y^2$, inside the cylinder $x^2+y^2=9$, and above the xy–plane.
19. The region under the sphere $x^2+y^2+z^2=16$, above the xy–plane, and inside the cylinder $x^2+y^2=4$.
20. The region in the first octant under the hyperbolic paraboloid $z = 2xy$ inside the cylinder $x^2+y^2=16$.
21. The region between the cone $z^2 = x^2+y^2$ and the paraboloid $z = x^2+y^2$.
22. The region between the paraboloids $z = x^2+y^2$ and $z = 8 - x^2 + y^2$.

23. The region under the paraboloid $z = \frac{1}{4}(x^2 + y^2)$ and inside the cylinder $x^2 + y^2 = 4x$.

24. The region bounded by the sphere $x^2 + y^2 + z^2 = 4a^2$ and the cylinder $x^2 + y^2 = 2ax$.

25. The region bounded by the spheres $x^2 + y^2 + z^2 = a^2$ and $x^2 + y^2 + (z - a)^2 = a^2$.

26. The region bounded by the sphere $x^2 + y^2 + z^2 = a^2$ and the planes $z = b$ and $z = c$, where $0 \le b < c \le a$.

27. The region bounded by the elliptic paraboloids $z = x^2 + 2y^2$ and $z = 12 - 2x^2 - y^2$.

28. The region bounded by the elliptic cone $z^2 = 2x^2 + 4y^2$ and the ellipsoid $4x^2 + 2y^2 + z^2 = 36$.

29. The solid ellipsoid $x^2/a^2 + y^2/a^2 + z^2/c^2 \le 1$.

30. The solid ellipsoid $x^2/a^2 + y^2/b^2 + z^2/b^2 \le 1$.

In Probs. 31–42 find the mass of the solid region with the given density σ. Use k for a proportionality constant, when appropriate.

31. The region in Prob. 17 if $\sigma(x, y, z) = \sqrt{x^2 + y^2}$.

32. The region in Prob. 18 if $\sigma(x, y, z) = |x|$.

33. The region in Prob. 19 if the density is proportional to the distance from the z–axis.

34. The region in Prob. 20 if the density is proportional to the distance from the yz–plane.

35. The region in Prob. 21 if the density is proportional to the distance from the xy–plane.

36. The region in Prob. 22 if the density is proportional to the distance from the z–axis.

37. The region in Prob. 23 if the density is $\sigma(x, y, z) = |xy|$.

38. The region in Prob. 24 if the density is $\sigma(x, y, z) = |xy|$.

39. The region in Prob. 25 if the density is proportional to the square of the distance from the z–axis.

40. The region in Prob. 26 if the density is proportional to the height above the xy–plane.

41. The region in Prob. 27 if the density is proportional to the product of the distances from the yz– and xz–planes.

42. The region in Prob. 28 if the density is proportional to the square of the distance from the origin.

43. Use reasoning similar to Ex. 7 to evaluate the integral

$$\int_0^\infty \int_0^\infty \frac{1}{(1 + x^2 + y^2)^2}\, dx\, dy.$$

The next problem helps you to establish the remarkable formula

$$\int_{-\infty}^{\infty} e^{-x^2} dx = \sqrt{\pi}.$$

Review Ex. 7 and Fig. 13 before you continue.

44. For $c > 0$, let S_c be the square in the xy–plane with $0 \leq x \leq c$ and $0 \leq y \leq c$. Let D_c be the quarter disk of points in the first quadrant with $x^2 + y^2 \leq c^2$. (a) Show that D_c lies inside S_c which lies inside $D_{\sqrt{2}c}$ and conclude that

$$\iint_{D_c} e^{-x^2-y^2} dA \leq \iint_{S_c} e^{-x^2-y^2} dA \leq \iint_{D_{\sqrt{2}c}} e^{-x^2-y^2} dA.$$

(b) Deduce that

$$\lim_{c \to \infty} \iint_{S_c} e^{-x^2-y^2} dA = \lim_{c \to \infty} \frac{1}{4} V_c = \frac{\pi}{4},$$

where V_c is the volume from Ex. 7.

(c) Show that

$$\iint_{S_c} e^{-x^2-y^2} dA = \left(\int_0^c e^{-x^2} dx \right)^2.$$

(d) Finally, conclude that $\int_{-\infty}^{\infty} e^{-x^2} dx = \sqrt{\pi}$.

The integral $\int_{-\infty}^{\infty} e^{-x^2} dx = \sqrt{\pi}$ helps evaluate several other integrals that come up in physical applications. Use it and an appropriate change of variable or integration by parts to evaluate the integrals in Probs. 45–48.

45. $\displaystyle\int_0^{\infty} \frac{e^{-x}}{\sqrt{x}} dx$

46. $\displaystyle\int_0^{\infty} \sqrt{x}\, e^{-x} dx$

47. $\displaystyle\int_0^{\infty} x^2 e^{-x^2} dx$

48. $\displaystyle\int_0^{1} \sqrt{|\ln x|}\, dx$

7.4 Triple Integrals in Spherical Coordinates

Solid regions with some degree of symmetry about the origin and integrals over such regions often are described most conveniently with spherical coordinates. Not surprisingly, triple integrals over such regions are usually easier to evaluate when expressed in spherical coordinates. We begin with a description of spherical coordinates and their relationships with rectangular and cylindrical coordinates. Then we show how to evaluate triple integrals by iterated integration using spherical coordinates.

Spherical Coordinates

The spherical coordinates (ρ, θ, φ) of the point P in Fig. 1 are given by

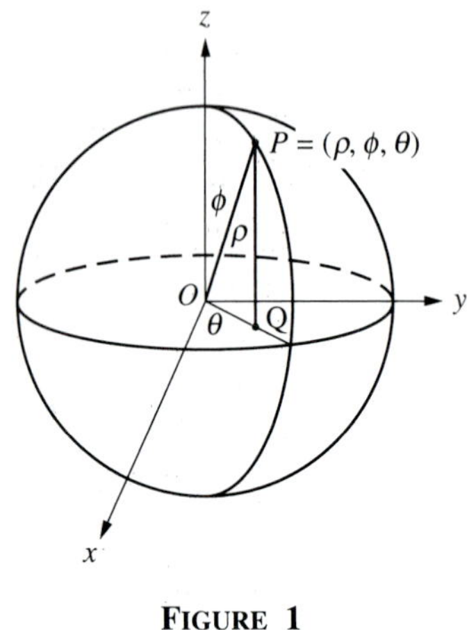

FIGURE 1

$$\rho = \text{distance from } P \text{ to the origin } O,$$
$$\varphi = \text{angle between } \overline{OP} \text{ and the } z\text{–axis},$$
$$\theta = \text{angle between } \overline{OQ} \text{ and the } x\text{–axis}.$$

where Q is the projection of P on the xy–plane and θ is the same as the cylindrical coordinate angle for P. The angle φ is restricted to $0 \leq \varphi \leq \pi$.

Figure 1 and a little imagination help us identify the following graphs.

$\rho = c$: sphere with center O and radius c.

$\varphi = c$: half-cone with vertex O and axis the z–axis.

$\theta = c$: half-plane with edge the z–axis.

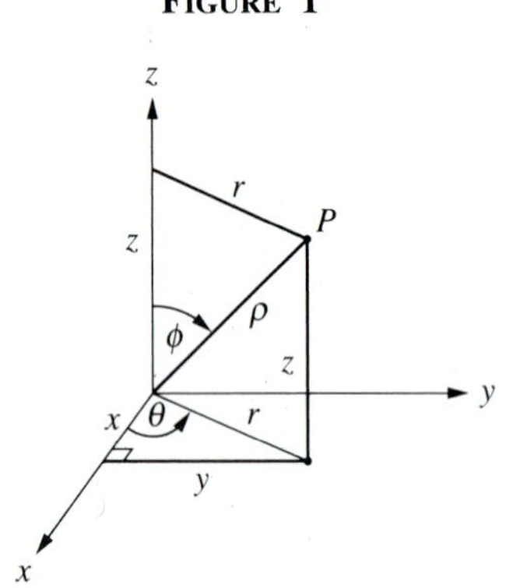

FIGURE 2

For example, $\rho = 2$ is the spherical coordinate equation for the sphere with radius 2 and center the origin; $\varphi = \pi/4$ is the equation for the half-cone with vertex at the origin obtained by rotating about the z–axis a ray emanating from the origin and inclined at a 45° angle to the positive z–axis; $\theta = \pi/2$ is the half of the yz–plane with $y \geq 0$.

Figure 2 shows the rectangular, cylindrical, and spherical coordinates of a point P. Thus, $P = (x, y, z)$ or (r, θ, z) or (ρ, φ, θ). The relations among the three coordinate systems in the following table can be read from Fig. 2 whenever you need them. Figure 2 and the table help us change coordinate systems.

rectangular coordinates (x, y, z)		cylindrical coordinates (r, θ, z)	spherical coordinates (ρ, φ, θ)
x	$=$	$r\cos\theta$	$= \rho\sin\varphi\cos\theta$
y	$=$	$r\sin\theta$	$= \rho\sin\varphi\sin\theta$
z	$=$	z	$= \rho\cos\varphi$
$x^2 + y^2 + z^2$	$=$	$r^2 + z^2$	$= \rho^2$
$\sqrt{x^2 + y^2}$	$=$	r	$= \rho\sin\varphi$
		r/z	$= \tan\varphi$
		θ	$= \theta$

EXAMPLE 1. If P in Fig. 2 has rectangular coordinates $(x, y, z) = (1, 1, 1)$, then

$$r = \sqrt{2}, \ \rho = \sqrt{3}, \ \theta = \pi/4, \quad \text{and} \quad \varphi = \arctan\sqrt{2} \approx 54.7°.$$

Consequently, P has cylindrical coordinates $(r, \theta, z) = (\sqrt{2}, \pi/4, 1)$ and spherical coordinates $(\rho, \varphi, \theta) = (\sqrt{3}, \arctan\sqrt{2}, \pi/4)$. □

EXAMPLE 2. If the point P in Fig. 2 has spherical coordinates $(\rho, \varphi, \theta) = (4, \pi/6, \pi/4)$, then

$$r = \rho \sin\varphi = 2, \; z = \rho\cos\varphi = 2\sqrt{3}, \quad \text{and} \quad x = y = r/\sqrt{2} = \sqrt{2}.$$

So P has cylindrical coordinates $(r, \theta, z) = (2, \pi/4, 2\sqrt{3})$ and rectangular coordinates $(x, y, z) = (\sqrt{2}, \sqrt{2}, 2\sqrt{3})$. □

The angles φ and θ are closely related to latitude and longitude on the earth. If the sphere in Fig. 1 is the surface of the earth with the xy–plane as the equatorial plane and Greenwich, England, in the xz–plane, then

$$\text{north latitude} = 90 - \varphi \text{ degrees,}$$
$$\text{west longitude} = 360 - \theta \text{ degrees,}$$

where north means north of the equator and west means west of Greenwich.

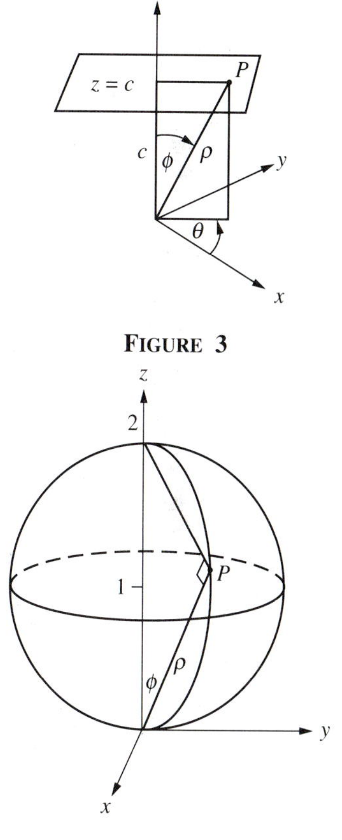

FIGURE 3

FIGURE 4

EXAMPLE 3. Express the horizontal plane $z = c$ in spherical coordinates.

Solution. The case with $c > 0$ is illustrated in Fig. 3. The point $P = (\rho, \varphi, \theta)$ lies in the plane $z = c$ if

$$\rho = c \sec\varphi.$$

This is the equation for the plane $z = c$ in spherical coordinates. The same equation is obtained by substituting $z = \rho\cos\varphi$ from Fig. 2 or the table into $z = c$ and solving for ρ. □

EXAMPLE 4. Express the sphere $x^2 + y^2 + (z-1)^2 = 1$ in spherical coordinates.

Solution. From Fig. 4, $P = (\rho, \varphi, \theta)$ lies on the sphere if

$$\rho = 2\cos\varphi.$$

This is the equation for the given sphere in spherical coordinates. As in Ex. 3, there is no restriction on θ, which could have been expected from the symmetry of the sphere about the z–axis. An algebraic derivation for the same equation is

$$x^2 + y^2 + (z-1)^2 = 1 \quad \Leftrightarrow \quad x^2 + y^2 + z^2 = 2z$$
$$\Leftrightarrow \quad \rho^2 = 2\rho\cos\varphi \quad \Leftrightarrow \quad \rho = 2\cos\varphi. \; \square$$

EXAMPLE 5. Describe the surface $\rho = 2\sin\varphi$.

Solution. Since there is no restriction on θ, the surface $\rho = 2\sin\varphi$ is a surface of revolution about the z–axis. First, let's find the trace of the surface in the half-plane $\theta = 0$, where $y = 0$ and $x \geq 0$. Thus, let $P = (x, 0, z)$ in rectangular coordinates with $x \geq 0$ and $P = (\rho, \varphi, 0)$ in spherical coordinates. From the table,

$$\rho = 2\sin\varphi \quad \Leftrightarrow \quad \rho^2 = 2\rho\sin\varphi \quad \Leftrightarrow \quad x^2 + z^2 = 2x \quad \Leftrightarrow \quad (x-1)^2 + z^2 = 1.$$

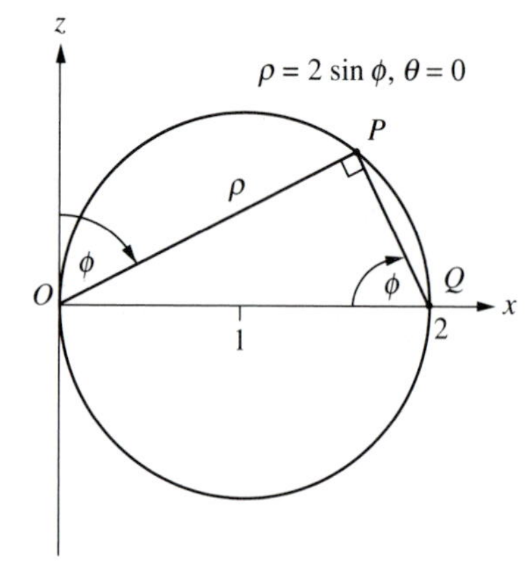

FIGURE 5

Therefore, P lies on the circle with center $(x, y, z) = (1, 0, 0)$ and radius 1 shown in Fig. 5. Since $\angle OPQ$ is a right angle, the two angles labeled φ are equal and the equation $\rho = 2\sin\varphi$ can be read from the figure. The graph of the surface $\rho = 2\sin\varphi$ with θ unrestricted is obtained by rotating the circle in Fig. 5 about the z–axis. The surface looks like a doughnut without a hole. □

Triple Integrals in Spherical Coordinates

Since the story for triple integrals in spherical coordinates is quite similar to the corresponding account for triple integrals in cylindrical coordinates, we shall be rather concise.

The building blocks for expressing triple integrals in spherical coordinates are **spherical blocks**. Such a spherical block D, shown in Fig. 6, consists of the points (ρ, φ, θ) with

$$0 \le a \le \rho \le b, \qquad 0 \le c \le \varphi \le d, \qquad p \le \theta \le q.$$

As before, $q \le p + 2\pi$. A solid sphere $\rho \le c$ is a special case of a spherical block. You will meet other examples as we go along.

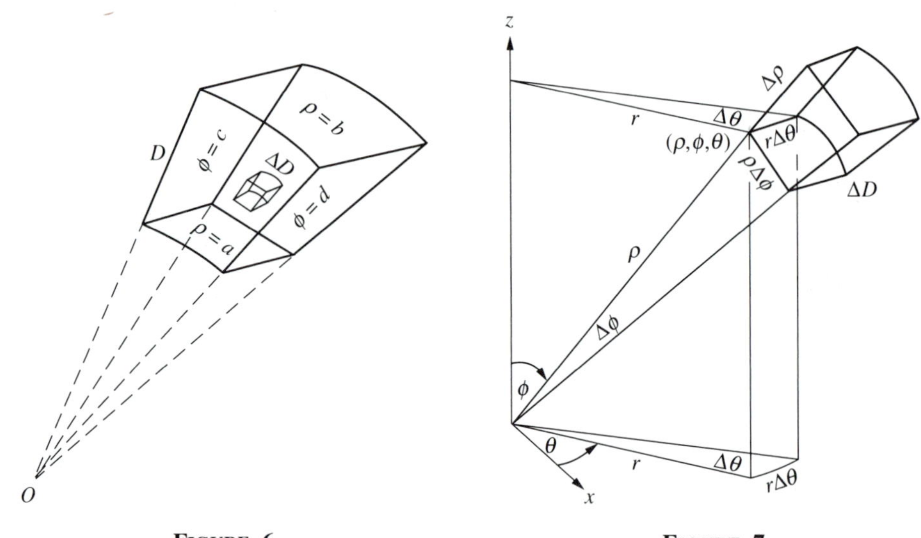

FIGURE 6 FIGURE 7

Imagine that D is partitioned into many small spherical blocks ΔD. A typical block ΔD is shown in Fig. 6 and enlarged in Fig. 7. If the dimensions $\Delta\rho$, $\rho\Delta\varphi$, and $r\Delta\theta = \rho\sin\varphi\Delta\theta$ in Fig. 7 are very small, then ΔD is nearly a rectangular box with volume ΔV given approximately by

$$\Delta V \approx \Delta\rho \cdot \rho\Delta\varphi \cdot \rho\sin\varphi\Delta\theta = \rho^2 \sin\varphi\Delta\rho\Delta\varphi\Delta\theta.$$

Let $f(\rho, \varphi, \theta)$ be a continuous function defined on D. Then

$$\iiint_D f dV = \lim \Sigma\,\Sigma\,\Sigma\, f(\rho, \varphi, \theta)\,\Delta V = \lim \Sigma\,\Sigma\,\Sigma\, f(\rho, \varphi, \theta)\,\rho^2 \sin\varphi\Delta\rho\Delta\varphi\Delta\theta.$$

This limit of Riemann sums can be expressed as an iterated integral in any order. In particular,

$$\iiint_D f dV = \int_p^q \int_c^d \int_a^b f(\rho, \varphi, \theta)\, \rho^2 \sin\varphi\, d\rho\, d\varphi\, d\theta.$$

The volume element dV is expressed in spherical coordinates by

$$dV = \rho^2 \sin\varphi\, d\rho\, d\varphi\, d\theta.$$

If $f(\rho, \varphi, \theta) = g(\rho)h(\varphi)k(\theta)$, then the foregoing triple integral of f can be expressed as a product of single integrals:

$$\iiint_D f dV = \int_a^b g(\rho)\rho^2 d\rho \cdot \int_c^d h(\varphi)\sin\varphi d\varphi \cdot \int_p^q k(\theta) d\theta.$$

EXAMPLE 6. Let D be the solid sphere $\rho \le a$ with radius a. Use spherical coordinates to verify that the volume of D is $V = \frac{4}{3}\pi a^3$.

Solution. The top half of D is the spherical block D^+ given by

$$0 \le \rho \le a, \qquad 0 \le \varphi \le \pi/2, \qquad 0 \le \theta \le 2\pi.$$

By symmetry,

$$\begin{aligned} V = 2\iiint_{D^+} 1 dV &= \int_0^a \int_0^{\pi/2} \int_0^{2\pi} \rho^2 \sin\varphi\, d\theta\, d\varphi\, d\rho \\ &= 2\int_0^a \rho^2 d\rho \cdot \int_0^{\pi/2} \sin\varphi\, d\varphi \cdot \int_0^{2\pi} d\theta \\ &= 2 \cdot \frac{1}{3}a^3 \cdot 1 \cdot 2\pi = \frac{4}{3}\pi a^3. \ \square \end{aligned}$$

EXAMPLE 7. Let D be the solid region inside the sphere $x^2 + y^2 + z^2 = 9$ and inside the top half of the cone $z^2 = \frac{1}{3}(x^2 + y^2)$. See Fig. 8. If D has density $\sigma = \sqrt{x^2 + y^2 + z^2}$, find its mass M. Use spherical coordinates (ρ, φ, θ).

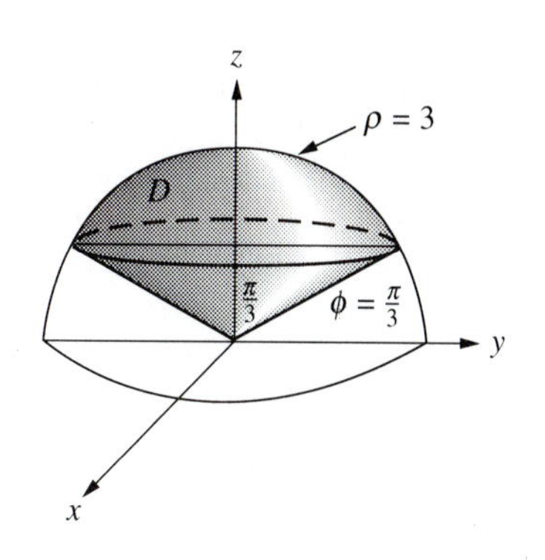

FIGURE 8

Solution. The sphere is given by $\rho = 3$ in spherical coordinates. Since $r^2 = x^2 + y^2$, the half-cone is expressed in cylindrical coordinates by $r = z\sqrt{3}$. Since $\tan\varphi = r/z$ (see Fig. 2), the half-cone is represented also by $\tan\varphi = \sqrt{3}$

and $\varphi = \pi/3$. Thus, the half-cone is expressed in spherical coordinates by $\varphi = \pi/3$. Consequently, the solid region D, shown in Fig. 8, is the spherical block

$$0 \le \rho \le 3, \qquad 0 \le \varphi \le \pi/3, \qquad 0 \le \theta \le 2\pi.$$

Since the density is $\sigma = \sqrt{x^2 + y^2 + z^2} = \rho$, the mass of D is

$$M = \iiint_D \sigma\, dV = \int_0^3 \int_0^{\pi/3} \int_0^{2\pi} \rho\rho^2 \sin\varphi\, d\theta\, d\varphi\, d\rho$$

$$= \int_0^3 \rho^3 d\rho \cdot \int_0^{\pi/3} \sin\varphi\, d\varphi \cdot \int_0^{2\pi} d\theta$$

$$= \frac{81}{4} \cdot \frac{1}{2} \cdot 2\pi = \frac{81\pi}{4} \approx 63.6. \;\square$$

EXAMPLE 8. Evaluate $\iiint_D e^{-\rho^3} dV$, where D is all of 3–space.

Solution. This is an improper integral. It is defined by

$$\iiint_D e^{-\rho^3} dV = \lim_{c \to \infty} \iiint_{D_c} e^{-\rho^3} dV,$$

where D_c is the solid sphere $\rho \le c$. Now,

$$\iiint_{D_c} e^{-\rho^3} dV = \int_0^{2\pi} \int_0^{\pi} \int_0^{c} e^{-\rho^3} \rho^2 \sin\varphi\, d\rho\, d\varphi\, d\theta$$

$$= \int_0^{2\pi} d\theta \cdot \int_0^{\pi} \sin\varphi\, d\varphi \cdot \int_0^{c} e^{-\rho^3} \rho^2 d\rho$$

$$= 4\pi \left[-\frac{1}{3} e^{-\rho^3} \right]_0^c = \frac{4\pi}{3}(1 - e^{-c^3}).$$

Let $c \to \infty$ to obtain

$$\iiint_D e^{-\rho^3} dV = \frac{4\pi}{3}. \;\square$$

Integration Over Spherically Simple Regions

We frequently meet triple integrals over regions that are more general than spherical blocks. Often these regions are spherically simple. Then the triple integral can be evaluated by iterated integration.

A region D in 3–space is **spherically simple** if it consists of the points (ρ, φ, θ) with

$$c \le \varphi \le d, \qquad p \le \theta \le q, \qquad \text{and} \qquad \rho_1(\varphi, \theta) \le \rho \le \rho_2(\varphi, \theta),$$

where $\rho_1(\varphi, \theta)$ and $\rho_2(\varphi, \theta)$ are continuous and c, d, p, q are restricted as before. See Fig. 9. Any spherical block is spherically simple. Let $f(\rho, \varphi, \theta)$ be a continuous function on D. Then

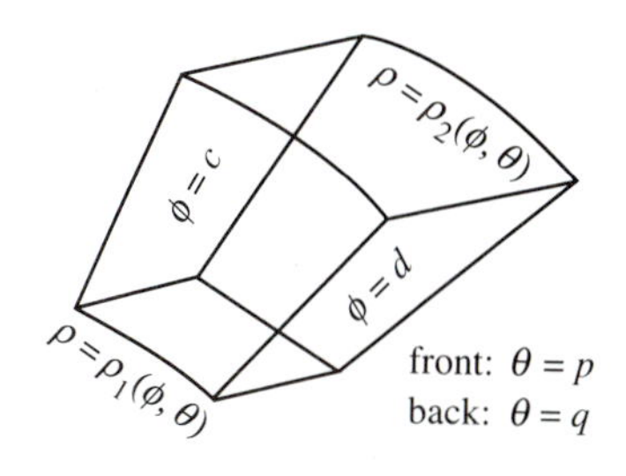

FIGURE 9

$$\iiint_D f dV = \int_p^q \int_c^d \int_{\rho_1(\varphi,\theta)}^{\rho_2(\varphi,\theta)} f(\rho, \varphi, \theta)\, \rho^2 \sin\varphi \, d\rho \, d\varphi \, d\theta.$$

The fact that the volume of a right circular cone with radius of base r and height h is $V = \frac{1}{3}\pi r^2 h$ is often established by the method of cross sections in one–variable calculus. The next problem uses spherical coordinates to confirm this result in a special case. The general case is left to the problems.

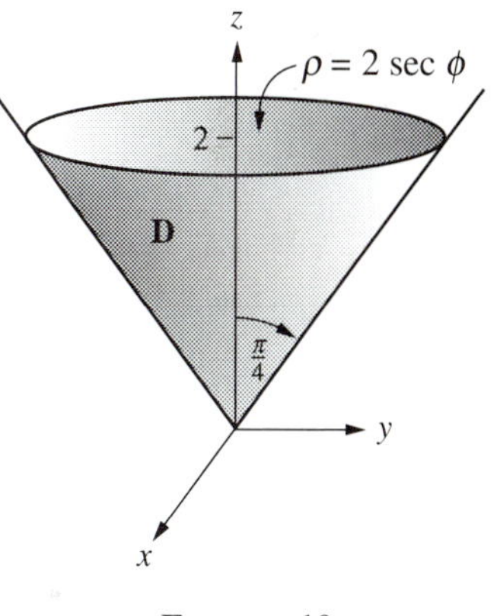

FIGURE 10

EXAMPLE 9. Find the volume of the solid cone D in Fig. 10. It is bounded by the plane $z = 2$ and the conical surface $z^2 = x^2 + y^2$.

Solution. In spherical coordinates, the plane is expressed by $\rho = 2\sec\varphi$ and the conical surface by $\varphi = \pi/4$. So D is the spherically simple region given by

$$0 \le \varphi \le \pi/4, \quad 0 \le \theta \le 2\pi, \quad \text{and} \quad 0 \le \rho \le 2\sec\varphi.$$

The volume of D is

$$V = \iiint_D 1 dV = \int_0^{2\pi} \int_0^{\pi/4} \int_0^{2\sec\varphi} \rho^2 \sin\varphi \, d\rho \, d\varphi \, d\theta$$

$$= 2\pi \int_0^{\pi/4} \left[\frac{1}{3}\rho^3\right]_{\rho=0}^{\rho=2\sec\varphi} \sin\varphi \, d\varphi = \frac{16\pi}{3} \int_0^{\pi/4} (\sec\varphi)^3 \sin\varphi \, d\varphi$$

$$= \frac{16\pi}{3} \int_0^{\pi/4} (\cos\varphi)^{-3} \sin\varphi \, d\varphi = \frac{16\pi}{3} \left[\frac{1}{2}(\cos\varphi)^{-2}\right]_0^{\pi/4} = \frac{8\pi}{3}.$$

Since the cone in Fig. 10 has height $h = 2$ and radius of base $r = 2$, the general formula $V = \frac{1}{3}\pi r^2 h$ gives the same value $V = 8\pi/3$ for the volume. □

EXAMPLE 10. Let D be the solid inside the surface $\rho = 2\sin\varphi$. This is the doughnut without a hole in Ex. 5. Given that D has density $\sigma = |z|$, find its mass.

Solution. By symmetry,

$$M = \iiint_D \sigma dV = 2\iiint_{D^+} z dV = 2\iiint_{D^+} \rho\cos\varphi dV,$$

where D^+ is the top half of D. From Fig. 5, D^+ is spherically simple with

$$0 \le \varphi \le \pi/2, \quad 0 \le \theta \le 2\pi, \quad \text{and} \quad 0 \le \rho \le 2\sin\varphi.$$

It follows that

$$\begin{aligned} M &= 2\int_0^{2\pi}\int_0^{\pi/2}\int_0^{2\sin\varphi} (\rho\cos\varphi)\,\rho^2\sin\varphi d\rho d\varphi d\theta \\ &= 4\pi\int_0^{\pi/2}\left[\frac{1}{4}\rho^4\right]_{\rho=0}^{\rho=2\sin\varphi} \sin\varphi\cos\varphi d\varphi = 16\pi\int_0^{\pi/2} \sin^5\varphi\cos\varphi d\varphi \\ &= 16\pi\left[\frac{1}{6}\sin^6\varphi\right]_0^{\pi/2} = \frac{8\pi}{3}. \ \square \end{aligned}$$

Newton's Law of Gravitation Revisited*

We close this section with another fundamental discovery of Isaac Newton.

> A homogeneous spherical body attracts an external particle as if the entire mass of the body were concentrated at its center.

This physical principle played an important role in Newton's reasoning that the gravitational forces near the Earth and those acting between heavenly bodies are essentially the same. See Sec. 13.5. The analogous result holds for a homogeneous charge distribution because charge distributions are governed by an inverse square law of the same form as the gravitational law.

We shall sketch a proof of Newton's discovery. To begin with, consider two point masses situated at $P = (x, y, z)$ and $P_0 = (x_0, y_0, z_0)$. The corresponding position vectors are $\mathbf{r} = x\mathbf{j} + y\mathbf{k} + z\mathbf{k}$ and $\mathbf{r}_0 = x_0\mathbf{i} + y_0\mathbf{j} + z_0\mathbf{k}$. According to Newton's law of gravitation, M exerts a force on m given by

$$\mathbf{F} = GmM\frac{\mathbf{r} - \mathbf{r}_0}{u^3}, \qquad u = \|\mathbf{r} - \mathbf{r}_0\|,$$

where G is the universal constant of gravitation. Now replace M by a solid body D with density $\sigma(\mathbf{r})$. By a Riemann sum and limit passage argument, D exerts a force on m given by

$$\mathbf{F} = Gm\iiint_D \frac{\mathbf{r} - \mathbf{r}_0}{u^3}\sigma(\mathbf{r})\,dV, \qquad u = \|\mathbf{r} - \mathbf{r}_0\|.$$

The vector integral is defined component by component:

$$\iiint_D \frac{\mathbf{r}-\mathbf{r}_0}{u^3}\,dV = \left(\iiint_D \frac{x-x_0}{u^3}\,dV\right)\mathbf{i} + \left(\iiint_D \frac{y-y_0}{u^3}\,dV\right)\mathbf{j} + \left(\iiint_D \frac{z-z_0}{u^3}\,dV\right)\mathbf{k}.$$

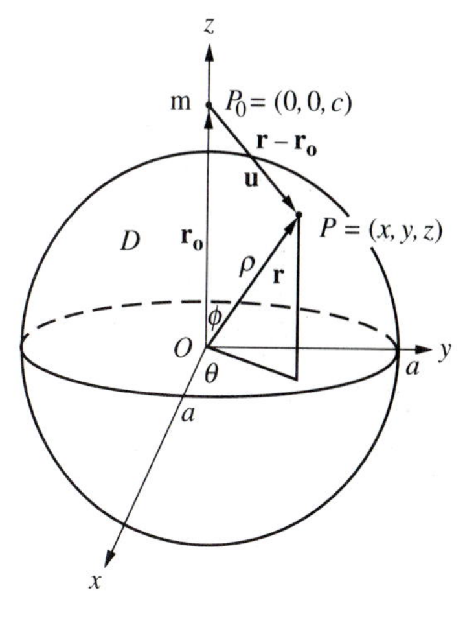

FIGURE 11

With this preparation, we are ready to address Newton's discovery. Figure 11 shows a solid sphere D with radius a and constant density σ, and a point mass m outside of D. Rectangular coordinates are set up with origin at the center O of D and so that m lies on the positive z–axis at $P_0 = (0, 0, c)$. The volume and mass of the spherical body are $V = \frac{4}{3}\pi a^3$ and $M = \frac{4}{3}\pi a^3 \sigma$. Let $P = (x, y, z)$ be any point in D. The position vectors for P and P_0 are $\mathbf{r} = x\mathbf{i} + y\mathbf{j} + z\mathbf{k}$ and $\mathbf{r}_0 = c\mathbf{k}$. Then

$$\mathbf{r} - \mathbf{r}_0 = x\,\mathbf{i} + y\mathbf{j} + (z - c)\,\mathbf{k}, \qquad u = \|\mathbf{r} - \mathbf{r}_0\| = [x^2 + y^2 + (z - c)^2]^{1/2}.$$

From the general results derived previously, the solid sphere D exerts a force on m given by

$$\mathbf{F} = Gm\sigma \iiint_D \frac{\mathbf{r} - \mathbf{r}_0}{u^3}\,dV,$$

where

$$\iiint_D \frac{\mathbf{r} - \mathbf{r}_0}{u^3}\,dV = \left(\iiint_D \frac{x}{u^3}\,dV\right)\mathbf{i} + \left(\iiint_D \frac{y}{u^3}\,dV\right)\mathbf{j} + \left(\iiint_D \frac{z - c}{u^3}\,dV\right)\mathbf{k}.$$

Since D is symmetric about all three coordinate planes and the quantities x/u^3 and y/u^3 take opposite values at symmetric points in x and y, respectively,

$$\iiint_D \frac{x}{u^3}\,dV = 0 \quad \text{and} \quad \iiint_D \frac{y}{u^3}\,dV = 0.$$

Therefore,

$$\mathbf{F} = GM\sigma\left(\iiint_D \frac{z - c}{u^3}\,dV\right)\mathbf{k}.$$

We shall use spherical coordinates to evaluate the integral on the right. First, we express u in spherical coordinates. From Fig. 11 and the law of cosines,

$$u = (\rho^2 + c^2 - 2c\,\rho\cos\varphi)^{1/2}.$$

Since $z = \rho \cos \varphi$,

$$\iiint_D \frac{z - c}{u^3} dV = \int_0^a \int_0^\pi \int_0^{2\pi} \frac{\rho \cos \varphi - c}{u^3} \rho^2 \sin \varphi d\theta d\varphi d\rho$$

$$= 2\pi \int_0^a \rho^2 \int_0^\pi \frac{\rho \cos \varphi - c}{u^3} \sin \varphi d\varphi d\rho.$$

Our first task is to evaluate the inner integral,

$$\int_0^\pi \frac{\rho \cos \varphi - c}{u^3} \sin \varphi d\varphi.$$

In this integral, ρ is held constant and u depends on φ. It turns out that the change of the variable of integration from φ to u greatly simplifies the integral. This change is a u–substitution. Since $u^2 = \rho^2 + c^2 - 2c\rho \cos \varphi$, we have

$$\rho \cos \varphi = \frac{1}{2c}(\rho^2 + c^2 - u^2), \qquad \rho \cos \varphi - c = \frac{1}{2c}(\rho^2 - c^2 - u^2),$$

$$2udu = 2c\rho \sin \varphi d\varphi, \qquad \sin \varphi d\varphi = \frac{u}{c\rho} du.$$

This takes care of the integrand in the preceding integral. Now for the limits of integration. From Fig. 11 and the fact that ρ is fixed for the moment, $0 \le \varphi \le \pi \quad \Leftrightarrow \quad c - \rho \le u \le c + \rho$. Therefore,

$$\int_0^\pi \frac{\rho \cos \varphi - c}{u^3} \sin \varphi d\varphi = \frac{1}{2c^2\rho} \int_{c-\rho}^{c+\rho} \frac{\rho^2 - c^2 - u^2}{u^2} du$$

$$= \frac{1}{2c^2\rho} \int_{c-\rho}^{c+\rho} \left(\frac{\rho^2 - c^2}{u^2} - 1 \right) du$$

$$= \frac{1}{2c^2\rho} \left[\frac{c^2 - \rho^2}{u} - u \right]_{c-\rho}^{c+\rho} = -\frac{2}{c^2}.$$

It follows that

$$\iiint_D \frac{z - c}{u^3} du = -\frac{4\pi}{c^2} \int_0^a \rho^2 d\rho = -\frac{4\pi a^3}{3c^2},$$

$$\mathbf{F} = Gm\sigma \left(-\frac{4\pi a^3}{3c^2} \right) \mathbf{k}.$$

Since $M = \frac{4}{3}\pi a^3 \sigma$, we finally arrive at

$$\mathbf{F} = -\frac{GMm}{c^2} \mathbf{k}.$$

This is the same force that would be exerted on the mass m if the entire mass M of the solid sphere were concentrated at its center. We have confirmed Newton's great discovery.

PROBLEMS

In Probs. 1–4, find the cylindrical coordinates (r, θ, z) and the rectangular coordinates (x, y, z) of the point with the given spherical coordinates (ρ, φ, θ).

1. $(2, \pi/4, \pi/6)$
2. $(2, \pi/2, 3\pi/2)$
3. $(\sqrt{3}, \pi/3, -\pi/6)$
4. $(4, 5\pi/6, 3\pi/4)$

In Probs. 5–8, find the cylindrical coordinates (r, θ, z) and spherical coordinates (ρ, φ, θ) of the point with the given rectangular coordinates (x, y, z).

5. $(1, 0, 0)$
6. $(1, 0, -1)$
7. $(2\sqrt{3}, 2, 4)$
8. $(2\sqrt{3}, -2, -4\sqrt{3})$

In Probs. 9–12, find the rectangular coordinates (x, y, z) and spherical coordinates (ρ, φ, θ) of the point with the given cylindrical coordinates (r, θ, z).

9. $(1, 3\pi/4, 1)$
10. $(1, -\pi, -1)$
11. $(4, -\pi/6, 4\sqrt{3})$
12. $(\sqrt{2}, \pi/6, -\sqrt{2})$

In Probs. 13–20, identify the graph of the spherical coordinate equation.

13. $\rho = 4 \sec \theta$
14. $\rho \cos \theta = -3$
15. $\rho = 5$
16. $\theta = 3\pi/4$
17. $\varphi = \pi/6$
18. $\varphi = \pi/2$
19. $\rho = 2 \cos \varphi$
20. $\rho \sin \varphi \cos \theta = 7$

In Probs. 21–26, identify the graph of the rectangular coordinate equation. Then express the equation in terms of (a) cylindrical and (b) spherical coordinates.

21. $x^2 + y^2 + z^2 = 4$
22. $x^2 + y^2 = 5$
23. $z^2 = 3(x^2 + y^2)$
24. $x^2 + y^2 = 2x$
25. $x^2 + y^2 + (z - 2)^2 = 4$
26. $x^2 + (y - 1)^2 + z^2 = 1$

In Probs. 27–30 express the triple integral as an iterated integral with three ordinary integrals using (a) rectangular, (b) cylindrical, and (c) spherical coordinates.

27. $\iiint_D (x^2 + y^2)\, dV$, where D is the ball $x^2 + y^2 + z^2 \leq 4$.
28. $\iiint_D x^2 dV$, where D is the region bounded by the cylinder $x^2 + y^2 = a^2$ and the planes $z = 0$ and $z = h$.
29. $\iiint_D z dV$, where D is the region in the first octant under the sphere

$x^2 + y^2 + z^2 = 36$, and inside the cylinder $x^2 + y^2 = 6x$. *Hint.* For (c) show first that the cylinder is given by $\rho \sin \varphi = 6 \cos \theta$ and that the sphere and cylinder intersect when $\varphi = (\pi/2) - \theta$.

30. $\iiint_D (x^2 + y^2 + z^2)\, dV$, where D is the region under the sphere $x^2 + y^2 + z^2 = a^2$ and above the plane $z = b$, where $0 \le b \le a$.

In Probs. 31–34, use spherical coordinates to evaluate the integral.

31. $\iiint_D (x^2 + y^2 + z^2)\, dV$, where D is the closed unit ball $x^2 + y^2 + z^2 \le 1$.
32. $\iiint_D x^2\, dV$, where D is the part of the ball $x^2 + y^2 + z^2 \le 16$ that lies in the first octant. (See Ex. 7 in Sec. 7.2.)
33. $\iiint_D xy\, dV$, where D is the region in the first octant in common to the ball $\rho \le 3$ and the solid half-cone $0 \le \varphi \le \pi/4$.
34. $\iiint_D e^{(x^2 + y^2 + z^2)^{3/2}}\, dV$, where D is the region in common to the ball $\rho \le 1$ and the solid half-cone $0 \le \varphi \le \pi/4$.

In Probs. 35–47 use any convenient coordinate system to evaluate the integrals that arise. If a proportionality constant is needed, call it k. You may wish to set up the problems in more than one coordinate system. This will help reinforce the advantage of a good choice of coordinates.

35. Find the mass of the ball $x^2 + y^2 + z^2 \le 4$ if the density is $\sigma(x, y, z) = |z|$.
36. A solid hemisphere has radius a and its density is proportional to the distance from its circular base. Find its mass.
37. A wedge-shaped region in the first octant is cut out of the ball $x^2 + y^2 + z^2 \le 9$ by the half-planes $\theta = \pi/6$ and $\theta = \pi/3$. Find the volume of the wedge.
38. Find the volume of the region bounded by the plane $z = 4$ and the cone $z^2 = x^2 + y^2$.
39. Find the volume of a solid right circular cone of height a and radius of base a. *Hint.* Choose coordinates so that the vertex of the cone is at the origin and the axis of the cone is along the positive z–axis.
40. Find the volume of the smaller region cut from the ball $x^2 + y^2 + z^2 = 4$ by the plane $z = 1$.
41. A spherical cap is the smaller region cut from a solid sphere by a plane. If the sphere has radius a and the (shortest) distance from the plane to the center of the sphere is $b < a$, find the volume of the cap.
42. The density of a right circular cone with height h and radius of base a is proportional to the distance from (a) its vertex (b) its base. Find the mass of the cone.
43. The density of the spherical cap in Prob. 41 is proportional to the distance from the center of the sphere. Find the mass of the cap if $a = 2$ and $b = 1$.
44. The region in common to the ball $\rho \le 2a \cos \varphi$ and the cone $0 \le \varphi \le \alpha$, where $a > 0$ and $0 < \alpha \le \pi/2$ are constants, has density proportional to the distance above the xy–plane. Find its mass.

45. Find the volume in common to the balls $\rho \leq a$ and $\rho \leq 2a\cos\varphi$.

46. Find the mass of the ball $\rho \leq a$ if the density is proportional to the distance to the surface of the ball.

47. Find the volume of the doughnut without a hole in Ex. 5.

48. In Fig. 7 regard (ρ, φ, θ) as *fixed.* Then the points $(\tilde{\rho}, \tilde{\varphi}, \tilde{\theta})$ of the spherical coordinate block ΔD satisfy

$$\rho \leq \tilde{\rho} \leq \rho + \Delta\rho, \qquad \varphi \leq \tilde{\varphi} \leq \varphi + \Delta\varphi, \qquad \theta \leq \tilde{\theta} \leq \theta + \Delta\theta.$$

(a) Use integration to show that the volume of ΔD is

$$\Delta V = \frac{1}{3}[(\rho + \Delta\rho)^3 - \rho^3] \cdot [\cos(\varphi + \Delta\varphi) - \cos\varphi] \cdot \Delta\theta.$$

(b) Show that

$$\lim \frac{\Delta V}{\rho^2 \sin\varphi\Delta\rho\Delta\varphi\Delta\theta} = 1,$$

where the limit is taken as $\Delta\rho$, $\Delta\varphi$, $\Delta\theta \to 0$. This conclusion serves as a comforting check on the assertion made in the text that $\Delta V \approx \rho^2 \sin\varphi\Delta$-$\rho\Delta\varphi\Delta\theta$ for small $\Delta\rho$, $\Delta\varphi$, and $\Delta\theta$.

Probs. 49–53 are about gravitational attraction. You may wish to review the corresponding discussion in the text before solving them.

49. The ice cream cone $0 \leq \varphi \leq \alpha$, $0 \leq \rho \leq a$, where $0 < \alpha \leq \pi/2$ and $a > 0$, has constant density σ_0. Find the gravitational force exerted by the cone on a point mass m placed at its vertex.

50. Replace the ice cream cone in the previous problem by the cone $0 \leq \varphi \leq \alpha$, $0 \leq \rho \leq a \sec\varphi$ with a flat top. Find the gravitational force exerted by the cone for a point mass m placed at its vertex.

51. Find the gravitational attraction of the top half of the ball $x^2 + y^2 + (z - a)^2 = a^2$ for a point mass m at the origin. Assume the ball has constant density σ_0.

52. Extend Newton's discovery about homogeneous spherical bodies to spherical bodies with radially symmetric (continuous) mass distributions $\sigma = \sigma(\rho)$. *Hint.* The reasoning is almost the same as for the case of a constant density.

53. A homogeneous ring of matter lies between the spheres $\rho = a$ and $\rho = b$ and a point mass m has rectangular coordinates $(0, 0, c)$ where $c < a < b$. Show that the gravitational force $\mathbf{F}$ of the homogeneous ring on the point mass m is $\mathbf{F} = \mathbf{0}$. (Notice that m can be located at any point inside the inner sphere. We just chose coordinates so that m lies on the z–axis.)

7.5 Further Applications of Double and Triple Integrals

Double and triple integrals occur in many applications, especially in motion problems for three–dimensional bodies. Such integrals help explain the balancing of a seesaw, the behavior of levers, the flight of a jet plane, and the launch of a space shuttle. We can only scratch the surface of this interesting and important circle of ideas. The first step toward analyzing the motion of real–world objects is to extend basic physical laws of force, equilibrium, and motion from idealized point masses to solid bodies which have physical extent. These extended laws are expressed in terms of quantities defined by double and triple integrals.

We shall restrict our attention to three main topics that are fundamental for analyzing the physical behavior of three–dimensional objects. First, centroids of plane or solid regions are introduced. As the name suggests, a centroid is a kind of center. For a disk or ball, the centroid is the center. The second topic concerns centers of mass, which are similar to centroids, but take mass distribution into account. The final theme, which is optional, concerns rotational properties of solid bodies. We shall formulate rotational analogues of Newton's second law. In the process, torque, angular momentum, and moments of inertia play key roles.

The topics just mentioned could easily fill a chapter or even an entire book. To keep the section down to a manageable size, we present many of the results with only brief indications of their derivations. Some of the details are given in the problems. Throughout the section we deal only with plane regions and solid bodies of the types described in Sec. 7.1 and Sec. 7.2. In particular, such regions have definite areas or volumes, depending on the dimension. All the functions that appear are assumed to be integrable, although this assumption will not be mentioned again.

Centroids

The centroid of a plane or solid region can be regarded as its geometric center. The coordinates of the centroid are average values of the coordinates of points in the region. So the discussion of centroids will be preceded by a brief introduction to average values, also known as mean values. We shall confine the consideration of mean values to their role in clarifying the idea of a centroid.

Mean Values

In one–variable calculus, the mean (or average) value of a function $f(x)$, $a \le x \le b$, is defined by

$$\overline{f} = \frac{1}{b-a}\int_a^b f(x)\,dx.$$

Mean values of functions of two or three variables are defined by similar formulas. The length $b - a$ of the interval of integration is replaced by the area or volume of the region of integration.

To begin with, suppose that $f(x,y)$ is an integrable function defined on a rectangle R with sides $a \le x \le b$, $c \le y \le d$, and area $A = (b-a)(d-c)$. The **mean** (or **average**) **value** of f on R is given by

$$\overline{f} = \frac{1}{A}\iint_R f dA.$$

The reason $\overline{f}$ is called a mean value is explained with the aid of Riemann sums. Partition R into smaller rectangles ΔR_{ij} for $i, j = 1, 2, \ldots, n$, with side lengths $\Delta x = (b-a)/n$ and $\Delta y = (d-c)/n$, as illustrated in Fig. 1. The rectangles ΔR_{ij} have equal areas $\Delta A = \Delta x \Delta y = A/n^2$. The upper right corner of ΔR_{ij} is the point (x_i, y_j) with $x_i = a + i\Delta x$ and $y_j = c + j\Delta y$.

The integral of f over R is a limit of Riemann sums:

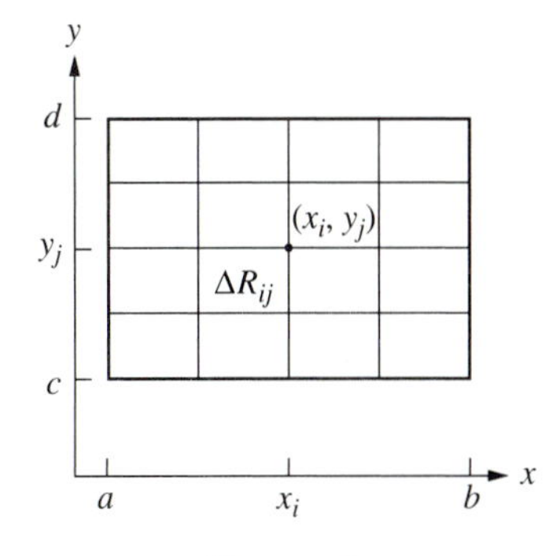

FIGURE 1

$$\iint_R f dA = \lim_{n\to\infty} \sum_{i=1}^{n} \sum_{j=1}^{n} f(x_i, y_j)\Delta A = A \lim_{n\to\infty} \frac{1}{n^2} \sum_{i=1}^{n} \sum_{j=1}^{n} f(x_i, y_j).$$

Divide by A to obtain

$$\overline{f} = \frac{1}{A} \iint_R f dA = \lim_{n\to\infty} \frac{1}{n^2} \sum_{i=1}^{n} \sum_{j=1}^{n} f(x_i, y_j).$$

Since the right member is a limit of average values of $f(x, y)$ over uniformly spaced points in R, it makes sense to call $\overline{f}$ a mean or average value of f on R.

The mean value $\overline{f}$ of a function f on a more general region R with area A is defined by the same double integral formula and can be justified in the same way, using inner Riemann sums.

The **mean** (or **average**) **value** of a function $f(x, y, z)$ on a solid region D with volume V is given by

$$\overline{f} = \frac{1}{V} \iiint_D f dV.$$

The mean value $\overline{f}$ is a limit of average values of f over uniformly spaced points in D.

Centroids for Plane Regions

The **centroid** of a plane region R with area A is the point $(\overline{x}, \overline{y})$ with coordinates

$$\overline{x} = \frac{1}{A} \iint_R x dA, \qquad \overline{y} = \frac{1}{A} \iint_R y dA.$$

Thus, $\overline{x}$ and $\overline{y}$ are the mean or average values of the functions $f(x, y) = x$ and $f(x, y) = y$ on R. In other words, $\overline{x}$ and $\overline{y}$ are the average values of the x–coordinates and the y–coordinates of points in R.

EXAMPLE 1. Let R be the disk with center $(0, 0)$ and radius a. Since R is symmetric in x and the integrand x of $\iint_R x dA$ takes on opposite values at symmetric points, $\iint_R x dA = 0$ and $\overline{x} = 0$. Similarly, $\overline{y} = 0$. Thus, the centroid of the disk R is its center $(\overline{x}, \overline{y}) = (0, 0)$. □

The centroid of a region R is a geometric property of the region. Consequently, it is unchanged if the coordinate axes are translated or rotated. Also, if

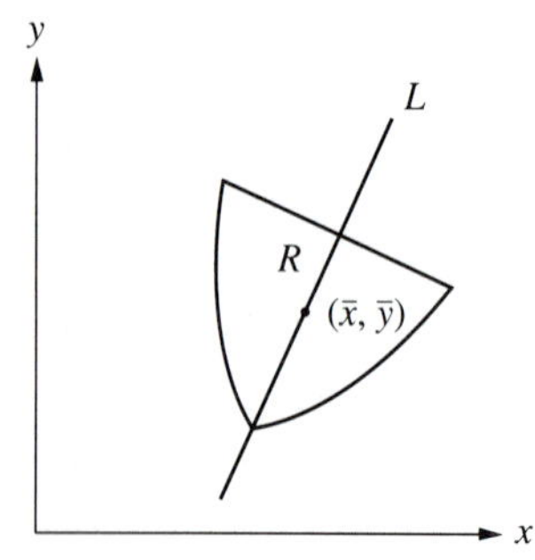

FIGURE 2

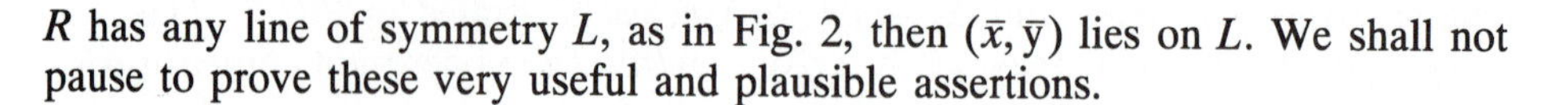
R has any line of symmetry L, as in Fig. 2, then $(\bar{x}, \bar{y})$ lies on L. We shall not pause to prove these very useful and plausible assertions.

EXAMPLE 2. Find the centroid of the top half R of the disk $(x-1)^2 + y^2 \leq 1$ in Fig. 3.

Solution. The area of R is $A = \pi/2$. Since the line $x = 1$ is a line of symmetry of R, the centroid lies on this vertical line. So $\bar{x} = 1$. It is convenient to set up a polar coordinate system with pole at (1,0). Then R is described in polar coordinates by $0 \leq \theta \leq \pi$ and $0 \leq r \leq 1$. Since $y = r \sin\theta$ and $dA = r\,dr\,d\theta$,

$$\iint_R y\,dA = \int_0^\pi \int_0^1 r^2 \sin\theta\,dr\,d\theta = \int_0^\pi \sin\theta\,d\theta \cdot \int_0^1 r^2\,dr = \frac{2}{3}.$$

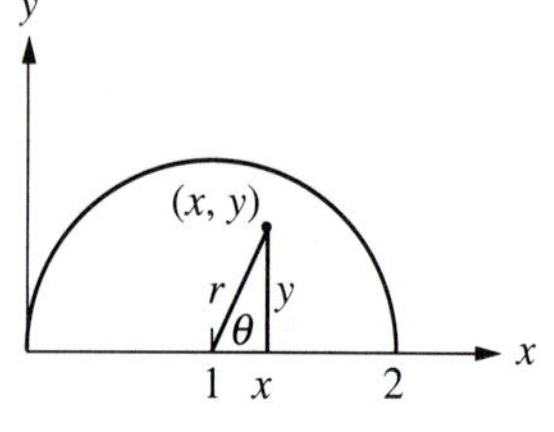

FIGURE 3

Hence,

$$\bar{y} = \frac{1}{A}\iint_R y\,dA = \frac{4}{3\pi} \approx 0.42.$$

The centroid of R is $(\bar{x}, \bar{y}) = (1, 4/3\pi)$. □

Centroids for Solid Bodies

The **centroid** of a solid body D with volume V is the point $(\bar{x}, \bar{y}, \bar{z})$ with

$$\bar{x} = \frac{1}{V}\iiint_D x\,dV, \qquad \bar{y} = \frac{1}{V}\iiint_D y\,dV, \qquad \bar{z} = \frac{1}{V}\iiint_D z\,dV.$$

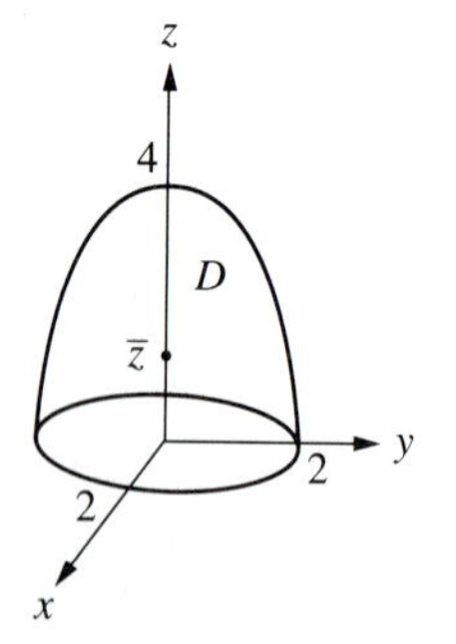

FIGURE 4

The coordinates of the centroid are the average values of x, y, z on D. The centroid lies on any plane of symmetry for D. It follows that the centroid of a ball is its center. Why?

EXAMPLE 3. Let D be the solid body bounded by the paraboloid $z = 4 - x^2 - y^2$ and the xy–plane. See Fig. 4. Find the centroid of D.

Solution. Since D is symmetric in x and y, $\bar{x} = 0$ and $\bar{y} = 0$. It remains to find $\bar{z}$. The solid D is described in cylindrical coordinates by

$$0 \leq r \leq 2, \qquad 0 \leq \theta \leq 2\pi, \qquad 0 \leq z \leq 4 - r^2.$$

Therefore,

$$\iiint_D z\,dV = \int_0^{2\pi}\int_0^2\int_0^{4-r^2} zr\,dz\,dr\,d\theta = 2\pi\int_0^2 r\left[\frac{1}{2}z^2\right]_{z=0}^{z=4-r^2} dr$$

$$= \pi\int_0^2 r(4-r^2)^2\,dr = \pi\left[-\frac{1}{6}(4-r^2)^3\right]_0^2 = \frac{32}{3}\pi.$$

From Ex. 5 in Sec. 7.3, the volume of D is $V = 8\pi$. Hence,

$$\bar{z} = \frac{1}{V}\iiint_D z\,dV = \frac{4}{3}.$$

The centroid of D is $(\bar{x}, \bar{y}, \bar{z}) = (0, 0, 4/3)$. □

Centers of Mass

As we mentioned earlier a center of mass of a plane or solid region is similar to a centroid, but it takes mass distribution into account. If the body is homogeneous, then the center of mass is the same as the centroid. Newton's second law of motion will be extended from a point mass to a solid body using center of mass. The extended law implies that the center of mass of the body moves under the influence of an external force, such as gravity, just as if its entire mass were concentrated at the center of mass. The center of mass will be defined first for a system of point masses and then extended to a solid body by a limit passage.

Center of Mass of a System of Point Masses

Let's start with a seesaw and two riders, as illustrated in Fig. 5. The two riders are regarded as points with masses m_1 and m_2. The corresponding weights are $w_1 = m_1 g$ and $w_2 = m_2 g$, which are downward forces. The weight of the seesaw is neglected. (See the problems for the case with the weight of the seesaw taken into account.) The point P in Fig. 5 is the *fulcrum* or *pivot point*. The distances of the riders from P are d_1 and d_2 in Fig. 5. Suppose that the riders just balance. Then P is the *balance point*. It is an experimental fact, attributed to Archimedes, that

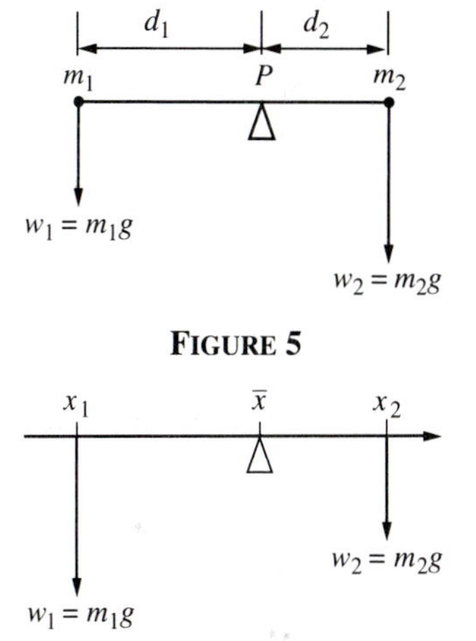

FIGURE 5

FIGURE 6

$$w_1 d_1 = w_2 d_2.$$

In modern language, $w_1 d_1$ and $w_2 d_2$ are *moments* of the forces w_1 and w_2 about P.

To calculate the balance point, we place the seesaw on the x–axis, as in Fig. 6. The riders are at x_1 and x_2. The balance point is $\bar{x}$. By Archimedes' principle,

$$w_1(\bar{x} - x_1) = w_2(x_2 - \bar{x}), \qquad (w_1 + w_2)\bar{x} = w_1 x_1 + w_2 x_2,$$

$$\bar{x} = \frac{w_1 x_1 + w_2 x_2}{w_1 + w_2}.$$

In this form, $\bar{x}$ is called the *center of gravity* of w_1 and w_2. Let $w_1 = m_1 g$ and $w_2 = m_2 g$ to obtain

$$\bar{x} = \frac{m_1 x_1 + m_2 x_2}{m_1 + m_2}.$$

In this form, $\bar{x}$ is called the *center of mass* of m_1 and m_2.

The center of mass is defined in a similar way for a system of point masses m_i located at points (x_i, y_i, z_i) for $i = 1, 2, \ldots, n$. The total mass of the system is

$$M = \sum_{i=1}^{n} m_i.$$

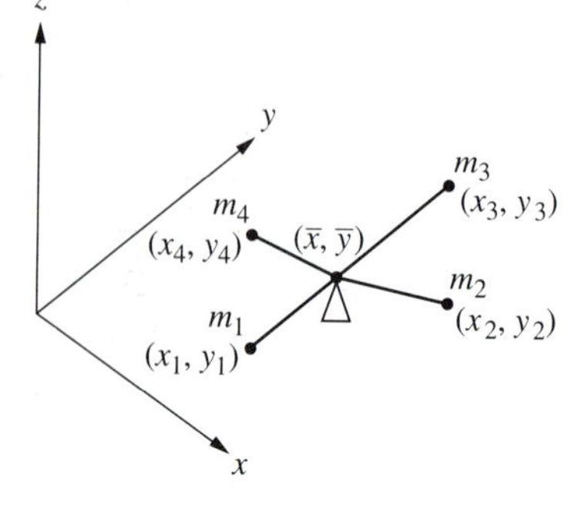

FIGURE 7

The **center of mass** of the system is the point $(\bar{x}, \bar{y}, \bar{z})$ with coordinates

$$\bar{x} = \frac{1}{M}\sum_{i=1}^{n} m_i x_i, \qquad \bar{y} = \frac{1}{M}\sum_{i=1}^{n} m_i y_i, \qquad \bar{z} = \frac{1}{M}\sum_{i=1}^{n} m_i z_i.$$

The same formulas serve for point masses on the x–axis or in the xy–plane; just ignore the unneeded coordinates. The seesaw illustrated in Fig. 6 is a special case. The plane case is illustrated in Fig. 7. Just like the seesaw, the system balances at the center of mass $(\bar{x}, \bar{y})$.

Denote the position vector for the point (x_i, y_i, z_i) by $\mathbf{r}_i = <x_i, y_i, z_i>$. Let $\mathbf{r}_{cm} = <\bar{x}, \bar{y}, \bar{z}>$ be the position vector for the center of mass. Then

$$\mathbf{r}_{cm} = \frac{1}{M}\sum_{i=1}^{n} m_i \mathbf{r}_i.$$

We shall refer to either $(\bar{x}, \bar{y}, \bar{z})$ or $\mathbf{r}_{cm} = <\bar{x}, \bar{y}, \bar{z}>$ as the center of mass.

Center of Mass of a Plane Region

Let R be a region in the xy–plane with density function $\sigma(x, y)$ in terms of mass per unit area. As we pointed out in Sec. 7.2, R can be regarded as an idealization of a thin plate. The **center of mass** of R is the point $(\bar{x}, \bar{y})$ with coordinates

$$\boxed{\bar{x} = \frac{1}{M}\iint_R \sigma x \, dA, \qquad \bar{y} = \frac{1}{M}\iint_R \sigma y \, dA.}$$

Let $\mathbf{r} = <x, y>$ and $\mathbf{r}_{cm} = <\bar{x}, \bar{y}>$. Then the center of mass is expressed in vector form by

$$\boxed{\mathbf{r}_{cm} = \frac{1}{M}\iint_R \sigma(\mathbf{r}) \mathbf{r} \, dA.}$$

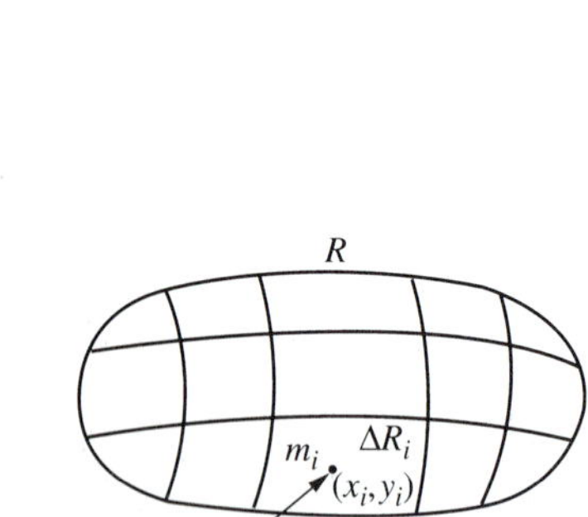

FIGURE 8

We call either $(\bar{x}, \bar{y})$ or $\mathbf{r}_{cm} = <\bar{x}, \bar{y}>$ the center of mass of R.

In Fig. 8, a plate R with mass M is approximated by a system of point masses m_i, $i = 1, 2, \ldots, n$, determined as follows. In the figure, R is partitioned into smaller pieces ΔR_i with areas ΔA_i. Let (x_i, y_i) be a point in ΔR_i with position vector $\mathbf{r}_i = <x_i, y_i>$. Let $m_i = \sigma(\mathbf{r}_i)\Delta A_i$ be a point mass located at (x_i, y_i). Then m_i is an approximation for the mass of ΔR_i and the region R is approximated by the system of point masses m_i, $i = 1, 2, \ldots, n$. The total mass and the center of mass of this system are given by

$$M_n = \sum_{i=1}^{n} m_i = \sum_{i=1}^{n} \sigma(\mathbf{r}_i)\Delta A_i,$$

$$(\mathbf{r}_{cm})_n = \frac{1}{M_n}\sum_{i=1}^{n} m_i \mathbf{r}_i = \frac{1}{M_n}\sum_{i=1}^{n} \sigma(\mathbf{r}_i)\mathbf{r}_i \Delta A_i.$$

The two sums on the right are Riemann sums for $M = \iint_R \sigma(\mathbf{r})\,dA$ and $\mathbf{r}_{cm} = \iint_R \sigma(\mathbf{r})_{\mathbf{r}}\,dA$. A limit passage gives

$$M_n \to M \quad \text{and} \quad (\mathbf{r}_{cm})_n \to \mathbf{r}_{cm} \quad \text{as} \quad n \to \infty.$$

Thus, the center of mass of R is the limit of the centers of mass of approximating systems of point masses. A consequence is that the region R, thought of as a thin plate, balances at its center of mass. See Fig. 9.

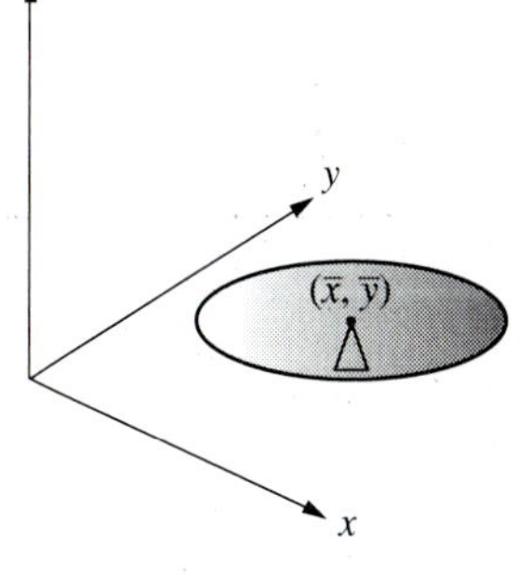

FIGURE 9

If the density σ is constant, then the mass of R is $M = \sigma A$, where A is the area of R, and

$$\bar{x} = \frac{1}{\sigma A}\iint_R \sigma x dA = \frac{1}{A}\iint_R x dA, \qquad \bar{y} = \frac{1}{\sigma A}\iint_R \sigma y dA = \frac{1}{A}\iint_R y dA.$$

Thus, the center of mass is the same as the centroid if R has constant density.

The remarks concerning symmetry for centroids carry over to centers of mass. If R has a line of symmetry L and the density is the same at symmetric points relative to L, then the center of mass lies on the line L.

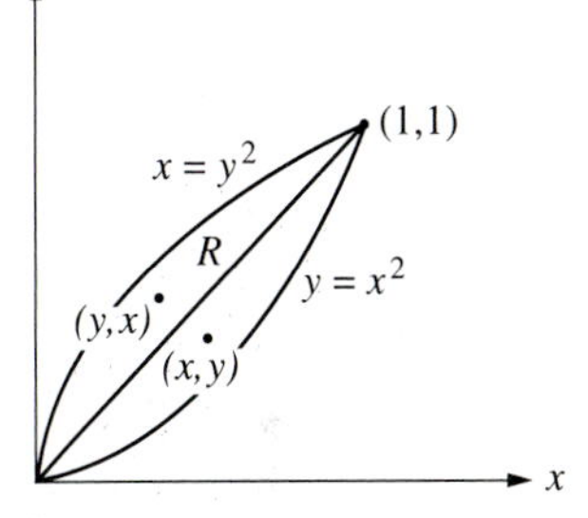

FIGURE 10

EXAMPLE 4. Let R be the region between the two parabolas $y = x^2$ and $x = y^2$ in Fig. 10. Given the density $\sigma(x, y) = xy$, find the center of mass of R.

Solution. It is clear from Fig. 10 that R is symmetric about the line L with equation $y = x$ and that the density is the same at the symmetric points (x, y) and (y, x) relative to L. So the center of mass lies on L and $\bar{y} = \bar{x}$. We found in Ex. 4 of Sec. 7.1 and in Ex. 9 of Sec. 7.2 that the region R is y–simple with $0 \le x \le 1$ and $x^2 \le y \le \sqrt{x}$, and that the mass of R is $M = 1/12$. Hence,

$$\bar{y} = \bar{x} = \frac{1}{M}\iint_R \sigma x dA = 12\int_0^1\int_{x^2}^{\sqrt{x}} x^2 y dy dx = 12\int_0^1 x^2\left[\frac{1}{2}y^2\right]_{y=x^2}^{y=\sqrt{x}} dx$$

$$= 6\int_0^1 x^2(x - x^4)\,dx = 6\left(\frac{1}{4} - \frac{1}{7}\right) = \frac{9}{14}.$$

The center of mass is $(\bar{x}, \bar{y}) = (9/14,\ 9/14)$. □

Center of Mass of a Solid Body

The story for the center of mass of a solid body is very similar to that for a plane region. Let D be a solid body with a continuous mass density function $\sigma(x, y, z)$. The total mass of D is

$$M = \iiint_D \sigma(\mathbf{r})dV.$$

The **center of mass** of D is the point $(\bar{x}, \bar{y}, \bar{z})$ with coordinates

$$\bar{x} = \frac{1}{M}\iiint_D \sigma x dV, \qquad \bar{y} = \frac{1}{M}\iiint_D \sigma y dV, \qquad \bar{z} = \frac{1}{M}\iiint_D \sigma z dV.$$

Let $\mathbf{r} = \langle x, y, z \rangle$ and let $\mathbf{r}_{cm} = \langle \bar{x}, \bar{y}, \bar{z} \rangle$ be the position vector of the center of mass. Then the center of mass is expressed in vector form by

$$\mathbf{r}_{cm} = \frac{1}{M}\iiint_D \sigma(\mathbf{r})\mathbf{r} dV.$$

We call either $(\bar{x}, \bar{y}, \bar{z})$ or $\mathbf{r}_{cm} = \langle \bar{x}, \bar{y}, \bar{z} \rangle$ the center of mass of D.

If the density σ is constant, then the center of mass is the same as the centroid of D. If D has a plane of symmetry and the density takes the same values at symmetric points, then the center of mass lies on the plane of symmetry.

EXAMPLE 5. Let D be the solid inside the sphere $x^2 + y^2 + z^2 = 9$ and also inside the top half of the cone $z^2 = \frac{1}{3}(x^2 + y^2)$. Given the density $\sigma = \sqrt{x^2 + y^2 + z^2}$, find the center of mass of D.

Solution. The same solid D with the same density σ appeared in Ex. 7 and Fig. 8 of Sec. 7.4. By symmetry, $\bar{x} = 0$ and $\bar{y} = 0$. To find $\bar{z}$ we shall use spherical coordinates (ρ, φ, θ). We found in Sec. 7.4 that D is described in spherical coordinates by

$$0 \le \rho \le 3, \qquad 0 \le \varphi \le \pi/3, \qquad 0 \le \theta \le 2\pi,$$

and the mass of D is $M = 81\pi/4$. Since $\sigma = \rho$ and $z = \rho \cos\varphi$,

$$\iiint_D \sigma z dV = \int_0^3 \int_0^{\pi/3} \int_0^{2\pi} \rho(\rho \cos\varphi)\rho^2 \sin\varphi d\theta d\varphi d\rho$$

$$= 2\pi \int_0^3 \rho^4 d\rho \cdot \int_0^{\pi/3} \sin\varphi \cos\varphi d\varphi$$

$$= 2\pi \left[\frac{1}{5}\rho^5\right]_0^3 \cdot \left[\frac{1}{2}\sin^2\varphi\right]_0^{\pi/3} = \frac{3^6\pi}{20}.$$

It follows that

$$\bar{z} = \frac{1}{M} \iiint_D \sigma z dV = \frac{4}{81\pi} \cdot \frac{3^6\pi}{20} = \frac{9}{5}.$$

The center of mass of D is $(0, 0, 9/5)$. □

Newton's Second Law of Motion

Newton's second law for the motion of a point mass m with position vector $\mathbf{r}(t)$ moving under the action of a force $\mathbf{F}(t)$ has the vector form $\mathbf{F}(t) = m\mathbf{r}''(t)$. The law has natural extensions to systems of point masses and to solid bodies. In either case, let M be the total mass, $\mathbf{r}_{cm}(t)$ the center of mass, and $\mathbf{F}(t)$ an applied force, such as gravity. We assume, as is commonly the case, that any internal forces cancel out. Then, as we shall see, Newton's second law takes the form

$$\boxed{\mathbf{F}(t) = M\,\mathbf{r}''_{cm}(t).}$$

This law implies a fundamental physical principle:

> The center of mass of a system of point masses or a solid body moves under the action of an external force as if the entire mass were concentrated at the center of mass.

For example, if you throw a handful of rocks into the air, then the center of mass moves in accordance with Newton's second law. For another example, throw a stick into the air. As it flies, the stick wobbles and turns. But viewed from some distance away, the stick looks like a point following a smooth trajectory. In fact, it is the center of mass that follows the smooth trajectory. If air resistance is neglected, the path is a parabola.

The derivation of $\mathbf{F}(t) = M\mathbf{r}''_{cm}(t)$ is not particularly difficult. First consider a system of point masses m_i, $i = 1, 2, \ldots, n$ that is in motion under the action of gravity or other forces. Let $\mathbf{r}_i(t)$ be the position vector for m_i at time t. The center of mass at time t is given by

$$\mathbf{r}_{cm}(t) = \frac{1}{M} \sum_{i=1}^{n} m_i \mathbf{r}_i(t).$$

Let $\mathbf{F}_i(t)$ be the external force acting on m_i. Under typical conditions, internal forces which the masses m_i exert on each other cancel out in pairs. Then the total force on the system is

$$\mathbf{F}(t) = \sum_{i=1}^{n} \mathbf{F}_i(t).$$

By Newton's second law applied to m_i, $\mathbf{F}_i(t) = m_i \mathbf{r}_i''(t)$ and

$$\sum_{i=1}^{n} \mathbf{F}_i(t) = \sum_{i=1}^{n} m_i \mathbf{r}_i''(t).$$

It follows that

$$\mathbf{F}(t) = M\,\mathbf{r}_{cm}''(t).$$

This is Newton's second law for a system of point masses with total mass M.

Newton's second law has the same form, $\mathbf{F}(t) = M\,\mathbf{r}_{cm}''(t)$, for a solid body D. The extension of the law to a solid body D is accomplished by approximating D by successive systems of point masses. The approximation procedure follows the same lines given earlier for obtaining the center of mass of a solid body as the limit of centers of mass of approximating systems of point masses. See the problems.

Rotational Properties of Solid Bodies*

Torque, angular momentum, and moments of inertia play important roles in the study of the rotation of a solid body about an axis. Torque was introduced in Sec. 3.3. Angular momentum and moments of inertia are introduced here. Then a rotational analogue of Newton's second law of motion is expressed in terms of these concepts.

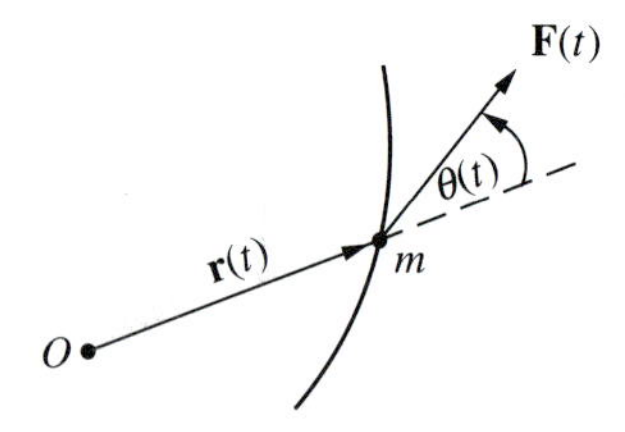

FIGURE 11

In Fig. 11 a point mass m with position vector $\mathbf{r}(t)$ is moving under the influence of a force $\mathbf{F}(t)$ in accordance with Newton's second law, $\mathbf{F}(t) = m\mathbf{r}''(t)$. The torque of $\mathbf{F}(t)$ about O is

$$\boldsymbol{\tau}(t) = \mathbf{r}(t) \times \mathbf{F}(t).$$

Torque measures the effect of the force $\mathbf{F}(t)$ to cause rotation about O. The magnitude of the torque,

$$\|\boldsymbol{\tau}(t)\| = \|\mathbf{r}(t)\|\,\|\mathbf{F}(t)\| \sin\theta(t),$$

is the **moment** of $\mathbf{F}(t)$ about O. For a central force, $\mathbf{F}(t)$ is parallel to $\mathbf{r}(t)$, so $\boldsymbol{\tau}(t) = 0$ and $\mathbf{F}(t)$ has no rotational effect about the center of force.

The momentum of a point mass m is $m\mathbf{r}'(t)$. The **angular momentum** of m is

$$\mathbf{L}(t) = \mathbf{r}(t) \times m\mathbf{r}'(t).$$

Differentiate $\mathbf{L}(t)$, using $\mathbf{r}' \times \mathbf{r}' = \mathbf{0}$ and $m\mathbf{r}'' = \mathbf{F}$, to obtain

$$\mathbf{L}'(t) = \boldsymbol{\tau}(t).$$

This is the basic dynamic equation for rotational motion. For a central force, $\mathbf{L}'(t) = \boldsymbol{\tau}(t) = \mathbf{0}$, which gives the **law of conservation of angular momentum:**

> For motion under a central force,
> the angular momentum is constant.

Since gravity is a central force, angular momentum is conserved during the motions of the planets about the sun or satellites about the earth.

The equation $\mathbf{L}'(t) = \boldsymbol{\tau}(t)$ extends to systems of point masses and to solid bodies in much the same manner as for Newton's second law. First consider a system of point masses m_i with position vectors $\mathbf{r}_i(t)$ and forces $\mathbf{F}_i(t)$ for $i = 1$, 2, ..., n. For convenience, t is suppressed in most of the following discussion. For each mass m_i,

$$\boldsymbol{\tau}_i = \mathbf{r}_i \times \mathbf{F}_i, \qquad \mathbf{L}_i = \mathbf{r}_i \times m_i\mathbf{r}_i', \qquad \mathbf{L}_i' = \boldsymbol{\tau}_i.$$

In general, $\mathbf{F}_i$ is composed of external and internal forces. Assume, as is very common, that the internal forces are central. Then their rotational effects cancel out (by an argument that we omit) and the **external torque** on the system is

$$\boldsymbol{\tau} = \sum_{i=1}^{n} \boldsymbol{\tau}_i = \sum_{i=1}^{n} \mathbf{r}_i \times \mathbf{F}_i.$$

The **angular momentum** of the system is

$$\mathbf{L} = \sum_{i=1}^{n} \mathbf{L}_i = \sum_{i=1}^{n} \mathbf{r}_i \times m_i \mathbf{r}_i'.$$

Differentiate with respect to time as before to obtain

$$\mathbf{L}'(t) = \boldsymbol{\tau}(t).$$

So the basic relationship $\mathbf{L}' = \boldsymbol{\tau}$ holds also for a system of point masses.

Now consider a solid body D with density $\sigma(x, y, z)$. By definition the **external torque** on D is

$$\boxed{\boldsymbol{\tau} = \iiint_D \mathbf{r} \times \mathbf{r}'' \sigma \, dV,}$$

and the **angular momentum** of D is

$$\boxed{\mathbf{L} = \iiint_D \mathbf{r} \times \mathbf{r}' \sigma \, dV.}$$

These definitions are motivated from approximations of D by systems of point masses and a limit passage. Similar arguments lead to the basic dynamic equation for rotation of a solid body D:

$$\boxed{\mathbf{L}'(t) = \boldsymbol{\tau}(t).}$$

Rotational Motion About a Fixed Axis

We specialize the foregoing general discussion to the important case of motion of a rigid body about a fixed axis. As we shall see, the angular momentum is conveniently expressed as a product of the moment of inertia of the body about the axis of rotation and its angular velocity. We also obtain the analogue of $\mathbf{F}(t) = M\,\mathbf{r}''_{cm}(t)$ for rotational motion about a fixed axis. As usual, we treat point masses first and then move on to a solid body.

Suppose that a point mass m is moving around the circle in the xy–plane with center at the origin O and fixed radius r. So m rotates about the z–axis. The quantity

$$I = mr^2$$

is called the **moment of inertia** of m about the z–axis. The position of the rotating mass m at time t can be expressed in polar coordinates by $(r, \theta(t))$ with constant r. As we know, the angular velocity of m is $\theta'(t)$. The vector $\boldsymbol{\omega}(t) = \theta'(t)\mathbf{k}$ describes both the angular velocity and the axis of rotation. By elementary calculations (see the problems) the angular momentum and torque of m are given by

$$\mathbf{L}(t) = I\,\theta'(t)\mathbf{k} \quad \text{and} \quad \boldsymbol{\tau}(t) = \mathbf{L}'(t) = I\,\theta''(t)\mathbf{k}.$$

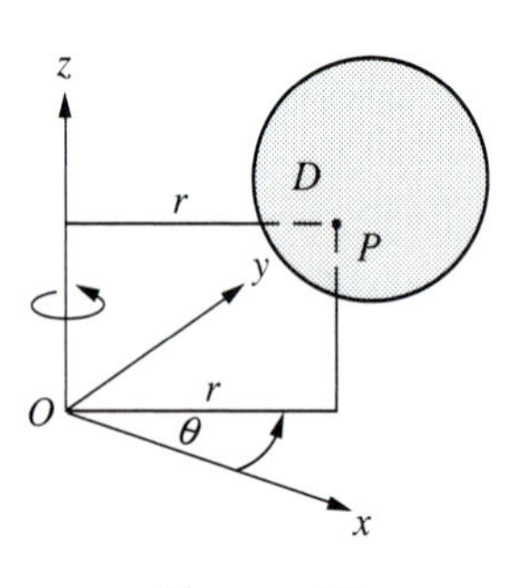

FIGURE 12

This equation for the torque is the rotational analogue of Newton's second law, $\mathbf{F}(t) = m\mathbf{r}''(t)$.

Next, we extend the foregoing ideas to the rotation of a solid body about a fixed axis. As a first step, we calculate the angular momentum of the body about the axis of rotation. Imagine that the solid D in Fig. 12 is rotating around the z–axis with angular velocity $\theta'(t)$. As before, let $\boldsymbol{\omega}(t) = \theta'(t)\mathbf{k}$. We shall use either rectangular or cylindrical coordinates according to the convenience of the moment. The point $P = (x, y, z) = (r, \theta, z)$ of the solid moves around a circle with fixed radius r and center on the z–axis. The position vector of P is

$$\mathbf{r} = x\mathbf{i} + y\mathbf{j} + z\mathbf{k} = r\cos\theta\mathbf{i} + \mathbf{r}\sin\theta\mathbf{j} + z\mathbf{k},$$

where r and z are independent of t. The velocity of P is

$$\mathbf{r}' = -(r\sin\theta)\theta'\mathbf{i} + (r\cos\theta)\theta'\mathbf{j} = -\theta' y\mathbf{i} + \theta' x\mathbf{j}.$$

The angular momentum of D is $\mathbf{L} = \iiint_D \mathbf{r} \times \mathbf{r}'\sigma dV$. The determinant formula for the cross product yields

$$\mathbf{r} \times \mathbf{r}' = \begin{vmatrix} \mathbf{i} & \mathbf{j} & \mathbf{k} \\ x & y & z \\ -\theta' y & \theta' x & 0 \end{vmatrix} = \theta'[-xz\mathbf{i} - yz\mathbf{j} + (x^2 + y^2)\mathbf{k}].$$

Substitute this expression into the triple integral for angular momentum to obtain

$$\mathbf{L}(t) = \theta'(t)\left\{ -\iiint_D xz\sigma dV\ \mathbf{i} - \iiint_D yz\sigma dV\ \mathbf{j} + \iiint_D (x^2 + y^2)\sigma\, dV\ \mathbf{k} \right\}.$$

The three integrals in the formula for $\mathbf{L}(t)$ are determined by the shape of D, its density σ, and its orientation relative to the z–axis. The third integral,

$$I = \iiint_D (x^2 + y^2)\sigma dV = \iiint_D r^2 \sigma dV,$$

is the **moment of inertia** about the z–axis. The first two integrals are called **products of inertia**. Very often, the products of inertia are zero for reasons of symmetry. Then

$$\mathbf{L}(t) = I\theta'(t)\mathbf{k} = I\,\boldsymbol{\omega}(t),$$

just as for a single point mass. Since $\mathbf{L}'(t) = \boldsymbol{\tau}(t)$, it follows that $\boldsymbol{\tau}(t) = I\theta''(t)\mathbf{k}$. Since the angular acceleration of the body about its axis of rotation is $\boldsymbol{\alpha}(t) = \theta''(t)\mathbf{k}$, we have

$$\boldsymbol{\tau}(t) = I\boldsymbol{\alpha}(t).$$

This result is the rotational analogue for a solid body of $\mathbf{F}(t) = M\mathbf{r}''_{cm}(t)$. Comparison of these two basic equations of motion shows that the moment of inertia I is a quantitative measure of the tendency of a body to resist rotational motion just as mass is a quantitative measure of the tendency of a body to resist rectilinear motion.

You can permute the variables x, y, and z in the foregoing formulas for $\mathbf{L}$ and I to obtain corresponding results for rotation about the x–axis or the y–axis.

EXAMPLE 6. The homogeneous solid cylinder D with constant density σ rotates around the z–axis in Fig. 13. Find the moment of inertia and the angular momentum.

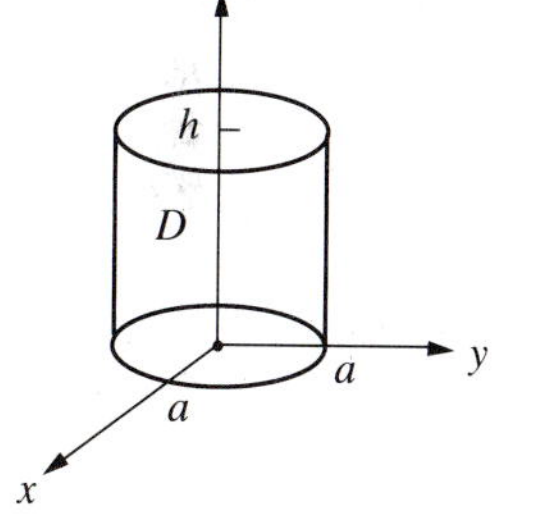

FIGURE 13

Solution. By symmetry, the products of inertia are zero. (See the problems.) Therefore, $\mathbf{L}(t) = I\boldsymbol{\omega}(t)$, where

$$I = \sigma \iiint_D r^2 dV.$$

The cylinder D is described in cylindrical coordinates by

$$0 \le \theta \le 2\pi, \qquad 0 \le r \le a, \qquad 0 \le z \le h.$$

Therefore,

$$I = \sigma \int_0^{2\pi}\int_0^a\int_0^h r^2 r dz dr d\theta = \frac{1}{2}\sigma\pi a^4 h.$$

Since the mass of the cylinder is $M = \sigma\pi a^2 h$,

$$I = \frac{1}{2}Ma^2 \quad \text{and} \quad \mathbf{L}(t) = \frac{1}{2}Ma^2\omega(t). \ \square$$

FIGURE 14

EXAMPLE 7. The cylinder D in Ex. 6 and Fig. 13 has been moved in Fig. 14. Now the axis of the cylinder is the vertical line through $(0, a, 0)$. Find the moment of inertia and the angular momentum of the cylinder about the z–axis.

Solution. By symmetry (see the problems) the xz–product of inertia is zero. Let P_{yz} denote the yz–product of inertia. Then $\mathbf{L}(t) = \theta'(t)[-P_{yz}\mathbf{j} + I\mathbf{k}]$. The cylinder D is described in cylindrical coordinates by

$$0 \le \theta \le \pi, \qquad 0 \le r \le 2a \sin\theta, \qquad 0 \le z \le h.$$

Therefore,

$$\begin{aligned} I &= \sigma \int_0^{\pi}\int_0^{2a\sin\theta}\int_0^{h} r^2 r\, dz\, dr\, d\theta \\ &= \sigma h \int_0^{\pi} \left[\frac{1}{4}r^4\right]_{r=0}^{r=2a\sin\theta} d\theta = 4\sigma h a^4 \int_0^{\pi} \sin^4\theta\, d\theta. \end{aligned}$$

By two applications of the identity $\sin^2 x = \frac{1}{2}(1 - \cos 2x)$ or by use of a CAS or an integral table, the value of the last integral is $3\pi/8$. Since the mass of the cylinder is $M = \sigma\pi a^2 h$,

$$I = \frac{3}{2}Ma^2.$$

We shall give an easier evaluation of I shortly.

Likewise,

$$\begin{aligned} P_{yz} &= \sigma \int_0^{\pi}\int_0^{2a\sin\theta}\int_0^{h} (r\sin\theta) z r\, dz\, dr\, d\theta \\ &= \frac{4\sigma h^2 a^3}{3}\int_0^{\pi} \sin^4\theta\, d\theta = \frac{\pi\sigma h^2 a^3}{2} = \frac{Mah}{2}, \end{aligned}$$

and, hence,

$$\mathbf{L}(t) = \theta'(t)[-P_{yz}\mathbf{j} + I\mathbf{k}] = \frac{1}{2}Ma\theta'(t)[3a\mathbf{k} - h\mathbf{j}]. \ \square$$

The triple integral for the moment of inertia about the z–axis, when properly interpreted, also defines the moment of inertia about any axis of rotation: The moment of inertia I of a solid body D about any axis of rotation A, as illustrated in Fig. 15, is defined by

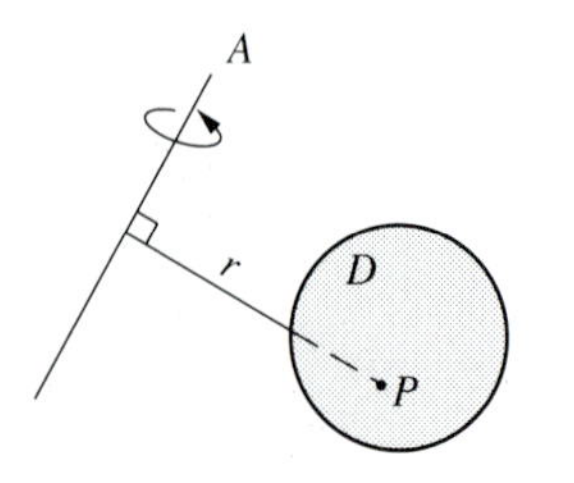

FIGURE 15

$$\boxed{I = \iiint_D r^2 \sigma\, dV,}$$

where (see Fig. 15) r is now the perpendicular distance from the axis of rotation to a variable point P in D. The moment of inertia is independent of the coordinate system. The moment of inertia depends only on the body D and its orientation relative to the axis of rotation. So the moment of inertia is unchanged if the body D and the axis of rotation A are translated or rotated without changing their relative positions.

Calculations of moments of inertia can be quite lengthy. Fortunately, there is a remarkable relationship between moments of inertia and centers of mass that can save a lot of work. Figure 16 will serve as a guide. In the figure, D is a solid body with total mass M. The axis A is arbitrary and the axis A_{cm} is parallel to A through the center of mass $(\bar{x}, \bar{y}, \bar{z})$ of D. The perpendicular distance between A and A_{cm} is denoted by r. Let I be the moment of inertia of D about A and I_{cm} the moment of inertia of D about A_{cm}. Finally, let I_M be the moment of inertia about A of a mass point M at $(\bar{x}, \bar{y}, \bar{z})$. Then $I_M = Mr^2$ and

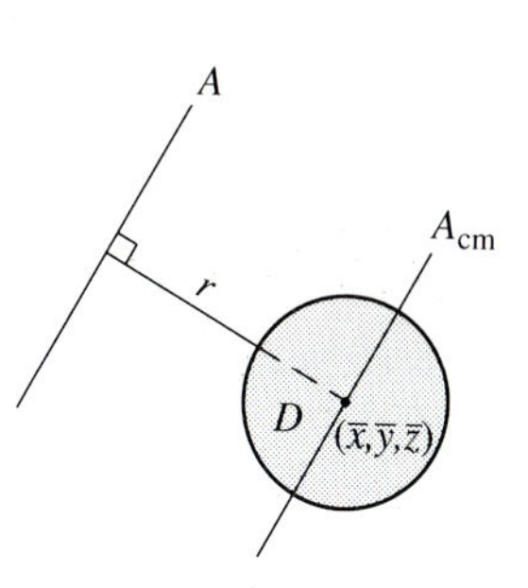

FIGURE 16

$$\boxed{I = I_{cm} + I_M = I_{cm} + Mr^2.}$$

This equation is called the **parallel axis theorem**. If you know the mass of D, its center of mass, and the moment of inertia of D about one axis, then you can easily calculate its moment of inertia about any parallel axis, using the parallel axis theorem.

EXAMPLE 8. Use the parallel axis theorem and Ex. 6 to find the moment of inertia of the cylinder in Ex. 7 and Fig. 14 about the z–axis.

Solution. Evidently, the center of mass of D is $(0, a, h/2)$. Let A denote the z–axis and A_{cm} the parallel axis through the center of mass. From Ex. 6, the moment of inertia of D about A_{cm} is $I_{cm} = \frac{1}{2}Ma^2$. The moment of inertia of a point mass M at $(0, a, h/2)$ about A (the z–axis) is $I_M = Ma^2$. By the parallel axis theorem, the moment of inertia of the cylinder about the z–axis is

$$I = I_{cm} + I_M = \frac{1}{2}Ma^2 + Ma^2 = \frac{3}{2}Ma^2.$$

This method of finding I is much easier than the direct approach used in Ex. 7, where several steps were suppressed. It is just as easy to calculate the moment of inertia of D about any other axis parallel to the z–axis. □

PROBLEMS

In Probs. 1–4, find the average value of the given function over the indicated region.

1. $f(x, y) = y \sin xy$
   $R: 1 \le x = 2, \quad 0 \le y \le \pi$

2. $f(x, y) = y \sin xy$
   $R: 1 \le x = 2, \quad 0 \le y \le \pi/2$

3. $f(x, y) = xy$
   $R: 1 \le x = 2, \quad \sqrt{x} \le y \le x^2$

4. $f(x, y) = xe^y/(4 - y)$
   $\mathrm{R}: 0 \le x = 2, \quad 0 \le y \le 4 - x^2$

Solve Probs. 5–45 using any convenient coordinate system. Take advantage of symmetry whenever possible. The letters a, b, c, and h denote positive constants. If a proportionality constant is needed, call it k. As usual σ denotes a density.

5. Find the centroid of the rectangular box $0 \le x \le 1$, $0 \le y \le 2$, $0 \le z \le 3$.

6. Find the centroid of the solid tetrahedron cut from the first octant by the plane $x/3 + y + z/2 = 1$.

7. Find the centroid of the solid below the paraboloid $z = 9 - x^2 - y^2$ and above the xy–plane.

8. Find the centroid of the part of the paraboloid in the previous problem that lies in the first octant.

9. Find the centroid of the half-ball $x^2 + y^2 + z^2 \le 4$, $z \ge 0$.

10. Find the centroid of the part of region in the previous problem that lies in the first octant.

11. Find the centroid of a right circular cylinder of height h and radius of base a.

12. Find the centroid of a right circular cone of height h and radius of base a.

13. Find the centroid of a solid hemisphere of radius a.

14. Find the centroid of the upper half of the solid ellipsoid $x^2/a^2 + y^2/a^2 + z^2/c^2 \le 1$.

15. Find the centroid of the solid bounded by the plane $z = h$ and the paraboloid $z = a(x^2 + y^2)$.

16. Find the centroid of the ice cream cone bounded by the sphere $x^2 + y^2 + z^2 = a^2$ and the top half of the cone $z^2 = b^2(x^2 + y^2)$.

17. Find the centroid of the region bounded by the plane $z = y$ and the paraboloid $z = x^2 + y^2$.

18. Find the centroid of the wedge in Ex. 8 in Sec. 7.1.

19. (a) Find the centroid of the region above the xy–plane and between the two spheres $\rho = a$ and $\rho = b$, where $a < b$. (b) Find the limiting value of the centroid as $b \to a$. What is a reasonable physical interpretation of this limit?

20. Find the centroid of the region above the xy–plane, below the plane $z = h$, and between the cones $z^2 = x^2 + y^2$ and $z^2 = 3(x^2 + y^2)$.

21. A wedge is cut from a ball of radius a by two half-planes that intersect along a diameter of the ball. If the angle between the half-planes is $0 < \gamma \le 2\pi$ find the centroid. (You may find it informative to consider the special cases $\gamma = \pi/2$, π, $3\pi/2$, 2π, and the limiting case $\gamma \to 0$. *Hint.* Choose spherical coordinates so that the wedge has description $0 \le \theta \le \gamma$, $0 \le \rho \le a$, $0 \le \varphi \le \pi$.

22. Find the center of mass of the region in Prob. 5 if $\sigma = y$.

23. Find the center of mass of the region in Prob. 6 if $\sigma = x + y + z$.

24. Find the center of mass of the region in Prob. 7 if $\sigma = \sqrt{x^2 + y^2}$.

25. Find the center of mass of the region in Prob. 8 if σ is constant.

26. Find the center of mass of the region in Prob. 9 if σ is constant.

27. Find the center of mass of the region in Prob. 10 if σ is constant.

28. A thin rectangular plate with length a and width b has density proportional to the square of the distance from one corner. Find the center of mass.

29. The rectangular plate in the previous problem has density proportional to the distance from one of the sides with length b. Find the center of mass.

30. A thin rectangular plate is bounded by the curves $y = ax^2$ and $x = ay^2$ and has density proportional to the square of the distance from the origin. Find the center of mass.

31. A thin semicircular plate with radius a has density proportional to the distance from its center. Find the center of mass.

32. A thin semicircular plate with radius a has density proportional to the distance from its straight side. Find the center of mass.

33. A circular plate with radius a has density proportional to the distance from a given point on its boundary. Find the center of mass. *Hint.* Choose coordinates so that the plate is tangent to the x–axis at the given point, which is the origin.

34. Find the center of mass of a cube with edge a if the density is proportional to the distance from one face.

35. Find the center of mass of a cube with edge a if the density is proportional to the square of the distance from one edge.

36. A right circular cone of height h and radius of base a has density proportional to the distance from its base. Find the center of mass.

37. Find the center of mass of the cone in the previous example if the density is proportional to the distance from its vertex.

38. Find the center of mass of the solid in Ex. 3 if the density is proportional to the distance above the xy–plane.

39. Find the center of mass of the solid hemisphere of radius a if the density is proportional to the distance from the plane of its base.

40. Find the mass of a solid hemisphere of radius a if the density is proportional to the distance from its center.

41. Start with the solid hemisphere $x^2 + y^2 + z^2 \leq a^2$, $z \geq 0$. Remove from the solid hemisphere the solid right circular cone whose base is the base of the hemisphere and whose vertex is $(0, 0, a)$. Find the center of mass of the remaining solid if the density is proportional to the distance from the xy–plane.

42. The region above the xy–plane, below the top half of the cone $z^2 = x^2 + y^2$, and inside the cylinder $x^2 + y^2 = 2ay$ has density proportional to the distance from the xz–plane. Find the center of mass.

43. The region inside the cylinder $x^2 + y^2 = 2ax$, below the cone $z^2 = x^2 + y^2$, and above the xy–plane has density proportional to the distance from the xy–plane. Find the center of mass.

44. The solid inside both the sphere $\rho = 2a \cos\varphi$ and the cone $\varphi = b$, where $b \leq \pi/2$, has density proportional to the distance from the *xy*–plane. Find the center of mass.

45. Find the center of mass of a ball of radius a if the density is proportional to the distance from a given point on its surface. *Hint.* Choose coordinates so that the ball sits on the *xy*–plane with the given point at the origin.

In Probs. 46–58 do the following: (a) Make a sketch of the solid D and then find the moment of inertia I about the given axis. (b) Express I as a multiple of the mass M of the solid. (Take advantage of the parallel axis theorem whenever you can.)

46. D is the homogeneous rectangular solid given by $-a/2 \leq x \leq a/2$, $0 \leq y \leq b$, and $-c/2 \leq z \leq c/2$. It has density σ_0 and the axis of rotation is perpendicular to the *xy*–plane and passes through the center of mass of the solid. (Perhaps the most interesting physical case is a long thin rectangular rod where $a, c \ll b$.)

47. D is the solid in the previous problem. The axis of rotation is the *z*–axis.

48. D is the solid in the previous problem. The axis of rotation is the *y*–axis.

49. D is the right circular cylinder $x^2 + z^2 \leq a^2$, $0 \leq y \leq h$ and has constant density σ_0. The axis of rotation is the *z*–axis. (Perhaps the most interesting physical case is a long thin rod where $a \ll h$.)

50. D is the cylinder in the previous problem. The axis of rotation is the *y*–axis.

51. D is a ball of radius a and constant density σ_0. The axis of rotation is a diameter of the ball.

52. D is a ball of radius a and density proportional to the square of the distance from the center. The axis of rotation is a diameter of the ball.

53. D is a homogeneous ball with density σ_0 and radius a. The axis of rotation is tangent to the ball.

54. D is a right circular cone with height h, radius of base a, and constant density σ_0. The axis of rotation is the axis of the cone.

55. D is the right circular cone of the previous problem. The axis of rotation passes through the vertex of the cone and is parallel to its base.

56. D is the region below the plane $z = h$ and above the paraboloid $z = \frac{h}{a^2}(x^2 + y^2)$ and has constant density σ_0. The axis of rotation is the axis of the paraboloid.

57. D is the homogeneous solid with density σ_0 inside a sphere of radius $2a$ and outside a right circular cylinder of radius a whose axis contains a diameter of the sphere. The axis of rotation is the axis of the cylinder.

58. D is the solid in the previous problem but with the density proportional to the distance from the center of the sphere.

In Probs. 59–68, find the angular momentum about the origin (or about the point specified) of the solid described in the indicated problem. The body rotates about the indicated axis.

59. Prob. 46

60. Prob. 47

61. Prob. 48
62. Prob. 49
63. Prob. 50
64. Prob. 51 (center of ball)
65. Prob. 52 (center of ball)
66. Prob. 53 (point of tangency)
67. Prob. 54 (vertex of cone)
68. Prob. 55 (vertex of cone)
69. Show that the products of inertia are zero for the body in Ex. 6.
70. Show that the xz–product of inertia is zero for the body in Ex. 7.
71. Show that the products of inertia are zero for any body that is symmetric in z.
72. Show that the products of inertia are zero for any body that is symmetric in both x and y.
73. Verify that $\mathbf{L}(t) = I\theta'(t)\mathbf{k}$ and $\tau(t) = \mathbf{L}'(t) = I\theta''(t)\mathbf{k}$, where $I = mr^2$, for a point mass m moving on the circle with radius r and center at the origin in the xy–plane.
74. A point P in a body D is located by its position vector $\mathbf{r}$. As the body moves in space, P and $\mathbf{r}$ vary in time. Start from the definition of the kinetic energy of a point mass and use a Riemann sum, limit passage argument to show that the kinetic energy of D should be defined by

$$\boxed{K = \frac{1}{2}\iiint_D ||\mathbf{v}||^2 \sigma dV,}$$

where $\sigma = \sigma(\mathbf{r})$ is the density and $\mathbf{v} = \mathbf{r}'$ is the velocity of the particle at $\mathbf{r}$.

75. Assume that the body D in the previous problem rotates about the z–axis as in Fig. 12. Show that the kinetic energy of a body rotating about the z–axis with angular velocity ω is

$$\boxed{K = \frac{1}{2}I\omega^2,}$$

where I is the moment of inertia of D about the axis of rotation.

76. Prove the parallel axis theorem by the following steps. Choose a coordinate system with the center of mass at the origin and the given axis parallel to the z–axis. Let the given axis cut the xy–plane at $(a, b, 0)$. Show:

(*a*) $I_{cm} = \displaystyle\iiint_D (x^2 + y^2)dV.$

(*b*) $I = \displaystyle\iiint_D [(x - a)^2 + (y - b)^2]dV.$

(c) Expand the formula for I and recall that $\bar{x} = \bar{y} = 0$ in the chosen coordinate system.

77. Given that the centroid of a plane region R is unchanged if the coordinate axes are translated or rotated, show that the centroid of R lies on any line of symmetry of R. *Hint.* Choose the line of symmetry to be the y–axis.

78. A body D has constant density. Justify briefly the following statements.
(a) If two planes of symmetry of D intersect, the center of mass of D lies on the line of intersection of the two planes.
(b) If three planes of symmetry of D intersect in a point, then that point is the center of mass of D.

79. What are the corresponding statements in the previous example if D has variable density? Explain briefly.

80. The **moment of inertia** about the z–axis of a plane region R in the xy–plane is defined by

$$I = \iint_R (x^2 + y^2)\sigma dA,$$

where $\sigma = \sigma(x, y)$ is the areal density of R. Let D be the solid right cylinder with base R, height h, and density $\sigma = \sigma(x, y, z)$ that is constant on each line perpendicular to the xy–plane. Show that the moment of inertia of D about the z–axis is the same as the moment of inertia of R if $\sigma(x, y) = \sigma(x, y, 0)h$. (This justifies treating solids with "vertically" constant density as plane regions for questions about moments or inertia about vertical axes.)

81. Extend $\mathbf{F} = M\mathbf{r}''_{cm}$ from a system of particles to a body D by the usual Riemann sum and limit passage reasoning. Assume that the external force acting on the body is given by a force density $\eta(\mathbf{r}) = lim_{\Delta V \to 0} \mathbf{F}/\Delta V$, where $\mathbf{F}$ is the external force acting on a small piece of the body with volume ΔV and the limit is defined in the same manner as when we defined mass density.

82. Provide the Riemann sum and limit passage argument that leads from $\mathbf{L} = \mathbf{r} \times m\mathbf{r}'$ for a particle to $\mathbf{L} = \iiint_D \sigma\mathbf{r} \times \mathbf{r}'dV$ for a solid body D with density $\sigma = \sigma(\mathbf{r})$.

83. A solid D is composed of two nonoverlapping parts D_1 and D_2. The respective masses and centers of mass are M, M_1, M_2, and $\mathbf{r}_{cm}$, $\mathbf{r}_{1,cm}$, $\mathbf{r}_{2,cm}$.
(a) Show that $\mathbf{r}_{cm} = \dfrac{M_1}{M}\mathbf{r}_{1,cm} + \dfrac{M_2}{M}\mathbf{r}_{2,cm}$.
(b) Express the result of (a) in terms of the coordinates $(\bar{x}, \bar{y}, \bar{z})$, $(\bar{x}_1, \bar{y}_1, \bar{z}_1)$, $(\bar{x}_2, \bar{y}_2, \bar{z}_2)$ of the respective centers of mass.
(c) How does (a) change if D is composed of three nonoverlapping parts?
(d) Make a sketch of the plane region R of points (x, y) such that either $0 \le x \le 2a$, $0 \le y \le b$ or $(x - a)^2 + (y - b)^2 \le a^2$, $y \ge b$. Use the analogue of the preceding results for plane regions to find the center of mass of R. Assume constant density.

84. Return to the seesaw in Fig. 5. This time assume the seesaw is a homogeneous board with (total) weight W. Now find the coordinate $\bar{x}$ of the balance point.

Chapter Highlights

Double and triple integrals are defined as limits of Riemann sums expressed symbolically by

$$\iint_R f dA = \lim \Sigma\Sigma f \Delta A, \qquad \iiint_D f dV = \lim \Sigma\Sigma\Sigma f \Delta V.$$

Usually the integrals are evaluated in iterated forms over simple regions of various types depending on the coordinate system being used. There is a similar pattern to all of these multiple integrals. In what follows assume all functions that appear are continuous.

Rectangular coordinates (x, y) **in 2–space.** Let R be a y–simple plane region

$$a \le x \le b \quad \text{and} \quad y_1(x) \le y \le y_2(x).$$

See Fig. 10 in Sec. 7.1. Then

$$\iint_R f dA = \int_a^b \int_{y_1(x)}^{y_2(x)} f(x, y)\, dy dx.$$

Rectangular coordinates (x, y, z) **in 3–space.** Let D be a z–simple solid region

$$(x, y) \text{ in } R \quad \text{and} \quad z_1(x, y) \le z \le z_2(x, y),$$

where R is x–simple or y–simple. See Fig. 3 in Sec. 7.2. Then

$$\iiint_D f dV = \iint_R \left[\int_{z_1(x,y)}^{z_2(x,y)} f(x, y, z)\, dz \right] dA.$$

Polar coordinates (r, θ) **in 2–space.** Let R be a radially simple plane region

$$c \le \theta \le d, \qquad g(\theta) \le r \le h(\theta).$$

See Fig. 14 in Sec. 7.3. Then

$$\iint_R f dA = \int_c^d \int_{g(\theta)}^{h(\theta)} f(r, \theta) r dr d\theta.$$

The area element in polar coordinates is $dA = r dr d\theta$.

Cylindrical coordinates (r, θ, z) **in 3–space.** Let D be a cylindrically simple solid region

$$(r,\theta) \text{ in } R \quad \text{and} \quad z_1(\theta) \le z \le z_2(\theta),$$

where R is radially simple. See Fig. 19 in Sec. 7.3. Then

$$\iiint_D f dV = \iint_R \left[\int_{z_1(r,\theta)}^{z_2(r,\theta)} f(r,\theta,z)\, dz \right] dA.$$

The volume element in cylindrical coordinates is $dV = r dr d\theta dz$.

Spherical coordinates $(\sigma, \varphi, \theta)$ **in 3–space.** Let D be a spherically simple solid region

$$c \le \varphi \le d, \qquad p \le \theta \le q, \qquad \rho_1(\varphi,\theta) \le \rho \le \rho_2(\varphi,\theta).$$

See Fig. 9 in Sec. 7.4. Then

$$\iiint_D f dV = \int_p^q \int_c^d \int_{\rho_1(\varphi,\theta)}^{\rho_2(\varphi,\theta)} f(\rho,\varphi,\theta)\, \rho^2 \sin\varphi d\rho d\varphi d\theta.$$

The volume element in spherical coordinates is $dV = \rho^2 \sin\varphi d\rho d\varphi d\theta$.

Many geometric and physical quantities are represented by double and triple integrals. The area of a plane region R is $A = \int\int_R 1 dA$. The volume of a solid region D is $V = \int\int_D 1 dV$. If D has mass density $\sigma(x,y,z)$, then the mass of D is

$$M = \iiint_D \sigma dV.$$

The center of mass of D is the point $(\bar{x},\bar{y},\bar{z})$ with

$$\bar{x} = \frac{1}{M}\iiint_D \sigma x dV, \qquad \bar{y} = \frac{1}{M}\iiint_D \sigma y dV, \qquad \bar{z} = \frac{1}{M}\iiint_D \sigma z dV.$$

In vector form, the center of mass of D is $\mathbf{r}_{cm} = <\bar{x},\bar{y},\bar{z}>$. If D is moving under the influence of an external force $\mathbf{F}(t)$ such as gravity, then the center of mass of D moves in accordance with Newton's second law of motion $\mathbf{F}(t) = M\mathbf{r}''_{cm}(t)$.

Other applications of double and triple integrals pertain to rotary motion.

The project for this chapter is a brief introduction to numerical approximation of multiple integrals. Just as for single integrals, many multiple integrals cannot be evaluated exactly. Numerical methods are required to approximate their values. The development of accurate, efficient numerical methods for multiple integrals is more challenging than for single integrals for several reasons. For example, typical numerical integration methods for a single integral are based on the fact that the interval of integration can be partitioned conveniently into many small subintervals. Corresponding partitions of a two–dimensional region of integration R are based on systems of small rectangles that usually do not "fit" R exactly. Special care must be taken to approximate the integral over the small rectangles that intersect the boundary of R. We do not have time to deal systematically with such issues in this project. Rather we concentrate on two useful ideas that underlie many numerical integration methods—the use of infinite series and the use of judiciously chosen double sums. The project illustrates these ideas for double integrals. You will find it convenient to use a scientific calculator, Maple, Mathematica, or other appropriate computer software in the project.

Chapter Project: Numerical Integration

Sometimes familiar power series expansions enable you to find accurate numerical approximations for double integrals in much the same way as for single integrals. The first two problems illustrate how this works in particular cases.

Problem 1 Make the change of variables $z = -x^2y^2$ in the power series for e^z and use term-by-term integration to obtain

$$\int_0^1\int_0^1 e^{-x^2y^2}dxdy = \sum_{n=0}^{\infty} \frac{(-1)^n}{(2n+1)^2 n!}.$$

Use this result to determine the value of the integral to within 10^{-4}.

Problem 2 Use power series methods, much as in Prob. 1, to approximate

$$\iint_R e^{-x^2y^2}dA$$

to within 10^{-4}, where R is the region in the xy–plane with $0 \le x \le 1$ and $0 \le y \le x^2$.

The rest of the project is concerned with the approximation of double integrals by double sums, with emphasis on a two–dimensional version of the trapezoidal rule. Let $f(x,y)$ be a continuous function on the rectangle R with $a \le x \le b$ and $c \le y \le d$, as in Fig. 1. The rectangle is subdivided into smaller rectangles R_{ij} for $i = 1, \ldots, m$ and $j = 1, \ldots, n$, with sides of length

$$h = \Delta x_i = \frac{b-a}{m} \quad \text{and} \quad k = \Delta y_j = \frac{d-c}{n}.$$

The area of R_{ij} is $\Delta x_i \Delta y_j = hk$. The upper right corner of R_{ij} is (x_i, y_j). The game plan is to approximate the integral of f over R_{ij} in a convenient fashion and then to sum the approximations over i and j to obtain an approximation for the integral of f over R. A straightforward numerical integration scheme of this type is given by

$$\iint_{R_{ij}} f(x,y)\,dA \approx hkf(x_i, y_j),$$

$$\iint_R f(x,y)\,dA \approx hk \sum_{i=1}^{m}\sum_{j=1}^{n} f(x_i, y_j). \qquad \text{(UR)}$$

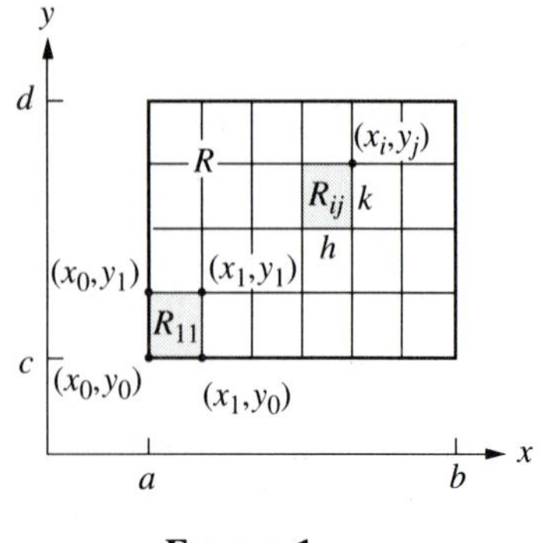

FIGURE 1

The double sum on the right may be called the *upper right rectangular rule*; see R_{ij} in Fig. 1.

Problem 3 There are three other natural rectangular rules that can be associated with the partition in Fig. 1 and may be referred to as (LR), (UL), and (LL). What are the formulas for the other three rules? Why do all four double sums converge to $\iint_R f(x,y)\,dA$ as $m \to \infty$ and $n \to \infty$?

The foregoing rectangular rules are seldom used in practice because you can do better with virtually no more effort by using the trapezoidal rule for double integrals. Focus your attention on the rectangle R_{11} in Fig. 1. Write out the integral of f over R_{11} as an iterated integral

$$\iint\limits_{R_{11}} f(x,y)\,dA = \int_{x_0}^{x_1}\int_{y_0}^{y_1} f(x,y)\,dy\,dx.$$

Equivalently,

$$\iint\limits_{R_{11}} f(x,y)\,dA = \int_{x_0}^{x_1} g(x)\,dx, \quad \text{where} \quad g(x) = \int_{y_0}^{y_1} f(x,y)\,dy.$$

Problem 4 Use the trapezoidal rule for single integrals to obtain the approximations

$$g(x) \approx \frac{k}{2}[f(x,y_0) + f(x,y_1)],$$

$$\int_{x_0}^{x_1} g(x)\,dx \approx \frac{h}{2}[g(x_0) + g(x_1)].$$

Deduce that

$$\iint\limits_{R_{11}} f(x,y)\,dA \approx \frac{hk}{4}[f(x_0,y_0) + f(x_1,y_0) + f(x_0,y_1) + f(x_1,y_1)].$$

Call the last approximation in Prob. 4 the basic trapezoidal rule approximation to f over the rectangle R_{11}.

Problem 5 Check that the basic trapezoidal rule approximation also is obtained by averaging the four rectangular rule approximations applied to R_{11} without further subdivision. What is the analogous result for single integrals?

Problem 6 Let $m = n = 2$ as in Fig. 2. Write out the basic trapezoidal rule approximations for the integrals of f over R_{12}, R_{21}, and R_{22}. Add these approximations to the approximation in Prob. 4 to obtain

$$\iint_R f(x,y)\,dA \approx \frac{hk}{4}[f(x_0,y_0) + f(x_2,y_0) + f(x_0,y_2) + f(x_2,y_2)$$

$$+ 2f(x_1,y_0) + 2f(x_0,y_1) + 2f(x_1,y_2) + 2f(x_2,y_1) + 4f(x_1,y_1)].$$

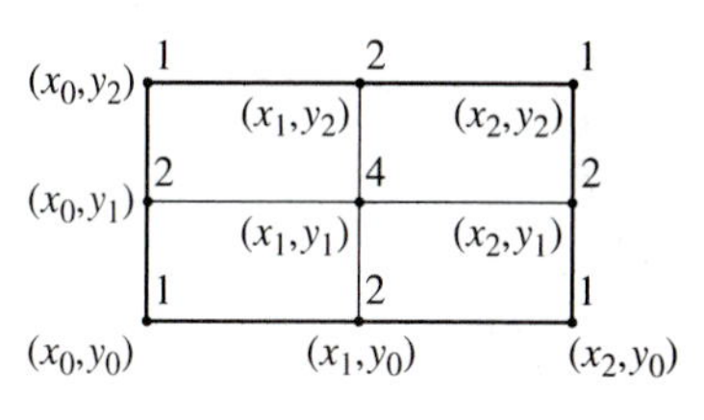

FIGURE 2

Explain how the coefficients 1, 2, 4 of $f(x_i,y_j)$ for $i, j = 0, 1, 2$ that are placed near the points (x_i,y_j) in Fig. 2 can be determined directly from the figure, before adding the four basic trapezoidal rule approximations together.
The **trapezoidal rule for double integrals** is given by

$$\iint_R f(x,y)\,dA \approx T_{mn}[f] = \frac{hk}{4}\sum_{i=0}^{m}\sum_{j=0}^{n} c_{ij} f(x_i,y_j),$$

where $c_{ij} = 1$ or 2 or 4 is the number of rectangles in the partition that share the vertex (x_i,y_j). Figure 3 illustrates the case $m = n = 4$.
Problem 7 Explain why $T_{mn}[f] \to \iint_R f(x,y)\,dA$ as $m \to \infty$ and $n \to \infty$.

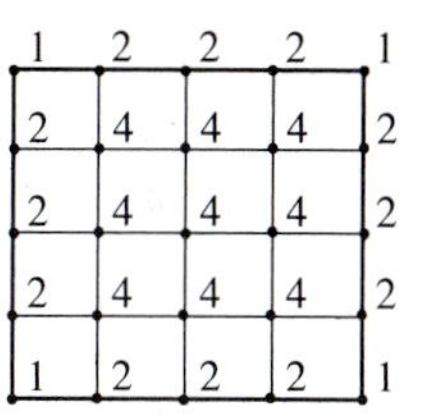

FIGURE 3

Problem 8 Use the trapezoidal rule with $m = n = 2, 4, 8, 16$ to approximate the double integral in Prob. 1.

Similar methods can be applied to double integrals over regions R that are not rectangles. For such regions, decisions must be made about how to deal with rectangles R_{ij} that lie partly inside and partly outside R. In Fig. 4, the dark polygonal line indicates one choice of the rectangles R_{ij} to be included in the calculation of $T_{mn}[f]$. The formula for $T_{mn}[f]$ has the same form as before except that the sum is only over the points (x_i,y_j) in the included rectangles. The coefficients c_{ij} are determined in the same way as before.

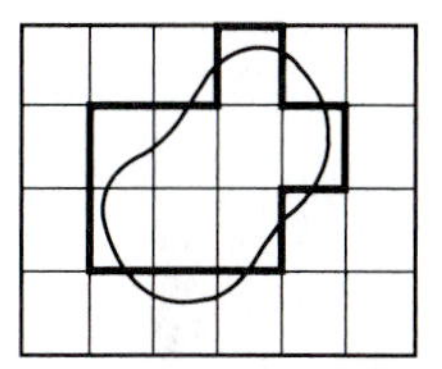
FIGURE 4

Problem 9 For $m = n = 2, 4, 8, 16$, use the trapezoidal rule to approximate the integral in Prob. 2. In the sum for $T_{mn}[f]$ include the points (x_i,y_j) in the rectangles that (a) lie entirely in R and (b) have at least one point in common with R. Average the results in (a) and (b) to obtain other approximations for the integral. Would you expect the averages to be better, worse, or about the same as approximations for the integral as compared to either of the corresponding approximations in (a) or (b)? Explain.

Another natural variant of the preceding ideas applies to y–simple regions. Let R be such a region given by $a \leq x \leq b$ and $\varphi(x) \leq y \leq \psi(x)$. Then

$$\iint_R f(x,y)\,dA = \int_a^b \left[\int_{\varphi(x)}^{\psi(x)} f(x,y)\,dy\right] dx = \int_a^b g(x)\,dx,$$

where $g(x)$ is the integral in square brackets.

Problem 10 Let R be y–simple as before. Apply the one–dimensional trapezoidal rule to $g(x)$ to obtain the approximation

$$\iint_R f(x,y)\,dA \approx h\left[\frac{1}{2}g_0 + g_1 + \cdots + g_{n-1} + \frac{1}{2}g_n\right],$$

where $g_i = g(a + ih)$ for $i = 0, 1, \ldots, n$, and $h = (b - a)/n$.

Usually each of the values g_i in Prob. 10 has to be evaluated by a one–dimensional numerical integration, perhaps using the trapezoidal rule with $n + 1$ points.

Problem 11 Let $f(x,y)$ and R be as in Prob. 2. Use the result of Prob. 10 with $n = 2, 4, 8, 16$ to approximate $\iint_R f(x,y)\,dA$. Use the same number of points and the trapezoidal rule when you evaluate the g_i.

This completes our brief introduction to the numerical evaluation of multiple integrals. A natural next step, which you may wish to carry out on your own, is to develop two–dimensional analogues of Simpson's rule by reasoning much as you have just done for the trapezoidal rule.

Chapter Review Problems

In Probs. 1–10, evaluate the iterated, the double, or the triple integral.

1. $\displaystyle\int_0^{\pi}\int_0^{x} \frac{2x\,\sin y}{\pi^2 - y^2}\,dy\,dx$

2. $\displaystyle\int_0^{\pi/4}\int_0^{y} \cos(x + y)\,dx\,dy$

3. $\displaystyle\int_0^{4}\int_{\sqrt{y}}^{2} \sqrt{1 + x^3}\,dx\,dy$

4. $\displaystyle\int_0^{1}\int_1^{e^x} \frac{1}{y}\,dy\,dx$

5. $\displaystyle\int_0^{1}\int_0^{z}\int_0^{y} (x + y + z)\,dx\,dy\,dz$

6. $\displaystyle\int_0^{1}\int_0^{y}\int_0^{y/(1+x)} dz\,dx\,dy$

7. $\displaystyle\iint_R \frac{1}{1 + x^2 + y^2}\,dA,$ R is the sector of the circle $x^2 + y^2 = 4$ between the lines $y = 0$ and $y = x$.

8. $\displaystyle\iiint_D dV,$ D is the region with $0 \le x \le \pi/4$, $0 \le y \le \sec x$, and $0 \le z \le \sec x$.

9. $\displaystyle\iint_R \sin\theta\,dA,$ R is the region inside the top half of the cardioid $r = 1 + \cos\theta$, $0 \le \theta \le \pi$.

10. $\displaystyle\iint_R xe^y\,dx\,dy,$ R is the region in the first quadrant bounded by the coordinate axes and the curve $y = 4 - x^2$.

11. Integrate $\sin(y^2)$ over the triangle with sides along the y–axis, the line $y = \sqrt{\pi}$, and the line $x = 3y$.

12. Explain why $\int_0^\infty\int_0^\infty[1/(1+x^2+y^2)^2]\,dy\,dx = \pi/4$.

13. Find the area in the first quadrant between the circle $r = 1$ and the parabola $r = 1/(1+\cos\theta)$.

14. Find the area between the spirals $r = \theta$ and $r = 2\theta$ for $0 \le \theta \le 2\pi$.

15. (a) Find the area of the region inside both the circle $x^2 + y^2 = 8$ and the parabola $y = x^2/2$. (b) Find the centroid of the region.

16. Find the centroid of the plane region bounded by the x–axis and the curve $y = \cos x$, $0 \le x \le 2$.

17. Find the average distance from a point in a solid ball of radius 4 to its center.

18. Find the average value of the areas of cross sections of a solid ball with radius 1 perpendicular to a diameter.

19. Find the average value of $f(x, y) = xy$ over the part of the unit disk centered at the origin that lies in the first quadrant.

20. Find the volume of the solid region bounded by the surfaces $y + z = 4$, $x^2 + y^2 = 4$, $y = 0$, and $z = 0$.

21. Find the volume of the solid D that lies above the xy–plane, below the half cone $z = \sqrt{x^2 + y^2}$, and inside the cylinder $x^2 + y^2 = 3x$.

22. Find the volume of the wedge cut from a solid ball with radius 3 by two planes that make a 30° angle at a diameter of the ball.

23. Find the volume of the solid region inside the sphere $x^2 + y^2 + z^2 = 9$ and inside the cylinder $x^2 + y^2 = 3x$.

24. Find the volume of the region bounded by the hyperboloid $z^2 = x^2 + y^2 + 1$ and the plane $z = 2$.

25. Find the volume of the solid in the first octant bounded by the cone $z = r$ and the cylinder $r = 3\cos\theta$.

26. Find the volume of the solid bounded above by the plane $z = 2 - x$, below by the xy–plane, and on the sides by the cylinder $x^2 + y^2 = 4$.

27. Find the volume of the ellipsoid $x^2 + y^2 + z^2/4 = 4$.

28. Find the volume of the solid region inside the sphere with center (0,0,1) and radius 1 and inside the cone $z^2 = 3(x^2 + y^2)$.

29. Find the volume of the solid region inside the sphere with center at the origin and radius 3 and inside the right circular cylinder with radius 3/2 and axis on the line with equations $x = 3$ and $y = 0$.

30. Find the volume of the solid region inside both of the cylinders $x^2 + z^2 = 9$ and $z^2 + y^2 = 9$.

31. (a) Find the volume above the xy–plane that is bounded by the paraboloid $z = 2 - x^2 - y^2$ and the cone $z^2 = x^2 + y^2$. (b) Find the centroid.

32. (a) Find the volume of the region above the xy–plane, below the plane $z = y$, and inside the cylinder $x^2 + y^2 = 4$. (b) Find the centroid.

33. A solid region D lies between the cone $\varphi = \pi/4$ and the sphere $\rho = 2\cos\varphi$. (a) Confirm that the volume of D is π. (b) Find the centroid.

34. Find the centroid of the solid region above the xy–plane, below the parabolic cylinder $z = 1 - y^2$, and between the planes $x = 0$ and $x = 1$.

In Probs. 35–51, the density of a lamina or solid region is described in terms of certain distances. It is understood that such a density is multiplied by a constant with magnitude 1 so that it has units of mass/area or mass/volume, as appropriate.

35. A lamina has the shape of the region inside the limaçon $r = 2 - \cos\theta$. The density at a point in the region is three times its distance from the origin. Find the mass of the region.

36. The lamina between the circles $r = 2$ and $r = 2\cos\theta$ has density r. Find (a) the mass and (b) the center of mass.

37. Let R be the region in the first quadrant bounded by the circle $x^2 + y^2 = 1$ and the lines $x = 1$ and $y = 1$. The density is $\sigma(x, y) = xy$. Find (a) the mass and (b) the center of mass of the R.

38. Let R be the triangular region in the first quadrant bounded by the coordinate axes and the line $x + y = 5$. The density is $\sigma(x, y) = 6\sqrt{x^2 + y^2}$. Find (a) the mass and (b) the center of mass of the region.

39. Find the mass of a solid ball with radius 1, center at the origin, and density (in spherical coordinates) given by e^{ρ^3}.

40. Find the mass of the solid bounded by the planes $x = 1$, $y = 0$, and $z = 0$, and also bounded by the cylinders $x = z^2$ and $y = x^2$. The density is $4xyz$.

41. The solid region between the spheres with center at the origin and radii 1 and 2 has density $15z^2$. Find the mass of the region.

42. Find the mass of the solid with density z and that is bounded by the xy–plane and the paraboloid $z = 1 - x^2 - y^2$.

43. Find the mass of the region inside the sphere with center at the origin and radius 2 if the density is z^2.

44. The solid region lying above the xy–plane and between the spheres with center at the origin and radii 4 and 5 has density $z/(x^2 + y^2 + z^2)$. Find its mass.

45. Let D be the top half of the solid ball with center at the origin and radius 2. Assume the density is $2\sqrt{x^2 + y^2 + z^2}$. Find (a) the mass and (b) the center of mass.

46. Let D be the part of the solid ball with center at the origin and radius 1 that lies in the first octant. Assume the density is $\sigma(x, y, z) = 1/(1 + x^2 + y^2 + z^2)$. Find (a) the mass and (b) the center of mass.

47. A solid D is bounded by the xy–plane, the xz–plane, and the paraboloid $z = 4 - x^2 - y^2$. The density is given by $\sigma(x, y, z) = \sqrt{x^2 + y^2}$. Find (a) the mass and (b) the y–coordinate of the center of mass.

48. Find the center of mass of the solid region inside the sphere $x^2 + y^2 + z^2 = 4$ and inside the cylinder $x^2 + y^2 = 2$ if the density is $\sigma(x, y, z) = 1 + z^4$.

49. The density of the solid sphere $\rho \leq 2\cos\varphi$ is $\sigma(\rho, \varphi, \theta) = 5\rho$. Find the moment of inertia about the z–axis.

50. The density of the solid region between the two spheres $\rho = 3$ and $\rho = 6$ is $\sigma = 1/8\rho$. Find the moment of inertia about the z–axis.

51. The solid rectangle with edges of lengths 3, 4, 5 has constant density 1. Find the moment of inertia about an edge of length 5.

The last few problems deal with angular momentum and assume the optional material in Sec. 4.5 on motion in polar coordinates.

52. Show that motion under a central force takes place in the plane through the origin that has the (constant) angular momentum vector for a normal.

53. An object moves in a plane and its location is given by polar coordinates (r, θ), where r and θ are functions of time along the path of motion. Assume the object has position $(r_0, 0)$ at time $t = 0$. (a) Use $\mathbf{r} = r\mathbf{u}_r$ and $\mathbf{r}' = r'\mathbf{u}_r + r\theta'\mathbf{u}_\theta$ to show that the angular momentum can be expressed as

$$\mathbf{L} = mr^2\theta'\mathbf{k}.$$

(b) Then show that

$$\mathbf{L} = 2mA'\mathbf{k}.$$

where $A = A(t)$ is the area swept out by the moving object up to time t. (c) Conclude that for motion in a plane, angular momentum is conserved if and only if the radius vector sweeps out equal areas in equal times.

54. An object moves with constant velocity. Show that its angular momentum is constant.

55. Verify that for uniform circular motion of a mass m with $r(t) = \rho$ and constant angular velocity $\theta'(t) = \omega$, the angular momentum is $\mathbf{L}(t) = m\rho^2\omega\mathbf{k}$.

CHAPTER 8
ELEMENTS OF VECTOR ANALYSIS

In previous chapters, we have met various scalar functions of two and three variables, such as density and temperature. We have also met vector functions, such as gravitational forces and normals to surfaces. In this chapter on vector analysis, we extend our knowledge of both scalar and vector functions and give a number of important physical applications, especially to fluid flow. In this setting, scalar and vector functions are called scalar fields and vector fields.

In Sec. 8.1, several important vector fields are described. Then the divergence and curl of a vector field are introduced and given physical interpretations. Line integrals are defined and applications to work and conservation of energy are given in Secs. 8.2 and 8.3. Next comes Green's theorem, which expresses a double integral over a plane region as an integral around its boundary. Surface area and surface integrals are introduced in Sec. 8.5. They are needed to extend the two-dimensional versions of the divergence theorem and Stokes' theorem covered in Sec. 8.4 to three dimensions. The rest of the chapter deals with two fundamental results of vector analysis, the divergence theorem and Stokes' theorem in 3–space. In the context of fluid flow, the divergence theorem tells us that the rate at which fluid leaves a solid is equal to the rate at which fluid crosses the surface in the outward direction. Stokes' theorem is concerned with the rotation of a fluid. Other applications to gravitational fields and electromagnetic fields are given.

8.1 Scalar and Vector Fields

The study of scalar and vector fields is motivated by a host of physical applications, including problems involving gravitation, fluid flow, heat conduction, and electromagnetic interactions. Since scalar fields are more familiar and easier to understand, we deal more briefly with them. Most of the section is devoted to vector fields. We begin with general material about scalar and vector fields. Then we focus on three especially important operations, the gradient, divergence, and curl, which produce new scalar or vector fields from given fields.

Definitions of Scalar and Vector Fields

A **scalar field** is simply another name for a real-valued function $f(x,y)$ or $f(x,y,z)$ defined on some region of 2–space or 3–space. A typical scalar field is given by the temperature $T(x,y,z)$ at any point in a room at a particular time. Another scalar field is given by the mass density $\sigma(x,y,z)$ of a solid body. In both examples, the word *field* is meant to convey the idea that a real number, temperature or density, is attached to each point in a region of interest. Standard notation for a scalar field is $f(\mathbf{r})$, where $\mathbf{r} = x\mathbf{i} + y\mathbf{j}$ or $\mathbf{r} = x\mathbf{i} + y\mathbf{j} + z\mathbf{k}$ is the position vector of a point (x, y) or (x, y, z) in the domain of f.

A **vector field** is a vector-valued function $\mathbf{F}(x, y)$ or $\mathbf{F}(x, y, z)$ with domain some region in 2–space or 3–space. A typical vector field is given by the velocity $\mathbf{v}(x, y, z)$ at each point in a river at a particular time. Figure 1 shows how it might look. Alternatively, Fig. 1 might represent a field of force vectors $\mathbf{F}(x, y, z)$. Whatever the interpretation, the picture illustrates the direction and magnitude of the field at each point. For example, the river in Fig. 1 flows more rapidly in the middle and more slowly near the banks. Much as for a scalar field, standard notation for a vector field is $\mathbf{F}(\mathbf{r})$, where $\mathbf{r} = x\mathbf{i} + y\mathbf{j}$ or $\mathbf{r} = x\mathbf{i} + y\mathbf{j} + z\mathbf{k}$ is the position vector of a point (x, y) or (x, y, z) in the domain of $\mathbf{F}$.

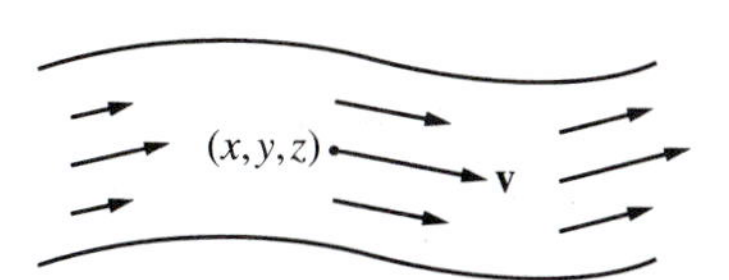

FIGURE 1

As with any function, the domain of a scalar or vector field must be explicitly given or understood from the context. If no instructions are given, the domain is assumed to be as large as possible.

A Few Basic Vector Fields

Next come several vector fields that occur in a number of applications and will serve as running examples. The abbreviations 2-d and 3-d stand for two-dimensional and three-dimensional.

EXAMPLE 1. The 2–d vector field

$$\mathbf{F}(\mathbf{r}) = \mathbf{r} = x\mathbf{i} + y\mathbf{j}$$

is illustrated in Fig. 2. Each vector $\mathbf{F}(\mathbf{r})$ is attached to the point (x, y) with position vector $\mathbf{r}$ and points radially away from the origin. Since $\|\mathbf{F}(\mathbf{r})\| = \|\mathbf{r}\| = \sqrt{x^2 + y^2}$, the length of $\mathbf{F}(\mathbf{r})$ is the distance of (x, y) from the origin. □

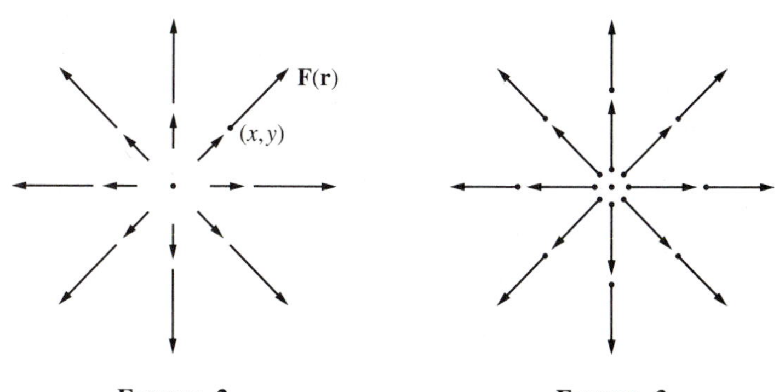

FIGURE 2

FIGURE 3

EXAMPLE 2. For each $\mathbf{r} \neq \mathbf{0}$, let $\mathbf{F}(\mathbf{r})$ be the unit vector

$$\mathbf{F}(\mathbf{r}) = \mathbf{u}_r = \frac{\mathbf{r}}{\|\mathbf{r}\|} = \frac{x\mathbf{i} + y\mathbf{j}}{\sqrt{x^2 + y^2}}.$$

The vectors $\mathbf{F}(\mathbf{r}) = \mathbf{u}_r$ point away from the origin in Fig. 3. These vectors were used in Sec. 4.5 to describe radial components of velocity and acceleration. □

EXAMPLE 3. The 2-d vector field

$$\mathbf{F}(\mathbf{r}) = -y\mathbf{i} + x\mathbf{j}$$

is illustrated in Fig. 4. Since $\mathbf{F}(\mathbf{r}) \cdot \mathbf{r} = 0$, each vector $\mathbf{F}(\mathbf{r})$ is perpendicular to the position vector $\mathbf{r}$. This vector field $\mathbf{F}(\mathbf{r})$ represents the velocity field of a wheel spinning counterclockwise with angular speed 1 rad/sec. □

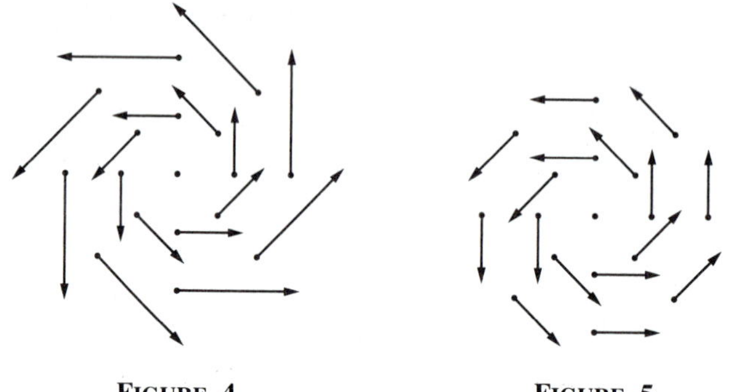

FIGURE 4 FIGURE 5

EXAMPLE 4. For each $\mathbf{r} \neq \mathbf{0}$, let $\mathbf{F}(\mathbf{r})$ be the unit vector

$$\mathbf{F}(\mathbf{r}) = \mathbf{u}_\theta = \frac{-y\mathbf{i} + x\mathbf{j}}{\sqrt{x^2 + y^2}}.$$

See Fig. 5. As in Ex. 3, $\mathbf{F}(\mathbf{r}) \cdot \mathbf{r} = 0$, so that $\mathbf{F}(\mathbf{r}) = \mathbf{u}_\theta$ is perpendicular to the position vector $\mathbf{r}$. These vectors were used in Sec. 4.5 to describe transverse components of velocity and acceleration. For another interpretation of Fig. 5, suppose that the z–axis points right at you. Imagine that it carries a constant electric current in the positive z–direction. Then the vector field in Fig. 5 models the direction of the magnetic field induced by the current. □

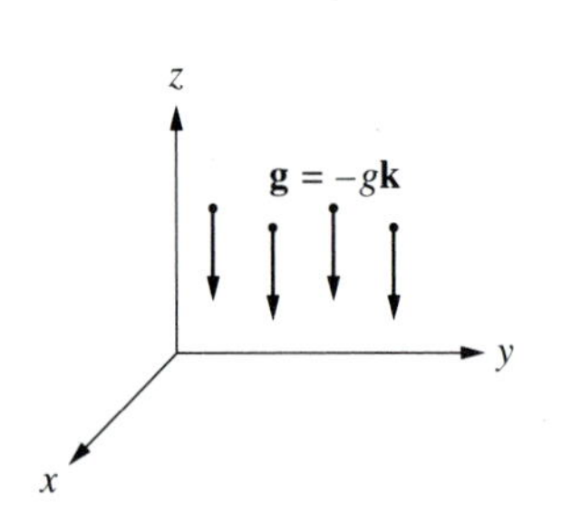

FIGURE 6

EXAMPLE 5. According to Galileo's law, the gravitational force near the earth's surface is given by the uniform (constant) vector field

$$\mathbf{g} = -g\mathbf{k}$$

illustrated in Fig. 6. Here, the xy–plane (near the origin) represents a part of the surface of the earth. □

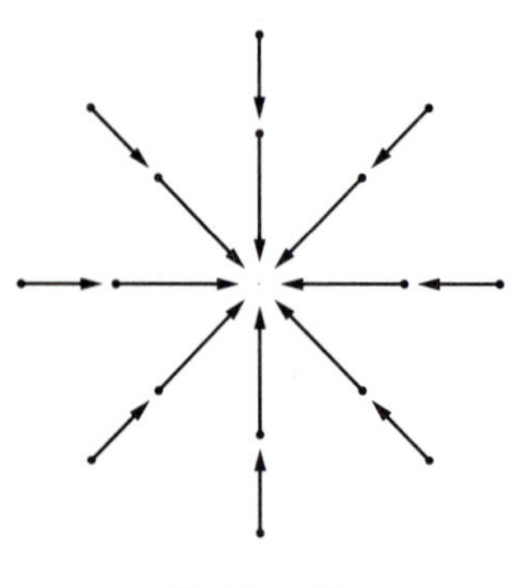

FIGURE 7

EXAMPLE 6. For $\mathbf{r} \neq \mathbf{0}$, let

$$\mathbf{F}(\mathbf{r}) = -\frac{\mathbf{r}}{||\mathbf{r}||^3} = -\frac{x\mathbf{i} + y\mathbf{j} + z\mathbf{k}}{(x^2 + y^2 + z^2)^{3/2}}.$$

This 3-d vector field plays a fundamental role in gravitational and electromagnetic problems. The vector $\mathbf{F}(\mathbf{r})$ points toward the origin and has magnitude $||\mathbf{F}(\mathbf{r})|| = 1/||\mathbf{r}||^2$. Such vector fields are called **inverse square fields**. Figure 7 shows typical vectors $\mathbf{F}(\mathbf{r})$ that lie in the xy–plane. The field looks the same in any other plane through the origin. □

Two inverse square fields deserve special mention. The **gravitational field strength G** due to a point mass M at the origin is

$$\mathbf{G} = -GM\frac{\mathbf{r}}{r^3},$$

where G is the universal gravitational constant. By Newton's law of gravitation, the gravitational force of M on a mass m at $\mathbf{r}$ is $m\mathbf{G}$. Similarly, the **electric field strength E** due to a point charge Q at the origin is

$$\mathbf{E} = \frac{Q}{4\pi\varepsilon_0}\frac{\mathbf{r}}{r^3},$$

where ε_0 is a costant, called the *primitivity of free space*. By Coulomb's law, the electric force that Q imparts to a charge q at $\mathbf{r}$ is $q\mathbf{E}$.

Gradient Fields

Gradient fields are fundamentally important both for their physical applications and from a mathematical point of view. Gradients were introduced in Sec. 5.6 before the concept of a vector field was defined. Now we emphasize that the gradient of a scalar field is a vector field.

Recall, paraphrasing the definition in Sec. 5.6, that the gradient of a differentiable scalar field $f(\mathbf{r})$ is the vector field given by

$$\nabla f = \text{grad } f = \frac{\partial f}{\partial x}\mathbf{i} + \frac{\partial f}{\partial y}\mathbf{j} \qquad \text{in 2–space,}$$

$$\nabla f = \text{grad } f = \frac{\partial f}{\partial x}\mathbf{i} + \frac{\partial f}{\partial y}\mathbf{j} + \frac{\partial f}{\partial z}\mathbf{k} \qquad \text{in 3–space.}$$

The **gradient operator**

$$\nabla = \frac{\partial}{\partial x}\mathbf{i} + \frac{\partial}{\partial y}\mathbf{j} \quad \text{or} \quad \nabla = \frac{\partial}{\partial x}\mathbf{i} + \frac{\partial}{\partial y}\mathbf{j} + \frac{\partial}{\partial z}\mathbf{k}$$

acts on a scalar field f to produce a vector field ∇f, called the **gradient field** of f. For example,

$$f(x,y) = xy \quad \Rightarrow \quad \nabla f = y\mathbf{i} + x\mathbf{j},$$

$$f(x,y,z) = e^{-x}yz^2 \quad \Rightarrow \quad \nabla f = -e^{-x}yz^2\mathbf{i} + e^{-x}z^2\mathbf{j} + 2e^{-x}yz\mathbf{k}.$$

A vector field $\mathbf{F}$ is **conservative** if $\mathbf{F} = \nabla f$ for some scalar field f with continuous first partials. Then f is called a **potential function** for $\mathbf{F}$. The terminology is justified in Sec. 8.3 where we show that a force field is conservative if and only if the total mechanical energy is conserved under the action of the force.

EXAMPLE 7. In the problems you are asked to show that the inverse square field

$$\mathbf{F}(\mathbf{r}) = -\mathbf{r}/\|\mathbf{r}\|^3$$

is a conservative vector field with potential function

$$f(\mathbf{r}) = 1/\|\mathbf{r}\|.$$

In particular, gravitational and Coulomb fields are conservative. □

Expressed in our current language, Th. 1 in Sec. 5.6 tells us that any two potential functions differ by an additive constant.

> **Theorem 1** *Uniqueness of Potentials*
> Assume that $\mathbf{F} = \nabla f$ in a connected open set where f has continuous first partials. Then the general potential function for $\mathbf{F}$ is given by
>
> $$f + C \text{ for any constant } C.$$

For example, all potential functions for the inverse square field $\mathbf{F}(\mathbf{r}) = -\mathbf{r}/\|\mathbf{r}\|^3$ in Ex. 7 are given by

$$f(\mathbf{r}) = \frac{1}{\|\mathbf{r}\|} + C \quad \text{for any constant} \quad C.$$

The following questions about conservative vector fields and potential functions are important both for practical and theoretical reasons. We shall return to them in Sec. 8.3.

Given a vector field $\mathbf{F}$, is $\mathbf{F}$ conservative?
If $\mathbf{F}$ is conservative, how can we find a potential function f?

Divergence of a Vector Field

The divergence operator acts on a vector field to produce a scalar field. It has important applications to fluid flow and many other fields. We shall concentrate on the interpretation of the divergence in the context of fluid flow.

We begin by introducing some standard notion used in vector analysis. The components of a general vector field $\mathbf{F}(\mathbf{r})$ are denoted by

$$\mathbf{F}(\mathbf{r}) = P(x,y)\,\mathbf{i} + Q(x,y)\,\mathbf{j} \quad \text{in 2–space,}$$
$$\mathbf{F}(\mathbf{r}) = P(x,y,z)\,\mathbf{i} + Q(x,y,z)\,\mathbf{j} + R(x,y,z)\,\mathbf{k} \text{ in 3–space.}$$

This notation is very common but, unfortunately, it conflicts with our previous use of P to denote a point. To avoid misunderstandings, we rarely use P for a point in this chapter. We assume throughout that P, Q, and R have partial derivatives with respect to each coordinate variable.

Definition *Divergence of a Vector Field*
The **divergence** of a vector field $\mathbf{F}$ is

$$\text{div}\,\mathbf{F} = \frac{\partial P}{\partial x} + \frac{\partial Q}{\partial y} \qquad \text{in 2–space,}$$

$$\text{div}\,\mathbf{F} = \frac{\partial P}{\partial x} + \frac{\partial Q}{\partial y} + \frac{\partial R}{\partial z} \quad \text{in 3–space.}$$

As we proceed, keep in mind that the divergence of a vector field is a scalar field.

EXAMPLE 8. Find the divergence of the vector field

$$\mathbf{F} = y \ln x\mathbf{i} + xe^{y^2}\mathbf{j} + y^2 \sin \pi z\mathbf{k}.$$

Solution. In this case, $P = y \ln x$, $Q = xe^{y^2}$, and $R = y^2 \sin \pi z$. So

$$\text{div}\,\mathbf{F} = \frac{\partial P}{\partial x} + \frac{\partial Q}{\partial y} + \frac{\partial R}{\partial z} = \frac{y}{x} + 2xye^{y^2} + \pi y^2 \cos \pi z \ \square.$$

The physical meaning of divergence is most easily understood in terms of a steady fluid flow. *Steady* refers to time and means that the flow pattern does not change as time passes. The vector field $\mathbf{v} = \mathbf{v}(x, y, z)$ that gives the velocity of a steady flow at a point (x, y, z) in the flow region is called the **velocity field** of the flow. Let $\sigma(x, y, z)$ be the mass density of the fluid at (x, y, z). Assume that $\mathbf{v}$ and σ are continuous functions. The vector field $\mathbf{F} = \sigma\mathbf{v}$ is called the **mass flow field**. For any (x, y, z) the vector $\mathbf{F}(x, y, z) = \sigma(x, y, z)\mathbf{v}(x, y, z)$ is the **mass flow vector** at (x, y, z).

We shall interpret divergence first in terms of a two-dimensional flow. Thus, suppose that the fluid velocity and mass density are independent of z. Then the mass flow vector is $\mathbf{F}(x, y) = \sigma(x, y)\mathbf{v}(x, y)$, where $\sigma(x, y)$ is the mass per unit area, as described in Sec. 7.2. Express the velocity field as $\mathbf{v}(\mathbf{r}) = v_1(\mathbf{r})\,\mathbf{i} + v_2(\mathbf{r})\,\mathbf{j}$ where $\mathbf{r} = x\mathbf{i} + y\mathbf{j}$. Then $v_1(\mathbf{r})$ and $v_2(\mathbf{r})$ are the horizontal and vertical components of the velocity field. The mass flow vector is given by

$$\mathbf{F} = P\mathbf{i} + Q\mathbf{j}, \text{ where } P = \sigma v_1 \text{ and } Q = \sigma v_2.$$

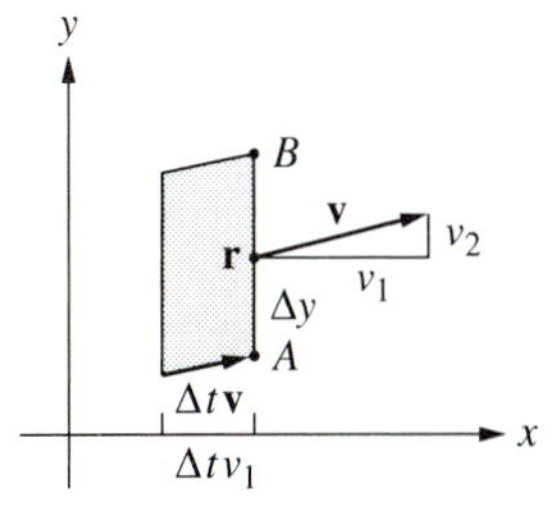

FIGURE 8

Figure 8 will lead to physical interpretations of P and Q.

In the figure, $\overline{AB}$ is a vertical line segment centered at $\mathbf{r}$ with small length Δy. During a short time interval Δt the area of fluid that crosses $\overline{AB}$ is approximately the area of the shaded parallelogram, which is $\Delta t v_1(\mathbf{r})\,\Delta y$. The mass crossing $\overline{AB}$ in time Δt is approximately $\sigma(\mathbf{r})\,v_1(\mathbf{r})\,\Delta t\Delta y = P(\mathbf{r})\,\Delta t\Delta y$. Divide by $\Delta t\Delta y$ and let $\Delta t \to 0$ and $\Delta y \to 0$ to find that

$P(\mathbf{r})$ = the net rate per unit time per unit length in the y–direction that fluid mass flows past $\mathbf{r}$ in the positive x–direction.

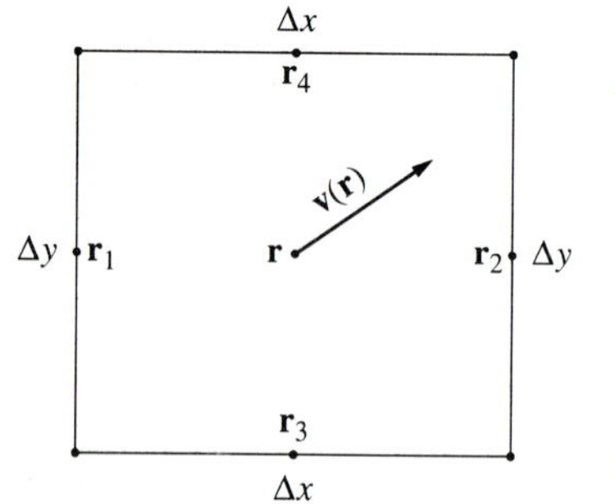

FIGURE 9

The fluid flows to the right at $\mathbf{r}$ if $P(\mathbf{r}) > 0$, as in Fig. 8, and to the left if $P(\mathbf{r}) < 0$. Similarly,

$Q(\mathbf{r})$ = the net rate per unit time per unit length in the x–direction that fluid mass flows past $\mathbf{r}$ in the positive y–direction.

Next we interpret $\partial P/\partial x$ and $\partial Q/\partial y$. To this end, Fig. 9 shows a rectangle with center $\mathbf{r} = <x, y>$, side lengths Δx and Δy, and area $\Delta A = \Delta x \Delta y$. Let

$$\mathbf{r}_1 = \left\langle x - \frac{\Delta x}{2}, y \right\rangle, \quad \mathbf{r}_2 = \left\langle x + \frac{\Delta x}{2}, y \right\rangle, \quad \mathbf{r}_3 = \left\langle x, y - \frac{\Delta y}{2} \right\rangle, \quad \mathbf{r}_4 = \left\langle x, y + \frac{\Delta y}{2} \right\rangle.$$

From Fig. 9 and the interpretation of $P(\mathbf{r})$ given previously, the net mass of fluid leaving the rectangle through the vertical sides per unit area per unit time is approximately

$$\frac{P(\mathbf{r}_2)\,\Delta t\,\Delta y - P(\mathbf{r}_1)\,\Delta t\,\Delta y}{\Delta t\,\Delta A} = \frac{P(x + \Delta x/2, y) - P(x - \Delta x/2, y)}{\Delta x} \approx \frac{\partial P}{\partial x}.$$

Similarly, the net mass of fluid leaving the rectangle through the horizontal sides per unit area per unit time is approximately

$$\frac{Q(\mathbf{r}_4)\,\Delta t\,\Delta x - Q(\mathbf{r}_3)\,\Delta t\,\Delta x}{\Delta t\,\Delta A} = \frac{Q(x, y + \Delta y/2) - Q(x, y - \Delta y/2)}{\Delta y} \approx \frac{\partial Q}{\partial y}.$$

The net mass of fluid leaving the rectangle through the four sides per unit area per unit time is approximately $\text{div}\,\mathbf{F} = \partial P/\partial x + \partial Q/\partial y$. The approximation becomes exact in the limit as $\Delta x \to 0$, $\Delta y \to 0$, and $\Delta t \to 0$. So we arrive at the following interpretation of divergence in terms of a two-dimensional flow.

div $\mathbf{F}(x, y)$	=	the net rate per unit area per unit time that fluid mass leaves the vicinity of (x, y).

Similarly, for a three-dimensional flow,

div $\mathbf{F}(x, y, z)$	=	the net rate per unit volume per unit time that fluid mass leaves the vicinity of (x, y, z).

A more complete discussion of the meaning of divergence will be given later in the chapter.

In view of the physical interpretation of the divergence for a mass flow field $\mathbf{F}$, if div $\mathbf{F}(x, y, z) > 0$, then (x, y, z) is called a **source** and fluid is injected into the flow at that point. If div $\mathbf{F}(x, y, z) < 0$, then (x, y, z) is called a **sink** and fluid is removed from the flow at that point. The same terminology is used for two-dimensional flows and also for a general vector field $\mathbf{F}$.

EXAMPLE 9. Find the divergence of the vector field in Ex. 2,

$$\mathbf{F} = \frac{x\mathbf{i} + y\mathbf{j}}{\sqrt{x^2 + y^2}}, \qquad (x,y) \neq (0,0).$$

Solution. In this case, $\mathbf{F} = P\mathbf{i} + Q\mathbf{j}$ with

$$P = x(x^2 + y^2)^{-1/2} \text{ and } Q = y(x^2 + y^2)^{-1/2}.$$

First,

$$\frac{\partial P}{\partial x} = x\left(-\frac{1}{2}\right)(x^2 + y^2)^{-3/2}(2x) + (x^2 + y^2)^{-1/2},$$

$$= \frac{-x^2}{(x^2 + y^2)^{3/2}} + \frac{x^2 + y^2}{(x^2 + y^2)^{3/2}} = \frac{y^2}{(x^2 + y^2)^{3/2}}.$$

By symmetry (meaning that P and Q differ only by the interchange of x and y),

$$\frac{\partial Q}{\partial y} = \frac{x^2}{(x^2 + y^2)^{3/2}}.$$

So the divergence of $\mathbf{F}$ is

$$\text{div } \mathbf{F} = \frac{\partial P}{\partial x} + \frac{\partial Q}{\partial y} = \frac{x^2 + y^2}{(x^2 + y^2)^{3/2}} = \frac{1}{(x^2 + y^2)^{1/2}}.$$

Since div $\mathbf{F}(x,y) > 0$, every point $(x,y) \neq (0,0)$ is a source for this vector field. □

EXAMPLE 10. Find the divergence of the vector field in Ex. 4,

$$\mathbf{F} = \frac{-y\mathbf{i} + x\mathbf{j}}{\sqrt{x^2 + y^2}}, \qquad (x,y) \neq (0,0).$$

Solution. By routine steps,

$$\frac{\partial}{\partial x}\left(\frac{-y}{\sqrt{x^2 + y^2}}\right) = \frac{2xy}{(x^2 + y^2)^{3/2}}, \qquad \frac{\partial}{\partial y}\left(\frac{x}{\sqrt{x^2 + y^2}}\right) = \frac{-2xy}{(x^2 + y^2)^{3/2}}.$$

Hence, div $\mathbf{F} = 0$. This vector field has no sources or sinks. □

EXAMPLE 11. Find the divergence of the inverse square field

$$\mathbf{F} = -\frac{\mathbf{r}}{\|\mathbf{r}\|^3} = -\frac{x\mathbf{i} + y\mathbf{j} + z\mathbf{k}}{(x^2 + y^2 + z^2)^{3/2}}, \qquad \mathbf{r} \neq \mathbf{0}.$$

Solution. By the quotient rule,

$$\frac{\partial}{\partial x}\left(-\frac{x}{(x^2+y^2+z^2)^{3/2}}\right)=\frac{2x^2-y^2-z^2}{(x^2+y^2+z^2)^{5/2}}.$$

By symmetry,

$$\frac{\partial}{\partial y}\left(-\frac{y}{(x^2+y^2+z^2)^{3/2}}\right)=\frac{2y^2-x^2-z^2}{(x^2+y^2+z^2)^{5/2}},$$

$$\frac{\partial}{\partial z}\left(-\frac{z}{(x^2+y^2+z^2)^{3/2}}\right)=\frac{2z^2-x^2-y^2}{(x^2+y^2+z^2)^{5/2}}.$$

Add these three partial derivatives to obtain

$$\text{div } \mathbf{F} = 0.$$

Thus, the divergence of a gravitational field or a Coulomb field is 0. Remember that $\mathbf{r} = \mathbf{0}$, which is the location of the point mass or point charge that creates the field, is not in the domain of $\mathbf{F}$. □

It is useful to think of div $\mathbf{F}$ as a symbolic dot product of the gradient operator ∇ with $\mathbf{F}$. For example,

$$\text{div } \mathbf{F} = \frac{\partial P}{\partial x}+\frac{\partial Q}{\partial y}=\left(\frac{\partial}{\partial x}\mathbf{i}+\frac{\partial}{\partial y}\mathbf{j}\right)\cdot(P\mathbf{i}+Q\mathbf{j}) \qquad \text{in 2–space,}$$

$$\text{div } \mathbf{F} = \frac{\partial P}{\partial x}+\frac{\partial Q}{\partial y}+\frac{\partial R}{\partial z}=\left(\frac{\partial}{\partial x}\mathbf{i}+\frac{\partial}{\partial y}\mathbf{j}+\frac{\partial}{\partial z}\mathbf{k}\right)\cdot(P\mathbf{i}+Q\mathbf{j}+R\mathbf{k}) \text{ in 3–space.}$$

In either case,

$$\boxed{\text{div } \mathbf{F} = \nabla\cdot\mathbf{F}.}$$

Both of these expressions for the divergence are in common use.

Curl of a Vector Field

The curl of a vector field is a measure of the rotation of the field about any point. In fact, an older term for curl is *rotation.* The concept arose in studies of the rotation of a solid body and the circulation of a fluid.

We start with the curl of a vector field in the xy–plane.

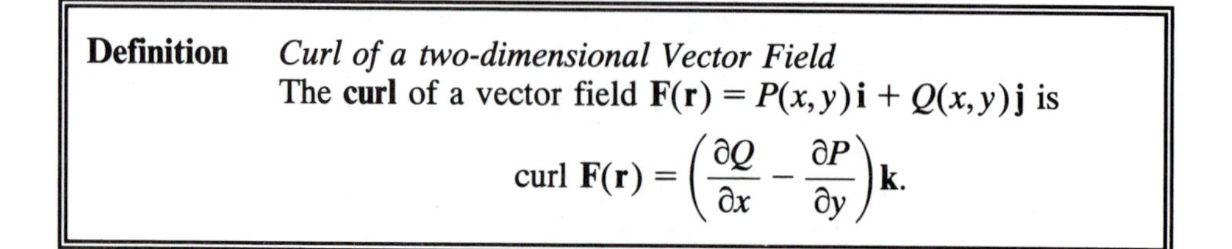

Definition *Curl of a two-dimensional Vector Field*
The **curl** of a vector field $\mathbf{F}(\mathbf{r}) = P(x,y)\mathbf{i} + Q(x,y)\mathbf{j}$ is

$$\text{curl } \mathbf{F}(\mathbf{r}) = \left(\frac{\partial Q}{\partial x}-\frac{\partial P}{\partial y}\right)\mathbf{k}.$$

Notice that the curl of a vector field is a vector field. The expression for curl may look strange at first. It will seem more reasonable after some discussion and a few examples.

Figure 10 will help explain how curl is related to rotational motion. It shows a thin plate in the xy–plane that rotates about the z–axis with (possibly variable) angular velocity $\omega = d\theta/dt$. The angular velocity vector is $\boldsymbol{\omega} = \omega\mathbf{k}$. (See Sec. 6.5.) Since each point $(x, y) = (r\cos\theta,\ r\sin\theta)$ moves on a circle with radius $r = \sqrt{x^2 + y^2}$,

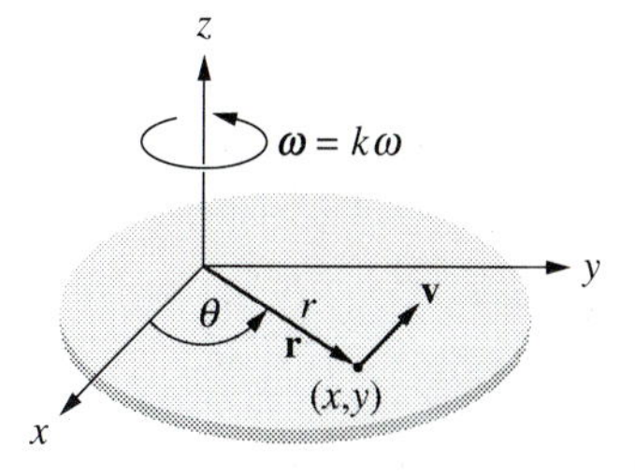

FIGURE 10

$$\mathbf{r} = x\mathbf{i} + y\mathbf{j} = r\cos\theta\,\mathbf{i} + r\sin\theta\,\mathbf{j}.$$

By the chain rule with $d\theta/dt = \omega$, the point (x, y) has velocity

$$\mathbf{v} = \frac{d\mathbf{r}}{dt} = (-r\sin\theta)\,\omega\mathbf{i} + (r\cos\theta)\,\omega\mathbf{j} = -\omega y\mathbf{i} + \omega x\mathbf{j}.$$

Equivalently,

$$\mathbf{v} = P\mathbf{i} + Q\mathbf{j} \text{ with } P = -\omega y \text{ and } Q = \omega x.$$

The curl of the velocity field is given by

$$\text{curl } \mathbf{v} = \left(\frac{\partial Q}{\partial x} - \frac{\partial P}{\partial y}\right)\mathbf{k} = 2\omega\mathbf{k} = 2\boldsymbol{\omega}.$$

Thus, apart from the factor 2, curl $\mathbf{v}$ is the angular velocity of the rotating plate. In the special case with $\omega = 1$, the plate rotates counterclockwise at 1 rad/sec and the velocity field $\mathbf{v}$ of the plate is the vector field in Ex. 3 and Fig. 4.

EXAMPLE 12. Find the curl of the vector field of Ex. 2 and Fig. 3,

$$\mathbf{F} = \frac{x\mathbf{i} + y\mathbf{j}}{\sqrt{x^2 + y^2}}, \qquad (x, y) \neq (0, 0).$$

Solution. Now $\mathbf{F} = P\mathbf{i} + Q\mathbf{j}$ with

$$P = \frac{x}{\sqrt{x^2 + y^2}} \quad \text{and} \quad Q = \frac{y}{\sqrt{x^2 + y^2}}.$$

Routine calculations give

$$\frac{\partial P}{\partial y} = \frac{\partial Q}{\partial x} = -\frac{2xy}{(x^2 + y^2)^{3/2}}.$$

Therefore, curl $\mathbf{F} = \mathbf{0}$. The vector field in Fig. 3 does not tend to rotate around any point. □

EXAMPLE 13. Find the curl of the vector field in Ex. 4,

$$\mathbf{F} = \frac{-y\mathbf{i} + x\mathbf{j}}{\sqrt{x^2 + y^2}}, \qquad (x, y) \neq (0, 0).$$

Solution. Here, $P = -y/\sqrt{x^2 + y^2}$ and $Q = x/\sqrt{x^2 + y^2}$. Then

$$\frac{\partial P}{\partial y} = -\frac{x^2}{(x^2 + y^2)^{3/2}}, \qquad \frac{\partial Q}{\partial x} = \frac{y^2}{(x^2 + y^2)^{3/2}},$$

and the curl of $\mathbf{F}$ is

$$\text{curl } \mathbf{F} = \frac{(x^2 + y^2)\mathbf{k}}{(x^2 + y^2)^{3/2}} = \frac{\mathbf{k}}{\sqrt{x^2 + y^2}} = \frac{\mathbf{k}}{\|\mathbf{r}\|}.$$

The field rotates rapidly about points near the origin and slowly about points far away from the origin. □

Now we turn to the curl of a vector field in 3–space.

Definition *Curl of a three-dimensional Vector Field*
The **curl** of a vector field
$\mathbf{F}(\mathbf{r}) = P(x, y, z)\mathbf{i} + Q(x, y, z)\mathbf{j} + R(x, y, z)\mathbf{k}$ is

$$\text{curl } \mathbf{F} = \left(\frac{\partial R}{\partial y} - \frac{\partial Q}{\partial z}\right)\mathbf{i} + \left(\frac{\partial P}{\partial z} - \frac{\partial R}{\partial x}\right)\mathbf{j} + \left(\frac{\partial Q}{\partial x} - \frac{\partial P}{\partial y}\right)\mathbf{k}.$$

One way to look at this formula for the curl is to regard it as a superposition of three 2–d curls. When $\mathbf{F} = P(x, y)\mathbf{i} + Q(x, y)\mathbf{j}$, we have $R = 0$ and

$$\text{curl } \mathbf{F} = \left(\frac{\partial Q}{\partial x} - \frac{\partial P}{\partial y}\right)\mathbf{k},$$

which is the expression for the curl in the *xy*–plane.

A symbolic formula for the curl in 3–space that is easier to remember is

$$\text{curl } \mathbf{F} = \nabla \times \mathbf{F} = \begin{vmatrix} \mathbf{i} & \mathbf{j} & \mathbf{k} \\ \dfrac{\partial}{\partial x} & \dfrac{\partial}{\partial y} & \dfrac{\partial}{\partial z} \\ P & Q & R \end{vmatrix}.$$

When evaluating the determinant on the right the *product* $(\partial/\partial x) \cdot Q$ is interpreted to mean $\partial Q/\partial x$, and so on. Expand the determinant to see that the correct formula for the curl is obtained.

EXAMPLE 14. Find the curl of the vector field $\mathbf{F} = xyz\mathbf{i} + ze^y\mathbf{j} + y\sin z\mathbf{k}$.

Solution. In this case,

$$\operatorname{curl}\mathbf{F} = \begin{vmatrix} \mathbf{i} & \mathbf{j} & \mathbf{k} \\ \dfrac{\partial}{\partial x} & \dfrac{\partial}{\partial y} & \dfrac{\partial}{\partial z} \\ xyz & ze^y & y\sin z \end{vmatrix} = \mathbf{i}(\sin z - e^y) - \mathbf{j}(0 - xy) + \mathbf{k}(0 - xz),$$

$$= (\sin z - e^y)\mathbf{i} + xy\mathbf{j} - xz\mathbf{k}. \ \square$$

The physical interpretations of curl for two-dimensional vector fields carry over to three-dimensional vector fields. In fact, the discussion accompanying Fig. 10 extends quite easily to a three-dimensional rotating solid as in Fig. 11. Suppose that the solid rotates about the axis shown in the figure with angular velocity vector $\boldsymbol{\omega}$. Then the scalar angular velocity is $\omega = ||\boldsymbol{\omega}||$ rad/sec. A point (x, y, z) in the body with position vector $\mathbf{r}$ moves with velocity $\mathbf{v} = \boldsymbol{\omega} \times \mathbf{r}$ and $\operatorname{curl}\mathbf{v} = \operatorname{curl}(\boldsymbol{\omega} \times \mathbf{r}) = 2\boldsymbol{\omega}$. We ask you to verify these facts in the problems.

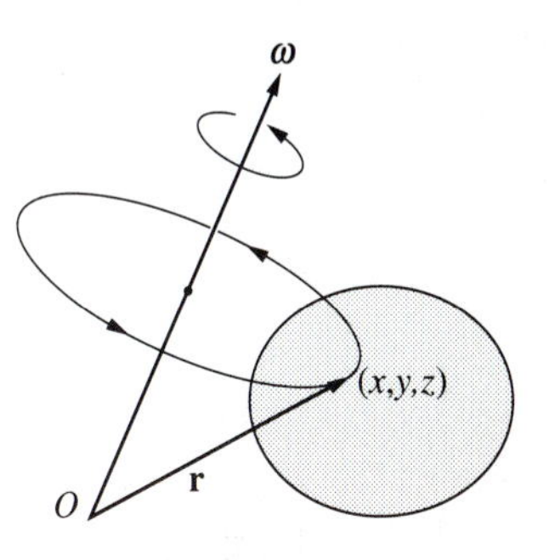

FIGURE 11

There are many useful identities involving gradient, divergence, and curl. We mention two of them here. Others appear in the problems. Let f and $\mathbf{F}$ be scalar and vector fields with continuous second-order partial derivatives. Then

$$\operatorname{curl}(\nabla f) = \mathbf{0},$$
$$\operatorname{div}(\operatorname{curl}\mathbf{F}) = 0.$$

In words, the curl of a gradient is zero and the divergence of a curl is zero. We leave the verifications to the problems.

EXAMPLE 15. Find the curl of the inverse square field,

$$\mathbf{F} = -\frac{\mathbf{r}}{||\mathbf{r}||} = -\frac{x\mathbf{i} + y\mathbf{j} + z\mathbf{k}}{(x^2 + y^2 + z^2)^{3/2}}.$$

Solution 1. In this case, $\mathbf{F} = P\mathbf{i} + Q\mathbf{j} + R\mathbf{k}$ with

$$P = \frac{-x}{(x^2 + y^2 + z^2)^{3/2}}, \qquad Q = \frac{-y}{(x^2 + y^2 + z^2)^{3/2}}, \qquad R = \frac{-z}{(x^2 + y^2 + z^2)^{3/2}}.$$

By symmetry or direct calculations,

$$\frac{\partial R}{\partial y} = \frac{\partial Q}{\partial z}, \qquad \frac{\partial R}{\partial x} = \frac{\partial P}{\partial z}, \qquad \frac{\partial Q}{\partial x} = \frac{\partial P}{\partial y}.$$

It follows that $\operatorname{curl}\mathbf{F} = \mathbf{0}$.

Solution 2. From Ex. 7, $\mathbf{F} = \nabla f$ for $f = 1/||\mathbf{r}||$. Hence,

$$\operatorname{curl}\mathbf{F} = \operatorname{curl}(\nabla f) = \mathbf{0}. \ \square$$

PROBLEMS

In Probs. 1–10 make a sketch of the vector field. Display enough vectors to illustrate the overall pattern of the field, as in Fig. 2–7.

1. $\mathbf{F}(x,y) = \mathbf{i} + \mathbf{j}$
2. $\mathbf{F}(x,y) = \mathbf{i} - 2\mathbf{j}$
3. $\mathbf{F}(x,y) = x\mathbf{i} + \mathbf{j}$
4. $\mathbf{F}(x,y) = \mathbf{i} - y\mathbf{j}$
5. $\mathbf{F}(x,y) = y\mathbf{i} + x\mathbf{j}$
6. $\mathbf{F}(x,y) = y\mathbf{i} - x\mathbf{j}$
7. $\mathbf{F}(x,y) = 2\mathbf{j}$
8. $\mathbf{F}(x,y) = \mathbf{j} + \mathbf{k}$
9. $\mathbf{F}(x,y,z) = x\mathbf{i} + y\mathbf{j} + z\mathbf{k}$
10. $\mathbf{F}(x,y,z) = (x\mathbf{i} + y\mathbf{j} + z\mathbf{k})/(x^2 + y^2 + z^2)$

In Probs. 11–18 find the vector field that has the given potential function.

11. $f(x,y) = x^2 - y^2$
12. $f(x,y) = \ln\sqrt{x^2 + y^2}$
13. $f(x,y) = \cos(x - y)$
14. $f(x,y) = e^{xy}$
15. $f(x,y,z) = xy^2z^3$
16. $f(x,y,z) = xy + yz + zx$
17. $f(x,y,z) = \ln(x + y + z)$
18. $f(x,y) = (x^2 + y^2)^{1/2}e^{-z}$

In Probs. 19–30 find (a) the divergence and (b) the curl of the vector field.

19. $\mathbf{F}(x,y) = x\mathbf{i} + y\mathbf{j}$
20. $\mathbf{F}(x,y) = \sqrt{x^2 + y^2}(x\mathbf{i} + y\mathbf{j})$
21. $\mathbf{F}(x,y) = e^x\cos y\mathbf{i} + e^x\sin y\mathbf{j}$
22. $\mathbf{F}(x,y) = e^{xy}\mathbf{i} - \ln(y/x)\mathbf{j}$
23. $\mathbf{F}(x,y) = (x^2 + y^2)^{-1}(x\mathbf{i} + y\mathbf{j})$
24. $\mathbf{F}(x,y) = (x^2 + y^2)^{-1}(y\mathbf{i} + x\mathbf{j})$
25. $\mathbf{F}(x,y,z) = x\mathbf{i} + y\mathbf{j} + z\mathbf{k}$
26. $\mathbf{F}(x,y,z) = xy\mathbf{i} + yz\mathbf{j} + zx\mathbf{k}$
27. $\mathbf{F}(x,y,z) = \sin(x+y)\mathbf{i} + e^{-x}\mathbf{j} + xz\mathbf{k}$
28. $\mathbf{F}(x,y,z) = (x+y+z)(x\mathbf{i} + y\mathbf{j} + z\mathbf{k})$
29. $\mathbf{F}(x,y,z) = f(x)\mathbf{i} + g(y)\mathbf{j} + h(z)\mathbf{k}$ where f, g, and h are differentiable functions.
30. $\mathbf{F}(x,y,z) = f(y,z)\mathbf{i} + g(x,z)\mathbf{j} + h(x,y)\mathbf{k}$ where f, g, and h have first-order partial derivatives.

The **streamlines** of a vector field $\mathbf{F}$ are the curves C whose tangent vectors at each point have the same direction as the field at that point. If $\mathbf{F} = \sigma\mathbf{v}$ is a mass flow field, the streamlines are the paths along which fluid particles move.

31. (a) Explain why the streamlines of a vector field $\mathbf{F}$ are given by the solution curves of the vector differential equation

$$\mathbf{r}'(t) = \mathbf{G}(\mathbf{r}(t))$$

where $\mathbf{G}$ is any convenient vector field that points in the same direction as $\mathbf{F}$ at each point. (b) If $\mathbf{G} = P\mathbf{i} + Q\mathbf{j}$ for a 2–d field, show that the vector equation in (a) is equivalent to the two scalar equations

$$x' = P(x,y), \qquad y' = Q(x,y),$$

where x and y are functions of t.

In Probs. 32–36 (a) guess the streamlines of the vector field and then (b) solve the differential equation in Prob. 31 to confirm your guess. *Hint.* In (b) you are free to choose any convenient $\mathbf{G}$ that points in the same direction as $\mathbf{F}$. Sometimes there are better choices than $\mathbf{G} = \mathbf{F}$.

32. $\mathbf{F}(x,y) = x\mathbf{i} + y\mathbf{j}$ from Ex. 1.
33. $\mathbf{F}(x,y) = \mathbf{u}_r$ from Ex. 2.
34. $\mathbf{F}(x,y) = \mathbf{u}_\theta$ from Ex. 4.
35. $\mathbf{F}(x,y) = \mathbf{g}$ from Ex. 5.
36. $\mathbf{F}(\mathbf{r}) = -\mathbf{r}/\|\mathbf{r}\|^3$ from Ex. 6.
37. Find the streamlines of the vector field

$$\mathbf{F}(x,y,z) = (y\mathbf{i} - x\mathbf{j} + \mathbf{k})/(x^2 + y^2 + z^2)^{3/2}.$$

Hint. The differential $x'' + x = 0$ has the general solution $x = x_0\cos(t + \varphi_0)$ where x_0 and φ_0 are arbitrary constants.

Problems 38–43 list a number of useful vector identities. Establish each identity by expanding the left member, simplifying, and combining as needed. Assume that the vector fields $\mathbf{F}$, $\mathbf{G}$ and the scalar fields f,g are functions of (x,y,z) with continuous partial derivatives. The letters a, b are constants.

38. $\operatorname{div}(a\mathbf{F} + b\mathbf{G}) = a \operatorname{div}\mathbf{F} + b \operatorname{div}\mathbf{G}$
39. $\operatorname{curl}(a\mathbf{F} + b\mathbf{G}) = a \operatorname{curl}\mathbf{F} + b \operatorname{curl}\mathbf{G}$
40. $\operatorname{div}(f\mathbf{F}) = f \operatorname{div}\mathbf{F} + \operatorname{grad} f \cdot \mathbf{F}$
41. $\operatorname{curl}(f\mathbf{F}) = f \operatorname{curl}\mathbf{F} + (\operatorname{grad} f) \times \mathbf{F}$
42. $\operatorname{div}(\operatorname{curl}\mathbf{F}) = 0$
43. $\operatorname{curl}(\operatorname{grad} f) = \mathbf{0}$
44. Write out each of the identities in Probs. 38–43 using the symbolic expressions grad $= \nabla$, div $= \nabla\,\cdot$, and curl $= \nabla\,\times$. Some of the identities are easier to remember in the symbolic forms.
45. Show that the inverse square field in Ex. 7 is the gradient of the given scalar field.
46. The **Laplacian** Δf of a scalar field $f(x,y)$ or $f(x,y,z)$ with continuous second-order partial derivatives is defined by

$$\Delta f = f_{xx} + f_{yy} + f_{zz},$$

where the last term is deleted in 2–d case. (a) Show that

$$\Delta f = \operatorname{div}(\operatorname{grad} f) = \nabla \cdot \nabla f.$$

(b) Show that $\Delta(fg) = f\Delta g + g\Delta f + 2\nabla f \cdot \nabla g$.

In Probs. 47–54, establish special identities for the position vector $\mathbf{r} = x\mathbf{i} + y\mathbf{j} + z\mathbf{k}$ and a constant vector $\mathbf{a}$. As usual, $r = \|\mathbf{r}\|$.

47. $\operatorname{div}\mathbf{r} = 3$
48. $\operatorname{curl}\mathbf{r} = \mathbf{0}$
49. $\operatorname{div}(\mathbf{r}/r^3) = 0$
50. $\operatorname{curl}(\mathbf{r}/r^3) = \mathbf{0}$
51. $\operatorname{div}(\mathbf{a} \times \mathbf{r}) = \mathbf{0}$
52. $\operatorname{curl}(\mathbf{a} \times \mathbf{r}) = 2\mathbf{a}$
53. $\operatorname{div}(r\mathbf{r}) = 4r$
54. $\operatorname{div}(\mathbf{r}/r^2) = 1/r^2$

55. Find the value(s) of n if any such that (a) $\mathrm{div}(r^n \mathbf{r}) = 0$ and (b) $\mathrm{curl}(r^n \mathbf{r}) = \mathbf{0}$.

56. Refer to the rotating solid in Fig. 11 and the accompanying text. Show that the point (x, y, z) with position vector $\mathbf{r}$ has velocity $\mathbf{v} = \boldsymbol{\omega} \times \mathbf{r}$. Then show that curl $\mathbf{v} = 2\boldsymbol{\omega}$.

8.2 Line Integrals

Line integrals are integrals along curves. They are natural generalizations of definite integrals. Line integrals come up in many applications. For example, the mass of a curved wire is the integral of its density (mass per unit length) with respect to length along the curve. If a force acts on an object as it moves along a curve, then the work done by the force is the integral of the tangential component of the force along the curve. We shall meet other applications of line integrals as we go along.

The terminology *line integral* is standard but rather misleading because the integrals are over curves (that may be lines). An older name is *curvilinear integral.*

A Review of Curves

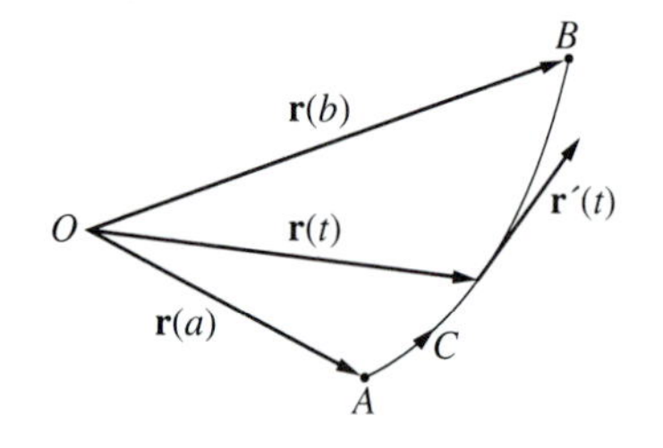

FIGURE 1

Since curves play an important role in the study of line integrals, we begin our discussion of line integrals with a brief review of curves that focuses on aspects of curves that are especially important for line integrals. See Ch. 4 for more details.

A *parametric curve* C, illustrated in Fig. 1, is represented in vector form by $\mathbf{r}(t)$, $a \le t \le b$, where $\mathbf{r}(t) = x(t)\mathbf{i} + y(t)\mathbf{j}$ in 2–space and $\mathbf{r}(t) = x(t)\mathbf{i} + y(t)\mathbf{j} + z(t)\mathbf{k}$ in 3–space. The parametric curve C is directed. The arrow in Fig. 1 indicates the direction on C, which is the direction that $\mathbf{r}(t)$ sweeps out the points of C as t increases. The point A with position vector $\mathbf{r}(a)$ is the *initial point* of C and the point B with position vector $\mathbf{r}(b)$ is the *terminal point* of C.

It is often useful to express curves in 2–space or 3–space as parametric curves. For example, the parabolic arc $y = x^2$ for $-2 \le x \le 2$ in Fig. 2 can be regarded as the parametric curve C in Fig. 3 given by $\mathbf{r}(t) = t\mathbf{i} + t^2\mathbf{j}$ for $-2 \le t \le 2$. Now the parabolic arc is directed. The same directed curve is given by the different parametric representation $\mathbf{r}(\theta) = \sin\theta\,\mathbf{i} + \sin^2\theta\,\mathbf{j}$ for $-\pi/2 \le \theta \le \pi/2$. Figure 4 shows the parabolic arc in Fig. 2 expressed as the parametric curve with $\mathbf{r}(t) = -t\mathbf{i} + t^2\mathbf{j}$ for $-2 \le t \le 2$. Now the direction is reversed from that of the curve C in Fig. 3. We denote the curve in Fig. 4 by $-C$. In general, for any parametric curve C the curve $-C$ is obtained from C by reversing its direction.

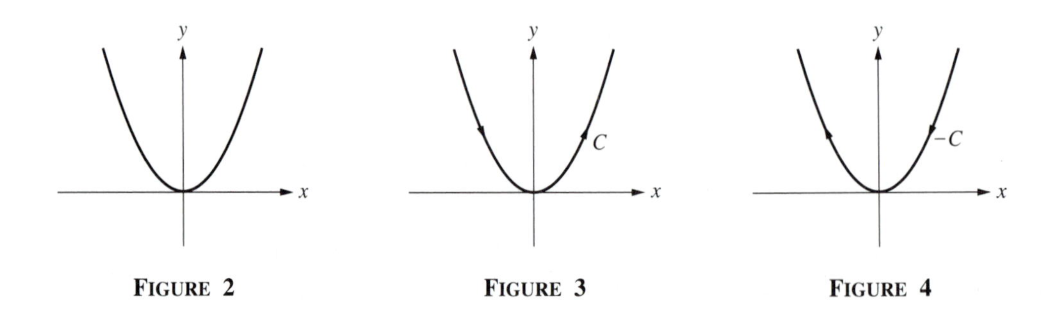

FIGURE 2 FIGURE 3 FIGURE 4

Recall that a parametric curve C represented by $\mathbf{r}(t)$, $a \le t \le b$, is *smooth* if $\mathbf{r}(t)$ is differentiable, $\mathbf{r}'(t)$ is continuous, and $\mathbf{r}'(t) \neq \mathbf{0}$ for $a \le x \le b$. Then $\mathbf{r}'(t)$ is tangent to the curve at $\mathbf{r}(t)$, as in Fig. 1. The length L of a smooth curve is

$$L = \int_a^b \|\mathbf{r}'(t)\| dt.$$

Length (or distance) along a smooth curve C is a geometric property of the curve. It does not depend on the direction or the particular parameterization used to describe C. In Fig. 5, $s = s(t)$ is the distance along C from $\mathbf{r}(a)$ to $\mathbf{r}(t)$:

$$s = s(t) = \int_a^t \|\mathbf{r}'(u)\| du, \qquad a \le t \le b.$$

We refer to s as *arc length* on C. We always measure arc length from the initial point of C. By the fundamental theorem of calculus,

$$\frac{ds}{dt} = \|\mathbf{r}'(t)\|,$$

which is often used in the differential form $ds = \|\mathbf{r}'(t)\| \, dt$. A smooth parametric curve C given by $\mathbf{r}(t)$ for $a \le t \le b$ also can be parameterized by its arc length s, as illustrated in Fig. 6. Then C is expressed by $\mathbf{r}(s)$ for $0 \le s \le L$, which is the *arc length parameterization* of C. The unit tangent vector to C at $\mathbf{r}(s)$ is $\mathbf{T}(s) = \mathbf{r}'(s)$.

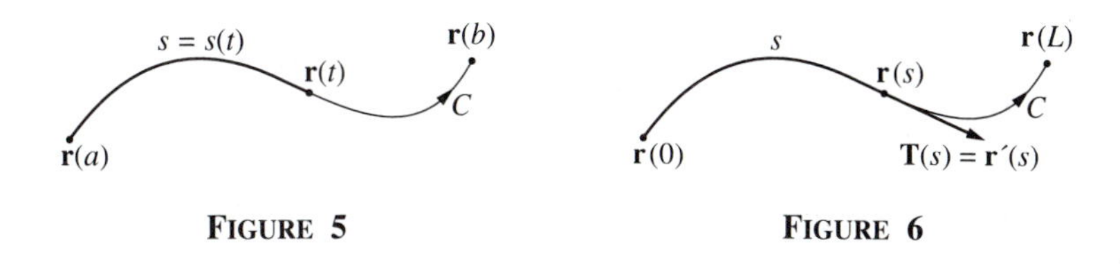

FIGURE 5 FIGURE 6

A **path**, or piecewise smooth directed curve, is made up of a finite number of smooth curves placed end to end. Figure 7 shows a path C, which consists of three smooth curves C_1, C_2, and C_3. We write

$$C = C_1 + C_2 + C_3.$$

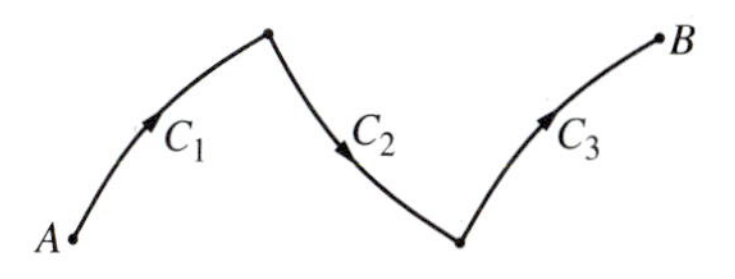

FIGURE 7

The direction on C induces a direction on C_1, C_2, and C_3. If a path C consists of n smooth curves C_1, C_2, ... C_n placed end to end, we write

$$C = C_1 + C_2 + \cdots + C_n.$$

We close our review of curves with a few comments about closed curves in 2–space. Recall that a curve is *closed* if its initial and terminal points coincide. Furthermore, a closed curve is *simple* (without self-intersections) if no points other than its initial and terminal points coincide. Figure 8 shows such a plane curve C. Apparently C divides the plane into two regions. One of these regions

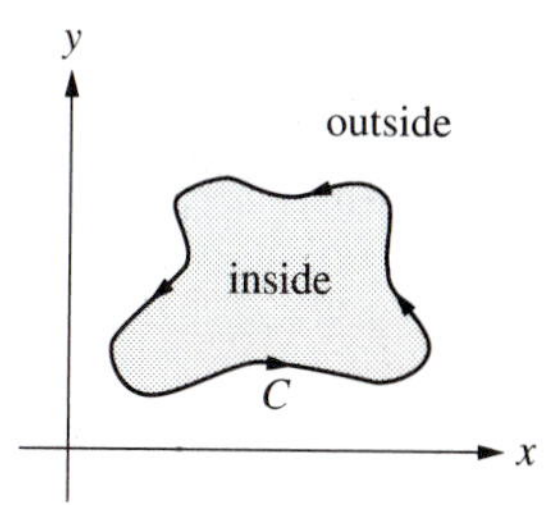

FIGURE 8

is inside C, the other is outside C, and C is the common boundary of the two regions. The direction of C shown in the figure is called the **positive direction** on C with respect to the region inside C. The positive direction may be described intuitively as the direction around C a person must walk to keep the inside of C always on the left. Although these properties of simple closed curves in a plane seem obvious, they are not easy to prove. The French mathematician Camille Jordan (1838–1922) was the first to formulate these properties in what is now called the Jordan curve theorem.

Line Integrals With Respect to Arc Length

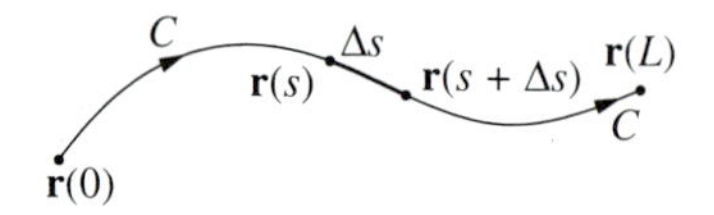

FIGURE 9

We begin with an example that motivates both the notation and the meaning of a line integral with respect to arc length. Suppose that a wire is represented by a smooth curve C parameterized by arc length s in Fig. 9. We assume that the density (mass/length) $\sigma(\mathbf{r})$ is a continuous function on C and inquire about the mass M of the wire.

The small segment of wire with length Δs shown in Fig. 9 has nearly constant density $\sigma(\mathbf{r}(s))$. Hence, the mass ΔM of the segment satisfies

$$\Delta M \approx \sigma(\mathbf{r}(s))\Delta s$$

and the total mass of the wire satisfies

$$M \approx \sum \sigma(\mathbf{r}(s))\Delta s,$$

where the approximate equality improves as Δs becomes smaller. Since the sum on the right is a Riemann sum, the usual limit passage argument leads us to define the mass of the wire by

$$M = \int_0^L \sigma(\mathbf{r}(s))ds.$$

It is natural and suggestive to express the mass M by $\int_C \sigma ds$, which conveys the idea that the mass of the wire should be thought of as the integral of its density with respect to length along the wire. Thus,

$$\boxed{M = \int_C \sigma ds = \int_0^L \sigma(\mathbf{r}(s))ds.}$$

We make the following general definition to prepare for other applications.

Definition *Line Integral With Respect to Arc Length*
Let C be a smooth curve parameterized by arc length, $\mathbf{r} = \mathbf{r}(s)$ for $0 \le s \le L$. Let $f(\mathbf{r})$ be continuous on C. The **line integral of f along C** (with respect to arc length) is

$$\int_C f ds = \int_0^L f(\mathbf{r}(s))ds.$$

Clearly the line integral inherits the linearity of the definite integral,

$$\int_C (af + bg)\,ds = a\int_C f\,ds + b\int_C g\,ds.$$

Another useful property of line integrals with respect to arc length is

$$\int_C f\,ds = \int_{-C} f\,ds.$$

If C is a curve from A to B, then s is measured from A in $\int_C f\,ds$ and s is measured from B in $\int_{-C} f\,ds$. The displayed equality should seem quite reasonable if you think of f as density along a wire C. The equality of the line integrals means that the mass of the wire does not depend on the direction along the wire.

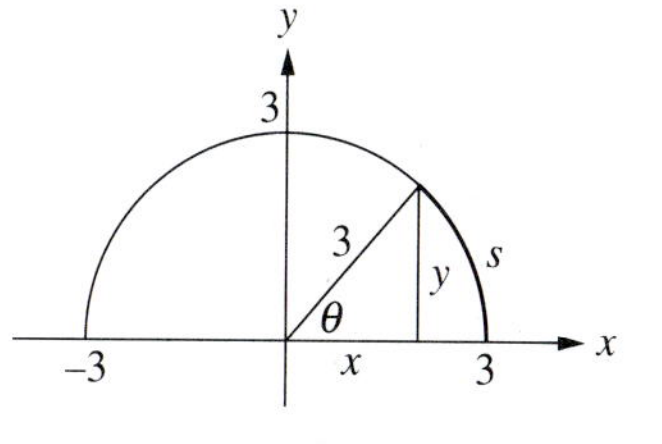

FIGURE 10

EXAMPLE 1. Let C be the upper semicircle with radius 3 in Fig. 10. Suppose that C is a wire with density $\sigma(\mathbf{r}) = y$ for $\mathbf{r} = x\mathbf{i} + y\mathbf{j}$ on C. Find the mass M of the wire.

Solution. In Fig. 10, $s = 3\theta$. Hence, C has length $L = 3\pi$ and arc length parameterization

$$\mathbf{r} = 3\cos\left(\frac{s}{3}\right)\mathbf{i} + 3\sin\left(\frac{s}{3}\right)\mathbf{j} \quad \text{for} \quad 0 \le s \le 3\pi.$$

Therefore, $\sigma(\mathbf{r}(s)) = 3\sin(s/3)$ and

$$M = \int_C \sigma\,ds = \int_0^{3\pi} 3\sin\left(\frac{s}{3}\right)ds = \left[-9\cos\left(\frac{s}{3}\right)\right]_0^{3\pi} = 18. \ \square$$

Although the definition of $\int_C f\,ds$ is very natural, it is usually not convenient for practical calculations because parametric curves are typically represented in the form $\mathbf{r}(t)$ for $a \le t \le b$ with t some parameter other than arc length, as in Fig. 5. Then how do we evaluate $\int_C f\,ds$? The answer is to make the change of variable

$$s = \int_a^t \|\mathbf{r}'(u)\|\,du, \qquad ds = \|\mathbf{r}'(t)\|\,dt.$$

Since $s = 0$ when $t = a$ and $s = L$ when $t = b$,

$$\int_C f\,ds = \int_0^L f(\mathbf{r}(s))\,ds = \int_a^b f(\mathbf{r}(t))\,\|\mathbf{r}'(t)\|\,dt.$$

In summary, if C is a smooth curve with parametric representation $\mathbf{r} = \mathbf{r}(t)$ for $a \le t \le b$ and f is continuous on C, then

$$\boxed{\int_C f\,ds = \int_a^b f(\mathbf{r}(t))\,\|\mathbf{r}'(t)\|\,dt.}$$

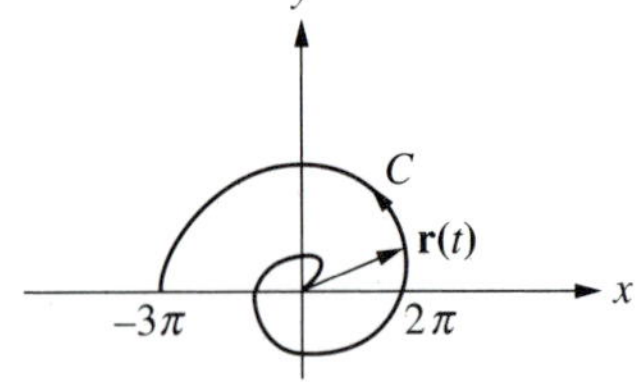

FIGURE 11

EXAMPLE 2. Let C be the spiral in Fig. 11 given by

$$\mathbf{r}(t) = t(\cos t\,\mathbf{i} + \sin t\,\mathbf{j}) \text{ for } 0 \le t \le 3\pi$$

and let $f(\mathbf{r}) = \|\mathbf{r}\|$ on C. Evaluate $\int_C f ds$.

Solution. We could regard C as a wire with density at a point numerically equal to its distance from the origin and the line integral as the mass of the wire. Whatever the interpretation, note that $\|\mathbf{r}(t)\| = t$, so that $f(\mathbf{r}(t)) = t$. Routine calculations yield

$$\mathbf{r}'(t) = (\cos t - t\sin t)\mathbf{i} + (\sin t + t\cos t)\mathbf{j},$$

$$\|\mathbf{r}'(t)\| = \sqrt{1 + t^2}.$$

Therefore,

$$\int_C f ds = \int_0^{3\pi} f(\mathbf{r}(t))\|\mathbf{r}'(t)\|dt = \int_0^{3\pi} t(1 + t^2)^{1/2}\,dt$$

$$= \frac{1}{3}[(1 + t^2)^{3/2}]_0^{3\pi} = \frac{1}{3}[(1 + 9\pi^2)^{3/2} - 1] \approx 283.45. \square$$

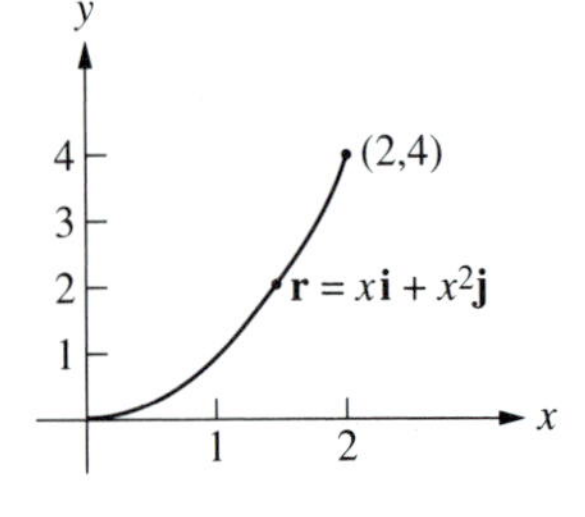

FIGURE 12

You will appreciate more fully the ease of the solution in Ex. 2 if you try to express the spiral in terms of arc length before evaluating the line integral.

EXAMPLE 3. Let C be the arc of the parabola $y = x^2$ from $(0,0)$ to $(2,4)$ in Fig. 12. Let $f(x,y) = x^{1/2}y^{1/4}$ on C. Evaluate $\int_C f ds$.

Solution. Let's use x as the parameter for C. Then

$$\mathbf{r}(x) = x\mathbf{i} + x^2\mathbf{j}, \qquad 0 \le x \le 2,$$

$$\mathbf{r}'(x) = \mathbf{i} + 2x\mathbf{j}, \qquad \|\mathbf{r}'(x)\| = \sqrt{1 + 4x^2},$$

$$f(\mathbf{r}(x)) = f(x, x^2) = x^{1/2}(x^2)^{1/4} = x.$$

Therefore,

$$\int_C f ds = \int_0^2 f(\mathbf{r}(x))\,\|\mathbf{r}'(x)\|\,dx = \int_0^2 x(1 + 4x^2)^{1/2}\,dx$$

$$= \left[\frac{1}{12}(1 + 4x^2)^{3/2}\right]_0^2 = \frac{1}{12}(17^{3/2} - 1) \approx 5.76. \square$$

EXAMPLE 4. Let C be the twisted cubic given by

$$\mathbf{r}(t) = 2t\mathbf{i} + t^2\mathbf{j} + \frac{1}{3}t^3\mathbf{k} \quad \text{for} \quad 0 \le t \le 2.$$

We first met this curve in Ex. 13 of Sec. 4.2. Let $f(x,y,z) = \sqrt{6xyz}$ on C. Evaluate $\int_C f ds$.

Solution. Straightforward calculations yield

$$\mathbf{r}'(t) = 2\mathbf{i} + 2t\mathbf{j} + t^2\mathbf{k}, \qquad \|\mathbf{r}'(t)\| = 2 + t^2,$$
$$f(\mathbf{r}(t)) = f(2t, t^2, t^3/3) = \sqrt{4t^6} = 2t^3.$$

Hence,

$$\int_C f ds = \int_0^2 f(\mathbf{r}(t))\,\|\mathbf{r}'(t)\|\,dt = \int_0^2 2t^3(2 + t^2)\,dt = \frac{112}{3}. \quad \square$$

Applications of line integrals often involve integration over paths. If $C = C_1 + C_2 + \cdots + C_n$ is a path and f is continuous on C, then the line integral of f along C (with respect to arc length) is defined by

$$\boxed{\int_C f ds = \int_{C_1} f ds + \int_{C_2} f ds + \cdots + \int_{C_n} f ds.}$$

Each line integral on the right is over a smooth curve and can be calculated using any convenient parameterization.

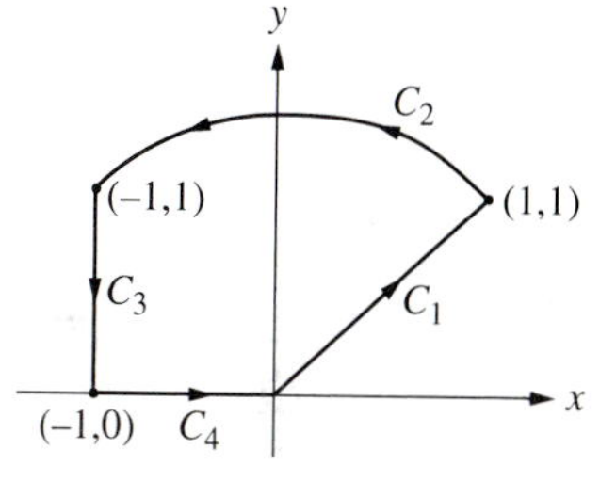

FIGURE 13

EXAMPLE 5. Evaluate $\int_C ye^{-x} ds$ for $C = C_1 + C_2 + C_3 + C_4$, the path in Fig. 13, where C_1, C_3, and C_4 are line segments and C_2 is a circular arc with center $(0,0)$.

Solution. On C_1 let's use x for the parameter. Then

$$y = x \quad \text{and} \quad 0 \le x \le 1,$$
$$\mathbf{r}(x) = x\mathbf{i} + x\mathbf{j}, \qquad \mathbf{r}'(x) = \mathbf{i} + \mathbf{j}, \qquad \|\mathbf{r}'(x)\| = \sqrt{2},$$
$$ds = \|\mathbf{r}'(x)\|\,dx = \sqrt{2}\,dx,$$
$$\int_{C_1} ye^{-x} ds = \int_0^1 xe^{-x}\sqrt{2}\,dx = \sqrt{2}\,[-xe^{-x} - e^{-x}]_0^1 = \sqrt{2}\left(1 - \frac{2}{e}\right).$$

On C_2, use the polar angle θ for the parameter. Then

$$\mathbf{r}(\theta) = \sqrt{2}(\cos\theta\,\mathbf{i} + \sin\theta\,\mathbf{j}) \quad \text{for} \quad \frac{\pi}{4} \le \theta \le \frac{3\pi}{4},$$
$$\mathbf{r}'(\theta) = \sqrt{2}(-\sin\theta\,\mathbf{i} + \cos\theta\,\mathbf{j}), \qquad \|\mathbf{r}'(\theta)\| = \sqrt{2},$$
$$x = \sqrt{2}\cos\theta, \quad y = \sqrt{2}\sin\theta, \quad ds = \sqrt{2}\,d\theta,$$
$$\int_{C_2} ye^{-x} ds = \int_{\pi/4}^{3\pi/4} \sqrt{2}\sin\theta e^{-\sqrt{2}\cos\theta}\sqrt{2}\,d\theta = \sqrt{2}\,[e^{-\sqrt{2}\cos\theta}]_{\pi/4}^{3\pi/4} = \sqrt{2}\left(e - \frac{1}{e}\right).$$

To evaluate the line integral along C_3, we use the general fact that $\int_C f ds = \int_{-C} f ds$ and the natural parameter y for $-C_3$. On $-C_3$,

$$x = -1 \quad \text{and} \quad 0 \leq y \leq 1,$$
$$\mathbf{r}(y) = -\mathbf{i} + y\mathbf{j}, \qquad \mathbf{r}'(y) = \mathbf{j},$$
$$\|\mathbf{r}'(y)\| = 1, \qquad ds = dy.$$

Hence,

$$\int_{C_3} ye^{-x} ds = \int_{-C_3} ye^{-x} ds = \int_0^1 ye \cdot 1 dy = \frac{e}{2}.$$

Finally, since $y = 0$ on C_4,

$$\int_{C_4} ye^{-x} ds = 0.$$

Addition of the four line integrals gives

$$\int_C ye^{-x} ds = \sqrt{2}\left(e - \frac{3}{e} + 1\right) + \frac{e}{2} \approx 5.06. \ \square$$

Return to the wire C with variable density σ in Fig. 9. Let M be the mass of the wire. Use of Riemann sum and limit passage arguments as in Sec. 7.5 leads us to define the **center of mass** of C by $(\bar{x}, \bar{y}, \bar{z})$ where

$$\bar{x} = \frac{1}{M}\int_C x\sigma ds, \qquad \bar{y} = \frac{1}{M}\int_C y\sigma ds, \qquad \bar{z} = \frac{1}{M}\int_C z\sigma ds.$$

Similarly, if $l(\mathbf{r})$ is the perpendicular distance from a point $\mathbf{r}$ on C to a given axis, then the **moment of inertia** of C about the axis is

$$I = \int_C l^2 \sigma ds.$$

EXAMPLE 6. As in Ex. 1, let C be a semicircular wire with radius 3 and density $\sigma(\mathbf{r}) = y$. Find the center of mass of C and its moment of inertia about the x–axis.

Solution. This time represent C by

$$\mathbf{r}(\theta) = 3(\cos\theta\mathbf{i} + \sin\theta\mathbf{j}) \text{ for } 0 \leq \theta \leq \pi.$$

Now $\mathbf{r}'(\theta) = 3(-\sin\theta\mathbf{i} + \cos\theta\mathbf{j})$ and $\|\mathbf{r}'(\theta)\| = 3$. By symmetry, $\bar{x} = 0$. Next,

$$\int_C y\sigma ds = \int_C y^2 ds = \int_0^\pi 9\sin^2\theta \cdot 3d\theta = \frac{27}{2}\int_0^\pi (1 - \cos 2\theta)\, d\theta$$
$$= \frac{27}{2}\left[\theta - \frac{1}{2}\sin 2\theta\right]_0^\pi = \frac{27\pi}{2}.$$

From Ex. 1, the mass of the wire is $M = 18$. So

$$\bar{y} = \frac{1}{M}\int_C y\sigma ds = \frac{1}{18}\cdot\frac{27\pi}{2} = \frac{3\pi}{4}$$

and the center of mass is $(\bar{x}, \bar{y}) = (0, 3\pi/4)$. Since the perpendicular distance from the point (x,y) on C to the x–axis is y, the moment of inertia of the semicircle about the x–axis is

$$I = \int_C y^2\sigma ds = \int_0^\pi 27\sin^3\theta \cdot 3d\theta = 81\int_0^\pi (1 - \cos^2\theta)\sin\theta d\theta$$
$$= 81\left[-\cos\theta + \frac{1}{3}\cos^3\theta\right]_0^\pi = 81\cdot\frac{4}{3} = 108. \square$$

Line Integrals With Respect to x, y, z

Line integrals with respect to x, y, and z, which are denoted by $\int_C f dx$, $\int_C f dy$, and $\int_C f dz$, are motivated, defined, and evaluated in virtually the same way as for line integrals with respect to arc length. Thus, we shall be brief.

If C is a smooth parametric curve given by $\mathbf{r}(t)$ for $a \le t \le b$ and f is continuous on C, then

$$\int_C f dx = \int_a^b f(\mathbf{r}(t))\frac{dx}{dt}dt,$$
$$\int_C f dy = \int_a^b f(\mathbf{r}(t))\frac{dy}{dt}dt,$$
$$\int_C f dz = \int_a^b f(\mathbf{r}(t))\frac{dz}{dt}dt.$$

Furthermore, if P, Q, and R are continuous on C, we set

$$\int_C P dx + Q dy + R dz = \int_C P dx + \int_C Q dy + \int_C R dz.$$

Consequently,

$$\int_C P dx + Q dy + R dz = \int_a^b \left[P(\mathbf{r}(t))\frac{dx}{dt} + Q(\mathbf{r}(t))\frac{dy}{dt} + R(\mathbf{r}(t))\frac{dz}{dt}\right]dt.$$

The formulas for $\int_C f dx, \int_C f dy, \int_C f dz$ express these line integrals in terms of a particular parameterization $\mathbf{r}(t)$ of C. However, the line integral is independent of the parameterization of C in the following sense. We can replace $\mathbf{r}(t)$ by any other smooth parametric representation of C that *preserves the direction* of C. This fact is a consequence of the change of variables theorem for definite integrals. We omit the details.

Line integrals with respect to x, y, and z over a path $C = C_1 + C_2 + \cdots + C_n$, with each C_i a smooth parametric curve, are defined in the expected manner. Thus,

$$\int_C f dx = \int_{C_1} f dx + \int_{C_2} f dx + \cdots + \int_{C_n} f dx.$$

There is one important difference between line integrals with respect to x, y, and z and line integrals with respect to arc length s:

$$\int_{-C} f dx = -\int_C f dx, \qquad \int_{-C} f dy = -\int_C f dy, \qquad \int_{-C} f dz = -\int_C f dz,$$

in contrast to $\int_{-C} f ds = \int_C f ds$. One way to confirm the foregoing results is to observe that if $\mathbf{r}(t)$ for $a \le t \le b$ is a parameterization of C where it is understood that t *increases* from a to b, then $\mathbf{r}(t)$ also parameterizes $-C$ when t *decreases* from b to a. Therefore,

$$\int_{-C} f dx = \int_b^a f(\mathbf{r}(t)) \frac{dx}{dt} dt = -\int_a^b f(\mathbf{r}(t)) \frac{dx}{dt} dt = -\int_C f dx.$$

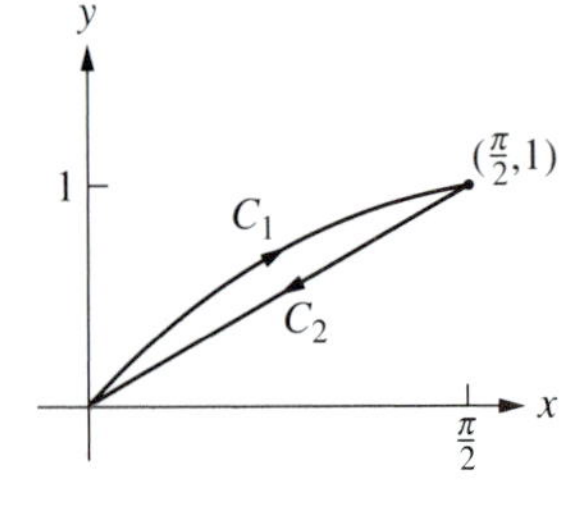

FIGURE 14

EXAMPLE 7. Let $C = C_1 + C_2$, where C_1 is the arc of the sine curve $y = \sin x$ for $0 \le x \le \pi/2$ and C_2 is the line segment from $(\pi/2, 1)$ to $(0, 0)$, as in Fig. 14. Evaluate $\int_C y dx + x dy$.

Solution. Parameterize C_1 by $x = t$, $y = \sin t$ for $0 \le t \le \pi/2$. Then

$$\int_{C_1} y dx + x dy = \int_0^{\pi/2} \left(y \frac{dx}{dt} + x \frac{dy}{dt} \right) dt = \int_0^{\pi/2} (\sin t + t \cos t) dt.$$

Integrate by parts to find that an antiderivative of $t \cos t$ is $t \sin t + \cos t$. Then

$$\int_{C_1} y dx + x dy = [-\cos t + t \sin t + \cos t]_0^{\pi/2} = \frac{\pi}{2}.$$

Since the line segment C_2 has slope $2/\pi$, we can parameterize C_2 by $x = t$, $y = 2t/\pi$ where t *decreases* from $\pi/2$ to 0. Then

$$\int_{C_2} y dx + x dy = \int_{\pi/2}^0 \left(y \frac{dx}{dt} + x \frac{dy}{dt} \right) dt = \int_{\pi/2}^0 \left(\frac{2t}{\pi} \cdot 1 + t \cdot \frac{2}{\pi} \right) dt = \left[\frac{2}{\pi} t^2 \right]_{\pi/2}^0 = -\frac{\pi}{2}.$$

Add the line integrals over C_1 and C_2 to obtain

$$\int_C xdy + ydx = \frac{\pi}{2} - \frac{\pi}{2} = 0. \ \square$$

Line Integrals $\int_C \mathbf{F} \cdot d\mathbf{r}$

We start with an example, involving work, that motivates both the notation and the meaning of the line integral $\int_C \mathbf{F} \cdot d\mathbf{r}$.

Recall from Sec. 3.2 that the work W done by a constant force $\mathbf{F}$ that acts on an object as it moves through a displacement vector $\mathbf{D}$, as shown in Fig. 15, is

$$W = \mathbf{F} \cdot \mathbf{D} = ||\mathbf{F}|| \ ||\mathbf{D}|| \cos\theta = (\text{comp}_{\mathbf{D}} \mathbf{F}) \, ||\mathbf{D}||.$$

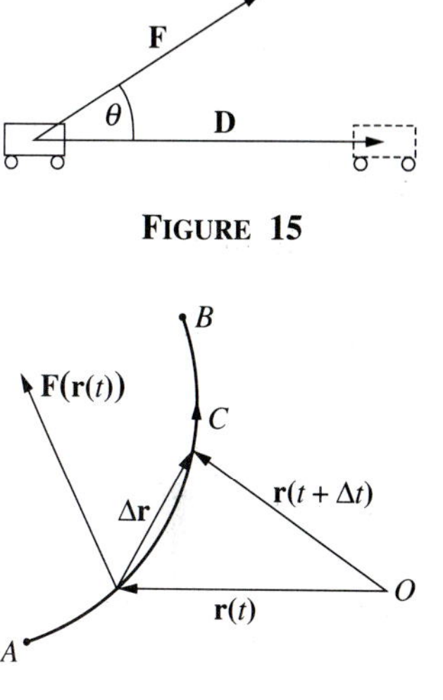

FIGURE 15

FIGURE 16

This formula provides the basis for a Riemann sum and limit passage argument which gives an integral for the work when the force is variable and the object moves along a curve.

Suppose that a constant or variable force $\mathbf{F} = \mathbf{F}(\mathbf{r})$ acts on an object as it moves from A to B along a smooth curve C in Fig. 16. Assume that $\mathbf{F}(\mathbf{r})$ is continuous on C and let $\mathbf{r}(t)$ for $a \le t \le b$ be any smooth parameterization for C. As the object moves from $\mathbf{r}(t)$ to $\mathbf{r}(t + \Delta t)$, with Δt small, its net displacement is

$$\Delta\mathbf{r} = \mathbf{r}(t + \Delta t) - \mathbf{r}(t)$$

and the work done by the force is

$$\Delta W \approx \mathbf{F}(\mathbf{r}(t)) \cdot \Delta\mathbf{r}.$$

The total work done as the object moves from A to B along C is

$$W_C \approx \Sigma \mathbf{F}(\mathbf{r}(t)) \cdot \Delta\mathbf{r}.$$

We shall examine this sum from two points of view. First, for small Δt,

$$\Delta\mathbf{r} \approx \mathbf{r}'(t)\Delta t, \qquad \Delta W \approx \mathbf{F}(\mathbf{r}(t)) \cdot \mathbf{r}'(t)\Delta t,$$
$$W_C \approx \Sigma \mathbf{F}(\mathbf{r}(t)) \cdot \mathbf{r}'(t)\Delta t.$$

This is a Riemann sum in our usual shorthand. A limit passage gives a definite integral for the work:

$$W_C = \int_a^b \mathbf{F}(\mathbf{r}(t)) \cdot \mathbf{r}'(t)\, dt.$$

On the other hand, since $W_C \approx \Sigma \mathbf{F}(\mathbf{r}(t)) \cdot \Delta\mathbf{r}$, it is natural to express the work by $\int_C \mathbf{F} \cdot d\mathbf{r}$. Thus,

$$\boxed{W_C = \int_C \mathbf{F} \cdot d\mathbf{r} = \int_a^b \mathbf{F}(\mathbf{r}(t)) \cdot \mathbf{r}'(t)\, dt.}$$

Now let $\mathbf{F}(\mathbf{r})$ be any continuous vector function on C. In applications, $\mathbf{F}(\mathbf{r})$ is often a vector field, such as a force field or electromagnetic field, defined in a region containing C.

Definition *Line Integral of* ***F*** *Over C*
Let C be a smooth curve parameterized by $\mathbf{r}(t)$ for $a \le t \le b$. Let $\mathbf{F} = \mathbf{F}(\mathbf{r})$ be a continuous vector function on C. Then

$$\int_C \mathbf{F} \cdot d\mathbf{r} = \int_a^b \mathbf{F}(\mathbf{r}(t)) \cdot \mathbf{r}'(t)\,dt.$$

It follows immediately from this definition and the earlier formula for $\int_C Pdx + Qdy + Rdz$ that

$$\int_C \mathbf{F} \cdot d\mathbf{r} = \int_C Pdx + Qdy + Rdz,$$

where $\mathbf{F} = P\mathbf{i} + Q\mathbf{j} + R\mathbf{k}$. Consequently, just as for the line integrals with respect to x, y, and z, the value of the line integral $\int_C \mathbf{F} \cdot d\mathbf{r}$ does not depend on the particular parameterization $\mathbf{r}(t)$ used to evaluate it in the sense that the same value is obtained for the line integral for all smooth parameterizations of C that give the same direction to C.

It is apparent from the definition of $\int_C \mathbf{F} \cdot d\mathbf{r}$ that

$$\int_C (a\mathbf{F} + b\mathbf{G}) \cdot d\mathbf{r} = a\int_C \mathbf{F} \cdot d\mathbf{r} + b\int_C \mathbf{G} \cdot d\mathbf{r}$$

for any continuous vector functions $\mathbf{F}$ and $\mathbf{G}$ on C and for any constants a and b. Also, by a change of variables argument that we omit,

$$\int_{-C} \mathbf{F} \cdot d\mathbf{r} = -\int_C \mathbf{F} \cdot d\mathbf{r}.$$

In the context of Fig. 16, the foregoing equality means that the work done by a force field when the object moves along $-C$ from B to A is the negative of the work done by the field when the object moves along C from A to B.

Let $t = s$ in the definition of $\int_C \mathbf{F} \cdot d\mathbf{r}$ and recall that the unit tangent vector to C at $\mathbf{r}(s)$ is $\mathbf{T}(s) = \mathbf{r}'(s)$ to obtain

$$\int_C \mathbf{F} \cdot d\mathbf{r} = \int_0^L \mathbf{F}(\mathbf{r}(s)) \cdot \mathbf{r}'(s)\,ds = \int_0^L \mathbf{F}(\mathbf{r}(s)) \cdot \mathbf{T}(s)\,ds.$$

The integral on the right is often expressed in the abbreviated form $\int_C \mathbf{F} \cdot \mathbf{T}ds$. Some authors prefer the notation $\int_C \mathbf{F} \cdot \mathbf{T}ds$ to $\int_C \mathbf{F} \cdot d\mathbf{r}$. Since $\mathbf{F} \cdot \mathbf{T} = \text{comp}_{\mathbf{T}}\mathbf{F}$, $\int_C \mathbf{F} \cdot d\mathbf{r} = \int_C \text{comp}_{\mathbf{T}}\mathbf{F}ds$. Consequently, if a force acts on an object as it moves along a curve, then the work done by the force is the integral (with respect to arc length) of the tangential component of the force along the curve, just as we mentioned earlier.

EXAMPLE 8. Find the work done by the force field $\mathbf{F}(\mathbf{r}) = 4xy\mathbf{i} + 3y^2\mathbf{j}$ when an object is moved from $(0,0)$ to $(2,4)$ along the curve C, which is the piece of the parabola $y = x^2$ in Fig. 12 and Ex. 3.

Solution. As in Ex. 3, we use x for the parameter on C:

$$\mathbf{r}(x) = x\mathbf{i} + x^2\mathbf{j}, \qquad \mathbf{r}'(x) = \mathbf{i} + 2x\mathbf{j}, \quad \text{for} \quad 0 \le x \le 2,$$
$$\mathbf{F}(\mathbf{r}(x)) = \mathbf{F}(x, x^2) = 4x^3\mathbf{i} + 3x^4\mathbf{j}.$$

Then

$$W_C = \int_C \mathbf{F} \cdot d\mathbf{r} = \int_0^2 \mathbf{F}(\mathbf{r}(x)) \cdot \mathbf{r}'(x)\,dx$$
$$= \int_0^2 (4x^3 + 6x^5)\,dx = [x^4 + x^6]_0^2 = 80 \text{ ft-lbs.} \ \square$$

If $C = C_1 + C_2 + \cdots + C_n$ is a path composed of the smooth curves C_i, then, as expected, we define

$$\boxed{\int_C \mathbf{F} \cdot d\mathbf{r} = \int_{C_1} \mathbf{F} \cdot d\mathbf{r} + \int_{C_2} \mathbf{F} \cdot d\mathbf{r} + \cdots + \int_{C_n} \mathbf{F} \cdot d\mathbf{r}.}$$

EXAMPLE 9. Let $C = C_1 + C_2 + C_3$ be the path in Fig. 17, where C_1 is the arc of the parabola $y = 4 - x^2$, $z = 0$, from $(2,0,0)$ to $(0,4,0)$, and C_2 and C_3 are the indicated line segments. Let $\mathbf{F}(\mathbf{r}) = xe^z\mathbf{i} + z\mathbf{j} + y\mathbf{k}$. Find $\int_C \mathbf{F} \cdot d\mathbf{r}$.

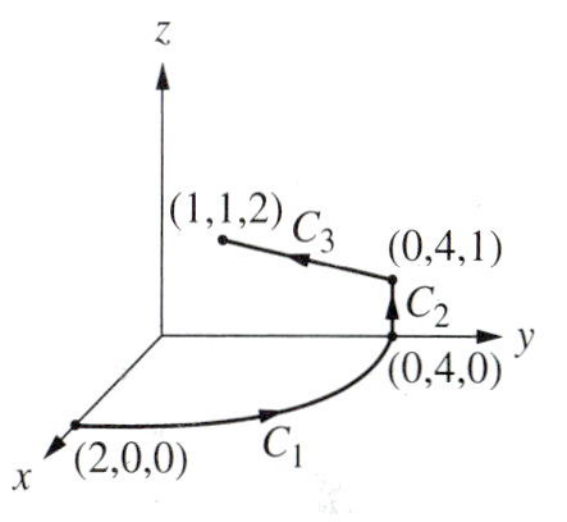

FIGURE 17

Solution. Instead of C_1, let's work with $-C_1$. A natural parameter for $-C_1$ is x. On $-C_1$,

$$\mathbf{r}(x) = x\mathbf{i} + (4 - x^2)\mathbf{j}, \qquad \mathbf{r}'(x) = \mathbf{i} - 2x\mathbf{j}, \quad \text{for} \quad 0 \le x \le 2,$$
$$\mathbf{F}(\mathbf{r}(x)) = \mathbf{F}(x, 4 - x^2, 0) = x\mathbf{i} + (4 - x^2)\mathbf{k},$$
$$\int_{C_1} \mathbf{F} \cdot d\mathbf{r} = -\int_{-C_1} \mathbf{F} \cdot d\mathbf{r} = -\int_0^2 \mathbf{F}(\mathbf{r}(x)) \cdot \mathbf{r}'(x)\,dx$$
$$= -\int_0^2 x\,dx = -\left[\frac{1}{2}x^2\right]_0^2 = -2.$$

On C_2, z is a natural parameter and $x = 0$. Then

$$\mathbf{r}(z) = 4\mathbf{j} + z\mathbf{k}, \qquad \mathbf{r}'(z) = \mathbf{k}, \quad \text{for} \quad 0 \le z \le 1,$$
$$\mathbf{F}(\mathbf{r}(z)) = \mathbf{F}(0, 4, z) = z\mathbf{j} + 4\mathbf{k},$$
$$\int_{C_2} \mathbf{F} \cdot d\mathbf{r} = \int_0^1 \mathbf{F}(\mathbf{r}(z)) \cdot \mathbf{r}'(z)\,dz = \int_0^1 4\,dz = 4.$$

On C_3 it is convenient to use the notation $\mathbf{r} = < x, y, z >$ for vectors. A direction vector for the line segment C_3 from $(0,4,1)$ to $(1,1,2)$ is $\mathbf{v} = < 1, -3, 1 >$. So a vector equation for C_3 is

$$\mathbf{r}(t) = < 0,4,1 > + t < 1, -3, 1 > = < t, 4 - 3t, 1 + t >, \qquad 0 \le t \le 1.$$

Then $\mathbf{r}'(t) = < 1, -3, 1 >$ and

$$\mathbf{F}(\mathbf{r}(t)) = \mathbf{F}(t, 4-3t, 1+t) = \langle te^{1+t}, 1+t, 4-3t \rangle,$$

$$\mathbf{F}(\mathbf{r}(t)) \cdot \mathbf{r}'(t) = te^{1+t} - 3(1+t) + (4-3t) = te^{1+t} - 6t + 1,$$

$$\int_{C_3} \mathbf{F} \cdot d\mathbf{r} = \int_0^1 \mathbf{F}(\mathbf{r}(t)) \cdot \mathbf{r}'(t)\, dt = \int_0^1 (ete^t - 6t + 1)\, dt = e + 4,$$

after integrating te^t by parts. Add the three integrals to obtain

$$\int_C \mathbf{F} \cdot d\mathbf{r} = \int_{C_1} \mathbf{F} \cdot d\mathbf{r} + \int_{C_2} \mathbf{F} \cdot d\mathbf{r} + \int_{C_3} \mathbf{F} \cdot d\mathbf{r} = e + 6. \ \square$$

Kinetic Energy and Work

Kinetic energy is energy associated with motion. Kinetic energy and work are closely related, as we are about to show. The story will continue in the next section when we discuss conservation of energy.

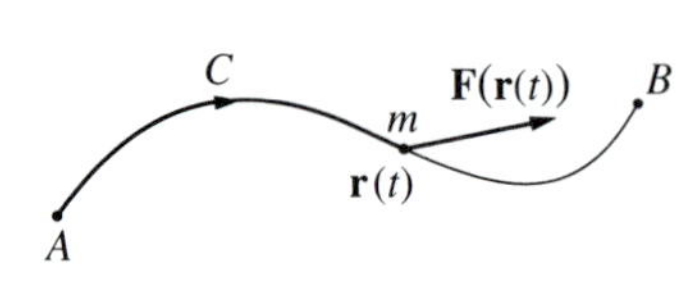

FIGURE 18

Figure 18 shows a particle of mass m with position vector $\mathbf{r}(t)$ at time t moving under the action of a continuous force field $\mathbf{F} = \mathbf{F}(\mathbf{r})$. The particle moves from A to B during the time interval $a \le t \le b$. According to Newton's second law, it will move on a trajectory C from A to B along which

$$m\mathbf{r}''(t) = \mathbf{F}(\mathbf{r}(t)).$$

At time t the particle has velocity $\mathbf{v}(t) = \mathbf{r}'(t)$ and **kinetic energy**

$$K(t) = \frac{1}{2} m \left(\frac{ds}{dt} \right)^2 = \frac{1}{2} m \, ||\mathbf{r}'(t)||^2.$$

Since $K(t) = \frac{1}{2} m\mathbf{r}'(t) \cdot \mathbf{r}'(t)$ and $m\mathbf{r}''(t) = \mathbf{F}(\mathbf{r}(t))$,

$$K'(t) = m\mathbf{r}''(t) \cdot \mathbf{r}'(t) = \mathbf{F}(\mathbf{r}(t)) \cdot \mathbf{r}'(t).$$

Consequently, the work done by $\mathbf{F}(\mathbf{r})$ in moving the mass m from A to B along the trajectory C is

$$W = \int_C \mathbf{F} \cdot d\mathbf{r} = \int_a^b \mathbf{F}(\mathbf{r}(t)) \cdot \mathbf{r}'(t)\, dt = \int_a^b K'(t)\, dt = K(b) - K(a).$$

Theorem 1 *Work and Kinetic Energy*
The work W done by a continuous force field $\mathbf{F}(\mathbf{r})$ in moving a particle with mass m from A to B is equal to the change in kinetic energy of the particle. In symbols,

$$W = K(B) - K(A) = \frac{1}{2} m \|\mathbf{v}(B)\|^2 - \frac{1}{2} m \|\mathbf{v}(A)\|^2.$$

This is a fundamental result. We shall have more to say about its physical significance in Sec. 8.3.

PROBLEMS

In Probs. 1–6, make a rough sketch of the path C and evaluate the given line integral.

1. $\int_C x\,ds, \quad C\colon\ \mathbf{r} = \cos 2t\mathbf{i} - \sin 2t\mathbf{j},\ 0 \le t \le \pi$

2. $\int_C (x^2 + y^2)\,ds, \quad C\colon\ x = e^{-t}\sin t,\ y = e^{-t}\cos t,\ 0 \le t \le 2\pi$

3. $\int_C (xy^2/z)\,ds, \quad C\colon\ x = t^3/3,\ y = t^2,\ z = 2t\ \ 0 \le t \le 2$

4. $\int_C \sqrt{x^2 + y^2 + z^2}\,ds, \quad C\colon\ \mathbf{r} = \cos\theta\mathbf{i} - \sin\theta\mathbf{j} + 2\theta\mathbf{k},\ 0 \le \theta \le 2\pi$

5. $\int_C y \sin(\pi x/2)\,ds$, C is the rectangle formed by joining the points $(0,0)$, $(1,0)$, $(1,2)$, $(0,2)$, and $(0,0)$ in that order.

6. $\int_C xe^y\,ds$, $C = C_1 + C_2 + C_3$, where C_1 is the line segment from $(1,0)$ to $(1,1)$, C_2 is the upper semicircle with center $(0,0)$ and radius $\sqrt{2}$ from $(1,1)$ to $(-1,1)$, and C_3 is the line segment from $(-1,1)$ to $(-1,0)$.

In Probs. 7–14, C is the upper semicircle with center the origin from $A = (a,0)$ to $B = (-a,0)$ in the xy–plane. C represents a wire and V denotes a variable point on C. Use k for the proportionality constant associated with the density.

7. Find the mass of C if the density is proportional to the square of the distance from V to A.

8. Find the mass of C if the density is proportional to the distance from V to B.

9. Find the mass of C if the density is proportional to the x–coordinate of V.

10. Find the mass of C if the density is proportional to the product of the x– and y–coordinates of V.

11. Find the center of mass of C if the density is proportional to the distance of V from A.

12. Find the center of mass of C if the density is proportional to the square of the y–coordinate of V.

13. Find the moment of inertia of C about the x–axis if the density is proportional to the distance of V from A.

14. Find the moment of inertia of C about the y–axis if the density is proportional to the distance of V from A.

In Probs. 15–20, a wire C has the shape of a circular helix given by $\mathbf{r}(t) = < a\cos t,\ a\sin t,\ bt >$ for $0 \le t \le 2\pi$ with a and b positive constants. Use k for the proportionality constant associated with the density. Let V denote a variable point on the helix.

15. Find the mass of C if the density is proportional to the height of V above the xy–plane.

16. Find the center of mass of C in Prob. 15.

17. Find the mass of C if the density is proportional to the distance of V from the origin.

18. Find the center of mass of C in Prob. 17.

19. Find the moment of inertia of C in Prob. 15 about the z–axis.

20. Find the moment of inertia of C in Prob. 17 about the x–axis.

21. A homogeneous wire C with density 1 (in suitable units) is shaped like the cycloid $\mathbf{r}(t) = a < 1 - \cos t,\ t - \sin t >$ for $0 \le t \le 2\pi$. Find (a) the mass and (b) the center of mass.

22. A homogeneous wire C with density 1 (in suitable units) is shaped like the quarter circle $\mathbf{r}(t) = a(\cos\theta\mathbf{i} + \sin\theta\mathbf{j})$ for $0 \le \theta \le \pi/2$. Find (a) the mass, (b) the center of mass, and (c) the moment of inertia of C about the axis $y = x$.

23. Evaluate $\int_C -ydx + xdy$ along the curve $y = \cos x$ for $0 \le x \le \pi$.

24. Evaluate $\int_C -ydx + xdy$ along (a) the line segment from $(4,0)$ to $(0,4)$ and (b) the quarter circle centered at the origin and extending from $(4,0)$ to $(0,4)$.

25. Evaluate $\int_C (x - y^2)\,dx + (2y + x^2)\,dy + xdz$ where C is the polygon formed by joining $(1,0,0)$ to $(1,2,0)$ to $(1,1,1)$ to $(1,0,0)$ in that order.

26. Evaluate $\int_C yzdx + xzdy + xydz$ along the twisted cubic $x = t$, $y = t^2$, $z = t^3$ for $0 \le t \le 1$.

In Probs. 27–32 make a rough sketch of C and evaluate $\int_C \mathbf{F} \cdot d\mathbf{r}$.

27. $\mathbf{F} = -y\mathbf{i} + x\mathbf{j}$, $\quad C$: $\mathbf{r} = t\mathbf{i} + 2t\mathbf{j}$, $0 \le t \le 2$

28. $\mathbf{F} = e^{x-y}(x\mathbf{i} + y\mathbf{j})$, $\quad C$: $\mathbf{r} = 3\cos\theta\mathbf{i} - 3\sin\theta\mathbf{j}$, $0 \le \theta \le \pi$

29. $\mathbf{F} = -y\mathbf{i} + x\mathbf{j}$, C is the arc of the parabola $y^2 = 4x$ extending from the origin to $(4,4)$.

30. $\mathbf{F} = (x^2 + y^2)\mathbf{i} + 2xy\mathbf{j}$, C is the square formed by joining the origin to $(2,0)$ to $(2,2)$ to $(0,2)$ to the origin in that order.

31. $\mathbf{F} = yz\mathbf{i} + xz\mathbf{j} + xy\mathbf{k}$, $\quad C$: $x = t$, $y = t^2$, $z = t^{-1}$, $1 \le t \le 2$

32. $\mathbf{F} = (x+y)\mathbf{i} + (y+z)\mathbf{j} + (x+y)\mathbf{k}$, C is the straight line segment from $(1,2,3)$ to $(3,-1,2)$.

33. The force $\mathbf{F} = (x^2+y^2)^{-1}(3x\mathbf{i} - 4y\mathbf{j})$ acts on an object as it moves from $(5,0)$ to $(0,5)$ to $(-5,0)$ along two straight line segments. Find the work done by the force.

34. The force $\mathbf{F}$ in the previous problem acts as an object moves from $(5,0)$ to $(-5,0)$ along the upper semicircle with radius 5 centered at the origin. Find the work done by the force.

35. A homogeneous line of charge, extending indefinitely in both directions along the z–axis, sets up an electric field $\mathbf{E}(x,y,z) = \sigma(x^2+y^2)^{-1}(x\mathbf{i} + y\mathbf{j})$ for any point (x,y,z) not on the z–axis. Here σ is a fixed constant. Find the work done by the field in moving a unit charge along the path in Prob. 33. *Hint.* The force on a unit positive charge is $\mathbf{E}$.

36. Repeat the previous problem for the path in Prob. 34.

37. A charged particle of mass m moves on the helix used in Probs. 15–20. Find the work done by gravity during the motion. Assume that the gravitational acceleration is the constant vector $\mathbf{g}$.

38. As a particle moves on the helix in Probs. 15–20, it is acted upon by a force directed toward the origin and with magnitude numerically equal to its distance from the origin. Find the work done by the force.

39. Find the work done by the force field $\mathbf{F} = -\frac{1}{2}(y\mathbf{i} - x\mathbf{j})$ as it moves an object once counterclockwise around (a) the circle $x^2+y^2 = a^2$ and (b) the ellipse $x^2/a^2 + y^2/b^2 = 1$.

40. An object of mass m has position vector $\mathbf{r} = t\mathbf{i} + t^2\mathbf{j} + t^3\mathbf{k}$ at any time t. Find (a) the total force acting on the object and (b) the work done by the force for $0 \le t \le 2$.

41. The force $\mathbf{F} = (x+2y)\mathbf{i} + xy\mathbf{j}$ acts as an object moves from $(0,0)$ to $(1,0)$ along the parabola $y = ax(1-x)$, where a is a constant. Find the parabolic arc along which the work done by $\mathbf{F}$ is as small as possible.

42. A central force field, such as Newton's gravitational field, has the form $\mathbf{F} = f(\mathbf{r})\mathbf{r}$ if the center of force is at the origin. Show that a central force field does no work on an object that moves on any path C that lies on the surface of the sphere $\|\mathbf{r}\| = a$, where a is any positive constant.

43. The force $\mathbf{F} = y^2\mathbf{i} + x\mathbf{j}$ acts as an object moves from $(0,0)$ to $(\pi,0)$ along the sine curve $y = a\sin x$, where a is a constant. Find the sine curve along which the work done by $\mathbf{F}$ is as large as possible.

44. A frictional force on a moving object has constant magnitude and is always directed opposite to the direction of motion. Show that the work done by friction along any path C is proportional to the length of C.

45. A constant force $\mathbf{F}$ acts as an object moves from position $\mathbf{r}_0$ to $\mathbf{r}_1$ along a path C. Show that the work done by $\mathbf{F}$ is $\mathbf{F} \cdot \mathbf{D}$ where $\mathbf{D} = \mathbf{r}_1 - \mathbf{r}_0$.

8.3 The Fundamental Theorem for Line Integrals

The fundamental theorem of calculus for functions of one variable leads to a corresponding fundamental theorem for line integrals of functions of two and three variables. This new fundamental theorem has a number of very important applications. It enables us to evaluate certain line integrals rather easily and to establish the conservation of mechanical energy, which is one of the most important principles in physics. Other applications will appear in the problems and later in the chapter.

Overview

The fundamental theorem for line integrals has several aspects not present in the fundamental theorem of calculus. So as not to lose the forest for the trees, we begin with an overview of the principal topics and results covered in this section. More complete statements and supporting discussion will follow later.

First, we adopt some standing notation. As usual, a vector field is expressed by

$$\mathbf{F}(\mathbf{r}) = P(x,y)\,\mathbf{i} + Q(x,y)\,\mathbf{j} \text{ in 2–space, and}$$
$$\mathbf{F}(\mathbf{r}) = P(x,y,z)\,\mathbf{i} + Q(x,y,z)\,\mathbf{j} + R(x,y,z)\,\mathbf{k} \text{ in 3–space.}$$

The domain of **F** is denoted by D. We always assume that the components of **F** have continuous partial derivatives and that D is open and connected. Recall that a set is open if every point in the set has a neighborhood that lies in the set. An open set is connected if for any two points A and B in D there is a smooth curve in D extending from A to B. In the discussion that follows all points and curves lie in the domain D of the vector field **F**.

For functions of one variable, the fundamental theorem of calculus has two forms

$$f(x) = f(a) + \int_a^x f'(t)\,dt,$$
$$\int_a^b f'(x)\,dx = f(b) - f(a).$$

The analogue of the second form for line integrals is

$$\int_C \nabla f \cdot d\mathbf{r} = f(B) - f(A),$$

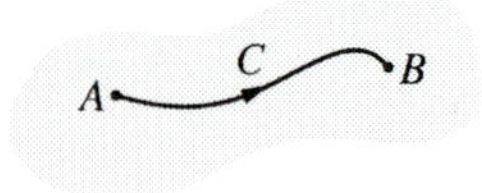

FIGURE 1

where $f(x,y)$ or $f(x,y,z)$ is a 2–d or 3–d scalar field and C is any path from a point A to a point B, as illustrated in Fig. 1. Notice that $f(B) - f(A)$ depends only on the endpoints of the path C and not on the particular path from A to B. We say that the line integral of ∇f is independent of the path from A to B. This fact makes it easy to evaluate line integrals of gradient fields. There is no need to introduce parametric representations and engage in sometimes laborious calculations. Since $\int_C \nabla f \cdot d\mathbf{r}$ is independent of the path, we often write

$$\int_C \nabla f \cdot d\mathbf{r} = \int_A^B \nabla f \cdot d\mathbf{r} = f(B) - f(A).$$

The analogue for line integrals of the first form of the fundamental theorem of calculus is obtained if we fix A and vary B in the preceding formula. Thus, in the two-variable case, fix $A = (x_0, y_0)$ and let $B = (x, y)$ vary. Then

$$f(x,y) = f(x_0, y_0) + \int_{(x_0,y_0)}^{(x,y)} \nabla f \cdot d\mathbf{r}.$$

Now we express the foregoing results a little differently. Let $\mathbf{F} = \mathbf{F}(\mathbf{r})$ be a given vector field with domain D. If $\mathbf{F}$ is a gradient field, that is, if $\mathbf{F} = \nabla f$ for some potential function f, then

$$\int_C \mathbf{F} \cdot d\mathbf{r} = \int_C \nabla f \cdot d\mathbf{r} = f(B) - f(A)$$

for any path C in D from a point A to a point B. Thus, the line integral of $\mathbf{F}$ is independent of the path. One part of the fundamental theorem for line integrals states that only gradient fields have this property:

$$\mathbf{F} = \nabla f \text{ for some } f \quad \Leftrightarrow \quad \int_C \mathbf{F} \cdot d\mathbf{r} \text{ is independent of the path in } D,$$

in which case

$$\int_C \mathbf{F} \cdot d\mathbf{r} = \int_A^B \mathbf{F} \cdot d\mathbf{r} = f(B) - f(A).$$

There is another description of independence of the path that is very useful:

$$\int_C \mathbf{F} \cdot d\mathbf{r} \text{ is independent of the path in } D$$

$$\Leftrightarrow$$

$$\int_C \mathbf{F} \cdot d\mathbf{r} = \mathbf{0} \text{ for every closed path in } D.$$

In order to take full advantage of the fundamental theorem for line integrals to simplify the calculation of $\int_C \mathbf{F} \cdot d\mathbf{r}$, we need to answer two questions:

Q1. How can we tell if $\mathbf{F}$ is a gradient field?
Q2. If $\mathbf{F}$ is a gradient field, how can we find a potential function f such that $\mathbf{F} = \nabla f$?

We defer the answer to Q2 until later in the section and concentrate here on Q1.

The answer to Q1 has some straightforward aspects and some subtle aspects. For instance, we know (Sec. 8.2) that the curl of a gradient is $\mathbf{0}$:

$$\mathbf{F} = \nabla f \quad \Rightarrow \quad \text{curl } \mathbf{F} = \mathbf{0}.$$

The curl condition must hold if $\mathbf{F}$ is a gradient field. Now we turn the tables and ask whether

$$\text{curl } \mathbf{F} = \mathbf{0} \quad \Rightarrow \quad \mathbf{F} = \nabla f \text{ for some } f.$$

It turns out that this implication is very often true. Surprisingly, its truth depends on a geometric property of the domain D of the vector field $\mathbf{F}$: The domain must be simply connected. We explain what this means next, beginning with the two-dimensional case.

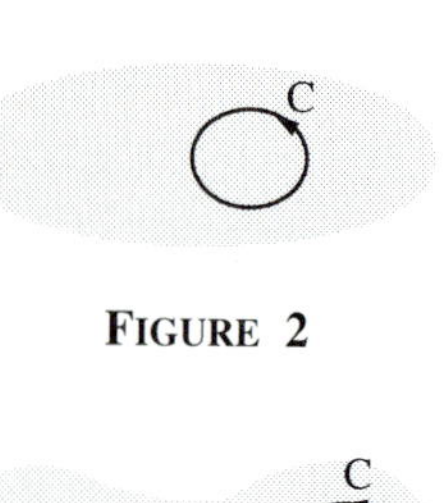

FIGURE 2

FIGURE 3

Intuitively, an open set D in 2–space is **simply connected** if it has no holes. For example, the entire xy–plane is simply connected and the region inside a circle, ellipse, or rectangle is simply connected. On the other hand, the annular region between two circles is not simply connected; it has a hole in it. Figures 2 and 3 illustrate typical sets in the xy–plane that are, respectively, simply connected and not simply connected.

The precise mathematical definition of *simply connected* is a little too involved to be given here. We shall describe only the general idea. If any continuous closed curve C in D can be deformed gradually into a single point in D without leaving D in the process, then D is simply connected. The region in Fig. 2 has this property and is simply connected. Since the curve C in Fig. 3 cannot be shrunk to a point in the required manner, the set D in Fig. 3 is not simply connected. The description of simply connected just given applies also to regions in 3–space. For example, all of 3–space is simply connected and the region inside a sphere, ellipsoid, or parallelepiped is simply connected. The region inside a torus (doughnut) is not simply connected.

Now we can give a reasonably satisfactory answer to Q1: In a simply connected domain D,

$$\mathbf{F} = \nabla f \text{ for some } f \quad \Leftrightarrow \quad \text{curl } \mathbf{F} = \mathbf{0}.$$

We have covered a lot of ground. Let's summarize some of the important conclusions. Let $\mathbf{F} = \mathbf{F}(\mathbf{r})$ be a vector field with components having continuous partial derivatives in a region D. If D is simply connected, then the following four properties of $\mathbf{F}$ are equivalent.

$\mathbf{F} = \nabla f$ for some potential function f in D.

$\int_C \mathbf{F} \cdot d\mathbf{r}$ is independent of the path in D.

$\int_C \mathbf{F} \cdot d\mathbf{r} = 0$ for every closed path in D.

curl $\mathbf{F} = \mathbf{0}$ in D.

Furthermore, if any one (and, hence, all four) of these properties holds, then

$$\int_C \mathbf{F} \cdot d\mathbf{r} = \int_A^B \mathbf{F} \cdot d\mathbf{r} = f(B) - f(A)$$

for any path C in D extending from A to B.

Now it's time to take a closer look at the properties summarized here.

The Fundamental Theorem and Independence of the Path

The fundamental theorem for line integrals establishes the equivalence of the first two properties mentioned in the overview and the convenient formula for evaluating line integrals of gradient fields. We begin by stating more carefully what independence of path means.

Definition *Independence of Path*
Let $\mathbf{F} = \mathbf{F}(\mathbf{r})$ be a continuous vector field defined on a connected open set D in 2–space or 3–space. The line integral $\int_C \mathbf{F} \cdot d\mathbf{r}$ is **independent of the path** in D if, for any two points A and B in D, the line integral $\int_C \mathbf{F} \cdot d\mathbf{r}$ has the same value for all paths C in D from A to B.

Since paths are composed of smooth curves, $\int_C \mathbf{F} \cdot d\mathbf{r}$ is independent of the path in D if $\int_C \mathbf{F} \cdot d\mathbf{r}$ has the same value for all smooth curves in D from A to B.

Theorem 1 *The Fundamental Theorem for Line Integrals*
Let $\mathbf{F} = \mathbf{F}(\mathbf{r})$ be a continuous vector field defined on a connected open set D in 2–space or 3–space. Then the following are equivalent:

(1) $\mathbf{F} = \nabla f$ in D for some potential function f.

(2) $\int_C \mathbf{F} \cdot d\mathbf{r}$ is independent of the path in D.

If (1) and hence (2) is true, then

$$\int_C \mathbf{F} \cdot d\mathbf{r} = \int_A^B \mathbf{F} \cdot d\mathbf{r} = f(B) - f(A) \tag{3}$$

for any path C in D from A to B.

Proof. Assume (1) holds. Let A, B be any two points in D and let C be any smooth curve from A to B with parameterization $\mathbf{r}(t)$ for $a \leq t \leq b$. By the vector form of the chain rule,

$$\frac{d}{dt} f(\mathbf{r}(t)) = \nabla f(\mathbf{r}(t)) \cdot \mathbf{r}'(t).$$

Therefore,

$$\begin{aligned}\int_C \mathbf{F} \cdot d\mathbf{r} &= \int_a^b \mathbf{F}(\mathbf{r}(t)) \cdot \mathbf{r}'(t)\,dt = \int_a^b \nabla f(\mathbf{r}(t)) \cdot \mathbf{r}'(t)\,dt \\ &= \int_a^b \left[\frac{d}{dt} f(\mathbf{r}(t))\right] dt = [f(\mathbf{r}(t))]_a^b = f(B) - f(A).\end{aligned}$$

This equality shows that $\int_C \mathbf{F} \cdot d\mathbf{r}$ depends only upon A and B, and not on the particular curve from A to B. Therefore, (1) $\Rightarrow$ (2) and (3).

It remains to prove that (2) $\Rightarrow$ (1). We give the details for the 2–d case. The 3–d case is handled in the same way. Assume that (2) holds. Then $\int_C \mathbf{F} \cdot d\mathbf{r}$ is independent of the path in D. To prove (1) we must manufacture a scalar field f such that $\nabla f = \mathbf{F}$. If there is such a scalar field f, then, since (1) $\Rightarrow$ (3),

$$\int_A^B \mathbf{F} \cdot d\mathbf{r} = f(B) - f(A)$$

and, consequently,

$$f(x,y) = f(x_0,y_0) + \int_{(x_0,y_0)}^{(x,y)} \mathbf{F} \cdot d\mathbf{r}.$$

We use this equation to define f. Fix any point (x_0,y_0) in D and choose any value (perhaps, 0) for $f(x_0,y_0)$. Define $f(x,y)$ by the preceding formula. We shall prove that $\nabla f = \mathbf{F}$ or, equating components, that

$$\frac{\partial f}{\partial x}(x,y) = P(x,y) \quad \text{and} \quad \frac{\partial f}{\partial y}(x,y) = Q(x,y).$$

Equivalently,

$$\lim_{h\to 0} \frac{f(x+h,y) - f(x,y)}{h} = P(x,y) \quad \text{and} \quad \lim_{h\to 0} \frac{f(x,y+h) - f(x,y)}{h} = Q(x,y).$$

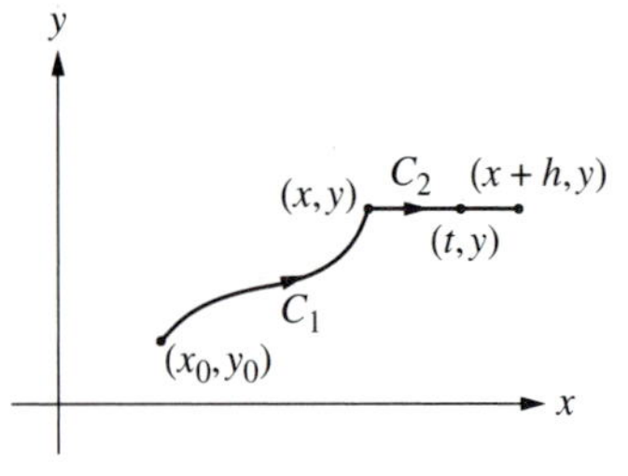

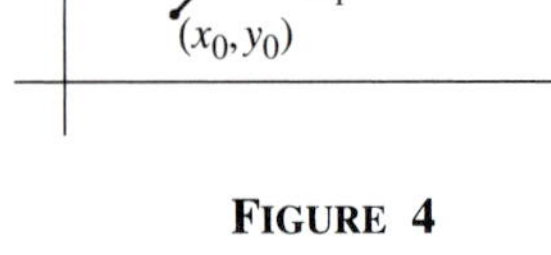

FIGURE 4

To verify the first of these limits, refer to Fig. 4, where $h > 0$. Now,

$$f(x,y) = f(x_0,y_0) + \int_{C_1} \mathbf{F} \cdot d\mathbf{r},$$

$$f(x+h,y) = f(x_0,y_0) + \int_{C_1} \mathbf{F} \cdot d\mathbf{r} + \int_{C_2} \mathbf{F} \cdot d\mathbf{r},$$

$$f(x+h,y) - f(x,y) = \int_{C_2} \mathbf{F} \cdot d\mathbf{r}.$$

On C_2,

$$\mathbf{r}(t) = t\mathbf{i} + y\mathbf{j}, \qquad \mathbf{r}'(t) = \mathbf{i}, \quad \text{for} \quad x \le t \le x+h,$$

$$\mathbf{F}(\mathbf{r}(t)) = P(t,y)\,\mathbf{i} + Q(t,y)\,\mathbf{j},$$

$$\int_{C_2} \mathbf{F} \cdot d\mathbf{r} = \int_x^{x+h} \mathbf{F}(\mathbf{r}(t)) \cdot \mathbf{r}'(t)\,dt = \int_x^{x+h} P(t,y)\,dt.$$

Consequently,

$$\frac{f(x+h,y) - f(x,y)}{h} = \frac{1}{h}\int_x^{x+h} P(t,y)\,dt.$$

There are several ways to determine the limit on the right as $h \to 0$ with x and y fixed. Perhaps the quickest way is to observe that the right member is an indeterminate form of type 0/0 as $h \to 0$. By l'Hôpital's rule, the fundamental theorem of calculus, and the continuity of P,

$$\lim_{h\to 0} \frac{f(x+h,y) - f(x,y)}{h} = \lim_{h\to 0} \frac{1}{h}\int_x^{x+h} P(t,y)\,dt = \lim_{h\to 0} \frac{P(x+h,y)}{1} = P(x,y).$$

It follows that

$$\frac{\partial f}{\partial x}(x,y) = P(x,y).$$

Likewise,

$$\frac{\partial f}{\partial y}(x,y) = Q(x,y).$$

Since P and Q are continuous, we see that f has continuous partial derivatives and that

$$\nabla f(x,y) = P(x,y)\mathbf{i} + Q(x,y)\mathbf{j} = \mathbf{F}(x,y),$$

which completes the proof that (2) $\Rightarrow$ (1). □

EXAMPLE 1. Let $\mathbf{F} = 2xe^{x^2}\sin\pi y\mathbf{i} + \pi e^{x^2}\cos\pi y\mathbf{j}$. Evaluate $\int_C \mathbf{F}\cdot d\mathbf{r}$ where C is the curve $y = x^3$, $0 \leq x \leq 1$, from $(0,0)$ to $(1,1)$.

Solution. A direct evaluation using the parameterization for C is arduous. If we notice that $\mathbf{F} = \nabla f$ for $f(x,y) = e^{x^2}\sin\pi y$, then the evaluation is easy from Th. 1:

$$\int_C \mathbf{F}\cdot d\mathbf{r} = f(1,1) - f(0,0) = 0 - 0 = 0. \ \square$$

A few additional remarks about Th. 1 are in order. Since we call a vector field $\mathbf{F}$ conservative if $\mathbf{F} = \nabla f$ for some f, the first part of Th. 1 can be expressed in the following way.

A vector field $\mathbf{F}$ is conservative in D $\Leftrightarrow$ $\int_C \mathbf{F}\cdot d\mathbf{r}$ is independent of the path in D.

Thus, conservative force fields are those for which the work $W = \int_A^B \mathbf{F}\cdot d\mathbf{r}$ done by the field as an object moves depends only on the initial and final locations of the object. Since both the Galilean (constant) and Newtonian (inverse square) gravitational fields are conservative, the work done by either field is independent of the path. This fact is needed to justify the somewhat bizarre procedure used in one-variable calculus to find the work done against gravity to pump liquid into or out of a tank. The reasoning is based on moving thin, horizontal slabs of liquid from one level to another. An actual pump moves the liquid quite differently, but the work done is the same because the work done is independent of the path. In the same way, you can calculate the work done against gravity to lift an astronaut into orbit by assuming the rocket rises straight up. A real rocket follows a very different path but the work to reach a given orbital height is the same, independent of the path used to reach that height.

Next we look at another very convenient way to describe conservative fields.

$\int_C \mathbf{F} \cdot d\mathbf{r}$ is independent of the path in D $\Leftrightarrow$ $\int_C \mathbf{F} \cdot d\mathbf{r} = 0$ for every closed path in D.

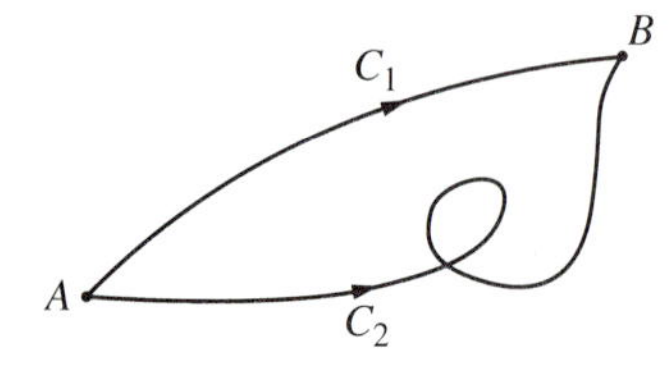

FIGURE 5

To understand why, consult Fig. 5 in which A and B are any two points in D and C_1 and C_2 are any two paths in D from A to B. Assume that $\int_C \mathbf{F} \cdot d\mathbf{r} = 0$ for every closed path in D. Since $C = C_1 - C_2 = C_1 + (-C_2)$ is such a path,

$$0 = \int_C \mathbf{F} \cdot d\mathbf{r} = \int_{C_1 - C_2} \mathbf{F} \cdot d\mathbf{r} = \int_{C_1} \mathbf{F} \cdot d\mathbf{r} - \int_{C_2} \mathbf{F} \cdot d\mathbf{r},$$

$$\int_{C_1} \mathbf{F} \cdot d\mathbf{r} = \int_{C_2} \mathbf{F} \cdot d\mathbf{r}.$$

So the line integral is independent of the path in D. Reverse the steps in the argument to show that $\int_C \mathbf{F} \cdot d\mathbf{r} = 0$ if the line integral is independent of the path in D. (See the problems.)

The Fundamental Theorem and Conservation of Energy

The fundamental theorem for line integrals enables us to establish the principle of conservation of mechanical energy for an object moving in a conservative (gradient) force field in two- and three-dimensions. As we mentioned earlier, this conservation law is a fundamental principle of physics.

Let $\mathbf{F} = \mathbf{F}(\mathbf{r})$ be a conservative force field. Thus, $\mathbf{F} = \nabla f$ for some f. In most physics and engineering books, it is customary to write $f = -V$ and call V a potential energy function for $\mathbf{F}$. That is, V is a **potential energy function** for $\mathbf{F}$ if $\mathbf{F} = -\nabla V$. Since V is unique up to an additive constant, we often refer to V as the **potential energy**. Often a convenient choice for the constant is suggested by the physical situation.

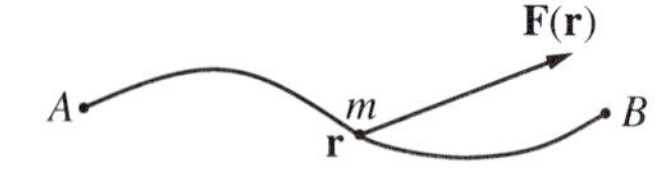

FIGURE 6

Suppose that the force field $\mathbf{F}(\mathbf{r})$ moves an object of mass m along a trajectory C from a point A to a point B, as shown in Fig. 6. Express the trajectory in parametric form by $\mathbf{r}(t)$ for $a \le t \le b$, where t is time. Then C is determined by Newton's second law $m\mathbf{r}'' = \mathbf{F}$. The kinetic energy at time t is $K(t) = \frac{1}{2}m\|\mathbf{r}'(t)\|^2$. By Th. 1 in Sec. 8.2, the work done by the force in moving the object from A to B is the change in kinetic energy,

$$W_C = \int_C \mathbf{F} \cdot d\mathbf{r} = K(b) - K(a).$$

On the other hand, by the fundamental theorem for line integrals, the work is expressed in terms of the potential energy by

$$W_C = \int_C \mathbf{F} \cdot d\mathbf{r} = \int_A^B -\nabla V \cdot d\mathbf{r} = V(A) - V(B).$$

Equate the two expressions for W_C to obtain

$$K(b) + V(B) = K(a) + V(A).$$

In words, the sum of the kinetic and potential energies is the same at any two points on a trajectory. The sum

$$E(t) = K(t) + V(\mathbf{r}(t))$$

is called the **total mechanical energy** of the object at time t. We have shown that $E(t)$ is constant. This fundamental result is summarized in the following theorem.

Theorem 2 *Conservation of Energy*
Let **F** be a conservative, continuous force field defined in a connected open set in 2–space or 3–space. Then the total mechanical energy of an object remains constant as it moves along a trajectory of the force field.

EXAMPLE 2. Verify directly that the total mechanical energy is conserved when a particle of mass m moves in the (essentially) constant gravitational field near the surface of the earth.

Solution. The gravitational force on the particle is $\mathbf{F} = -mg\mathbf{k}$ when the z–axis is directed upward. The force field $\mathbf{F}$ is conservative with potential energy $V = mgz$ because $-\nabla V = -mg\mathbf{k} = \mathbf{F}$. Let $\mathbf{r}(0) = \mathbf{r}_0$ and $\mathbf{r}'(0) = \mathbf{v}_0$. From Newton's second law,

$$m\mathbf{r}''(t) = -mg\mathbf{k}, \qquad \mathbf{r}'(t) = -gt\mathbf{k} + \mathbf{v}_0, \qquad \mathbf{r}(t) = -\frac{1}{2}gt^2\mathbf{k} + t\mathbf{v}_0 + \mathbf{r}_0.$$

So the kinetic energy is

$$K(t) = \frac{1}{2}m\,||\mathbf{r}'(t)||^2 = \frac{1}{2}m\mathbf{r}'(t) \cdot \mathbf{r}'(t) = \frac{1}{2}m(g^2t^2 - 2gt\mathbf{v}_0 \cdot \mathbf{k} + ||\mathbf{v}_0||^2)$$

and the potential energy is

$$V(t) = mgz(t) = mg\mathbf{r}(t) \cdot \mathbf{k} = mg\left(-\frac{1}{2}gt^2 + t\mathbf{v}_0 \cdot \mathbf{k} + \mathbf{r}_0 \cdot \mathbf{k}\right).$$

Add to obtain the total mechanical energy,

$$\begin{aligned} E(t) = K(t) + V(t) &= \frac{1}{2}m\,||\mathbf{v}_0||^2 + mg\mathbf{r}_0 \cdot \mathbf{k} \\ &= K(0) + mgz(0) = E(0). \end{aligned}$$

Thus, $E(t) = E(0)$ for all t and the total mechanical energy is constant, just as guaranteed by Th. 2. □

Finding Potential Functions

To make full use of Ths. 1 and 2 we need to know when a given vector field **F** is conservative ($\mathbf{F} = \nabla f$ for some f) and if so, how to find a potential function f for **F**. These questions were raised in the preview of this section, where we discussed the first question. Here is a formal statement of what we announced.

Theorem 3 *Characterization of Conservative Vector Fields*
Let $\mathbf{F}(\mathbf{r})$ be a vector field defined in a simply connected domain. Assume that the components of **F** have continuous first partials. Then

$$\mathbf{F} = \nabla f \text{ for some } f \quad \Leftrightarrow \quad \text{curl } \mathbf{F} = \mathbf{0}.$$

For a two-dimensional field $\mathbf{F}(\mathbf{r}) = P(x,y)\mathbf{i} + Q(x,y)\mathbf{j}$,

$$\text{curl } \mathbf{F} = \left(\frac{\partial Q}{\partial x} - \frac{\partial P}{\partial y}\right)\mathbf{k}.$$

Consequently,

$$\text{curl } \mathbf{F} = \mathbf{0} \quad \Leftrightarrow \quad \frac{\partial P}{\partial y} = \frac{\partial Q}{\partial x}.$$

We shall sketch a proof of Th. 3 a little later, after we give some examples.

EXAMPLE 3. Let $\mathbf{F}(\mathbf{r}) = y^2\mathbf{i} + (2xy - 1)\mathbf{j}$. (a) Show that **F** is a conservative field. (b) Find a potential function f for **F**.

Solution. (a) In this case, $\mathbf{F} = P\mathbf{i} + Q\mathbf{j}$ with

$$P = y^2, \qquad Q = 2xy - 1.$$

Since the domain of **F** is the entire xy–plane, which is simply connected, and

$$\frac{\partial P}{\partial y} = 2y = \frac{\partial Q}{\partial x},$$

F is a gradient field (hence, conservative) by Th. 3.
(b) Since $\mathbf{F} = P\mathbf{i} + Q\mathbf{j}$ and $\nabla f = f_x\mathbf{i} + f_y\mathbf{j}$,

$$\mathbf{F} = \nabla f \quad \Leftrightarrow \quad \frac{\partial f}{\partial x} = P = y^2, \qquad \frac{\partial f}{\partial y} = Q = 2xy - 1.$$

We must solve the equations on the right for f. Antidifferentiate $\partial f/\partial x = y^2$ with respect to x holding y constant to find that

$$f(x,y) = xy^2 + \varphi(y),$$

where $\varphi(y)$ is any function of y. In other words, $\varphi(y)$ is constant with respect to x. No matter what choice is made for $\varphi(y)$, $\partial f/\partial x = y^2$, as required. Since $\partial f/\partial y = 2xy - 1$, we must also have

$$\frac{\partial f}{\partial y} = 2xy + \varphi'(y) = 2xy - 1.$$

Therefore, $\varphi'(y) = -1$ and $\varphi(y) = -y + C$ with any constant C. Then

$$f(x,y) = xy^2 - y + C.$$

Since we seek only one potential function f, we can take $C = 0$. Then $f(x,y) = xy^2 - y$. It is easy to check that $\nabla f = \mathbf{F}$. □

The reasoning in (b) of Ex. 3 illustrates a general procedure for finding a potential function for a conservative vector field. It can be used for both two- and three-dimensional fields.

EXAMPLE 4. As in Ex. 1, let $\mathbf{F} = 2xe^{x^2} \sin \pi y \mathbf{i} + \pi e^{x^2} \cos \pi y \mathbf{j}$. (a) Show that $\mathbf{F}$ is a conservative vector field. (b) Find a potential function f for $\mathbf{F}$.

Solution. Here $\mathbf{F} = P\mathbf{i} + Q\mathbf{j}$ with

$$P = 2xe^{x^2} \sin \pi y, \qquad Q = \pi e^{x^2} \cos \pi y.$$

(a) Since the domain of $\mathbf{F}$ is the entire xy–plane and

$$\frac{\partial P}{\partial y} = 2\pi x e^{x^2} \cos \pi y = \frac{\partial Q}{\partial x},$$

$\mathbf{F} = \nabla f$ for some f by Th. 3.
(b) Now

$$\mathbf{F} = \nabla f \quad \Leftrightarrow \quad \frac{\partial f}{\partial x} = P = 2xe^{x^2} \sin \pi y, \; \frac{\partial f}{\partial y} = Q = \pi e^{x^2} \cos \pi y.$$

Antidifferentiate $\partial f/\partial y$ with respect to y holding x fixed to obtain

$$f(x,y) = e^{x^2} \sin \pi y + \varphi(x)$$

for some $\varphi(x)$. Then

$$\frac{\partial f}{\partial x} = 2xe^{x^2} \sin \pi y + \varphi'(x).$$

On the other hand, since $\partial f/\partial x = 2xe^{x^2} \sin \pi y$ from our earlier work, we must have $\varphi'(x) = 0$. Let's take $\varphi(x) = 0$. Then

$$f(x,y) = e^{x^2} \sin \pi y.$$

This is the same potential function we "pulled out of the hat" in Ex. 1. □

EXAMPLE 5. Let $\mathbf{F} = y^2z^3\mathbf{i} + 2xyz^3\mathbf{j} + (3xy^2z^2 + z)\mathbf{k}$. (a) Show that $\mathbf{F}$ is a conservative vector field. (b) Find a potential function f for $\mathbf{F}$.

Solution. In this case $\mathbf{F} = P\mathbf{i} + Q\mathbf{j} + R\mathbf{k}$ with

$$P = y^2z^3, \qquad Q = 2xyz^3, \qquad R = 3xy^2z^2 + z.$$

(a) The domain of $\mathbf{F}$ is all of 3–space and by a routine calculation

$$\text{curl } \mathbf{F} = \begin{vmatrix} \mathbf{i} & \mathbf{j} & \mathbf{k} \\ \dfrac{\partial}{\partial x} & \dfrac{\partial}{\partial y} & \dfrac{\partial}{\partial z} \\ y^2z^3 & 2xyz^3 & 3xy^2z^2 + z \end{vmatrix} = \mathbf{0}.$$

By Th. 3, $\mathbf{F} = \nabla f$ for some f.
(b) To find f, we must solve the equations

$$\frac{\partial f}{\partial x} = P = y^2z^3, \qquad \frac{\partial f}{\partial y} = Q = 2xyz^3, \qquad \frac{\partial f}{\partial z} = R = 3xy^2z^2 + z.$$

Antidifferentiate the first equation with respect to x to get

$$f(x,y,z) = xy^2z^3 + \varphi(y,z)$$

for some function $\varphi(y,z)$. Since $\partial f/\partial y = Q$, we find that

$$\frac{\partial f}{\partial y} = 2xyz^3 + \frac{\partial}{\partial y}\varphi(y,z) = 2xyz^3, \qquad \frac{\partial \varphi}{\partial y} = 0.$$

Thus, φ depends only upon z. Since $\partial f/\partial z = R$,

$$\frac{\partial f}{\partial z} = 3xy^2z^2 + \frac{\partial \varphi}{\partial z} = 3xy^2z^2 + z, \qquad \frac{\partial \varphi}{\partial z} = z.$$

Since φ depends only on z, an obvious choice is $\varphi = z^2/2$. (As usual no additive constant is necessary.) Then

$$f(x,y,z) = xy^2z^3 + \frac{1}{2}z^2.$$

Finally, evaluate ∇f to confirm that $\nabla f = \mathbf{F}$. □

Proof of Theorem 3

We shall sketch the proof of Th. 3 only in the two-dimensional case and with the added restriction that the domain of the vector field $\mathbf{F(r)}$ is the entire xy–plane. This case illustrates the key ideas needed for the more general case of a simply connected domain and avoids technical details.

If $\mathbf{F} = \nabla f$, then we know that curl $\mathbf{F} = \mathbf{0}$. For the converse, assume that curl $\mathbf{F} = \mathbf{0}$. Equivalently,

$$\frac{\partial P}{\partial y} = \frac{\partial Q}{\partial x}.$$

We seek a scalar function f such that $\mathbf{F} = \nabla f$. If there is such an f, then, as in the proof of Th. 1, it satisfies

$$f(x,y) = f(x_0, y_0) + \int_{(x_0,y_0)}^{(x,y)} \mathbf{F} \cdot d\mathbf{r}$$

for any (x_0,y_0) and (x,y). Since we do not yet know that $\mathbf{F} = \nabla f$, the value of the line integral from (x_0,y_0) to (x,y) might depend on the path along which we integrate. To ensure that $f(x,y)$ is uniquely defined, we choose a specific path $C = C_1 + C_2$ from (x_0,y_0) to (x,y), where C_1 is a horizontal line segment and C_2 is a vertical line segment. The case with $x > x_0$ and $y > y_0$ is illustrated in Fig. 7.

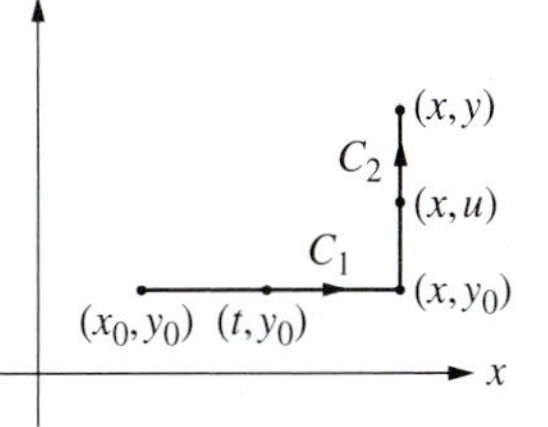

FIGURE 7

Now we define $f(x,y)$ by

$$f(x,y) = f(x_0,y_0) + \int_{C_1} \mathbf{F} \cdot d\mathbf{r} + \int_{C_2} \mathbf{F} \cdot d\mathbf{r},$$

where (x_0,y_0) is any fixed point and $f(x_0,y_0)$ may be assigned any particular value. Parameterize C_1 and C_2 by

$$\mathbf{r}(t) = t\mathbf{i} + y_0\mathbf{j} \quad \text{for} \quad x_0 \le t \le x \quad \text{on} \quad C_1,$$
$$\mathbf{r}(u) = x\mathbf{i} + u\mathbf{j} \quad \text{for} \quad y_0 \le u \le y \quad \text{on} \quad C_2,$$

as shown in Fig. 7. Then

$$f(x,y) = f(x_0,y_0) + \int_{x_0}^{x} P(t,y_0)\,dt + \int_{y_0}^{y} Q(x,u)\,du.$$

By the fundamental theorem of calculus and $\partial P/\partial y = \partial Q/\partial x$,

$$\begin{aligned}
\frac{\partial f}{\partial x} &= \frac{\partial}{\partial x}\int_{x_0}^{x} P(t,y_0)\,dt + \frac{\partial}{\partial x}\int_{y_0}^{y} Q(x,u)\,du, \\
&= P(x,y_0) + \int_{y_0}^{y} \frac{\partial}{\partial x} Q(x,u)\,du, \\
&= P(x,y_0) + \int_{y_0}^{y} \frac{\partial}{\partial y} P(x,u)\,du, \\
&= P(x,y_0) + [P(x,y) - P(x,y_0)] = P(x,y).
\end{aligned}$$

Thus, $\partial f/\partial x = P(x,y)$. A very similar calculation gives $\partial f/\partial y = Q(x,y)$. Hence,

$$\text{curl } \mathbf{F} = \mathbf{0} \quad \Rightarrow \quad \mathbf{F} = \nabla f \text{ for some } f,$$

which completes our sketch of the proof of Th. 3.

PROBLEMS

In Probs. 1–8 (a) use Th. 3 to verify that the vector field **F** is conservative; (b) find a potential for it; and (c) evaluate the line integral $\int_C \mathbf{F} \cdot d\mathbf{r}$ using Th. 1.

1. $\mathbf{F} = (3x - 2y)\mathbf{i} + (3y - 2x)\mathbf{j}$
   C: $x = 4\cos t$, $y = 4\sin t$, $0 \leq t \leq \pi$

2. $\mathbf{F} = (2xy - y^2)\mathbf{i} + (x^2 - 2xy)\mathbf{j}$
   C: $y = x^3$, $0 \leq x \leq 2$

3. $\mathbf{F} = 2xe^{x^2}\cos y\mathbf{i} + -e^{x^2}\sin y\mathbf{j}$
   C: $\mathbf{r} = t\cos t\mathbf{i} + \mathbf{t}\sin t\mathbf{j}$, $0 \leq t \leq \pi/2$

4. $\mathbf{F} = (e^y + y\sin x)\mathbf{i} + (xe^y - \cos x + 2)\mathbf{j}$
   C: $x = \pi t$, $y = 3 - 3t$, $0 \leq t = 1$

5. $\mathbf{F} = < \cos xy - xy\sin xy, -x^2\sin xy >$
   C: $\mathbf{r} = 3\cos\theta\mathbf{i} - \mathbf{3}\sin\theta\mathbf{j}$, $0 \leq \theta \leq 3\pi/2$

6. $\mathbf{F} = < \cos x\cosh y + y\sin xy, \sin x\sinh y + x\sin xy >$
   C: $x = 1 - \cos t$, $y = t - \sin t$, $0 \leq t \leq 2\pi$

7. $\mathbf{F} = (yz + 1)\mathbf{i} + (xz + 2)\mathbf{j} + (xy + 3)\mathbf{k}$
   C: $\mathbf{r} = < \cos 2\theta, \sin 2\theta, 4\theta/\pi >$, $0 \leq \theta \leq \pi/2$

8. $\mathbf{F} = < yz^2 - y^2z, xz^2 - 2xyz, 2xyz - xy^2 >$
   C: is the polygonal line with successive vertices $(1,0,0)$, $(1,1,0)$, $(1,1,1)$, $(0,0,1)$

In Probs. 9–16 evaluate the line integral $\int_C \mathbf{F} \cdot d\mathbf{r}$.

9. $\mathbf{F} = (x^2 + y^2)\mathbf{i} + (2xy + \cos y)\mathbf{j}$
   C: $y = \sin(\pi x/2)$, $0 \leq x \leq 1$

10. $\mathbf{F} = (2x - 2y)e^{x^2 - 2xy}\mathbf{i} - 2xe^{x^2 - 2xy}\mathbf{j}$
    C: $\mathbf{r} = t^2\mathbf{i} - t^3\mathbf{j}$, $0 \leq t \leq 2$

11. $\mathbf{F} = < x^3y, x - y >$
    C: $y = x^2$, $-1 \leq x \leq 2$

12. $\mathbf{F} = < xy, y^2 >$,
    C: $x = \cos\theta$, $y = \sin\theta$, $0 \leq \theta \leq 2\pi$

13. $\mathbf{F} = y\sin xy\mathbf{i} + x\sin xy\mathbf{j}$
    C: $x = t$, $y = 4 - t^2$, $0 \leq t \leq 2$

14. $\mathbf{F} = (1 + x^2y^2)^{-1} < y, x >$
    C: $y = x^2$, $-1 \leq x \leq 1$

15. $\mathbf{F} = \sin y\mathbf{i} + (x + x\cos y)\mathbf{j}$
    C: $\mathbf{r} = < \cos\theta, \sin\theta >$, $0 \leq \theta \leq 2\pi$

    Hint. Write $\mathbf{F} = \mathbf{F}_1 + \mathbf{F}_2$ where $\mathbf{F}_1$ is conservative.

16. $\mathbf{F} = (ye^{xy} - y)\mathbf{i} + (xe^{xy} + x)\mathbf{j}$
    C: $x = 4\cos t$, $y = 3\sin t$, $0 \leq t \leq 2\pi$

In Probs. 17–20 you are asked to take advantage of the fact that line integrals of conservative fields are independent of the path to evaluate some specific integrals without first finding a potential function. Recall that each of the vector fields in Probs. 1–8 is conservative.

17. Evaluate the line integral in Prob. 1(c) by explicitly calculating $\int_L \mathbf{F} \cdot d\mathbf{r}$ where L is the line segment joining the endpoints of C.

18. Evaluate the line integral in Prob. 2(c) by explicitly calculating $\int_L \mathbf{F} \cdot d\mathbf{r}$ where L is the line segment joining the endpoints of C.

19. Evaluate the line integral in Prob. 5(c) by explicitly calculating $\int_L \mathbf{F} \cdot d\mathbf{r}$ where L is the polygonal line joining (3,0) to (0,0) to (0,3).

20. Evaluate the line integral in Prob. 8(c) by explicitly calculating $\int_L \mathbf{F} \cdot d\mathbf{r}$ where L is the line segment joining the endpoints of C.

21. Let $\mathbf{F} = \mathbf{F}(\mathbf{r})$ be a conservative force field with potential energy V. Show that the work done *against* the field when an object is moved from $\mathbf{r}_0$ to $\mathbf{r}_1$ is $V(\mathbf{r}_1) - V(\mathbf{r}_0)$.

22. Assume the Galilean model of constant gravitational acceleration g. Show that the work done against gravity to move an object from (x_0, y_0, z_0) to (x_1, y_1, z_1) along any path is mgh where $h = z_1 - z_0$. (As usual, the positive z–axis points vertically upward).

23. The Newtonian gravitational force of the earth on a object with mass m is $\mathbf{F} = -GMm\mathbf{r}/r^3$ where M is the mass of the earth, m is located at $\mathbf{r}$ and $r = \|\mathbf{r}\| \geq R$, the radius of the earth. The origin is at the center of the earth. (a) Show that the work done against gravity to move m from position $\mathbf{r}_0$ to position $\mathbf{r}_1$ along any path is $GMm(1/r_1 - 1/r_0)$.

24. Find the work done against the gravitational field of the earth when the moon moves from its closest distance of 356,400 km from the earth to its greatest distance of 407,000 km from the earth. The mass of the moon is 1/81 the mass of the earth, which is 5.97×10^{24} kg, and the universal gravitational constant is $G \approx 6.67 \times 10^{-11}$ nt–m^2/kg^2.

25. Let n be a fixed constant and $r = \|\mathbf{r}\|$ where $\mathbf{r}$ is the position vector of a point (x, y) or (x, y, z). Show that the central force field $\mathbf{F} = (k/r^n)\mathbf{r}$, k constant, is conservative by finding a potential function for it. (Be careful. There are two cases: $n = 2$ and $n \neq 2$.)

26. Show that the line integral $\int_C \mathbf{r} \cdot d\mathbf{r}$ is independent of the path and find its value for any path joining $\mathbf{r}_0$ to $\mathbf{r}_1$.

27. Show that the line integral $\int_C r^n\mathbf{r} \cdot d\mathbf{r}$ is independent of the path for any $n \geq 0$ and find its value for any path joining $\mathbf{r}_0$ to r_1.

28. A general central force field with center of force at the origin has the form $\mathbf{F} = g(r)(\mathbf{r}/r)$ where $\mathbf{r} = \langle x, y \rangle$ or $\langle x, y, z \rangle$, $r = \|\mathbf{r}\|$, and $g(r)$ is continuous. Let h be an antiderivative of g so that $h'(r) = g(r)$. (a) Show that $h(\mathbf{r})$ is a potential for $\mathbf{F(r)}$. (b) Use (a) to find a potential function for the Newtonian gravitational field.

29. An object with mass m is attached to a spring and moves along the x–axis. Assume the spring obeys Hooke's law so that the spring force is $\mathbf{F} = -kx\mathbf{i}$ where $k > 0$ is the spring constant. (a) Show that $\mathbf{F}$ is conservative and find a potential energy function for it. (b) Find a formula for the total mechanical energy of the mass m attached to the spring. (c) Check directly, much as in

Ex. 2, that energy is conserved, which verifies Th. 2 in another particular case. (d) Let $x(t)$ be the position of m at time t. If $x(0) = 3$ and $x'(0) = -5$, find the maximum speed and maximum displacement of the mass. Use only results of this section.

The **equipotential surfaces** (**curves** in 2–d) of a conservative force field are the surfaces (curves) along which the potential energy is constant.

30. Let $\mathbf{F}(\mathbf{r})$ be conservative force field with a potential whose first partials are continuous. (a) Show that $\mathbf{F}$ is perpendicular to an equipotential surface at each of its points. (b) Find the equipotential surfaces of the Newtonian gravitational field and then confirm the statement in (a) directly for the Newtonian field.
31. An object moves on a trajectory of a conservative force field. If the trajectory lies in an equipotential surface, show that the object has constant speed.
32. (a) Find the equipotential curves of $\mathbf{F} = x\mathbf{i} + y\mathbf{j}$. (b) Check directly that $\mathbf{F}$ is perpendicular to any equipotential curve at each of its points.

Problems 33–36 shed further light on the subtle interplay among the conservative nature of a vector field $\mathbf{F}$, the geometry of the domain of $\mathbf{F}$, and the condition curl $\mathbf{F} = \mathbf{0}$.

33. Let $\mathbf{F} = (x^2 + y^2)^{-1/2}(-y\mathbf{i} + x\mathbf{j})$ for $(x,y) \neq (0,0)$, which is the vector field in Ex. 4 and Fig. 5 of Sec. 7.1.
    (a) Show that $\int_C \mathbf{F} \cdot d\mathbf{r} = 2\pi$ where C is the unit circle $\mathbf{r} = \cos\theta\mathbf{i} + \sin\theta\mathbf{j}$ for $0 \leq \theta \leq 2\pi$.
    (b) Show that $\operatorname{curl} \mathbf{F} = \mathbf{0}$.
    (c) Conclude from (a) that $\mathbf{F}$ is not a conservative field. Then explain why (b) and Th. 3 are not in contradiction with the previous sentence.
34. Is the vector field $\mathbf{F}$ in Prob. 33 a gradient field in the annulus $1 < x^2 + y^2 < 2$? Explain carefully why or why not.
35. Is the vector field $\mathbf{F}$ in Prob. 33 a gradient field in the region bounded by the circle with center $(1, 2)$ and radius 2? Explain carefully why or why not.
36. Is the vector field $\mathbf{F}$ in Prob. 33 a gradient field in the annulus $2 < (x-3)^2 + (y-4)^2 < 4$? Explain carefully why or why not.
37. Let $\mathbf{F}$ be a continuous vector field with domain D. If $\int_C \mathbf{F} \cdot d\mathbf{r}$ is independent of the path in D, then show that $\int_C \mathbf{F} \cdot d\mathbf{r} = 0$ for every *closed* path in D by elaborating on the "reverse the steps" argument mentioned in the text.

8.4 Green's Theorem and Applications

Green's theorem expresses a double integral over a region D in the xy–plane as a line integral over the boundary of D. The theorem first appeared in *Essay on the Application of Mathematical Analysis to the Theories of Electricity and Magnetism.* This monograph was authored by George Green (1793–1841) and published at his own expense. Thus, the result now known as Green's theorem had its origin in electricity and magnetism. We shall express Green's theorem first in a scalar form. Then we reformulate it in two revealing vector forms, called the divergence theorem and Stokes' theorem in the plane. These vector forms of Green's theorem will help us better understand divergence and curl and they prepare the way for the divergence theorem and Stokes' theorem in 3–space. Green's theorem and a number of closely related results, now called Green's identities, have come to play a fundamental role in many areas of mathematics, physics, and engineering. We shall indicate some of these applications as we go along.

Green's Theorem

The setting for Green's theorem is a region D in the xy–plane that is bounded by a simple closed path C, as in Fig. 1. The theorem asserts that

$$\iint_D \left(\frac{\partial Q}{\partial x} - \frac{\partial P}{\partial y}\right) dA = \oint_C Pdx + Qdy$$

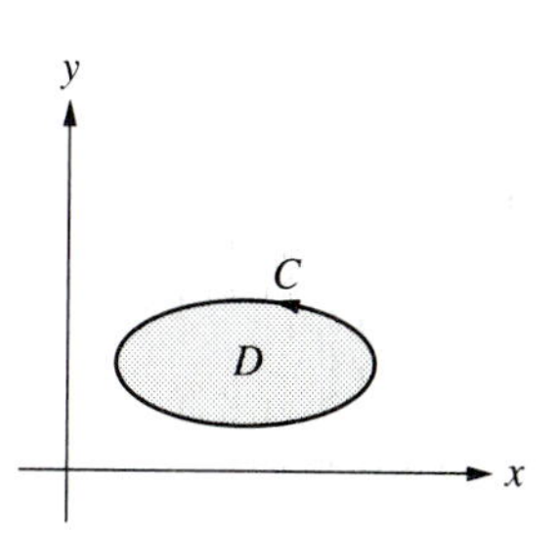

FIGURE 1

under reasonable conditions on P and Q. The circle on the line integral means that the path C is positively directed relative to D. (See Sec. 8.2.) Thus, D is to the left as C is traversed in the positive sense. Although it is not apparent from the formula, Green's theorem can be viewed as a two-dimensional extension of the fundamental theorem of calculus, expressed in the form

$$\int_a^b f'(x)\,dx = f(b) - f(a).$$

Here the boundary of the interval $[a,b]$ consists of the two points a and b.

> **Theorem 1** *Green's Theorem*
> Let C be a simple closed path in the xy–plane and let D be the closed region bounded by C. Assume that $P(x,y)$ and $Q(x,y)$ have continuous first partials in D. Then
>
> $$\iint_D \left(\frac{\partial Q}{\partial x} - \frac{\partial P}{\partial y}\right) dA = \oint_C Pdx + Qdy.$$

The basic idea of the proof of Green's theorem is informative and not particularly difficult. Nevertheless, we prefer to defer the proof to the end of the section after we have given a number of examples and applications.

EXAMPLE 1. Evaluate $\oint_C (e^{x2} - y)\,dx + (x + \sin\sqrt{y})\,dy$, where C is the circle with center (0,3) and radius 2.

Solution. A direct evaluation of this line integral would be formidable. Let's use Green's theorem instead. In this case,

$$P = e^{x2} - y, \qquad Q = x + \sin\sqrt{y},$$

$$\frac{\partial P}{\partial y} = -1, \qquad \frac{\partial Q}{\partial x} = 1,$$

for (x, y) in the closed disk D bounded by the given circle C. So the hypotheses of Green's theorem are satisfied and, since D has area 4π,

$$\oint_C (e^{x^2} - y)\,dx + (x + \sin\sqrt{y})\,dy = \iint_D 2\,dA = 2 \cdot 4\pi = 8\pi. \quad \square$$

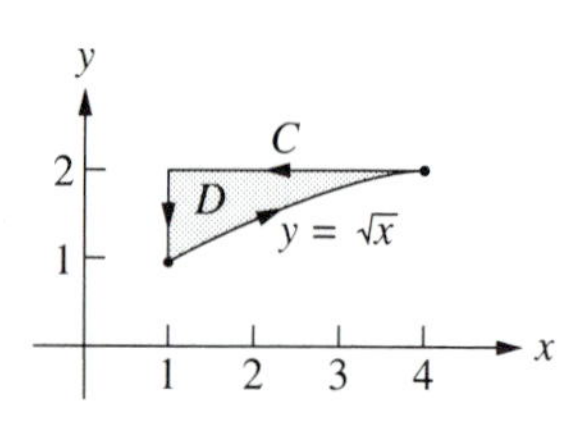

FIGURE 2

EXAMPLE 2. Evaluate $\oint_C (y^2 + \sin x)\,dx + (xy + \ln x)\,dy$, where C is the simple closed path around the region D in Fig. 2.

Solution. Again, a direct evaluation of the line integral is tedious at best. Now,

$$P = y^2 + \sin x, \qquad Q = xy + \ln x,$$

$$\frac{\partial P}{\partial y} = 2y, \qquad \frac{\partial Q}{\partial x} = y,$$

and D is described by $1 \le x \le 4$, $\sqrt{x} \le y \le 2$. By Green's theorem,

$$\oint_c (y^2 + \sin x)\,dx + (xy \ln x)\,dy = \iint_D -y\,dA = \int_1^4 \int_{\sqrt{x}}^2 -y\,dy\,dx$$

$$= \int_1^4 \left[-\frac{1}{2}y^2\right]_{\sqrt{x}}^2 dx = \int_1^4 \left(\frac{1}{2}x - 2\right) dx = \left[\frac{1}{4}x^2 - 2x\right]_1^4 = -\frac{9}{4}. \quad \square$$

It may surprise you to learn that line integrals can be used to find areas enclosed by curves. Green's theorem provides the connection. Let $P = 0$ and $Q = x$ in Th. 1 to obtain the area A of the region D bounded by the curve C in Fig. 1:

$$A = \iint_D 1\,dA = \oint_C x\,dy.$$

Likewise, let $P = -y$ and $Q = 0$ to find

$$A = \iint_D 1\,dA = -\oint_C y\,dx.$$

Average these two formulas for the area to obtain

$$A = \frac{1}{2}\oint_C x\,dy - y\,dx.$$

The following theorem summarizes our findings.

Theorem 2 *Areas by Line Integrals*
Let D be a plane region bounded by a simple closed path C. Then the area A of D is given by

$$A = \oint_C x\,dy = -\oint_C y\,dx = \frac{1}{2}\oint_C x\,dy - y\,dx.$$

EXAMPLE 3. Find the area bounded by the ellipse $x^2/a^2 + y^2/b^2 = 1$.

Solution. Parametric equations for the ellipse are

$$x = a\cos t, \qquad y = b\sin t, \quad 0 \le t \le 2\pi.$$

See Sec. 4.1. By the third area formula in Th. 2,

$$\begin{aligned} A = \frac{1}{2}\oint_C x\,dy - y\,dx &= \frac{1}{2}\int_0^{2\pi} [x(t)y'(t) - y(t)x'(t)]dt \\ &= \frac{1}{2}\int_0^{2\pi} [(a\cos t)(b\cos t) - (b\sin t)(-a\sin t)]\,dt \\ &= \frac{ab}{2}\int_0^{2\pi} (\cos^2 t + \sin^2 t)\,dt = \frac{ab}{2}\cdot 2\pi = \pi ab. \end{aligned}$$

You are invited to try either of the other formulas for the area in Th. 2. You will find that, even though the other formulas for the area are shorter, more work is required (in this example). □

Theorem 2 is the mathematical basis for a planimeter, which is an instrument used to measure the area enclosed by a curve. You roll the planimeter once around the boundary curve and the planimeter mechanically approximates a line integral for the area enclosed.

The closed paths in Exs. 1, 2, and 3 are all simple; however, Green's theorem applies to many closed paths that are not simple. For example, it applies to the figure–eight–like path C in Fig. 3 with the positive (counterclockwise) direction shown. Likewise, Green's theorem applies to closed paths with a finite number of self–intersections. In fact, it applies to any closed path that traverses the region it surrounds in the positive direction while only "winding around the region once." We shall not attempt to be more precise about what this means because in most applications of Green's theorem, the path C is simple or, at worst, has a finite number of self–intersections.

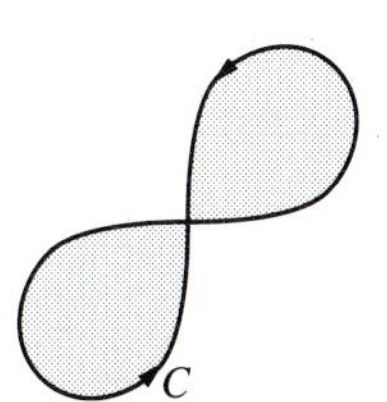

FIGURE 3

Green's Theorem for Regions with Holes

An extension of Green's theorem applies to regions with holes, that is, regions that are not simply connected. In fact, the most important applications of Green's theorem are made for such regions. Often the region has a single hole, which is a point where the functions P and Q become infinite.

To prepare the way for Green's theorem with holes and for the proof of Th.

1 later in the section, we begin with a useful general observation about Green's theorem:

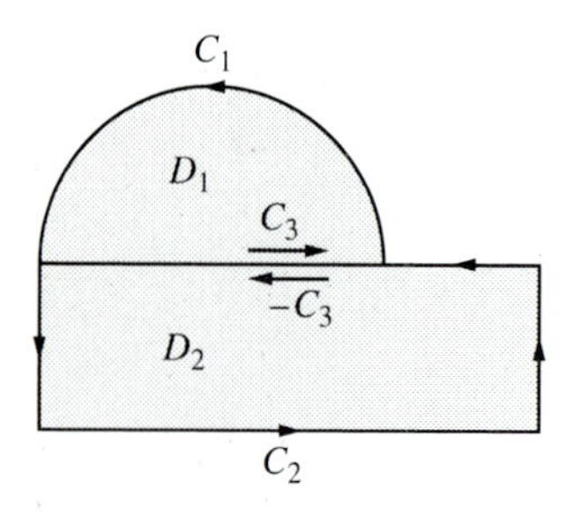

FIGURE 4

> If Green's theorem is valid for each of two nonoverlapping regions whose boundaries share a particular path in common, then Green's theorem is valid for the region consisting of the union of the two regions.

To understand why this statement is true, consult Fig. 4. In the figure, the positively directed boundary of the region D_1 is $C_1 + C_3$ where C_3 is the diameter of the semicircle; the positively directed boundary of the region D_2 is $C_2 + (-C_3)$, which we write as $C_2 - C_3$; and the positively directed boundary of D, the union of D_1 and D_2, is $C = C_1 + C_2$. Theorem 1 applies to both D_1 and D_2 and gives

$$\iint_{D_1}\left(\frac{\partial Q}{\partial x} - \frac{\partial P}{\partial y}\right)dA = \int_{C_1} Pdx + Qdy + \int_{C_3} Pdx + Qdy,$$

$$\iint_{D_2}\left(\frac{\partial Q}{\partial x} - \frac{\partial P}{\partial y}\right)dA = \int_{C_2} Pdx + Qdy - \int_{C_3} Pdx + Qdy.$$

Add to obtain

$$\iint_{D}\left(\frac{\partial Q}{\partial x} - \frac{\partial P}{\partial y}\right)dA = \oint_{C} Pdx + Qdy,$$

which is the conclusion of Th. 1 for the entire region D.

Figure 5 shows a region D with one hole. The boundary path of D consists of two parts, C and C_1, traversed in the positive sense relative to D. For C, the direction is as expected. But for C_1, the direction indicated in Fig. 5 may be unexpected. Remember the rule: The region D should be to the left as the boundary path is traversed. For the type of region in Fig. 5, the conclusion of Green's theorem is

$$\iint_{D}\left(\frac{\partial Q}{\partial x} - \frac{\partial P}{\partial y}\right)dA = \int_{C} Pdx + Qdy + \int_{C_1} Pdx + Qdy.$$

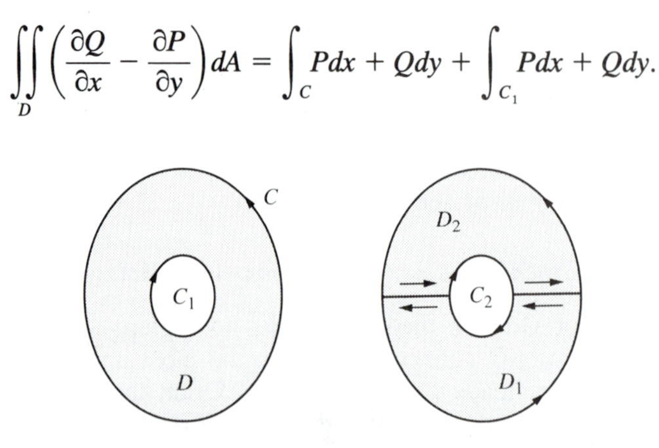

FIGURE 5 FIGURE 6

The derivation of the foregoing equation is based on Fig. 6. It shows the same region D as in Fig. 5 but with two so-called **cross-cuts** (often line segments) that separate D into two nonoverlapping subregions D_1 and D_2 with positively directed boundary paths indicated in Fig. 6. Apply Green's theorem to D_1 and to D_2 and add the results, taking into account the cancellation of the line integrals along the cross-cuts, to obtain Green's theorem for the region D with one hole.

Similar reasoning applies to a region D with outer bounding path C and with any finite number of holes. Suppose that there are n holes with bounding paths $C_1, C_2, \ldots, C_n$, directed positively with respect to D. The case with $n = 3$ is illustrated in Fig. 7. For a region D with n holes, we obtain

FIGURE 7

$$\iint_D \left(\frac{\partial Q}{\partial x} - \frac{\partial P}{\partial y}\right) dA = \int_C P dx + Q dy + \sum_{i=1}^{n} \int_{C_i} P dx + Q dy.$$

EXAMPLE 4. Evaluate

$$\oint_C \frac{x dy - y dx}{x^2 + y^2},$$

where C is the ellipse $x^2/4 + y^2/9 = 1$ in Figure 8.

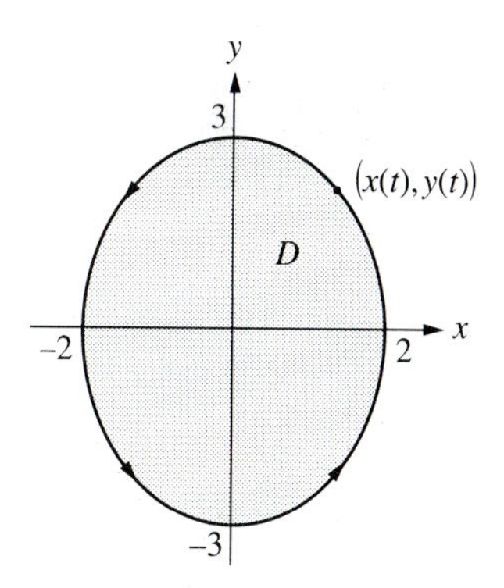

FIGURE 8

Solution 1. The ellipse has the parametric representation

$$x(t) = 2\cos t, \qquad y(t) = 3\sin t, \quad 0 \le t \le 2\pi.$$

Hence,

$$\begin{aligned}\oint_c \frac{x dy - y dx}{x^2 + y^2} &= \int_0^{2\pi} \frac{(2\cos t)(3\cos t\, dt) - (3\sin t)(-2\sin t\, dt)}{4\cos^2 t + 9\sin^2 t} \\ &= 6\int_0^{2\pi} \frac{dt}{4\cos^2 t + 9\sin^2 t} = 6\int_0^{2\pi} \frac{dt}{4 + 5\sin^2 t}.\end{aligned}$$

We cannot evaluate this integral by means we have studied so far. Let's abandon this attempt.

Solution 2. Suppose we try to apply Green's theorem. Now

$$\oint_C \frac{x dy - y dx}{x^2 + y^2} = \oint_C P dx + Q dy$$

with

$$P = \frac{-y}{x^2 + y^2}, \qquad Q = \frac{x}{x^2 + y^2}.$$

By the quotient rule,

$$\frac{\partial P}{\partial y} = \frac{y^2 - x^2}{(x^2 + y^2)^2}, \qquad \frac{\partial Q}{\partial x} = \frac{y^2 - x^2}{(x^2 + y^2)^2}.$$

Substitution into Green's theorem gives

$$\oint_C \frac{xdy - ydx}{x^2 + y^2} = \oint_C Pdx + Qdy = \iint_D \left(\frac{\partial Q}{\partial x} - \frac{\partial P}{\partial y}\right) dA = 0.$$

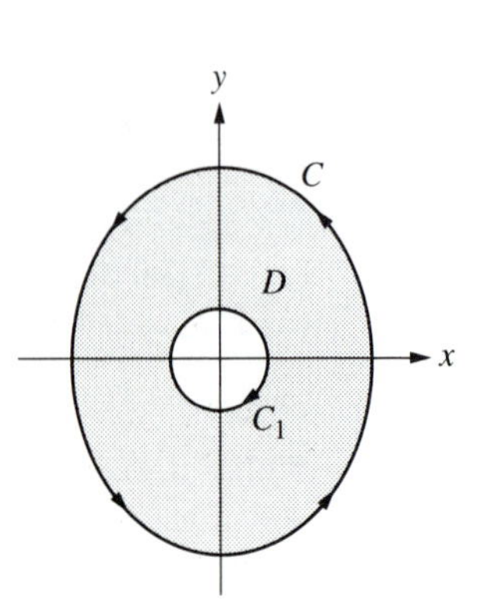

FIGURE 9

This evaluation of the line integral is WRONG! Why? Because the hypotheses of Green's theorem are not satisfied. The functions $P(x,y)$ and $Q(x,y)$ do not have continuous derivatives at *all* points inside and on the ellipse C. Indeed, $P(x,y)$ and $Q(x,y)$ become infinite as $(x,y) \to (0,0)$.

Solution 3. We can overcome the problem at $(0,0)$ by applying the extended form of Green's theorem to the closed region D in Fig. 9 between the ellipse C and the unit circle C_1, directed clockwise as in Fig. 9. Now $P(x,y)$ and $Q(x,y)$ do have continuous derivatives in D. Since our previous calculations give $\partial Q/\partial x = \partial P/\partial y$ in D, the extended form of Green's theorem yields

$$\oint_C \frac{xdy - ydx}{x^2 + y^2} + \oint_{C_1} \frac{xdy - ydx}{x^2 + y^2} = \iint_D \left(\frac{\partial Q}{\partial x} - \frac{\partial P}{\partial y}\right) dA = 0,$$

$$\oint_C \frac{xdy - ydx}{x^2 + y^2} = -\oint_{C_1} \frac{xdy - ydx}{x^2 + y^2} = \oint_{-C_1} \frac{xdy - ydx}{x^2 + y^2},$$

where $-C_1$ is directed counterclockwise. On $-C_1$,

$$x = \cos\theta, \qquad y = \sin\theta, \qquad 0 \le \theta \le 2\pi.$$

Hence,

$$\oint_C \frac{xdy - ydx}{x^2 + y^2} = \int_0^{2\pi} \frac{(\cos\theta)(\cos\theta d\theta) - (\sin\theta)(-\sin\theta d\theta)}{\cos^2\theta + \sin^2\theta} = \int_0^{2\pi} d\theta = 2\pi.$$

This is the correct value of the line integral. Virtually the same reasoning shows that the value of the foregoing line integral along any simple closed path C going once around the origin in the counterclockwise sense is 2π. Just replace the ellipse with the more general path C and the unit circle by any convenient circle that lies inside C. □

The Divergence Theorem in the Plane

There is an important vector formulation of Green's theorem that involves the divergence of a plane vector field. We describe this version of Green's theorem next. It will lead us to the concept of flux of a vector field and to a better understanding of divergence.

We assume throughout that $\mathbf{F}(\mathbf{r}) = P(x,y)\mathbf{i} + Q(x,y)\mathbf{j}$ where $\mathbf{r} = x\mathbf{i} + y\mathbf{j}$ and that P and Q have continuous partial derivatives in the regions that come up. Then

$$\operatorname{div}\mathbf{F} = \frac{\partial P}{\partial x} + \frac{\partial Q}{\partial y}.$$

In Green's theorem, replace Q by P and P by $-Q$ to obtain

$$\iint_D \operatorname{div} \mathbf{F}\, dA = \iint_D \left(\frac{\partial P}{\partial x} + \frac{\partial Q}{\partial y} \right) dA = \oint_C P dy - Q dx.$$

The line integral on the right will be expressed more informatively in terms of arc length.

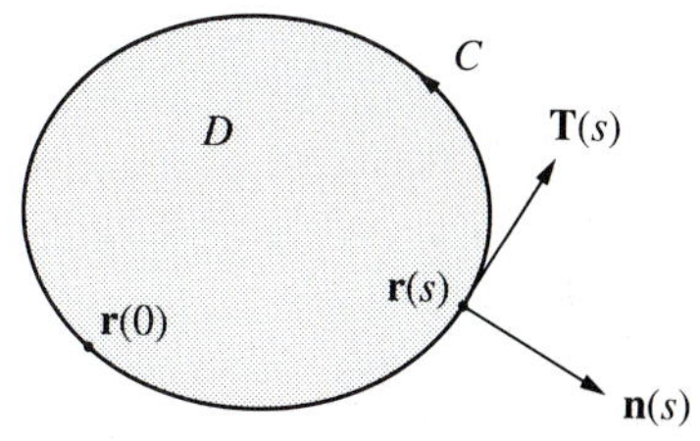

In Fig. 10, C is parameterized by arc length s measured from any convenient point on C. Let L be the length of C. Then C is represented by $\mathbf{r} = \mathbf{r}(s)$ for $0 \leq s \leq L$. The unit tangent vector to C at $\mathbf{r}(s)$ is given by

$$\mathbf{T}(s) = \mathbf{r}'(s) = x'(s)\mathbf{i} + y'(s)\mathbf{j}.$$

FIGURE 10

The **outer unit normal** to C at $\mathbf{r}(s)$ is

$$\mathbf{n}(s) = y'(s)\mathbf{i} - x'(s)\,\mathbf{j}.$$

It is obtained by rotating $\mathbf{T}(s)$ 90° clockwise. In Fig. 10, $\mathbf{n}(s)$ points away from the region D bounded by C. Note that $\|\mathbf{n}\| = \|\mathbf{T}\| = 1$ and $\mathbf{n} \cdot \mathbf{T} = 0$.

The line integral $\oint_C P dy - Q dx$ is expressed in terms of arc length by

$$\begin{aligned} \oint_C P dy - Q dx &= \int_0^L [P(\mathbf{r}(s))y'(s) - Q(\mathbf{r}(s))x'(s)]\, ds \\ &= \int_0^L \mathbf{F}(\mathbf{r}(s)) \cdot \mathbf{n}(s)\, ds = \oint_C \mathbf{F} \cdot \mathbf{n} ds. \end{aligned}$$

On the other hand, we showed earlier that the same line integral is equal to the integral of div $\mathbf{F}$ over D. Thus,

$$\boxed{\iint_D \operatorname{div} \mathbf{F} dA = \oint_C \mathbf{F} \cdot \mathbf{n} ds.}$$

This vector formulation of Green's theorem is called the **divergence theorem in the plane**.

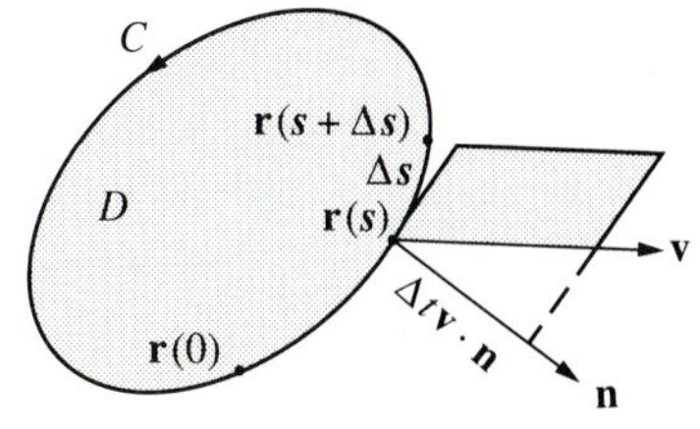

Next we relate $\oint_C \mathbf{F} \cdot \mathbf{n} ds$ to the "flow" of the vector field $\mathbf{F}$. To begin with, suppose that $\mathbf{F}$ is the mass flow vector for a two-dimensional steady fluid flow. Then $\mathbf{F} = \sigma \mathbf{v}$, where $\sigma$is the density (mass/area) and $\mathbf{v}$ is the velocity field of the fluid. Figure 11 shows a simple closed path C, which is not a barrier to the flow. We are interested in the flow of fluid across C. Focus your attention on a small arc of C with length Δs from $\mathbf{r}(s)$ to $\mathbf{r}(s + \Delta s)$. The fluid particles that cross Δs all have nearly the same velocity $\mathbf{v} = \mathbf{v}(\mathbf{r}(s))$. First, suppose that $\mathbf{v} \cdot \mathbf{n} > 0$, as in Fig. 11. Then fluid leaves D across Δs. During a short time interval Δt, the fluid crossing Δs nearly fills the shaded region, which is approximately a parallelogram with base Δs, height $\Delta t \mathbf{v} \cdot \mathbf{n}$, and area $\Delta s \Delta t \mathbf{v} \cdot \mathbf{n}$. So the mass of fluid crossing Δs in time Δt is approximately

FIGURE 11

$$\sigma \Delta s \Delta t \mathbf{v} \cdot \mathbf{n} = \mathbf{F} \cdot \mathbf{n} \Delta s\, \Delta t.$$

The mass per unit time leaving D across Δs is approximately

$$\sigma \Delta s \mathbf{v} \cdot \mathbf{n} = \mathbf{F} \cdot \mathbf{n} \Delta s.$$

Here, $\mathbf{F} \cdot \mathbf{n} > 0$ because $\mathbf{v} \cdot \mathbf{n} > 0$. If $\mathbf{v} \cdot \mathbf{n} < 0$, then fluid enters D across Δs and $\mathbf{F} \cdot \mathbf{n} < 0$. In either case, $\mathbf{F} \cdot \mathbf{n} \Delta s$ measures the net flow rate across Δs in the outward direction. The usual Riemann sum and limit passage argument gives the net flow rate across C:

$$\oint_C \mathbf{F} \cdot \mathbf{n} ds = \lim \sum \mathbf{F} \cdot \mathbf{n} \Delta s.$$

In summary, if $\mathbf{F} = \sigma \mathbf{v}$ is the mass flow vector for a steady two-dimensional fluid flow, then $\oint_C \mathbf{F} \cdot \mathbf{n} ds$ gives the (net) mass per unit time that flows outward across a simple closed path C in the flow region.

These ideas extend to other vector fields. For a general vector field, the word *flux* is used in place of *flow*. It comes from the Latin word for *flow*.

Definition *Flux Across a Simple Closed Path in the Plane*
Let $\mathbf{F} = P\mathbf{i} + Q\mathbf{j}$ be a continuous vector field defined on a simple closed path C with outer normal $\mathbf{n}$. The **flux of F outward across C** is

$$\oint_C \mathbf{F} \cdot \mathbf{n} ds.$$

The flux across any path with respect to a chosen normal direction along the path can be defined in the same way. The path does not have to be closed.

The divergence theorem and the physical interpretation of the flux integral for a mass flow vector $\mathbf{F} = \sigma \mathbf{v}$ lead us, by another route, to the physical interpretation of divergence discussed in Sec. 8.1. Assume that div $\mathbf{F}$ is continuous. In Fig. 12, (x, y) is a fixed point in the flow plane and C is a simple closed curve around (x, y). Imagine that C shrinks to (x, y) in the sense that C ultimately lies inside any circle centered at (x, y). Then the area ΔA of the region D enclosed by C tends to zero. By the mean value theorem for double integrals,

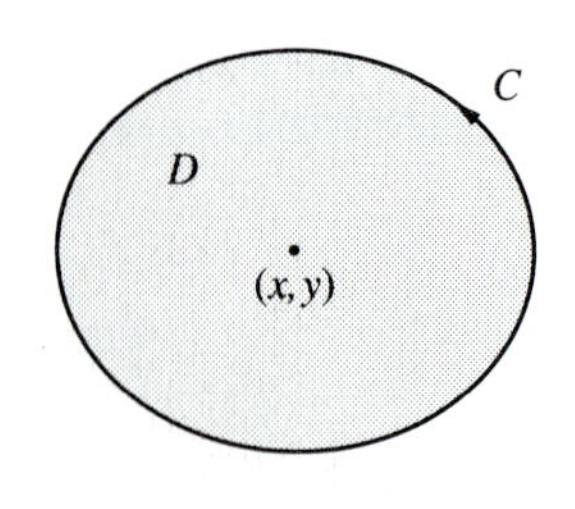

FIGURE 12

$$\oint_C \mathbf{F} \cdot \mathbf{n} ds = \iint_D \operatorname{div} \mathbf{F} dA = \operatorname{div} \mathbf{F}(\bar{x}, \bar{y}) \Delta A,$$

$$\operatorname{div} \mathbf{F}(\bar{x}, \bar{y}) = \frac{1}{\Delta A} \oint_C \mathbf{F} \cdot \mathbf{n} ds,$$

for some point $(\bar{x}, \bar{y})$ in D. Since $(\bar{x}, \bar{y}) \to (x, y)$ as C shrinks to (x, y) and div $\mathbf{F}$ is continuous,

$$\operatorname{div} \mathbf{F}(x, y) = \lim_{\Delta A \to 0} \frac{1}{\Delta A} \oint_C \mathbf{F} \cdot \mathbf{n} ds.$$

In words,

$\operatorname{div}\mathbf{F}(x,y)$ is the net rate per unit area per unit time that fluid leaves the vicinity of (x,y).

This interpretation of $\operatorname{div}\mathbf{F}$ will reveal the physical significance of the divergence theorem,

$$\iint_D \operatorname{div}\mathbf{F}\,dA = \oint_C \mathbf{F}\cdot\mathbf{n}\,ds.$$

The divergence integral $\iint_D \operatorname{div}\mathbf{F}dA$ is a limit of Riemann sums $\sum \operatorname{div}\mathbf{F}\,\Delta A$. Each term in the sum is approximately the net rate per unit time that fluid leaves ΔA. So $\iint_D \operatorname{div}\mathbf{F}\,dA$ is the net rate per unit time that fluid leaves the region D. The integral $\oint_C \mathbf{F}\cdot\mathbf{n}\,ds$ is the net rate per unit time that fluid crosses C in the outward direction. Thus, the divergence theorem tells us that the rate at which fluid mass leaves D is equal to the rate at which fluid mass crosses the boundary C in the outward direction. From this point of view, the divergence theorem is a conservation law.

Stokes' Theorem in the Plane

Another important vector formulation of Green's theorem involves the curl of a plane vector field. After describing this version of Green's theorem, we use it to obtain two physical interpretations of the curl of a vector field. Finally, we give an alternative derivation of the characterization of a conservative vector field defined on a simply connected domain in terms of curl.

We assume throughout that $\mathbf{F}(\mathbf{r}) = P(x,y)\mathbf{i} + Q(x,y)\mathbf{j}$ where P and Q have continuous partial derivatives in the region of interest. Then

$$\operatorname{curl}\mathbf{F} = \left(\frac{\partial Q}{\partial x} - \frac{\partial P}{\partial y}\right)\mathbf{k}, \qquad \frac{\partial Q}{\partial x} - \frac{\partial P}{\partial y} = \operatorname{curl}\mathbf{F}\cdot\mathbf{k},$$

and Green's theorem can be written as

$$\iint_D \operatorname{curl}\mathbf{F}\cdot\mathbf{k}\,dA = \oint_C P dx + Q dy = \oint_C \mathbf{F}\cdot d\mathbf{r}.$$

Thus

$$\oint_C \mathbf{F}\cdot d\mathbf{r} = \iint_D \operatorname{curl}\mathbf{F}\cdot\mathbf{k}\,dA.$$

This vector formulation of Green's theorem is called **Stokes' theorem in the plane**. It will be extended to 3–space in Sec. 8.7.

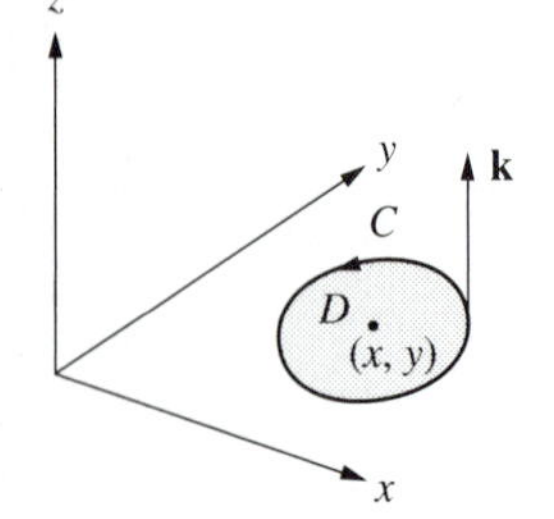

FIGURE 13

Stokes' theorem leads to two important physical interpretations of the curl of a vector field, one related to the work done by a force field and the other related to circulation in fluid flow. Figure 13 will serve as a guide to both results. In the figure, (x, y) is a fixed point surrounded by a simple closed path C that encloses a region D with area ΔA. We plan to let C shrink to (x, y) in the sense described earlier in the treatment of divergence.

We begin with a force field $\mathbf{F}$ and work. The work done by the force field to move an object once around C in the positive direction is

$$W_C = \oint_C \mathbf{F} \cdot d\mathbf{r} = \iint_D \operatorname{curl} \mathbf{F} \cdot \mathbf{k} dA = \operatorname{curl} \mathbf{F}(\bar{x}, \bar{y}) \cdot \mathbf{k} \Delta A$$

for some $(\bar{x}, \bar{y})$ in D. Thus,

$$\operatorname{curl} \mathbf{F}(\bar{x}, \bar{y}) \cdot \mathbf{k} = \frac{W_C}{\Delta A}.$$

Now let C shrink to (x, y). Then $(\bar{x}, \bar{y}) \to (x, y)$, $\operatorname{curl} \mathbf{F}(\bar{x}, \bar{y}) \to \operatorname{curl} \mathbf{F}(x, y)$ and, hence,

$$\operatorname{curl} \mathbf{F}(x, y) \cdot \mathbf{k} = \lim_{\Delta A \to 0} \frac{W_C}{\Delta A}.$$

In words,

$\operatorname{curl} \mathbf{F}(x, y) \cdot \mathbf{k}$ is the limiting value of the work per unit area done by $\mathbf{F}$ to move an object once around (x, y) in the positive direction.

Next consider fluid flow. In Stokes' theorem let $\mathbf{F} = \mathbf{v}$, the velocity field of a steady two-dimensional fluid flow, to obtain

$$\iint_D \operatorname{curl} \mathbf{v} \cdot \mathbf{k}\, dA = \oint_C \mathbf{v} \cdot d\mathbf{r}.$$

The line integral $\oint_C \mathbf{v} \cdot d\mathbf{r}$ is called the **circulation of the flow about C.** It measures the tendency of the fluid to flow around the closed path C. In the problems you are asked to confirm this statement by expressing $\oint_C \mathbf{v} \cdot d\mathbf{r}$ as a limit of Riemann sums. An argument, much like the preceding one about work, gives

$$\operatorname{curl} \mathbf{v}(x, y) \cdot \mathbf{k} = \lim_{\Delta A \to 0} \frac{1}{\Delta A} \oint_C \mathbf{v} \cdot d\mathbf{r},$$

in the context of Fig. 13. In words, $\operatorname{curl} \mathbf{v} \cdot \mathbf{k}$ is the limiting value of the circulation per unit area about (x, y).

Finally, we use Stokes' theorem to sketch an argument that, in a simply connected domain D, a force field $\mathbf{F}$ is conservative if $\operatorname{curl} \mathbf{F} = \mathbf{0}$. Assume that curl $\mathbf{F} = \mathbf{0}$ in D. Let C be any closed path in D surrounding a closed region D_1. Since D is simply connected, D_1 lies entirely in D. Consequently, curl $\mathbf{F} = \mathbf{0}$ in D_1 and, by Stokes' theorem,

$$\oint_C \mathbf{F} \cdot d\mathbf{r} = \iint_{D_1} \operatorname{curl} \mathbf{F} \cdot \mathbf{k}\, dA = 0.$$

Therefore, $\oint_C \mathbf{F} \cdot dr = 0$ for any simple closed path in D. It follows, by an argument which we omit, that $\oint_C F \cdot d\mathbf{r} = \mathbf{0}$ for any closed path in D. We learned in Sec. 7.3 that the last condition implies $\mathbf{F}$ is conservative.

Proof of Green's Theorem

Now that we have a much better idea of the importance of Green's theorem, we close this section with a proof of the theorem for the types of regions you are likely to meet.

We begin with a simplifying remark. Two special cases of Green's theorem, one with $Q = 0$ and the other with $P = 0$, are

$$\iint_D \frac{\partial P}{\partial y} dA = -\oint_C P dx \quad \text{and} \quad \iint_D \frac{\partial Q}{\partial x} dA = \oint_C Q dy.$$

Conversely, if both of these equations are established, then subtraction yields Green's theorem:

$$\iint_D \left(\frac{\partial Q}{\partial x} - \frac{\partial P}{\partial y} \right) dA = \oint_C P dx + Q dy.$$

Consequently, it is enough to establish Green's theorem for $P(x,y)$ with $Q = 0$ and for $Q(x,y)$ with $P = 0$. We proceed with the proof, which has four cases, depending on the type of region D.

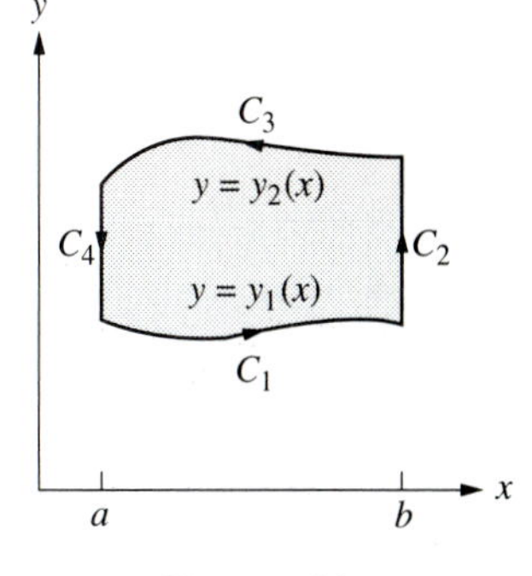

FIGURE 14

Case 1. $\iint_D (\partial P/\partial y)\, dA = -\oint_C P dx$ if D is y–simple, as in Fig. 14, where $C = C_1 + C_2 + C_3 + C_4$ and the line segments C_2 and C_4 may reduce to single points.

To establish Case 1 of Green's theorem, observe that a point (x,y) lies in D if $y_1(x) \le y \le y_2(x)$ and $a \le x \le b$. Consequently,

$$\iint_D \frac{\partial P}{\partial y} dA = \int_a^b \int_{y_1(x)}^{y_2(x)} \frac{\partial P}{\partial y} dA = \int_a^b [P(x,y)]_{y=y_1(x)}^{y=y_2(x)} dx$$

$$= \int_a^b P(x, y_2(x))\, dx - \int_a^b P(x, y_1(x))\, dx$$

$$= -\int_{C_3} P dx - \int_{C_1} P dx$$

Since $dx = 0$ on C_2 and C_4,

$$\int_{C_2} Pdx = 0 \quad \text{and} \quad \int_{C_4} Pdx = 0.$$

Combine these results to obtain

$$\iint_D \frac{\partial P}{\partial y}\, dA = -\oint_C Pdx.$$

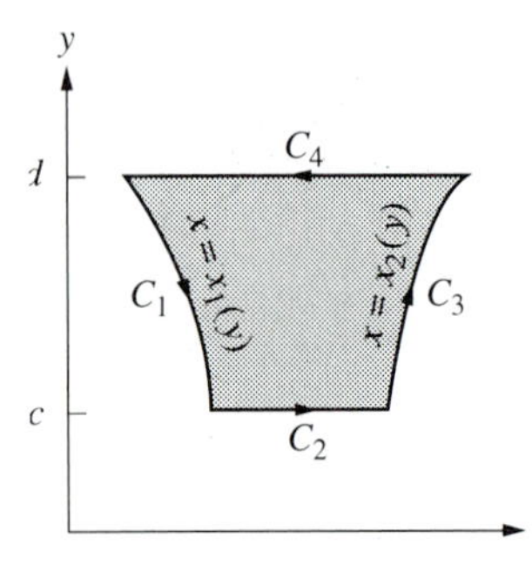

FIGURE 15

Case 2. $\iint_D (\partial Q/\partial x)\, dA = \oint_C Qdy$ if D is x–simple, as in Fig. 15.

The proof is almost the same as in Case 1. See the problems.

Case 3. $\iint_D(\partial Q/\partial x - \partial P/\partial y)\, dA = \oint_C Pdx + Qdy$ if D is both x–simple and y–simple.

Just add the results in Cases 1 and 2.

Case 4. $\iint_D(\partial Q/\partial x - \partial P/\partial y)\, dA = \oint_C Pdx + Qdy$ if D can be decomposed into a finite number of nonoverlapping subregions $D_1, D_2, \ldots, D_n$, each of which is both x–simple and y–simple.

The key idea behind the proof was used earlier in Fig. 4 and the accompanying discussion in the text. It is illustrated again in Fig. 16. Cross-cuts are introduced in order to decompose D into subregions that are both x–simple and y–simple. Apply Green's theorem to each subregion and add the results to obtain Green's theorem for the original region D. As before, integrals along the cross-cuts cancel out.

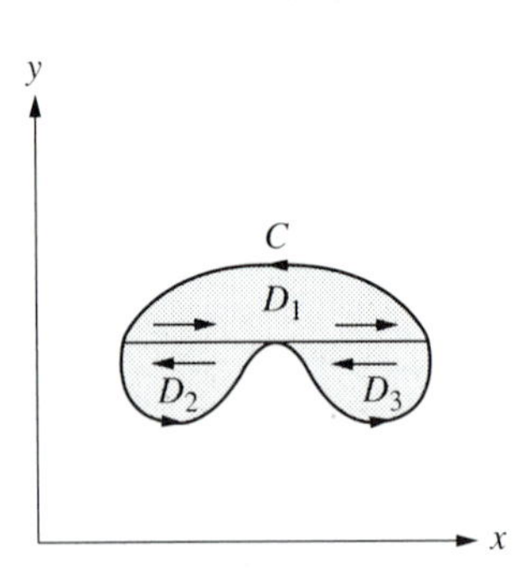

FIGURE 16

Cases 1–4 establish Green's theorem for any region that can be decomposed into a finite number of nonoverlapping regions that are both x–simple and y–simple. Such regions cover virtually all applications of Green's theorem.

PROBLEMS

In Probs. 1–4 evaluate the line integral in two ways: (a) directly and (b) by Green's theorem. Assume the curve is positively directed and traversed once.

1. $\oint_C -ydx + xdy$, $\quad C$: $x^2 + y^2 = 4$
2. $\oint_C xy^2dx + x^2ydy$, $\quad C$: $x^2 + y^2 = 4$
3. $\oint_C ydx + x^2dy$, C the boundary of the bounded region between $y = x^2$ and $y = \sqrt{x}$
4. $\oint_C xydy$, $\quad C$ the rectangle with vertices $(0,0)$, $(2,0)$, $(2,3)$, $(0,3)$

In Probs. 5–12 evaluate the given line integral. Assume the curve is positively directed and traversed once.

5. $\oint_C y^2dx - xydy$, $\quad C$ the square with vertices $(0,0)$, $(1,0)$, $(1,1)$, and $(0,1)$
6. $\oint_C (2x + y)\, dx + (2x - y)\, dy$, C the boundary of the bounded region between $y = x$ and $y = x^2$
7. $\oint_C (x^2 - y^2)\, dx + 2xydy$, $\quad C$ the ellipse $x^2/4^2 + y^2/3^2 = 1$
8. $\oint_C \sin y\, dx + (x + x\cos y)\, dy$, $\quad C$ the circle $x^2 + y^2 = 9$

9. $\oint_C e^x \cos y\, dx + e^x \sin y\, dy$, C the rectangle with vertices $(0,0)$, $(\ln 4, 0)$, $(\ln 4, \pi/2)$, $(0, \pi/2)$

10. $\oint_C e^{-x} \cos y\, dx + e^{-x} \sin y\, dy$, C the quarter disk in the first quadrant bounded by $x = 0$, $y = 0$, and $x^2 + y^2 = 1$

11. $\oint_C (y - \arctan\sqrt{x})\, dx + (x + \ln y)\, dy$, C the ellipse $x = 4 + 2\sin t$, $y = 4 + 3\cos t$, $0 \le t \le 2\pi$

12. $\oint_C (e^x + y^2)\, dx + (e^y + x^2)\, dy$, C the boundary of the region enclosed by $y = 0$, $x = \pi/2$, and $y = \sin x$.

In Probs. 13–16 (a) sketch the area in question and (b) use line integrals to evaluate it. The number a is a positive constant.

13. The area between one arch of the cycloid $x = a(t - \sin t)$, $y = a(1 - \cos t)$, and the x–axis.

14. The area enclosed by the astroid $x^{2/3} + y^{2/3} = a^{2/3}$. *Hint.* Find a parameterization with $x = a\cos^3 t$.

15. The area enclosed by the curve $x = a\cos t$, $y = a \sin 2t$, $-\pi/2 \le t \le \pi/2$.

16. The area of the triangle with vertices (x_0, y_0), (x_1, y_1), (x_2, y_2). Assume the triangle is traversed in the positive sense when the vertices are encountered in the given order.

In Probs. 17–18 C is a closed path in the xy–plane that bounds a region D with area A.

17. Show that the centroid $(\bar{x}, \bar{y})$ of D is given by

$$\bar{x} = \frac{1}{2A}\oint_C x^2 dy, \qquad \bar{y} = -\frac{1}{2A}\oint_C y^2 dx.$$

Recall that the centroid of a plane region is its center of mass when regarded as a thin body with constant density.

18. Assume that D has constant density σ. Show that the moments of inertia I_x, I_y, and I_z of D about the x–axis, y–axis, and z–axis are, respectively,

$$\text{(a) } I_x = -\frac{\sigma}{3}\oint_C y^3 dx, \quad \text{(b) } I_y = \frac{\sigma}{3}\oint_C x^3 dy, \quad \text{(c) } I_z = \frac{\sigma}{3}\oint_C x^3\, dy - y^3 dx.$$

19. Use Prob. 17 and symmetry to find the centroid of the top half of the ellipse $x^2/a^2 + y^2/b^2 = 1$.

20. Use Prob. 17 and symmetry to find the centroid of the part of the asteroid $x^{2/3} + y^{2/3} = a^{2/3}$ that lies in the first quadrant.

21. Use Prob. 18 to find the moment of inertia (a) I_x, (b) I_y, and (c) I_z for the ellipse $x^2/a^2 + y^2/b^2 = 1$ with constant density σ. It is informative to express the answer using the mass $M = \sigma(\pi ab)$ of the ellipse.

22. Use Prob. 18 to find the moment of inertia (a) I_x, (b) I_y, and (c) I_z for the semidisk $x^2 + y^2 \le a^2$ and $y \ge 0$ with constant density σ. It is informative to express the answer using the mass $M = \sigma(\pi a^2)/2$ of the semidisk.

23. Let $y = f(x)$ for $a \leq x \leq b$ and assume that $f(x) \geq 0$. Then the area A of the region D between the graph of f and the x–axis is $A = \int_a^b f(x)\,dx$. Let C be the positively directed boundary of D. Then $A = -\oint_C y\,dx$ by Th. 2. Show by a direct evaluation of the line integral that $-\oint_C y\,dx = \int_b^a f(x)\,dx$.

24. If $f(x)$ and $g(y)$ have continuous derivatives for all x, y and C is any closed path, evaluate $\oint_C f(x)\,dx + g(y)\,dy$.

In Probs. 25–28 find the work done by the given force field **F**, which acts as an object moves once in the positive direction around the path C.

25. $\mathbf{F} = (x^2 + y^2)\mathbf{i} + xy^2\,\mathbf{j}$ and C is the path bounding the region enclosed by $y = \sqrt{x}$ and $y = x$.

26. $\mathbf{F} = \langle xy, \sin y \rangle$, and C is the boundary of the quarter disk $x^2 + y^2 \leq 4$, $x \geq 0$, $y \geq 0$.

27. $\mathbf{F} = -y\mathbf{i} + x\mathbf{j}$ and C is the asteroid $x^{2/3} + y^{2/3} = 1$. *Hint.* See Prob. 14.

28. $\mathbf{F} = \langle xy, x + y \rangle$ and C is the boundary of the semiannulus $1 \leq x^2 + y^2 \leq 4$, $y \geq 0$.

In Probs. 29–32 find the flux of **F** outward across the given closed path.

29. $\mathbf{F} = xy^2\mathbf{i} + (x^2 + y^2)\mathbf{j}$ and C is the path bounding the region enclosed by $y = \sqrt{x}$ and $y = x^2$.

30. $\mathbf{F} = \langle x\sin y, x\cos y \rangle$ and C is the rectangle with vertices $(0,0)$, $(1,0)$, $(1,\pi)$, $(0,\pi)$.

31. $\mathbf{F} = \langle -y, x \rangle$ and C is the ellipse $x^2/4 + y^2/9 = 1$.

32. $\mathbf{F} = (e^y + xy)\mathbf{i} + (xy - e^{-x})\mathbf{j}$ and C is the boundary of the semidisk $x^2 + y^2 = 4$, $y \geq 0$.

Problems 33–36 deal with ideas related to Ex. 4.

33. Let $\mathbf{F} = (x^2 + y^2)^{-1}(-y\mathbf{i} + x\mathbf{j})$ so that P and Q are as in Ex. 4. Show directly that $\oint_{-C_a} \mathbf{F} \cdot d\mathbf{r} = 2\pi$ where C_a is the circle, traversed clockwise, with center at the origin and radius a. Conclude that C_1 in the third solution of Ex. 4 could have been replaced by C_a for any $a < 2$.

34. Evaluate $\oint_C \mathbf{F} \cdot d\mathbf{r}$ with **F** from Prob. 33 and C any closed path that surrounds the origin once and is traversed counterclockwise.

35. Evaluate $\oint_C \mathbf{F} \cdot d\mathbf{r}$ with **F** from Prob. 33 and C the ellipse $(x - 5)^2/4 + y^2/9 = 1$, which looks like the ellipse in Fig. 8 moved 5 units to the right.

36. Evaluate $\oint_C \mathbf{F} \cdot d\mathbf{r}$ with **F** from Prob. 33 and C any closed path such that the origin is outside C.

37. Evaluate $\oint_C \mathbf{F} \cdot d\mathbf{r}$ where $\mathbf{F} = \mathbf{r}/r^2$ and C is any positively directed closed path with (a) the origin inside C or (b) the origin outside C.

38. Suppose that **F** is a vector field whose components have continuous first partials in the region with three holes, such as in Fig. 7. Assume that $\operatorname{curl} \mathbf{F} = \mathbf{0}$ in the region. What relation exists among $\oint_C \mathbf{F} \cdot d\mathbf{r}$, $\oint_{C_1} \mathbf{F} \cdot d\mathbf{r}$, $\oint_{C_2} \mathbf{F} \cdot d\mathbf{r}$, and $\oint_{C_3} \mathbf{F} \cdot d\mathbf{r}$, where C is the outer bounding curve and C_1, C_2, and C_3 bound the three holes?

39. Let $\mathbf{v}$ be the velocity field of a steady fluid flow with constant density σ Use a Riemann sum and limit passage argument to explain why $\oint_C \mathbf{v} \cdot d\mathbf{r}$ is a reasonable measure of the tendency of the fluid to circulate around a smooth closed curve C. *Hint.* If C is parameterized by arc length s, then $\mathbf{T} = \mathbf{r}'(s)$ is the unit tangent vector and $d\mathbf{r} = \mathbf{T}ds$.

40. (a) Show that the symmetric form of the area formula in Th. 2 can be expressed as

$$A = \frac{1}{2}\oint_C \mathbf{r} \times d\mathbf{r}$$

where $\mathbf{r}$ is the position vector to a point (x,y) on C. (b) Now develop this formula from scratch using a Riemann sum and limit passage argument. *Hint.* Write $A = \sum \Delta A$ where the ΔA are chosen much like the area elements used to develop the area formula for regions bounded by polar curves.

41. Establish the equality in Case 2 in the proof of Green's theorem.

8.5 Surface Area and Surface Integrals

In Sec. 4.2 we used Riemann sum and limit passage reasoning to define lengths of curves as definite integrals. We use similar reasoning now to define surface areas and surface integrals. Among other applications, surface integrals will be needed in order to extend the divergence theorem and Stokes' theorem in the plane to 3–space.

Surfaces and Normals

As you will see, normals to surfaces play an important role in the study of surface areas and surface integrals. So we begin with a brief review of surfaces and their normals. For further details see Ch. 5.

We shall deal mainly with surfaces S that are given either explicitly by an equation of the form $z = f(x,y)$ or implicitly by $F(x,y,z) = c$ for some constant c. For example, $z = x^2 + y^2$ is an explicit equation for a paraboloid and $x^2 + y^2 + z^2 = 4$ is an implicit equation for a sphere.

A surface S given in implicit form $F(x,y,z) = c$ is **smooth** if F has continuous first partials on S and the gradient vector ∇F is nonzero on S. Then, as we learned in Sec. 5.6,

$$\mathbf{N} = \nabla F = \langle F_x, F_y, F_z \rangle \text{ is normal to } S \text{ at } (x,y,z),$$

where ∇F is evaluated at the point (x,y,z) on S. A corresponding unit normal is

$$\mathbf{n} = \frac{\mathbf{N}}{||\mathbf{N}||} = \frac{\langle F_x, F_y, F_z \rangle}{\sqrt{F_x^2 + F_y^2 + F_z^2}}.$$

The vector fields $\mathbf{N}$ and $\mathbf{n}$ vary continuously on S, as suggested in Fig. 1.

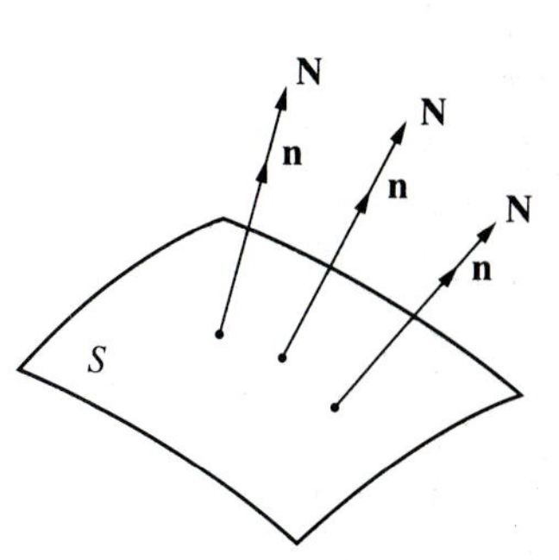

FIGURE 1

EXAMPLE 1. The sphere $x^2 + y^2 + z^2 = 4$ has normal vector field

$$\mathbf{N} = \nabla(x^2 + y^2 + z^2) = \langle 2x, 2y, 2z \rangle = 2\mathbf{r},$$

where $\mathbf{r} = <x,y,z>$ is the position vector for the point (x,y,z) on the sphere. The corresponding unit normal vector field is

$$\mathbf{n} = \frac{\mathbf{N}}{\|\mathbf{N}\|} = \frac{\mathbf{r}}{\|\mathbf{r}\|} = \frac{\mathbf{r}}{2}.$$

Since $\mathbf{N}$ and $\mathbf{n}$ point outward from the center of the sphere, they are called **outward normals** to the sphere. □

Now let S be a surface given explicitly by $z = f(x,y)$. An equivalent implicit equation for S is $F(x,y,z) = z - f(x,y) = 0$. It follows that S is smooth if f has continuous first partials on S. Then

$$\mathbf{N} = \nabla F = <-f_x, -f_y, 1>$$

is a nonzero normal vector field on S. The corresponding unit normal vector field is

$$\mathbf{n} = \frac{\mathbf{N}}{\|\mathbf{N}\|} = \frac{\langle -f_x, -f_y, 1\rangle}{\sqrt{f_x^2 + f_y^2 + 1}}.$$

Both $\mathbf{N}$ and $\mathbf{n}$ are **upward normals** in the sense that their z–components are positive.

EXAMPLE 2. The paraboloid $z = x^2 + y^2$ has normal vector field $\mathbf{N} = \nabla(z - x^2 - y^2) = <-2x, -2y, 1>$ and corresponding unit normal

$$\mathbf{n} = \frac{<-2x, -2y, 1>}{\sqrt{4x^2 + 4y^2 + 1}}.$$

Both $\mathbf{N}$ and $\mathbf{n}$ are upward normals. □

Piecewise Smooth Surfaces

Practical applications of surface area and surface integrals often involve surfaces that are not smooth, but are piecewise smooth. In general, a surface S is **piecewise smooth** if it consists of a finite number of nonoverlapping smooth pieces, $S_1, S_2, \ldots, S_m$. For example, a soup can (circular cylinder with top and bottom) is a piecewise smooth surface made up of three smooth pieces. Outward unit normals $\mathbf{n}$ to the three pieces are shown in Fig. 2. On the curved surface, $\mathbf{n}$ points in a radial direction. On the top, $\mathbf{n} = \mathbf{k}$. On the bottom, $\mathbf{n} = -\mathbf{k}$.

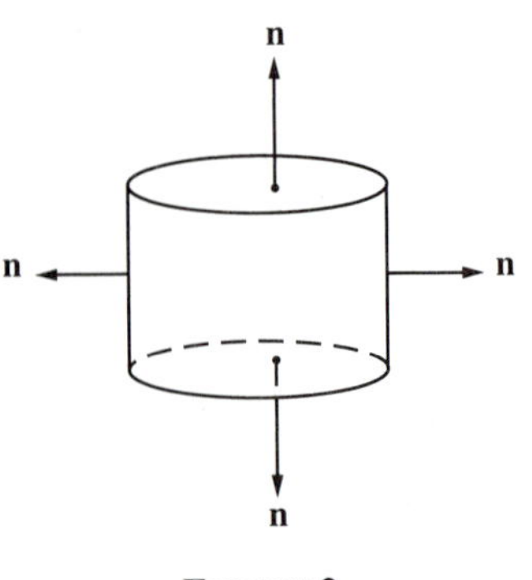

FIGURE 2

Surface Area

Now we are ready to define the area of a surface and develop practical means for calculating surface areas. In this subsection, we break with our standing convention and let P, Q, and R denote points in space rather than components of a vector. We start with a simple situation that will carry us further than you might think. We seek the area of a parallelogram in 3–space that lies in a nonvertical plane Π. Suppose that Π passes through the point $P = (x_0, y_0, z_0)$ and has the point–slope equation

$$z - z_0 = a(x - x_0) + b(y - y_0).$$

The parallelogram we are interested in is shown in Fig. 3. It lies in the plane Π directly above a rectangle in the xy–plane with sides Δx, Δy and area $\Delta A = \Delta x\,\Delta y$. The area of the parallelogram is

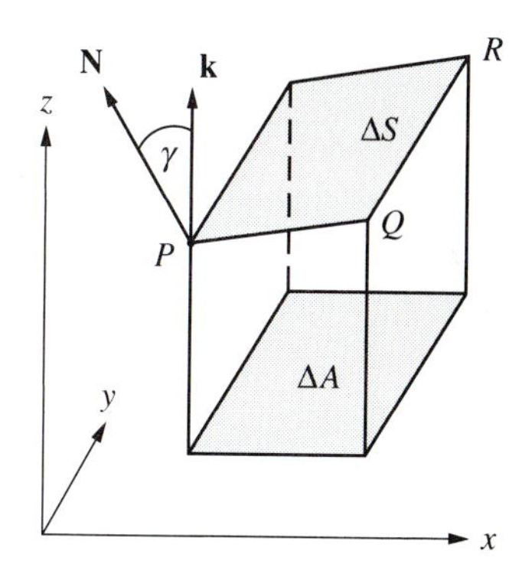

FIGURE 3

$$\Delta S = \sqrt{a^2 + b^2 + 1}\,\Delta A.$$

The verification is not difficult. From Fig. 3 and the point–slope equation for Π, three corners of the parallelogram are the points

$$P=(x_0, y_0, z_0),\ Q= (x_0+\Delta x, y_0, z_0+a\Delta x),\ R = (x_0 + \Delta x, y_0 + \Delta y, z_0 + a\Delta x + b\Delta y).$$

Then

$$\overrightarrow{PQ} < \Delta x, 0, a\Delta x >, \quad \overrightarrow{QR} = < 0, \Delta y, b\Delta y > .$$

Recall that the cross product $\overrightarrow{PQ} \times \overrightarrow{QR}$ is perpendicular to Π and $||\overrightarrow{PQ} \times \overrightarrow{QR}||$ is the area of the parallelogram with sides $\overrightarrow{PQ}$ and $\overrightarrow{QR}$. Thus, $\Delta S = ||\overrightarrow{PQ} \times \overrightarrow{QR}||$. By a routine calculation,

$$\overrightarrow{PQ} \times \overrightarrow{QR} = \Delta x\,\Delta y < -a, -b, 1 > = \Delta A < -a, -b, 1 > .$$

Therefore, as announced, $\Delta S = \sqrt{a^2 + b^2 + 1}\,\Delta A$.

It is informative to express ΔS in other ways. From the point–slope equation, the plane Π has upward normal $\mathbf{N} = < -a, -b, 1 >$. Since $||\mathbf{N}|| = \sqrt{a^2 + b^2 + 1}$, another formula for ΔS is

$$\Delta S = ||\mathbf{N}||\,\Delta A.$$

Here $\mathbf{N}$ is the upward normal to S with z–component 1. Yet another formula for ΔS involves the angle γ between $\mathbf{N} = < -a, -b, 1 >$ and $\mathbf{k}$ in Fig. 3. From

$$\cos\gamma = \frac{\mathbf{N}\cdot\mathbf{k}}{||\mathbf{N}||\,||\mathbf{k}||} = \frac{1}{||\mathbf{N}||}, \qquad \sec\gamma = ||\mathbf{N}||,$$

we obtain

$$\Delta S = \sec\gamma\,\Delta A.$$

Since $\sec\gamma$ increases as the acute angle γ increases, the area ΔS above ΔA increases as the plane Π becomes more nearly vertical. This behavior should agree with your intuition.

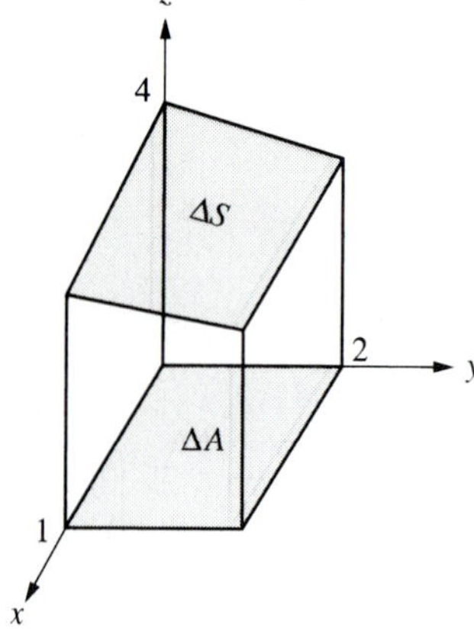

FIGURE 4

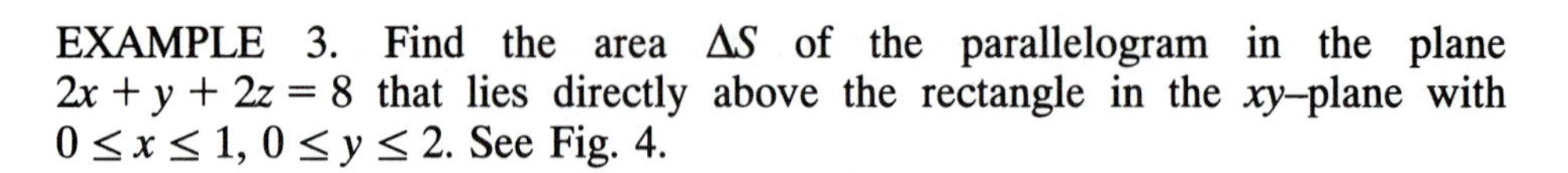
EXAMPLE 3. Find the area ΔS of the parallelogram in the plane $2x + y + 2z = 8$ that lies directly above the rectangle in the xy–plane with $0 \le x \le 1$, $0 \le y \le 2$. See Fig. 4.

Solution. The point–slope equation for the plane is $z - 4 = -x - \frac{1}{2}y$. So the upward normal to S with z–component 1 is $\mathbf{N} = <1, \frac{1}{2}, 1>$ and $\sec\gamma = ||\mathbf{N}|| = 3/2$. Since $\Delta A = 1 \cdot 2 = 2$,

$$\Delta S = \sec\gamma\,\Delta A = \frac{3}{2} \cdot 2 = 3. \;\square$$

The formula $\Delta S = \sec\gamma\Delta A$ extends directly to other flat surfaces besides parallelograms. Figure 5 shows a flat surface S in a nonvertical plane Π that is directly above a region D in the xy–plane with area $\mathcal{A}(D)$. The area of S is given by

$$\mathcal{A}(S) = \sec\gamma\mathcal{A}(D).$$

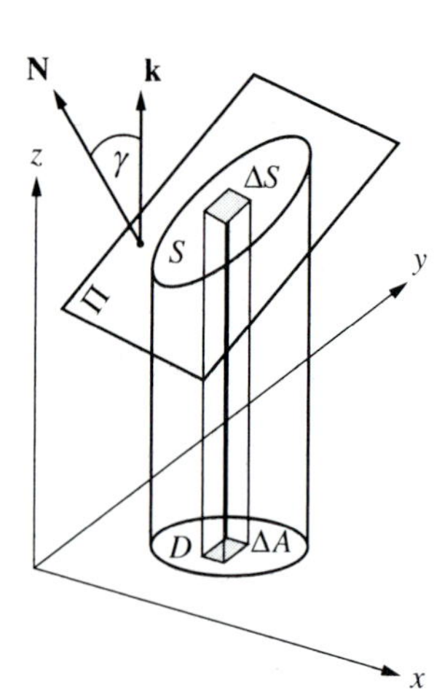

FIGURE 5

The verification is suggested in Fig. 5. Partition D into rectangles with sides parallel to the coordinate axes. A typical rectangle in D, with area ΔA, is shown in Fig. 5. Let ΔS be the area of the part of the surface S that lies directly above ΔA. Then $\Delta S = \sec\gamma\Delta A$ and, hence,

$$\sum\sum \Delta S = \sum\sum \sec\gamma\Delta A = \sec\gamma \sum\sum \Delta A.$$

A limit passage leads us to define the area of S by $\mathcal{A}(S) = \sec\gamma\,\mathcal{A}(D)$. The same formula is valid if S lies below the xy–plane or intersects the xy–plane. In all cases, D is the vertical projection of S onto the xy–plane.

EXAMPLE 4. Find the area of the elliptical surface S cut from the solid cylinder $x^2 + y^2 \le 1$ by the plane $2x + y + 2z = 8$. A sketch of S is shown in Fig. 6, which is not drawn to scale.

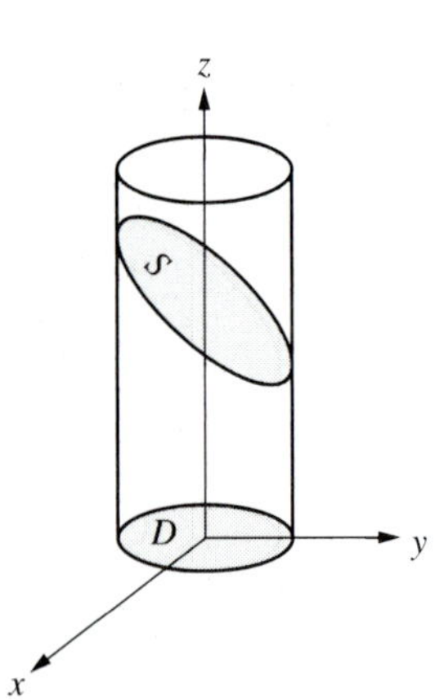

FIGURE 6

Solution. In this case, D is the unit disk in the xy–plane, with area $\mathcal{A}(D) = \pi$. From Ex. 3, $\sec\gamma = 3/2$. So the surface S has area

$$\mathcal{A}(S) = \sec\gamma\,\mathcal{A}(D) = \frac{3}{2}\pi. \;\square$$

Areas of curved surfaces can be found by an argument similar to that used for flat surfaces. In Fig. 7, S is a smooth surface given by $z = f(x,y)$ for (x,y) in a domain D with area $\mathcal{A}(D)$. As we learned earlier,

$$\mathbf{N} = < -f_x, -f_y, 1 >$$

is an upward normal to S. The angle γ between $\mathbf{N}$ and $\mathbf{k}$ satisfies

$$\cos\gamma = \frac{\mathbf{N} \cdot \mathbf{k}}{||\mathbf{N}||\;||\mathbf{k}||} = \frac{1}{||\mathbf{N}||},$$

$$\sec\gamma = ||\mathbf{N}|| = \sqrt{f_x^2 + f_y^2 + 1}.$$

Since S is smooth, $\sec\gamma$ is a continuous function of (x, y) on S.

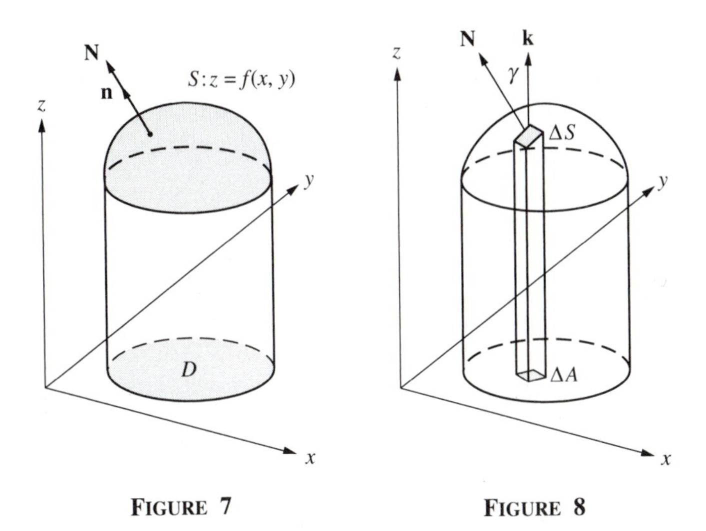

FIGURE 7 FIGURE 8

Partition D into rectangles as indicated in Fig. 8. Approximate the piece of S over ΔA by a parallelogram in the tangent plane to S at any convenient point on S over ΔA. The piece of S over ΔA has the approximate area $\Delta S = \sec\gamma\Delta A$. So the area of S is given approximately by

$$\mathcal{A}(S) \approx \Sigma\Sigma\Delta S = \Sigma\Sigma\sec\gamma\Delta A.$$

This is a Riemann sum in shorthand. A limit passage gives

$$\mathcal{A}(S) = \iint_D \sec\gamma dA.$$

Actually, $\mathcal{A}(S)$ is defined by this integral, which was motivated by the foregoing discussion.

Definition *Surface Area*
Let S be a smooth surface given by $z = f(x, y)$ for (x, y) in a closed domain D inside and on a simple closed path. The **area** of S is

$$\mathcal{A}(S) = \iint_D \sec\gamma dA,$$

where γ is the angle between an upward normal to S and **k**.

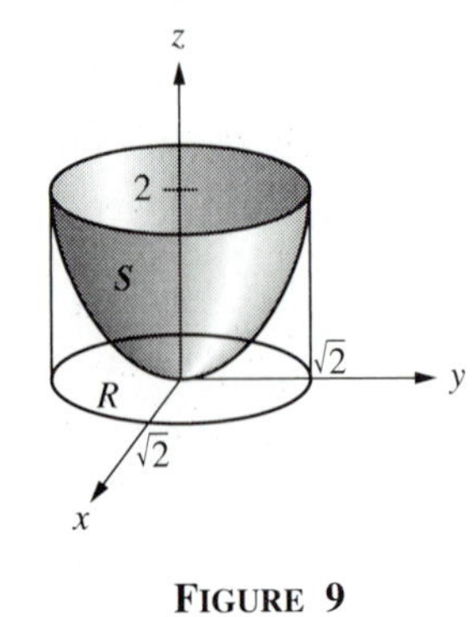

FIGURE 9

Convenient formulas for $\sec\gamma$ are

$$\boxed{\sec\gamma = \sqrt{f_x^2 + f_y^2 + 1} = \sqrt{z_x^2 + z_y^2 + 1}.}$$

EXAMPLE 5. Find the area of the part S of the paraboloid $z = x^2 + y^2$ with $z \leq 2$. See Fig. 9.

Solution. Let D be the closed disk $x^2 + y^2 \leq 2$ with radius $\sqrt{2}$. Since $z_x = 2x$, $z_y = 2y$, and $\sec\gamma = \sqrt{4x^2 + 4y^2 + 1}$, the area of S is

$$\mathcal{A}(S) = \iint_D \sqrt{4x^2 + 4y^2 + 1}\, dA.$$

Since the paraboloid has cylindrical symmetry, let's use cylindrical coordinates r, θ, z to evaluate the double integral. Then $x^2 + y^2 = r^2$, $dA = r\,dr\,d\theta$, and

$$\mathcal{A}(S) = \int_0^{2\pi}\int_0^{\sqrt{2}} \sqrt{4r^2 + 1}\, r\,dr\,d\theta = 2\pi \int_0^{\sqrt{2}} (4r^2 + 1)^{1/2}\, r\,dr,$$

$$= 2\pi\left[\frac{1}{12}(4r^2 + 1)^{3/2}\right]_0^{\sqrt{2}} = \frac{\pi}{6}[27 - 1] = \frac{13\pi}{3}. \ \square$$

Because partial or full cylindrical symmetry often occurs, it is worthwhile to derive formulas for $\sec\gamma$ and $\mathcal{A}(S)$ in terms of cylindrical coordinates r, θ, z. From the chain rule and $x = r\cos\theta$, $y = r\sin\theta$,

$$z_r = \frac{\partial z}{\partial r} = \frac{\partial z}{\partial x}\frac{\partial x}{\partial r} + \frac{\partial z}{\partial y}\frac{\partial y}{\partial r} = z_x \cos\theta + z_y \sin\theta,$$

$$z_\theta = \frac{\partial z}{\partial \theta} = \frac{\partial z}{\partial x}\frac{\partial x}{\partial \theta} + \frac{\partial z}{\partial y}\frac{\partial y}{\partial \theta} = r(-z_x \sin\theta + z_y \cos\theta).$$

A little algebra yields

$$z_r^2 + \frac{1}{r^2} z_\theta^2 = z_x^2 + z_y^2.$$

Therefore,

$$\boxed{\sec\gamma = \sqrt{z_r^2 + \frac{1}{r^2} z_\theta^2 + 1}.}$$

Since $dA = rdrd\theta$ in cylindrical coordinates,

$$\mathcal{A}(S) = \iint_D \sqrt{z_r^2 + \frac{1}{r^2} z_\theta^2 + 1}\, rdrd\theta.$$

The displayed formulas simplify further if S has cylindrical symmetry because then z is a function of r alone and $z_\theta = 0$. For instance, the surface in Ex. 5 and Fig. 9 is represented in cylindrical coordinates by $z = r^2$ for $0 \le r \le \sqrt{2}$ and $0 \le \theta \le 2\pi$. So $z_r = 2r$, $z_\theta = 0$, and

$$\mathcal{A}(S) = \int_0^{2\pi} \int_0^{\sqrt{2}} \sqrt{4r^2 + 1}\, rdrd\theta,$$

the same as before.

Next we derive an area formula that you may have learned long ago but probably never proved.

EXAMPLE 6. Show that the surface area of a sphere of radius a is $4\pi a^2$.

Solution. By symmetry, the area of the sphere is twice the area of the upper hemisphere

$$z = (a^2 - x^2 - y^2)^{1/2} \quad \text{for} \quad x^2 + y^2 \le a^2.$$

Let's change to cylindrical coordinates. Then

$$z = (a^2 - r^2)^{1/2} \quad \text{for} \quad 0 \le r \le a, \qquad 0 \le \theta \le 2\pi,$$

$$z_r = \frac{-r}{(a^2 - r^2)^{1/2}}, \quad z_\theta = 0,$$

$$z_r^2 + \frac{1}{r^2} z_\theta^2 + 1 = \frac{r^2}{a^2 - r^2} + 1 = \frac{a^2}{a^2 - r^2}.$$

Hence, the area of the hemisphere is

$$a \int_0^{2\pi} \int_0^a (a^2 - r^2)^{-1/2} rdrd\theta = 2\pi \int_0^a (a^2 - r^2)^{-1/2} rdr$$

$$= 2\pi a [-(a^2 - r^2)^{1/2}]_0^a = 2\pi a^2,$$

and the area of the sphere is $4\pi a^2$. □

Did you notice a questionable step in the calculation of the area of the hemisphere? The r–integral is improper because the integrand becomes infinite as $r \to a$. The trouble comes from the fact that the normal vectors to the sphere on the "equator" with $r = a$ and $z = 0$ are horizontal. On the equator, the angle γ is 90° so $\cos \gamma = 0$ and $\sec \gamma$ is undefined. The use of the improper integral can be justified by finding the area of the part of the upper hemisphere with $x^2 + y^2 \leq b^2$ for $b < a$ and then taking the limit as $b \to a$. You will encounter areas of other surfaces that involve improper integrals in the problems.

Areas of piecewise smooth surfaces present no difficulties. If S is piecewise smooth and consists of the nonoverlapping smooth pieces $S_1, S_2, \ldots, S_m$, then the area of S is defined by

$$\mathcal{A}(S) = \mathcal{A}(S_1) + \mathcal{A}(S_2) + \cdots + \mathcal{A}(S_m).$$

Surface Integrals

We begin with a physical situation that leads naturally to the definition of a surface integral and reasonable notation for such integrals.

Again consider the surface in Fig. 7. Imagine that S is a thin sheet of metal with a continuous mass density function $\sigma = \sigma(x, y, z)$ in units of mass per unit area. In Fig. 8, the piece of the surface S lying above a small rectangle with area ΔA has approximate area $\Delta S = \sec\gamma\Delta A$ and approximate mass $\sigma\Delta S = \sigma\sec\gamma\Delta A$. So the total mass of S is approximately

$$\sum\sum\sigma\Delta S = \sum\sum\sigma\sec\gamma\Delta A.$$

The usual limit passage argument lead us to conclude that the total mass M of S should be given by $M = \int\int_S \sigma dS = \int\int_D \sigma\sec\gamma dA$. The preceding discussion motivates the definition of a surface integral.

Definition *Surface Integral*
Let S be a smooth surface given by $z = f(x, y)$ for (x, y) in a closed domain D inside and on a simple closed path. Let $h(x, y, z)$ be continuous on S. The **surface integral** of h over S is

$$\iint_S h dS = \iint_D h\sec\gamma dA,$$

where γ is the angle between an upward normal to S and **k**.

In the integral over S, $h = h(x, y, z)$ with (x, y, z) on S. In the integral over D, $h = h(x, y, z)$ with (x, y) in D and $z = f(x, y)$. Notice that if $h = 1$ on S, then

$$\mathcal{A}(S) = \iint_D \sec\gamma dA = \iint_S 1 dS.$$

The expressions for $\sec\gamma$ given earlier,

$$\sec\gamma = \sqrt{z_x^2 + z_y^2 + 1} \quad \text{in rectangular coordinates,}$$

$$\sec\gamma = \sqrt{z_r^2 + \frac{1}{r^2}z_\theta^2 + 1} \quad \text{in cylindrical coordinates,}$$

help with the evaluation of surface integrals, as we illustrate in a moment.

As we have anticipated, the **mass** of a surface S with a continuous density function σ is defined by

$$M = \iint_S \sigma dS.$$

The **center of mass** of S is the point $(\bar{x}, \bar{y}, \bar{z})$ with

$$\bar{x} = \frac{1}{M}\iint_S x\sigma dS, \qquad \bar{y} = \frac{1}{M}\iint_S y\,\sigma dS, \quad \bar{z} = \frac{1}{M}\iint_S z\,\sigma dS.$$

If $\sigma = 1$, the center of mass is also called the **centroid** of S.

EXAMPLE 7. A sheet of mass has the shape of the cone $z = \frac{3}{4}\sqrt{x^2 + y^2}$ for $x^2 + y^2 \leq 16$ and has density $\sigma = \sqrt{x^2 + y^2}$ in suitable units. Find the mass and center of mass of the cone. See Fig. 10. For convenience, a larger scale is used on the z–axis.

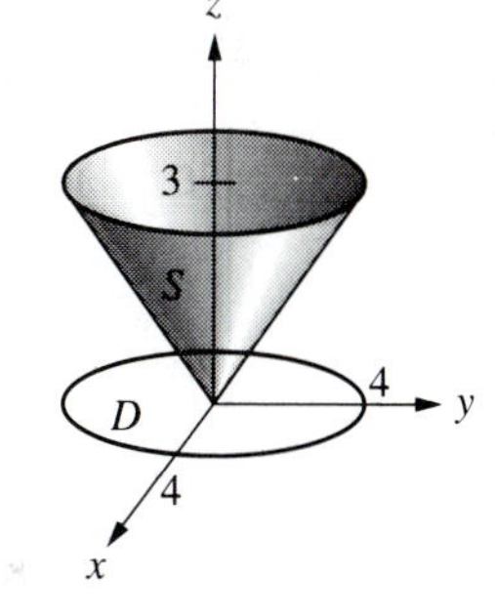

FIGURE 10

Solution. Since the cone and the density have cylindrical symmetry, it is advantageous to use cylindrical coordinates. Then the cone is described by

$$z = \frac{3}{4}r \quad \text{for} \quad 0 \leq r \leq 4, \qquad 0 \leq \theta \leq 2\pi,$$

and the density is $\sigma = r$. Since $z_r = 3/4$ and $z_\theta = 0$,

$$\sec\gamma = \sqrt{z_r^2 + \frac{1}{r^2}z_\theta^2 + 1} = \sqrt{\frac{9}{16} + 1} = \frac{5}{4}$$

and

$$M = \iint_S \sigma dS = \iint_D \sigma \sec\gamma dA = \int_0^{2\pi}\int_0^4 r \cdot \frac{5}{4} r dr d\theta$$

$$= \frac{5\pi}{2}\int_0^4 r^2 dr = \frac{5\pi}{2} \cdot \frac{64}{3} = \frac{160\,\pi}{3}.$$

It remains to find the center of mass. By symmetry $\bar{x} = 0$ and $\bar{y} = 0$. Next, since $z = 3r/4$ on S,

$$\iint_S z\,\sigma\, dS = \int_0^{2\pi}\int_0^4 \frac{3}{4} r \cdot r \cdot \frac{5}{4} r dr d\theta = \frac{15\pi}{8}\int_0^4 r^4 dr = \frac{15\pi}{8} \cdot 64 = 120\,\pi.$$

Therefore,

$$\bar{z} = \frac{1}{M}\iint_S z\sigma\, dS = \frac{3}{160\pi}\cdot 120\,\pi = \frac{9}{4}\pi,$$

and the center of mass is $(\bar{x},\bar{y},\bar{z}) = (0,0,9\,\pi/4)$. □

The formulas highlighted earlier for the evaluation of surface integrals are for a smooth surface given explicitly by $z = f(x,y)$ for (x,y) in D. By permuting the roles of x, y, and z in those formulas we obtain corresponding formulas for other explicitly given smooth surfaces:

$$S:\ \ y = f(x,z) \text{ for } (x,z) \text{ in } D \quad \Rightarrow \quad \iint_S h dS = \iint_D h\sqrt{f_x^2 + f_z^2 + 1}\, dA,$$

$$S:\ \ x = f(y,z) \text{ for } (y,z) \text{ in } D \quad \Rightarrow \quad \iint_S h dS = \iint_D h\sqrt{f_y^2 + f_z^2 + 1}\, dA.$$

When $h = 1$ these surface integrals give the area of S. In the problems, we ask you to develop corresponding formulas for surface integrals of smooth surfaces given implicitly.

As for surface area, the extension of surface integrals to piecewise smooth surfaces presents no difficulty. If S is piecewise smooth and consists of the nonoverlapping smooth pieces S_1, S_2, ..., S_m, then the surface integral of a continuous function h over S is defined by

$$\iint_S h dS = \iint_{S_1} h dS + \iint_{S_2} h dS + \ldots + \iint_{S_m} h dS.$$

PROBLEMS

In Probs. 1–14 find the area of the indicated surface. In these problems, a, b, c, and h are positive constants.

1. The part of the plane $2x - y + 2z = 5$ that lies vertically over the rectangle $0 \le x \le 2,\ -1 \le y \le 2$.
2. The part of the plane $2x + y + 2z = 8$ that lies inside the cylinder $x^2 + y^2 = 9$.
3. The part of the plane $x + y + z = a$ that lies inside the cylinder $x^2 + y^2 = b^2$.
4. The part of the plane $x/a + y/b + z/c = 1$ that lies in the first octant.
5. The part of the cone $z^2 = x^2 + y^2$ above the xy–plane and below the plane $z = 9$.
6. A right circular cone with radius of base a and height h. *Hint.* The cone in question is part of $z^2 = c^2\,(x^2 + y^2)$ for an appropriate choice of c.

7. The part of the sphere $x^2 + y^2 + z^2 = 25$ that lies between the planes $z = 0$ and $z = 4$.

8. Let $0 \le b < c \le a$. The part of the sphere, called a **zone**, between the planes $z = b$ and $z = c$. (Observe that the area depends only upon the difference $c - b$. Does this surprise you?)

9. The part of the cylinder $x^2 + z^2 = 4$ in the first octant and to the left of the plane $y = 8$.

10. A right circular cylinder with radius of base a and height h. *Hint.* Use symmetry and a cylinder much as in the previous problem.

11. The surface cut from the top half of the cone $z^2 = x^2 + y^2$ by the cylinder $x^2 + y^2 = 2ax$.

12. The surface cut from the cylinder $x^2 + z^2 = a^2$ by the cylinder $x^2 + y^2 = a^2$. *Hint.* Use symmetry to restrict to the first octant.

13. The part of the sphere $x^2 + y^2 + z^2 = 2z$ that is within the paraboloid $z = (2/3)(x^2 + y^2)$.

14. The surface that bounds the region inside both the sphere $x^2 + y^2 + z^2 = a^2$ and the cone $z = \sqrt{x^2 + y^2}$.

15. Let S be a smooth surface with equation $z = f(x,y)$ for (x,y) in D, as in the definition of surface area. Let $\mathbf{N}$ be *any* upward normal that varies continuously over S. Show that $\sec\gamma = \|\mathbf{N}\|/\mathbf{N}\cdot\mathbf{k}$ and that

$$\mathcal{A}(S) = \iint_D \frac{\|\mathbf{N}\|}{\mathbf{N}\cdot\mathbf{k}}\, dA.$$

16. A smooth surface S is given by $F(x,y,z) = c$. Assume that $F_z \neq 0$ on S. Suppose that distinct points on S project onto distinct points in the xy–plane and that the projection of S onto the xy–plane is D. Show that

$$\mathcal{A}(S) = \iint_D \frac{\sqrt{F_x^2 + F_y^2 + F_z^2}}{|F_z|}\, dA.$$

It may be helpful first to treat the case $F_z > 0$.

17. Use Prob. 15 or 16 to find the surface area of the upper hemisphere of $x^2 + y^2 + z^2 = a^2$.

18. Use Prob. 15 or 16 to find the area of the top half of the cone $z^2 = x^2 + y^2$ that lies between the spheres $x^2 + y^2 + z^2 = 32$ and $x^2 + y^2 + z^2 = 50$.

In Probs. 19–26, find (a) the mass and (b) the center of mass of the given surface. In these problems, a and h are positive constants. Also, let $k > 0$ be the proportionality constant.

19. The parallelogram in Fig. 4 if the density is proportional to the distance of a point on the parallelogram from the xy–plane.

20. The solid ellipse in Fig. 6 if the density is proportional to the distance of a point on the surface to the xy–plane.

21. The part of the paraboloid in Fig. 9 if the density is constant.

22. The part of the paraboloid $z = 4 - x^2 - y^2$ with $z \geq 0$ if the density is proportional to the distance of a point on the paraboloid from the plane $z = 4$.

23. The part of the cone $z^2 = x^2 + y^2$ between the planes $z = 0$ and $z = h$ if the density is proportional to the distance of a point on the cone above the plane $z = 0$.

24. The hemisphere $x^2 + y^2 + z^2 = a^2$, $z \geq 0$ with density of a point on the hemisphere proportional to its distance from the z–axis.

25. The part of the cylinder $x^2 + z^2 = a^2$ in the first octant and between the xz–plane and the plane $y = x$ if the density of a point on the surface is proportional to the square of its distance from the xy–plane.

26. The part of the cylinder $x^2 + z^2 = a^2$ in the first octant that lies inside the cylinder $x^2 + y^2 = a^2$ if the density of a point on the first cylinder is proportional to its distance from the xz–plane.

In Probs. 27–30, find the centroid of the given surface. The numbers a, b, and c are positive.

27. The hemisphere $x^2 + y^2 + z^2 = a^2$, $z \geq 0$.

28. The part of the plane $x/a + y/b + z/c = 1$ that lies in the first octant.

29. The part of the sphere $x^2 + y^2 + z^2 = a^2$ that lies within the cone $z = \sqrt{x^2 + y^2}$.

30. The part of the sphere $x^2 + y^2 + z^2 = 16$ that lies above the xy–plane and inside the cylinder $x^2 + y^2 = 4x$. *Hint.* The intersection of the cylinder with the xy–plane has a simple polar equation.

8.6 The Divergence Theorem (Gauss' Theorem) and Applications

The divergence theorem for a plane region was established in Sec. 8.4. Now we extend the theorem to a solid region in 3–space and apply it to physical situations, particularly fluid flow fields and inverse square fields for gravitational and electric forces.

The divergence theorem in 2–space asserts that

$$\iint_D \operatorname{div} \mathbf{F}\, dA = \oint_C \mathbf{F} \cdot \mathbf{n} ds,$$

where C is a path around a closed region D and $\mathbf{n}$ is the outer unit normal to C, as in Fig. 1.

The divergence theorem in 3–space has a similar appearance:

$$\iiint_{\Omega} \operatorname{div} \mathbf{F}\, dV = \iint_{S} \mathbf{F} \cdot \mathbf{n} dS,$$

where S is the boundary of a solid region Ω and $\mathbf{n}$ is the outer unit normal to S, as in Fig. 2. A more formal statement of the divergence theorem will be made a little later, after a preliminary discussion of oriented surfaces and flux integrals.

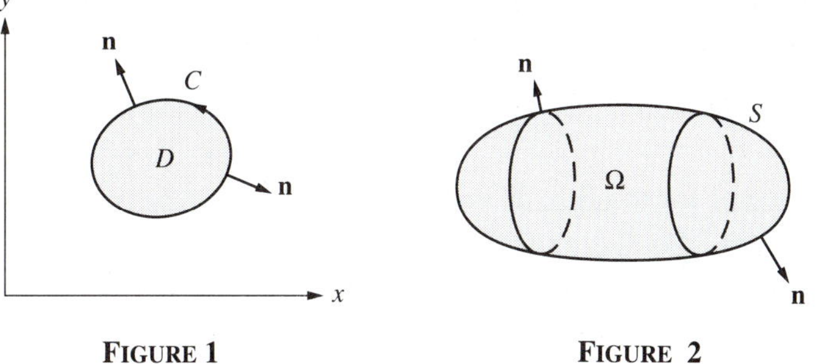

FIGURE 1

FIGURE 2

Oriented Surfaces

In fluid flow problems, and other problems as well, we are interested in flow across a surface S. For this purpose, we need to distinguish a direction of flow across S. A continuous unit normal field $\mathbf{n} = \mathbf{n}(x,y,z)$ on S, as illustrated in Fig. 3, determines a **positive direction** across S. Then S is said to be **oriented** with the **orientation** $\mathbf{n} = \mathbf{n}(x,y,z)$. The unit vectors $-\mathbf{n}$ provide another orientation for S, with positive direction opposite to that of $\mathbf{n}$. A surface S is said to be **orientable** if there exists an orientation for S. Practically every surface you will ever meet is orientable. An exception is the Moebius strip in Prob. 25.

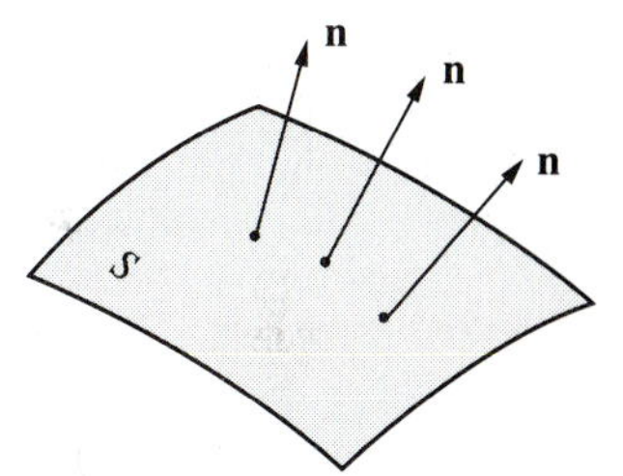

FIGURE 3

EXAMPLE 1. The sphere $x^2 + y^2 + z^2 = a^2$ is orientable. One orientation is given by the outer unit normals $\mathbf{n} = \mathbf{r}/||\mathbf{r}||$. Another orientation is by the inner unit normals $-\mathbf{n} = -\mathbf{r}/||\mathbf{r}||$. □

All smooth surfaces are orientable. Recall that a surface S given by $F(x,y,z) = c$ is smooth if F has continuous first partials on S and the gradient $\nabla F = < F_x, F_y, F_z >$ is nonzero on S. Then, as we learned in Sec. 8.5, S has the normal vector field $\mathbf{N} = \nabla F$ and the orientation $\mathbf{n} = \mathbf{N}/||\mathbf{N}||$. The sphere in Ex. 1 is a case in point.

Often, as in Fig. 4, a smooth surface S is given in explicit form by $z = f(x,y)$ for (x,y) in a domain D and where f has continuous first partials on S. Then S is smooth and the vectors $\mathbf{N} = < -f_x, -f_y, 1 >$ are upward normals to S. The corresponding upward unit normals

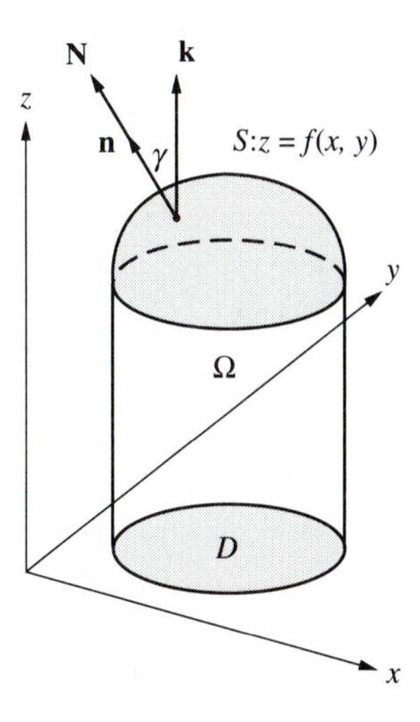

FIGURE 4

$$\mathbf{n} = \frac{\mathbf{N}}{||\mathbf{N}||} = \frac{< -f_x, -f_y, 1 >}{\sqrt{f_x^2 + f_y^2 + 1}}$$

provide an orientation for S. For later convenience, Fig. 4 also shows the angle γ between **k** and **N** (or **n**). Recall that

$$\cos\gamma = \frac{\mathbf{N}\cdot\mathbf{k}}{||\mathbf{N}||\ ||\mathbf{k}||} = \frac{1}{||\mathbf{N}||}, \qquad \sec\gamma = ||\mathbf{N}||.$$

The extension of orientation to piecewise smooth surfaces is straightforward. Let S be a piecewise smooth surface that consists of the nonoverlapping smooth pieces $S_1, S_2, \dots, S_m$. We write $S = S_1 + S_2 + \dots + S_m$. If each of the smooth surfaces S_i has orientation $\mathbf{n}_i$, then an **orientation** of S is comprised of the unit normal vectors $\mathbf{n}_1, \mathbf{n}_2, \dots, \mathbf{n}_m$. For example, the surface $S = S_1 + S_2$ with S_1 the upper hemisphere $x^2 + y^2 + z^2 = 1, z \geq 0$, and S_2 the disk with $x^2 + y^2 \leq 1, z = 0$, has orientation comprised of $\mathbf{n}_1 = \mathbf{r}$ on S_1 and $\mathbf{n}_2 = -\mathbf{k}$ on S_2.

Flux Integrals

A line integral for the flux of a plane vector field across a curve in 2–space was introduced in Sec. 8.4 in connection with fluid flow. The corresponding surface integral for flux in 3–space is similar.

Definition *Flux of a Vector Field*
Let S be a piecewise smooth oriented surface with orientation $\mathbf{n} = \mathbf{n}(x,y,z)$. Let $\mathbf{F} = \mathbf{F}(x,y,z)$ be a continuous vector field defined on S. Then the **flux** of **F** across S (in the positive direction determined by **n**) is

$$\iint_S \mathbf{F}\cdot\mathbf{n}\,dS.$$

We concentrate first on the computation of flux integrals. Then we discuss the physical significance of flux, especially for fluid flow.

Flux integrals are particular cases of surface integrals. As in Fig. 4, suppose that S is a smooth surface given by $z = f(x,y)$ for (x,y) in a domain D. Let $\mathbf{N} = \langle -f_x, -f_y, 1\rangle$ and $\mathbf{n} = \mathbf{N}/||\mathbf{N}||$. From Sec. 8.5, the integral of a continuous function $h(x,y,z)$ over S is given by

$$\iint_S h\,dS = \iint_D h\sec\gamma\,dA,$$

where γ is the angle between **N** and **k** in Fig. 4. Let $h = \mathbf{F}\cdot\mathbf{n}$ to obtain

$$\iint_S \mathbf{F}\cdot\mathbf{n}\,dS = \iint_D \mathbf{F}\cdot\mathbf{n}\sec\gamma\,dA.$$

The integral on the right simplifies remarkably. Since $\mathbf{n} = \mathbf{N}/||\mathbf{N}||$ and $\sec\gamma = ||\mathbf{N}||$,

$$\mathbf{F} \cdot \mathbf{n} \sec\gamma = \mathbf{F} \cdot \left(\frac{\mathbf{N}}{||\mathbf{N}||}\right) ||\mathbf{N}|| = \mathbf{F} \cdot \mathbf{N}$$

and

$$\boxed{\iint\limits_S \mathbf{F} \cdot \mathbf{n}\, dS = \iint\limits_D \mathbf{F} \cdot \mathbf{N}\, dA.}$$

In this formula, $\mathbf{N}$ is the upward normal to S with z–component 1 and $\mathbf{n} = \mathbf{N}/||\mathbf{N}||$.

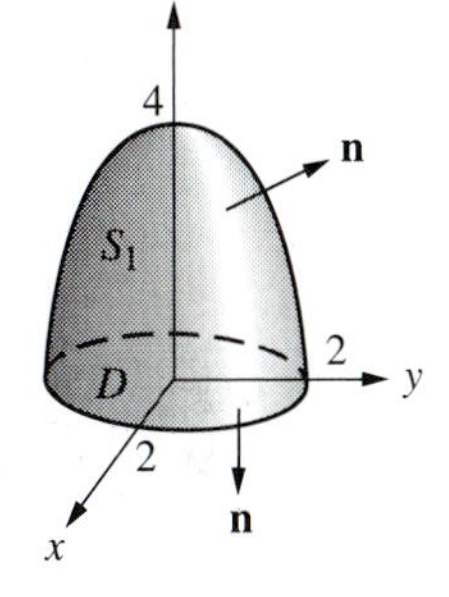

FIGURE 5

EXAMPLE 2. Let S be the piecewise smooth surface in Fig. 5 consisting of the smooth pieces S_1 and D, where D is the disk $x^2 + y^2 \leq 4$ in the xy–plane and S_1 is the surface $z = 4 - x^2 - y^2$ for (x, y) in D. Let $\mathbf{F} = 2x\mathbf{i} + 2y\mathbf{j} + z\mathbf{k}$. Find the flux of $\mathbf{F}$ across S in the outward direction.

Solution. The outward normal to S_1 with z–component 1 is $\mathbf{N} = < 2x, 2y, 1 >$. Since $z = 4 - x^2 - y^2$ on S_1,

$$\mathbf{F} \cdot \mathbf{N} = 4x^2 + 4y^2 + z = 3x^2 + 3y^2 + 4 \text{ on } S_1.$$

In cylindrical coordinates,

$$\mathbf{F} \cdot \mathbf{N} = 3r^2 + 4.$$

Therefore,

$$\iint\limits_S \mathbf{F} \cdot \mathbf{n} dS = \iint\limits_D \mathbf{F} \cdot \mathbf{N} dA = \int_0^{2\pi} \int_0^2 (3r^2 + 4)\, r\, dr\, d\theta$$

$$= 2\pi \int_0^2 (3r^3 + 4r)\, dr = 2\pi \left[\frac{3}{4} r^4 + 2r^2\right]_0^2 = 40\pi.$$

On D the outer normal is $\mathbf{n} = -\mathbf{k}$ and $\mathbf{F} \cdot \mathbf{n} = -z$. Since $z = 0$ on D, $\mathbf{F} \cdot \mathbf{n} = 0$ on D and $\iint_D \mathbf{F} \cdot \mathbf{n}\, dS = 0$. Add the fluxes for S_1 and D to obtain

$$\iint\limits_S \mathbf{F} \cdot \mathbf{n} dS = 40\pi$$

for the flux of $\mathbf{F}$ across the surface S. □

Now we turn to the physical interpretation of flux for fluid flow. Since the story is essentially the same as in the 2–d case, we merely sketch the ideas.

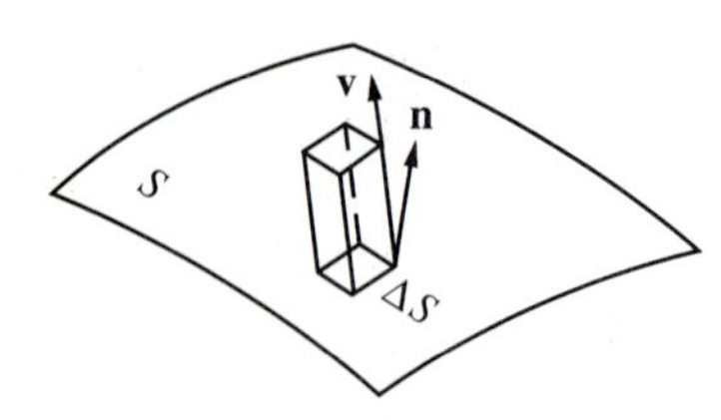

FIGURE 6

Suppose that $\mathbf{F} = \sigma \mathbf{v}$, the mass flow vector for a steady fluid flow. Let S be an oriented surface in the flow field with orientation $\mathbf{n} = \mathbf{n}(x, y, z)$, as illustrated in Fig. 6.

The flux integral is a limit of Riemann sums:

$$\iint_S \mathbf{F} \cdot \mathbf{n}\, dS = \lim \sum\sum \mathbf{F} \cdot \mathbf{n}\Delta S.$$

A piece of the surface with area ΔS is shown in Fig. 6. First, suppose that $\mathbf{v} \cdot \mathbf{n} > 0$ on ΔS. Then the flow is in the positive direction across ΔS. The fluid that crosses ΔS in time Δt nearly fills out a parallelepiped with base ΔS, height $\Delta t\, \mathbf{v} \cdot \mathbf{n}$, volume $\Delta S\, \Delta t \mathbf{v} \cdot \mathbf{n}$, and mass $\sigma \Delta S \Delta t \mathbf{v} \cdot \mathbf{n} = \mathbf{F} \cdot \mathbf{n} \Delta S \Delta t$. So $\mathbf{F} \cdot \mathbf{n}\Delta S$ approximates the mass flow rate (per unit time) across ΔS. For $\mathbf{v} \cdot \mathbf{n}$ positive or negative, $\mathbf{F} \cdot \mathbf{n}\Delta S$ approximates the net mass flow rate across ΔS and, hence, $\sum\sum \mathbf{F} \cdot \mathbf{n}\Delta S$ approximates the net mass flow rate across S. A limit passage gives

$$\iint_S \mathbf{F} \cdot \mathbf{n} dS = \text{the net rate that fluid crosses } S \text{ in the positive direction.}$$

An obvious but very useful property of flux integrals is that if the orientation of a surface S is reversed (that is, if $\mathbf{n}$ is replaced by $-\mathbf{n}$), then the flux across S changes sign.

The Divergence Theorem

With this preparation, we are ready to make a formal statement of the divergence theorem.

Theorem 1 *The Divergence Theorem (Gauss' Theorem)*
Let Ω be a closed and bounded region in 3–space with a piecewise smooth boundary surface S and outer unit normal n to S. Let $\mathbf{F} = \mathbf{F}(x, y, z)$ be a vector field with components having continuous first partials for (x, y, z) in Ω. Then

$$\iiint_\Omega \operatorname{div} \mathbf{F} dV = \iint_S \mathbf{F} \cdot \mathbf{n} dS.$$

For steady fluid flow, the flux integral in Th. 1 is the net rate that fluid crosses S in the outward (positive) direction. As we shall see in a moment,

$$\iiint_\Omega \operatorname{div} \mathbf{F} dV \quad = \quad \text{the net rate that fluid leaves } \Omega.$$

Thus, the divergence theorem tells us that the net rate that fluid leaves Ω is equal to the net rate that fluid crosses S in the outward direction. With this interpretation, the divergence theorem is a conservation law.

The divergence integral is a limit of Riemann sums:

$$\iiint_{\Omega} \operatorname{div} \mathbf{F} dV = \lim \sum\sum\sum \operatorname{div} \mathbf{F} \Delta V.$$

Recall from Sec. 8.1 that $\operatorname{div} \mathbf{F}(x,y,z)$ is the net rate per unit volume per unit time that fluid leaves the vicinity of (x,y,z). So $\operatorname{div} \mathbf{F} \Delta V$ approximates the net rate (per unit time) that fluid leaves ΔV and $\sum\sum\sum \operatorname{div} \mathbf{F} \Delta V$ approximates the net rate that fluid leaves Ω. A limit passage gives the interpretation of the divergence integral displayed previously.

EXAMPLE 3. Use the divergence theorem to find the flux of $\mathbf{F} = 2x\mathbf{i} + 2y\mathbf{j} + z\mathbf{k}$ across the surface S in Ex. 2 and Fig. 5.

Solution. First, note that $\operatorname{div} \mathbf{F} = 5$. Let Ω be the solid region bounded by S. Then

$$\iint_{S} \mathbf{F} \cdot \mathbf{n} dS = \iiint_{\Omega} \operatorname{div} \mathbf{F} dV = \iiint_{\Omega} 5 dV = 5V,$$

where V is the volume of Ω. The upper surface S_1 in Fig. 5 is expressed in cylindrical coordinates by $z = 4 - r^2$ for $0 \le r \le 2$ and $0 \le \theta \le 2\pi$. Therefore,

$$V = \int_0^{2\pi} \int_0^2 (4 - r^2) r dr d\theta = 2\pi \left[2r^2 - \frac{r^4}{4} \right]_0^2 = 8\pi.$$

So the outward flux across S is 40π, which agrees with what we found in Ex. 2. □

We shall sketch a proof of the divergence theorem for types of regions you are apt to meet. The strategy is to start with a relatively simple case and build up to more complicated situations. As usual,

$$\mathbf{F}(x,y,z) = P(x,y,z)\mathbf{i} + Q(x,y,z)\mathbf{j} + R(x,y,z)\mathbf{k}.$$

Then

$$\operatorname{div} \mathbf{F} = \frac{\partial P}{\partial x} + \frac{\partial Q}{\partial y} + \frac{\partial R}{\partial z}.$$

Throughout our considerations, Ω is a solid region with a piecewise smooth boundary.

First let $\mathbf{F} = R\mathbf{k}$ and let Ω be the region in Fig. 4. We must prove that

$$\iiint_{\Omega} \frac{\partial R}{\partial z} dV = \iint_{S} R\mathbf{k} \cdot \mathbf{n} dS,$$

where now S is the complete boundary of Ω. It consists of three pieces, the upper surface S_1(labeled S in Fig. 4), the lateral surface S_2, and the lower surface $S_3 = D$. Consider the integral of $R\mathbf{k} \cdot \mathbf{n}$ separately over S_1, S_2, and S_3. Let $\mathbf{F} = R\ \mathbf{k}$ in the general formula $\int\int_{S_1} \mathbf{F} \cdot \mathbf{n}dS = \int\int_D \mathbf{F} \cdot \mathbf{N}dA$, where $\mathbf{N} = < -f_x, -f_y, 1 >$, to obtain

$$\iint_{S_1} R\mathbf{k} \cdot \mathbf{n}dS = \iint_D R\mathbf{k} \cdot \mathbf{N}dA = \iint_D R(x,y,f(x,y))\,dA.$$

On the lateral surface S_2, the unit normal $\mathbf{n}$ is horizontal. So $\mathbf{k} \cdot \mathbf{n} = 0$ and

$$\iint_{S_2} R\mathbf{k} \cdot \mathbf{n}\,dS = 0.$$

On $S_3 = D$, $z = 0$, and $\mathbf{n} = -\mathbf{k}$. Hence, $\mathbf{k} \cdot \mathbf{n} = -1$ and

$$\iint_{S_3} R\mathbf{k} \cdot \mathbf{n}dS = -\iint_D R(x,y,0)\,dA.$$

Combine the three flux integrals over S_1, S_2, and S_3 to obtain

$$\iint_S R\mathbf{k} \cdot \mathbf{n}dS = \iint_D [R(x,y,f(x,y)) - R(x,y,0))]\,dA.$$

Now consider the volume integral:

$$\iiint_\Omega \frac{\partial R}{\partial z}dV = \iint_D \left[\int_0^{f(x,y)} \frac{\partial R}{\partial z}dz\right]dA = \iint_D [R(x,y,f(x,y)) - R(x,y,0)]\,dA.$$

Thus,

$$\iiint_\Omega \frac{\partial R}{\partial z}dV = \iint_S R\mathbf{k} \cdot \mathbf{n}dS$$

and the divergence theorem is valid for $\mathbf{F} = R\mathbf{k}$ and Ω the region in Fig. 4.

It follows easily that the divergence theorem is valid for $\mathbf{F} = R\mathbf{k}$ on the z–simple region in Fig. 7 determined by $g(x,y) \le z \le h(x,y)$ for (x,y) in D. Just apply the theorem to $\mathbf{F} = R\mathbf{k}$ on the regions with $0 \le z \le h(x,y)$ and $0 \le z \le g(x,y)$. Then subtract. (Minor adjustments cover the case when $g(x,y)$ may assume negative values.)

By essentially the same reasoning, the divergence theorem is valid for $\mathbf{F} = P\mathbf{i}$ if Ω is x–simple and for $\mathbf{F} = Q\mathbf{j}$ if Ω is y–simple.

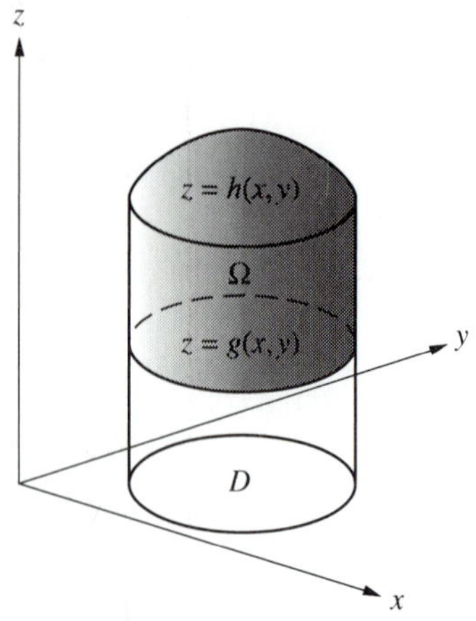

z
z = h(x, y)
Ω
z = g(x, y)
y
D
x

FIGURE 7

For brevity, we say that a region Ω is **simple** if it is x–simple, y–simple, and z–simple. For example, a solid ball or box is simple. If Ω is simple, then the divergence theorem is valid for $P\mathbf{i}$, $Q\mathbf{j}$, $R\mathbf{k}$ and, by addition, for their sum, $\mathbf{F} = P\mathbf{i} + Q\mathbf{j} + R\mathbf{k}$.

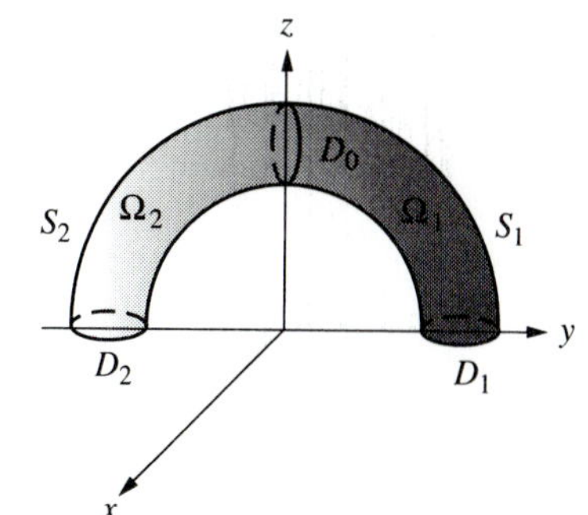

FIGURE 8

Finally, the divergence theorem is valid for a region Ω that can be decomposed into a finite number of nonoverlapping simple regions with piecewise smooth boundaries. To see why this is so, consider the "tubular arch" Ω in Fig. 8, which is composed of two simple regions Ω_1 and Ω_2 with $y \geq 0$ and $y \leq 0$. The disk D_0 that separates Ω_1 and Ω_2 is analogous to a cross-cut in 2–space. Let $\mathbf{n}_1$ and $\mathbf{n}_2$ be the outward unit normals to the surfaces of Ω_1 and Ω_2. Then

$$\iiint_{\Omega_1} \operatorname{div} \mathbf{F} dV = \iint_{D_0} \mathbf{F} \cdot \mathbf{n}_1 dS + \iint_{D_1} \mathbf{F} \cdot \mathbf{n}_1 \, dS + \iint_{S_1} \mathbf{F} \cdot \mathbf{n}_1 \, dS,$$

$$\iiint_{\Omega_2} \operatorname{div} \mathbf{F} dV = \iint_{D_0} \mathbf{F} \cdot \mathbf{n}_2 dS + \iint_{D_2} \mathbf{F} \cdot \mathbf{n}_2 dS + \iint_{S_2} \mathbf{F} \cdot \mathbf{n}_2 dS.$$

Since $\mathbf{n}_2 = -\mathbf{n}_1 = \mathbf{j}$ on D_0,

$$\iint_{D_0} \mathbf{F} \cdot \mathbf{n}_1 dS + \iint_{D_0} \mathbf{F} \cdot \mathbf{n}_2 dS = 0.$$

Add the two previous equations for the divergence integrals to obtain

$$\iiint_{\Omega} \mathbf{F} dV = \iint_{D_1} \mathbf{F} \cdot \mathbf{n} dS + \iint_{D_2} \mathbf{F} \cdot \mathbf{n} dS + \iint_{S_1} \mathbf{F} \cdot \mathbf{n} dS + \iint_{S_2} \mathbf{F} \cdot \mathbf{n} dS,$$

where $\mathbf{n}$ is the outer unit normal to the surface S of Ω. The sum of the four flux integrals is $\iint_S \mathbf{F} \cdot \mathbf{n} dS$. Thus, the divergence theorem is valid for the tubular region Ω.

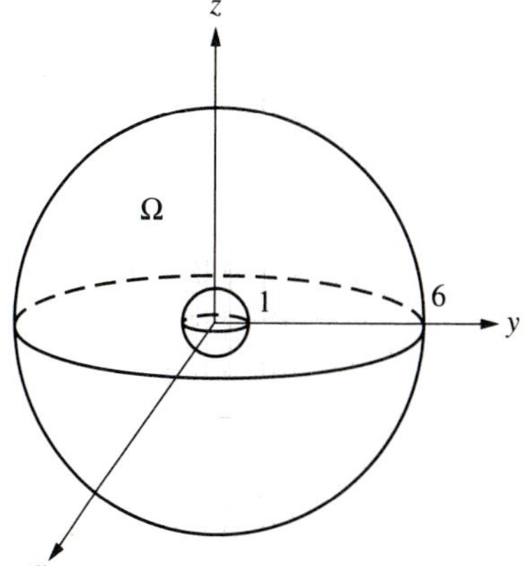

FIGURE 9

For another example, let Ω be the region between the concentric spheres $x^2 + y^2 + z^2 = 1$ and $x^2 + y^2 + z^2 = 36$ in Fig. 9. The boundary S of Ω consists of the two spheres, each oriented outwardly from Ω. This region Ω can be decomposed into eight simple regions Ω_i, one in each of the eight octants. Apply the divergence theorem to each subregion Ω_i and add the results to obtain

$$\iiint_{\Omega} \operatorname{div} \mathbf{F} dV = \iint_{S} \mathbf{F} \cdot \mathbf{n} dS.$$

The flux integrals over the surfaces that separate the regions Ω_i cancel out in pairs because the outer normals on these faces are oppositely directed.

Applications

The divergence theorem has many applications, the most important of which establish general principles and characteristics of physical and mathematical systems. The theorem can be used also for practical calculations.

EXAMPLE 4. A fluid has mass flow vector $\mathbf{F} = x^2\mathbf{i} + y^3\mathbf{j} + z^4\mathbf{k}$. Find the flux outward from the unit cube Ω: $0 \le x \le 1$, $0 \le y \le 1$, $0 \le z \le 1$.

Solution. Instead of calculating six flux integrals, one for each face of the cube, we use the divergence theorem. The flux across the surface S of the cube is

$$\iint_S \mathbf{F} \cdot \mathbf{n} dS = \iiint_\Omega \operatorname{div} \mathbf{F} dV = \int_0^1\int_0^1\int_0^1 (2x+3y^2+4z^3)\, dxdydz = 1+1+1=3. \; \square$$

Earlier, the divergence of a mass flow vector $\mathbf{F}(x,y,z)$ was interpreted as the net rate per unit volume that fluid leaves the vicinity of (x,y,z). Now that the divergence theorem has been established, we can use it to derive a more general result for any vector field $\mathbf{F}$ with components having continuous first partials. Fix a point (x,y,z) in the domain of $\mathbf{F}$. Let Ω be a simple region containing (x,y,z) with a piecewise smooth boundary S and volume V. By the divergence theorem and the mean value theorem,

$$\iint_S \mathbf{F} \cdot \mathbf{n} dS = \iiint_\Omega \operatorname{div} \mathbf{F} dV = \operatorname{div} \mathbf{F}(\bar{x},\bar{y},\bar{z}) \; V$$

for some $(\bar{x},\bar{y},\bar{z})$ in Ω. Then

$$\operatorname{div} \mathbf{F}(\bar{x},\bar{y},\bar{z}) = \frac{1}{V}\iint_S \mathbf{F} \cdot \mathbf{n} dS.$$

Imagine that Ω shrinks to (x,y,z) in the sense that Ω ultimately lies in any ball centered at (x,y,z). Then $V \to 0$, $(\bar{x},\bar{y},\bar{z}) \to (x,y,z)$ and $\mathbf{F}(\bar{x},\bar{y},\bar{z}) \to \mathbf{F}(x,y,z)$. Hence,

$$\operatorname{div} \mathbf{F}(x,y,z) = \lim_{V\to 0} \frac{1}{V}\iint_S \mathbf{F} \cdot \mathbf{n} dS.$$

In words,

$\operatorname{div} \mathbf{F}(x,y,z)$	is the flux per unit volume leaving the vicinity of (x,y,z).

EXAMPLE 5. A point mass or charge at the origin sets up an inverse square field

$$\mathbf{F} = \mathbf{F}(\mathbf{r}) = k\frac{\mathbf{r}}{r^3},$$

where $\mathbf{r} = <x, y, z>$, $r = ||\mathbf{r}||$, and k is a constant. Find the flux of the field outward across the ellipsoid $x^2/4 + y^2/3 + z^2/5 = 1$.

Solution. We could try to apply the divergence theorem to the solid bounded by the ellipsoid, but that would be wrong because $\mathbf{F}$ is undefined at the origin and the hypotheses of Th. 1 are not satisfied. Instead, we apply the divergence theorem to the solid region Ω inside the ellipsoid S and outside the unit sphere S_1. By the divergence theorem,

$$\iiint_\Omega \operatorname{div} \mathbf{F}\, dV = \iint_S \mathbf{F} \cdot \mathbf{n} dS + \iint_{S_1} \mathbf{F} \cdot \mathbf{n} dS.$$

By a routine calculation (see Ex. 11 in Sec. 8.1), $\operatorname{div} \mathbf{F}(\mathbf{r}) = 0$ for $\mathbf{r} \neq \mathbf{0}$. Therefore, $\iiint_\Omega \operatorname{div} \mathbf{F} dV = 0$ and

$$\iint_S \mathbf{F} \cdot \mathbf{n} dS = -\iint_{S_1} \mathbf{F} \cdot \mathbf{n} dS.$$

On S_1, $r = ||\mathbf{r}|| = 1$, $\mathbf{n} = -\mathbf{r}$, $\mathbf{F} = k\mathbf{r}$, $\mathbf{F} \cdot \mathbf{n} = -k$, and

$$\iint_{S_1} \mathbf{F} \cdot \mathbf{n} dS = -k \iint_{S_1} dS = -4\pi k.$$

Finally,

$$\iint_S \mathbf{F} \cdot \mathbf{n} dS = 4\pi k\ .\square$$

The argument in the solution to Ex. 5 works if the ellipsoid is replaced by any piecewise smooth surface that is the boundary of a simple region Ω containing the origin in its interior. We still get

$$\iint_S \mathbf{F} \cdot \mathbf{n} dS = 4\pi k.$$

The point mass or charge can be located at any point $\mathbf{r}_0$. Then

$$\mathbf{F} = \frac{k(\mathbf{r} - \mathbf{r}_0)}{||\mathbf{r} - \mathbf{r}_0||^3}$$

and, once again, we get the flux law $\iint_S \mathbf{F} \cdot \mathbf{n} dS = 4\pi k$ for S, the boundary of the simple region Ω that contains $\mathbf{r}_0$.

The two most important applications of the flux equation $\iint_S \mathbf{F} \cdot \mathbf{n} dS = 4\pi k$ are to gravitational and electric fields. As we learned in Sec. 8.1, the *gravitational field strength* $\mathbf{G}$ due to a point mass M at the origin is

$$\mathbf{G} = -GM\frac{\mathbf{r}}{r^3},$$

where G is the universal gravitational constant. By Newton's law of gravitation, the gravitational force of M on a mass m at $\mathbf{r}$ is $m\mathbf{G}$. Similarly, the *electrical field strength* $\mathbf{E}$ due to a point charge Q at the origin is

$$\mathbf{E} = \frac{Q}{4\pi\varepsilon_0}\frac{\mathbf{r}}{r^3},$$

where ε_0 is a constant, called the *primitivity of free space.* By Coulomb's law, the electric force that Q imparts to a charge q at $\mathbf{r}$ is $q\mathbf{E}$. Apply the general flux law for inverse square fields to obtain, for the gravitational and electric fields,

$$\iint_S \mathbf{G} \cdot \mathbf{n} dS = -4\pi GM, \qquad \iint_S \mathbf{E} \cdot \mathbf{n} dS = \frac{Q}{\varepsilon_0}.$$

These flux laws, especially the latter, are referred to as **Gauss' law**. They are named after the great German mathematician Karl Friedrich Gauss (1777–1855), who was called the prince of mathematicians. The two laws assert that the flux of the gravitational or electric field across a closed surface is proportional to the mass or charge enclosed by the surface. Although we have derived the laws only for point masses and charges, they are valid for any mass or charge distribution inside the surface.

To convey the significance and fundamental nature of Gauss' law, we use it and a reasonable symmetry assumption to derive Newton's inverse square law for a gravitational field. Let M be a point mass located for convenience at the origin. We assume that M sets up a gravitational field $\mathbf{G}(\mathbf{r})$ that obeys Gauss' law, that $\mathbf{G}(\mathbf{r})$ is radially directed, and that $\|\mathbf{G}(\mathbf{r})\|$ is constant on any sphere $\|\mathbf{r}\| = a$. In formulas, these conditions mean that

$$\mathbf{G}(\mathbf{r}) = f(r)\frac{\mathbf{r}}{r}, \qquad r = \|\mathbf{r}\|,$$

where $|f(r)| = \|\mathbf{G}(\mathbf{r})\|$. Apply Gauss' law to the sphere S_a with center at the origin and radius a. On S_a,

$$\mathbf{G} = f(a)\frac{\mathbf{r}}{a}, \qquad \mathbf{n} = \frac{\mathbf{r}}{a}, \qquad \mathbf{G} \cdot \mathbf{n} = f(a).$$

Hence, by Gauss' law,

$$-4\pi GM = \iint_{S_a} \mathbf{G} \cdot \mathbf{n} dS = \iint_{S_a} f(a)\, dS = f(a) \cdot 4\pi a^2,$$

$$f(a) = -\frac{GM}{a^2}.$$

Since a is arbitrary, $f(r) = -GM/r^2$ and

$$\mathbf{G}(\mathbf{r}) = f(r)\frac{\mathbf{r}}{r} = -GM\frac{\mathbf{r}}{r^3},$$

which is Newton's inverse square law for gravitation.

PROBLEMS

1. Find the flux of $\mathbf{F} = < 1, -1, 1 >$ across the part of the plane $2x + 6y + 3z = 6$ in the first octant oriented by its upward normal.
2. Find the flux of $\mathbf{F} = \mathbf{i} - \mathbf{j} + \mathbf{k}$ across the hemisphere $x^2 + y^2 + z^2 = 4$, $z \le 0$ oriented by its upward normal.
3. Find the flux of the electric field $\mathbf{E} = z\mathbf{k}$ across the hemisphere $x^2 + y^2 + z^2 = 9$, $z \ge 0$ oriented by its outer normal.
4. The mass flow vector of a steady flow is $\mathbf{F} = -y\mathbf{i} + x\mathbf{j} + z\mathbf{k}$. Find the net amount of mass/sec moving across the parabolic surface $z = x^2 + y^2$, $z \le 16$ oriented by its downward (negative z–component) unit normal.

In Probs. 5–8 verify the divergence theorem by calculating both integrals separately for the given vector field $\mathbf{F}$ and region Ω.

5. $\mathbf{F} = x\mathbf{i} + y\mathbf{j} + z\mathbf{k}$ and Ω is the unit ball $x^2 + y^2 + z^2 \le 1$.
6. $\mathbf{F} = < xy, yz, zx >$ and Ω is the unit cube $0 \le x, y, z \le 1$.
7. $\mathbf{F} = z\mathbf{k}$ and Ω is the semiball $x^2 + y^2 + z^2 \le 4$, $z \le 0$.
8. $\mathbf{F} = < x^2 - y^2, 2xy, y^2 - xy >$ and Ω is the solid tetrahedron with vertices $(0,0,0)$, $(1,0,0)$, $(0,1,0)$, $(0,0,1)$.
9. Find the flux of the electric field $\mathbf{E} = < x, -y, z^2 - 1 >$ outward across the region bounded by the planes $z = 0$, $z = 1$, and the cylinder $x^2 + y^2 = 4$.
10. The mass flow vector of a steady flow is $\mathbf{F} = -y\mathbf{i} + x\mathbf{j} + z\mathbf{k}$. Find the net amount of mass/sec moving outward across the surface of the region bounded by the paraboloids $z = x^2 + y^2$ and $z = 32 - x^2 - y^2$.
11. The mass flow vector of a steady flow is $\mathbf{F} = < x^3 + \sin\pi z, x^2y + \cos\pi z, \cos\pi xy >$. Find the net mass/sec moving outward across the surface of the region bounded by the coordinate planes and $x + 2y + 3z = 6$.
12. Find the flux of the electric field $\mathbf{E} = (x^2 - ye^z)\mathbf{i} + (y^2 - xe^z)\mathbf{j} + (z^2 + ye^x)\mathbf{k}$ outward across the region inside the cylinder $x^2 + y^2 = 16$, below the plane $x + 2y + 3z = 12$, and above the plane $z = -3$.

According to *Fourier's law,* the flow of heat (energy) from one region of a body to another is governed by the *heat flow vector* $\mathbf{q} = -k\nabla u$, where $u = u(x,y,z)$ is the temperature at (x,y,z) and $k > 0$ is the *thermal conductivity* of the body. The heat (energy) per second crossing an oriented surface S is the flux of the vector field $\mathbf{q}$ across S.

13. A body has the shape of the solid cone $\sqrt{x^2+y^2} \le z \le 2$. The temperature in the body is $u = x^2 + 3y^2 + 5z^2$ and its thermal conductivity is 2. Find the rate at which heat energy crosses outward over the surface of the body.

14. The solid in the first octant bounded by the coordinate planes, the plane $y = h$, and the cylinder $x^2 + z^2 = a^2$ lies in a region of space where the temperature is $u = xe^{-y^2-z^2}$ and the thermal conductivity is 1. Find the rate of heat flow outward across the surface of the solid.

In Probs. 15–22, assume that the bounded domain Ω, its bounding surface S, and the vector field S satisfy the hypotheses in the divergence theorem.

15. Theorem 2 in Sec. 8.4 expresses the area enclosed by a curve as a line integral along the curve. Use the divergence theorem to express the volume of a region Ω as a surface integral over its bounding surface S.

16. The vector field $\mathbf{F}$ is tangent to the surface S at every point on it. Evaluate $\iiint_\Omega \operatorname{div} \mathbf{F} dV$.

17. Evaluate the flux of a constant vector field outward across a closed surface.

18. Show that $\iint_S \operatorname{curl} \mathbf{F} \cdot \mathbf{n} dS = 0$, if the components of $\mathbf{F}$ have continuous second partials.

Let $f(x,y,z)$ be a scalar field with continuous first partials in a region of interest. Let $\mathbf{n}$ be a unit normal vector field to a surface S. The directional derivative $D_\mathbf{n} f(x,y,z)$ is called the **normal derivative** of f on S. Usually this normal derivative is denoted by $\partial f / \partial n$:

$$\boxed{\frac{\partial f}{\partial n} = D_\mathbf{n} f.}$$

The next few problems give several important identities. Assume the scalar fields f and g have continuous second partials in the regions of interest.

19. Show that

$$\iint_S \frac{\partial f}{\partial n} dS = \iiint_\Omega \Delta f dV,$$

where $\Delta f = f_{xx} + f_{yy} + f_{zz}$ is the Laplacian of f.

20. Show that

$$\iint_S f\frac{\partial f}{\partial n}\,dS = \iiint_\Omega |\nabla f|^2\,dV.$$

21. (a) Prove **Green's first identity**

$$\iint_S f\frac{\partial g}{\partial n}\,dS = \iiint_\Omega (f\Delta g + \nabla f \cdot \nabla g)\,dV.$$

Hint. Use $\mathbf{F} = f\nabla g$ in Th. 1.

(b) Then interchange f and g in (a) to obtain **Green's second identity**

$$\iint_S \left(f\frac{\partial g}{\partial n} - g\frac{\partial f}{\partial n}\right)dS = \iiint_\Omega (f\Delta g - g\Delta f)\,dV.$$

22. Show that

$$\iint_S f\mathbf{n}\,dS = \iiint_\Omega \nabla f\,dV.$$

Integrals of vector functions are defined by componentwise integration. *Hint.* $f\mathbf{n} = < f\mathbf{n}\cdot\mathbf{i}, f\mathbf{n}\cdot\mathbf{j}, f\mathbf{n}\cdot\mathbf{k} >$.

Probs. 23–24 illustrate a typical application of the divergence theorem in developing physical principles.

23. Let $\mathbf{B}$ denote the buoyant force exerted by a fluid (like water) on a submerged body Ω with boundary S and outer unit normal $\mathbf{n}$. Let $p = p(x,y,z)$ be the hydrostatic pressure (force per unit area) at (x,y,z). Then the buoyant force $\Delta\mathbf{B}$ on a surface element ΔS of the submerged surface S is $\Delta\mathbf{B} \approx -p\mathbf{n}\Delta S$. (The minus sign means that the buoyant force on ΔS is directed into Ω.) Use a Riemann sum and limit passage argument to justify the statement that the total buoyant force on Ω is

$$\mathbf{B} = -\iint_S p\mathbf{n}\,dS.$$

24. On physical grounds, the buoyant force does not depend on the nature of the submerged body. So, for conceptual purposes, we can assume the body is composed of material whose density is the same at each point as the fluid that it displaces. Use the previous two problems; the *hydrostatic pressure law* $\nabla p = \sigma\mathbf{g}$, where $\sigma(x,y,z)$ is the density of the fluid and $\mathbf{g}$ is the constant gravitational acceleration near the earth; and the divergence theorem to obtain

$$\mathbf{B} = -\iiint_V \sigma\mathbf{g}\,dV = -\mathbf{W}$$

where $\mathbf{W}$ is the weight of fluid displaced by the submerged body. The equation $\mathbf{B} = -\mathbf{W}$ is **Archimedes' principle**. In words, the buoyant force on a submerged object is the negative of the weight of the fluid displaced by the object. (The principle extends to partially submerged objects.)

25. *(A Nonorientable Surface)*. If you follow the unit normals of an orientable surface around any smooth closed curve on the surface, the normals vary continuously and you end up with the same unit normal that you started with. Here is a surface that does not have this property and, hence, is not orientable. Cut a strip of paper about 1 inch wide and 11 inches long. Draw an 11-inch line down the center of one side of the strip and mark a reference point on the line. Give the strip a single twist and tape the far ends of the twisted strip together to obtain a **Moebius strip**. Use a pencil or short stick to represent a unit normal $\mathbf{n}$ to the surface. Start at the reference point on the line you drew and move the normal $\mathbf{n}$ once around the Moebius strip. Note that $\mathbf{n}$ points in the opposite direction when it returns to the reference point. This should convince you that there is no way to set up a continuous unit normal vector field on the Moebius strip.

8.7 Stokes' Theorem and Applications

Stokes' theorem in the xy–plane was established in Sec. 8.4. Now we extend the theorem to 3–space and give several applications.

Stokes' Theorem

The story begins with Green's theorem in the xy–plane,

$$\iint_D \left(\frac{\partial Q}{\partial x} - \frac{\partial P}{\partial y} \right) dA = \oint_C P dx + Q dy,$$

where $P(x,y)$ and $Q(x,y)$ have continuous first partials in a domain D and C is a simple closed path around D traversed in the positive direction. As we learned in Sec. 8.4, Stokes' theorem in the plane is a vector reformulation of Green's theorem. It is helpful to review that reformulation. Let $\mathbf{F}(x,y) = P(x,y)\,\mathbf{i} + Q(x,y)\,\mathbf{j}$. Then

$$\operatorname{curl} \mathbf{F} = \left(\frac{\partial Q}{\partial x} - \frac{\partial P}{\partial y} \right) \mathbf{k}, \qquad \operatorname{curl} \mathbf{F} \cdot \mathbf{k} = \frac{\partial Q}{\partial x} - \frac{\partial P}{\partial y},$$

$$\mathbf{F} \cdot d\mathbf{r} = P dx + Q dy.$$

So Green's theorem can be expressed as

$$\iint_D \operatorname{curl} \mathbf{F} \cdot \mathbf{k} \, dA = \oint_C \mathbf{F} \cdot d\mathbf{r}.$$

This is Stokes' theorem in the xy–plane. See Fig. 1.

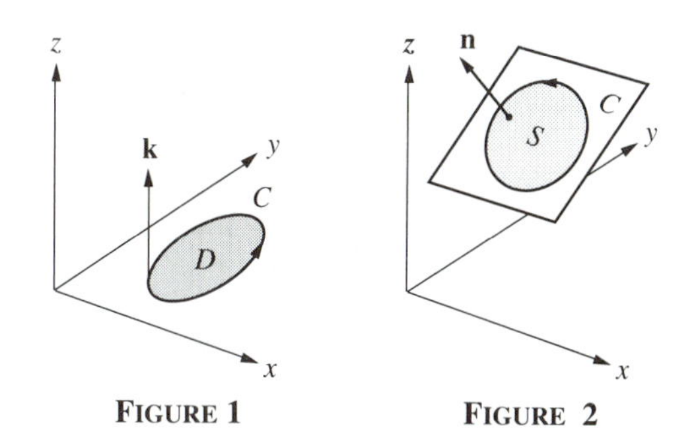

FIGURE 1 FIGURE 2

Stokes' theorem holds in the slightly modified form

$$\iint_S \operatorname{curl} \mathbf{F} \cdot \mathbf{n} dA = \oint_C \mathbf{F} \cdot d\mathbf{r}$$

for a 3–d vector field **F** and a flat oriented surface S with orientation **n**, as in Fig. 2. The bounding curve C in Fig. 2 is **positively directed** relative to S (or **n**). The intuitive idea is that if you walk around C with your head in the direction of **n**, while keeping S to your left, then you are walking in the positive direction along C. For example, in Fig. 1, D is oriented by the normal vector $\mathbf{n} = \mathbf{k}$ and the curve C is positively directed relative to D (or **k**).

With this preparation, we are ready to state the general version of Stokes' theorem. It pertains to an oriented surface S in 3–space with a positively directed boundary curve C defined as before. See Fig. 3.

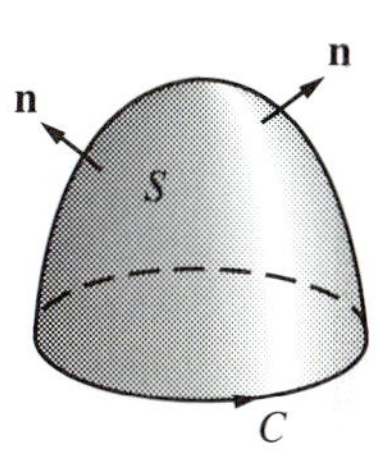

FIGURE 3

Theorem 1 *Stokes' Theorem*
Let S be a piecewise smooth surface with orientation **n** bounded by a positively directed simple closed path C. Let **F** be a vector field with components having continuous first partials on S. Then

$$\iint_S \operatorname{curl} \mathbf{F} \cdot \mathbf{n} dS = \oint_C \mathbf{F} \cdot d\mathbf{r}.$$

Stokes' theorem is valid when the boundary curve C of the surface S consists of more than one simple closed path. For example, Stokes' theorem applies to the surface S in Fig. 4. The positively directed boundary of S is $C = C_1 + C_2$. Note that C_1 and C_2 are positively directed relative to S, as described earlier.

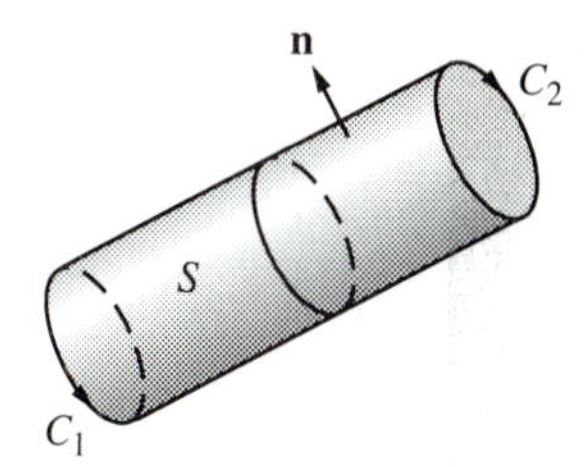

FIGURE 4

The proof of Stokes' theorem in full generality is beyond the scope of a first course in calculus. Some indications of the proof are given at the end of the section. In the meanwhile, we illustrate how the theorem is used.

Applications

Stokes' theorem is used much like the divergence theorem. Its most important applications establish general principles and characteristics of physical and

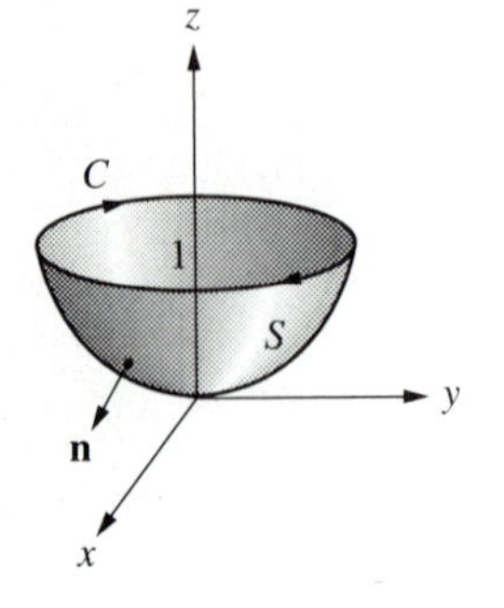

FIGURE 5

mathematical systems. The theorem also simplifies certain practical computations. We begin with the practical use of Stokes' theorem.

EXAMPLE 1. Let S be the hemisphere in Fig. 5 given by $x^2 + y^2 + (z-1)^2 = 1$ for $z \leq 1$, oriented by the outward unit normal $\mathbf{n}$ to S. Evaluate $\iint_S \operatorname{curl} \mathbf{F} \cdot \mathbf{n}\, dS$ where $\mathbf{F} = z^2\mathbf{i} + 3x\mathbf{j} - 2y^3\mathbf{k}$.

Solution. You are asked in the problems to evaluate the surface integral directly. Here we use Stokes' theorem,

$$\iint_S \operatorname{curl} \mathbf{F} \cdot \mathbf{n}\, dS = \oint_C \mathbf{F} \cdot d\mathbf{r},$$

where C is the unit circle $x^2 + y^2 = 1$, $z = 1$ directed positively with respect to S, as shown in Fig. 5. Viewed from above, C is traversed in the clockwise direction. A convenient parameterization for C is $x = \cos\theta$, $y = \sin\theta$, $z = 1$ or

$$\mathbf{r} = \cos\theta\mathbf{i} + \sin\theta\mathbf{j} + \mathbf{k}\,,$$

where θ decreases from 2π to 0. Then

$$\frac{d\mathbf{r}}{d\theta} = -\sin\theta\mathbf{i} + \cos\theta\mathbf{j}, \qquad \mathbf{F} = \mathbf{i} + 3\cos\theta\mathbf{j} - 2\sin^3\theta\mathbf{k},$$

and

$$\iint_S \operatorname{curl} \mathbf{F} \cdot \mathbf{n}\, dS = \oint_C \mathbf{F} \cdot d\mathbf{r} = \int_{2\pi}^{0} \mathbf{F} \cdot \frac{d\mathbf{r}}{d\theta}\, d\theta$$

$$= \int_{2\pi}^{0} (-\sin\theta + 3\cos^2\theta)\, d\theta = -3\pi,$$

after a short calculation. □

EXAMPLE 2. Let C be the ellipse obtained by intersecting the plane $6x + 4y + 3z = 12$ and the cylinder $x^2 + y^2 = 1$. Find the work done by the force field $\mathbf{F} = 3z\mathbf{i} + 5x\mathbf{j} - 2y\mathbf{k}$ as it moves an object once counterclockwise around C as viewed from above.

Solution 1. See Fig. 6 in Sec. 8.5 for a sketch of the situation. The work done is $W = \oint_C \mathbf{F} \cdot d\mathbf{r}$. We shall find W with the aid of Stokes' theorem. Let S be the flat elliptical surface enclosed by C. The upward normal to S with z–component 1 is $\mathbf{N} = \frac{1}{3}(6\mathbf{i} + 4\mathbf{j} + 3\mathbf{k})$. The corresponding upward unit normal is $\mathbf{n} = \mathbf{N}/\|\mathbf{N}\|$. The counterclockwise direction on C is the positive direction relative to $\mathbf{n}$. By Stokes' theorem and the general formula for evaluating flux integrals,

$$W = \oint_C \mathbf{F} \cdot d\mathbf{r} = \iint_S \operatorname{curl} \mathbf{F} \cdot \mathbf{n}\, dS = \iint_D \operatorname{curl} \mathbf{F} \cdot \mathbf{N}\, dA,$$

where D is the disk $x^2 + y^2 \leq 1$. Now,

$$\operatorname{curl}\mathbf{F} = \begin{vmatrix} \mathbf{i} & \mathbf{j} & \mathbf{k} \\ \dfrac{\partial}{\partial x} & \dfrac{\partial}{\partial y} & \dfrac{\partial}{\partial z} \\ 3z & 5x & -2y \end{vmatrix} = -2\mathbf{i} + 3\mathbf{j} + 5\mathbf{k},$$

$$\operatorname{curl}\mathbf{F} \cdot \mathbf{N} = \frac{1}{3}(-12 + 12 + 15) = 5.$$

Since the area of D is π,

$$W = \iint_D 5\,dA = 5\pi.$$

Solution 2. Let's evaluate $W = \oint_C \mathbf{F} \cdot d\mathbf{r}$ directly. (We are going to have to work harder.) Now we need a parametric representation for C. Since C is the intersection of the plane and the cylinder in Ex. 2, equations for C are

$$z = 4 - 2x - \frac{4}{3}y, \qquad x^2 + y^2 = 1.$$

In cylindrical coordinates, C is given by

$$x = \cos\theta, \quad y = \sin\theta, \quad z = 4 - 2\cos\theta - \frac{4}{3}\sin\theta.$$

A vector equation for C is

$$\mathbf{r} = \cos\theta\mathbf{i} + \sin\theta\mathbf{j} + \left(4 - 2\cos\theta - \frac{4}{3}\sin\theta\right)\mathbf{k}, \qquad 0 \leq \theta \leq 2\pi.$$

Then

$$\frac{d\mathbf{r}}{d\theta} = -\sin\theta\mathbf{i} + \cos\theta\mathbf{j} + \left(2\sin\theta - \frac{4}{3}\cos\theta\right)\mathbf{k},$$

$$\mathbf{F} = 3\left(4 - 2\cos\theta - \frac{4}{3}\sin\theta\right)\mathbf{i} + 5\cos\theta\mathbf{j} - 2\sin\theta\mathbf{k} \text{ on } C.$$

Finally, leaving out several steps,

$$\begin{aligned} W &= \oint_C \mathbf{F} \cdot d\mathbf{r} = \int_0^{2\pi} \mathbf{F} \cdot \frac{d\mathbf{r}}{d\theta}\,d\theta \\ &= \int_0^{2\pi} \left(5\cos^2\theta + \frac{26}{3}\sin\theta\cos\theta - 12\sin\theta\right) d\theta = 5\pi. \end{aligned}$$

Clearly, the first solution using Stokes' theorem is more efficient. □

Circulation of a Vector Field

In Sec. 8.4 we used Stokes' theorem in the plane to give two physical interpretations of the curl of a 2–d vector field. One interpretation involved work and the other concerned the circulation of a fluid. Both interpretations extend to 3–d fields. We develop the 3–d interpretation for circulation here. The work interpretation is left for the problems.

Let C be a directed closed path in space and $\mathbf{F}$ a continuous vector field on C. Then $\oint_C \mathbf{F} \cdot d\mathbf{r}$ is called the **circulation of F around** C. The terminology comes from fluid flow problems where $\mathbf{F} = \mathbf{v}$ or $\mathbf{F} = \sigma\mathbf{v}$. We discussed circulation for a 2–d flow in Sec. 8.4. We shall apply the reasoning used there, but in a 3–d context and to a general vector field, to obtain more insight into the curl. In the discussion that follows, $\mathbf{F}$ is a vector field whose components have continuous partial derivatives.

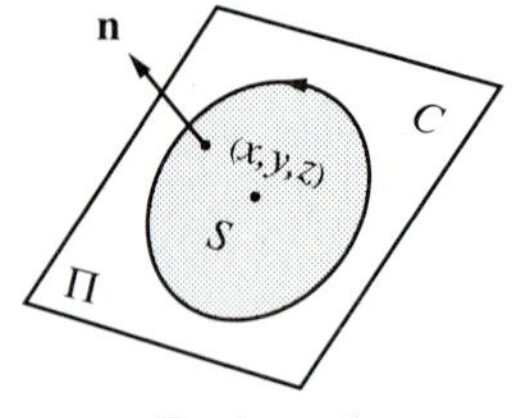

FIGURE 6

Figure 6 will serve as a guide. In the figure, (x, y, z) is a fixed point in space and Π is a plane through (x, y, z) with unit normal $\mathbf{n}$. A simple closed path C encloses a plane surface S with area ΔA. We plan to let C shrink to (x, y, z) in the plane Π in the sense that C is ultimately inside any circle in the plane Π with center (x, y, z). By Stokes' theorem and the mean value theorem for integrals,

$$\oint_C \mathbf{F} \cdot d\mathbf{r} = \iint_S \operatorname{curl} \mathbf{F} \cdot \mathbf{n} dA = \operatorname{curl} \mathbf{F}(\bar{x}, \bar{y}, \bar{z}) \cdot \mathbf{n} \Delta A$$

for some $(\bar{x}, \bar{y}, \bar{z})$ in S. Thus,

$$\operatorname{curl} \mathbf{F}(\bar{x}, \bar{y}, \bar{z}) \cdot \mathbf{n} = \frac{1}{\Delta A} \oint_C \mathbf{F} \cdot d\mathbf{r}.$$

Now let C shrink to (x, y, z). Then $(\bar{x}, \bar{y}, \bar{z}) \to (x, y, z)$, $\operatorname{curl} \mathbf{F}(\bar{x}, \bar{y}, \bar{z}) \to \operatorname{curl} \mathbf{F}(x, y, z)$ and, hence,

$$\operatorname{curl} \mathbf{F}(x, y, z) \cdot \mathbf{n} = \lim_{\Delta A \to 0} \frac{1}{\Delta A} \oint_C \mathbf{F} \cdot d\mathbf{r}.$$

In words,

$\operatorname{curl} \mathbf{F} \cdot \mathbf{n}$ is the limiting value of the circulation per unit area about (x, y, z) of the vector field $\mathbf{F}$ about a curve C in the plane Π with unit normal $\mathbf{n}$.

Consequently, for a surface S in Fig. 6 with very small area ΔA,

$$\oint_C \mathbf{F} \cdot d\mathbf{r} \approx \Delta A \operatorname{curl} \mathbf{F}(x, y, z) \cdot \mathbf{n}.$$

Now imagine that $\mathbf{n}$ varies in direction so the plane Π through (x, y, z) in Fig. 6 changes its orientation in space. Since $\operatorname{curl} \mathbf{F}(x, y, z) \cdot \mathbf{n}$ assumes its maximum value when $\mathbf{n}$ is in the same direction as $\operatorname{curl} \mathbf{F}$, we conclude that (to within the given approximation) the circulation of the vector field $\mathbf{F}$ at (x, y, z) is maximum in the plane through (x, y, z) with normal in the direction of $\operatorname{curl} \mathbf{F}$. If $\mathbf{F} = \sigma \mathbf{v}$ is the mass flow vector of a steady fluid, this means that the fluid circulates about (x, y, z) most rapidly in the plane with normal vector $\operatorname{curl} \mathbf{F}$.

Ampère's Law

The basic equations of electromagnetism are called Maxwell's equations. For steady (time independent) currents, Maxwell's fourth equation,

$$\operatorname{curl} \mathbf{B} = \mu_0 \mathbf{J},$$

relates the magnetic field strength $\mathbf{B}$ of a magnetic field and the *current density* $\mathbf{J}$. The positive constant μ_0 is the *permeability of free space.* The current density $\mathbf{J}$ is analogous to the mass flow vector $\mathbf{F} = \sigma \mathbf{v}$ of a steady fluid flow. In the case of a fluid, $\iint_S \mathbf{F} \cdot \mathbf{n} \, dS$ is the mass/sec crossing S in the positive direction, while

$$I = \iint_S \mathbf{J} \cdot \mathbf{n} \, dS$$

is the current (charge/sec) crossing S. Since $\operatorname{curl} \mathbf{B} = \mu_0 \mathbf{J}$, the current and magnetic field strength are related by

$$\iint_S \operatorname{curl} \mathbf{B} \cdot \mathbf{n} \, dS = \mu_0 I.$$

We shall use Stokes' theorem to relate the current and circulation of the magnetic field. Let C be a simple closed path in a steady electromagnetic field and let S be any piecewise smooth oriented surface that has C for its positively directed boundary. By Stokes' theorem,

$$\iint_S \operatorname{curl} \mathbf{B} \cdot \mathbf{n} \, dS = \oint_C \mathbf{B} \cdot d\mathbf{r}.$$

Therefore,

$$\oint_C \mathbf{B} \cdot d\mathbf{r} = \mu_0 I,$$

which is *Ampère's law.* In words, the circulation of the magnetic field around C is μ_0 times the current crossing any surface bounded by C.

Conservative Vector Fields

Remember that a vector field $\mathbf{F}(\mathbf{r})$ is conservative in a domain Ω if $\mathbf{F} = \nabla f$ for some scalar function f with continuous first partials in Ω. In Th. 3 of Sec. 8.3 we learned the following important test for conservative fields.

Let $\mathbf{F}(\mathbf{r})$ be a vector field defined in a simply connected domain.

Assume that the components of $\mathbf{F}$ have continuous first partials. Then

$$\mathbf{F} = \nabla f \text{ for some } f \quad \Leftrightarrow \quad \operatorname{curl} \mathbf{F} = \mathbf{0}.$$

The emphasis in Sec. 8.3 was on 2-d fields. We only proved Th. 3 for a 2-d field defined on a rectangle. Stokes' theorem enables us to establish the general result. Since the curl of a gradient is zero, the forward implication above is elementary and does not require Ω to be simply connected. The reverse implication,

$$\operatorname{curl} \mathbf{F} = \mathbf{0} \text{ in } \Omega \quad \Rightarrow \quad \mathbf{F} \text{ is conservative in } \Omega,$$

is more difficult. We outline a proof based on Stokes' theorem. The argument will help clarify why the domain Ω needs to be simply connected. The key idea of the proof is easy. We learned in Sec. 8.3 that $\mathbf{F}$ is conservative in Ω if and only if $\oint_C \mathbf{F} \cdot d\mathbf{r} = 0$ for every closed path C in Ω. Let C be a simple closed path in Ω. Then, as illustrated in Fig. 7, there is a smooth oriented surface S that lies in Ω and has C as for its boundary curve. (In fact, there are a great many such surfaces S.) By Stokes' theorem,

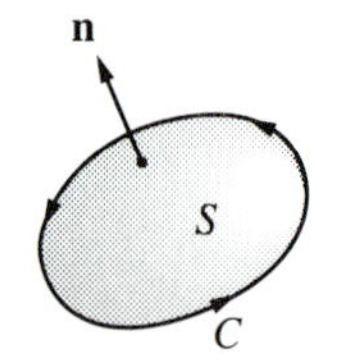

FIGURE 7

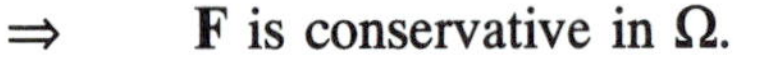

$$\operatorname{curl} \mathbf{F} = \mathbf{0} \text{ in } \Omega \quad \Rightarrow \quad \oint_C \mathbf{F} \cdot d\mathbf{r} = \iint_S \operatorname{curl} \mathbf{F} \cdot \mathbf{n} \, dS = 0,$$

$$\Rightarrow \quad \mathbf{F} \text{ is conservative in } \Omega.$$

Although it should seem plausible that C is the boundary of a smooth oriented surface S that lies in Ω, how can we be sure? To answer this question, recall that a connected set is simply connected if every simple closed path in Ω can be deformed to a point in Ω without leaving Ω in the process. With some effort it can be shown that the curves in the deformation process form such a surface S, which enables us to use Stokes' theorem. A final point needs to be clarified. The reasoning just outlined, based on Stokes' theorem, gives $\oint_C \mathbf{F} \cdot d\mathbf{r} = 0$ for every simple closed path in Ω. This implies, by an argument that we omit, that $\oint_C \mathbf{F} \cdot d\mathbf{r} = 0$ for every closed path in Ω. We used this fact implicitly when we claimed earlier that $\mathbf{F}$ was conservative.

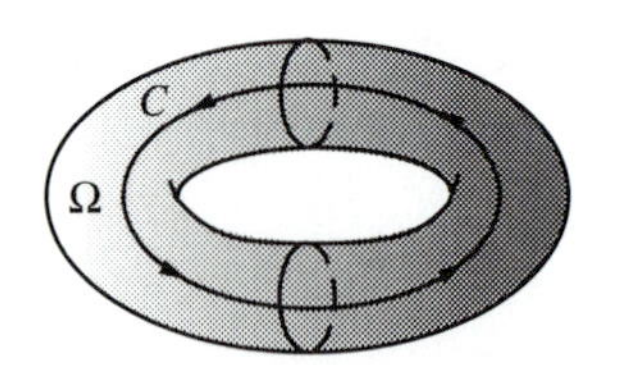

FIGURE 8

Figure 8 shows why it is essential that Ω be simply connected in order to obtain the surface S. The torus Ω in Fig. 8 is not simply connected. Clearly, the curve C shown cannot be deformed to a point without leaving Ω, and C is not the boundary curve of a surface S lying in Ω.

Sketch of a Proof of Stokes' Theorem*

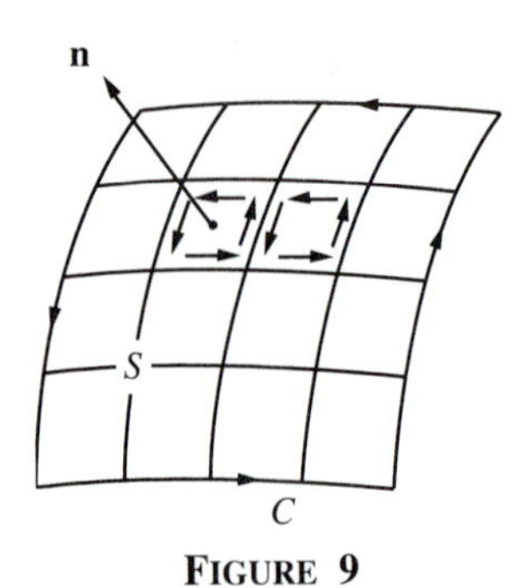

FIGURE 9

In the sketch of the proof of Stokes' theorem that follows, assume that S, C, and $\mathbf{F} = P\mathbf{i} + Q\mathbf{j} + R\mathbf{k}$ satisfy the hypotheses in Th. 1. As you will see, a key step in the proof reduces Stokes' theorem in space to Stokes' theorem in the plane.

Figure 9 shows a surface S partitioned into nonoverlapping surface elements

S_i, $i = 1, 2, \ldots, m$. Let C_i be the positively directed boundary of S_i relative to the unit normal that orients S. Suppose that Stokes' theorem holds for each S_i:

$$\iint_{S_i} \operatorname{curl} \mathbf{F} \cdot \mathbf{n}\, dS = \oint_{C_i} \mathbf{F} \cdot d\mathbf{r}, \qquad i = 1, 2, \ldots, m.$$

Add these equations, taking account of the pairwise cancellation of line integrals along shared boundary curves, to obtain

$$\iint_S \operatorname{curl} \mathbf{F} \cdot \mathbf{n}\, dS = \oint_C \mathbf{F} \cdot d\mathbf{r}.$$

Thus, Stokes' theorem holds for any oriented surface S that can be decomposed into a finite number of nonoverlapping surface elements for which the theorem is known to be valid.

Next we prove Stokes' theorem for a simple surface (defined later). Then Stokes' theorem holds for any surface S that can be decomposed into a finite number of simple surfaces as in Fig. 9. This line of reasoning establishes Stokes' theorem for any oriented surface you are likely to meet in practice.

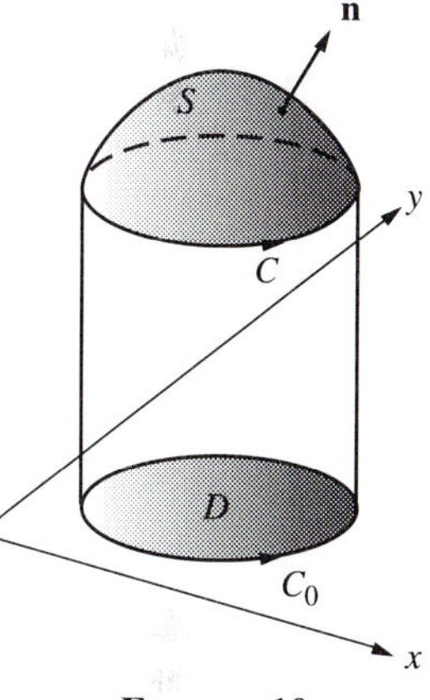

FIGURE 10

A surface S is **z–simple** if it is the graph of a function $z = f(x, y)$ with continuous first partials defined on a domain D bounded by a simple closed path C_0, as in Fig. 10. Likewise, a surface S is **x–simple** if it is the graph of $x = g(y, z)$ and **y–simple** if it is the graph of $y = h(x, z)$. A surface is **simple** if it is x–simple, y–simple, and z–simple. We claim that

$$S \ \ x\text{–simple} \ \Rightarrow \ \iint_S \operatorname{curl} R\mathbf{k} \cdot \mathbf{n}\, dS = \oint_C R\mathbf{k} \cdot d\mathbf{r},$$

$$S \ \ y\text{–simple} \ \Rightarrow \ \iint_S \operatorname{curl} Q\mathbf{j} \cdot \mathbf{n}\, dS = \oint_C Q\mathbf{j} \cdot d\mathbf{r},$$

$$S \ \ z\text{–simple} \ \Rightarrow \ \iint_S \operatorname{curl} P\mathbf{i} \cdot \mathbf{n}\, dS = \oint_C P\mathbf{i} \cdot d\mathbf{r},$$

Notice that the equalities on the right are the special cases of Stokes' theorem with $\mathbf{F} = P\mathbf{i}$, $Q\mathbf{i}$, and $R\mathbf{k}$. If S is simple, then all three special cases hold and by addition

$$\iint_S \operatorname{curl} \mathbf{F} \cdot \mathbf{n}\, dS = \oint_C \mathbf{F} \cdot d\mathbf{r}$$

for $\mathbf{F} = P\mathbf{i} + Q\mathbf{j} + R\mathbf{k}$. Thus, Stokes' theorem holds for any simple surface.

The proofs of the three special cases of Stokes' theorem for x–simple, y–simple, and z–simple surfaces all use the same reasoning. We give the proof for a z–simple surface. So let S be the graph of a function $z = z(x, y)$ with continuous first partials in a domain D bounded by a simple closed path C_0, as in Fig. 10. The vector $\mathbf{N} = \langle -z_x, -z_y, 1 \rangle$ is an upward normal to S. Orient S by the upward unit normal $\mathbf{n} = \mathbf{N}/\|\mathbf{N}\|$. In Fig. 10, the boundary C of S is positively directed relative to $\mathbf{n}$. Note that C_0, the projection of C onto the xy–plane, is positively directed relative to D. We must prove that

$$\iint_S \operatorname{curl} P\mathbf{i} \cdot \mathbf{n}\, dS = \oint_C P\mathbf{i} \cdot d\mathbf{r}.$$

To this end, let $\mathbf{F}_0$ be the vector field in the xy–plane defined by

$$\mathbf{F}_0 = P_0(x, y)\mathbf{i} \quad \text{with} \quad P_0(x, y) = P(x, y, z(x, y)).$$

We shall show that

(a)
$$\iint_S \operatorname{curl} P\mathbf{i} \cdot \mathbf{n}\, dS = \iint_D \operatorname{curl} \mathbf{F}_0 \cdot \mathbf{k}\, dA,$$

(b)
$$\oint_C P\mathbf{i} \cdot d\mathbf{r} = \oint_{C_0} \mathbf{F}_0 \cdot d\mathbf{r}.$$

Since the right members of (a) and (b) are equal by Stokes' theorem in the xy–plane,

$$\iint_S \operatorname{curl} P\mathbf{i} \cdot \mathbf{n}\, dS = \oint_C P\mathbf{i} \cdot d\mathbf{r},$$

as claimed.

It remains to establish (a) and (b) to complete our sketch of a proof of Stokes' theorem. As the point $(x, y, 0)$ traverses C_0 once in the positive direction in Fig. 10, the point $(x, y, z(x, y))$ traverses C once in the positive direction. Since $\mathbf{F}_0 = P_0(x, y)\mathbf{i} = P(x, y, z(x, y))\mathbf{i}$, and $dz = 0$ on C because it lies in a horizontal plane, it is easy to verify (b); we leave the details to the problems. We shall establish (a), which is a little harder. First, note that

$$\operatorname{curl} P\mathbf{i} = \frac{\partial P}{\partial z}\mathbf{j} - \frac{\partial P}{\partial y}\mathbf{k}.$$

Let $\mathbf{F} = \operatorname{curl} P\mathbf{i}$ and $\mathbf{N} = \langle -z_x, -z_y, 1 \rangle$ in the general formula $\iint_S \mathbf{F} \cdot \mathbf{n}\, dS = \iint_D \mathbf{F} \cdot \mathbf{N}\, dA$ to obtain

$$\iint_S \operatorname{curl} P\mathbf{i} \cdot \mathbf{n}\, dS = \iint_D \operatorname{curl} P\mathbf{i} \cdot \mathbf{N}\, dA = \iint_D -\left(\frac{\partial P}{\partial y} + \frac{\partial P}{\partial z}\frac{\partial z}{\partial y}\right) dA.$$

Since $\operatorname{curl}\mathbf{F}_0 = -(\partial P_0/\partial y)\mathbf{k}$ and $P_0 = P(x, y, z(x, y))$,

$$\operatorname{curl}\mathbf{F}_0 \cdot \mathbf{k} = -\frac{\partial P_0}{\partial y} = -\left(\frac{\partial P}{\partial y} + \frac{\partial P}{\partial z}\frac{\partial z}{\partial y}\right)$$

by the chain rule. Consequently,

$$\iint_S \operatorname{curl} P\mathbf{i} \cdot \mathbf{n}\,dS = \iint_D \operatorname{curl}\mathbf{F}_0 \cdot \mathbf{k}\,dA,$$

which is (a).

As we have already observed, this discussion establishes Stokes' theorem for simple surfaces and, by the earlier decomposition argument, for surfaces composed of a finite number of simple surface elements. The general version of Stokes' theorem stated in Th. 1 can be proved by similar means.

PROBLEMS

1. Evaluate $\iint_S \operatorname{curl}\mathbf{F} \cdot \mathbf{n}\,dS$ directly for the vector field and surface in Ex. 1. This calculation, together with the solution to Ex. 1, verifies Stokes' theorem in a particular case.

In Probs. 2 and 3 verify Stokes' theorem for the given vector field and given oriented surface.

2. $\mathbf{F} = \langle y, z, x \rangle$ and S is the lower hemisphere of the sphere $x^2 + y^2 + z^2 = 4$ oriented with its upper unit normal.

3. $\mathbf{F} = \langle x, z, -y \rangle$ and S is the part of the cone $z^2 = x^2 + y^2$ between the planes $z = 1$ and $z = 25$ and is oriented by its downward unit normal.

In Probs. 4–7 use Stokes' theorem to find the flux of $\operatorname{curl}\mathbf{F}$ across the given oriented surface S.

4. $\mathbf{F} = \langle z^2, -3x, 2y^2 \rangle$ and S is the hemisphere $x^2 + y^2 + z^2 = 1$, $z \le 0$ oriented by its upward unit normal.

5. $\mathbf{F} = yz\mathbf{i} + xz\mathbf{j} + xy\mathbf{k}$ and S consists of the five faces of the solid unit cube $0 \le x, y, z \le 1$ that do not lie entirely in the xy–plane. Orient S by the outer unit normals on the cube.

6. $\mathbf{F} = z\mathbf{i} - x\mathbf{j} + y\mathbf{k}$ and S is given by $z = x^2 + y^2$ for $z \le 4$ and is oriented by its downward unit normal.

7. $\mathbf{F} = \langle x^2 - y^2, y^2 - z^2, x^2 - z^2 \rangle$ and S is the triangle cut from the first octant by the plane $2x + y + 3z = 12$. Orient S with its upward unit normal.

In Probs. 8–11 use Stokes' theorem to find the circulation of the given vector field $\mathbf{F}$ about the given curve C.

8. $\mathbf{F} = y^2\mathbf{i} + x\mathbf{j} + z^2\mathbf{k}$ and C is the circle $x^2 + y^2 = 1$ and $z = 2$ traversed clockwise when viewed from above.

9. $\mathbf{F} = < -3y, x, 2z >$ and C is the ellipse in which the cylinder $x^2 + y^2 = 4$ meets the plane $z = y$.

10. $\mathbf{F} = < ye^{xy} \sin z,\ xe^{xy} \sin z,\ e^{xy} \cos z >$ and C is the parametric curve with $x = \sin \pi t$, $y = \cos \pi t$, $z = t^2 - 2t$ for $0 \leq t \leq 2$.

11. $\mathbf{F} = < y + y^2 z^3, z + 2xyz^3, x + 3xy^2 z^2 >$ and C is the triangle with vertices $(1,0,0)$, $(0,2,0)$, $(0,0,3)$ traversed in the given order.

12. Let S be the unit sphere $x^2 + y^2 + z^2 = 1$. Apply Stokes' theorem to the upper and lower hemispheres with $z \geq 0$ and $z \leq 0$ to deduce that $\int\int_S \operatorname{curl} \mathbf{F} \cdot \mathbf{n} dS = 0$ for any vector field with continuously differentiable components on S.

13. Repeat the last problem for the ellipse $x^2/a^2 + y^2/b^2 + z^2/c^2 = 1$.

14. Use Stokes' theorem to show that $\int\int_S \operatorname{curl} \mathbf{F} \cdot \mathbf{n} dS = 0$ for any vector field $\mathbf{F}$ and closed surface S that satisfy the hypotheses in Stokes' theorem.

15. Let S_1 and S_2 be surfaces and $\mathbf{F}$ a vector field as in Stokes' theorem. Assume S_1 and S_2 have the same positively directed boundary curve C. Show that the flux of $\operatorname{curl} \mathbf{F}$ across S_1 and the flux of $\operatorname{curl} \mathbf{F}$ across S_2 are equal.

16. Let S and $\mathbf{F}$ be as in Prob. 4. Use Prob. 15 with $S_1 = S$ and a well-chosen surface S_2 to evaluate the flux of $\operatorname{curl} \mathbf{F}$ across S by evaluating the flux of $\operatorname{curl} \mathbf{F}$ across S_2 directly.

17. Let S be the part of the paraboloid $z = 1 - x^2 - y^2$ with $z \geq 0$ oriented by its upward unit normal and $\mathbf{F} = < \cos yz,\ \sin xz,\ e^{xy} >$. Use Prob. 15 with $S_1 = S$ and a well-chosen surface S_2 to evaluate the flux of $\operatorname{curl} \mathbf{F}$ across S by evaluating the flux of $\operatorname{curl} \mathbf{F}$ across S_2 directly.

18. Maxwell's third equation, $\operatorname{curl} \mathbf{E} = -\,\partial \mathbf{B}/\partial t$, relates the electric field $\mathbf{E}$ and the magnetic field $\mathbf{B}$ in a region of space. The line integral $\oint_C \mathbf{E} \cdot d\mathbf{r}$ gives the voltage drop around a closed directed path C. Let S be any piecewise smooth oriented surface that has C for its positively directed boundary curve. Show that the voltage drop around C is equal to the negative of the time rate of change of the flux of the magnetic field through S, a result called *Faraday's law.*

19. Review the discussion in Sec. 8.4 which interprets $\operatorname{curl} \mathbf{F} \cdot \mathbf{k}$ as work per unit area done by a force field $\mathbf{F}$ in the xy–plane. Then develop the corresponding interpretation for $\operatorname{curl} \mathbf{F} \cdot \mathbf{n}$ where $\mathbf{F}$ is a 3–d force field and $\mathbf{n}$ is a unit normal to a plane Π in space.

The next three problems extend the discussion of buoyant forces begun in Probs. 23 and 24 in Sec. 8.6. The notation introduced and assumptions made in those problems remain in force.

20. Under the hypotheses in the divergence theorem, prove that

$$\iint_S \mathbf{n} \times \mathbf{F} dS = \iiint_\Omega \operatorname{curl} \mathbf{F} dV.$$

Hint. Apply the divergence theorem to the vector field $\mathbf{F} \times \mathbf{a}$ where $\mathbf{a}$ is a constant vector. Then choose $\mathbf{a} = \mathbf{i}, \mathbf{j}, \mathbf{k}$ to confirm that the preceding vector–valued integrals have the same components.

21. The submerged body from Probs. 23 and 24 of Sec. 8.6 occupies the region of space Ω. If fluid were to occupy the region Ω, then the center of mass of that fluid region would be

$$\bar{\mathbf{r}} = \frac{1}{M} \iiint_{\Omega} \sigma \mathbf{r}\, dV,$$

where M is the mass of region Ω when filled with fluid, σ is the fluid density, and $\mathbf{r} = \langle x, y, z \rangle$. Use a Riemann sum and limit passage argument to show that the total torque τ about $\bar{\mathbf{r}}$ of the buoyant forces acting on the surface S is

$$\tau = \iint_{S} (\mathbf{r} - \bar{\mathbf{r}}) \times (-p\mathbf{n})\, dS.$$

22. Use Probs. 20, 21, and the hydrostatic pressure law $\nabla p = \sigma \mathbf{g}$ to show that $\tau = 0$. (It follows that the buoyant force $\mathbf{B}$ in Sec. 8.6 acts along the vertical line through the center of mass of the region Ω when it is filled with fluid. For most fluids and most practical situations, the fluid density can be assumed constant. Then the center of mass $\bar{\mathbf{r}}$ is also the centroid of the submerged solid. For example, the buoyant force on a submarine acts vertically upward through the centroid of the submarine.)

Chapter Highlights

Vector analysis is devoted largely to the interplay between vector and scalar fields, which are other names for vector and scalar functions $\mathbf{F}(\mathbf{r})$ and $f(\mathbf{r})$ defined on regions in 2–space or 3–space. For vector functions, we use the notation $\mathbf{F}(\mathbf{r}) = P\mathbf{i} + Q\mathbf{j}$ in 2–space and $\mathbf{F}(\mathbf{r}) = P\mathbf{i} + Q\mathbf{j} + R\mathbf{k}$ in 3–space, where P, Q and R are functions of position.

The gradient of a scalar field f is the vector field $\mathbf{F} = \nabla f$. A vector field $\mathbf{F}$ is conservative if it is the gradient of a scalar field f, in which case f is called a potential function for $\mathbf{F}$. Gravitational force fields are conservative.

The divergence of a vector field $\mathbf{F}$ is the scalar field defined by

$$\operatorname{div}\mathbf{F} = \frac{\partial P}{\partial x} + \frac{\partial Q}{\partial y} \qquad \text{in 2–space,}$$

$$\operatorname{div}\mathbf{F} = \frac{\partial P}{\partial x} + \frac{\partial Q}{\partial y} + \frac{\partial R}{\partial z} \qquad \text{in 3–space.}$$

In the context of fluid flow, divergence measures the rate that fluid leaves the vicinity of a point.

The curl of a vector field $\mathbf{F}$ is the vector field defined by

$$\operatorname{curl}\mathbf{F} = \left(\frac{\partial Q}{\partial x} - \frac{\partial P}{\partial y}\right)\mathbf{k} \qquad \text{in 2–space,}$$

$$\operatorname{curl}\mathbf{F} = \left(\frac{\partial R}{\partial y} - \frac{\partial Q}{\partial z}\right)\mathbf{i} + \left(\frac{\partial P}{\partial z} - \frac{\partial R}{\partial x}\right)\mathbf{j} + \left(\frac{\partial Q}{\partial x} - \frac{\partial P}{\partial y}\right)\mathbf{k} \qquad \text{in 3–space.}$$

Curl measures the rotation of a solid body and the circulation of a fluid.

Line integrals are integrals along curves. Particular line integrals represent mass, work, and other physical quantities. Three types of line integrals we have met are

$$\int_C f ds = \int_a^b f(\mathbf{r}(t)) \|\mathbf{r}'(t)\| dt,$$

$$\int_C f dx = \int_a^b f(\mathbf{r}(t)) x'(t) dt,$$

$$\int_C \mathbf{F} \cdot d\mathbf{r} = \int_a^b \mathbf{F}(\mathbf{r}(t)) \cdot \mathbf{r}'(t) dt.$$

The fundamental theorem of calculus for line integrals,

$$\int_C \nabla f \cdot d\mathbf{r} = \int_A^B \nabla f \cdot d\mathbf{r} = f(B) - f(A),$$

has many important physical applications, including conservation of energy. If $\mathbf{F}$ is conservative, so that $\mathbf{F} = \nabla f$ for some f, then $\int_C \mathbf{F} \cdot d\mathbf{r} = f(b) - f(a)$. In a simply connected domain, the following properties of a vector field $\mathbf{F}$ are equivalent:

$\mathbf{F} = \nabla f$ in D.

$\int_C \mathbf{F} \cdot d\mathbf{r}$ is independent of the path in D.

$\int_C \mathbf{F} \cdot d\mathbf{r} = 0$ for any closed path in D.

curl $\mathbf{F} = \mathbf{0}$ in D.

Green's theorem in the plane,

$$\iint_D \left(\frac{\partial Q}{\partial x} - \frac{\partial P}{\partial y} \right) dA = \oint_C P dx + Q dy$$

has many geometric and physical applications. For example, it gives areas enclosed by curves and the flux of a fluid across a curve. The vector forms of Green's theorem,

$$\iint_D \operatorname{div} \mathbf{F} dA = \oint_C \mathbf{F} \cdot \mathbf{n} ds \quad \text{and} \quad \iint_D \operatorname{curl} \mathbf{F} \cdot \mathbf{k} dA = \oint_C \mathbf{F} \cdot d\mathbf{r},$$

are called the divergence theorem and Stokes' theorem in the plane.

Surface areas and surface integrals are given by the double integrals

$$A(S) = \iint_S 1 dS = \iint_D \sec \gamma dA,$$

$$\iint_S h dS = \iint_D h \sec \gamma dA,$$

where γ is the angle between the upward normal to the surface and $\mathbf{k}$. If the surface S is given by $z = f(x,y)$ for (x,y) in D, then

$$\sec\gamma = \sqrt{1 + z_x^2 + z_y^2},$$

while if S is given by $z = f(r,\theta)$ for (r,θ) in D, then

$$\sec\gamma = \sqrt{1 + z_r^2 + \frac{1}{r^2} z_\theta^2}.$$

The geometric and physical applications of surface integrals are natural analogues of applications of line integrals.

The flux of a vector field $\mathbf{F}$ across an oriented surface S with orientation $\mathbf{n}$ is

$$\iint_S \mathbf{F} \cdot \mathbf{n}\, dS.$$

The divergence theorem,

$$\iiint_\Omega \operatorname{div} \mathbf{F}\, dV = \iint_S \mathbf{F} \cdot \mathbf{n}\, dS,$$

which expresses the flux across S as a volume integral, is one of the most important theorems in vector analysis. From a physical point of view, the divergence theorem is a conservation law. Consequences of the divergence theorem include Gauss' law for gravitational and electric fields.

Stokes' theorem,

$$\iint_S \operatorname{curl} \mathbf{F} \cdot \mathbf{n}\, dS = \oint_C \mathbf{F} \cdot d\mathbf{r},$$

is another cornerstone of vector analysis with important applications. Stokes' theorem was used in the text to show that a vector field $\mathbf{F}$ in a simply connected domain Ω is conservative in Ω if and only if curl $\mathbf{F} = \mathbf{0}$ in Ω.

The project for this chapter develops basic equations that model heat conduction in a solid. The lines of reasoning that follow typify how the methods of vector calculus and, in particular, the divergence theorem and Stokes' theorem are used to translate basic physical principles such as conservation of mass, energy, and momentum into differential equations that describe the evolution of many physical systems. For example, instead of heat conduction, similar projects could deal with the transport of pollutants, aerodynamics, acoustics, and general questions of fluid flow and gas dynamics. Other applications of the divergence theorem and Stokes' theorem lead to representations of the solutions to many physical problems as integrals. An important numerical method, the boundary element method, provides effective means for evaluating such integrals.

Chapter Project: Heat Conduction

Heat is a form of energy. Heat flow is energy in transition from one part of a region to another as a result of temperature differences in the region. The goal of this project is to describe heat flow in a solid by means of a fundamental equation, the heat equation, that expresses how temperature varies in a solid. For example, the solid may be a pie baking in an oven, the outside of a building as it heats and cools throughout the day, or a metal ingot as it heats or cools.

For simplicity, assume that the solid, call it Ω, is a rigid body at rest so that the only energy present is thermal energy. Assume further that the density (mass/volume) of the solid, say $\sigma(x,y,z)$, depends only on position and not on the time t or the temperature $u = u(x,y,z,t)$ in the solid. Finally, assume that both the density and temperature are continuous functions. All of these assumptions are valid for many practical problems in heat conduction in which temperature variations are not too extreme. The heat content of such a solid Ω is described by its *specific internal energy* (function) $e = e(x,y,z,t)$. This function is an energy density with units of energy/mass.

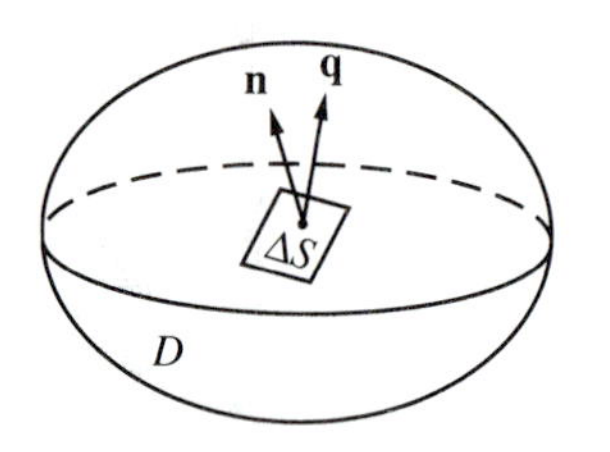

FIGURE 1

The flow of heat from one part of Ω to another is described by a **heat flow vector** $\mathbf{q} = \mathbf{q}(x,y,z,t)$ (analogous to the mass flow vector discussed in the text). See Fig. 1. Let ΔS be the area of a small planar surface with unit normal $\mathbf{n}$. Then the net heat energy per second crossing ΔS in the direction $\mathbf{n}$ is approximately $\mathbf{q} \cdot \mathbf{n}\Delta S$. This approximation is exact if $\mathbf{q}$ is constant on ΔS.

The following experimental results establish a connection between the internal energy, the heat flow vector, and the temperature. These results hold for many materials and over a wide range of temperatures.

1. The internal energy is proportional to the temperature; hence, the time rate of change of the internal energy is proportional to the rate of change of temperature.
2. Heat flows from hot to cold and the rate of flow in a particular direction is proportional to the rate of change of temperature in that direction.

Problem 1 Explain why these experimental properties are expressed concisely by the equations

$$e_t = cu_t \qquad \text{and} \qquad \mathbf{q} \cdot \mathbf{n} = -k\nabla u \cdot \mathbf{n}$$

where $\mathbf{n}$ is any unit vector and $c = c(x,y,z)$, the *specific heat*, and $k = k(x,y,z)$, the *thermal conductivity*, are positive functions determined by the solid.

Problem 2 Deduce from Prob. 1 that $\mathbf{q} = -k\nabla u$, which is Fourier's law of heat conduction.

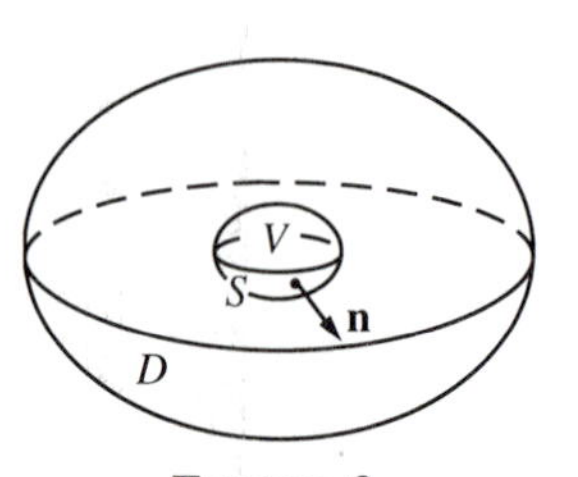

FIGURE 2

The derivation of the heat equation you are about to carry out uses "test" regions V in the solid Ω. A test region V is a reasonably shaped, closed, bounded part of the solid Ω. Often test regions are small neighborhoods of points of interest. Test regions are used to analyze how heat energy moves for one part of Ω to another. Figure 2 shows a test typical test region V with boundary S and unit *outer* normal $\mathbf{n}$.

Problem 3 If V is a test region in Ω, explain why the total heat energy in V at time t is $\iiint_V e\sigma dV$. Conclude that the rate of change of thermal energy in V is

$$\frac{d}{dt}\iiint_V e\sigma dV = \iiint_V \sigma e_t\, dV,$$

assuming that differentiation under the integral is valid. (It is if e_t and σ are continuous.)

Problem 4 Refer to the test region in Fig. 2. Use a Riemann sum and limit passage argument to explain why $\iint_S \mathbf{q}\cdot\mathbf{n}dS$ gives the net heat per second *leaving* V across its surface S.

The solid Ω may contain sources and sinks of heat energy. For example, a wire running through Ω carrying an electric current will produce joule heating in Ω. Such sources and sinks are assumed known and are described by a density function $f(x,y,z,t)$ that gives the heat energy per second per unit volume generated by the sources and sinks in Ω.

Problem 5 Show that the rate of change of heat energy in a test region V in Ω due to the sources and sinks in V is given by $\iiint_V f dV$.

Problem 6 Justify the following equation on the basis of conservation of energy. For any test region V in Ω,

$$\iiint_V \sigma e_t dV = -\iint_S \mathbf{q}\cdot\mathbf{n}dS + \iiint_V f dV.$$

Be sure to explain why there is a minus sign before the surface integral.

Problem 7 Deduce from Prob. 6 and the divergence theorem that

$$\iiint_V (\sigma e_t + \operatorname{div}\mathbf{q} - f)dV = 0$$

for any test region V in Ω.

The result in Prob. 7 holds at any time t. Next comes a very useful and rather evident mathematical fact.

Problem 8 Let $g(x,y,z,t)$ be a continuous function for (x,y,z) in Ω and for all t in some interval of interest. Show: if $\iiint_V g dV = 0$ for any test region V in Ω, then $g(x,y,z,t) = 0$ for all (x,y,z) in Ω and all t of interest. *Hint.* If $g(P_0) \neq 0$ at some point P_0 in Ω, then $g(P)$ is always positive or always negative in a small enough closed ball centered at P_0. Why?

Assume henceforth that the rate of change of specific internal energy e_t, the heat flow vector $\mathbf{q}$, and the source term f are all continuous.

Problem 9 Deduce that

$$\sigma e_t + \operatorname{div}\mathbf{q} - f = 0$$

holds at each point (x, y, z) of Ω and at each time t of interest.

The equation in Prob. 9 expresses conservation of heat energy in a convenient mathematical form. At this point, you have one equation and four unknowns, e and the three components of $\mathbf{q}$. The final key step is to express the foregoing equation in terms of the temperature.

Problem 10 Show that the temperature u in Ω satisfies the **heat equation**

$$\sigma c u_t = \operatorname{div}(k\nabla u) + f$$

at each point in Ω. Expand to obtain

$$\sigma c \frac{\partial u}{\partial t} = \frac{\partial}{\partial x}\left(k\frac{\partial u}{\partial x}\right) + \frac{\partial}{\partial y}\left(k\frac{\partial u}{\partial y}\right) + \frac{\partial}{\partial z}\left(k\frac{\partial u}{\partial z}\right) + f(x, y, z, t).$$

Problem 11 If the solid is homogeneous, so that all the thermal coefficients are constant, show that the heat equation reduces to

$$u_t = a\Delta u + \frac{1}{\sigma c} f$$

where $\Delta u = u_{xx} + u_{yy} + u_{zz}$ is the Laplacian of u and $a = k/\sigma c$ is the *thermal diffusivity* of the solid.

Very often the sources and sinks described by f are independent of the time t and the surroundings of Ω are maintained at a time-independent temperature. Then it is an experimental fact that the temperature in the solid tends to a time–independent state, say, $u(x, y, z, t) \to U(x, y, z)$ as $t \to \infty$. The limit $U(x, y, z)$ is called a *steady–state temperature distribution.*

Problem 12 Make a formal passage to the limit in the heat equation in Prob. 11 to infer that U should satisfy the *Poisson equation*

$$\Delta U = -f/k \text{ in } \Omega.$$

If there are no sources or sinks in Ω, conclude that the steady–state temperature distribution U satisfies the *Laplace equation*

$$\Delta U = 0 \text{ in } \Omega.$$

We have just scratched the surface of the important topic of heat conduction. Further work is needed to describe how heat flows back and forth between Ω and its surroundings. This leads to boundary conditions for the heat equation.

Once appropriate boundary conditions are given, it is natural to anticipate on physical grounds that the temperature $u(x,y,z,t)$ is uniquely determined by the heat equation and the boundary conditions. All of this and more is a story for another day and a course in applied mathematics.

Chapter Review Problems

1. Let $\mathbf{F} = (2y/x)\mathbf{i} - (1/x)\mathbf{j}$. Find div$\mathbf{F}$ and curl$\mathbf{F}$.
2. Find div$\mathbf{F}$ and curl $\mathbf{F}$ for the vector field $\mathbf{F} = (x/y)\mathbf{i} + (y/z)\mathbf{j} + (z/x)\mathbf{k}$.
3. (a) Show that $\mathbf{F}(x,y) = < y \cos xy,\ x \cos xy >$ is the gradient of a scalar field. (b) Evaluate $\int_C \mathbf{F} \cdot d\mathbf{r}$ where C is a path from $(0,0)$ to $(1,\pi/2)$.
4. Verify that $\mathbf{F} = (1 + 4x^4y^2)^{-1}\ (4xy\mathbf{i} + 2x^2\mathbf{j})$ is conservative and find a potential function.
5. Is $\mathbf{F} = < -x/(x^2 + y^2)^{3/2}, -y/(x^2 + y^2)^{3/2} >$ conservative? If so, find a potential function.
6. Show that $\mathbf{F}(x,y,z) = (2xy + z^2)\mathbf{i} + (x^2 + 1)\mathbf{j} + 2xz\mathbf{k}$ is conservative and find a potential function.
7. (a) Is $\mathbf{F}(x,y) = (xe^{2y} + 1)\mathbf{i} + (x^2e^{2y} - y)\mathbf{j}$ the gradient of a scalar field $f(x,y)$? (b) If it is, find $f(x,y)$.
8. Show that $\mathbf{F}(x,y,z) = (2x \cos z - 3)\mathbf{i} - (2yz - 2)\mathbf{j} - (x^2 \sin z + y^2)\mathbf{k}$ is conservative and find a potential function.
9. (a) Sketch the vector field $\mathbf{F} = x^2\mathbf{i} + y^2\mathbf{j}$ in the first quadrant. (b) Find the streamlines of the field.
10. Let C be the path in the xy–plane consisting of the line segments from $(0,0)$ to $(1,0)$, from $(1,0)$ to $(1,\pi/2)$, and from $(1,\pi/2)$ to $(0,\pi/2)$. Describe two different way to evaluate $\int_C e^x \sin y\,dx + e^x \cos y\,dy$ and carry out the evaluation both ways.
11. Evaluate the line integral $\int_C y \sec^2 x\,dx + \tan x\,dy$ along any path from $(-2,0)$ to $(4,\pi/4)$.
12. Evaluate (a) $\oint_C \mathbf{F} \cdot \mathbf{T}ds$ and (b) $\oint_C \mathbf{F} \cdot \mathbf{n}ds$ where $\mathbf{F} = y^2\mathbf{i} + x^2\mathbf{j}$ and C is the unit square with opposite corners $(0,0)$ and $(1,1)$.
13. Find $\int_C \mathbf{F} \cdot d\mathbf{r}$, where $\mathbf{F} = e^y\mathbf{i} + xe^y\mathbf{j}$ and C consists of the line segments from $(0,0)$ to $(5,0)$ and from $(5,0)$ to $(5,7)$.
14. Find $\int_C \mathbf{F} \cdot d\mathbf{r}$, where $\mathbf{F}(x,y) = < xy,\ 2x - y >$ and C is the arc of the parabola $y = x^2/2$ from $(0,0)$ to $(2,2)$.
15. Evaluate the line integral
$$\int_C \left(\frac{2}{y} + \frac{z}{x^2}\right)dx + \left(\frac{1}{z} - \frac{2x}{y^2}\right)dy - \left(\frac{1}{x} + \frac{y}{z^2}\right)dz$$
along any convenient path from $(1,1,1)$ to $(2,2,2)$ with $x > 0$, $y > 0$, and $z > 0$.
16. Evaluate $\int_C (1/y)\,dx + (1/x)\,dy$ where C is the path from $(1,2)$ to $(2,1)$ along (a) the line $x + y = 3$ and (b) the hyperbola $xy = 2$.

17. Evaluate the line integral $\int_C (y^2 - x)\,dx + 2xy\,dy$ where C is the line segment from $(3,0)$ to $(9,3)$.

18. Evaluate $\int_C y\,dx - x\,dy + xyz^2 dz$, where C is the curve given by $\mathbf{r}(t) = e^{-t}\mathbf{i} + e^t\mathbf{j} + t\mathbf{k}$ for $0 \le t \le 1$.

19. Find $\oint_C (y + 2z)\,ds$ where C is the triangle with vertices $(-1,0,0)$, $(0,1,0)$, and $(0,0,1)$.

20. Find $\int_C (1 + xy)\,ds$, where (a) C is the circular arc $x^2 + y^2 = 4$ from $(0,-2)$ to $(0,2)$ with $x \ge 0$, (b) C is the line segment from $(0,2)$ to $(0,-2)$, and (c) C is the sum of the paths in (a) and (b).

21. Let $\mathbf{F}(\mathbf{r}) = \langle e^y,\ xe^y + y^3 \rangle$ for $\mathbf{r} = \langle x, y \rangle$. (a) Find $f(\mathbf{r})$ such that $\mathbf{F} = \nabla f$. (b) Evaluate $W = \int_C \mathbf{F} \cdot d\mathbf{r}$ for C the unit circle with center $(0,0)$. (c) Evaluate $W = \int_C \mathbf{F} \cdot d\mathbf{r}$ for C the arc of the parabola $y = 1 - x^2$ from $(-1,0)$ to $(1,0)$.

22. Let $\mathbf{F}(\mathbf{r}) = \langle 1/y,\ \ln z,\ e^x \rangle$ for $\mathbf{r} = \langle x, y, z \rangle$. Evaluate $W = \int_C \mathbf{F} \cdot d\mathbf{r}$ when (a) C is the straight line from $(1,2,1)$ to $(3,6,1)$, (b) C is the arc of the circle in the xy–plane with center $(0,0)$ and radius $\sqrt{2}$ from $(-1,1)$ to $(1,1)$.

23. A force field $\mathbf{F}(x,y) = \langle 2x^2 + y,\ 2xy \rangle$ moves a particle along the parabola $y = x^2$ from $(-1,1)$ to $(1,1)$. Find the work done.

24. A force field $\mathbf{F}(\mathbf{r}) = x^3y\mathbf{i} + (x - y)\mathbf{j}$ moves a particle along the curve $\mathbf{r}(t) = \langle t, t^2 \rangle$ from $(-2,4)$ to $(1,1)$. Find the work done.

25. Find the work done when the force field $\mathbf{F}(x,y,z) = \langle 4y/x,\ 4\ln 3x + 1/y \rangle$ moves a particle once in the clockwise direction around the circle with center $(2,2)$ and radius 1.

26. A force $\mathbf{F}(\mathbf{r}) = -3\mathbf{r}/\|\mathbf{r}\|^3$ for $\mathbf{r} = \langle x,y,z \rangle \ne \langle 0,0,0 \rangle$ moves a particle from $(5,0,12)$ to $(0,3,4)$. How much work is done?

In Probs. 27–33, use Green's theorem to evaluate the given integral or find the indicated quantity.

27. $\oint_C (3y + \tanh x)\,dx + (5x - \sec y)\,dy$, where C is the circle $x^2 + y^2 = 4$.

28. $\oint_C (-y^2 + \arctan x)\,dx + \ln y\ dy$, where C is the closed curve formed by $y = x^2$ and $x = y^2$.

29. $\oint_C (\arctan y\,dx - xy^2(y^2 + 1)^{-1}dy$, where C is the ellipse $x^2/4 + y^2/25 = 1$.

30. $\oint_C (x^2 - y^3)\,dx + (y^2 + x^3)\,dy$, where C is the unit circle with center at $(0,0)$.

31. The area of the (bounded) region enclosed by the curves $y = x^2$ and $x = y^2$.

32. The area bounded by the astroid $x = 2\cos^3 t$, $y = 2\sin^3 t$ for $0 \le t \le 2\pi$.

33. Find the center of mass of the part of the astroid $x = 5\cos^3 t$, $y = 5\sin^3 t$ that lies in the first quadrant. Assume constant density.

34. Find the area of the part of the plane $2x + 2y + z = 4$ in the first octant.

35. Find the area of the surface $z = 2\sqrt{2x} + y^{3/2}$, $1 \le x \le 2$, $0 \le y \le 1$.

36. Find the area of the surface cut out from the half-cone $z = 2\sqrt{x^2 + y^2}$ by the cylinder $x^2 + (y - 1)^2 = 1$.

37. Find the area of the surface cut from the hemisphere $x^2 + y^2 + z^2 = 4$, $z \geq 0$, by the cylinder $(x - 1)^2 + y^2 = 1$.

38. Evaluate the surface integral $\int\int_S xyz\,dS$ where S is the part of the cylinder $x^2 + y^2 = 4$ with $1 \leq z \leq 3$.

39. Find the mass of the surface $z = 9 - r^2$, $0 \leq z \leq 9$ with density (mass/area) $1/\sqrt{4r^2 + 1}$.

40. Find the mass of the hemisphere $x^2 + y^2 + z^2 = 4$ with $z \geq 0$ if the density is $\delta(x, y, z) = 3z^2$.

41. Find the flux across the conical surface $z = \sqrt{x^2 + y^2}$, $0 \leq z \leq 1$ if the mass flow vector is $\mathbf{F} = x^2\mathbf{i} + y^2\mathbf{j} + z\mathbf{k}$.

42. A fluid has mass flow vector $\mathbf{F} = y\mathbf{i} - x\mathbf{j} + 2\mathbf{k}$. Find the flux outward across the part of the sphere $x^2 + y^2 + z^2 = 9$ with $z \geq \sqrt{5}$.

43. Evaluate $\int\int_S \mathbf{F} \cdot \mathbf{n}\,dS$ where $\mathbf{F}(x, y, z) = < 1/x, 1/y, 1/z >$, S is the unit sphere with center $(0, 0, 0)$, and $\mathbf{n}$ is the outer normal.

44. Evaluate the surface integral $\int\int_S x^2\,dS$ over the hemisphere $x^2 + y^2 + z^2 = 9$ with $y \geq 0$.

45. Evaluate $\int\int_S \mathbf{F} \cdot \mathbf{n}\,dS$ where $\mathbf{F} = < x^2, y^2, -xy >$, S is the surface of the cube with opposite corners $(0, 0, 0)$ and $(1, 1, 1)$, and $\mathbf{n}$ is the outer normal.

46. Verify the divergence theorem for $\mathbf{F}(\mathbf{r}) = \mathbf{r}$ and the solid cylinder with $x^2 + y^2 \leq 1$ and $0 \leq z \leq 1$ by calculating both integrals.

47. Let S be the surface of the solid region Ω in the first octant bounded by the coordinate planes and the paraboloid $z = 1 - x^2 - y^2$. Evaluate (a) $\int\int\int_\Omega \nabla \cdot \mathbf{F}\,dV$ and (b) $\int\int_S \mathbf{F} \cdot \mathbf{n}\,dS$ when $\mathbf{F}(\mathbf{r}) = < y, -z, xz >$.

48. Verify Stokes' theorem for $\mathbf{F} = < y^2, x, z^2 >$ and the part of the paraboloid $z = x^2 + y^2$, $z \leq 1$ by evaluating both integrals.

49. Verify Stokes' theorem for $\mathbf{F} = < x + y, y + z, z + x >$ and the elliptical region $x^2 + y^2/4 = 1$ in the xy–plane by calculating both integrals.

50. (a) Sketch the surface $z = y/\sqrt{x^2 + y^2}$ for $x^2 + y^2 \leq 1$, $x > 0$, $y > 0$. What is the boundary of the surface? *Hint.* Cylindrical coordinates. (b) Verify Stokes' theorem for $\mathbf{F} = -yz\mathbf{i} + xz\mathbf{j}$ and the surface in (a) by evaluating both integrals.

51. Use Stokes' theorem to evaluate the line integral

$$\oint_C y^2x^3dx + 2xyz^3dy + 3xy^2z^2dz$$

where C is given by $\mathbf{r}(t) = 2\cos t\,\mathbf{i} + 3\mathbf{j} + 2\sin t\,\mathbf{k}$, $0 \leq t \leq 2\pi$.

APPENDIX A

RADIUS OF CONVERGENCE OF A POWER SERIES

Every power series has a finite or infinite radius of convergence. Theorem 1 tells the story.

Theorem 1 *Radius of Convergence*
Every power series $\sum a_n x^n$ has a *radius of convergence* r with $0 \leq r \leq \infty$ such that

(a) $\sum a_n x^n$ converges absolutely for $|x| < r$;

(b) $\sum a_n x^n$ diverges for $|x| > r$.

A slightly different form of this theorem appears in Sec. 2.2.

The proof of Th. 1 is based on two lemmas. The first lemma reveals that the set where the power series converges absolutely is an interval containing the origin.

Lemma 1

Let A be the set of all x for which $\sum a_n x^n$ converges absolutely. Then

$$A = [0,0] \text{ or } (-r, r) \text{ or } [-r, r] \text{ or } (-\infty, \infty)$$

for some $r > 0$.

Proof. If x is in A, then $\sum |a_n x^n|$ converges by the meaning of absolute convergence. Since $|a_n(-x)^n| = |a_n x^n|$, x is in A if and only if $-x$ is in A. So A is symmetric about 0. Let A^+ be the right half of A. It consists of all $x \geq 0$ for which $\sum |a_n x^n| = \sum |a_n| x^n$ converges. Evidently, 0 is in A^+. Observe that

$$0 \leq t \leq x \quad \Rightarrow \quad \sum |a_n| t^n \leq \sum |a_n| x^n.$$

It follows that if x is in A^+, then t is in A^+ for $0 \leq t \leq x$. A little thought leads to the conclusion that A^+ is an interval with left endpoint 0 and right endpoint the least upper bound r of A^+. Thus, there are four possibilities for A^+:

$$A^+ = [0,0] \text{ or } [0, r) \text{ or } [0, r] \text{ or } [0, \infty)$$

with $0 < r < \infty$. Since A is symmetric about 0, the lemma follows. □

In Lemma 1, let $r = 0$ if $A = [0, 0]$ and $r = \infty$ if $A = (-\infty, \infty)$. Then, as we shall see presently, r is the radius of convergence of $\sum a_n x^n$.

Our second lemma relates convergence and absolute convergence.

Lemma 2

For some $x \neq 0$, suppose that $\sum a_n x^n$ converges. Then $\sum a_n t^n$ converges absolutely for $|t| < |x|$.

Proof. Since $\sum a_n x^n$ converges, $a_n x^n \to 0$ as $n \to \infty$, and, hence,

$$|a_n x^n| \leq 1 \text{ for } n \text{ sufficiently large.}$$

For such n and any t with $|t| < |x|$,

$$|a_n t^n| = |a_n x^n| \left|\frac{t}{x}\right|^n \leq \left|\frac{t}{x}\right|^n$$

Since $|t/x| < 1$, the geometric series $\sum |t/x|^n$ converges and, by comparison, $\sum |a_n t^n|$ converges. That is, $\sum a_n t^n$ is absolutely convergent. □

Now, it is just a matter of putting the pieces together to prove (a) and (b) in Th. 1. Let r be determined as in Lemma 1. Then Lemma 1 gives us (a). To prove (b), let $|x| > r$. Choose and fix any t with $|x| > t > r$. By Lemma 1, $\sum a_n t^n$ does not converge absolutely. Consequently, by Lemma 2, $\sum a_n x^n$ diverges for $|x| > r$, which is (b). This completes the proof of Th. 1.

APPENDIX B
DIFFERENTIATION AND INTEGRATION OF POWER SERIES

A power series can be differentiated or integrated term-by-term inside its interval of convergence. We begin with term-by-term integration.

Theorem 1 *Integration of Power Series*
Let $f(x) = \sum_{n=0}^{\infty} a_n x^n$ with radius of convergence $r > 0$. Then

$$\int_0^x f(t)\,dt = \sum_{n=0}^{\infty} a_n \frac{x^{n+1}}{n+1} \quad \text{for} \quad -r < x < r.$$

Proof. Express $f(x)$ as

$$f(x) = \sum_{n=0}^{N} a_n x^n + R_N(x), \qquad R_N(x) = \sum_{n=N+1}^{\infty} a_n x^n.$$

By Th. 1 of Sec. 2.3, $f(x)$ is continuous for $-r < x < r$. Therefore, $f(x)$ is integrable and

$$\int_0^x f(t)\,dt = \sum_{n=0}^{N} a_n \frac{x^{n+1}}{n+1} + \int_0^x R_N(t)\,dt.$$

We must show that

$$\int_0^x R_N(t)\,dt \to 0 \quad \text{as} \quad N \to \infty \quad \text{for} \quad -r < x < r.$$

Fix x with $0 \le x < r$. Choose any c with $x < c < r$. Since $\sum a_n c^n$ converges, $a_n c^n \to 0$ as $n \to \infty$ and, hence,

$$|a_n c^n| \le 1 \text{ for } n \text{ sufficiently large.}$$

Then, for such n and any t with $0 \le t \le x$,

$$|a_n t^n| \le |a_n x^n| = |a_n c^n| \left(\frac{x}{c}\right)^n \le \left(\frac{x}{c}\right)^n \quad \text{and} \quad 0 \le \frac{x}{c} < 1.$$

Consequently, for large enough N and $0 \le t \le x$,

$$|R_N(t)| \le \sum_{n=N+1}^{\infty} |a_n t^n| \le \sum_{n=N+1}^{\infty} \left(\frac{x}{c}\right)^n = \left(\frac{x}{c}\right)^{N+1} \frac{1}{1-(x/c)} = \left(\frac{x}{c}\right)^{N+1} \frac{c}{c-x},$$

$$\left|\int_0^x R_N(t)\,dt\right| \le \left(\frac{x}{c}\right)^{N+1} \frac{cx}{c-x}.$$

Since $0 < x/c < 1$, $(x/c)^{N+1} \to 0$ as $N \to \infty$ and, hence,

$$\int_0^x R_N(t) \to 0 \quad \text{as} \quad N \to \infty \quad \text{for} \quad 0 \le x < r.$$

The reasoning for $-r < x \le 0$ is almost the same. It follows that

$$\int_0^x f(t)\,dt = \sum_{n=0}^{\infty} a_n \frac{x^{n+1}}{n+1} \quad \text{for} \quad -r < x < r. \quad \square$$

Now consider term-by-term differentiation.

Theorem 2 *Differentiation of Power Series*
Let $f(x) = \sum_{n=0}^{\infty} a_n x^n$ with radius of convergence $r > 0$. Then $f(x)$ is differentiable and

$$f'(x) = \sum_{n=1}^{\infty} a_n n x^{n-1} \quad \text{for} \quad -r < x < r.$$

Proof. By the ratio test, $\sum n t^{n-1}$ converges for $|t| < 1$. We shall make use of this series in some comparisons. Let $0 \le x < r$. Choose any c with $x < c < r$. As above,

$$|a_n c^n| \le 1 \quad \text{for all large } n.$$

Then

$$|a_n n x^{n-1}| = |a_n c^n| \frac{n}{c}\left(\frac{x}{c}\right)^{n-1} \le \frac{n}{c}\left(\frac{x}{c}\right)^{n-1} \quad \text{and} \quad 0 \le \frac{x}{c} < 1.$$

Since $0 \le x/c < 1$, the series $\sum a_n n (x/c)^{n-1}$ converges. By comparison, $\sum a_n n x^{n-1}$ converges for $0 \le x < r$. The convergence of $\sum a_n n x^{n-1}$ for $-r < x \le 0$ is established in the same way. Let.

$$g(x) = \sum_{n=1}^{\infty} a_n n x^{n-1} \quad \text{for} \quad -r < x < r.$$

By Th. 1, we can integrate term-by-term to obtain

$$\int_0^x g(t)\,dt = \sum_{n=1}^{\infty} a_n x^n = f(x) - a_0 \quad \text{for} \quad -r < x < r.$$

By the fundamental theorem of calculus,

$$g(x) = \frac{d}{dx}\int_0^x g(t)\,dt = f'(x).$$

Thus, $f(x)$ is differentiable and

$$f'(x) = \sum_{n=1}^{\infty} a_n n x^{n-1} \quad \text{for} \quad -r < x < r. \ \Box$$

It follows from Ths. 1 and 2 that

The differentiated and integrated series have the same radius of convergence as the original series.

There is not much to the verification. Let r_1 be the radius of convergence of the series for $f'(x)$. By Th. 2, $r_1 \geq r$. Since term-by-term integration of the power series for $f'(x)$ yields the power series for $f(x)$, $r \geq r_1$ by Th. 1. So $r_1 = r$. Likewise, if r_2 is the radius of convergence of the series $\int_0^x f(t)\,dt$, then $r_2 = r$. $\Box$

APPENDIX C
ANSWERS TO SELECTED ODD-NUMBERED PROBLEMS

This appendix contains the answers to most of the odd-numbered problems in the book. Exceptions are as follows. If a problem contains its own answer, that answer is not repeated here. Such problems are usually statements of results to prove or formulas to verify. In certain multipart problems, some parts are primarily hints for reaching conclusions in other parts or follow immediately from a prior part. In such cases, only answers to the primary part or parts of the problem are given.

Section 1.1

1. converges, limit 1
3. diverges, no limit
5. diverges, no limit
7. converges, limit 1
9. diverges, no limit
11. converges, limit 1
13. diverges, no limit
15. converges, limit 0
17. converges, limit 0
19. converges, limit 2
21. converges, limit 1
23. converges, limit 0
25. diverges to ∞
27. diverges to ∞
29. converges, limit 1
31. converges, limit 1
33. converges, limit 0
35. converges, limit 0
37. converges, limit 0
39. converges, limit 1
41. converges, limit $\sin(1)$
43. converges, limit $\sin(2)$
45. converges, limit 0
47. converges, limit $e^{x/2}$
49. diverges to ∞
51. converges, limit $\pi/2$
53. converges, limit $1/e$
55. converges, limit e^2
57. converges, limit 1/2
59. converges, limit 1
61. converges, limit 0
63. converges, limit $\ln(1/2)$
65. converges, limit 0
67. converges, limit 1
69. converges, limit 0
71. limit 0, $N = 31$
73. limit 0, $N = 9$
75. limit 0, $N = 29$
77. limit $\pi/2$, $N = 999$

Section 1.2

In problems 1–19, l means greatest lower bound, u means least upper bound, and L means limit.

1. decreasing, $L = l = 1/2$, $u = 5/2$
3. increasing, $l = 0$, $L = u = 1/3$
5. ultimately increasing, $l = \tan\dfrac{9\pi - 1}{13}$, $u = \tan\dfrac{4\pi - 1}{8}$, $L = 0$
7. ultimately decreasing, $L = 0$, $l = 0$, $u = \dfrac{10^9}{9!}$
9. ultimately decreasing, $L = l = 0$, $u = 1/2$
11. ultimately decreasing, $L = 0$, $l = \sin(6)$, $u = \sin(3/2)$
13. increasing, $l = \sin(1)$, $L = u = 1$
15. decreasing, $L = l = 0$, $u = \ln(3)$

17. ultimately decreasing, $L = l = 0$, $u = \dfrac{\ln(7)^2}{7}$

19. increasing, $l = \sin(1)$, $L = u = 1$

21. increasing, concave up on $[2,3]$; the Newton sequence with $x_0 = 3$ will decrease to r; $n = 3$ and $x_4 = 2.23607$ correctly rounded

23. decreasing, concave up on $[1,e]$; the Newton sequence with $x_0 = 1$ will increase to r; $n = 3$ and $x_4 = 1.30980$ correctly rounded

25. increasing, concave up on $[0,1]$; the Newton sequence with $x_0 = 1$ will decrease to r; $n = 3$ and $x_4 = 0.76538$ correctly rounded

27. $n = 6$, $x_7 = 0.45019$

29. $n = 16$, $x_{17} = 0.56712$

31. $n = 4$, $x_5 = 2.99501$

37. $L = (1 + \sqrt{1 + 4c})/2$

39. (a) If L exists, $L = (-1 \pm \sqrt{5})/2$; (b) no limit exists

41. (a) $L = \pm 2\sqrt{-1}$; (b) the sequence diverges

Section 1.3

1. converges, sum 4/9
3. converges, sum 3
5. converges, sum 3/2
7. converges, sum 4/15
9. converges, sum 15
11. diverges to ∞
13. converges, sum $1/(1 + e^{-\pi})$
15. diverges to ∞
17. diverges to ∞
19. diverges, no sum
21. diverges to ∞
23. converges, sum $-1/2$
25. converges, sum 1/3
27. diverges to ∞
29. 43/99
31. 101/999
37. $\sum_{n=1}^{\infty} 2^{n-1}x^n$, $|x| < 1/2$
39. $\sum_{n=1}^{\infty} 4^{n-1}x^{2n}$, $|x| < 1/2$
41. $-\sum_{n=2}^{\infty} x^n$, $|x| < 1$
43. $x^3/(1-x)$, $|x| < 1$
45. $x^{-1}/(1-x^3)$, $|x| < 1$
47. $2/(2 + x^2)$, $|x| < \sqrt{2}$
49. $-1/2(x-1)$, $0 < x < 1$
51. $1/(1 - 2\sin x)$, for $-\dfrac{\pi}{6} < x < \dfrac{\pi}{6}$ and translates of this interval by any integer multiple of π
53. $\sum_{n=0}^{\infty} t^n/2^n = 2/(2-t)$, $|t| < 2$
55. $\sum_{n=0}^{\infty} (2\cos t)^n = 1/(1 - 2\cos t)$ for $\pi/3 < t < 2\pi/3$ and translates of this interval by any integer multiple of π
57. $\sum_{n=0}^{\infty} \left(\dfrac{2t}{1-t}\right)^n = \dfrac{1-t}{1-3t}, -1 < t < \dfrac{1}{3}$
59. $\sum_{n=0}^{\infty} e^{-nt} = \dfrac{e^t}{e^t - 1}$, $t > 0$
61. 6/11
63. 8
65. $10r/(1-r)$ million dollars

Section 1.4

1. diverges
3. converges
5. diverges
7. converges
9. converges
11. converges
13. converges
15. diverges
17. diverges
19. converges
21. converges
23. diverges
25. diverges
27. diverges
29. converges
31. converges
33. converges
35. diverges
37. converges
39. converges
41. converges

43. converges
45. converges
47. converges $\Leftrightarrow$ $p > 1$
49. converges $\Leftrightarrow$ $p > 1$
51. (b) $a_n = 1/n$
53. $T_3 = 1.0363$
55. $T_2 = 0.40457$

Section 1.5

1. converges conditionally
3. converges absolutely
5. converges conditionally
7. converges conditionally
9. converges conditionally
11. converges conditionally
13. converges absolutely
15. converges absolutely
17. converges absolutely
19. converges conditionally
21. converges conditionally
23. converges absolutely for $|x| \leq 1$, diverges for $|x| > 1$
25. converges absolutely for $|x| \leq 2$, diverges for $|x| > 2$
27. $S_{10} = 0.9011$
29. $S_{31} = 0.6731$
31. a_n not decreasing

Section 1.6

If a series has positive terms and it converges, it also converges absolutely; in such cases, only convergence is mentioned in the answers.

1. converges
3. diverges
5. converges
7. converges
9. converges
11. diverges
13. diverges
15. converges absolutely
17. converges
19. converges
21. converges absolutely
23. converges
25. diverges
27. converges
29. diverges
31. diverges
33. diverges
35. converges
37. diverges
39. converges
41. (b) converges
45. Use the root test to find $n = 25$ and $S_{25} = 2.0000$ correctly rounded.
47. Use the root test to find $n = 10$ and $S_{10} = 0.6492$ correctly rounded.
49. Use the ratio test to find $n = 14$ and $S_{14} = 7.3891$ correctly rounded.

Chapter 1 Review Problems

1. diverges to ∞
3. diverges, no limit
5. converges, limit 1
7. converges, limit 1
9. converges, limit 0
11. converges, limit 1
13. converges, limit $1/e$
15. converges, limit 0
17. $L = 0$
19. $L = 2$
21. decreasing, $0 \leq a_n \leq 1$
23. diverges

In Probs. 23–49, if a series with positive terms converges, it also converges absolutely.

25. converges
27. diverges
29. diverges
31. diverges
33. converges conditionally
35. converges
37. converges
39. converges conditionally
41. converges

43. diverges
45. converges
47. converges conditionally
49. converges absolutely
51. converges conditionally
53. converges absolutely for $|x| < 1$, diverges otherwise
55. converges for $x = 0$, diverges otherwise
57. converges absolutely for $|x| < 1$, diverges for $|x| \geq 1$ (Assume the results for $|x| = 1$)
59. converges absolutely for $-2 < x < 0$, diverges otherwise
61. $1/(1 - e^{-1})$
63. 3/4
65. 0.36788 rounded
67. Use the integral test methods, $T_{31} = 2.0762$ rounded

Section 2.1

1. $1 + 2x + (3/2)x^2 + (2/3)x^3$
3. $3 - x^2 + 3x^3 - x^4$
5. $50 - 30t + 8t^2$
7. $1 - 3t - (\sin(1))t^2$
9. (a) $(1/8)t^2 - (1/96)t^3$; (b) $(1/8)t^2 - (1/96)t^3 - (95/1536)t^4$
11. $\sum_{n=0}^{6} \dfrac{2^n}{n!}x^n$
13. $1 - x^3$
15. $1-\dfrac{1}{2}x+\dfrac{3}{8}x^2-\dfrac{5}{16}x^3+\dfrac{35}{128}x^4-\dfrac{63}{256}x^5$
17. $2x - \dfrac{4}{3}x^3 + \dfrac{4}{15}x^5$
19. $x^3 - 2x^2 + 5x - 1$
21. $x^3 + 27$
23. $1 + \dfrac{1}{2}x^2 + \dfrac{5}{24}x^4$
25. $x - \dfrac{1}{3}x^3 + \dfrac{1}{5}x^5$
27. $\sum_{k=0}^{n} (-1)^k x^k$
29. $\sum_{k=0}^{n} \dfrac{(-1)^k 2^k}{k!}x^k$
31. $1 + \sum_{k=1}^{n} (-1)^k \dfrac{1 \cdot 3 \cdot \cdots \cdot (2k-1)}{2^k k!}x^k$
33. $P_0(x) = P_1(x) = 0$
35. $\sum_{k=1}^{n} \dfrac{(-1)^{k+1}}{(k-1)!}x^k$
37. $2\sum_{\substack{k=1 \\ k \text{ odd}}}^{n} \dfrac{x^k}{k}$
39. $\sum_{\substack{k=0 \\ k \text{ even}}}^{n} \dfrac{x^k}{k!}$
41. error $\leq \dfrac{e}{4!} \approx 0.113$
43. error $\leq \dfrac{3}{8}\dfrac{(1/4)^3}{3!} \approx 0.001$
45. $n =$ 40, not reasonable
47. $P_8(x) = P_9(x)$
49. (b) $\sum_{k=0}^{n} e^a \dfrac{(x-a)^k}{k!}$
51. $$\frac{\sqrt{2}}{2}\left[1 - \left(x - \frac{\pi}{4}\right) - \frac{\left(x - \frac{\pi}{4}\right)^2}{2!} + \frac{\left(x - \frac{\pi}{4}\right)^3}{3!} + \frac{\left(x - \frac{\pi}{4}\right)^4}{4!} - - + + \ldots \pm \frac{\left(x - \frac{\pi}{4}\right)^n}{n!}\right]$$

Section 2.2

1. $\sum_{n=0}^{\infty} (-1)^n \dfrac{x^n}{n!}$
3. $\sum_{n=0}^{\infty} (-1)^n x^n$
5. $\sum_{n=0}^{\infty} \dfrac{x^{n+1}}{n!}$
7. $\sum_{n=0}^{\infty} (-1)^n 2^{2n} \dfrac{x^{2n}}{(2n)!}$
9. $\sum_{n=0}^{\infty} (-1)^n x^{n+1}$
11. $\sum_{n=0}^{\infty} \dfrac{x^{2n+1}}{(2n+1)!}$
13. $1 + \dfrac{1}{2}x + \sum_{n=2}^{\infty} \dfrac{(-1)^{n-1} 1 \cdot 3 \cdot 5 \cdot \cdots \cdot (2n-3)}{2^n n!}x^n$

29. $r = 1, \quad [-1, 1)$
31. $r = 1, \quad [-1, 1]$
33. $r = 1, \quad [-1, 1]$
35. $r = 1, (-1, 1)$
37. $r = 1, \quad (-1, 1)$
39. $r = 2, [-2, 2)$
41. $r = 1, \quad [-1, 1)$
43. $r = \frac{1}{\sqrt{2}}, \left[\frac{-1}{\sqrt{2}}, \frac{1}{\sqrt{2}}\right]$
45. $r = 0, [0, 0]$
47. $r = \ln 2, \quad (-\ln 2, \ln 2)$
49. $r = 2, \quad [-2, 2)$
51. $r = 1, \quad (1, 3]$
53. $r = \infty, (-\infty, \infty)$
55. $r = 4$
57. $r = \sqrt{2}$
59. $r = 1$
61. (a) $\sum_{n=0}^{\infty} e^2 \frac{(x-2)^n}{n!}$
63. (a) $\sum_{n=0}^{\infty} e^a \frac{(x-a)^n}{n!}$

Section 2.3

1. $\sum_{n=0}^{\infty} \frac{(-1)^n x^{n+2}}{n!}, |x| < \infty$
3. $\frac{1}{2} \sum_{n=2}^{\infty} n(n-1)x^{n-2}, |x| < 1$
5. $\sqrt{x} \sum_{n=0}^{\infty} \frac{(-1)^n x^n}{(2n+1)!}, \quad |x| < \infty$
7. $\sum_{n=0}^{\infty} \frac{(-1)^n x^n}{n+1}, \quad |x| < 1$
9. $1 + \sum_{n=1}^{\infty} \frac{(-1)^n 2^{2n-1} x^{2n}}{(2n)!}, \quad |x| < \infty$
11. $\sum_{n=0}^{\infty} \frac{x^{2n}}{(2n)!}, \quad |x| < \infty$
13. $\sum_{n=0}^{\infty} \frac{(-1)^n x^{2n+1}}{(2n+1)(2n+1)!}, \quad |x| < \infty$
15. $\sum_{n=0}^{\infty} \frac{(-1)^n}{(3n+1)2^{3n+1}}$
17. $\sum_{n=0}^{\infty} \frac{x^{2n+1}}{2n+1}, \quad |x| < 1$
19. $x/(1-x^2)$
21. $x + (1-x)\ln(1-x)$
23. 3/2
25. $-1 + 2\ln(2)$
27. $\sum_{n=0}^{2} \frac{(-1)^2}{(2n+1)(2n+1)!} \approx 0.9461$
29. $\sum_{n=0}^{2} \frac{(-1)^2}{(4n+3)(2n+1)!} \approx 0.3103$
31. $\sum_{n=1}^{6} (1/nn!) \approx 1.3179$
33. (a) $b \geq 13$, (b) 0.8862
35. 1/2
37. 0
39. 2
43. at least $\frac{10^6 - 1}{2} \approx 5 \times 10^5$ terms
45. 4 terms of each series

Section 2.4

5. (a) $xy' = -\ln(1-x)$
7. $\sum_{n=0}^{\infty} \frac{(-1)^n 2^n}{n!} x^n = e^{-2x}$
9. $3\sum_{n=0}^{\infty} \frac{(-1)^n}{2^n n!} x^{2n} = 3e^{-x^2/2}$
11. $\sum_{n=0}^{\infty} 5 \frac{(-1)^n 4^n}{(2n)!} x^{2n} + (-8) \frac{(-1)^n 4^n}{(2n+1)!} x^{2n+1} = 5\cos 2x - 4\sin 2x$
13. $1 + \sum_{n=1}^{\infty} \frac{\left(-\frac{2}{3}\right)\left(-\frac{2}{3}-1\right)\left(-\frac{2}{3}-2\right) \cdots \left(-\frac{2}{3}-n+1\right)}{n!} 4^n x^n, \quad r = \frac{1}{4}$
15. $1 + \sum_{n=1}^{\infty} \frac{\left(-\frac{1}{2}\right)\left(-\frac{1}{2}-1\right)\left(-\frac{1}{2}-2\right) \cdots \left(-\frac{1}{2}-n+1\right)}{n!} x^{2n}, \quad r = 1$
17. $\frac{1}{2} + \frac{1}{2 \cdot 5}\left(\frac{1}{2}\right)^5 = \frac{161}{320} \approx 0.503$

19. $\frac{1}{3(5/2)} + \frac{1}{2}\frac{1}{3^3}\frac{1}{3(9/2)} = \frac{491}{3645} \approx 0.13471$

21. $L = 2\pi a\left(1 - \frac{k^2}{4} - \sum_{n=2}^{\infty}\left[\frac{1 \cdot 3 \cdot \cdots \cdot (2n-3)}{2^n n!}\right]^2 (2n-1)k^{2n}\right)$

Chapter 2 Review Problems

1. (a) $1 - 2(x-1) + 3(x-1)^2$
3. $x + (1/3)x^3 + (1/5)x^5$
5. $\sum_{n=0}^{\infty}(-1)^n(n+1)x^n$
7. $20 + 80t + 6t^2$
9. $(\sin x)/x$
11. $e^{-x/2}$
13. $r = 1, \quad I = [-1,1)$
15. $r = 2, \quad I = (-2,2]$
17. $r = e^{-2}, \quad I = (-e^{-2}, e^{-2})$
19. $r = 2, \quad I = (-2,2)$
21. $r = 0, \quad I = [0,0]$
23. 3/8
25. $\frac{1}{2}\sum_{n=1}^{\infty}(-1)^{n+1}\frac{2^{2n}x^{2n}}{(2n)!}$
27. 1/6
29. $\frac{2}{\sqrt{\pi}}\left(1 - \frac{1}{3} + \frac{1}{10} - \frac{1}{42}\right) \approx 0.838$
31. $\sum_{n=0}^{\infty}\frac{x^{2n+1}}{2n+1}$
33. 0.262 within 0.01
39. $(x-1) + \frac{3}{2}(x-1)^2 + \sum_{n=3}^{\infty}\frac{(-1)^{n+1}2(x-1)^n}{n(n-1)(n-2)}$
41. $|R_3(x)| \leq 2/3$ both ways
43. $\sum_{n=0}^{\infty}(1 - \frac{1}{2^n})x^n$
45. $y = \cos x$

Section 3.1

1. 10
3. 3
5. $z = 5$
7. $x = 3$
9. $x^2 + y^2 + z^2 = 16$
11. $(x-2)^2+(y+5)^2+(z-3)^2=25$
13. $P_0 = (2, -3, 0)$ and $r = 4$

Section 3.2

1. $\langle -5, -4\rangle, \sqrt{5}, \left\langle \frac{3}{5}, \frac{4}{5}\right\rangle$
3. $\langle 43,12\rangle, \quad \sqrt{193}, \quad \left\langle -\frac{5}{13}, \frac{12}{13}\right\rangle$
5. $\langle -7, -4, -24\rangle, \quad \sqrt{14}, \quad \left\langle \frac{5}{13}, 0, \frac{12}{13}\right\rangle$
7. $\langle 13, -5, -14\rangle, \quad \sqrt{6}, \quad \left\langle -\frac{3}{\sqrt{22}}, \frac{3}{\sqrt{22}}, \frac{2}{\sqrt{22}}\right\rangle$
9. $\langle 2,2\rangle = 2\mathbf{i} + 2\mathbf{j}$
11. $\langle -12,0\rangle = -12\mathbf{i}$
13. $\langle 3,1,9\rangle = 3\mathbf{i} + \mathbf{j} + 9\mathbf{k}$
15. $\langle -4,2,4\rangle = -4\mathbf{i} + 2\mathbf{j} + 4\mathbf{k}$
17. All four vectors are parallel; **u**, **v**, **w** have the same direction; **r** opposite direction to other three.
19. Proceed as in Fig. 8.
21. Proceed as in Fig. 8 but in the plane $y = 1$.
23. (a) $\frac{3}{5}; \ -\frac{4}{5}$
25. (a) $-\frac{3}{13}; \ \frac{4}{13}; \ \frac{12}{13}$
27. $\frac{2}{3}; \ \frac{1}{3}; \ \frac{2}{3}$
29. $-\frac{12}{13}; \ -\frac{5}{13}$
31. $\pm\left\langle \frac{2}{\sqrt{5}}, \frac{4}{\sqrt{5}}\right\rangle; \ \left\langle \frac{2}{\sqrt{5}}, \frac{4}{\sqrt{5}}\right\rangle; \ -\left\langle \frac{2}{\sqrt{5}}, \frac{4}{\sqrt{5}}\right\rangle$

33. $\lambda = -1$; none; $\lambda = -1$

35. The diagonal vector extending from the common initial point of $\mathbf{v}$ and $\mathbf{w}$ is $\mathbf{v} + \mathbf{w}$. The diagonal vector with initial point at $\mathbf{w}$ and tip at $\mathbf{v}$ is $\mathbf{v} - \mathbf{w}$.

39. $\|\mathbf{T}\| = 500\sqrt{2}$; $\|\mathbf{R}\| = 500$

Section 3.3

1. 11

3. -11

5. 0

7. 10

9. $\arccos \dfrac{1}{\sqrt{5}} \approx 63.4°$

11. $\arccos \dfrac{\sqrt{3}}{2} = 30°$

13. $\arccos \dfrac{1}{2} = 60°$

15. $\arccos\left(-\dfrac{1}{5}\right) \approx 101.5°$

17. $\arccos \dfrac{2}{\sqrt{5}} \approx 26.6°$; $\arccos \dfrac{1}{\sqrt{5}} \approx 63.4°$

19. $\arccos \dfrac{1}{\sqrt{3}} \approx 54.7°$

25. 1; $\left\langle -\dfrac{3}{5}, \dfrac{4}{5} \right\rangle$

27. $\sqrt{3}$; $\dfrac{1}{2}(\sqrt{3}\,\mathbf{i} + 3\mathbf{j})$

29. $\dfrac{5}{2}\sqrt{14}$; $\dfrac{1}{2}\langle -10, 13, -9 \rangle$

31. $-\dfrac{1}{5}\sqrt{3}$; $-\dfrac{1}{25}(5\mathbf{i} - \mathbf{j} + 7\mathbf{k})$

33. $\langle 2,3,0 \rangle$ and $\langle 0,4,2 \rangle$; also linear combinations of these vectors.

35. $\left\langle \dfrac{20}{9}, -\dfrac{40}{9}, \dfrac{49}{9} \right\rangle$

37. $\langle b, -a \rangle$

45. $\mathbf{v} = 20\sqrt{2}\langle -\sqrt{3}, 1, 2 \rangle$; horizontal and vertical projections are $20\sqrt{2}\langle -\sqrt{3}, 1, 0 \rangle$ and $40\sqrt{2}\langle 0,0,1 \rangle$.

47. $W = 400 \cos 20° \approx 376$; effective force is $8 \cos 20° \mathbf{i}$.

49. $\arccos \dfrac{1}{6} \approx 80.4°$

51. $\langle 290\sqrt{2}, -210\sqrt{2} \rangle$; $20\sqrt{641}$; $\arccos \dfrac{21}{90} \approx 43.6°$ south of east

Section 3.4

1. $\langle -5, 2, 3 \rangle$

3. $\langle -2, -8, -1 \rangle$

5. 1

7. $2\mathbf{k}$

9. $\mathbf{0}$

11. $\dfrac{1}{5\sqrt{5}}\langle -6, -5, -8 \rangle$

13. $\langle 2,1,1 \rangle$

15. $\dfrac{1}{2}\sqrt{3}$

17. $2\sqrt{3}$

19. $\sqrt{3}$

21. $4\sqrt{5}$

23. not coplanar

25. not coplanar

41. $5\mathbf{k}$; 5

Section 3.5

1. (a) $\mathbf{r} = \langle 3 + 3t, 4 + 2t \rangle$; (b) $x = 3 + 3t$, $y = 4 + 2t$; (c) slope 2/3 intercept 2

3. (a) $\mathbf{r} = \langle 1 - 3t, 1 + 2t \rangle$; (b) $x = 1 - 3t$, $y = 1 + 2t$; (c) slope $-2/3$ intercept 5/3

5. (a) $\mathbf{r} = \langle 3t, 1+t, -2-2t \rangle$, (b) $x = 3t,\ y = 1+t,\ z = -2-2t$

7. (a) $\mathbf{r} = \langle 3+2t, 1+t, -1-3t \rangle$, (b) $x = 3+2t,\ y = 1+t,\ z = -1-3t$

9. (a) $\mathbf{r} = \langle 1+t, 2, 3 \rangle$, (b) $x = 1+t,\ y = 2,\ z = 3$

11. (a) $\langle -3, 2, -1 \rangle$; (b) $\mathbf{r} = \langle 4-3t, 2t, 3-t \rangle$; (c) $(-5, 6, 0)$; (d) $m = \tan\varphi = -\dfrac{1}{\sqrt{13}}$; $\varphi = \arctan\left(-\dfrac{1}{\sqrt{13}}\right) \approx -15.5°$

13. (a) $\langle 1, 1, -1 \rangle$; (b) $\mathbf{r} = \langle 1+t, t, 2-t \rangle$; (c) $(3, 2, 0)$; (d) $m = \tan\varphi = -\dfrac{1}{\sqrt{2}}$; $\varphi = \arctan\left(-\dfrac{1}{\sqrt{2}}\right) \approx -35.3°$

15. (a) $(1, 2, 3)$; (b) $\arccos(1/\sqrt{28}) \approx 79.1°$

17. no intersection

19. $\dfrac{3}{7}\sqrt{42}$

21. $2(x-1) - (y-2) + 4(z-3) = 0$

23. $x = 3$

25. $x - 2y + 3z = 4$

27. $2x - 6y + 2z = -8$

29. $-7x + 10y + 11z = 46$

31. $\pi/2$

33. $\arccos\left(\dfrac{1}{\sqrt{3}}\right) \approx 54.7°$

35. $\mathbf{r} = \left\langle \dfrac{8}{11} - 4t, \dfrac{29}{11} - 9t, 11t \right\rangle$, $x = \dfrac{8}{11} - 4t,\ y = \dfrac{29}{11} - 9t,\ z = 11t$

37. $x/2 = (1-y)/3 = z/2$

39. $(x-2)/2 = (7-z)/6,\ y = 5$

41. $1/\sqrt{29}$

47. 1

Chapter 3 Review Problems

1. (a) $\langle 0, 7 \rangle$; (b) $\langle 14, 7 \rangle$

3. (a) $\langle -8, 10, 11 \rangle$; (b) -11

5. (a) $\langle 3, 2, 6 \rangle$; (b) 7

7. 4/21

9. $\mathbf{u} = (25x^2 + z^2)^{-1/2}\langle 3x, -4x, z \rangle$ for $(x, z) \neq (0,0)$

11. (a) $\mathbf{v} = 15\langle -\sqrt{2}, -\sqrt{2}, 2\sqrt{3} \rangle$; (b) $15\langle -\sqrt{2}, -\sqrt{2}, 0 \rangle$; $\langle 0, 0, 30\sqrt{3} \rangle$

13. $\langle 2, 5, -3 \rangle$

15. $\mathbf{r} = \langle 3+6t, 2t, 2-3t \rangle$

17. $\mathbf{r} = \langle 4, 1, -3+t \rangle$

19. $\mathbf{r} = \langle 1+t, 1-4t, 1+2t \rangle$

21. $(5-x)/4 = y+3 = (1-z)/2$

23. $\sqrt{2}$

25. coplanar

27. $(1, 3, -5)$

31. (a) 2; (b) $\sqrt{5}$; (c) $(2/13)\langle 3, 4, 12 \rangle$; (d) $(1/13)\langle 20, 18, -11 \rangle$

33. (a) $\mathbf{v}_0 = \langle 6, 8, 0 \rangle$; $\mathbf{v}_1 = \langle 0, 0, 5 \rangle$; (b) 1/2; (c) $\arctan(1/2)$

35. 2/39

37. $\mathbf{r} = t\langle 1, 1, 1 \rangle$ and $x = t,\ y = t,\ z = t$

39. $(3/2, 1, 5/2)$

41. $\mathbf{n} = \langle 8, 4, 4 \rangle$

43. $5x + 4y + z = 3$

Section 4.1

1. $y = 3 - x$; the straight line is traversed from left to right.

3. $x = (2-y)^2 + 1$; the parabola opens to the right with vertex (1,2) and axis the line $y = 2$; it is traversed in a counterclockwise sense.

5. $y = (x+1)^2 + 1$; the parabola opens up with vertex $(-1,1)$ and axis $x = -1$; it is traversed from left to right.

7. $y = -2x^2$, $x > 0$; the parabola opens down with vertex (0,0) and axis the y–axis; it is traversed from left to right.

9. $y = x^2, \quad x \geq 0$; the right half of the standard parabola is traversed from left to right.

11. $x^2 + y^2 = 16$; the circle with center (0,0) and radius 4 is traversed repeatedly counterclockwise.

13. $x^2 + y^2 = 16$; the circle with center (0,0) and radius 4 is traversed repeatedly clockwise.

15. $x = 1 - 2y^2, \quad -1 \leq y \leq 1$; the graph is an arc on a parabola that opens to the left with vertex (1,0) and axis the x–axis; it is traversed repeatedly back and forth between $(-1, 1)$ and $(-1, -1)$.

17. 0   19. $-e^{-1}$ 21. -1

23. $x = 3\cos t, \ y = -3\sin t, \ 0 \leq t \leq 2\pi$ 25. $x = 3\cos t, y = 4\sin t, \ 0 \leq t \leq 2\pi$

27. The curve is the parabolic arc $y = x^2$, $0 \leq x \leq 1$ traversed from left to right.

29. (a) circle; (b) an object moves counterclockwise on the circle with constant angular speed ω.

31. (a) the projections on the coordinate plane are lines; (b) the curve is a straight line in space.

33. (a) projection on xy–plane is a quarter circle with center at the origin and radius 4 traversed counterclockwise; projection on the yz–plane is the first-quadrant part of the ellipse $y^2/4^2 + z^2/3^2 = 1$ traversed clockwise; projection on the xz–plane is the line segment $z = 3x/4$ for $0 \leq x \leq 4$; (b) the curve lies in the first octant directly over the quarter circle and is traversed downward from $(4, 0, 3)$ to $(0, 4, 0)$.

35. (a) circle with center $(0, 0)$ and radius 1 in the xy–plane, which is traversed clockwise starting from $(0, 1)$; (b) one turn on a circular helix descending from the xy–plane

37. (a) the parabolic arc $y = \sqrt{x}$ for $0 \leq x \leq \pi$ traversed from left to right; (b) the curve lies vertically above the arc and descends from $(0, 0, 4)$ to $(\pi, \sqrt{\pi}, -4)$, crossing the arc at $(\pi/2, \sqrt{\pi/2}, 0)$.

41. $\left(\pm 1, \pm \dfrac{\sqrt{2}}{2}\right)$ four points   43. $2\ln 4$ 45. 1

47. (a) $A = a^2\left(\dfrac{1}{2}\sin^2\theta_0 - \dfrac{1}{6}\theta_0^3 - \dfrac{1}{2}\theta_0^2\sin\theta_0 + \sin\theta_0 - \theta_0\cos\theta_0 + \dfrac{1}{24}\pi^3 + \dfrac{1}{4}\pi^2 - 3\right)$; (b) $\theta_0 \approx 4.4934$; $A \approx a^2(2.6799)$

Section 4.2

1. (a) $\langle 1, -4\rangle$; $\langle 0, 2\rangle$; (b) -4 3. (a) $e\mathbf{i} - 4e^2\mathbf{j}$; $e\mathbf{i} - 8e^2\mathbf{j}$; (b) $-4e$ 5. (a) $\langle 0, 0\rangle$; $\langle 2, -2\rangle$; (b) -1

7. (a) $\langle 0, 4\rangle$; $\langle -4, 0\rangle$; (b) no slope   9. (a) $\langle 0, -1, 4\rangle$; $\langle -4, 0, 0\rangle$; (c) $\langle 0, -1, 0\rangle$; $\langle 0, 0, 4\rangle$; $\arctan 4 \approx 76°$

11. (a) $-8\mathbf{j} - 3\mathbf{k}$; $16\mathbf{i}$; (c) $-8\mathbf{j}$; $-3\mathbf{k}$; $-\arctan 3/8 \approx -20.6°$ 13. (a) $\langle 1, 0\rangle$; $\langle 1, 2\rangle$; (b) 0

15. (a) $4\mathbf{i} + 2\mathbf{j}$; $2\mathbf{i}$; (b) 1/2   17. (a) $\langle -8, 3, 0\rangle$; $\langle 0, 0, -16\rangle$; (c) $\langle -8, 3, 0\rangle$; $\langle 0, 0, 0\rangle$; 0°

19. (a) $\mathbf{j} + \mathbf{k}$; $\mathbf{i} - \mathbf{k}$; (c) $\mathbf{j}$; $\mathbf{k}$; 45°   21. $\mathbf{r}_{\tan} = (1 + t)\mathbf{i} - (2 + 4t)\mathbf{k}$ 23. $\mathbf{r}_{\tan} = \langle -4t, 4\rangle$

25. $\mathbf{r}_{\tan} = \langle -1, t, (\pi/2) - t\rangle$ 27. $\mathbf{r}_{\tan} = \left\langle \dfrac{\pi}{2} + t, \sqrt{\dfrac{\pi}{2}} + \dfrac{t}{\sqrt{2\pi}}, -4t \right\rangle$ 29. 90°

31. (a) circle, center $(0, 0)$, radius a, traversed once counterclockwise from $(a, 0)$; (b) circle, center $(0, 0)$, radius a, traversed counterclockwise from $(a, 0)$ to $(a\cos 2\pi\omega, a\sin 2\pi\omega)$; (c) circle, center $(0, 0)$, radius a, traversed once clockwise from $(0, a)$

33. $\langle 0, \pi a/2\rangle$; $\langle -a\pi, 0\rangle$ 35. $(13/27)\sqrt{13} - 8/27$ 37. $\sqrt{17} + (1/8)\ln(33 + 8\sqrt{17})$

39. $\sqrt{10} - \sqrt{2} + \dfrac{1}{2}\ln\left(\dfrac{(\sqrt{10} - 1)(\sqrt{2} + 1)}{(\sqrt{10} + 1)(\sqrt{2} - 1)}\right)$

41. $e - e^{-1}$ 43. $\dfrac{15}{8}\sqrt{2}$

Section 4.3

1. (a) $\mathbf{r}' = \mathbf{i} + 2t\mathbf{j}$, $\|\mathbf{r}'\| = \sqrt{1+4t^2}$, $\mathbf{r}'' = 2\mathbf{j}$; (b) left to right on the parabola $y = x^2$

3. (a) $\mathbf{r}' = \langle -2\sin t, 3, 2\cos t\rangle$, $\|\mathbf{r}'\| = \sqrt{13}$, $\mathbf{r}'' = \langle -2\cos t, 0, -2\sin t\rangle$; (b) moves with increasing y along a circular helix with axis the y–axis

5. (a) $\mathbf{r}' = \mathbf{i} + 2t\mathbf{j} + 3t^2\mathbf{k}$, $\|\mathbf{r}'\| = \sqrt{1+4t^2+9t^4}$, $\mathbf{r}'' = 2\mathbf{j} + 6t\mathbf{k}$; (b) upward on the standard twisted cubic

7. (a) $\mathbf{r}' = \langle 1, 2t, 1/t\rangle$, $\|\mathbf{r}'\| = \sqrt{1+4t^2+1/t^2}$, $\mathbf{r}' = \langle 0, 2, -1/t^2\rangle$; (b) moves with increasing x–coordinate vertically above the parabola $y = x^2$ with elevation function $z = \ln x$

9. $t^2\mathbf{i} + e^{-t}\mathbf{j} + \mathbf{c}$

11. $e^t\mathbf{i} + ((1/2)t + (1/4)\sin 2t)\mathbf{j} + (1/2)\cos 2t\,\mathbf{k} + \mathbf{c}$

13. $\langle \ln t, t\ln t - t, -(1/\pi)\ln|\cos \pi t|\rangle + \mathbf{c}$

15. (a) $\mathbf{r}' = -a\omega\sin\omega t\,\mathbf{i} + a\omega\cos\omega t\,\mathbf{j} + b\mathbf{k}$; $\mathbf{r}'' = -a\omega^2\cos\omega t\,\mathbf{i} - a\omega^2\sin\omega t\,\mathbf{j}$;

(b) $\arccos(b^2t[(a^2+b^2t^2)(a^2\omega^2+b^2)]^{-1/2})$; (c) $\pi/2$; (d) a circular helix

17. The maximum speed is 2 units/sec when $t = (2k+1)\pi$ and the minimum 0 when $t = 2\pi k$ for k any integer.

19. (b) $\mathbf{r}' = a\omega^2 t\langle \cos\omega t, \sin\omega t\rangle$; $\|\mathbf{r}'\| = a\omega^2 t$; (c) $\mathbf{r}'' = a\omega^2\langle \cos\omega t - \omega t\sin\omega t, \sin\omega t + \omega t\cos\omega t\rangle$;

$\|\mathbf{r}''\| = a\omega^2\sqrt{\omega^2t^2+1}$; (d) $\arccos(1/\sqrt{\omega^2t^2+1})$

23. $\mathbf{r}(2\pi) = \langle 4, 0, 6\pi\rangle$; $\mathbf{r}'(2\pi) = \langle 0, 4, 3\rangle$; $\|\mathbf{r}'(2\pi)\| = 5$; $\mathbf{r}''(2\pi) = \langle -4, 0, 0\rangle$

25. (a) $\dfrac{313}{4} = 78.25$ ft (b) $11\sqrt{3}\left(\dfrac{11+\sqrt{313}}{2}\right) \approx 273$ ft

27. (b) $\dfrac{v_0^2}{2g}\sin^2\alpha$; (c) $\dfrac{v_0^2}{g}\sin 2\alpha$; (d) 45°; (e) parabolic arc

29. (a) y max ≈ 113.75 ft (b) maximum range approximately 99.8 ft

33. $t(s) = s/\sqrt{2}$

35. $t(s) = \ln((s/3)+1)$

37. $t(s) = \sqrt{2s}$

Section 4.4

1. both 3

3. both 1

5. both $\dfrac{3\cos^2 t\sin^2 t}{\cos^6 t + \sin^6 t}$

7. 1

9. 3/25

11. $2^{-3/2}(1+\sin\theta)^{-1/2}$

13. $\dfrac{|\sin x|}{(1+\cos^2 x)^{3/2}}$

15. $\dfrac{x}{(x^2+1)^{3/2}}$

17. $\kappa = \dfrac{1}{4a|\sin(t/2)|}$; $\kappa_{\min} = \dfrac{1}{4}a$; the graph of $\kappa(t)$ is roughly U-shaped with vertical asymptotes as $t \to 0^+$ or $(2\pi)^-$

19. $\kappa_{\max}$ occurs at $x = \dfrac{1}{2}\ln 2$; $\kappa(x) \to 0$ as $x \to \pm\infty$; the graph has the x–axis as a horizontal asymptote and has a roughly bell-shaped profile.

21. $\dfrac{t+2t^3}{|t|\sqrt{1+t^2}}$; $\dfrac{|t|}{\sqrt{1+t^2}}$

23. 0; 4

25. $\sqrt{2}\sinh t$; 1

27. $\dfrac{-4\sin 4t}{\sqrt{1+4\cos^2 2t}}$; $\sqrt{\dfrac{17-12\cos^2 2t}{1+4\cos^2 2t}}$

29. ellipse $\dfrac{x^2}{a^2}+\dfrac{y^2}{b^2}=1$; $a_T=\dfrac{1}{\sqrt{2}}\dfrac{a^2-b^2}{(a^2+b^2)^{1/2}}$; $a_N=\sqrt{2}\dfrac{ab}{(a^2+b^2)^{1/2}}$;

$\kappa=\dfrac{2^{3/2}ab}{(a^2+b^2)^{3/2}}$; $\mathbf{T}=\dfrac{-a\mathbf{i}+b\mathbf{j}}{(a^2+b^2)^{1/2}}$; $\mathbf{N}=-\dfrac{b\mathbf{i}+a\mathbf{j}}{(a^2+b^2)^{1/2}}$

31. $a_T=0$; $a_N=\rho\omega^2$; $\kappa=\dfrac{\rho\omega^2}{\rho^2\omega^2+b^2}$; $\mathbf{T}=\dfrac{-\rho\omega\sin\omega t\,\mathbf{i}+\rho\omega\cos\omega t\,\mathbf{j}+b\mathbf{k}}{\sqrt{\rho^2\omega^2+b^2}}$; $\mathbf{N}=-\cos\omega t\,\mathbf{i}-\sin\omega t\,\mathbf{j}$

33. $a_T=e^t-e^{-t}$; $a_N=\sqrt{2}$; $\kappa=\dfrac{\sqrt{2}}{(e^t+e^{-t})^2}$; $\mathbf{T}=\dfrac{\langle e^t,-e^{-t},\sqrt{2}\rangle}{(e^t+e^{-t})}$; $\mathbf{N}=\dfrac{\langle\sqrt{2},\sqrt{2},e^{-t}-e^t\rangle}{(e^t+e^{-t})}$

39. $\rho=\dfrac{5^{3/2}}{2}$ $\left(-4,\dfrac{7}{2}\right)$

41. $\rho=2^{3/2}$; $(3,-2)$

Section 4.5

1. (a) $\mathbf{r}'(t)=4\mathbf{u}_\theta$; $\mathbf{r}''(t)=-4\mathbf{u}_r$; (b) $\mathbf{r}'(0)=4\mathbf{j}$; $\mathbf{r}''(0)=-4\mathbf{i}$; (c) circle, center (0,0), radius 4

3. (a) $\mathbf{r}'(t)=2\pi(\cos\pi t)\mathbf{u}_r+2\pi(\sin\pi t)\mathbf{u}_\theta$; $\mathbf{r}''(t)=-4\pi^2(\sin\pi t)\mathbf{u}_r+4\pi^2(\cos\pi t)\mathbf{u}_\theta$; (b) $\mathbf{r}'(1)=2\pi\mathbf{i}$; $\mathbf{r}''(1)=4\pi^2\mathbf{j}$; (c) circle, center (0, 1), radius 1

5. (a) $\mathbf{r}'(t)=\dfrac{1}{4}e^{t/4}\mathbf{u}_r+\dfrac{1}{4}e^{t/4}\mathbf{u}_\theta$; $\mathbf{r}''(t)=\dfrac{1}{8}e^{t/4}\mathbf{u}_\theta$; (b) $\mathbf{r}'(0)=\dfrac{1}{4}(\mathbf{i}+\mathbf{j})$; $\mathbf{r}''(0)=\dfrac{1}{8}\mathbf{j}$; (c) spiral turning counterclockwise about the origin

7. (a) $\mathbf{r}'(t)=\dfrac{-4\cos 2t}{(1+\sin 2t)^2}\mathbf{u}_r+\dfrac{4}{1+\sin 2t}\mathbf{u}_\theta$; $\mathbf{r}''(t)=8\dfrac{\cos^2 2t-\sin 2t-\sin^2 2t}{(1+\sin 2t)^3}\mathbf{u}_r-\dfrac{16\cos 2t}{(1+\sin 2t)^2}\mathbf{u}_\theta$;
(b) $\mathbf{r}'(\pi/4)=-2\mathbf{i}$; $\mathbf{r}''(\pi/4)=-2\mathbf{j}$; (c) parabola opening down, vertex (0,1), axis the y–axis

9. (a) $\dfrac{a\omega\sin\omega t}{(1+\cos\omega t)^2}$; $\dfrac{a\omega}{1+\cos\omega t}$; (b) $a\omega^2\dfrac{2\sin^2\omega t-\cos\omega t-1}{(1+\cos\omega t)^3}$; $\dfrac{2a\omega^2\sin\omega t}{(1+\cos\omega t)^2}$;
(c) parabola opening left, vertex (a/2,0), axis the x–axis; (d) no, acceleration is not central.

11. (a) 0; $\rho\omega$; $-\rho\omega^2$; 0; (b) velocity vector tangent to the circle and acceleration vector directed toward the center; (c) yes, acceleration is central.

13. $\mathbf{F}=-m\omega^2\mathbf{r}(t)$; (b) yes, force is central; (c) ellipse, center (0, 0), semiaxes a and b

17. $(7/12)\pi^3a^2$

19. $8a$

21. $\sqrt{2}$

23. (b) 94.148 million miles; (c) 3278.3 million miles

25. 12,800 miles approx.

Chapter 4 Review Problems

1. $y=\dfrac{3}{2}x-\dfrac{17}{2}$; line

3. $\dfrac{x^2}{4}+\dfrac{y^2}{9}=1$; ellipse

5. $\dfrac{1}{3}$

7. 1

9. $t=2$

11. (a) $x=\dfrac{4}{9}y^2-2$; $-3\le y\le 3$; (b) parabolic segment, vertex $(-2,0)$, opens right, axis the x–axis; particle oscillates back and forth.

13. (a) $xy=1$, $x,y>0$, $z=0$, first quadrant part of a hyperbola; (b) $z=2-y$ at (x,y); (c) curve ascends from (1,1,1) toward level (height) 2 while remaining vertically above the hyperbola in (a).

15. $A=4$

17. $t=\sqrt{2}$

19. (a) $r=e^\theta$; (b) 45°

21. 2

23. (a) 24/25; (b) 4/3; (c) 3/4

25. $\sqrt{6} + \ln(\sqrt{2} + \sqrt{3})$

27. 15

29. $\sqrt{2}$

31. $s(t) = (1+t)^{3/2} - 1$

33. $\operatorname{sech}^2 x$

35. 0

37. $(\sin t)\mathbf{i} - (\cos t)\mathbf{j}$; $(\cos t)\mathbf{i} + (\sin t)\mathbf{j}$

39. (a) $y = x + 1$, $-1 \le x \le 3$; line segment traversed back and forth from (3,4) to $(-1,0)$ to (3,4); (b) $a_T = \pm 2\sqrt{2}$; $a_N = 0$

41. (a) $s'(t) = \sqrt{2\cosh 2t}$; (b) $s'(0) = \sqrt{2}$; (c) $a_T = 2\sinh 2t/\sqrt{2\cosh 2t}$; (d) $a_N = \sqrt{2\,\operatorname{sech} 2t}$

43. $\mathbf{T} = (1/3)\langle 2, -1, 2\rangle$; $\mathbf{N} = (1/3)\langle 2, 2, -1\rangle$

45. (a) cardioid; (b) $\mathbf{r}'(t) = (2\cos t)\mathbf{u}_r + 2(1+\sin t)\mathbf{u}_\theta$; $\mathbf{r}''(t) = -(2 + 4\sin t)\mathbf{u}_r + (4\cos t)\mathbf{u}_\theta$; (c) $\mathbf{r}'(\pi/2) = -4\mathbf{i}$; $\mathbf{r}''(\pi/2) = -6\mathbf{j}$

47. $-(5/4)\sin t$ and $\cos t$

Section 5.1

1. (a) all (x,y); (b) $(-\infty,\infty)$

3. (a) (x,y) with $x^2 + y^2 \le 4$; (b) $[0,2]$

5. (a) all (x,y) with $x \ne 0$; (b) $(-\pi/2, \pi/2)$

7. (a) all $(x,y) \ne (0,0)$; (b) $(-\pi/2, \pi/2)$

9. (a) all $(x,y,z) \ne (0,0,0)$; (b) $(0,1)$

The surfaces in Probs. 11–33 are of the types described in the text and illustrated in Figs. 4–13.

11. ellipsoid

13. elliptic paraboloid

15. circular paraboloid

17. circular cone

19. elliptic cone

21. hyperbolic paraboloid

23. hyperboloid of two sheets

25. hyperboloid of one sheet

27. circular cylinder

29. hyperbolic cylinder

31. cylinder

33. cylinder

35. ellipses

37. hyperbolas and the x– and y–axis

39. lines through the origin with $(0,0)$ removed

41. lines $x - y = a + 2\pi k$, $x - y = -a + 2(k+1)\pi$, $-\pi/2 \le a \le \pi/2$

43. parabolas $y = x^2 + k$, $k > -1$

45. paraboloids $z = x^2 + y^2 - c$, any c

47. ellipsoids $4x^2 + 9y^2 + 25z^2 = c$, any $c > 0$ and $(0,0,0)$

49. sphere, center $(1,-2,3)$, radius 4

51. hyperboloid of one sheet with center $(0,3,-1)$

53. circular paraboloid

55. $z = 1/\sqrt{x^2 + y^2}$

57. yes; a function cannot have two values at the same point.

Section 5.2

1. (a) $x^2 + y^2 < 4$; (b) all (x,y) inside the circle $x^2 + y^2 = 4$; (c) open; (d) bounded; (e) connected

3. (a) $x^2 + y^2 > 1$ and $4x^2 + 9y^2 \le 36$; (b) all (x,y) outside the circle $x^2 + y^2 = 1$ and inside or on the ellipse $4x^2 + 9y^2 = 36$; (c) neither; (d) bounded; (e) connected

5. (a) $x > 0$ and $y > 0$; (b) first quadrant without the axes; (c) open; (d) unbounded; (e) connected

7. (a) $y/x > 0$; (b) first and third quadrants without axes; (c) open; (d) unbounded; (e) disconnected

9. (a) $y \le e^x$; (b) all (x,y) on or below the graph of $y = e^x$; (c) closed; (d) unbounded; (e) connected

11. all (x,y,z) with $y > x$

13. $xy > 0$ and $z \ne 0$

15. (x,y,z,t), $z \ge 0$, $t > -1$

17. $-6/13$

19. $\sqrt{5/7}$

21. 0

23. 1

25. 0

31. possible, $f(0,0) = 0$

Section 5.3

1. (a) $f_x = 3x^2 + 8xy$, $f_y = 4x^2$; (b, c) $f_x(1,2) = 19$, $f_y(1,2) = 4$

3. (a) $f_x = -2x/(x^2+y^2)^2$, $f_y = -2y/(x^2+y^2)^2$; (b, c) $f_x(2,3) = -4/169$, $f_y(2,3) = -6/169$

5. (a) $f_x = -y/(x^2+y^2)$, $f_y = x/(x^2+y^2)$; (b, c) $f_x(3,4) = -4/25$, $f_y(3,4) = 3/25$

7. (a) $f_x = (x^2 \sin y + 2x)e^{x \sin y}$, $f_y = x^3 e^{x \sin y} \cos y$; (b, c) $f_x(2,0) = 4$, $f_y(2,0) = 8$

9. $z_x = e^{x-y}$, $z_y = -e^{x-y}$

11. $z_x = e^x \cos y$, $z_y = -e^x \sin y$

13. $z_x = \cosh(x+3y)$, $z_y = 3\cosh(x+3y)$

15. $f_x = y + z$, $f_y = x + z$, $f_z = x + y$

17. $w_x = yze^{xyz}$, $w_y = xze^{xyz}$, $w_z = xye^{xyz}$

19. $w_x = \dfrac{-wx}{2t}$, $w_y = \dfrac{-wy}{2t}$, $w_z = -\dfrac{wz}{2t}$, $w_t = w\left[\dfrac{(x^2+y^2+z^2)}{4t^2} - \dfrac{3}{2t}\right]$

21. (a) $z_x = \dfrac{4x}{z}$, $z_y = \dfrac{3y}{z}$; (b) $z_x = \pm 2$, $z_y = 0$ when $(x,y) = (2,0)$ and $z = \pm 4$

23. (a) $z_x = \dfrac{4z^2 - 2x}{5z^4 - 8(x+y)z}$, $z_y = \dfrac{4z^4 + 2y}{5z^4 - 8(x+y)z}$; (b) $z_x = \dfrac{7}{24}$, $z_y = \dfrac{3}{8}$ when $(x,y,z) = (1,1,2)$

25. (a) $\dfrac{\partial x}{\partial r} = \cos\theta$, $\dfrac{\partial x}{\partial \theta} = -r\sin\theta$, $\dfrac{\partial y}{\partial r} = \sin\theta$, $\dfrac{\partial y}{\partial \theta} = r\cos\theta$;
(b) $\dfrac{\partial r}{\partial x} = \dfrac{x}{\sqrt{x^2+y^2}}$, $\dfrac{\partial r}{\partial y} = \dfrac{y}{\sqrt{x^2+y^2}}$, $\dfrac{\partial \theta}{\partial x} = -\dfrac{y}{x^2+y^2}$, $\dfrac{\partial \theta}{\partial y} = \dfrac{x}{x^2+y^2}$

27. (a) $\dfrac{\partial r}{\partial x} = \dfrac{x}{x^2+y^2}$, $\dfrac{\partial \theta}{\partial x} = -\dfrac{y}{x^2+y^2}$; (b) $\dfrac{\partial r}{\partial y} = \dfrac{y}{x^2+y^2}$, $\dfrac{\partial \theta}{\partial y} = \dfrac{x}{x^2+y^2}$

31. $z_x = -9/2\sqrt{23}$

33. (a) $T_x = (1/2)e^{-1/4}\cos 2$; (b) $T_t = -(1/4)e^{-1/4}\sin 2$

39. $(nR - c\gamma PT^{\gamma-1})^2/nR$

Section 5.4

1. $3x + 2y - z = 6$

3. $3x + 4z = 25$

5. $4x + 3y + 2z = 24$

7. $2x + 5y - 3z = 0$

9. $4x + 2y - z = -84$

15. $3x + 4y - 5z = 0$ common tangent plane

17. $F(x,y) = 3 + 3(x-1) - 3(y+1)$

19. $F(x,y) = \dfrac{\pi}{3} - \dfrac{\sqrt{3}}{4}(x-1) + \dfrac{1}{4}(y - \sqrt{3})$

21. $F(x,y,z) = 25 + 10(x-1) + 13(y-2) + 13(z-3)$

27. $df=\left(\frac{1}{y}-\frac{y}{x^2}\right)dx+\left(\frac{1}{x}-\frac{x}{y^2}\right)dy$ 29. $dz=(y\cos xy)\,dx+(x\cos xy)\,dy$

31. $dw=e^{x^2+y^2-2z^2}(2xdx+2ydy-2dz)$

33. (a) $dV=-47/8$; (b) $\frac{dV}{V}=-\frac{47}{480}$ 35. (a) $dT=\frac{\pi}{\sqrt{Lg}}dL-\frac{\pi\sqrt{L}}{g^{3/2}}dg$; (b) $\left|\frac{\Delta T}{T}\right|\le 0.00275$

37. (a) $\frac{\Delta F}{F}\approx 0.04$; (b) equally significant

Section 5.5

1. $\frac{dz}{dt}=0$ 3. $\frac{dz}{dt}=4(t-t^{-3})e^{2(t^2+t^{-2})}$ 5. $\frac{dw}{dt}=\frac{18t}{1+9t^2}$

7. The ellipse $x^2/3^2+y^2/4^2=1$ traversed counterclockwise from (3,0); (a) $7\sqrt{3}/2\sqrt{43}$; (b) $-7\sqrt{2}/10$

9. $3\pi\ \text{ft}^3/\text{yr}$

11. The temperature oscillates between $T_{\min}$ and $T_{\max}$; $T_{\min}=35$ at $(\sqrt{2},1/\sqrt{2})$ and $(-\sqrt{2},-1/\sqrt{2})$; $T_{\max}=37$ at $(-\sqrt{2},1/\sqrt{2})$ and $(\sqrt{2},-1/\sqrt{2})$.

13. $z_s=4-\sqrt{3};\quad z_t=-2$ 15. $z_s=1;\quad z_t=0$ 17. -2

33. (a) $-\frac{\sqrt{3}+2}{6}$; (b) $\frac{\sqrt{3}+2}{6}$ 35. $\frac{5}{2}$ 41. $z_x=-\frac{6}{11}$; $z_y=-\frac{4}{11}$

Section 5.6

1. $\left\langle 4,\frac{2}{27}\right\rangle$ 3. $e^2\langle 2,0\rangle$ 5. $\frac{1}{2}\langle 1,1\rangle$

7. $\langle 1,1,0\rangle$ 9. $\left\langle -\frac{15}{16},\frac{9}{8},\frac{9}{16}\right\rangle$ 15. $\frac{55}{27}\sqrt{2}$

17. $e^2\sqrt{3}$ 19. 1   21. (a) $\frac{36}{\sqrt{14}}$; (b) $\frac{1}{\sqrt{6}}\langle 1,1,2\rangle$; $4\sqrt{6}$

25. $-3/2\pi$ 27. (a) $-72e^{-1}$; (b) $-\langle 24e^{-1},0\rangle$ 29. $-3\sqrt{3}$

31. $\mathbf{N}=\langle -1,2\rangle$; $2y-x=2$ 33. $3x+5y-7z=-41$ 35. $x+y+z=\pm 11$

Chapter 5 Review Problems

1. (a) all (x,y) with $xy>0$; (b) $y=x/e$ 3. $y=\pm(4/3)x,\ x\neq 0$

5. elliptic cone; two lines $z=0$, $y=\pm 3x$; two lines $x=0$, $y=\pm 4z$; ellipse $y=1$, $9x^2+16z^2=1$

7. 3   9. $F(x,y,z)=\frac{1}{3}(x+y+z)-8$ 11. $z_x=-\frac{y+z^2}{y+2xz}$; $z_y=-\frac{x+z}{y+2xz}$

13. $z=4x-3,\quad y=1$ 15. $\sqrt{2}$

17. $df=(\tan y+y\sec x\tan x)\,dx+(x\sec^2 y+\sec x)\,dy$

19. $dz=0.28$ 21. $(-2,0,5)$ 23. $3x+4y-9z=22$

25. $z=\frac{1}{\sqrt{3}}(y-x)$ 27. $x+4y+4z=44$ 29. $\frac{21}{2}$

31. $\pm\frac{1}{\sqrt{6}}\langle 1,-1,2\rangle$

33. $\pi-\frac{\pi^2}{2}$

35. (a) $-\frac{2}{25}\langle 3,4\rangle$; (b) $-\frac{48}{125}$; (c) $\frac{2}{5}$

37. 0

39. $-2/\sqrt{6}$

41. $-16/81$

43. 2

45. 1.7

Section 6.1

1. global min at $(1,-2)$

3. neither max nor min at $(0,0)$

5. no critical points

7. global max at $(0,0)$

9. global min at $(1,-2,3)$

11. $f_{max}=8$; $f_{min}=-9$

13. $f_{max}=0$; $f_{min}=-9$

15. $f_{max}=1$; $f_{min}=-1$

17. $f_{max}=3\sqrt{3}/2$; $f_{min}=-3\sqrt{3}/2$

19. $f_{max}=\frac{1+\sqrt{3}}{2}$; $f_{min}=\frac{1-\sqrt{3}}{2}$

21. (a) $f_{max}=1$

23. (a) $f_{min}=2$

25. (a) $f_{max}=2$

27. $V_{max}=8$

29. $V_{max}=18^2\cdot 36$

31. $2\sqrt[3]{3}$ by $2\sqrt[3]{3}$ by $3\sqrt[3]{3}$

33. $V_{max}=\frac{1}{2}\left(\frac{S}{3}\right)^{3/2}$; S fixed

35. $V_{max}=\frac{abc}{27}$

37. $V_{max}=\frac{4}{3}\pi\frac{S^3}{27}$; S fixed sum

39. $f_{max}=\frac{3\sqrt{3}}{2}$; no min

41. $\frac{x}{3}+\frac{y}{6}+\frac{z}{6}=1$

43. $V_{max}=\frac{4}{9}\sqrt{3}a^3$

49. no global extrema

Section 6.2

21. $(0,0)$, local min

23. $(1,1)$, local min

25. $(0,0)$, saddle; $(1,1)$, $(-1,-1)$, local minima

27. $(0,0)$, saddle; $(1,1)$ local max

29. $(-1,0)$, $(1,0)$ local maxima

31. $(0,0)$, saddle; $\left(\frac{1}{2},\frac{1}{\sqrt{2}}\right)$, $\left(-\frac{1}{2},-\frac{1}{\sqrt{2}}\right)$ local maxima; $\left(\frac{1}{2},-\frac{1}{\sqrt{2}}\right)$, $\left(-\frac{1}{2},\frac{1}{\sqrt{2}}\right)$ local minima

33. critical points $(m\pi,n\pi)$, m and n any integers; saddle when both are even or both odd; minima when m is even and n is odd; maxima when m is odd and n is even.

41. critical points $(m\pi,n\pi)$, m and n any integers; saddle when one is even and the other odd; minima when both are odd; maxima when both are even.

Section 6.3

1. $f_{max}=2$; $f_{min}=-2$

3. $f_{max}=3/2$; $f_{min}=1/2$

5. $f_{max}=3^{3/4}$; $f_{min}=-3^{3/4}$

7. no global extrema

9. $f_{max}=3/2$; $f_{min}=0$

11. $f_{max}=6$; $f_{min}=0$

13. square

15. cube

17. $r=4\pi^{-1/3}$; $h=16\pi^{-1/3}$

19. $2\cdot 3^{1/3}$ by $2\cdot 3^{1/3}$ by $3^{4/3}$

21. base and sides 2/3, angle between base and a side $2\pi/3$

23. $a=b=c=S/3$; S fixed sum

25. $f_{max}=\frac{3\sqrt{3}}{2}$

27. $\frac{x}{3}+\frac{y}{6}+\frac{z}{6}=1$

29. $\frac{2}{\sqrt{3}}$ by $\frac{2\sqrt{2}}{\sqrt{3}}$

31. $\frac{2a}{\sqrt{3}}$ by $\frac{2\sqrt{2}b}{\sqrt{3}}$

33. (a) $f_{max}=\frac{a}{3}$

35. $3a^{2/3}$

Chapter 6 Review Problems

1. $\left(\frac{1}{2},\frac{1}{2}\right)$, $\left(-\frac{1}{2},-\frac{1}{2}\right)$ local maxima; $(0,0)$, $(\pm 1,0)$, $(0,\pm 1)$ saddle points; $\left(-\frac{1}{2},\frac{1}{2}\right)$, $\left(\frac{1}{2},-\frac{1}{2}\right)$ local minima

3. $(0,0)$ saddle; $(2,1)$ local min

5. $(2,-2)$, $\left(-2,-\frac{2}{3}\right)$ saddles; $(0,0)$ local min

7. $(0,0)$ test fails; $(3,3)$ saddle

9. (a) $(0,0)$ loc max; $\left(\frac{1}{2},\pm\frac{1}{2}\right)$, $\left(-\frac{1}{2},\pm\frac{1}{2}\right)$ saddles; (b) $f_{\max}=0$, $f_{\min}=-1$

11. $f_{\max}=6$ 13. $f_{\min}=3$ 15. $\sqrt{6}$

17. (a) $V_{\max}=4$; (b) $V \to 0$ as any dimension tends to zero

19. $(1,0,1)$; $\sqrt{5}$ 21. $3\sqrt{2}/8$

23. 2 25. 6 27. $V_{\max}=1000$

29. (a) $f_{\max}=3/2$; (b) $f \to -1$ when any two angles tend to 0

Section 7.1

1. 5/6 3. $(1/2)(e^4-1)(e-e^{-1})$ 5. 12

7. 41 9. $(e^2-3-e^{-1})/2$ 11. $2(3^{5/2}-2^{5/2}-1)/15$

13. (b) 5/4 15. (b) $1-\cos 1$ 17. (b) 4/5

19. (b) $(1/4)(e-1)$ 21. (b) 8π 23. (b) 28/3

25. (b) 36π 27. (b) $(3/10)a^3$ 29. (b) 8π

31. (a) $0\le y\le 1$, $y\le x\le 1$ or $0\le x\le 1$, $0\le y\le x$; (b) $1-\cos 1$

33. (a) $0\le y\le 1$, $y^2\le x\le 1$ or $0\le x\le 1$, $0\le y\le x^{1/2}$; (b) $(e-1)/4$

35. (a) $0\le x\le 2$, $x/2\le y\le 1$ or $0\le y\le 1$, $0\le x\le 2y$; (b) $e-1$

37. (a) $1\le x\le 2$, $0\le y\le \pi$; (b) 0 45. 1

47. 0 49. 2 51. 0

Section 7.2

1. 3/2 3. 96 5. 1

7. 11/84 9. $-(3/4)e^2+2e-3/4$ 11. 12

13. $\frac{23}{120}$ 15. $\frac{151}{60}$ 17. $\frac{5}{24}a^4$

19. $\frac{2}{3}\sqrt{3}\left(36-\frac{216}{5}+\frac{108}{7}\right)$ 21. (a) $\frac{a^3bc}{60}$ (b) $\frac{abc}{60}(a^2+b^2)$ 23. $\frac{48}{5}$

25. $\frac{1}{2}$ 27. $\frac{2}{3}$ 29. 512π

31. 32 33. $\frac{16}{15}$ 35. 1

37. $\frac{2048}{135}\sqrt{2}$

39. $\frac{\pi-2}{8}$

41. $\frac{9}{2}$

Section 7.3

1. circular cylinder, axis the z–axis, radius 5
3. half plane with edge the z–axis making a 45° angle with the xz–plane
5. plane $x = 4$
7. circular half-cone, axis the positive z–axis, with apex angle 30°
9. circular cylinder, axis the line $x = 0$, $y = 1$, any z, radius 1

11. $\frac{\pi}{3}$

13. $\frac{\pi^2}{4}$

15. $\ln(\sqrt{2}+1)+\sqrt{2}-1$

17. 4π

19. $\frac{16}{3}\pi(8-3\sqrt{3})$

21. $\frac{\pi}{6}$

23. 6π

25. $\frac{5\pi a^3}{12}$

27. 24π

29. $\frac{4}{3}\pi a^2 c$

31. $\frac{32\pi}{5}$

33. $\frac{8k\pi}{3}(4\pi-3\sqrt{3})$

35. $\frac{\pi k}{12}$

37. $\frac{128}{3}$

39. $\frac{53}{480}\pi k a^5$

41. $32k$

43. $\frac{\pi}{4}$

45. $\sqrt{\pi}$

47. $(1/4)\sqrt{\pi}$

Section 7.4

1. $(r,\theta,z)=\left(\sqrt{2},\frac{\pi}{6},\sqrt{2}\right),(x,y,z)=\left(\frac{\sqrt{6}}{2},\frac{\sqrt{2}}{2},\sqrt{2}\right)$

3. $(r,\theta,z)=\left(\frac{3}{2},-\frac{\pi}{6},\frac{\sqrt{3}}{2}\right),\quad (x,y,z)=\left(3\frac{\sqrt{3}}{4},-\frac{3}{4},\frac{\sqrt{3}}{2}\right)$

5. $(r,\theta,z)=(1,0,0),\quad (\rho,\varphi,\theta)=\left(1,\frac{\pi}{2},0\right)$

7. $(r,\theta,z)=\left(4,\frac{\pi}{6},4\right),\quad (\rho,\varphi,\theta)=\left(4\sqrt{2},\frac{\pi}{4},\frac{\pi}{6}\right)$

9. $(x,y,z)=\left(-\frac{\sqrt{2}}{2},\frac{\sqrt{2}}{2},1\right),\quad (\rho,\varphi,\theta)=\left(\sqrt{2},\frac{\pi}{4},\frac{3\pi}{4}\right)$

11. $(x,y,z)=(2\sqrt{3},-2,4\sqrt{3}),\quad (\rho,\varphi,\theta)=\left(8,\frac{\pi}{6},-\frac{\pi}{6}\right)$

13. horizontal plane $z = 4$

15. sphere, center 0, radius 5

17. half-cone, vertex 0, axis the positive z–axis

19. sphere, center (0,0,1), radius 1

21. sphere, center 0, radius 2; $r^2 + z^2 = 4$; $\rho = 2$

23. cone; $z = \pm\sqrt{3}r$; $\varphi = \frac{\pi}{6}$ and $\varphi = \frac{5\pi}{6}$

25. sphere, center (0,0,2), radius 2; $r^2 + (z-2)^2 = 4$; $\rho = 4\cos\varphi$

27. (a) $\int_{-2}^{2}\int_{\sqrt{4-x^2}}^{\sqrt{4-x^2}}\int_{-\sqrt{4-x^2-y^2}}^{\sqrt{4-x^2-y^2}}(x^2+y^2)\,dzdydx$; (b) $\int_0^{2\pi}\int_0^{2}\int_{-\sqrt{4-r^2}}^{\sqrt{4-r^2}} r^3\,dzdrd\theta$; (c) $\int_0^{2\pi}\int_0^{\pi}\int_0^{2}\rho^4\sin^3\varphi d\rho d\varphi d\theta$

29. (a) $\int_0^6\int_0^{\sqrt{6x-x^2}}\int_0^{\sqrt{36x-x^2-y^2}} zdzdydx$; (b) $\int_0^{\pi/2}\int_0^{6\cos\theta}\int_0^{\sqrt{36-r^2}} zrdzdrd\theta$;

(c) $\left(\int_0^{\frac{\pi}{2}}\int_0^{\frac{\pi}{2}-\theta}\int_0^{6} + \int_0^{\frac{\pi}{2}}\int_{\frac{\pi}{2}-\theta}^{\frac{\pi}{2}}\int_0^{6\cos\theta/\sin\varphi}\right)\rho^3\sin\varphi\cos\varphi d\rho d\varphi d\theta$

31. $\frac{4\pi}{5}$

33. $\frac{81}{40}(8-5\sqrt{2})$

35. 8π

37. $\frac{3\pi}{2}$

39. $\frac{1}{3}\pi a^3$

41. $\frac{\pi}{3}(a-b)^2(2a+b)$

43. $\frac{17\pi k}{6}$

45. $\frac{5\pi a^3}{12}$

47. $2\pi^2$

49. $\mathbf{F} = Gm\sigma_0\pi a\sin^2\alpha\mathbf{k}$

51. $\mathbf{F} = \frac{2}{3}Gm\sigma_0\pi a(\sqrt{2}-1)\mathbf{k}$

Section 7.5

1. 0

3. $\frac{9+4\sqrt{2}}{4}$

5. $\left(\frac{1}{2}, 1, \frac{3}{2}\right)$

7. $(0,0,3)$

9. $\left(0,0,\frac{3}{4}\right)$

11. on axis midway between bases

13. $\frac{3}{8}a$ from center on radius perpendicular to base

15. $\left(0,0,\frac{2}{3}h\right)$

17. $\left(0,\frac{1}{2},\frac{5}{12}\right)$

19. (a) $\left(0,0,\frac{3(b^4-a^4)}{8(b^3-a^3)}\right)$; (b) $\left(0,0,\frac{a}{2}\right)$, centroid of spherical surface

21. $\left(\frac{3}{16}\pi a\frac{\sin\gamma}{\gamma}, \frac{3}{16}\pi a\frac{1-\cos\gamma}{\gamma}, 0\right)$

23. $\left(\frac{9}{10},\frac{7}{30},\frac{8}{15}\right)$

25. $\left(\frac{16}{5\pi},\frac{16}{5\pi},3\right)$

27. $\left(\frac{3}{4},\frac{3}{4},\frac{3}{4}\right)$

29. $\left(\frac{2}{3}a,\frac{1}{2}b\right)$ if rectangle has vertices $(0,0)$, $(a,0)$, (a,b), $(0,b)$

31. $\frac{3a}{2\pi}$ from center on radius perpendicular to base

33. $\left(0,\frac{6}{5}a\right)$

35. $\left(\frac{5}{8}a,\frac{5}{8}a,\frac{1}{2}a\right)$ for cube in first octant, three faces in the coordinate planes, one vertex at $(0,0,0)$, and "one edge" the z–axis

37. $\left(0,0,\frac{4}{5}h\right)$ cone with vertex $(0,0,0)$, axis $0 \le z \le h$

39. $\dfrac{8a}{15}$ from center on radius perpendicular to base

41. $\left(0,0,\dfrac{3}{5}a\right)$

43. $\left(\dfrac{4}{3}a,0,\dfrac{2048}{675\pi}a\right)$

45. $\left(0,0,\dfrac{8}{7}a\right)$

47. (b) $I=\dfrac{1}{12}(a^2+4b^2)M$ where $M=\sigma_0\,abc$

49. (b) $I=\dfrac{1}{12}(3a^2+4h^2)M$ where $M=\sigma_0\,\pi a^2 h$

51. (b) $I=\dfrac{2}{5}a^2\,M$ where $M=\sigma_0\,\dfrac{4}{3}\pi a^3$

53. (b) $I=\dfrac{7}{5}a^2\,M$ where $M=\sigma_0\,\dfrac{4}{3}\,\pi a^3$

55. (b) $I=\dfrac{3}{20}(a^2+4h^2)M$ where $M=\sigma_0\dfrac{1}{3}\pi a^2 h$

57. $I=\dfrac{11}{5}a^2M$ where $M=4\,\pi\sigma_0\sqrt{3}\,a^3$

59. $\mathbf{L}=\theta'(t)\dfrac{1}{12}(a^2+b^2)M\mathbf{k}$ where $M=\sigma_0\,abc$

61. $\mathbf{L}=\theta'(t)\dfrac{1}{12}(a^2+c^2)M\mathbf{j}$ where $M=\sigma_0 abc$

63. $\mathbf{L}=\theta'(t)\dfrac{1}{2}\,a^2M\mathbf{j}$ where $M=\sigma_0\pi a^2 h$

65. $\mathbf{L}=\theta'(t)\dfrac{10}{21}\,a^2M\mathbf{k}$ where $M=\dfrac{4}{5}\,\pi k a^5$

67. $\mathbf{L}=\theta'(t)\dfrac{3}{10}\,a^2M\mathbf{k}$ where $M=\sigma_0\,\dfrac{1}{3}\,\pi a^2 h$

79. The center of mass lies in any plane of symmetry of D if the density takes the same value at symmetric points.

83. (d) $\mathbf{r}_{cm}=\left\langle a,\dfrac{6b^2+3ab\pi+4a^2}{3(4b+a\pi)}\right\rangle$

Chapter 7 Review Problems

1. 2

3. 52/9

5. 1/4

7. $\dfrac{\pi}{8}\ln 5$

9. $\dfrac{4}{3}$

11. 3

13. $\dfrac{\pi}{4}-\dfrac{1}{3}$

15. (a) $\dfrac{4+6\pi}{3}$; (b) $\left(0,\dfrac{88}{5(2+3\pi)}\right)$

17. 3

19. $\dfrac{1}{2\pi}$

21. 12

23. $6(3\pi-4)$

25. 6

27. $32\pi/3$

29. $6(3\pi-4)$

31. (a) $\dfrac{5\pi}{6}$; (b) $\left(0,0,\dfrac{11}{10}\right)$

33. (b) $\left(0,0,\dfrac{7}{6}\right)$

35. 22π

37. (a) $\dfrac{1}{8}$; (b) $\left(\dfrac{4}{5},\dfrac{4}{5}\right)$

39. $\dfrac{4\pi}{3}(e-1)$

41. 124π

43. $\dfrac{128\pi}{15}$

45. (a) 16π; (b) $\left(0,0,\dfrac{4}{5}\right)$

47. (a) $\dfrac{64\pi}{15}$; (b) $\dfrac{5}{2\pi}$

49. $\dfrac{1280\pi}{189}$

51. 500

Section 8.1

1. Vectors from (x,y) to $(x+1,\ y+1)$
3. Vectors from (x,y) to $(2x,\ y+1)$
5. Vectors from (x,y) to $(x+y,\ x+y)$
7. Vectors from (x,y) to $(x,\ y+2)$
9. Vectors from (x,y,z) to $(2x, 2y, 2z)$
11. $\mathbf{F}(x,y) = 2x\mathbf{i} - 2y\mathbf{j}$
13. $\mathbf{F}(x,y) = \sin(x-y)(\mathbf{j}-\mathbf{i})$
15. $\mathbf{F}(x,y,z) = y^2z^3\mathbf{i} + 2xyz^3\mathbf{j} + 3xy^2z^2\mathbf{k}$
17. $\mathbf{F}(x,y,z) = \dfrac{\mathbf{i}+\mathbf{j}+\mathbf{k}}{(x+y+z)}$
19. $\operatorname{div}\mathbf{F} = 2, \quad \operatorname{curl}\mathbf{F} = \mathbf{0}$
21. $\operatorname{div}\mathbf{F} = 2e^x\cos y; \quad \operatorname{curl}\mathbf{F} = 2e^x\sin y\,\mathbf{k}$
23. $\operatorname{div}\mathbf{F} = 0; \quad \operatorname{curl}\mathbf{F} = \mathbf{0}$
25. $\operatorname{div}\mathbf{F} = 3; \quad \operatorname{curl}\mathbf{F} = \mathbf{0}$
27. $\operatorname{div}\mathbf{F} = x + \cos(x+y);$ $\operatorname{curl}\mathbf{F} = -z\mathbf{j} - [e^{-x} + \cos(x+y)]\mathbf{k}$
29. $\operatorname{div}\mathbf{F} = f'(x) + g'(y) + h'(z); \quad \operatorname{curl}\mathbf{F} = \mathbf{0}$
33. (a) lines through the origin
35. (a) lines $x = c$, a constant
37. circular helices on the cylinder with axis the z–axis
55. (a) $n = -3$; (b) all n

Section 8.2

1. 0
3. $\dfrac{2944}{189}$
5. $2 - \dfrac{4}{\pi}$
7. $2ka^3\pi$
9. $2ka^2$
11. $\left(-\dfrac{a}{3}, \dfrac{2a}{3}\right)$
13. $\dfrac{32ka^4}{15}$
15. $2kb\sqrt{a^2+b^2}\,\pi^2$
17. $\dfrac{2}{3}\pi k(3a^2 + 4\pi^2b^2)\sqrt{a^2+b^2}$
19. $2\pi^2ka^2b\sqrt{a^2+b^2}$
21. (a) $8a$; (b) $(4a/3, \pi a)$
23. $-\pi$
25. 0
27. 0
29. $-16/3$
31. 3
33. 0
35. 0
37. $W = -2\pi bmg$
39. (a) $W = \pi a^2$; (b) $W = \pi ab$
41. $y = 10x(1-x)$
43. $(2/\pi)\sin x$

Section 8.3

1. (b) $f(x,y) = \dfrac{3}{2}x^2 - 2xy + \dfrac{3}{2}y^2$; (c) 0
3. (b) $f(x,y) = e^{x^2}\cos y$; (c) –1
5. (b) $f(x,y) = x\cos xy$; (c) -3
7. (b) $f(x,y,z) = xyz + x + 2y + 3z$; (c) 4
9. $\dfrac{4}{3} + \sin 1$
11. 9
13. 0
15. π
17. 0
19. -3
25. $\dfrac{k}{2-n}r^{2-n}$, $n \neq 2$; $k\ln r$, $n = 2$
27. $\dfrac{\|\mathbf{r}_1\|^{n+2} - \|\mathbf{r}_0\|^{n+2}}{n+2}$
29. (a) $\dfrac{1}{2}kx^2$; (b) $E = \dfrac{1}{2}kx(t)^2 + \dfrac{1}{2}mx'(t)^2$; (d) $\max x = \sqrt{\dfrac{9k+25m}{k}}, \quad \max x' = \sqrt{\dfrac{9k+25m}{m}}$

33. (c) **F** does not have continuous partials on and inside C.

35. Yes; Th. 3 applies.

Section 8.4

1. 8π
3. $-1/30$
5. $-3/2$
7. 0
9. 6
11. 0
13. (b) $3\pi a^2$
15. (b) $4a^2/3$
19. $(0, 4b/3\pi)$
21. (*a*) $Mb^2/4$; (*b*) $Ma^2/4$; (*c*) $M(a^2+b^2)/4$
25. $-7/60$
27. $3\pi/4$
29. 27/70
31. 0
35. 0
37. (a) 0; (b) 0

Section 8.5

1. 9
3. $\pi b^2\sqrt{3}$
5. $81\pi\sqrt{2}$
7. 40π
9. 8π
11. $\sqrt{2}\,\pi a^2$
13. 3π
17. $2\pi a^2$
19. (a) $9k$; (b) $\left(\frac{17}{36}, \frac{17}{18}, \frac{55}{18}\right)$
21. (a) $\frac{13}{3}k\pi$; (b) $\left(0, 0, \frac{149}{130}\right)$
23. (a) $\frac{2}{3}kh^3\pi\sqrt{2}$; (b) $\left(0, 0, \frac{3}{4}h\right)$
25. (a) $\frac{1}{3}ka^4$; (b) $\left(\frac{3}{16}\pi a, \frac{3}{32}\pi a, \frac{3}{4}a\right)$
27. $\left(0, 0, \frac{a}{2}\right)$
29. $\left(0, 0, \frac{a}{2(2-\sqrt{2})}\right)$

Section 8.6

1. $-1/2$
3. 18π
5. 4π
7. $16\pi/3$
9. 4π
11. 432/5
13. -96π
15. $V = \frac{1}{3}\int\int_S \mathbf{r}\cdot\mathbf{n}dS$, or $V = \int\int_S x\mathbf{i}\cdot\mathbf{n}dS$, or $V = \int\int_S y\mathbf{i}\cdot\mathbf{n}dS$, or $V = \int\int_S z\mathbf{i}\cdot\mathbf{n}dS$
17. 0

Section 8.7

1. -3π
3. both integrals are 0.
5. 0
7. 304
9. 16π
11. $-11/2$
13. 0
17. 0

Chapter 8 Review Problems

1. $\operatorname{div}\mathbf{F} = -\frac{2y}{x^2}$; $\operatorname{curl}\mathbf{F} = \frac{1-2x}{x^2}\mathbf{k}$
3. (b) 1
5. Yes; $f(x, y) = \frac{1}{(x^2+y^2)^{1/2}}$
7. (a) yes; (b) $f(x, y) = \frac{1}{2}x^2e^{2y} + x - \frac{1}{2}y^2 + c$
9. (b) $\frac{1}{y} = \frac{1}{2x^2} + c$
11. 4
13. $5e^7$
15. 0

17. 45

19. $3\sqrt{2}$

21. (a) $f(x,y) = xe^y + \dfrac{y^4}{4} + c$; (b) 0; (c) 2

23. $\dfrac{18}{5}$

25. 0

27. 8π

29. 10π

31. 1/3

33. $(2,2)$

35. $5\sqrt{5} - 8$

37. $4\pi - 8$

39. 9π

41. $2\pi/3$

43. 12π

45. 2

47. (a) 2/15; (b) 2/15

49. both integrals -2π

51. 108π

Index

A

acceleration, 220, 222
  normal component, 241
  radial component, 248
  tangential component, 241
  transverse component, 248
alternating harmonic series, 62
alternating series test, 64
angle of inclination, 161
angular momentum, 454, 455
  conservation of, 454
angular velocity, 234
aphelion, 258
Archimedes' principle of buoyancy, 558
arc length parameterization, 229
area
  of a plane region, 396
  of a surface, 537
area element, 390

B

basic comparison test, 46
Bessel function, 138
binomial series, 130
Biot-Savat law, 233
boundary, 282
boundary point, 282
bounded, 282
bounded sequences, 17
box, 281
  closed, 281
  open, 281

C

Cauchy-Riemann equations, 326
center of gravity, 449
center of mass
  of a curve, 494
  of a plane region, 447
  of a solid, 452
  of a surface, 541
central force, 245
  attractive, 245
  repulsive, 245
centrifugal force, 227
centroid
  of a plane region, 447
  of a solid, 448
  of a surface, 541
centripetal acceleration, 223
centripetal force, 227
chain rule
  for partial derivatives, 312, 314, 315, 317
charge
  on a plane region, 417
  on a solid, 413
circle of curvature, 244
circulation, 494, 528, 562
closed ball, 281
closed disk, 281
completeness, 18
component
  along a vector, 157
  orthogonal to a vector, 159
cone
  circular, 272
  elliptic, 272
conservation of energy, 511
conservative vector fields, 477
  characterization of, 512
continuity, 210, 286
convex, 342
coordinate planes, 140
coordinate system
  right-handed, 139
convergence
  absolute, 59
  conditional, 60
Cramer's rule, 166
critical point
  function of several variables, 353
cross product, 168
curl, 482, 484
current density, 563
curvature, 236
curve length, 221
cycloid, 199
cylinder, 274
  axis of, 274
  circular, 274
  directrix of, 274
  quadric, 274
cylindrical block, 427
cylindrical coordinates, 421

D

decreasing sequence, 18
determinants
  2×2, 164
  3×3, 166
  permutation identity, 167
differentiable
  function of several variables, 330
differential, 305, 307
direction angles, 146, 149
direction cosines, 146, 149
directional derivative
  function of three variables, 321
  function of two variables, 319
  second order, 368
displacement vector, 160
divergence, 479
divergence theorem
  in the plane, 525
  in space, 548
domain, 268
dot product, 153
double integral
  in polar coordinates, 423

over a rectangle, 391
over a region, 395
double sums, 390

E

electric field strength, 477
ellipsoid, 270
equation of state, 297
equipotential
curves, 518
surfaces, 518
error function, 137
Euler's constant, 58

F

factorial, 13
Faraday's law, 568
flux
across a curve, 526
across a surface, 546
Fourier's law, 556

G

Gauss' law, 554
geometric series, 35
common ratio, 36
global extremum
function of several variables, 351
global maximum
function of several variables, 351
global minimum
function of several variables, 351
gradient, 328
gradient operator, 477
gradient vector field, 477
graph
of a function, 269
of an equation, 269
gravitational field strength, 477
Green's first identity, 577
Green's second identity, 577
Green's theorem, 519

H

harmonic series, 32
heat equation
one-dimensional, 363
three-dimensional, 574
two-dimensional, 371
heat flow vector, 572
helix
circular, 208
hyperbolic paraboloid, 273
hyperboloid
of one sheet, 273
of two sheets, 273

I

ideal gas, 297
constant, 297
implicit function theorem
for three variables, 324
for two variables, 323
implicit partial differentiation, 271
impulse, 233
increasing sequence, 18
integrable, 391, 395, 408
integral test, 51
interior, 281
interior point, 281
iterated integrals, 392, 398, 409
iterated sums, 390

K

Kepler's laws, 252
kinetic energy, 500

L

Lagrange equations, 377
Lagrange multipliers, 376
Lagrange's identity, 170
lamina, 415
Laplace's equation
two-dimensional, 371
Laplacian, 487
level curve, 275, 336
level surface, 276, 338
limit, 283, 286
of a vector function, 209
limit comparison test, 50
line
directed, 177
direction vector of, 177
normal equation of, 188
parametric equations for, 178
symmetric equations for, 179
vector equation of, 177
line integral, 488
fundamental theorem for, 507
independent of the path, 507
related to work, 497
with respect to arc length, 490
with respect to x, y, or z, 495
linear approximation, 301, 304
error in, 301, 304
linear function
of three variables, 304
of two variables, 292
local extremum
function of several variables, 350
local maximum
function of several variables, 350
local minimum
function of several variables, 350

M

mass
of a curve, 490
of a plane region, 416
of a solid body, 412
of a surface, 541
mass flow field, 479
mass flow vector, 479
mean value, 446

mechanical energy, 511
mesh, 391
mixed partial derivatives
  equality of, 362
Moebius strip, 558
moment of a force, 174, 454
moment of inertia, 456, 457, 494
momentum, 228
monotone sequences, 20

N

neighborhood, 281
Newton's law of gravitation
  vector form, 151, 440
Newton's method, 21
Newton's second law, 227, 228, 453
normal
  outward on a sphere, 534
  upward, 534
normal derivative, 556

O

octant, 140
open, 282
open ball, 281
open disk, 281
open rectangle, 281
orientable surface, 545
orientation, 545, 546
outer unit normal, 524

P

paraboloid
  circular, 271
  elliptic, 272
parallel axis theorem, 459
parallelepiped, 172
parallelogram law, 163
parameter, 178, 195
parameterization
  by arc length, 229
parametric curve
  closed, 209
  continuous, 195, 211
  differentiable, 195, 211
  elevation function, 202
  in a plane, 195
  in space, 201
  initial point, 208
  positive direction, 195
  projection on xy -plane, 201
  simple, 209
  smooth, 212
  terminal point, 208
parametric equations, 195
partial derivative, 290
  as rates of change, 294
  as slopes, 292
  mixed, 362
  second order, 361
path, 489
permeability of free space, 563
plane
  normal vector to, 182
  point-slope equation, 293
  vector equation of, 182
planes
  angle between, 183
polar coordinates
  radial unit vector, 247
  transverse unit vector, 247
polar parametric equations, 246
polar rectangle, 422
position vector, 143
positive direction
  across a surface, 545
  along a curve, 490
positively directed curve
  relative to a surface, 599
potential energy, 510
potential energy function, 510
potential function, 477, 512
primitivity of free space, 477
products of inertia, 457
projection
  along a vector, 158
  orthogonal to a vector, 159
p-series, 46, 53
power series
  interval of convergence, 106, 110
  radius of convergence, 106, 110

Q

quadric surface, 274

R

radius of curvature, 244
radius vector, 248
range, 268
ratio test, 71, 72
  for sequences, 12
rectangular coordinates, 139
resultant, 145
Riemann sum, 391, 395, 407
  inner, 395, 408
root test, 74, 75

S

saddle point, 353
scalar, 142
scalar field, 474
scalar multiple, 144, 148
scalar triple product, 171
Schwarz inequality, 163
second partials test, 365
sequence, 2
  convergent, 4
  converges, 4
  diverge, 6
  divergent, 6
  diverges to infinity, 6
  diverges to minus infinity, 6
  infinite limits, 6
  limit of, 4
  terms of, 2
set
  bounded, 282

closed, 282
connected, 283
open, 282
series
alternating, 62
converges, 30
diverges, 30
infinite, 30
*n*th partial sum, 30
*n*th term, 30
with nonnegative terms, 44
simple plane regions
radially, 426
x-, 398
y-, 400
simple solid regions
cylindrically, 429
spherically, 439
x-, 411
y-, 411
z-, 409
simply connected, 506
sink, 480
source, 480
spherical block, 436
spherical coordinates, 434
steepest ascent, 344
steepest descent, 344
Stokes' theorem,
in the plane, 527
in space, 559
streamlines, 486
successive approximations, 21
successive substitutions, 23
surface
of revolution, 271
oriented surface, 545
piecewise smooth, 534
smooth, 533
tangent plane of a, 299
trace of a, 270
x-, *y*-, *z*- simple, 565
surface integral, 540
symmetric
about the origin, 405
in *x*, 402, 414
in *y*, 403, 414
in *z*, 413

T

tangent vector, 212
Taylor polynomial
about 0, 92
about *a*, 97
remainder, 94
Taylor series,
about 0, 103
about *c*, 111
telescoping sum, 39
thermal conductivity, 556
torque, 174
trapezoidal rule
for double integrals, 469
triple integrals
in cylindrical coordinates, 427
in rectangular coordinates, 406
in spherical coordinates, 433
twisted cubic, 201

U

unit normal vector, 242
unit tangent vector, 230
unit vector, 147, 149
as a direction, 149
universal gravitational constant, 260

V

Van der Waal, 297
vector
2-dimensional, 142
3-dimensional, 147
components of, 142, 147
head of, 143
horizontal projection of a, 161
in standard position, 143
magnitude or length, 142, 147
scalar multiple of, 144, 148
slope relative to xy-plane, 161
tail of, 143
vertical projection of a, 161
vectors
angle between, 155
basic unit, 150
difference of, 145, 149
orthogonal, 155
parallel, 144, 149
sum of, 145, 149
vector addition
parallelogram law of, 145
triangle law of, 145
vector field, 475
conservative, 477, 563
sink, 480
source, 480
vector function
antiderivatives of, 225
continuity of, 210
derivative of, 225
differentiable, 211
integral of, 226
vector triple product, 176
velocity, 220
radial component, 248
transverse component, 248
velocity field, 479

W

wave equation
one-dimensional, 371
Witch of Agnesi, 206
work, 160

Z

zero vector, 142
zone, 543